Protein Folding Handbook

Edited by J. Buchner and T. Kiefhaber

Further Titles of Interest

K. H. Nierhaus, D. N. Wilson (eds.)

Protein Biosynthesis and Ribosome Structure

ISBN 3-527-30638-2

R. J. Mayer, A. J. Ciechanover, M. Rechsteiner (eds.)

Protein Degradation

ISBN 3-527-30837-7 (Vol. 1)
ISBN 3-527-31130-0 (Vol. 2)

G. Cesareni, M. Gimona, M. Sudol, M. Yaffe (eds.)

Modular Protein Domains

ISBN 3-527-30813-X

S. Brakmann, A. Schwienhorst (eds.)

Evolutionary Methods in Biotechnology

ISBN 3-527-30799-0

Protein Folding Handbook

Edited by Johannes Buchner and Thomas Kiefhaber

WILEY-VCH Verlag GmbH & Co. KGaA

Editors

Prof. Dr. Johannes Buchner
Institut für Organische Chemie und
Biochemie
Technische Universität München
Lichtenbergstrasse 4
85747 Garching
Germany
johannes.buchner@ch.tum.de

Prof. Dr. Thomas Kiefhaber
Biozentrum der Universität Basel
Division of Biophysical Chemistry
Klingelbergstrasse 70
4056 Basel
Switzerland
t.kiefhaber@unibas.ch

Cover
Artwork by Prof. Erich Gohl, Regensburg

■ This book was carefully produced. Nevertheless, authors, editors and publisher do not warrant the information contained therein to be free of errors. Readers are advised to keep in mind that statements, data, illustrations, procedural details or other items may inadvertently be inaccurate.

Library of Congress Card No.: applied for
A catalogue record for this book is available from the British Library.
Bibliographic information published by Die Deutsche Bibliothek
Die Deutsche Bibliothek lists this publication in the Deutsche Nationalbibliografie; detailed bibliographic data is available in the Internet at http://dnb.ddb.de

© 2005 WILEY-VCH Verlag GmbH & Co. KGaA, Weinheim
All rights reserved (including those of translation in other languages). No part of this book may be reproduced in any form – by photoprinting, microfilm, or any other means – nor transmitted or translated into machine language without written permission from the publishers. Registered names, trademarks, etc. used in this book, even when not specifically marked as such, are not to be considered unprotected by law.

Printed in the Federal Republic of Germany.
Printed on acid-free paper.

Typesetting   Asco Typesetters, Hong Kong
Printing   betz-druck gmbh, Darmstadt
Bookbinding   Litges & Dopf Buchbinderei GmbH, Heppenheim

ISBN-13 978-3-527-30784-5
ISBN-10 3-527-30784-2

Contents

Part I, Volume 1

Preface   *LVIII*

Contributors of Part I   *LX*

I/1	**Principles of Protein Stability and Design** *1*	
1	**Early Days of Studying the Mechanism of Protein Folding** *3*	
	Robert L. Baldwin	
1.1	Introduction *3*	
1.2	Two-state Folding *4*	
1.3	Levinthal's Paradox *5*	
1.4	The Domain as a Unit of Folding *6*	
1.5	Detection of Folding Intermediates and Initial Work on the Kinetic Mechanism of Folding *7*	
1.6	Two Unfolded Forms of RNase A and Explanation by Proline Isomerization *9*	
1.7	Covalent Intermediates in the Coupled Processes of Disulfide Bond Formation and Folding *11*	
1.8	Early Stages of Folding Detected by Antibodies and by Hydrogen Exchange *12*	
1.9	Molten Globule Folding Intermediates *14*	
1.10	Structures of Peptide Models for Folding Intermediates *15*	
	Acknowledgments *16*	
	References *16*	
2	**Spectroscopic Techniques to Study Protein Folding and Stability** *22*	
	Franz Schmid	
2.1	Introduction *22*	
2.2	Absorbance *23*	
2.2.1	Absorbance of Proteins *23*	
2.2.2	Practical Considerations for the Measurement of Protein Absorbance *27*	

Protein Folding Handbook. Part I. Edited by J. Buchner and T. Kiefhaber
Copyright © 2005 WILEY-VCH Verlag GmbH & Co. KGaA, Weinheim
ISBN: 3-527-30784-2

2.2.3	Data Interpretation 29
2.3	Fluorescence 29
2.3.1	The Fluorescence of Proteins 30
2.3.2	Energy Transfer and Fluorescence Quenching in a Protein: Barnase 31
2.3.3	Protein Unfolding Monitored by Fluorescence 33
2.3.4	Environmental Effects on Tyrosine and Tryptophan Emission 36
2.3.5	Practical Considerations 37
2.4	Circular Dichroism 38
2.4.1	CD Spectra of Native and Unfolded Proteins 38
2.4.2	Measurement of Circular Dichroism 41
2.4.3	Evaluation of CD Data 42
	References 43
3	**Denaturation of Proteins by Urea and Guanidine Hydrochloride 45**
	C. Nick Pace, Gerald R. Grimsley, and J. Martin Scholtz
3.1	Historical Perspective 45
3.2	How Urea Denatures Proteins 45
3.3	Linear Extrapolation Method 48
3.4	$\Delta G(H_2O)$ 50
3.5	*m*-Values 55
3.6	Concluding Remarks 58
3.7	Experimental Protocols 59
3.7.1	How to Choose the Best Denaturant for your Study 59
3.7.2	How to Prepare Denaturant Solutions 59
3.7.3	How to Determine Solvent Denaturation Curves 60
3.7.3.1	Determining a Urea or GdmCl Denaturation Curve 62
3.7.3.2	How to Analyze Urea or GdmCl Denaturant Curves 63
3.7.4	Determining Differences in Stability 64
	Acknowledgments 65
	References 65
4	**Thermal Unfolding of Proteins Studied by Calorimetry 70**
	George I. Makhatadze
4.1	Introduction 70
4.2	Two-state Unfolding 71
4.3	Cold Denaturation 76
4.4	Mechanisms of Thermostabilization 77
4.5	Thermodynamic Dissection of Forces Contributing to Protein Stability 79
4.5.1	Heat Capacity Changes, ΔC_p 81
4.5.2	Enthalpy of Unfolding, ΔH 81
4.5.3	Entropy of Unfolding, ΔS 83
4.6	Multistate Transitions 84
4.6.1	Two-state Dimeric Model 85

4.6.2	Two-state Multimeric Model	86
4.6.3	Three-state Dimeric Model	86
4.6.4	Two-state Model with Ligand Binding	88
4.6.5	Four-state (Two-domain Protein) Model	90
4.7	Experimental Protocols	92
4.7.1	How to Prepare for DSC Experiments	92
4.7.2	How to Choose Appropriate Conditions	94
4.7.3	Critical Factors in Running DSC Experiments	94
	References	95

5	**Pressure–Temperature Phase Diagrams of Proteins**	**99**
	Wolfgang Doster and Josef Friedrich	
5.1	Introduction	99
5.2	Basic Aspects of Phase Diagrams of Proteins and Early Experiments	100
5.3	Thermodynamics of Pressure–Temperature Phase Diagrams	103
5.4	Measuring Phase Stability Boundaries with Optical Techniques	110
5.4.1	Fluorescence Experiments with Cytochrome c	110
5.4.2	Results	112
5.5	What Do We Learn from the Stability Diagram?	116
5.5.1	Thermodynamics	116
5.5.2	Determination of the Equilibrium Constant of Denaturation	117
5.5.3	Microscopic Aspects	120
5.5.4	Structural Features of the Pressure-denatured State	122
5.6	Conclusions and Outlook	123
	Acknowledgment	124
	References	124

6	**Weak Interactions in Protein Folding: Hydrophobic Free Energy, van der Waals Interactions, Peptide Hydrogen Bonds, and Peptide Solvation**	**127**
	Robert L. Baldwin	
6.1	Introduction	127
6.2	Hydrophobic Free Energy, Burial of Nonpolar Surface and van der Waals Interactions	128
6.2.1	History	128
6.2.2	Liquid–Liquid Transfer Model	128
6.2.3	Relation between Hydrophobic Free Energy and Molecular Surface Area	130
6.2.4	Quasi-experimental Estimates of the Work of Making a Cavity in Water or in Liquid Alkane	131
6.2.5	Molecular Dynamics Simulations of the Work of Making Cavities in Water	133
6.2.6	Dependence of Transfer Free Energy on the Volume of the Solute	134
6.2.7	Molecular Nature of Hydrophobic Free Energy	136

6.2.8	Simulation of Hydrophobic Clusters 137
6.2.9	ΔC_p and the Temperature-dependent Thermodynamics of Hydrophobic Free Energy 137
6.2.10	Modeling Formation of the Hydrophobic Core from Solvation Free Energy and van der Waals Interactions between Nonpolar Residues 142
6.2.11	Evidence Supporting a Role for van der Waals Interactions in Forming the Hydrophobic Core 144
6.3	Peptide Solvation and the Peptide Hydrogen Bond 145
6.3.1	History 145
6.3.2	Solvation Free Energies of Amides 147
6.3.3	Test of the Hydrogen-Bond Inventory 149
6.3.4	The Born Equation 150
6.3.5	Prediction of Solvation Free Energies of Polar Molecules by an Electrostatic Algorithm 150
6.3.6	Prediction of the Solvation Free Energies of Peptide Groups in Different Backbone Conformations 151
6.3.7	Predicted Desolvation Penalty for Burial of a Peptide H-bond 153
6.3.8	Gas–Liquid Transfer Model 154
	Acknowledgments 156
	References 156

7   Electrostatics of Proteins: Principles, Models and Applications   163
Sonja Braun-Sand and Arieh Warshel

7.1	Introduction 163
7.2	Historical Perspectives 163
7.3	Electrostatic Models: From Microscopic to Macroscopic Models 166
7.3.1	All-Atom Models 166
7.3.2	Dipolar Lattice Models and the PDLD Approach 168
7.3.3	The PDLD/S-LRA Model 170
7.3.4	Continuum (Poisson-Boltzmann) and Related Approaches 171
7.3.5	Effective Dielectric Constant for Charge–Charge Interactions and the GB Model 172
7.4	The Meaning and Use of the Protein Dielectric Constant 173
7.5	Validation Studies 176
7.6	Systems Studied 178
7.6.1	Solvation Energies of Small Molecules 178
7.6.2	Calculation of pK_a Values of Ionizable Residues 179
7.6.3	Redox and Electron Transport Processes 180
7.6.4	Ligand Binding 181
7.6.5	Enzyme Catalysis 182
7.6.6	Ion Pairs 183
7.6.7	Protein–Protein Interactions 184
7.6.8	Ion Channels 185
7.6.9	Helix Macrodipoles versus Localized Molecular Dipoles 185
7.6.10	Folding and Stability 186
7.7	Concluding Remarks 189

Acknowledgments *190*
References *190*

8 Protein Conformational Transitions as Seen from the Solvent: Magnetic Relaxation Dispersion Studies of Water, Co-solvent, and Denaturant Interactions with Nonnative Proteins *201*
Bertil Halle, Vladimir P. Denisov, Kristofer Modig, and Monika Davidovic

8.1 The Role of the Solvent in Protein Folding and Stability *201*
8.2 Information Content of Magnetic Relaxation Dispersion *202*
8.3 Thermal Perturbations *205*
8.3.1 Heat Denaturation *205*
8.3.2 Cold Denaturation *209*
8.4 Electrostatic Perturbations *213*
8.5 Solvent Perturbations *218*
8.5.1 Denaturation Induced by Urea *219*
8.5.2 Denaturation Induced by Guanidinium Chloride *225*
8.5.3 Conformational Transitions Induced by Co-solvents *228*
8.6 Outlook *233*
8.7 Experimental Protocols and Data Analysis *233*
8.7.1 Experimental Methodology *233*
8.7.1.1 Multiple-field MRD *234*
8.7.1.2 Field-cycling MRD *234*
8.7.1.3 Choice of Nuclear Isotope *235*
8.7.2 Data Analysis *236*
8.7.2.1 Exchange Averaging *236*
8.7.2.2 Spectral Density Function *237*
8.7.2.3 Residence Time *239*
8.7.2.4 ^{19}F Relaxation *240*
8.7.2.5 Coexisting Protein Species *241*
8.7.2.6 Preferential Solvation *241*
References *242*

9 Stability and Design of α-Helices *247*
Andrew J. Doig, Neil Errington, and Teuku M. Iqbalsyah

9.1 Introduction *247*
9.2 Structure of the α-Helix *247*
9.2.1 Capping Motifs *248*
9.2.2 Metal Binding *250*
9.2.3 The 3_{10}-Helix *251*
9.2.4 The π-Helix *251*
9.3 Design of Peptide Helices *252*
9.3.1 Host–Guest Studies *253*
9.3.2 Helix Lengths *253*
9.3.3 The Helix Dipole *253*
9.3.4 Acetylation and Amidation *254*
9.3.5 Side Chain Spacings *255*

9.3.6	Solubility	*256*
9.3.7	Concentration Determination	*257*
9.3.8	Design of Peptides to Measure Helix Parameters	*257*
9.3.9	Helix Templates	*259*
9.3.10	Design of 3_{10}-Helices	*259*
9.3.11	Design of π-helices	*261*
9.4	Helix Coil Theory	*261*
9.4.1	Zimm-Bragg Model	*261*
9.4.2	Lifson-Roig Model	*262*
9.4.3	The Unfolded State and Polyproline II Helix	*265*
9.4.4	Single Sequence Approximation	*265*
9.4.5	N- and C-Caps	*266*
9.4.6	Capping Boxes	*266*
9.4.7	Side-chain Interactions	*266*
9.4.8	N1, N2, and N3 Preferences	*267*
9.4.9	Helix Dipole	*267*
9.4.10	3_{10}- and π-Helices	*268*
9.4.11	AGADIR	*268*
9.4.12	Lomize-Mosberg Model	*269*
9.4.13	Extension of the Zimm-Bragg Model	*270*
9.4.14	Availability of Helix/Coil Programs	*270*
9.5	Forces Affecting α-Helix Stability	*270*
9.5.1	Helix Interior	*270*
9.5.2	Caps	*273*
9.5.3	Phosphorylation	*276*
9.5.4	Noncovalent Side-chain Interactions	*276*
9.5.5	Covalent Side-chain interactions	*277*
9.5.6	Capping Motifs	*277*
9.5.7	Ionic Strength	*279*
9.5.8	Temperature	*279*
9.5.9	Trifluoroethanol	*279*
9.5.10	pK_a Values	*280*
9.5.11	Relevance to Proteins	*281*
9.6	Experimental Protocols and Strategies	*281*
9.6.1	Solid Phase Peptide Synthesis (SPPS) Based on the Fmoc Strategy	*281*
9.6.1.1	Equipment and Reagents	*281*
9.6.1.2	Fmoc Deprotection and Coupling	*283*
9.6.1.3	Kaiser Test	*284*
9.6.1.4	Acetylation and Cleavage	*285*
9.6.1.5	Peptide Precipitation	*286*
9.6.2	Peptide Purification	*286*
9.6.2.1	Equipment and Reagents	*286*
9.6.2.2	Method	*286*
9.6.3	Circular Dichroism	*287*
9.6.4	Acquisition of Spectra	*288*

9.6.4.1	Instrumental Considerations	288
9.6.5	Data Manipulation and Analysis	289
9.6.5.1	Protocol for CD Measurement of Helix Content	291
9.6.6	Aggregation Test for Helical Peptides	291
9.6.6.1	Equipment and Reagents	291
9.6.6.2	Method	292
9.6.7	Vibrational Circular Dichroism	292
9.6.8	NMR Spectroscopy	292
9.6.8.1	Nuclear Overhauser Effect	293
9.6.8.2	Amide Proton Exchange Rates	294
9.6.8.3	^{13}C NMR	294
9.6.9	Fourier Transform Infrared Spectroscopy	295
9.6.9.1	Secondary Structure	295
9.6.10	Raman Spectroscopy and Raman Optical Activity	296
9.6.11	pH Titrations	298
9.6.11.1	Equipment and Reagents	298
9.6.11.2	Method	298
	Acknowledgments	299
	References	299
10	**Design and Stability of Peptide β-Sheets**	**314**
	Mark S. Searle	
10.1	Introduction	314
10.2	β-Hairpins Derived from Native Protein Sequences	315
10.3	Role of β-Turns in Nucleating β-Hairpin Folding	316
10.4	Intrinsic ϕ, ψ Propensities of Amino Acids	319
10.5	Side-chain Interactions and β-Hairpin Stability	321
10.5.1	Aromatic Clusters Stabilize β-Hairpins	322
10.5.2	Salt Bridges Enhance Hairpin Stability	325
10.6	Cooperative Interactions in β-Sheet Peptides: Kinetic Barriers to Folding	330
10.7	Quantitative Analysis of Peptide Folding	331
10.8	Thermodynamics of β-Hairpin Folding	332
10.9	Multistranded Antiparallel β-Sheet Peptides	334
10.10	Concluding Remarks: Weak Interactions and Stabilization of Peptide β-Sheets	339
	References	340
11	**Predicting Free Energy Changes of Mutations in Proteins**	**343**
	Raphael Guerois, Joaquim Mendes, and Luis Serrano	
11.1	Physical Forces that Determine Protein Conformational Stability	343
11.1.1	Protein Conformational Stability [1]	343
11.1.2	Structures of the N and D States [2–6]	344
11.1.3	Studies Aimed at Understanding the Physical Forces that Determine Protein Conformational Stability [1, 2, 8, 19–26]	346
11.1.4	Forces Determining Conformational Stability [1, 2, 8, 19–27]	346

11.1.5	Intramolecular Interactions	347
11.1.5.1	van der Waals Interactions	347
11.1.5.2	Electrostatic Interactions	347
11.1.5.3	Conformational Strain	349
11.1.6	Solvation	350
11.1.7	Intramolecular Interactions and Solvation Taken Together	350
11.1.8	Entropy	351
11.1.9	Cavity Formation	352
11.1.10	Summary	353
11.2	Methods for the Prediction of the Effect of Point Mutations on in vitro Protein Stability	353
11.2.1	General Considerations on Protein Plasticity upon Mutation	353
11.2.2	Predictive Strategies	355
11.2.3	Methods	356
11.2.3.1	From Sequence and Multiple Sequence Alignment Analysis	356
11.2.3.2	Statistical Analysis of the Structure Databases	356
11.2.3.3	Helix/Coil Transition Model	357
11.2.3.4	Physicochemical Method Based on Protein Engineering Experiments	359
11.2.3.5	Methods Based only on the Basic Principles of Physics and Thermodynamics	364
11.3	Mutation Effects on in vivo Stability	366
11.3.1	The N-terminal Rule	366
11.3.2	The C-terminal Rule	367
11.3.3	PEST Signals	368
11.4	Mutation Effects on Aggregation	368
	References	369

I/2 Dynamics and Mechanisms of Protein Folding Reactions 377

12.1 Kinetic Mechanisms in Protein Folding 379
Annett Bachmann and Thomas Kiefhaber

12.1.1	Introduction	379
12.1.2	Analysis of Protein Folding Reactions using Simple Kinetic Models	379
12.1.2.1	General Treatment of Kinetic Data	380
12.1.2.2	Two-state Protein Folding	380
12.1.2.3	Complex Folding Kinetics	384
12.1.2.3.1	Heterogeneity in the Unfolded State	384
12.1.2.3.2	Folding through Intermediates	388
12.1.2.3.3	Rapid Pre-equilibria	391
12.1.2.3.4	Folding through an On-pathway High-energy Intermediate	393
12.1.3	A Case Study: the Mechanism of Lysozyme Folding	394
12.1.3.1	Lysozyme Folding at pH 5.2 and Low Salt Concentrations	394
12.1.3.2	Lysozyme Folding at pH 9.2 or at High Salt Concentrations	398
12.1.4	Non-exponential Kinetics	401

12.1.5	Conclusions and Outlook *401*
12.1.6	Protocols – Analytical Solutions of Three-state Protein Folding Models *402*
12.1.6.1	Triangular Mechanism *402*
12.1.6.2	On-pathway Intermediate *403*
12.1.6.3	Off-pathway Mechanism *404*
12.1.6.4	Folding Through an On-pathway High-Energy Intermediate *404*
	Acknowledgments *406*
	References *406*

12.2	**Characterization of Protein Folding Barriers with Rate Equilibrium Free Energy Relationships** *411*
	Thomas Kiefhaber, Ignacio E. Sánchez, and Annett Bachmann
12.2.1	Introduction *411*
12.2.2	Rate Equilibrium Free Energy Relationships *411*
12.2.2.1	Linear Rate Equilibrium Free Energy Relationships in Protein Folding *414*
12.2.2.2	Properties of Protein Folding Transition States Derived from Linear REFERs *418*
12.2.3	Nonlinear Rate Equilibrium Free Energy Relationships in Protein Folding *420*
12.2.3.1	Self-Interaction and Cross-Interaction Parameters *420*
12.2.3.2	Hammond and Anti-Hammond Behavior *424*
12.2.3.3	Sequential and Parallel Transition States *425*
12.2.3.4	Ground State Effects *428*
12.2.4	Experimental Results on the Shape of Free Energy Barriers in Protein Folding *432*
12.2.4.1	Broadness of Free Energy Barriers *432*
12.2.4.2	Parallel Pathways *437*
12.2.5	Folding in the Absence of Enthalpy Barriers *438*
12.2.6	Conclusions and Outlook *438*
	Acknowledgments *439*
	References *439*

13	**A Guide to Measuring and Interpreting ϕ-values** *445*
	Nicholas R. Guydosh and Alan R. Fersht
13.1	Introduction *445*
13.2	Basic Concept of ϕ-Value Analysis *445*
13.3	Further Interpretation of ϕ *448*
13.4	Techniques *450*
13.5	Conclusions *452*
	References *452*

14	**Fast Relaxation Methods** *454*
	Martin Gruebele
14.1	Introduction *454*

14.2	Techniques *455*	
14.2.1	Fast Pressure-Jump Experiments *455*	
14.2.2	Fast Resistive Heating Experiments *456*	
14.2.3	Fast Laser-induced Relaxation Experiments *457*	
14.2.3.1	Laser Photolysis *457*	
14.2.3.2	Electrochemical Jumps *458*	
14.2.3.3	Laser-induced pH Jumps *458*	
14.2.3.4	Covalent Bond Dissociation *459*	
14.2.3.5	Chromophore Excitation *460*	
14.2.3.6	Laser Temperature Jumps *460*	
14.2.4	Multichannel Detection Techniques for Relaxation Studies *461*	
14.2.4.1	Small Angle X-ray Scattering or Light Scattering *462*	
14.2.4.2	Direct Absorption Techniques *463*	
14.2.4.3	Circular Dichroism and Optical Rotatory Dispersion *464*	
14.2.4.4	Raman and Resonance Raman Scattering *464*	
14.2.4.5	Intrinsic Fluorescence *465*	
14.2.4.6	Extrinsic Fluorescence *465*	
14.3	Protein Folding by Relaxation *466*	
14.3.1	Transition State Theory, Energy Landscapes, and Fast Folding *466*	
14.3.2	Viscosity Dependence of Folding Motions *470*	
14.3.3	Resolving Burst Phases *471*	
14.3.4	Fast Folding and Unfolded Proteins *472*	
14.3.5	Experiment and Simulation *472*	
14.4	Summary *474*	
14.5	Experimental Protocols *475*	
14.5.1	Design Criteria for Laser Temperature Jumps *475*	
14.5.2	Design Criteria for Fast Single-Shot Detection Systems *476*	
14.5.3	Designing Proteins for Fast Relaxation Experiments *477*	
14.5.4	Linear Kinetic, Nonlinear Kinetic, and Generalized Kinetic Analysis of Fast Relaxation *477*	
14.5.4.1	The Reaction $D \rightleftharpoons F$ in the Presence of a Barrier *477*	
14.5.4.2	The Reaction $2A \rightleftharpoons A_2$ in the Presence of a Barrier *478*	
14.5.4.3	The Reaction $D \rightleftharpoons F$ at Short Times or over Low Barriers *479*	
14.5.5	Relaxation Data Analysis by Linear Decomposition *480*	
14.5.5.1	Singular Value Decomposition (SVD) *480*	
14.5.5.2	χ-Analysis *481*	
	Acknowledgments *481*	
	References *482*	
15	**Early Events in Protein Folding Explored by Rapid Mixing Methods** *491*	
	Heinrich Roder, Kosuke Maki, Ramil F. Latypov, Hong Cheng, and M. C. Ramachandra Shastry	
15.1	Importance of Kinetics for Understanding Protein Folding *491*	
15.2	Burst-phase Signals in Stopped-flow Experiments *492*	
15.3	Turbulent Mixing *494*	

15.4	Detection Methods *495*	
15.4.1	Tryptophan Fluorescence *495*	
15.4.2	ANS Fluorescence *498*	
15.4.3	FRET *499*	
15.4.4	Continuous-flow Absorbance *501*	
15.4.5	Other Detection Methods used in Ultrafast Folding Studies *502*	
15.5	A Quenched-Flow Method for H-D Exchange Labeling Studies on the Microsecond Time Scale *502*	
15.6	Evidence for Accumulation of Early Folding Intermediates in Small Proteins *505*	
15.6.1	B1 Domain of Protein G *505*	
15.6.2	Ubiquitin *508*	
15.6.3	Cytochrome *c* *512*	
15.7	Significance of Early Folding Events *515*	
15.7.1	Barrier-limited Folding vs. Chain Diffusion *515*	
15.7.2	Chain Compaction: Random Collapse vs. Specific Folding *516*	
15.7.3	Kinetic Role of Early Folding Intermediates *517*	
15.7.4	Broader Implications *520*	
	Appendix *521*	
A1	Design and Calibration of Rapid Mixing Instruments *521*	
A1.1	Stopped-flow Equipment *521*	
A1.2	Continuous-flow Instrumentation *524*	
	Acknowledgments *528*	
	References *528*	
16	**Kinetic Protein Folding Studies using NMR Spectroscopy** *536*	
	Markus Zeeb and Jochen Balbach	
16.1	Introduction *536*	
16.2	Following Slow Protein Folding Reactions in Real Time *538*	
16.3	Two-dimensional Real-time NMR Spectroscopy *545*	
16.4	Dynamic and Spin Relaxation NMR for Quantifying Microsecond-to-Millisecond Folding Rates *550*	
16.5	Conclusions and Future Directions *555*	
16.6	Experimental Protocols *556*	
16.6.1	How to Record and Analyze 1D Real-time NMR Spectra *556*	
16.6.1.1	Acquisition *556*	
16.6.1.2	Processing *557*	
16.6.1.3	Analysis *557*	
16.6.1.4	Analysis of 1D Real-time Diffusion Experiments *558*	
16.6.2	How to Extract Folding Rates from 1D Spectra by Line Shape Analysis *559*	
16.6.2.1	Acquisition *560*	
16.6.2.2	Processing *560*	
16.6.2.3	Analysis *561*	
16.6.3	How to Extract Folding Rates from 2D Real-time NMR Spectra *562*	

16.6.3.1	Acquisition 563
16.6.3.2	Processing 563
16.6.3.3	Analysis 563
16.6.4	How to Analyze Heteronuclear NMR Relaxation and Exchange Data 565
16.6.4.1	Acquisition 566
16.6.4.2	Processing 567
16.6.4.3	Analysis 567
	Acknowledgments 569
	References 569

Part I, Volume 2

17	Fluorescence Resonance Energy Transfer (FRET) and Single Molecule Fluorescence Detection Studies of the Mechanism of Protein Folding and Unfolding 573
	Elisha Haas
	Abbreviations 573
17.1	Introduction 573
17.2	What are the Main Aspects of the Protein Folding Problem that can be Addressed by Methods Based on FRET Measurements? 574
17.2.1	The Three Protein Folding Problems 574
17.2.1.1	The Chain Entropy Problem 574
17.2.1.2	The Function Problem: Conformational Fluctuations 575
17.3	Theoretical Background 576
17.3.1	Nonradiative Excitation Energy Transfer 576
17.3.2	What is FRET? The Singlet–Singlet Excitation Transfer 577
17.3.3	Rate of Nonradiative Excitation Energy Transfer within a Donor–Acceptor Pair 578
17.3.4	The Orientation Factor 583
17.3.5	How to Determine and Control the Value of R_o? 584
17.3.6	Index of Refraction n 584
17.3.7	The Donor Quantum Yield Φ_D^o 586
17.3.8	The Spectral Overlap Integral J 586
17.4	Determination of Intramolecular Distances in Protein Molecules using FRET Measurements 586
17.4.1	Single Distance between Donor and Acceptor 587
17.4.1.1	Method 1: Steady State Determination of Decrease of Donor Emission 587
17.4.1.2	Method 2: Acceptor Excitation Spectroscopy 588
17.4.2	Time-resolved Methods 588
17.4.3	Determination of E from Donor Fluorescence Decay Rates 589
17.4.4	Determination of Acceptor Fluorescence Lifetime 589
17.4.5	Determination of Intramolecular Distance Distributions 590

17.4.6	Evaluation of the Effect of Fast Conformational Fluctuations and Determination of Intramolecular Diffusion Coefficients	*592*
17.5	Experimental Challenges in the Implementation of FRET Folding Experiments	*594*
17.5.1	Optimized Design and Preparation of Labeled Protein Samples for FRET Folding Experiments	*594*
17.5.2	Strategies for Site-specific Double Labeling of Proteins	*595*
17.5.3	Preparation of Double-labeled Mutants Using Engineered Cysteine Residues (strategy 4)	*596*
17.5.4	Possible Pitfalls Associated with the Preparation of Labeled Protein Samples for FRET Folding Experiments	*599*
17.6	Experimental Aspects of Folding Studies by Distance Determination Based on FRET Measurements	*600*
17.6.1	Steady State Determination of Transfer Efficiency	*600*
17.6.1.1	Donor Emission	*600*
17.6.1.2	Acceptor Excitation Spectroscopy	*601*
17.6.2	Time-resolved Measurements	*601*
17.7	Data Analysis	*603*
17.7.1	Rigorous Error Analysis	*606*
17.7.2	Elimination of Systematic Errors	*606*
17.8	Applications of trFRET for Characterization of Unfolded and Partially Folded Conformations of Globular Proteins under Equilibrium Conditions	*607*
17.8.1	Bovine Pancreatic Trypsin Inhibitor	*607*
17.8.2	The Loop Hypothesis	*608*
17.8.3	RNase A	*609*
17.8.4	Staphylococcal Nuclease	*611*
17.9	Unfolding Transition via Continuum of Native-like Forms	*611*
17.10	The Third Folding Problem: Domain Motions and Conformational Fluctuations of Enzyme Molecules	*611*
17.11	Single Molecule FRET-detected Folding Experiments	*613*
17.12	Principles of Applications of Single Molecule FRET Spectroscopy in Folding Studies	*615*
17.12.1	Design and Analysis of Single Molecule FRET Experiments	*615*
17.12.1.1	How is Single Molecule FRET Efficiency Determined?	*615*
17.12.1.2	The Challenge of Extending the Length of the Time Trajectories	*617*
17.12.2	Distance and Time Resolution of the Single Molecule FRET Folding Experiments	*618*
17.13	Folding Kinetics	*619*
17.13.1	Steady State and trFRET-detected Folding Kinetics Experiments	*619*
17.13.2	Steady State Detection	*619*
17.13.3	Time-resolved FRET Detection of Rapid Folding Kinetics: the "Double Kinetics" Experiment	*621*
17.13.4	Multiple Probes Analysis of the Folding Transition	*622*
17.14	Concluding Remarks	*625*

Acknowledgments 626
References 627

18 Application of Hydrogen Exchange Kinetics to Studies of Protein Folding 634
Kaare Teilum, Birthe B. Kragelund, and Flemming M. Poulsen
18.1 Introduction 634
18.2 The Hydrogen Exchange Reaction 638
18.2.1 Calculating the Intrinsic Hydrogen Exchange Rate Constant, k_{int} 638
18.3 Protein Dynamics by Hydrogen Exchange in Native and Denaturing Conditions 641
18.3.1 Mechanisms of Exchange 642
18.3.2 Local Opening and Closing Rates from Hydrogen Exchange Kinetics 642
18.3.2.1 The General Amide Exchange Rate Expression – the Linderstrøm-Lang Equation 643
18.3.2.2 Limits to the General Rate Expression – EX1 and EX2 644
18.3.2.3 The Range between the EX1 and EX2 Limits 646
18.3.2.4 Identification of Exchange Limit 646
18.3.2.5 Global Opening and Closing Rates and Protein Folding 647
18.3.3 The "Native State Hydrogen Exchange" Strategy 648
18.3.3.1 Localization of Partially Unfolded States, PUFs 650
18.4 Hydrogen Exchange as a Structural Probe in Kinetic Folding Experiments 651
18.4.1 Protein Folding/Hydrogen Exchange Competition 652
18.4.2 Hydrogen Exchange Pulse Labeling 656
18.4.3 Protection Factors in Folding Intermediates 657
18.4.4 Kinetic Intermediate Structures Characterized by Hydrogen Exchange 659
18.5 Experimental Protocols 661
18.5.1 How to Determine Hydrogen Exchange Kinetics at Equilibrium 661
18.5.1.1 Equilibrium Hydrogen Exchange Experiments 661
18.5.1.2 Determination of Segmental Opening and Closing Rates, k_{op} and k_{cl} 662
18.5.1.3 Determination of ΔG_{fluc}, m, and $\Delta G°_{unf}$ 662
18.5.2 Planning a Hydrogen Exchange Folding Experiment 662
18.5.2.1 Determine a Combination of t_{pulse} and pH_{pulse} 662
18.5.2.2 Setup Quench Flow Apparatus 662
18.5.2.3 Prepare Deuterated Protein and Chemicals 663
18.5.2.4 Prepare Buffers and Unfolded Protein 663
18.5.2.5 Check pH in the Mixing Steps 664
18.5.2.6 Sample Mixing and Preparation 664
18.5.3 Data Analysis 664
Acknowledgments 665
References 665

19 Studying Protein Folding and Aggregation by Laser Light Scattering *673*
Klaus Gast and Andreas J. Modler

19.1 Introduction *673*
19.2 Basic Principles of Laser Light Scattering *674*
19.2.1 Light Scattering by Macromolecular Solutions *674*
19.2.2 Molecular Parameters Obtained from Static Light Scattering (SLS) *676*
19.2.3 Molecular Parameters Obtained from Dynamic Light Scattering (DLS) *678*
19.2.4 Advantages of Combined SLS and DLS Experiments *680*
19.3 Laser Light Scattering of Proteins in Different Conformational States – Equilibrium Folding/Unfolding Transitions *680*
19.3.1 General Considerations, Hydrodynamic Dimensions in the Natively Folded State *680*
19.3.2 Changes in the Hydrodynamic Dimensions during Heat-induced Unfolding *682*
19.3.3 Changes in the Hydrodynamic Dimensions upon Cold Denaturation *683*
19.3.4 Denaturant-induced Changes of the Hydrodynamic Dimensions *684*
19.3.5 Acid-induced Changes of the Hydrodynamic Dimensions *685*
19.3.6 Dimensions in Partially Folded States – Molten Globules and Fluoroalcohol-induced States *686*
19.3.7 Comparison of the Dimensions of Proteins in Different Conformational States *687*
19.3.8 Scaling Laws for the Native and Highly Unfolded States, Hydrodynamic Modeling *687*
19.4 Studying Folding Kinetics by Laser Light Scattering *689*
19.4.1 General Considerations, Attainable Time Regions *689*
19.4.2 Hydrodynamic Dimensions of the Kinetic Molten Globule of Bovine α-Lactalbumin *690*
19.4.3 RNase A is Only Weakly Collapsed During the Burst Phase of Folding *691*
19.5 Misfolding and Aggregation Studied by Laser Light Scattering *692*
19.5.1 Overview: Some Typical Light Scattering Studies of Protein Aggregation *692*
19.5.2 Studying Misfolding and Amyloid Formation by Laser Light Scattering *693*
19.5.2.1 Overview: Initial States, Critical Oligomers, Protofibrils, Fibrils *693*
19.5.2.2 Aggregation Kinetics of Aβ Peptides *694*
19.5.2.3 Kinetics of Oligomer and Fibril Formation of PGK and Recombinant Hamster Prion Protein *695*
19.5.2.4 Mechanisms of Misfolding and Misassembly, Some General Remarks *698*
19.6 Experimental Protocols *698*
19.6.1 Laser Light Scattering Instrumentation *698*

19.6.1.1	Basic Experimental Set-up, General Requirements	698
19.6.1.2	Supplementary Measurements and Useful Options	700
19.6.1.3	Commercially Available Light Scattering Instrumentation	701
19.6.2	Experimental Protocols for the Determination of Molecular Mass and Stokes Radius of a Protein in a Particular Conformational State	701
	Protocol 1 702	
	Protocol 2 704	
	Acknowledgments 704	
	References 704	

20 Conformational Properties of Unfolded Proteins 710
Patrick J. Fleming and George D. Rose

20.1	Introduction	710
20.1.1	Unfolded vs. Denatured Proteins	710
20.2	Early History	711
20.3	The Random Coil	712
20.3.1	The Random Coil – Theory	713
20.3.1.1	The Random Coil Model Prompts Three Questions	716
20.3.1.2	The Folding Funnel	716
20.3.1.3	Transition State Theory	717
20.3.1.4	Other Examples	717
20.3.1.5	Implicit Assumptions from the Random Coil Model	718
20.3.2	The Random Coil – Experiment	718
20.3.2.1	Intrinsic Viscosity	719
20.3.2.2	SAXS and SANS	720
20.4	Questions about the Random Coil Model	721
20.4.1	Questions from Theory	722
20.4.1.1	The Flory Isolated-pair Hypothesis	722
20.4.1.2	Structure vs. Energy Duality	724
20.4.1.3	The "Rediscovery" of Polyproline II Conformation	724
20.4.1.4	P_{II} in Unfolded Peptides and Proteins	726
20.4.2	Questions from Experiment	727
20.4.2.1	Residual Structure in Denatured Proteins and Peptides	727
20.4.3	The Reconciliation Problem	728
20.4.4	Organization in the Unfolded State – the Entropic Conjecture	728
20.4.4.1	Steric Restrictions beyond the Dipeptide	729
20.5	Future Directions	730
	Acknowledgments 731	
	References 731	

21 Conformation and Dynamics of Nonnative States of Proteins studied by NMR Spectroscopy 737
Julia Wirmer, Christian Schlörb, and Harald Schwalbe

21.1	Introduction	737
21.1.1	Structural Diversity of Polypeptide Chains	737

21.1.2	Intrinsically Unstructured and Natively Unfolded Proteins	739
21.2	Prerequisites: NMR Resonance Assignment	740
21.3	NMR Parameters	744
21.3.1	Chemical shifts δ	745
21.3.1.1	Conformational Dependence of Chemical Shifts	745
21.3.1.2	Interpretation of Chemical Shifts in the Presence of Conformational Averaging	746
21.3.2	J Coupling Constants	748
21.3.2.1	Conformational Dependence of J Coupling Constants	748
21.3.2.2	Interpretation of J Coupling Constants in the Presence of Conformational Averaging	750
21.3.3	Relaxation: Homonuclear NOEs	750
21.3.3.1	Distance Dependence of Homonuclear NOEs	750
21.3.3.2	Interpretation of Homonuclear NOEs in the Presence of Conformational Averaging	754
21.3.4	Heteronuclear Relaxation (^{15}N R_1, R_2, hetNOE)	757
21.3.4.1	Correlation Time Dependence of Heteronuclear Relaxation Parameters	757
21.3.4.2	Dependence on Internal Motions of Heteronuclear Relaxation Parameters	759
21.3.5	Residual Dipolar Couplings	760
21.3.5.1	Conformational Dependence of Residual Dipolar Couplings	760
21.3.5.2	Interpretation of Residual Dipolar Couplings in the Presence of Conformational Averaging	763
21.3.6	Diffusion	765
21.3.7	Paramagnetic Spin Labels	766
21.3.8	H/D Exchange	767
21.3.9	Photo-CIDNP	767
21.4	Model for the Random Coil State of a Protein	768
21.5	Nonnative States of Proteins: Examples from Lysozyme, α-Lactalbumin, and Ubiquitin	771
21.5.1	Backbone Conformation	772
21.5.1.1	Interpretation of Chemical Shifts	772
21.5.1.2	Interpretation of NOEs	774
21.5.1.3	Interpretation of J Coupling Constants	780
21.5.2	Side-chain Conformation	784
21.5.2.1	Interpretation of J Coupling Constants	784
21.5.3	Backbone Dynamics	786
21.5.3.1	Interpretation of ^{15}N Relaxation Rates	786
21.6	Summary and Outlook	793
	Acknowledgments	794
	References	794
22	**Dynamics of Unfolded Polypeptide Chains**	**809**
	Beat Fierz and Thomas Kiefhaber	
22.1	Introduction	809

22.2	Equilibrium Properties of Chain Molecules	809
22.2.1	The Freely Jointed Chain	810
22.2.2	Chain Stiffness	810
22.2.3	Polypeptide Chains	811
22.2.4	Excluded Volume Effects	812
22.3	Theory of Polymer Dynamics	813
22.3.1	The Langevin Equation	813
22.3.2	Rouse Model and Zimm Model	814
22.3.3	Dynamics of Loop Closure and the Szabo-Schulten-Schulten Theory	815
22.4	Experimental Studies on the Dynamics in Unfolded Polypeptide Chains	816
22.4.1	Experimental Systems for the Study of Intrachain Diffusion	816
22.4.1.1	Early Experimental Studies	816
22.4.1.2	Triplet Transfer and Triplet Quenching Studies	821
22.4.1.3	Fluorescence Quenching	825
22.4.2	Experimental Results on Dynamic Properties of Unfolded Polypeptide Chains	825
22.4.2.1	Kinetics of Intrachain Diffusion	826
22.4.2.2	Effect of Loop Size on the Dynamics in Flexible Polypeptide Chains	826
22.4.2.3	Effect of Amino Acid Sequence on Chain Dynamics	829
22.4.2.4	Effect of the Solvent on Intrachain Diffusion	831
22.4.2.5	Effect of Solvent Viscosity on Intrachain Diffusion	833
22.4.2.6	End-to-end Diffusion vs. Intrachain Diffusion	834
22.4.2.7	Chain Diffusion in Natural Protein Sequences	834
22.5	Implications for Protein Folding Kinetics	837
22.5.1	Rate of Contact Formation during the Earliest Steps in Protein Folding	837
22.5.2	The Speed Limit of Protein Folding vs. the Pre-exponential Factor	839
22.5.3	Contributions of Chain Dynamics to Rate- and Equilibrium Constants for Protein Folding Reactions	840
22.6	Conclusions and Outlook	844
22.7	Experimental Protocols and Instrumentation	844
22.7.1	Properties of the Electron Transfer Probes and Treatment of the Transfer Kinetics	845
22.7.2	Test for Diffusion-controlled Reactions	847
22.7.2.1	Determination of Bimolecular Quenching or Transfer Rate Constants	847
22.7.2.2	Testing the Viscosity Dependence	848
22.7.2.3	Determination of Activation Energy	848
22.7.3	Instrumentation	849
	Acknowledgments	849
	References	849

23	Equilibrium and Kinetically Observed Molten Globule States 856
	Kosuke Maki, Kiyoto Kamagata, and Kunihiro Kuwajima
23.1	Introduction 856
23.2	Equilibrium Molten Globule State 858
23.2.1	Structural Characteristics of the Molten Globule State 858
23.2.2	Typical Examples of the Equilibrium Molten Globule State 859
23.2.3	Thermodynamic Properties of the Molten Globule State 860
23.3	The Kinetically Observed Molten Globule State 862
23.3.1	Observation and Identification of the Molten Globule State in Kinetic Refolding 862
23.3.2	Kinetics of Formation of the Early Folding Intermediates 863
23.3.3	Late Folding Intermediates and Structural Diversity 864
23.3.4	Evidence for the On-pathway Folding Intermediate 865
23.4	Two-stage Hierarchical Folding Funnel 866
23.5	Unification of the Folding Mechanism between Non-two-state and Two-state Proteins 867
23.5.1	Statistical Analysis of the Folding Data of Non-two-state and Two-state Proteins 868
23.5.2	A Unified Mechanism of Protein Folding: Hierarchy 870
23.5.3	Hidden Folding Intermediates in Two-state Proteins 871
23.6	Practical Aspects of the Experimental Study of Molten Globules 872
23.6.1	Observation of the Equilibrium Molten Globule State 872
23.6.1.1	Two-state Unfolding Transition 872
23.6.1.2	Multi-state (Three-state) Unfolding Transition 874
23.6.2	Burst-phase Intermediate Accumulated during the Dead Time of Refolding Kinetics 876
23.6.3	Testing the Identity of the Molten Globule State with the Burst-Phase Intermediate 877
	References 879
24	**Alcohol- and Salt-induced Partially Folded Intermediates** 884
	Daizo Hamada and Yuji Goto
24.1	Introduction 884
24.2	Alcohol-induced Intermediates of Proteins and Peptides 886
24.2.1	Formation of Secondary Structures by Alcohols 888
24.2.2	Alcohol-induced Denaturation of Proteins 888
24.2.3	Formation of Compact Molten Globule States 889
24.2.4	Example: β-Lactoglobulin 890
24.3	Mechanism of Alcohol-induced Conformational Change 893
24.4	Effects of Alcohols on Folding Kinetics 896
24.5	Salt-induced Formation of the Intermediate States 899
24.5.1	Acid-denatured Proteins 899
24.5.2	Acid-induced Unfolding and Refolding Transitions 900
24.6	Mechanism of Salt-induced Conformational Change 904
24.7	Generality of the Salt Effects 906

24.8	Conclusion 907
	References 908

25 Prolyl Isomerization in Protein Folding 916
Franz Schmid

25.1	Introduction 916
25.2	Prolyl Peptide Bonds 917
25.3	Prolyl Isomerizations as Rate-determining Steps of Protein Folding 918
25.3.1	The Discovery of Fast and Slow Refolding Species 918
25.3.2	Detection of Proline-limited Folding Processes 919
25.3.3	Proline-limited Folding Reactions 921
25.3.4	Interrelation between Prolyl Isomerization and Conformational Folding 923
25.4	Examples of Proline-limited Folding Reactions 924
25.4.1	Ribonuclease A 924
25.4.2	Ribonuclease T1 926
25.4.3	The Structure of a Folding Intermediate with an Incorrect Prolyl Isomer 928
25.5	Native-state Prolyl Isomerizations 929
25.6	Nonprolyl Isomerizations in Protein Folding 930
25.7	Catalysis of Protein Folding by Prolyl Isomerases 932
25.7.1	Prolyl Isomerases as Tools for Identifying Proline-limited Folding Steps 932
25.7.2	Specificity of Prolyl Isomerases 933
25.7.3	The Trigger Factor 934
25.7.4	Catalysis of Prolyl Isomerization During de novo Protein Folding 935
25.8	Concluding Remarks 936
25.9	Experimental Protocols 936
25.9.1	Slow Refolding Assays ("Double Jumps") to Measure Prolyl Isomerizations in an Unfolded Protein 936
25.9.1.1	Guidelines for the Design of Double Jump Experiments 937
25.9.1.2	Formation of U_S Species after Unfolding of RNase A 938
25.9.2	Slow Unfolding Assays for Detecting and Measuring Prolyl Isomerizations in Refolding 938
25.9.2.1	Practical Considerations 939
25.9.2.2	Kinetics of the Formation of Fully Folded IIHY-G3P* Molecules 939
	References 939

26 Folding and Disulfide Formation 946
Margherita Ruoppolo, Piero Pucci, and Gennaro Marino

26.1	Chemistry of the Disulfide Bond 946
26.2	Trapping Protein Disulfides 947
26.3	Mass Spectrometric Analysis of Folding Intermediates 948
26.4	Mechanism(s) of Oxidative Folding so Far – Early and Late Folding Steps 949

26.5	Emerging Concepts from Mass Spectrometric Studies	950
26.5.1	Three-fingered Toxins	951
26.5.2	RNase A	953
26.5.3	Antibody Fragments	955
26.5.4	Human Nerve Growth Factor	956
26.6	Unanswered Questions	956
26.7	Concluding Remarks	957
26.8	Experimental Protocols	957
26.8.1	How to Prepare Folding Solutions	957
26.8.2	How to Carry Out Folding Reactions	958
26.8.3	How to Choose the Best Mass Spectrometric Equipment for Your Study	959
26.8.4	How to Perform Electrospray (ES)MS Analysis	959
26.8.5	How to Perform Matrix-assisted Laser Desorption Ionization (MALDI) MS Analysis	960
	References	961

27	**Concurrent Association and Folding of Small Oligomeric Proteins**	965
	Hans Rudolf Bosshard	
27.1	Introduction	965
27.2	Experimental Methods Used to Follow the Folding of Oligomeric Proteins	966
27.2.1	Equilibrium Methods	966
27.2.2	Kinetic Methods	968
27.3	Dimeric Proteins	969
27.3.1	Two-state Folding of Dimeric Proteins	970
27.3.1.1	Examples of Dimeric Proteins Obeying Two-state Folding	971
27.3.2	Folding of Dimeric Proteins through Intermediate States	978
27.4	Trimeric and Tetrameric Proteins	983
27.5	Concluding Remarks	986
	Appendix – Concurrent Association and Folding of Small Oligomeric Proteins	987
A1	Equilibrium Constants for Two-state Folding	988
A1.1	Homooligomeric Protein	988
A1.2	Heterooligomeric Protein	989
A2	Calculation of Thermodynamic Parameters from Equilibrium Constants	990
A2.1	Basic Thermodynamic Relationships	990
A2.2	Linear Extrapolation of Denaturant Unfolding Curves of Two-state Reaction	990
A2.3	Calculation of the van't Hoff Enthalpy Change from Thermal Unfolding Data	990
A2.4	Calculation of the van't Hoff Enthalpy Change from the Concentration-dependence of T_m	991
A2.5	Extrapolation of Thermodynamic Parameters to Different Temperatures: Gibbs-Helmholtz Equation	991

A3	Kinetics of Reversible Two-state Folding and Unfolding: Integrated Rate Equations *992*	
A3.1	Two-state Folding of Dimeric Protein *992*	
A3.2	Two-state Unfolding of Dimeric Protein *992*	
A3.3	Reversible Two-state Folding and Unfolding *993*	
A3.3.1	Homodimeric protein *993*	
A3.3.2	Heterodimeric protein *993*	
A4	Kinetics of Reversible Two-state Folding: Relaxation after Disturbance of a Pre-existing Equilibrium (Method of Bernasconi) *994*	
	Acknowledgments *995*	
	References *995*	

28	**Folding of Membrane Proteins** *998*	
	Lukas K. Tamm and Heedeok Hong	
28.1	Introduction *998*	
28.2	Thermodyamics of Residue Partitioning into Lipid Bilayers *1000*	
28.3	Stability of β-Barrel Proteins *1001*	
28.4	Stability of Helical Membrane Proteins *1009*	
28.5	Helix and Other Lateral Interactions in Membrane Proteins *1010*	
28.6	The Membrane Interface as an Important Contributor to Membrane Protein Folding *1012*	
28.7	Membrane Toxins as Models for Helical Membrane Protein Insertion *1013*	
28.8	Mechanisms of β-Barrel Membrane Protein Folding *1015*	
28.9	Experimental Protocols *1016*	
28.9.1	SDS Gel Shift Assay for Heat-modifiable Membrane Proteins *1016*	
28.9.1.1	Reversible Folding and Unfolding Protocol Using OmpA as an Example *1016*	
28.9.2	Tryptophan Fluorescence and Time-resolved Distance Determination by Tryptophan Fluorescence Quenching *1018*	
28.9.2.1	TDFQ Protocol for Monitoring the Translocation of Tryptophans across Membranes *1019*	
28.9.3	Circular Dichroism Spectroscopy *1020*	
28.9.4	Fourier Transform Infrared Spectroscopy *1022*	
28.9.4.1	Protocol for Obtaining Conformation and Orientation of Membrane Proteins and Peptides by Polarized ATR-FTIR Spectroscopy *1023*	
	Acknowledgments *1025*	
	References *1025*	

29	**Protein Folding Catalysis by Pro-domains** *1032*	
	Philip N. Bryan	
29.1	Introduction *1032*	
29.2	Bimolecular Folding Mechanisms *1033*	
29.3	Structures of Reactants and Products *1033*	
29.3.1	Structure of Free SBT *1033*	

29.3.2	Structure of SBT/Pro-domain Complex	*1036*
29.3.3	Structure of Free ALP	*1037*
29.3.4	Structure of the ALP/Pro-domain Complex	*1037*
29.4	Stability of the Mature Protease	*1039*
29.4.1	Stability of ALP	*1039*
29.4.2	Stability of Subtilisin	*1040*
29.5	Analysis of Pro-domain Binding to the Folded Protease	*1042*
29.6	Analysis of Folding Steps	*1043*
29.7	Why are Pro-domains Required for Folding?	*1046*
29.8	What is the Origin of High Cooperativity?	*1047*
29.9	How Does the Pro-domain Accelerate Folding?	*1048*
29.10	Are High Kinetic Stability and Facile Folding Mutually Exclusive?	*1049*
29.11	Experimental Protocols for Studying SBT Folding	*1049*
29.11.1	Fermentation and Purification of Active Subtilisin	*1049*
29.11.2	Fermentation and Purification of Facile-folding Ala221 Subtilisin from E. coli	*1050*
29.11.3	Mutagenesis and Protein Expression of Pro-domain Mutants	*1051*
29.11.4	Purification of Pro-domain	*1052*
29.11.5	Kinetics of Pro-domain Binding to Native SBT	*1052*
29.11.6	Kinetic Analysis of Pro-domain Facilitated Subtilisin Folding	*1052*
29.11.6.1	Single Mixing	*1052*
29.11.6.2	Double Jump: Renaturation–Denaturation	*1053*
29.11.6.3	Double Jump: Denaturation–Renaturation	*1053*
29.11.6.4	Triple Jump: Denaturation–Renaturation–Denaturation	*1054*
	References	*1054*
30	**The Thermodynamics and Kinetics of Collagen Folding**	*1059*
	Hans Peter Bächinger and Jürgen Engel	
30.1	Introduction	*1059*
30.1.1	The Collagen Family	*1059*
30.1.2	Biosynthesis of Collagens	*1060*
30.1.3	The Triple Helical Domain in Collagens and Other Proteins	*1061*
30.1.4	N- and C-Propeptide, Telopeptides, Flanking Coiled-Coil Domains	*1061*
30.1.5	Why is the Folding of the Triple Helix of Interest?	*1061*
30.2	Thermodynamics of Collagen Folding	*1062*
30.2.1	Stability of the Triple Helix	*1062*
30.2.2	The Role of Posttranslational Modifications	*1063*
30.2.3	Energies Involved in the Stability of the Triple Helix	*1063*
30.2.4	Model Peptides Forming the Collagen Triple Helix	*1066*
30.2.4.1	Type of Peptides	*1066*
30.2.4.2	The All-or-none Transition of Short Model Peptides	*1066*
30.2.4.3	Thermodynamic Parameters for Different Model Systems	*1069*
30.2.4.4	Contribution of Different Tripeptide Units to Stability	*1075*

30.2.4.5	Crystal and NMR Structures of Triple Helices	*1076*
30.2.4.6	Conformation of the Randomly Coiled Chains	*1077*
30.2.4.7	Model Studies with Isomers of Hydroxyproline and Fluoroproline	*1078*
30.2.4.8	*Cis* ⇌ *trans* Equilibria of Peptide Bonds	*1079*
30.2.4.9	Interpretations of Stabilities on a Molecular Level	*1080*
30.3	Kinetics of Triple Helix Formation	*1081*
30.3.1	Properties of Collagen Triple Helices that Influence Kinetics	*1081*
30.3.2	Folding of Triple Helices from Single Chains	*1082*
30.3.2.1	Early Work	*1082*
30.3.2.2	Concentration Dependence of the Folding of $(PPG)_{10}$ and $(POG)_{10}$	*1082*
30.3.2.3	Model Mechanism of the Folding Kinetics	*1085*
30.3.2.4	Rate Constants of Nucleation and Propagation	*1087*
30.3.2.5	Host–guest Peptides and an Alternative Kinetics Model	*1088*
30.3.3	Triple Helix Formation from Linked Chains	*1089*
30.3.3.1	The Short N-terminal Triple Helix of Collagen III in Fragment Col1–3	*1089*
30.3.3.2	Folding of the Central Long Triple Helix of Collagen III	*1090*
30.3.3.3	The Zipper Model	*1092*
30.3.4	Designed Collagen Models with Chains Connected by a Disulfide Knot or by Trimerizing Domains	*1097*
30.3.4.1	Disulfide-linked Model Peptides	*1097*
30.3.4.2	Model Peptides Linked by a Foldon Domain	*1098*
30.3.4.3	Collagen Triple Helix Formation can be Nucleated at either End	*1098*
30.3.4.4	Hysteresis of Triple Helix Formation	*1099*
30.3.5	Influence of *cis–trans* Isomerase and Chaperones	*1100*
30.3.6	Mutations in Collagen Triple Helices Affect Proper Folding	*1101*
	References	*1101*
31	**Unfolding Induced by Mechanical Force**	*1111*
	Jane Clarke and Phil M. Williams	
31.1	Introduction	*1111*
31.2	Experimental Basics	*1112*
31.2.1	Instrumentation	*1112*
31.2.2	Sample Preparation	*1113*
31.2.3	Collecting Data	*1114*
31.2.4	Anatomy of a Force Trace	*1115*
31.2.5	Detecting Intermediates in a Force Trace	*1115*
31.2.6	Analyzing the Force Trace	*1116*
31.3	Analysis of Force Data	*1117*
31.3.1	Basic Theory behind Dynamic Force Spectroscopy	*1117*
31.3.2	The Ramp of Force Experiment	*1119*
31.3.3	The Golden Equation of DFS	*1121*
31.3.4	Nonlinear Loading	*1122*

31.3.4.1	The Worm-line Chain (WLC)	*1123*
31.3.5	Experiments under Constant Force	*1124*
31.3.6	Effect of Tandem Repeats on Kinetics	*1125*
31.3.7	Determining the Modal Force	*1126*
31.3.8	Comparing Behavior	*1127*
31.3.9	Fitting the Data	*1127*
31.4	Use of Complementary Techniques	*1129*
31.4.1	Protein Engineering	*1130*
31.4.1.1	Choosing Mutants	*1130*
31.4.1.2	Determining $\Delta\Delta G_{D-N}$	*1131*
31.4.1.3	Determining $\Delta\Delta G_{TS-N}$	*1131*
31.4.1.4	Interpreting the Φ-values	*1132*
31.4.2	Computer Simulation	*1133*
31.5	Titin I27: A Case Study	*1134*
31.5.1	The Protein System	*1134*
31.5.2	The Unfolding Intermediate	*1135*
31.5.3	The Transition State	*1136*
31.5.4	The Relationship Between the Native and Transition States	*1137*
31.5.5	The Energy Landscape under Force	*1139*
31.6	Conclusions – the Future	*1139*
	References	*1139*

32 Molecular Dynamics Simulations to Study Protein Folding and Unfolding *1143*
Amedeo Caflisch and Emanuele Paci

32.1	Introduction	*1143*
32.2	Molecular Dynamics Simulations of Peptides and Proteins	*1144*
32.2.1	Folding of Structured Peptides	*1144*
32.2.1.1	Reversible Folding and Free Energy Surfaces	*1144*
32.2.1.2	Non-Arrhenius Temperature Dependence of the Folding Rate	*1147*
32.2.1.3	Denatured State and Levinthal Paradox	*1148*
32.2.1.4	Folding Events of Trp-cage	*1149*
32.2.2	Unfolding Simulations of Proteins	*1150*
32.2.2.1	High-temperature Simulations	*1150*
32.2.2.2	Biased Unfolding	*1150*
32.2.2.3	Forced Unfolding	*1151*
32.2.3	Determination of the Transition State Ensemble	*1153*
32.3	MD Techniques and Protocols	*1155*
32.3.1	Techniques to Improve Sampling	*1155*
32.3.1.1	Replica Exchange Molecular Dynamics	*1155*
32.3.1.2	Methods Based on Path Sampling	*1157*
32.3.2	MD with Restraints	*1157*
32.3.3	Distributed Computing Approach	*1158*
32.3.4	Implicit Solvent Models versus Explicit Water	*1160*
32.4	Conclusion	*1162*
	References	*1162*

33	**Molecular Dynamics Simulations of Proteins and Peptides: Problems, Achievements, and Perspectives** *1170*	
	Paul Tavan, Heiko Carstens, and Gerald Mathias	
33.1	Introduction *1170*	
33.2	Basic Physics of Protein Structure and Dynamics *1171*	
33.2.1	Protein Electrostatics *1172*	
33.2.2	Relaxation Times and Spatial Scales *1172*	
33.2.3	Solvent Environment *1173*	
33.2.4	Water *1174*	
33.2.5	Polarizability of the Peptide Groups and of Other Protein Components *1175*	
33.3	State of the Art *1177*	
33.3.1	Control of Thermodynamic Conditions *1177*	
33.3.2	Long-range Electrostatics *1177*	
33.3.3	Polarizability *1179*	
33.3.4	Higher Multipole Moments of the Molecular Components *1180*	
33.3.5	MM Models of Water *1181*	
33.3.6	Complexity of Protein–Solvent Systems and Consequences for MM-MD *1182*	
33.3.7	What about Successes of MD Methods? *1182*	
33.3.8	Accessible Time Scales and Accuracy Issues *1184*	
33.3.9	Continuum Solvent Models *1185*	
33.3.10	Are there Further Problems beyond Electrostatics and Structure Prediction? *1187*	
33.4	Conformational Dynamics of a Light-switchable Model Peptide *1187*	
33.4.1	Computational Methods *1188*	
33.4.2	Results and Discussion *1190*	
	Summary *1194*	
	Acknowledgments *1194*	
	References *1194*	

Part II, Volume 1

Contributors of Part II *LVIII*

1	**Paradigm Changes from "Unboiling an Egg" to "Synthesizing a Rabbit"** *3*	
	Rainer Jaenicke	
1.1	Protein Structure, Stability, and Self-organization *3*	
1.2	Autonomous and Assisted Folding and Association *6*	
1.3	Native, Intermediate, and Denatured States *11*	
1.4	Folding and Merging of Domains – Association of Subunits *13*	
1.5	Limits of Reconstitution *19*	
1.6	In Vitro Denaturation-Renaturation vs. Folding in Vivo *21*	

1.7	Perspectives 24
	Acknowledgements 26
	References 26

2	**Folding and Association of Multi-domain and Oligomeric Proteins** 32
	Hauke Lilie and Robert Seckler
2.1	Introduction 32
2.2	Folding of Multi-domain Proteins 33
2.2.1	Domain Architecture 33
2.2.2	γ-Crystallin as a Model for a Two-domain Protein 35
2.2.3	The Giant Protein Titin 39
2.3	Folding and Association of Oligomeric Proteins 41
2.3.1	Why Oligomers? 41
2.3.2	Inter-subunit Interfaces 42
2.3.3	Domain Swapping 44
2.3.4	Stability of Oligomeric Proteins 45
2.3.5	Methods Probing Folding/Association 47
2.3.5.1	Chemical Cross-linking 47
2.3.5.2	Analytical Gel Filtration Chromatography 47
2.3.5.3	Scattering Methods 48
2.3.5.4	Fluorescence Resonance Energy Transfer 48
2.3.5.5	Hybrid Formation 48
2.3.6	Kinetics of Folding and Association 49
2.3.6.1	General Considerations 49
2.3.6.2	Reconstitution Intermediates 50
2.3.6.3	Rates of Association 52
2.3.6.4	Homo- Versus Heterodimerization 52
2.4	Renaturation versus Aggregation 54
2.5	Case Studies on Protein Folding and Association 54
2.5.1	Antibody Fragments 54
2.5.2	Trimeric Tail Spike Protein of Bacteriophage P22 59
2.6	Experimental Protocols 62
	References 65

3	**Studying Protein Folding in Vivo** 73
	I. Marije Liscaljet, Bertrand Kleizen, and Ineke Braakman
3.1	Introduction 73
3.2	General Features in Folding Proteins Amenable to in Vivo Study 73
3.2.1	Increasing Compactness 76
3.2.2	Decreasing Accessibility to Different Reagents 76
3.2.3	Changes in Conformation 77
3.2.4	Assistance During Folding 78
3.3	Location-specific Features in Protein Folding 79
3.3.1	Translocation and Signal Peptide Cleavage 79
3.3.2	Glycosylation 80

3.3.3	Disulfide Bond Formation in the ER	81
3.3.4	Degradation	82
3.3.5	Transport from ER to Golgi and Plasma Membrane	83
3.4	How to Manipulate Protein Folding	84
3.4.1	Pharmacological Intervention (Low-molecular-weight Reagents)	84
3.4.1.1	Reducing and Oxidizing Agents	84
3.4.1.2	Calcium Depletion	84
3.4.1.3	ATP Depletion	85
3.4.1.4	Cross-linking	85
3.4.1.5	Glycosylation Inhibitors	85
3.4.2	Genetic Modifications (High-molecular-weight Manipulations)	86
3.4.2.1	Substrate Protein Mutants	86
3.4.2.2	Changing the Concentration or Activity of Folding Enzymes and Chaperones	87
3.5	Experimental Protocols	88
3.5.1	Protein-labeling Protocols	88
3.5.1.1	Basic Protocol Pulse Chase: Adherent Cells	88
3.5.1.2	Pulse Chase in Suspension Cells	91
3.5.2	(Co)-immunoprecipitation and Accessory Protocols	93
3.5.2.1	Immunoprecipitation	93
3.5.2.2	Co-precipitation with Calnexin ([84]; adapted from Ou et al. [85])	94
3.5.2.3	Co-immunoprecipitation with Other Chaperones	95
3.5.2.4	Protease Resistance	95
3.5.2.5	Endo H Resistance	96
3.5.2.6	Cell Surface Expression Tested by Protease	96
3.5.3	SDS-PAGE [13]	97
	Acknowledgements	98
	References	98
4	**Characterization of ATPase Cycles of Molecular Chaperones by Fluorescence and Transient Kinetic Methods**	**105**
	Sandra Schlee and Jochen Reinstein	
4.1	Introduction	105
4.1.1	Characterization of ATPase Cycles of Energy-transducing Systems	105
4.1.2	The Use of Fluorescent Nucleotide Analogues	106
4.1.2.1	Fluorescent Modifications of Nucleotides	106
4.1.2.2	How to Find a Suitable Analogue for a Specific Protein	108
4.2	Characterization of ATPase Cycles of Molecular Chaperones	109
4.2.1	Biased View	109
4.2.2	The ATPase Cycle of DnaK	109
4.2.3	The ATPase Cycle of the Chaperone Hsp90	109
4.2.4	The ATPase Cycle of the Chaperone ClpB	111
4.2.4.1	ClpB, an Oligomeric ATPase With Two AAA Modules Per Protomer	111

4.2.4.2	Nucleotide-binding Properties of NBD1 and NBD2	*111*
4.2.4.3	Cooperativity of ATP Hydrolysis and Interdomain Communication	*114*
4.3	Experimental Protocols	*116*
4.3.1	Synthesis of Fluorescent Nucleotide Analogues	*116*
4.3.1.1	Synthesis and Characterization of (P_β)MABA-ADP and (P_γ)MABA-ATP	*116*
4.3.1.2	Synthesis and Characterization of N8-MABA Nucleotides	*119*
4.3.1.3	Synthesis of MANT Nucleotides	*120*
4.3.2	Preparation of Nucleotides and Proteins	*121*
4.3.2.1	Assessment of Quality of Nucleotide Stock Solution	*121*
4.3.2.2	Determination of the Nucleotide Content of Proteins	*122*
4.3.2.3	Nucleotide Depletion Methods	*123*
4.3.3	Steady-state ATPase Assays	*124*
4.3.3.1	Coupled Enzymatic Assay	*124*
4.3.3.2	Assays Based on $[\alpha\text{-}^{32}P]$-ATP and TLC	*125*
4.3.3.3	Assays Based on Released P_i	*125*
4.3.4	Single-turnover ATPase Assays	*126*
4.3.4.1	Manual Mixing Procedures	*126*
4.3.4.2	Quenched Flow	*127*
4.3.5	Nucleotide-binding Measurements	*127*
4.3.5.1	Isothermal Titration Calorimetry	*127*
4.3.5.2	Equilibrium Dialysis	*129*
4.3.5.3	Filter Binding	*129*
4.3.5.4	Equilibrium Fluorescence Titration	*130*
4.3.5.5	Competition Experiments	*132*
4.3.6	Analytical Solutions of Equilibrium Systems	*133*
4.3.6.1	Quadratic Equation	*133*
4.3.6.2	Cubic Equation	*134*
4.3.6.3	Iterative Solutions	*138*
4.3.7	Time-resolved Binding Measurements	*141*
4.3.7.1	Introduction	*141*
4.3.7.2	One-step Irreversible Process	*142*
4.3.7.3	One-step Reversible Process	*143*
4.3.7.4	Reversible Second Order Reduced to Pseudo-first Order	*144*
4.3.7.5	Two Simultaneous Irreversible Pathways – Partitioning	*146*
4.3.7.6	Two-step Consecutive (Sequential) Reaction	*148*
4.3.7.7	Two-step Binding Reactions	*150*
	References	*152*
5	**Analysis of Chaperone Function in Vitro**	*162*
	Johannes Buchner and Stefan Walter	
5.1	Introduction	*162*
5.2	Basic Functional Principles of Molecular Chaperones	*164*
5.2.1	Recognition of Nonnative Proteins	*166*

5.2.2	Induction of Conformational Changes in the Substrate	167
5.2.3	Energy Consumption and Regulation of Chaperone Function	169
5.3	Limits and Extensions of the Chaperone Concept	170
5.3.1	Co-chaperones	171
5.3.2	Specific Chaperones	171
5.4	Working with Molecular Chaperones	172
5.4.1	Natural versus Artificial Substrate Proteins	172
5.4.2	Stability of Chaperones	172
5.5	Assays to Assess and Characterize Chaperone Function	174
5.5.1	Generating Nonnative Conformations of Proteins	174
5.5.2	Aggregation Assays	174
5.5.3	Detection of Complexes Between Chaperone and Substrate	175
5.5.4	Refolding of Denatured Substrates	175
5.5.5	ATPase Activity and Effect of Substrate and Cofactors	176
5.6	Experimental Protocols	176
5.6.1	General Considerations	176
5.6.1.1	Analysis of Chaperone Stability	176
5.6.1.2	Generation of Nonnative Proteins	177
5.6.1.3	Model Substrates for Chaperone Assays	177
5.6.2	Suppression of Aggregation	179
5.6.3	Complex Formation between Chaperones and Polypeptide Substrates	183
5.6.4	Identification of Chaperone-binding Sites	184
5.6.5	Chaperone-mediated Refolding of Test Proteins	186
5.6.6	ATPase Activity	188
	Acknowledgments	188
	References	189
6	**Physical Methods for Studies of Fiber Formation and Structure**	**197**
	Thomas Scheibel and Louise Serpell	
6.1	Introduction	197
6.2	Overview: Protein Fibers Formed in Vivo	198
6.2.1	Amyloid Fibers	198
6.2.2	Silks	199
6.2.3	Collagens	199
6.2.4	Actin, Myosin, and Tropomyosin Filaments	200
6.2.5	Intermediate Filaments/Nuclear Lamina	202
6.2.6	Fibrinogen/Fibrin	203
6.2.7	Microtubules	203
6.2.8	Elastic Fibers	204
6.2.9	Flagella and Pili	204
6.2.10	Filamentary Structures in Rod-like Viruses	205
6.2.11	Protein Fibers Used by Viruses and Bacteriophages to Bind to Their Hosts	206
6.3	Overview: Fiber Structures	206

6.3.1	Study of the Structure of β-sheet-containing Proteins	*207*
6.3.1.1	Amyloid *207*	
6.3.1.2	Paired Helical Filaments *207*	
6.3.1.3	β-Silks *207*	
6.3.1.4	β-Sheet-containing Viral Fibers *208*	
6.3.2	α-Helix-containing Protein Fibers *209*	
6.3.2.1	Collagen *209*	
6.3.2.2	Tropomyosin *210*	
6.3.2.3	Intermediate Filaments *210*	
6.3.3	Protein Polymers Consisting of a Mixture of Secondary Structure *211*	
6.3.3.1	Tubulin *211*	
6.3.3.2	Actin and Myosin Filaments *212*	
6.4	Methods to Study Fiber Assembly *213*	
6.4.1	Circular Dichroism Measurements for Monitoring Structural Changes Upon Fiber Assembly *213*	
6.4.1.1	Theory of CD *213*	
6.4.1.2	Experimental Guide to Measure CD Spectra and Structural Transition Kinetics *214*	
6.4.2	Intrinsic Fluorescence Measurements to Analyze Structural Changes *215*	
6.4.2.1	Theory of Protein Fluorescence *215*	
6.4.2.2	Experimental Guide to Measure Trp Fluorescence *216*	
6.4.3	Covalent Fluorescent Labeling to Determine Structural Changes of Proteins with Environmentally Sensitive Fluorophores *217*	
6.4.3.1	Theory on Environmental Sensitivity of Fluorophores *217*	
6.4.3.2	Experimental Guide to Labeling Proteins With Fluorophores *218*	
6.4.4	1-Anilino-8-Naphthalensulfonate (ANS) Binding to Investigate Fiber Assembly *219*	
6.4.4.1	Theory on Using ANS Fluorescence for Detecting Conformational Changes in Proteins *219*	
6.4.4.2	Experimental Guide to Using ANS for Monitoring Protein Fiber Assembly *220*	
6.4.5	Light Scattering to Monitor Particle Growth *220*	
6.4.5.1	Theory of Classical Light Scattering *221*	
6.4.5.2	Theory of Dynamic Light Scattering *221*	
6.4.5.3	Experimental Guide to Analyzing Fiber Assembly Using DLS *222*	
6.4.6	Field-flow Fractionation to Monitor Particle Growth *222*	
6.4.6.1	Theory of FFF *222*	
6.4.6.2	Experimental Guide to Using FFF for Monitoring Fiber Assembly *223*	
6.4.7	Fiber Growth-rate Analysis Using Surface Plasmon Resonance *223*	
6.4.7.1	Theory of SPR *223*	
6.4.7.2	Experimental Guide to Using SPR for Fiber-growth Analysis *224*	
6.4.8	Single-fiber Growth Imaging Using Atomic Force Microscopy *225*	

6.4.8.1	Theory of Atomic Force Microscopy	225
6.4.8.2	Experimental Guide for Using AFM to Investigate Fiber Growth	225
6.4.9	Dyes Specific for Detecting Amyloid Fibers	226
6.4.9.1	Theory on Congo Red and Thioflavin T Binding to Amyloid	226
6.4.9.2	Experimental Guide to Detecting Amyloid Fibers with CR and Thioflavin Binding	227
6.5	Methods to Study Fiber Morphology and Structure	228
6.5.1	Scanning Electron Microscopy for Examining the Low-resolution Morphology of a Fiber Specimen	228
6.5.1.1	Theory of SEM	228
6.5.1.2	Experimental Guide to Examining Fibers by SEM	229
6.5.2	Transmission Electron Microscopy for Examining Fiber Morphology and Structure	230
6.5.2.1	Theory of TEM	230
6.5.2.2	Experimental Guide to Examining Fiber Samples by TEM	231
6.5.3	Cryo-electron Microscopy for Examination of the Structure of Fibrous Proteins	232
6.5.3.1	Theory of Cryo-electron Microscopy	232
6.5.3.2	Experimental Guide to Preparing Proteins for Cryo-electron Microscopy	233
6.5.3.3	Structural Analysis from Electron Micrographs	233
6.5.4	Atomic Force Microscopy for Examining the Structure and Morphology of Fibrous Proteins	234
6.5.4.1	Experimental Guide for Using AFM to Monitor Fiber Morphology	234
6.5.5	Use of X-ray Diffraction for Examining the Structure of Fibrous Proteins	236
6.5.5.1	Theory of X-Ray Fiber Diffraction	236
6.5.5.2	Experimental Guide to X-Ray Fiber Diffraction	237
6.5.6	Fourier Transformed Infrared Spectroscopy	239
6.5.6.1	Theory of FTIR	239
6.5.6.2	Experimental Guide to Determining Protein Conformation by FTIR	240
6.6	Concluding Remarks	241
	Acknowledgements	242
	References	242
7	**Protein Unfolding in the Cell**	**254**
	Prakash Koodathingal, Neil E. Jaffe, and Andreas Matouschek	
7.1	Introduction	254
7.2	Protein Translocation Across Membranes	254
7.2.1	Compartmentalization and Unfolding	254
7.2.2	Mitochondria Actively Unfold Precursor Proteins	256
7.2.3	The Protein Import Machinery of Mitochondria	257
7.2.4	Specificity of Unfolding	259

7.2.5	Protein Import into Other Cellular Compartments	259
7.3	Protein Unfolding and Degradation by ATP-dependent Proteases	260
7.3.1	Structural Considerations of Unfoldases Associated With Degradation	260
7.3.2	Unfolding Is Required for Degradation by ATP-dependent Proteases	261
7.3.3	The Role of ATP and Models of Protein Unfolding	262
7.3.4	Proteins Are Unfolded Sequentially and Processively	263
7.3.5	The Influence of Substrate Structure on the Degradation Process	264
7.3.6	Unfolding by Pulling	264
7.3.7	Specificity of Degradation	265
7.4	Conclusions	266
7.5	Experimental Protocols	266
7.5.1	Size of Import Channels in the Outer and Inner Membranes of Mitochondria	266
7.5.2	Structure of Precursor Proteins During Import into Mitochondria	266
7.5.3	Import of Barnase Mutants	267
7.5.4	Protein Degradation by ATP-dependent Proteases	267
7.5.5	Use of Multi-domain Substrates	268
7.5.6	Studies Using Circular Permutants	268
	References	269

8	**Natively Disordered Proteins**	**275**
	Gary W. Daughdrill, Gary J. Pielak, Vladimir N. Uversky, Marc S. Cortese, and A. Keith Dunker	
8.1	Introduction	275
8.1.1	The Protein Structure-Function Paradigm	275
8.1.2	Natively Disordered Proteins	277
8.1.3	A New Protein Structure-Function Paradigm	280
8.2	Methods Used to Characterize Natively Disordered Proteins	281
8.2.1	NMR Spectroscopy	281
8.2.1.1	Chemical Shifts Measure the Presence of Transient Secondary Structure	282
8.2.1.2	Pulsed Field Gradient Methods to Measure Translational Diffusion	284
8.2.1.3	NMR Relaxation and Protein Flexibility	284
8.2.1.4	Using the Model-free Analysis of Relaxation Data to Estimate Internal Mobility and Rotational Correlation Time	285
8.2.1.5	Using Reduced Spectral Density Mapping to Assess the Amplitude and Frequencies of Intramolecular Motion	286
8.2.1.6	Characterization of the Dynamic Structures of Natively Disordered Proteins Using NMR	287
8.2.2	X-ray Crystallography	288
8.2.3	Small Angle X-ray Diffraction and Hydrodynamic Measurements	293

8.2.4	Circular Dichroism Spectropolarimetry 297
8.2.5	Infrared and Raman Spectroscopy 299
8.2.6	Fluorescence Methods 301
8.2.6.1	Intrinsic Fluorescence of Proteins 301
8.2.6.2	Dynamic Quenching of Fluorescence 302
8.2.6.3	Fluorescence Polarization and Anisotropy 303
8.2.6.4	Fluorescence Resonance Energy Transfer 303
8.2.6.5	ANS Fluorescence 305
8.2.7	Conformational Stability 308
8.2.7.1	Effect of Temperature on Proteins with Extended Disorder 309
8.2.7.2	Effect of pH on Proteins with Extended Disorder 309
8.2.8	Mass Spectrometry-based High-resolution Hydrogen-Deuterium Exchange 309
8.2.9	Protease Sensitivity 311
8.2.10	Prediction from Sequence 313
8.2.11	Advantage of Multiple Methods 314
8.3	Do Natively Disordered Proteins Exist Inside Cells? 315
8.3.1	Evolution of Ordered and Disordered Proteins Is Fundamentally Different 315
8.3.1.1	The Evolution of Natively Disordered Proteins 315
8.3.1.2	Adaptive Evolution and Protein Flexibility 317
8.3.1.3	Phylogeny Reconstruction and Protein Structure 318
8.3.2	Direct Measurement by NMR 320
8.4	Functional Repertoire 322
8.4.1	Molecular Recognition 322
8.4.1.1	The Coupling of Folding and Binding 322
8.4.1.2	Structural Plasticity for the Purpose of Functional Plasticity 323
8.4.1.3	Systems Where Disorder Increases Upon Binding 323
8.4.2	Assembly/Disassembly 325
8.4.3	Highly Entropic Chains 325
8.4.4	Protein Modification 327
8.5	Importance of Disorder for Protein Folding 328
8.6	Experimental Protocols 331
8.6.1	NMR Spectroscopy 331
8.6.1.1	General Requirements 331
8.6.1.2	Measuring Transient Secondary Structure in Secondary Chemical Shifts 332
8.6.1.3	Measuring the Translational Diffusion Coefficient Using Pulsed Field Gradient Diffusion Experiments 332
8.6.1.4	Relaxation Experiments 332
8.6.1.5	Relaxation Data Analysis Using Reduced Spectral Density Mapping 333
8.6.1.6	In-cell NMR 334
8.6.2	X-ray Crystallography 334
8.6.3	Circular Dichroism Spectropolarimetry 336

Acknowledgements *337*
References *337*

9 **The Catalysis of Disulfide Bond Formation in Prokaryotes** *358*
Jean-Francois Collet and James C. Bardwell
9.1 Introduction *358*
9.2 Disulfide Bond Formation in the *E. coli* Periplasm *358*
9.2.1 A Small Bond, a Big Effect *358*
9.2.2 Disulfide Bond Formation Is a Catalyzed Process *359*
9.2.3 DsbA, a Protein-folding Catalyst *359*
9.2.4 How is DsbA Re-oxidized? *361*
9.2.5 From Where Does the Oxidative Power of DsbB Originate? *361*
9.2.6 How Are Disulfide Bonds Transferred From DsbB to DsbA? *362*
9.2.7 How Can DsbB Generate Disulfide by Quinone Reduction? *364*
9.3 Disulfide Bond Isomerization *365*
9.3.1 The Protein Disulfide Isomerases DsbC and DsbG *365*
9.3.2 Dimerization of DsbC and DsbG Is Important for Isomerase and Chaperone Activity *366*
9.3.3 Dimerization Protects from DsbB Oxidation *367*
9.3.4 Import of Electrons from the Cytoplasm: DsbD *367*
9.3.5 Conclusions *369*
9.4 Experimental Protocols *369*
9.4.1 Oxidation-reduction of a Protein Sample *369*
9.4.2 Determination of the Free Thiol Content of a Protein *370*
9.4.3 Separation by HPLC *371*
9.4.4 Tryptophan Fluorescence *372*
9.4.5 Assay of Disulfide Oxidase Activity *372*
References *373*

10 **Catalysis of Peptidyl-prolyl *cis/trans* Isomerization by Enzymes** *377*
Gunter Fischer
10.1 Introduction *377*
10.2 Peptidyl-prolyl *cis/trans* Isomerization *379*
10.3 Monitoring Peptidyl-prolyl *cis/trans* Isomerase Activity *383*
10.4 Prototypical Peptidyl-prolyl *cis/trans* Isomerases *388*
10.4.1 General Considerations *388*
10.4.2 Prototypic Cyclophilins *390*
10.4.3 Prototypic FK506-binding Proteins *394*
10.4.4 Prototypic Parvulins *397*
10.5 Concluding Remarks *399*
10.6 Experimental Protocols *399*
10.6.1 PPIase Assays: Materials *399*
10.6.2 PPIase Assays: Equipment *400*
10.6.3 Assaying Procedure: Protease-coupled Spectrophotometric Assay *400*

10.6.4	Assaying Procedure: Protease-free Spectrophotometric Assay	401
	References 401	

11 Secondary Amide Peptide Bond *cis/trans* Isomerization in Polypeptide Backbone Restructuring: Implications for Catalysis 415
Cordelia Schiene-Fischer and Christian Lücke

11.1	Introduction 415	
11.2	Monitoring Secondary Amide Peptide Bond *cis/trans* Isomerization 416	
11.3	Kinetics and Thermodynamics of Secondary Amide Peptide Bond *cis/trans* Isomerization 418	
11.4	Principles of DnaK Catalysis 420	
11.5	Concluding Remarks 423	
11.6	Experimental Protocols 424	
11.6.1	Stopped-flow Measurements of Peptide Bond *cis/trans* Isomerization 424	
11.6.2	Two-dimensional ^{1}H-NMR Exchange Experiments 425	
	References 426	

12 Ribosome-associated Proteins Acting on Newly Synthesized Polypeptide Chains 429
Sabine Rospert, Matthias Gautschi, Magdalena Rakwalska, and Uta Raue

12.1	Introduction 429
12.2	Signal Recognition Particle, Nascent Polypeptide–associated Complex, and Trigger Factor 432
12.2.1	Signal Recognition Particle 432
12.2.2	An Interplay between Eukaryotic SRP and Nascent Polypeptide–associated Complex? 435
12.2.3	Interplay between Bacterial SRP and Trigger Factor? 435
12.2.4	Functional Redundancy: TF and the Bacterial Hsp70 Homologue DnaK 436
12.3	Chaperones Bound to the Eukaryotic Ribosome: Hsp70 and Hsp40 Systems 436
12.3.1	Sis1p and Ssa1p: an Hsp70/Hsp40 System Involved in Translation Initiation? 437
12.3.2	Ssb1/2p, an Hsp70 Homologue Distributed Between Ribosomes and Cytosol 438
12.3.3	Function of Ssb1/2p in Degradation and Protein Folding 439
12.3.4	Zuotin and Ssz1p: a Stable Chaperone Complex Bound to the Yeast Ribosome 440
12.3.5	A Functional Chaperone Triad Consisting of Ssb1/2p, Ssz1p, and Zuotin 440
12.3.6	Effects of Ribosome-bound Chaperones on the Yeast Prion $[PSI^+]$ 442
12.4	Enzymes Acting on Nascent Polypeptide Chains 443

12.4.1	Methionine Aminopeptidases	443
12.4.2	N^{α}-acetyltransferases	444
12.5	A Complex Arrangement at the Yeast Ribosomal Tunnel Exit	445
12.6	Experimental Protocols	446
12.6.1	Purification of Ribosome-associated Protein Complexes from Yeast	446
12.6.2	Growth of Yeast and Preparation of Ribosome-associated Proteins by High-salt Treatment of Ribosomes	447
12.6.3	Purification of NAC and RAC	448
	References	449

Part II, Volume 2

13 The Role of Trigger Factor in Folding of Newly Synthesized Proteins 459
Elke Deuerling, Thomas Rauch, Holger Patzelt, and Bernd Bukau

13.1	Introduction	459
13.2	In Vivo Function of Trigger Factor	459
13.2.1	Discovery	459
13.2.2	Trigger Factor Cooperates With the DnaK Chaperone in the Folding of Newly Synthesized Cytosolic Proteins	460
13.2.3	In Vivo Substrates of Trigger Factor and DnaK	461
13.2.4	Substrate Specificity of Trigger Factor	463
13.3	Structure–Function Analysis of Trigger Factor	465
13.3.1	Domain Structure and Conservation	465
13.3.2	Quaternary Structure	468
13.3.3	PPIase and Chaperone Activity of Trigger Factor	469
13.3.4	Importance of Ribosome Association	470
13.4	Models of the Trigger Factor Mechanism	471
13.5	Experimental Protocols	473
13.5.1	Trigger Factor Purification	473
13.5.2	GAPDH Trigger Factor Activity Assay	475
13.5.3	Modular Cell-free *E. coli* Transcription/Translation System	475
13.5.4	Isolation of Ribosomes and Add-back Experiments	483
13.5.5	Cross-linking Techniques	485
	References	485

14 Cellular Functions of Hsp70 Chaperones 490
Elizabeth A. Craig and Peggy Huang

14.1	Introduction	490
14.2	"Soluble" Hsp70s/J-proteins Function in General Protein Folding	492
14.2.1	The Soluble Hsp70 of *E. coli*, DnaK	492
14.2.2	Soluble Hsp70s of Major Eukaryotic Cellular Compartments	493
14.2.2.1	Eukaryotic Cytosol	493
14.2.2.2	Matrix of Mitochondria	494
14.2.2.3	Lumen of the Endoplasmic Reticulum	494

14.3	"Tethered" Hsp70s/J-proteins: Roles in Protein Folding on the Ribosome and in Protein Translocation 495	
14.3.1	Membrane-tethered Hsp70/J-protein 495	
14.3.2	Ribosome-associated Hsp70/J-proteins 496	
14.4	Modulating of Protein Conformation by Hsp70s/J-proteins 498	
14.4.1	Assembly of Fe/S Centers 499	
14.4.2	Uncoating of Clathrin-coated Vesicles 500	
14.4.3	Regulation of the Heat Shock Response 501	
14.4.4	Regulation of Activity of DNA Replication-initiator Proteins 502	
14.5	Cases of a Single Hsp70 Functioning With Multiple J-Proteins 504	
14.6	Hsp70s/J-proteins – When an Hsp70 Maybe Isn't Really a Chaperone 504	
14.6.1	The Ribosome-associated "Hsp70" Ssz1 505	
14.6.2	Mitochondrial Hsp70 as the Regulatory Subunit of an Endonuclease 506	
14.7	Emerging Concepts and Unanswered Questions 507	
	References 507	
15	**Regulation of Hsp70 Chaperones by Co-chaperones** 516	
	Matthias P. Mayer and Bernd Bukau	
15.1	Introduction 516	
15.2	Hsp70 Proteins 517	
15.2.1	Structure and Conservation 517	
15.2.2	ATPase Cycle 519	
15.2.3	Structural Investigations 521	
15.2.4	Interactions With Substrates 522	
15.3	J-domain Protein Family 526	
15.3.1	Structure and Conservation 526	
15.3.2	Interaction With Hsp70s 530	
15.3.3	Interactions with Substrates 532	
15.4	Nucleotide Exchange Factors 534	
15.4.1	GrpE: Structure and Interaction with DnaK 534	
15.4.2	Nucleotide Exchange Reaction 535	
15.4.3	Bag Family: Structure and Interaction With Hsp70 536	
15.4.4	Relevance of Regulated Nucleotide Exchange for Hsp70s 538	
15.5	TPR Motifs Containing Co-chaperones of Hsp70 540	
15.5.1	Hip 541	
15.5.2	Hop 542	
15.5.3	Chip 543	
15.6	Concluding Remarks 544	
15.7	Experimental Protocols 544	
15.7.1	Hsp70s 544	
15.7.2	J-Domain Proteins 545	
15.7.3	GrpE 546	
15.7.4	Bag-1 547	

15.7.5	Hip	*548*
15.7.6	Hop	*549*
15.7.7	Chip	*549*
	References	*550*

16 Protein Folding in the Endoplasmic Reticulum Via the Hsp70 Family *563*
Ying Shen, Kyung Tae Chung, and Linda M. Hendershot

16.1	Introduction *563*	
16.2	BiP Interactions with Unfolded Proteins *564*	
16.3	ER-localized DnaJ Homologues *567*	
16.4	ER-localized Nucleotide-exchange/releasing Factors *571*	
16.5	Organization and Relative Levels of Chaperones in the ER *572*	
16.6	Regulation of ER Chaperone Levels *573*	
16.7	Disposal of BiP-associated Proteins That Fail to Fold or Assemble *575*	
16.8	Other Roles of BiP in the ER *576*	
16.9	Concluding Comments *576*	
16.10	Experimental Protocols *577*	
16.10.1	Production of Recombinant ER Proteins *577*	
16.10.1.1	General Concerns *577*	
16.10.1.2	Bacterial Expression *578*	
16.10.1.3	Yeast Expression *580*	
16.10.1.4	Baculovirus *581*	
16.10.1.5	Mammalian Cells *583*	
16.10.2	Yeast Two-hybrid Screen for Identifying Interacting Partners of ER Proteins *586*	
16.10.3	Methods for Determining Subcellular Localization, Topology, and Orientation of Proteins *588*	
16.10.3.1	Sequence Predictions *588*	
16.10.3.2	Immunofluorescence Staining *589*	
16.10.3.3	Subcellular Fractionation *589*	
16.10.3.4	Determination of Topology *590*	
16.10.3.5	*N*-linked Glycosylation *592*	
16.10.4	Nucleotide Binding, Hydrolysis, and Exchange Assays *594*	
16.10.4.1	Nucleotide-binding Assays *594*	
16.10.4.2	ATP Hydrolysis Assays *596*	
16.10.4.3	Nucleotide Exchange Assays *597*	
16.10.5	Assays for Protein–Protein Interactions in Vitro/in Vivo *599*	
16.10.5.1	In Vitro GST Pull-down Assay *599*	
16.10.5.2	Co-immunoprecipitation *600*	
16.10.5.3	Chemical Cross-linking *600*	
16.10.5.4	Yeast Two-hybrid System *601*	
16.10.6	In Vivo Folding, Assembly, and Chaperone-binding Assays *601*	
16.10.6.1	Monitoring Oxidation of Intrachain Disulfide Bonds *601*	
16.10.6.2	Detection of Chaperone Binding *602*	

Acknowledgements 603
References 603

17	**Quality Control In Glycoprotein Folding** 617	
	E. Sergio Trombetta and Armando J. Parodi	
17.1	Introduction 617	
17.2	ER N-glycan Processing Reactions 617	
17.3	The UDP-Glc:Glycoprotein Glucosyltransferase 619	
17.4	Protein Folding in the ER 621	
17.5	Unconventional Chaperones (Lectins) Are Present in the ER Lumen 621	
17.6	In Vivo Glycoprotein-CNX/CRT Interaction 623	
17.7	Effect of CNX/CRT Binding on Glycoprotein Folding and ER Retention 624	
17.8	Glycoprotein-CNX/CRT Interaction Is Not Essential for Unicellular Organisms and Cells in Culture 627	
17.9	Diversion of Misfolded Glycoproteins to Proteasomal Degradation 629	
17.10	Unfolding Irreparably Misfolded Glycoproteins to Facilitate Proteasomal Degradation 632	
17.11	Summary and Future Directions 633	
17.12	Characterization of N-glycans from Glycoproteins 634	
17.12.1	Characterization of N-glycans Present in Immunoprecipitated Samples 634	
17.12.2	Analysis of Radio-labeled N-glycans 636	
17.12.3	Extraction and Analysis of Protein-bound N-glycans 636	
17.12.4	GII and GT Assays 637	
17.12.4.1	Assay for GII 637	
17.12.4.2	Assay for GT 638	
17.12.5	Purification of GII and GT from Rat Liver 639	
	References 641	
18	**Procollagen Biosynthesis in Mammalian Cells** 649	
	Mohammed Tasab and Neil J. Bulleid	
18.1	Introduction 649	
18.1.1	Variety and Complexity of Collagen Proteins 649	
18.1.2	Fibrillar Procollagen 650	
18.1.3	Expression of Fibrillar Collagens 650	
18.2	The Procollagen Biosynthetic Process: An Overview 651	
18.3	Disulfide Bonding in Procollagen Assembly 653	
18.4	The Influence of Primary Amino Acid Sequence on Intracellular Procollagen Folding 654	
18.4.1	Chain Recognition and Type-specific Assembly 654	
18.4.2	Assembly of Multi-subunit Proteins 654	
18.4.3	Coordination of Type-specific Procollagen Assembly and Chain Selection 655	

18.4.4	Hypervariable Motifs: Components of a Recognition Mechanism That Distinguishes Between Procollagen Chains? *656*
18.4.5	Modeling the C-propeptide *657*
18.4.6	Chain Association *657*
18.5	Posttranslational Modifications That Affect Procollagen Folding *658*
18.5.1	Hydroxylation and Triple-helix Stability *658*
18.6	Procollagen Chaperones *658*
18.6.1	Prolyl 4-Hydroxylase *658*
18.6.2	Protein Disulfide Isomerase *659*
18.6.3	Hsp47 *660*
18.6.4	PPI and BiP *661*
18.7	Analysis of Procollagen Folding *662*
18.8	Experimental Part *663*
18.8.1	Materials Required *663*
18.8.2	Experimental Protocols *664*
	References *668*

19 Redox Regulation of Chaperones *677*
Jörg H. Hoffmann and Ursula Jakob

19.1	Introduction *677*
19.2	Disulfide Bonds as Redox-Switches *677*
19.2.1	Functionality of Disulfide Bonds *677*
19.2.2	Regulatory Disulfide Bonds as Functional Switches *679*
19.2.3	Redox Regulation of Chaperone Activity *680*
19.3	Prokaryotic Hsp33: A Chaperone Activated by Oxidation *680*
19.3.1	Identification of a Redox-regulated Chaperone *680*
19.3.2	Activation Mechanism of Hsp33 *681*
19.3.3	The Crystal Structure of Active Hsp33 *682*
19.3.4	The Active Hsp33-Dimer: An Efficient Chaperone Holdase *683*
19.3.5	Hsp33 is Part of a Sophisticated Multi-chaperone Network *684*
19.4	Eukaryotic Protein Disulfide Isomerase (PDI): Redox Shuffling in the ER *685*
19.4.1	PDI, A Multifunctional Enzyme in Eukaryotes *685*
19.4.2	PDI and Redox Regulation *687*
19.5	Concluding Remarks and Outlook *688*
19.6	Appendix – Experimental Protocols *688*
19.6.1	How to Work With Redox-regulated Chaperones in Vitro *689*
19.6.1.1	Preparation of the Reduced Protein Species *689*
19.6.1.2	Preparation of the Oxidized Protein Species *690*
19.6.1.3	In Vitro Thiol Trapping to Monitor the Redox State of Proteins *691*
19.6.2	Thiol Coordinating Zinc Centers as Redox Switches *691*
19.6.2.1	PAR-PMPS Assay to Quantify Zinc *691*
19.6.2.2	Determination of Zinc-binding Constants *692*
19.6.3	Functional Analysis of Redox-regulated Chaperones in Vitro/in Vivo *693*
19.6.3.1	Chaperone Activity Assays *693*

19.6.3.2	Manipulating and Analyzing Redox Conditions in Vivo	694
	Acknowledgements 694	
	References 694	
20	**The *E. coli* GroE Chaperone** 699	
	Steven G. Burston and Stefan Walter	
20.1	Introduction 699	
20.2	The Structure of GroEL 699	
20.3	The Structure of GroEL-ATP 700	
20.4	The Structure of GroES and its Interaction with GroEL 701	
20.5	The Interaction Between GroEL and Substrate Polypeptides 702	
20.6	GroEL is a Complex Allosteric Macromolecule 703	
20.7	The Reaction Cycle of the GroE Chaperone 705	
20.8	The Effect of GroE on Protein-folding Pathways 708	
20.9	Future Perspectives 710	
20.10	Experimental Protocols 710	
	Acknowledgments 719	
	References 719	
21	**Structure and Function of the Cytosolic Chaperonin CCT** 725	
	José M. Valpuesta, José L. Carrascosa, and Keith R. Willison	
21.1	Introduction 725	
21.2	Structure and Composition of CCT 726	
21.3	Regulation of CCT Expression 729	
21.4	Functional Cycle of CCT 730	
21.5	Folding Mechanism of CCT 731	
21.6	Substrates of CCT 735	
21.7	Co-chaperones of CCT 739	
21.8	Evolution of CCT 741	
21.9	Concluding Remarks 743	
21.10	Experimental Protocols 743	
21.10.1	Purification 743	
21.10.2	ATP Hydrolysis Measurements 744	
21.10.3	CCT Substrate-binding and Folding Assays 744	
21.10.4	Electron Microscopy and Image Processing 744	
	References 747	
22	**Structure and Function of GimC/Prefoldin** 756	
	Katja Siegers, Andreas Bracher, and Ulrich Hartl	
22.1	Introduction 756	
22.2	Evolutionary Distribution of GimC/Prefoldin 757	
22.3	Structure of the Archaeal GimC/Prefoldin 757	
22.4	Complexity of the Eukaryotic/Archaeal GimC/Prefoldin 759	
22.5	Functional Cooperation of GimC/Prefoldin With the Eukaryotic Chaperonin TRiC/CCT 761	

22.6	Experimental Protocols 764	
22.6.1	Actin-folding Kinetics 764	
22.6.2	Prevention of Aggregation (Light-scattering) Assay 765	
22.6.3	Actin-binding Assay 765	
	Acknowledgements 766	
	References 766	
23	**Hsp90: From Dispensable Heat Shock Protein to Global Player 768**	
	Klaus Richter, Birgit Meinlschmidt, and Johannes Buchner	
23.1	Introduction 768	
23.2	The Hsp90 Family in Vivo 768	
23.2.1	Evolutionary Relationships within the Hsp90 Gene Family 768	
23.2.2	In Vivo Functions of Hsp90 769	
23.2.3	Regulation of Hsp90 Expression and Posttranscriptional Activation 772	
23.2.4	Chemical Inhibition of Hsp90 773	
23.2.5	Identification of Natural Hsp90 Substrates 774	
23.3	In Vitro Investigation of the Chaperone Hsp90 775	
23.3.1	Hsp90: A Special Kind of ATPase 775	
23.3.2	The ATPase Cycle of Hsp90 780	
23.3.3	Interaction of Hsp90 with Model Substrate Proteins 781	
23.3.4	Investigating Hsp90 Substrate Interactions Using Native Substrates 783	
23.4	Partner Proteins: Does Complexity Lead to Specificity? 784	
23.4.1	Hop, p23, and PPIases: The Chaperone Cycle of Hsp90 784	
23.4.2	Hop/Sti1: Interactions Mediated by TPR Domains 787	
23.4.3	p23/Sba1: Nucleotide-specific Interaction with Hsp90 789	
23.4.4	Large PPIases: Conferring Specificity to Substrate Localization? 790	
23.4.5	Pp5: Facilitating Dephosphorylation 791	
23.4.6	Cdc37: Building Complexes with Kinases 792	
23.4.7	Tom70: Chaperoning Mitochondrial Import 793	
23.4.8	CHIP and Sgt1: Multiple Connections to Protein Degradation 793	
23.4.9	Aha1 and Hch1: Just Stimulating the ATPase? 794	
23.4.10	Cns1, Sgt2, and Xap2: Is a TPR Enough to Become an Hsp90 Partner? 796	
23.5	Outlook 796	
23.6	Appendix – Experimental Protocols 797	
23.6.1	Calculation of Phylogenetic Trees Based on Protein Sequences 797	
23.6.2	Investigating the in Vivo Effect of Hsp90 Mutations in *S. cerevisiae* 797	
23.6.3	Well-characterized Hsp90 Mutants 798	
23.6.4	Investigating Activation of Heterologously Expressed Src Kinase in *S. cerevisiae* 800	
23.6.5	Investigation of Heterologously Expressed Glucocorticoid Receptor in *S. cerevisiae* 800	

23.6.6	Investigation of Chaperone Activity	*801*
23.6.7	Analysis of the ATPase Activity of Hsp90	*802*
23.6.8	Detecting Specific Influences on Hsp90 ATPase Activity	*803*
23.6.9	Investigation of the Quaternary Structure by SEC-HPLC	*804*
23.6.10	Investigation of Binding Events Using Changes of the Intrinsic Fluorescence	*806*
23.6.11	Investigation of Binding Events Using Isothermal Titration Calorimetry	*807*
23.6.12	Investigation of Protein-Protein Interactions Using Cross-linking	*807*
23.6.13	Investigation of Protein-Protein Interactions Using Surface Plasmon Resonance Spectroscopy	*808*
	Acknowledgements	*810*
	References	*810*

24	**Small Heat Shock Proteins: Dynamic Players in the Folding Game**	*830*
	Franz Narberhaus and Martin Haslbeck	
24.1	Introduction	*830*
24.2	α-Crystallins and the Small Heat Shock Protein Family: Diverse Yet Similar	*830*
24.3	Cellular Functions of α-Hsps	*831*
24.3.1	Chaperone Activity in Vitro	*831*
24.3.2	Chaperone Function in Vivo	*835*
24.3.3	Other Functions	*836*
24.4	The Oligomeric Structure of α-Hsps	*837*
24.5	Dynamic Structures as Key to Chaperone Activity	*839*
24.6	Experimental Protocols	*840*
24.6.1	Purification of sHsps	*840*
24.6.2	Chaperone Assays	*843*
24.6.3	Monitoring Dynamics of sHsps	*846*
	Acknowledgements	*847*
	References	*848*

25	**Alpha-crystallin: Its Involvement in Suppression of Protein Aggregation and Protein Folding**	*858*
	Joseph Horwitz	
25.1	Introduction	*858*
25.2	Distribution of Alpha-crystallin in the Various Tissues	*858*
25.3	Structure	*859*
25.4	Phosphorylation and Other Posttranslation Modification	*860*
25.5	Binding of Target Proteins to Alpha-crystallin	*861*
25.6	The Function of Alpha-crystallin	*863*
25.7	Experimental Protocols	*863*
25.7.1	Preparation of Alpha-crystallin	*863*
	Acknowledgements	*870*
	References	*870*

26	**Transmembrane Domains in Membrane Protein Folding, Oligomerization, and Function** *876*
	Anja Ridder and Dieter Langosch
26.1	Introduction *876*
26.1.1	Structure of Transmembrane Domains *876*
26.1.2	The Biosynthetic Route towards Folded and Oligomeric Integral Membrane Proteins *877*
26.1.3	Structure and Stability of TMSs *878*
26.1.3.1	Amino Acid Composition of TMSs and Flanking Regions *878*
26.1.3.2	Stability of Transmembrane Helices *879*
26.2	The Nature of Transmembrane Helix-Helix Interactions *880*
26.2.1	General Considerations *880*
26.2.1.1	Attractive Forces within Lipid Bilayers *880*
26.2.1.2	Forces between Transmembrane Helices *881*
26.2.1.3	Entropic Factors Influencing Transmembrane Helix–Helix Interactions *882*
26.2.2	Lessons from Sequence Analyses and High-resolution Structures *883*
26.2.3	Lessons from Bitopic Membrane Proteins *886*
26.2.3.1	Transmembrane Segments Forming Right-handed Pairs *886*
26.2.3.2	Transmembrane Segments Forming Left-handed Assemblies *889*
26.2.4	Selection of Self-interacting TMSs from Combinatorial Libraries *892*
26.2.5	Role of Lipids in Packing/Assembly of Membrane Proteins *893*
26.3	Conformational Flexibility of Transmembrane Segments *895*
26.4	Experimental Protocols *897*
26.4.1	Biochemical and Biophysical Techniques *897*
26.4.1.1	Visualization of Oligomeric States by Electrophoretic Techniques *898*
26.4.1.2	Hydrodynamic Methods *899*
26.4.1.3	Fluorescence Resonance Transfer *900*
26.4.2	Genetic Assays *901*
26.4.2.1	The ToxR System *901*
26.4.2.2	Other Genetic Assays *902*
26.4.3	Identification of TMS-TMS Interfaces by Mutational Analysis *903*
	References *904*

Part II, Volume 3

27	**SecB** *919*
	Arnold J. M. Driessen, Janny de Wit, and Nico Nouwen
27.1	Introduction *919*
27.2	Selective Binding of Preproteins by SecB *920*
27.3	SecA-SecB Interaction *925*
27.4	Preprotein Transfer from SecB to SecA *928*
27.5	Concluding Remarks *929*
27.6	Experimental Protocols *930*
27.6.1	How to Analyze SecB-Preprotein Interactions *930*

27.6.2	How to Analyze SecB-SecA Interaction	931
	Acknowledgements	932
	References	933

28 Protein Folding in the Periplasm and Outer Membrane of *E. coli* 938
Michael Ehrmann

28.1	Introduction	938
28.2	Individual Cellular Factors	940
28.2.1	The Proline Isomerases FkpA, PpiA, SurA, and PpiD	941
28.2.1.1	FkpA	942
28.2.1.2	PpiA	942
28.2.1.3	SurA	943
28.2.1.4	PpiD	943
28.2.2	Skp	944
28.2.3	Proteases and Protease/Chaperone Machines	945
28.2.3.1	The HtrA Family of Serine Proteases	946
28.2.3.2	*E. coli* HtrAs	946
28.2.3.3	DegP and DegQ	946
28.2.3.4	DegS	947
28.2.3.5	The Structure of HtrA	947
28.2.3.6	Other Proteases	948
28.3	Organization of Folding Factors into Pathways and Networks	950
28.3.1	Synthetic Lethality and Extragenic High-copy Suppressors	950
28.3.2	Reconstituted in Vitro Systems	951
28.4	Regulation	951
28.4.1	The Sigma E Pathway	951
28.4.2	The Cpx Pathway	952
28.4.3	The Bae Pathway	953
28.5	Future Perspectives	953
28.6	Experimental Protocols	954
28.6.1	Pulse Chase Immunoprecipitation	954
	Acknowledgements	957
	References	957

29 Formation of Adhesive Pili by the Chaperone-Usher Pathway 965
Michael Vetsch and Rudi Glockshuber

29.1	Basic Properties of Bacterial, Adhesive Surface Organelles	965
29.2	Structure and Function of Pilus Chaperones	970
29.3	Structure and Folding of Pilus Subunits	971
29.4	Structure and Function of Pilus Ushers	973
29.5	Conclusions and Outlook	976
29.6	Experimental Protocols	977
29.6.1	Test for the Presence of Type 1 Piliated *E. coli* Cells	977
29.6.2	Functional Expression of Pilus Subunits in the *E. coli* Periplasm	977
29.6.3	Purification of Pilus Subunits from the *E. coli* Periplasm	978

29.6.4	Preparation of Ushers 979	
	Acknowledgements 979	
	References 980	

30 Unfolding of Proteins During Import into Mitochondria   987
Walter Neupert, Michael Brunner, and Kai Hell
30.1 Introduction   987
30.2 Translocation Machineries and Pathways of the Mitochondrial Protein Import System   988
30.2.1 Import of Proteins Destined for the Mitochondrial Matrix   990
30.3 Import into Mitochondria Requires Protein Unfolding   993
30.4 Mechanisms of Unfolding by the Mitochondrial Import Motor   995
30.4.1 Targeted Brownian Ratchet   995
30.4.2 Power-stroke Model   995
30.5 Studies to Discriminate between the Models   996
30.5.1 Studies on the Unfolding of Preproteins   996
30.5.1.1 Comparison of the Import of Folded and Unfolded Proteins   996
30.5.1.2 Import of Preproteins With Different Presequence Lengths   999
30.5.1.3 Import of Titin Domains   1000
30.5.1.4 Unfolding by the Mitochondrial Membrane Potential $\Delta\Psi$ 1000
30.5.2 Mechanistic Studies of the Import Motor   1000
30.5.2.1 Brownian Movement of the Polypeptide Within the Import Channel   1000
30.5.2.2 Recruitment of mtHsp70 by Tim44   1001
30.5.2.3 Import Without Recruitment of mtHsp70 by Tim44   1002
30.5.2.4 MtHsp70 Function in the Import Motor   1003
30.6 Discussion and Perspectives   1004
30.7 Experimental Protocols   1006
30.7.1 Protein Import Into Mitochondria in Vitro   1006
30.7.2 Stabilization of the DHFR Domain by Methotrexate   1008
30.7.3 Import of Precursor Proteins Unfolded With Urea   1009
30.7.4 Kinetic Analysis of the Unfolding Reaction by Trapping of Intermediates   1009
References   1011

31 The Chaperone System of Mitochondria   1020
Wolfgang Voos and Nikolaus Pfanner
31.1 Introduction   1020
31.2 Membrane Translocation and the Hsp70 Import Motor   1020
31.3 Folding of Newly Imported Proteins Catalyzed by the Hsp70 and Hsp60 Systems   1026
31.4 Mitochondrial Protein Synthesis and the Assembly Problem   1030
31.5 Aggregation versus Degradation: Chaperone Functions Under Stress Conditions   1033
31.6 Experimental Protocols   1034

31.6.1	Chaperone Functions Characterized With Yeast Mutants 1034
31.6.2	Interaction of Imported Proteins With Matrix Chaperones 1036
31.6.3	Folding of Imported Model Proteins 1037
31.6.4	Assaying Mitochondrial Degradation of Imported Proteins 1038
31.6.5	Aggregation of Proteins in the Mitochondrial Matrix 1038
	References 1039

32 Chaperone Systems in Chloroplasts 1047
Thomas Becker, Jürgen Soll, and Enrico Schleiff

32.1	Introduction 1047
32.2	Chaperone Systems within Chloroplasts 1048
32.2.1	The Hsp70 System of Chloroplasts 1048
32.2.1.1	The Chloroplast Hsp70s 1049
32.2.1.2	The Co-chaperones of Chloroplastic Hsp70s 1051
32.2.2	The Chaperonins 1052
32.2.3	The HSP100/Clp Protein Family in Chloroplasts 1056
32.2.4	The Small Heat Shock Proteins 1058
32.2.5	Hsp90 Proteins of Chloroplasts 1061
32.2.6	Chaperone-like Proteins 1062
32.2.6.1	The Protein Disulfide Isomerase (PDI) 1062
32.2.6.2	The Peptidyl-prolyl *cis* Isomerase (PPIase) 1063
32.3	The Functional Chaperone Pathways in Chloroplasts 1065
32.3.1	Chaperones Involved in Protein Translocation 1065
32.3.2	Protein Transport Inside of Plastids 1070
32.3.3	Protein Folding and Complex Assembly Within Chloroplasts 1071
32.3.4	Chloroplast Chaperones Involved in Proteolysis 1072
32.3.5	Protein Storage Within Plastids 1073
32.3.6	Protein Protection and Repair 1074
32.4	Experimental Protocols 1075
32.4.1	Characterization of Cpn60 Binding to the Large Subunit of Rubisco via Native PAGE (adopted from Ref. [6]) 1075
32.4.2	Purification of Chloroplast Cpn60 From Young Pea Plants (adopted from Ref. [203]) 1076
32.4.3	Purification of Chloroplast Hsp21 From Pea (*Pisum sativum*) (adopted from [90]) 1077
32.4.4	Light-scattering Assays for Determination of the Chaperone Activity Using Citrate Synthase as Substrate (adopted from [196]) 1078
32.4.5	The Use Of *Bis*-ANS to Assess Surface Exposure of Hydrophobic Domains of Hsp17 of *Synechocystis* (adopted from [202]) 1079
32.4.6	Determination of Hsp17 Binding to Lipids (adopted from Refs. [204, 205]) 1079
	References 1081

33 An Overview of Protein Misfolding Diseases 1093
Christopher M. Dobson

| 33.1 | Introduction   1093 |

33.2	Protein Misfolding and Its Consequences for Disease 1094
33.3	The Structure and Mechanism of Amyloid Formation 1097
33.4	A Generic Description of Amyloid Formation 1101
33.5	The Fundamental Origins of Amyloid Disease 1104
33.6	Approaches to Therapeutic Intervention in Amyloid Disease 1106
33.7	Concluding Remarks 1108
	Acknowledgements 1108
	References 1109

34 Biochemistry and Structural Biology of Mammalian Prion Disease   1114
Rudi Glockshuber

34.1	Introduction 1114
34.1.1	Prions and the "Protein-Only" Hypothesis 1114
34.1.2	Models of PrP^{Sc} Propagation 1115
34.2	Properties of PrP^C and PrP^{Sc} 1117
34.3	Three-dimensional Structure and Folding of Recombinant PrP 1120
34.3.1	Expression of the Recombinant Prion Protein for Structural and Biophysical Studies 1120
34.3.2	Three-dimensional Structures of Recombinant Prion Proteins from Different Species and Their Implications for the Species Barrier of Prion Transmission 1120
34.3.2.1	Solution Structure of Murine PrP 1120
34.3.2.2	Comparison of Mammalian Prion Protein Structures and the Species Barrier of Prion Transmission 1124
34.3.3	Biophysical Characterization of the Recombinant Prion Protein 1125
34.3.3.1	Folding and Stability of Recombinant PrP 1125
34.3.3.2	Role of the Disulfide Bond in PrP 1127
34.3.3.3	Influence of Point Mutations Linked With Inherited TSEs on the Stability of Recombinant PrP 1129
34.4	Generation of Infectious Prions in Vitro: Principal Difficulties in Proving the Protein-Only Hypothesis 1131
34.5	Understanding the Strain Phenomenon in the Context of the Protein-Only Hypothesis: Are Prions Crystals? 1132
34.6	Conclusions and Outlook 1135
34.7	Experimental Protocols 1136
34.7.1	Protocol 1 [53, 55] 1136
34.7.2	Protocol 2 [54] 1137
	References 1138

35 Insights into the Nature of Yeast Prions   1144
Lev Z. Osherovich and Jonathan S. Weissman

35.1	Introduction 1144
35.2	Prions as Heritable Amyloidoses 1145
35.3	Prion Strains and Species Barriers: Universal Features of Amyloid-based Prion Elements 1149

35.4	Prediction and Identification of Novel Prion Elements	*1151*
35.5	Requirements for Prion Inheritance beyond Amyloid-mediated Growth	*1154*
35.6	Chaperones and Prion Replication	*1157*
35.7	The Structure of Prion Particles	*1158*
35.8	Prion-like Structures as Protein Interaction Modules	*1159*
35.9	Experimental Protocols	*1160*
35.9.1	Generation of Sup35 Amyloid Fibers in Vitro	*1160*
35.9.2	Thioflavin T–based Amyloid Seeding Efficacy Assay (Adapted from Chien et al. 2003)	*1161*
35.9.3	AFM-based Single-fiber Growth Assay	*1162*
35.9.4	Prion Infection Protocol (Adapted from Tanaka et al. 2004)	*1164*
35.9.5	Preparation of Lyticase	*1165*
35.9.6	Protocol for Counting Heritable Prion Units (Adapted from Cox et al. 2003)	*1166*
	Acknowledgements	*1167*
	References	*1168*
36	**Polyglutamine Aggregates as a Model for Protein-misfolding Diseases** *1175*	
	Soojin Kim, James F. Morley, Anat Ben-Zvi, and Richard I. Morimoto	
36.1	Introduction	*1175*
36.2	Polyglutamine Diseases	*1175*
36.2.1	Genetics	*1175*
36.2.2	Polyglutamine Diseases Involve a Toxic Gain of Function	*1176*
36.3	Polyglutamine Aggregates	*1176*
36.3.1	Presence of the Expanded Polyglutamine Is Sufficient to Induce Aggregation in Vivo	*1176*
36.3.2	Length of the Polyglutamine Dictates the Rate of Aggregate Formation	*1177*
36.3.3	Polyglutamine Aggregates Exhibit Features Characteristic of Amyloids	*1179*
36.3.4	Characterization of Protein Aggregates in Vivo Using Dynamic Imaging Methods	*1180*
36.4	A Role for Oligomeric Intermediates in Toxicity	*1181*
36.5	Consequences of Misfolded Proteins and Aggregates on Protein Homeostasis	*1181*
36.6	Modulators of Polyglutamine Aggregation and Toxicity	*1184*
36.6.1	Protein Context	*1184*
36.6.2	Molecular Chaperones	*1185*
36.6.3	Proteasomes	*1188*
36.6.4	The Protein-folding "Buffer" and Aging	*1188*
36.6.5	Summary	*1189*
36.7	Experimental Protocols	*1190*
36.7.1	FRAP Analysis	*1190*
	References	*1192*

37	Protein Folding and Aggregation in the Expanded Polyglutamine Repeat Diseases *1200*
	Ronald Wetzel
37.1	Introduction *1200*
37.2	Key Features of the Polyglutamine Diseases *1201*
37.2.1	The Variety of Expanded PolyGln Diseases *1201*
37.2.2	Clinical Features *1201*
37.2.2.1	Repeat Expansions and Repeat Length *1202*
37.2.3	The Role of PolyGln and PolyGln Aggregates *1203*
37.3	PolyGln Peptides in Studies of the Molecular Basis of Expanded Polyglutamine Diseases *1205*
37.3.1	Conformational Studies *1205*
37.3.2	Preliminary in Vitro Aggregation Studies *1206*
37.3.3	In Vivo Aggregation Studies *1206*
37.4	Analyzing Polyglutamine Behavior With Synthetic Peptides: Practical Aspects *1207*
37.4.1	Disaggregation of Synthetic Polyglutamine Peptides *1209*
37.4.2	Growing and Manipulating Aggregates *1210*
37.4.2.1	Polyglutamine Aggregation by Freeze Concentration *1210*
37.4.2.2	Preparing Small Aggregates *1211*
37.5	In vitro Studies of PolyGln Aggregation *1212*
37.5.1	The Universe of Protein Aggregation Mechanisms *1212*
37.5.2	Basic Studies on Spontaneous Aggregation *1213*
37.5.3	Nucleation Kinetics of PolyGln *1215*
37.5.4	Elongation Kinetics *1218*
37.5.4.1	Microtiter Plate Assay for Elongation Kinetics *1219*
37.5.4.2	Repeat-length and Aggregate-size Dependence of Elongation Rates *1220*
37.6	The Structure of PolyGln Aggregates *1221*
37.6.1	Electron Microscopy Analysis *1222*
37.6.2	Analysis with Amyloid Dyes Thioflavin T and Congo Red *1222*
37.6.3	Circular Dichroism Analysis *1224*
37.6.4	Presence of a Generic Amyloid Epitope in PolyGln Aggregates *1225*
37.6.5	Proline Mutagenesis to Dissect the Polyglutamine Fold Within the Aggregate *1225*
37.7	Polyglutamine Aggregates and Cytotoxicity *1227*
37.7.1	Direct Cytotoxicity of PolyGln Aggregates *1228*
37.7.1.1	Delivery of Aggregates into Cells and Cellular Compartments *1229*
37.7.1.2	Cell Killing by Nuclear-targeted PolyGln Aggregates *1229*
37.7.2	Visualization of Functional, Recruitment-positive Aggregation Foci *1230*
37.8	Inhibitors of polyGln Aggregation *1231*
37.8.1	Designed Peptide Inhibitors *1231*
37.8.2	Screening for Inhibitors of PolyGln Elongation *1231*
37.9	Concluding Remarks *1232*
37.10	Experimental Protocols *1233*

37.10.1	Disaggregation of Synthetic PolyGln Peptides *1233*
37.10.2	Determining the Concentration of Low-molecular-weight PolyGln Peptides by HPLC *1235*
	Acknowledgements *1237*
	References *1238*

38   **Production of Recombinant Proteins for Therapy, Diagnostics, and Industrial Research by in Vitro Folding** *1245*
Christian Lange and Rainer Rudolph

38.1	Introduction *1245*
38.1.1	The Inclusion Body Problem *1245*
38.1.2	Cost and Scale Limitations in Industrial Protein Folding *1248*
38.2	Treatment of Inclusion Bodies *1250*
38.2.1	Isolation of Inclusion Bodies *1250*
38.2.2	Solubilization of Inclusion Bodies *1250*
38.3	Refolding in Solution *1252*
38.3.1	Protein Design Considerations *1252*
38.3.2	Oxidative Refolding With Disulfide Bond Formation *1253*
38.3.3	Transfer of the Unfolded Proteins Into Refolding Buffer *1255*
38.3.4	Refolding Additives *1257*
38.3.5	Cofactors in Protein Folding *1260*
38.3.6	Chaperones and Folding-helper Proteins *1261*
38.3.7	An Artificial Chaperone System *1261*
38.3.8	Pressure-induced Folding *1262*
38.3.9	Temperature-leap Techniques *1263*
38.3.10	Recycling of Aggregates *1264*
38.4	Alternative Refolding Techniques *1264*
38.4.1	Matrix-assisted Refolding *1264*
38.4.2	Folding by Gel Filtration *1266*
38.4.3	Direct Refolding of Inclusion Body Material *1267*
38.5	Conclusions *1268*
38.6	Experimental Protocols *1268*
38.6.1	Protocol 1: Isolation of Inclusion Bodies *1268*
38.6.2	Protocol 2: Solubilization of Inclusion Bodies *1269*
38.6.3	Protocol 3: Refolding of Proteins *1270*
	Acknowledgements *1271*
	References *1271*

39   **Engineering Proteins for Stability and Efficient Folding** *1281*
Bernhard Schimmele and Andreas Plückthun

39.1	Introduction *1281*
39.2	Kinetic and Thermodynamic Aspects of Natural Proteins *1281*
39.2.1	The Stability of Natural Proteins *1281*
39.2.2	Different Kinds of "Stability" *1282*
39.2.2.1	Thermodynamic Stability *1283*

39.2.2.2	Kinetic Stability	*1285*
39.2.2.3	Folding Efficiency	*1287*
39.3	The Engineering Approach	*1288*
39.3.1	Consensus Strategies	*1288*
39.3.1.1	Principles	*1288*
39.3.1.2	Examples	*1291*
39.3.2	Structure-based Engineering	*1292*
39.3.2.1	Entropic Stabilization	*1294*
39.3.2.2	Hydrophobic Core Packing	*1296*
39.3.2.3	Charge Interactions	*1297*
39.3.2.4	Hydrogen Bonding	*1298*
39.3.2.5	Disallowed Phi-Psi Angles	*1298*
39.3.2.6	Local Secondary Structure Propensities	*1299*
39.3.2.7	Exposed Hydrophobic Side Chains	*1299*
39.3.2.8	Inter-domain Interactions	*1300*
39.3.3	Case Study: Combining Consensus Design and Rational Engineering to Yield Antibodies with Favorable Biophysical Properties	*1300*
39.4	The Selection and Evolution Approach	*1305*
39.4.1	Principles	*1305*
39.4.2	Screening and Selection Technologies Available for Improving Biophysical Properties	*1311*
39.4.2.1	In Vitro Display Technologies	*1313*
39.4.2.2	Partial in Vitro Display Technologies	*1314*
39.4.2.3	In Vivo Selection Technologies	*1315*
39.4.3	Selection for Enhanced Biophysical Properties	*1316*
39.4.3.1	Selection for Solubility	*1316*
39.4.3.2	Selection for Protein Display Rates	*1317*
39.4.3.3	Selection on the Basis of Cellular Quality Control	*1318*
39.4.4	Selection for Increased Stability	*1319*
39.4.4.1	General Strategies	*1319*
39.4.4.2	Protein Destabilization	*1319*
39.4.4.3	Selections Based on Elevated Temperature	*1321*
39.4.4.4	Selections Based on Destabilizing Agents	*1322*
39.4.4.5	Selection for Proteolytic Stability	*1323*
39.5	Conclusions and Perspectives	*1324*
	Acknowledgements	*1326*
	References	*1326*

Index   *1334*

17
Fluorescence Resonance Energy Transfer (FRET) and Single Molecule Fluorescence Detection Studies of the Mechanism of Protein Folding and Unfolding

Elisha Haas

Abbreviations

AMP, adenosine 5′-monophosphate; ANS, anilino naphthalene sulfonate; ATP, adenosine triphosphate; AK, *E. Coli* adenylate kinase (EC 2.7.4.3); BPTI, bovine pancreatic trypsin inhibitor; (1-n)BPTI, N^aMNA-Arg^1-N^e-DA-coum-lys^n-BPTI; DA, donor and acceptor, D, donor; DA-coum, 7-(dimethylamino-(-coumarin-4-yl-acetyl; DTT, dithiothriethol; *E*, transfer efficiency (%); EED, end-to-end distance; FRET, resonance energy transfer; FWHM, full width at half maximum; GdmCl, guanidinium hydrochloride; LI, local interaction; LID, a domain in AK; MNA, 2-methoxy-naphthyl-1-methylenyl; NLI, nonlocal interaction; PGK, phosphoglycerate kinase; PMT, photomultiplier tube; RNase A, ribonuclease A; R(1-n)BPTI, reduced (1-n)BPTI; R-BPTI, reduced BPTI; Tris, tris-(hydroxymethyl)aminomethane.

17.1
Introduction

An ideal folding experiment would yield the distributions of the coordinates of each atom or residue in a protein molecule at each time interval during the transition from any ensemble of nonnative conformers to the ensemble of native conformers. The ideal time resolution (i.e., the length of the time interval) should be defined by the rates of motions of the chain segments, within the nanosecond to microsecond time regime. While such an experiment is not feasible with present-day techniques, methods for the determination of intramolecular distances with subnanosecond time resolution provide significant though partial information. This determination can be achieved through the development of methods based on distance-dependent interactions between selected main-chain or side-chain atoms or dipoles.

Multidimensional NMR spectroscopy of isotopically labeled protein samples has the potential for the rapid determination of solution structures of partially folded protein molecules (see Chapter 16). Although the first method of choice, multidimensional NMR spectroscopy has three primary limitations that restrict the applicability of 2D and 3D NMR spectroscopy in folding research: limited sensitivity (millimolar concentrations are required while unfolded and partially folded proteins are very insoluble); limited time resolution and the short interaction length which is limited to less than 5 Å. The second method of choice is the optical analog of the measurements of the magnetic dipole–dipole interactions. Measurements of dynamic nonradiative excitation energy transfer (FRET) [1–6], which is based on the distance-dependent interactions between excited state dipoles, can be applied for the determination of long-range distances between amino-acyl residues in partially folded proteins with the following capabilities: close to the ideal time resolution; very high sensitivity up to the single molecule detection, over distances of molecular dimensions (10–100 Å); and the ability to recover distributions of intramolecular distances in transient ensembles of refolding protein molecules within the fast folding transitions. With these capabilities, FRET measurements can help answer some of the central questions in the problem of protein folding as described in the next section.

17.2
What are the Main Aspects of the Protein Folding Problem that can be Addressed by Methods Based on FRET Measurements?

17.2.1
The Three Protein Folding Problems

The term "protein folding" covers three different general phenomena: (a) the transition of the backbone to its native fold, known as the *chain entropy problem*; (b) the stabilization of the native fold, that is *the stability problem*; and (c) the refolding of the protein molecule to different conformations in response to small changes of the solution conditions, ligand binding, membrane insertion, or protein–protein interactions. This is known as *the function problem*.

17.2.1.1 The Chain Entropy Problem
The first protein folding problem refers primarily to the reduction of the conformational entropy of the main chain. According to the experimental data [7], the backbone entropy loss is a factor of almost 3 times greater than the factor of side-chain packing. Therefore protein folding is largely a backbone conformational transition process, rather than the process of side-chain ordering and packing.

A central difficulty of the chain entropy problem lies in the unknown nature of the 1D to 3D information translation in transferring amino-acyl sequence information to the native protein conformation. The information processing cannot occur in a sequential symbol by symbol fashion, but must operate simultaneously using remote parts of the sequence, and hence is essentially a nonlocal, collective process rather than the trivial translation of a message [8]. Moreover all along the folding pathways, the folding intermediates and the transient, partially ordered states of protein molecules form ensembles that can be characterized by distributions of intramolecular distances. The following parameters best characterize the conformational transitions: the means; the widths; the number of subpopulations; and the shapes of the distributions of the intramolecular distances. Time-resolved FRET (trFRET) measurements of double-labeled protein samples can yield these parameters. Both ensemble and single molecule FRET experiments were applied in studies targeted to the chain entropy problem.

Three common stages compose the folding transition of most proteins. These stages can occur sequentially or in parallel. The hydrophobic collapse is the first stage of folding. In this stage, a question of much interest is asked: is the collapse a general nonspecific solvent exclusion process which reduces the volume and hence the conformational space available for the protein molecules, or is the collapse a specifically directed transition in which specific subdomain structures are formed? The second stage of folding is the formation of secondary and tertiary structures according to the balance of local and nonlocal, native and nonnative interactions (see Chapters 6 and 7) [9]. Most proteins are very complex systems made of multiple structural elements. The search for the timing of formation of such native or nonnative structural elements during the first two folding stages either as a function of time or as a function of the solution conditions is another major challenge of experimental folding research. It calls for methods for determination of distributions of intramolecular distances in the protein molecule either in the kinetic or equilibrium transitions. The third stage of folding is final packing of side chains and secondary structure elements.

In the unfolded or partially folded states, protein molecules undergo rapid conformational fluctuations. Changes in these parameters are another characteristic that reflect the progress of the folding of chain elements. The lifetime of the excited states of donor probes in trFRET experiments (nanoseconds) define time windows that enable the detection of rapid fluctuations of intramolecular distances both in the ensemble and in the single molecule modes. Slower fluctuations can be detected by single molecule FRET spectroscopy or autocorrelation analysis of intensity fluctuations of double-labeled protein samples.

FRET experiments can address also the question of the extent of randomness of protein molecule structures in the presumably unfolded or denatured states (see Chapter 20). Small bias from statistical coil distributions of intramolecular distances can serve as a sensitive measure of subdomain structures in otherwise unfolded protein molecules. Such bias is of particular interest in the context of the so-called "natively unfolded" proteins (see Chapter 8 in Part II). These proteins appear to be unfolded by some measurements, but might demonstrate partial order that is difficult to detect by other measurements.

This chapter will focus mainly on the first problem with some reference to the third problem.

17.2.1.2 The Function Problem: Conformational Fluctuations

A protein molecule may be regarded as a "system" in the thermodynamic sense, immersed in a solvent "bath" [10]. The number of degrees of freedom of a protein and its volume are relatively small (although still amounting to several thousands).

The protein molecule, may therefore be regarded as a "small system" [10]. In small systems, fluctuations are an inherent characteristic, and such systems may spend a substantial amount of time in states quite different from the average conformation.

The extent of fluctuations was estimated, at least qualitatively, from measurable thermodynamic properties. For example, the heat capacity of a system is linked to fluctuations in energy through the following relation [10]:

$$\langle \partial E^2 \rangle = k_B T^2 C_V \tag{1}$$

Where $\langle \partial E^2 \rangle$ is the mean squared energy fluctuation, related to the second moment of energy distribution, k_B is the Boltzmann constant, T is the absolute temperature, and C_V is the heat capacity at constant volume. A typical heat capacity for proteins is approximately 0.3 cal K^{-1} g^{-1} [11], and for a 25 000 Da protein this leads to a root-mean-squared fluctuation of 35 kcal mol^{-1}. This number should be compared to the common average free energy of stabilization of the folded states of proteins, ~10–15 kcal mol^{-1} [12].

The dynamics of protein molecules do not have one typical time scale, and there is a hierarchy of time scales for motions of different structural elements in a protein. The fastest motions are those of side chains that occur on the picosecond time scale, while slow, large scale conformational changes may take microseconds to be completed [13]. The investigation of the functional properties of many proteins, especially enzymes, still relies on relatively slow kinetic techniques [14]. Ensemble trFRET and single molecule FRET spectroscopy are best posed to detect and characterize such fluctuations.

In short, the sensitivity, the time and spatial resolution, the specificity of measurements of structural elements, the option of measurements of distributions of intramolecular distances and their rapid fluctuations by the ensemble trFRET and single molecule FRET experiments, makes FRET a powerful tool in protein folding research. The currently limited extent of applications of this approach is probably a result of the complex procedure involved in the biochemistry and chemistry of site specific labeling of the protein samples. The following sections contain the theoretical background of the FRET experiments, the principles of their applications in folding research, and the essential control experiments that should support the data analyses of these experiments.

17.3
Theoretical Background

17.3.1
Nonradiative Excitation Energy Transfer

Nonradiative transfer of excitation energy requires some interaction between a donor molecule and an acceptor molecule. This transfer can occur if at least some vibronic levels in the donor have the same energy as the corresponding transitions in the acceptor. Energy transfer can result from different interaction mechanisms. The interactions may be Coulombic and/or due to intermolecular orbital overlap. The Coulombic interactions consist of long-range dipole–dipole interactions (Förster's mechanism) and short-range multipolar interactions. In the energy transfer process, the initially excited electron on the donor, D, returns to ground state orbital on D, while simultaneously an electron on the acceptor, A, is promoted to an excited state. For permitted transitions on D and A, the Coulombic interaction is predominant, even at short distances. This is the case of the singlet-singlet transfer:

$$^1D^* + {}^1A \rightarrow {}^1D + {}^1A^*$$

which is effective over a long interaction range (up to 80–100 Å). For forbidden transitions on D or A, the Coulombic interaction is negligible and the exchange mechanism is dominant. Here, the interaction is effective only at short distances (< 10 Å) because the interaction requires the overlap of the molecular orbitals. This is the case of triplet–triplet transfer

$$^3D^* + {}^1A \rightarrow {}^1D + {}^3A^* \quad \text{and} \quad {}^3D^* + {}^1A \rightarrow {}^1D + {}^1A^*$$

where the transitions $T_1 \rightarrow S_o$ in D and $S_o \rightarrow T_1$ on A are forbidden (Figure 17.1).

Two main mechanisms are relevant in the context of the application of energy transfer in protein folding research: The Förster mechanism for long-range interactions, and the triplet–triplet transfer for the detection of short-range interactions.

17.3.2
What is FRET? The Singlet–Singlet Excitation Transfer

The term fluorescence resonance energy transfer (FRET) is commonly used to describe singlet–singlet energy transfer via a mechanism based on long-range dipole–dipole resonance coupling[1] [1].

Classically, it is possible to approximate the donor and the acceptor molecules by idealized oscillating dipoles (or higher order multipoles). The electric field surrounding the emitting oscillating dipole is expressed by equation that includes several terms [15] (page 158 in ref. [15]). At distances as large as the wavelength, the term that contributes the radiation of energy dominates. At shorter distances the other terms that describe the electric field surrounding the oscillating dipole dominate. An acceptor chromophore positioned close to the donor will therefore

---

1) The acronym FRET denoting fluorescence resonance energy transfer is incorrect because the transfer does not involve any fluorescence but the electronic excitation energy of the donor. This mechanism does not depend on the fluorescence properties of the acceptor. The correct term should be EET, representing "electronic energy transfer" or "excitation energy transfer" or RET for "resonance energy transfer." However, since the literature is saturated with the term FRET, we will use it.

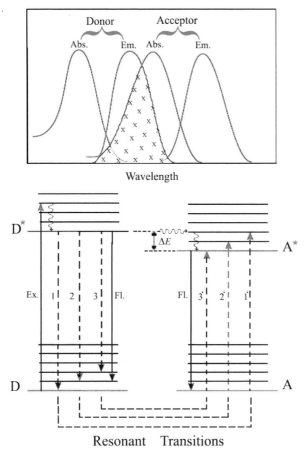

Fig. 17.1. Energy level diagram demonstrating the mechanism of resonance coupling of the nonradiative transitions in the donor and acceptor probes under conditions where the vibrational relaxation is faster than the energy transfer (very weak coupling). The nonradiative transitions are shown in dashed lines and the radiative transitions are shown as continuous vertical lines. The upper inset shows a scheme of the corresponding absorption and emission spectra of the probes and the overlap integral.

be affected by the components of the field that do not contribute to the radiation of energy.

17.3.3
Rate of Nonradiative Excitation Energy Transfer within a Donor–Acceptor Pair

A molecule in an electronically excited state (donor) can transfer its excitation energy to another molecule (acceptor) [1] provided that the pair fulfills several conditions [20]: (a) The energy donor must be luminescent; (b) the emission spectrum of the donor should have some overlap with the absorption spectrum of the acceptor; and (c) the distance between the two probes should not exceed an upper limit (usually within the range of up to 80 Å). The transfer is readily observed when two different probes are involved, but transfer within a homogeneous population of chromophores can occur. This transfer was detected already in 1927 by Perrin by the loss of polarization of the emitted light [16]. The theory for the mechanism of long-range nonradiative energy transfer was developed by Förster with both classical and quantum mechanical approaches. In both approaches, the interaction is assumed to be of the dipole–dipole type between the energy donor and acceptor [1]. Förster considered two neighboring, electrically charged classical mechanical oscillators coupled through weak electrostatic interactions. This coupling is analogous to two resonating pendulums that are weakly coupled mechanically and exchange oscillation energy back and forth. Förster assumed that only the donor oscillator is initially vibrating. Classically, the donor oscillator will lose energy via radiation and nonradiative exchange with the acceptor provided that they are close enough. The two dipoles, of dipole moment μ (an approximation) interact over distance R with interaction energy

$$E_{int} \approx \kappa \mu^2 / n^2 R^3 \qquad (2)$$

where κ is the orientation factor (see below) and n is the refractive index of the medium. Förster calculated the distance between the two oscillators where the time for radiative emission of a classical oscillator $\tau_{rad} \approx 3hc^2/\mu^2\omega^2$ and the time for exchange of energy between the two oscillators $\tau_{int} \approx h/E_{int} = hn^2 R^3/\kappa\mu^2$ are equal. Here $R \equiv R_o$ when $\tau_{rad} = \tau_{int}$. R_o, also known as the Förster critical distance, is the distance between the donor and the acceptor, where half of the energy that would have otherwise been emitted by the donor is transferred to the acceptor. Spectral broadening by very rapid dephasing of the molecular excitation energies lengthens the time considerably for the exchange of energy between the weakly coupled oscillators, and this leads to smaller R_o values. Förster calculated the probability of full overlap of the spectral regions of the two oscillators and w, the probability that the energies of both oscillators are simultaneously nearly the same within the small interaction energy E_{int}. This probability is $w = (\Omega'/\Omega)(E_{int}/h\Omega)$, where Ω and Ω' are respectively the spectral width (assumed to be the same for both probes) and the overlap spectral region (the second term is the probability that frequency within the broadband frequencies, Ω, will fall within the relatively narrow bandwidth of E_{int}/h). The rate of transfer calculated without this dephasing correction $(1/\tau_{int} \approx \kappa\mu^2/hn^2 R^3)$ is then multiplied by w to give the rate of transfer with real systems:

$$K_T \approx 1/\tau'_{int} = w/\tau_{int} = \mu^4 \kappa^2 \Omega' / (h^2 n^4 R^6 \Omega^2) \qquad (3)$$

Here, we see that the correction for the spectral broadening in real systems gives the famous R^6 dependence of the rate of excitation energy transfer on the interchromophoric distance.

By solving the distance R_o where $\tau_{rad} = \tau'_{int}$, Förster calculated the dependence of R_o on the characteristics of the two probes:

$$R_o^6 = 3(c^3/\omega^3)\mu^2\kappa^2(\Omega'/\Omega^2)(1/hn^4) = 9(\lambda^6/2\pi(\kappa^2/n^4)(\Omega'/\Omega^2)(1/\tau_{rad})) \tag{4}$$

where $\tau_{rad} = 3hc^3/\mu^2\omega^3$ replaces μ^2 and $c = \lambda\omega/2\pi$. The most relevant feature of this derivation in the context of application of FRET methods in folding research is the $1/R^6$ dependence of the FRET efficiency. This strong distance dependence derives from the distance dependence of the interaction energy of a dipole on $1/R^3$ which is multiplied by the correction for the dephasing of the two spectra. The strict requirement for exact resonance between the two interacting dipoles means that overlapping narrower (line) spectra can exchange energy over larger distances. Förster emphasized that although the transfer mechanism depends on exchange of excitation energy quanta, it is essentially a resonance mechanism and hence the classical approximation gives satisfactory results. Förster published his rigorous quantum mechanical derivation in 1948 and showed that the original classical derivation included all the essentials of process [1].

An alternative classical derivation by Steinberg et al. [17] also assumes an interaction between two idealized oscillating dipoles. Let us denote the donor dipole by $f = f_o \sin 2\pi v t$ where f_o is the amplitude, t is time, and v is the frequency of oscillations. The fields surrounding such an oscillating dipole are given in polar coordinates as follows:

$$E_r = 2\left[\frac{f_o}{D_e}\cos\theta\right]\left[\frac{2\pi v}{r^2 c/n}\cos 2\pi v\left(t - \frac{r}{c/n}\right) + \frac{1}{r^3}\sin 2\pi v\left(t - \frac{r}{c/n}\right)\right]$$

$$E_\theta = \left[\frac{f_o}{D_e}\sin\theta\right]\left[-\frac{(2\pi v)^2}{r(c/n)^2}\sin 2\pi v\left(t - \frac{r}{c/n}\right) + \frac{2\pi v}{r^2 c/n}\cos 2\pi v\left(t - \frac{r}{c/n}\right)\right.$$

$$\left. + \frac{1}{r^3}\sin 2\pi v\left(t - \frac{r}{c/n}\right)\right]$$

$$H_\phi = \left[\frac{f_o}{(D_e\mu)^{1/2}}\sin\theta\right]\left[-\frac{(2\pi v)^2}{r(c/n)^2}\sin 2\pi v\left(t - \frac{r}{c/n}\right) + \frac{2\pi v}{r^2 c/n}\cos 2\pi v\left(t - \frac{r}{c/n}\right)\right]$$

$$E_\phi = H_r = H_\theta = 0 \tag{5}$$

where E and H are the electric and magnetic fields, respectively. The subscripts r, θ and ϕ denote components along the r, θ and ϕ spherical coordinates, respectively; D_e and μ are respectively, the dielectric constants and magnetic permeability of the medium; n is the refractive index of the medium, and c is the velocity of light in vacuum. Only components whose Poynting's vector $[(c/4\pi)(E \times H)]$ is nonzero carry radiative energy flux. Only the first terms of E_θ and H_ϕ, which show $1/r$ dependence contribute to the energy flux radiated from the oscillating dipole over complete cycle of oscillation. The other terms carry energy from the dipole to the field and back in an oscillatory way, with zero net radiative flux. At distances shorter than the wavelength of the emitted radiation, the $1/r^3$ terms in E_θ and E_r are the dominant ones in magnitude, but they do not carry a net flux of energy. An acceptor molecule positioned at such short distances from the oscillating dipole is affected predominantly by the field components which do not contribute to the radiation field. Therefore the excitation transfer has no effect on the shape of the emission spectrum of the donor. The transferred energy is drawn from the field components that exchange energy with the oscillating dipole.

The dipole–dipole interaction that leads to the transfer of excitation energy is very weak, usually of the order of ~2–4 cm^{-1}, while the spectroscopic energies that are transferred are much higher, ~15 000–40 000 cm^{-1}.

Based on the above discussion, which identifies the short-range components of the dipole field, the rate of energy transfer can be calculated. Several treatments lead to the same results. Förster's theory describe the rate of energy transfer for an isolated pair of chromophores which fulfill the requirements for energy transfer by the dipole–dipole interaction to be

$$k_T = \frac{9(\ln 10)\kappa^2 \Phi_D^o}{128\pi^5 N_A' r^6 \tau_D^o} \int_0^\infty f(\lambda)\varepsilon(\lambda)\lambda^4 \, d\lambda \tag{6}$$

$$k_T = \frac{1}{\tau_D^o}\left(\frac{R_o}{r}\right)^6 \tag{6a}$$

where $\kappa = \cos\theta_{DA} - 3\cos\theta_D \cos\theta_A = \sin\theta_D \sin\theta_A \cos\varphi - 2\cos\theta_D \cos\theta_A$ (θ_{DA} is the angle between the donor and the acceptor dipoles; θ_D and θ_A are respectively the angles between the donor and the acceptor dipoles and the line joining their centers and φ is the angle between the projections of the transition moments on the plane perpendicular to the line through the centers (Figure 17.2). κ^2 can in principle assume values from 0 (perpendicular transition moments) to 4 (collinear

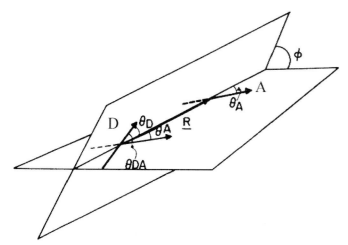

Fig. 17.2. The angles involved in the definition of the orientation factor κ^2.

transition moments), or 1 when the transition moments are parallel)); Φ_D^o and τ_D^o are respectively the quantum yield and the fluorescence lifetime of the donor in the absence of an acceptor; r is the distance between the centers of the two dipoles (donor and acceptor); N_A' is Avogadro's number per mmole, $f(\lambda)$ is the fluorescence intensity of the donor in the range λ to $\lambda + d\lambda$ normalized so that $\int_0^\infty f(\lambda)\, d\lambda = 1$ and $\varepsilon(\lambda)$ is the absorption coefficient of the acceptor at the wavelength λ. In Eq. (6), r and λ are in centimeters, $\varepsilon_A(\lambda)$ in cm^2 mol^{-1}, and $J(\lambda)$ in cm^6 mol^{-1}. When those units are used R_o is defined by Eq. (7) and is given by:

$$R_o^6 = 8.8 \times 10^{-28} \Phi_D^o \kappa^2 n^{-4} J \tag{7}$$

where J is the overlap integral in Eq. (6). The energy transfer process competes with the spontaneous decay of the excited state of the donor, characterized by the rate constant $k_D^o = 1/\tau_D^o$. If $\varepsilon_A(\lambda)$ is given in cm^{-1} mol^{-1} L units, then $R_o^6 = 8.8 \times 10^{-25} \Phi_D^o \kappa^2 n^{-4} J$. Thus the probability ρ for the donor to retain its excitation energy during the time t after excitation is given by:

$$-\left(\frac{1}{\rho}\right)\frac{d\rho}{dt} = \frac{1}{\tau_D^o} + \frac{1}{\tau_D^o}\left(\frac{R_o}{r}\right)^6 \tag{8}$$

and the efficiency of E of energy transfer is expressed by:

$$E = \frac{R_o^6}{R_o^6 + r^6} \tag{9}$$

R_o is thus the inter-dipole distance at which the transfer efficiency (and the donor lifetime) is reduced to 50% and has the strongest dependence on changes of that distance (Figure 17.3).

There is strong experimental support for Förster's equations (Eqs (6)–(9)) (reviewed by Steinberg [4]. Weber and Teal [18], and Latt et al. [19] showed the dependence on the overlap integral. The R^{-6} dependence was demonstrated by Stryer and Haugland [20].

Equation (9) shows that the distance between a donor and an acceptor can be determined by measuring the efficiency of transfer, provided that r is not too different from R_o and that all molecules in the sampled ensemble share the same intramolecular distance. If this is not the case, an average distance r_{av} that would be extracted from the measured transfer efficiency does not correspond to any simple average distance.

The efficiency of energy transfer is independent of the value of the lifetime of the excited state of the donor. Efficient energy transfer can occur even for "long life" excited states (e.g., for phosphorescence emission) provided that the quantum yield of emission is reasonably high. Such transfer phenomena has been observed. Equations (8) and (9) thus define the characteristic time and distance ranges ("windows") in which the transfer efficiency, E, is most sensitive to conformational transitions and their rates (Figure 17.3). Typical values available for probes that are used in protein chemistry are in the range of 10–80 Å and down to picosecond

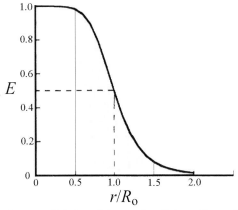

Fig. 17.3. Variation of the transfer efficiency, E, as a function of the distance between the donor and the acceptor probes. The shadowed range represents the limits of the distance range where reliable measurements of distances are possible.

time intervals. Many conformational changes and processes in proteins occur in these intervals. Hence it is in the hands of the researcher to design the experiments with the time and distance resolutions most suitable for each molecular question.

17.3.4
The Orientation Factor

Distance determination by measurements of transfer efficiencies is complicated by the strong dependence of the probability for energy transfer on the orientation of the interacting dipoles. Therefore knowledge of the orientation factor, κ^2, is essential for applications of FRET in studies of protein folding. When the two dipoles undergo rapid orientational averaging during the lifetime of the excited state of the donor, the orientation factor can be averaged to yield the numerical value of 2/3.

The large span of possible κ^2 values, between 0 and 4, makes it the primary source of uncertainty in distance determination by FRET measurements.

The main uncertainties that should to be eliminated in FRET-based folding experiments are (a) the question whether changes in transfer efficiencies during folding are caused by changes of intramolecular distances or merely by changes of the orientation of the probes and (b) in time-resolved FRET experiments, multi-exponential decay of the donor emission can result from the distribution of either the distances or the orientations. Various approaches were proposed for the reduction of these uncertainties and the determination of the range of possible error in distance determination. Dale et al. [21–23] reported the most extensive studies in which they utilized the partial rotations of the chromophores (as revealed by their emission anisotropy) to estimate limits for the range that κ^2 can assume. In cases where the distance between the donor and the acceptor does not change appreciably during the time that the transfer takes place, it is possible to derive a $\langle \kappa^2 \rangle_{min}$ and a $\langle \kappa^2 \rangle_{max}$ from anisotropy measurements. This procedure is based on assumptions that the two probes can be represented by single oscillating dipoles whose orientations fluctuate (e.g., wobble) within a cone and that the orientation of the axis of the cone also fluctuates. κ^2 values can be averaged over the corresponding range of orientations. Rapid reorientation during the lifetime of the excited state of the donor is reflected in the anisotropy measurements and therefore can be estimated experimentally for each preparation. The average κ^2 will depend on "depolarization factors" that are (averages of) second-rank Legendre polynomials of a cosine of a polar angle θ between 0 and π. The depolarization factors have the form:

$$d_\theta = \frac{1}{2} \cos^2 \theta - \frac{1}{2} \Rightarrow -\frac{1}{2} \leq d_\theta \leq 1 \tag{10}$$

or

$$\langle d_\theta \rangle = \left\langle \frac{1}{2} \cos^2 \theta - \frac{1}{2} \right\rangle \Rightarrow -\frac{1}{2} \leq \langle d_\theta \rangle \leq 1 \tag{10a}$$

where the brackets denote a weighted average over the range of angles covered by θ during times that are brief as compared to the average transfer time. The depolarization factors are derived from the anisotropy values under each experimental setup and solution conditions (for details of all the angles taken into account see Dale et al. [22, 23] and Van der Meer [6], pp. 55–83).

An ideal case, where full averaging is an intrinsic characteristic of the system, is achieved with atom probes, such as lanthanides and chelated transitions metal ions [5, 24–26]. An alternative approach to this problem is based on the fact that many chromophores show mixed polarizations in their spectral transitions (i.e., their absorption and emission across the relevant spectral range of overlap involves a combination of two or more incoherent dipole moments). The physical basis for this phenomenon has been discussed by Albrecht [27]. Its manifestation is well known in the observation that the limiting anisotropy of many organic probes used in FRET experiments is lower than the theoretically predicted upper limit. Haas et al. [26] showed that the occurrence of mixed polarizations in the energy donor and acceptor may markedly limit the range of possible values that κ^2 can assume and thus in favorable conditions alleviate the problem of the orientation factor in FRET experiments. The fact that any one of the two probes has two or more transition dipole moments involved in the transfer mechanism eliminates the extreme values that κ^2 can assume. It is equivalent to ultrafast rotation of an ideal dipole over the very wide range of orientations and hence to fast averaging. This effect, when considered in planning FRET experiments by using probes that show a high degree of internal depolarization, can reduce the κ^2-related uncertainties to the range of the other common experimental uncertainties.

A table was prepared of the possible range of values of κ^2 corresponding to the experimentally observed anisotropy values for the donor and the acceptor. An improved approximation of the value of κ^2 is obtained when the contribution of the rotation of the whole molecule to the emission anisotropy of the probes is first subtracted from the observed anisotropy. This procedure can be achieved when the anisotropy decay is determined and the analysis shows the relative amplitude associated with the decay component corresponding to the molecular rotational diffusion.

This procedure produces a distribution $P(\kappa^2)d(\kappa^2)$ which is the probability that the orientation factor assumes a value between κ^2 and $\kappa^2 + d(\kappa^2)$ for a frozen system of randomly oriented donor–acceptor pairs. As expected, the extreme values of κ^2 are eliminated even for moderate values of emission anisotropy of the probes. When the emission anisotropy of both the donor and the acceptor is relatively low, $P(\kappa^2)$ attains its maximal value at κ^2 close to $2/3$. This is the case for many common probes which have high levels of mixed polarization, such as aromatic probes and derivatives of naphthalene.

The uncertainty in r can be further reduced by employing two independent FRET experiments. In these experiments different pairs of probes are used for FRET determination of the distance between one pair of sites in a protein molecule.

The tables given in Ref. [26] can be used for the evaluation of the probable range of uncertainty in the distances estimated from FRET experiments for pairs of probes based on measurements of their anisotropies.

17.3.5
How to Determine and Control the Value of R_o?

The distance range for the statistically significant determination of distances for each pair is defined by the characteristics of the probes and is given by the R_o value. Equation (9) shows that the limits of that range are $r = R_o \pm 0.5 R_o$ (Figure 17.3). Maximal sensitivity of the FRET effect to small changes in r is observed when $r = R_o$. Therefore optimal design of FRET experiments includes a selection of a pair of probes that have an R_o value close to the expected r value.

R_o depends on four parameters which are characteristic of each pair of probes (Eq. (6)). These parameters were discussed critically by Eisinger, Feuer, and Lamola [28]. Modulation of the value of R_o of any specific pair in order to adjust it to the expected r value, is limited to a narrow range due to the sixth power of relation of R_o and the spectroscopic constants of the probes. Equation (6) shows that the donor quantum yield Φ_D^o and the overlap integral parameters are the most readily available for manipulation to modulate the R_o value.

17.3.6
Index of Refraction n

The values used for n to calculate R_o for a pair of probes attached to a protein molecule vary widely, from 1.33 to 1.6. Eisinger, Feuer, and Lamola [28] recommended using the value $n = 1.5$ in the region of spectral overlap of the aromatic amino acids in proteins.

17.3.7
The Donor Quantum Yield Φ_D^o

Φ_D^o should be determined for each probe when attached to the protein molecule at the same site used for the FRET experiment. Many experiments showed that the Φ_D^o is affected by both the site of attachment of the probes and by modifications at other sites in the protein molecule; by attachment of an acceptor at second sites; by changes of solvent conditions; and by ligand binding or folding/unfolding transitions. Therefore R_o should be determined repeatedly after changes in the conditions to avoid misinterpretation of changes of R_o as changes of r. The R_o depends on the sixth root of Φ_D^o and is therefore not very sensitive to uncertainties in Φ_D^o. Nevertheless, this same dependence limits the range that R_o can be manipulated. For instance, a twofold decrease Φ_D^o would result in ~9% decrease of R_o. Increasing Φ_D^o without a change of probes usually requires change of solvent and therefore is not very useful in folding studies; but Φ_D^o can be decreased by adding quenching solute without a change of solvent.

Frequently bound donor probes may have few modes of interactions with the macromolecular environment in the ground or in the electronically excited state. In such a case, each subpopulation which has a different lifetime also has a different R_o, as indicated in Eqs (6) and (7).

17.3.8
The Spectral Overlap Integral J

The most effective variation of R_o is available by selecting probes according to the overlap of their spectra. The value of J depends on the local environment of the probes. As a rule, absorption spectra vary relatively little with a change of solvent, temperature, or local environment, but emission spectra may be very sensitive to such changes of the local environment of the probes. An example of this dependence is the fluorescence of indole derivatives [29]. The selection of probes with narrow emission and absorption spectra and with high extinction coefficients of the acceptor can yield high R_o values.

17.4
Determination of Intramolecular Distances in Protein Molecules using FRET Measurements

Clear presentation of theoretical background for the methods described below may be found in the monographs by Valeur [30], Lakowicz [31], Van Der Mear [6], and numerous reviews such as Refs [4, 32–35] published in the past 40 years.

17.4.1
Single Distance between Donor and Acceptor

The Förster resonance energy transfer can be used as a spectroscopic ruler in the range 10–100 Å. Measurements can be conducted under equilibrium conditions or in the kinetics mode during conformational transitions. Distances shorter than 8–10 Å should be avoided since short-range multipoles contribute interactions which have different distance dependence.

E can be defined analogously to the quantum efficiency in terms of the rates of elementary competing processes of de-excitation of the donor excited states:

$$E = k_{\text{ret}} / \left(k_{\text{ret}} + k_{\text{fl}} + \sum k_i \right) \tag{11}$$

where $k_{\text{ret}}, k_{\text{fl}}$, and $\sum k_i$ are respectively the rate constants of FRET, fluorescence, and all other competing modes of de-excitation of the donor's first singlet excited state. E can be effectively monitored by observing the effect of the acceptor on either the lifetime of the donor excited state or on emission quantum efficiency. To evaluate E one must compare τ_D^o or Φ_D^o (in the absence of an acceptor) and the donor lifetime τ_D or quantum efficiency Φ_D in the presence of an acceptor.

Determination of distances via determination of FRET efficiency, E, is possible via a determination of the decrease of donor emission or an enhancement of the acceptor emission by steady state and time-resolved methods.

$$r = \left(\frac{1}{E} - 1 \right)^{1/6} R_o \tag{12}$$

Due to the comparative nature of this mode of determination of E, the concentrations of the probes (and hence the labeled protein molecules) and their microenvironments must be the same for all samples in every set of measurements.

17.4.1.1 Method 1: Steady State Determination of Decrease of Donor Emission

The competition of the FRET process results in decreased quantum yield of the donor emission:

$$E = 1 - \frac{\Phi_D}{\Phi_D^o} \tag{13}$$

since only relative quantum yields are to be determined, Eq. (13) can be written directly in terms of single wavelength donor emission intensities at wavelengths where the acceptor emission intensity is negligible:

$$E = 1 - \frac{A(\lambda_D)}{A_D(\lambda_D)} \frac{I_D(\lambda_D, \lambda_D^{\text{em}})}{I_D^o(\lambda_D, \lambda_D^{\text{em}})} \tag{14}$$

The factor A/A_D corrects for the absorption by the acceptor. This method can be readily applied at the single molecule level.

17.4.1.2 Method 2: Acceptor Excitation Spectroscopy

The most direct mode of determination of E, independent of correction factors or differences in concentrations, can be achieved via measurements of acceptor excitation spectra. In this procedure, the acceptor excitation spectra of three samples should be measured under the same solution conditions. These spectra are: (a) The acceptor excitation spectrum of the double-labeled protein (labeled by both the donor and the acceptor), $I_A(\lambda_D, \lambda_A^{em})$; (b) the excitation spectrum of the acceptor in the absence of a donor, using protein sample labeled by the acceptor alone, $I_{ref\,0}(\lambda_D, \lambda_A^{em})$ (a reference for $E = 0$); and (c) the excitation spectrum of the acceptor attached to a double-labeled (donor and acceptor) model compound where E is well known under the conditions of the measurements $I_{ref\,1}(\lambda_D, \lambda_A^{em})$. The acceptor excitation spectra of the three samples monitored at an acceptor emission wavelength (with negligible donor emission contribution) are then normalized at the acceptor excitation wavelength. E can then be obtained with high accuracy by

$$E = \frac{I_A(\lambda_D, \lambda_A^{em}) - I_{ref\,0}(\lambda_D, \lambda_A^{em})}{I_{ref\,1}(\lambda_D, \lambda_A^{em}) - I_{ref\,0}(\lambda_D, \lambda_A^{em})} \tag{15}$$

There is no need for additional independent measurements and exact knowledge of the concentrations or instrumental corrections. (Further reduced noise can be achieved by using the ratio of the area under the excitation spectra in the donor absorption range.) Therefore, this is the preferred method for the determination of E by steady state methods. Non-FRET mechanisms of quenching of the donor might affect the determination of E by method 1. Control experiments should be performed that confirm that the missing donor emission intensity is observed in the acceptor emission. Such non-FRET change of the donor emission does not affect the determination of E by method 2. Both steady state methods might be affected by the inner filter effect and care must be taken to account for changes in emission intensity changes due to that effect.

17.4.2
Time-resolved Methods

Time-resolved emission of the donor or the acceptor fluorescence provides direct information on the transfer rates independent of the concentrations, as this information is in the shape of the fluorescence decay curves, and not in the amplitudes. The donor and the acceptor fluorescence decay curves contain additional information not available from the steady state measurements. This information, which can be extracted directly by proper analytical procedures, is of particular interest for folding studies. An analysis of time-resolved FRET experiments can resolve conformational subpopulations and distributions of distances and fast conformational changes on the nanosecond time scales.

17.4.3
Determination of from Donor Fluorescence Decay Rates

Equation (11) can be rewritten:

$$1/\tau_D = 1/\tau_D^o + k_{ret} \tag{16}$$

and Eq. (13) can be replaced by

$$E = 1 - \frac{\tau_D}{\tau_D^o} \tag{17}$$

and

$$r = \frac{R_o}{(\tau_D/\tau_D^o - 1)^{1/6}} \tag{18}$$

when a nonexponential decay of the donor alone (τ_D^o) is observed, a common situation due to heterogeneity of microenvironments, average lifetime values can be used to calculate E, provided that the deviation from monoexponentiality is moderate:

$$E = 1 - \frac{\langle \tau_D \rangle}{\langle \tau_D^o \rangle} \quad \text{where} \quad \langle \tau \rangle = \sum_i \alpha_i \tau_i \Big/ \sum_i \alpha_i \tag{19}$$

when the heterogeneity of the decay rates is enhanced by the FRET effect, then the use of Eq. (18) might yield only an approximation of an average intramolecular distance. In such a case, distribution analysis accompanied by appropriate control experiments should be applied.

17.4.4
Determination of Acceptor Fluorescence Lifetime

The concentration of excited acceptor probes, $A^*(t)$, after a δ-pulse excitation at the donor excitation wavelength, obeys the following rate law:

$$\frac{dA^*(t)}{dt} = k_{ret} D^*(t) - (1/\tau_A^o) A^*(t) \tag{20}$$

where $D^*(t)$ is the time dependent concentration of excited donor probes. Under ideal conditions, where there is no direct excitation of the acceptor, and the donor decay is single-exponential, the time dependent excited acceptor concentration, $A^*(t)$, shows a "growing in" component (marked by the negative pre-exponent). Under the initial condition that $D^*(t=0) = D^*(0)$, the time dependence of the concentration of excited acceptor probes follows a nonexponential decay law,

$$A^*(t) = \frac{D^*(0)k_{\text{ret}}}{1/\tau_D - 1/\tau_A^o}[e^{-t/\tau_A^o} - e^{-t/\tau_D}] \tag{21}$$

This expression is typical of an excited state reaction where the second exponent is equal to the decay of the donor excited state that produces the acceptor excited states. The pre-exponential factor of the donor excited state is the same as for the first exponent, but is negative. Therefore the measurement of the acceptor decay under excitation of the donor is an important control experiment which confirms that any reduction of the donor excited state lifetime is due to the FRET mechanism, and that the distance determination is valid.

In practice, direct excitation of the acceptor to create a subpopulation, $A^*(0) = A^*(t=0)$, of directly excited acceptor probes at $t=0$, cannot be avoided in most cases. Consequently, Eq. (21) should be rewritten to add the decay of this subpopulation:

$$A^*(t) = \left[\frac{D^*(0)k_{\text{ret}}}{1/\tau_D - 1/\tau_A^o} + A^*(0)\right]e^{-t/\tau_A^o} - \frac{D^*(0)k_{\text{ret}}}{1/\tau_D - 1/\tau_A^o}e^{-t/\tau_D} \tag{22}$$

17.4.5
Determination of Intramolecular Distance Distributions

Distance determination by means of steady state or time-resolved measurements using Eqs (13) or (18) is valid for single molecule experiments or when all molecules in an ensemble share the same D–A distance during the time of the data collection. However, a situation where all molecules in an ensemble of partially folded molecules share the same intramolecular distances is quite rare, and an equilibrium distribution of intramolecular distances, $N_o(r)$, with finite FWHM is the rule. In most folding experiments, the FRET efficiency determined by steady state measurements is averaged over the efficiencies of all fractions in the distribution,

$$\langle E \rangle = \int_0^\infty N_o(r) \frac{R_o^6}{R_o^6 + r^6} dr \tag{23}$$

where $N_o(r)\,dr$ is the fraction of molecules with D–A pairs at distance r to $r + dr$. It is not possible to evaluate $N_o(r)$ from single steady state measurement. Cantor and Pechukas [36] suggested the use of a series of measurements where the same ensemble would be labeled by different pairs of probes, characterized by different R_o values for each experiment. A single time-resolved experiment contains all the information needed for the determination of $N_o(r)$.

Consider the time dependence of a sample of the labeled molecules in which the donor has been excited at time $t=0$ by a very short pulse of excitation light. At the moment of excitation, the intramolecular distance probability distribution (IDD) of

17.4 Determination of Intramolecular Distances in Protein Molecules using FRET Measurements

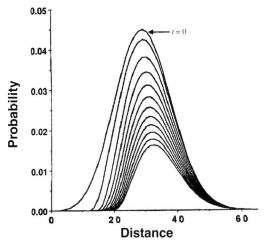

Fig. 17.4. Demonstration of the perturbation of the shape of the intramolecular distance distribution by the energy transfer effect. Simulation of the time dependence of an intramolecular distance distribution of the excited state donor subpopulation, $N^*(r, t)$, starting at time $t = 0$ with subsequent time intervals every 0.4 ns ($R_o = 27$ Å, $\tau = 6.0$ ns) without diffusion. The shift of the mean of the distribution to longer distances and the depletion of the short distance fractions as a function of time is demonstrated.

the sample of molecules with excited state donor probe, which is selected at random by the excitation pulse, $N^*(r, 0)$, is congruent with the equilibrium distance distribution of the whole population, $N_o(r)$ (within a proportionality factor). The rate of decay of the donor excited state in the fraction of labeled protein molecules which are characterized by a short D–A distance is nonlinearly accelerated relative to the corresponding rate observed for labeled molecules with larger D–A distances. This is due to the $1/r^6$ dependence of the dipole–dipole interactions. Figure 17.4 shows the rapid depletion of the lower end fractions of $N^*(r, t)$.

The net effect is (a) the generation of a virtual concentration gradient (within the original Gaussian boundaries); (b) a time-dependent shift of the mean of the time-dependent distance distribution of the sample of protein molecules with excited donor probe, $N^*(r, t)$, to larger intramolecular distances; and (c) an enhanced rate of decay of the excited state of the donor probe. This effect can be illustrated as an equivalent of optical "hole burning" in the distance probability distribution. When the intramolecular distances are unchanged during the lifetime of the excited state of the donor, the decay kinetics of the donor emission of the entire population of molecules with excited donor is multiexponential,

$$i(t) = k \int_0^{\infty} N_o(r) \exp\left[-\frac{t}{\tau_D^o} \left(1 + \left(\frac{R_o}{r^6} \right)^6 \right) \right] dr \tag{24}$$

where k is a proportionality factor. Knowledge of R_o, and τ_D^o enables the extraction of $N_o(r)$ from the multiexponential decay curve, $i(t)$.

17.4.6
Evaluation of the Effect of Fast Conformational Fluctuations and Determination of Intramolecular Diffusion Coefficients

As a result of the optically generated "concentration gradient" within the envelope of the equilibrium distribution, the random Brownian motion of the labeled segments causes a net transfer of molecules with excited donor probe to the short distance fractions. This rapid exchange between distance fractions has no effect on the shape of the equilibrium distance distribution, $N_o(r)$, and the molecules transferred to the shorter distances thus have enhanced transferred probability. This phenomenon can be viewed as a process of diffusion under the conformational force field which governs the equilibrium IDD [17, 37, 38]. Within the ensemble of labeled protein molecules whose donor was excited at the time $t = 0$, the fast conformational fluctuations act towards restoration of the equilibrium distribution, an effect that is well visible in the simulation shown in Figure 17.5. The Brownian fluctuations of segments incorporated in a complex molecule such as a globular protein, or even an unfolded polypeptide, are probably not purely random walks. Yet, in the absence of specific knowledge of the modes of motion, one can approximate these motions using a Fick equation. This is the basis for derivation of Eq. (25) which considers both the energy transfer reaction term, $k(r)$, and the restoration force by the diffusion term. Equation (25) is the basis for the experimental data analysis [37]:

$$\frac{\partial \bar{N}_i(r,t)}{\partial t} = \frac{D}{N_0(r)} \times \frac{\partial}{\partial r}\left[N_0(r)\frac{\partial}{\partial r}\bar{N}_i(r,t)\right] - k_i(r)\bar{N}_i(r,t) \qquad (25)$$

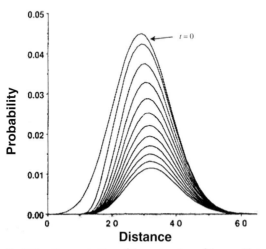

Fig. 17.5. Demonstration of the restoration of the equilibrium intramolecular distance distribution by the diffusion effect. The simulation shown in Figure 17.4 was repeated with nonzero value of the diffusion coefficient, $D = 20 \times 10^{-7}$ cm^2 s^{-1}. The replenishment of the short distance fractions is demonstrated.

where i denotes the ith species of the donor fluorescence decay (in the absence of an acceptor), $N^*_i(r,0) = N_0(r)$ (the equilibrium distance distribution); $N^*_i(r,t)\,dr$ is the probability for molecules with an excited state species i of the donor to have an intramolecular distance in the range r to $r+dr$ and $\bar{N}_i(r,t) = N^*_i(r,t)/N_0(r)$. D is defined as the intramolecular diffusion coefficient of the segments carrying the two probes; τ_i and α_i are respectively the lifetime and the normalized pre-exponential factor of the excited state of the ith species of the donor in the absence of an acceptor. $k_i(r)$, the reaction term, includes the spontaneous emission rate and the Förster energy transfer rate:

$$k_i(r) = \frac{1}{\tau_i} + \frac{1}{\tau_r}\left(\frac{8.79 \times 10^{-25} n^{-4} \kappa^2 J}{r^6}\right) \quad (26)$$

τ_r is the radiative lifetime of the donor, and the other terms were defined in Eqs (6) and (6a). Equation (26) considers the variation of R_o due to variations of the donor quantum yield Φ_D^0. This is represented by the ratio of the donor lifetime components and the radiative lifetime, which is assumed to be independent of local environmental effects.

The solution of Eq. (25) is obtained by numerical methods. Since the boundaries are chosen in such a way that the distribution function is very small at both the upper and lower extreme distances, reflective boundary conditions are arbitrarily used [39]:

$$[\delta\bar{N}(r,t)/\delta r]_{r=r_{min},\,r_{max}} = 0 \quad (27)$$

$i_c^D(t)$, the calculated fluorescence decay curve of the donor is readily obtained from $\bar{N}(r,t)$ by the relationship [37]

$$i_c^D = m \int_{r_{min}}^{r_{max}} N_0(r) \left[\sum_i \alpha_i \bar{N}_i(r,t)\right] dr \quad (28)$$

where m is a constant and α_i is the pre-exponential factor of the ith excited state species of the donor in the absence of an acceptor.

The ideal case is such that the fluorescence decay of the donor in the absence of an acceptor is monoexponential. In such a case, all molecules share one value of R_o and the sum in Eq. (28) reduces to one term and the reactive term, $k_i(r)$ (Eq. (26)), can be replaced by a simpler expression:

$$k(r) = -1/\tau[1 + (R_o/r)^6] \quad (29)$$

In a case where the diffusion rate is known to be negligible on the time scale of the lifetime of the excited state of the donor, the Fick term in Eq. (25) is zero and $N_o(r)$ can be derived directly from an analysis of the donor decay curve. In such a case, Eq. (25) can be replaced by [17]:

$$i_c^D(t) = k \int_0^\infty N_o(r)(\exp -k(r))\,dr \quad (30)$$

17.5
Experimental Challenges in the Implementation of FRET Folding Experiments

Practical implementation of the FRET approach in protein folding experiments either in the ensemble or the single molecule modes involves three very different steps:

1. Design and preparation of pure homogenous structurally and functionally intact double-labeled protein samples.
2. Spectroscopic determination of transfer efficiencies and fluorescence decay profiles of the probes and control experiments.
3. Data analysis.

17.5.1
Optimized Design and Preparation of Labeled Protein Samples for FRET Folding Experiments

A minimum of at least two, and more often three labeled protein species should be prepared for each well controlled, folding FRET experiment. These are (a) a double-labeled (by both donor and acceptor) mutant; (b) a single donor-labeled mutant (at the same site as in (a)); and (c) a single acceptor-labeled mutant (at the same site as in the double-labeled mutant) (optional). An ideal folding FRET experiment would employ three homogeneously ($>$ 97%) labeled samples of the model protein where the donor and the acceptor probes would be selectively attached to selected residues. In this ideal experiment, the following requirements should be satisfied: (a) the distance between the two selected sites should be sensitive to changes most relevant to the working hypothesis to be tested by the planned experiments; (b) the structure and function of the labeled protein derivatives should be unperturbed or minimally perturbed; (c) the mechanism of folding (kinetics and equilibrium transitions) should be undisturbed; (d) the probes should be of small size and preferably close to the main chain; and (e) the probes should have ideal spectroscopic characteristics that would make the detection and data analysis free of corrections factors.

Like everything in life, ideal experimental systems are good for dreams only, proteins are rarely totally unaffected by chemical or mutational modifications; the sites that are most desirable for insertion of the probes are in many cases also critical for folding mechanism and the spectroscopically ideal probes are usually also very large. Therefore, the design process is a matter of optimization, not maximization.

The compositional and structural homogeneity as well as the extent of perturbation of the folding mechanism; the structure; the function (activity) and stability of each labeled preparation should be monitored by standard methods. The extent of perturbation is an important factor in the evaluation of the FRET results. Occasionally modified protein molecules are found to be unsuitable for planned experiments due to very strong perturbations. Consequently, an alternative modification design should be attempted. The concern for purity cannot be overemphasized since background or impurity fluorescence can drastically distort the results of distance determinations.

Incomplete labeling of the two reference single-labeled mutants in each set should not prohibit the use of the preparation in FRET experiments, provided that the extent of labeling is known. Even incomplete labeling of the double-labeled mutant can be tolerated in ensemble steady state or time-resolved detected experiments, provided that the extent of labeling is known, and (in some cases) does not exceed 15–20% of single-labeled molecules. The strict requirements of complete labeling and accurate knowledge of the extent of labeling is relaxed in single molecule FRET experiments. In these measurements, it is possible to discard single molecule records that do not show both the donor and the acceptor emission during data processing.

The main spectral characteristics that should be considered in the design of FRET experiments are as follows: (1) R_o of the selected pair should be as close as possible to the expected mean distance between the two labeled sites (including the linker of the probes); (2) the orientation of the probes should be averaged during the lifetime of the fluorescence of the excited state of the donor; (3) in case of time-resolved ensemble experiments, the ratio $\tau_D/\tau_A \geq 3$; (4) photostability; (5) minimal environmental sensitivity of the spectral characteristics; (6) large Stokes shifts of the acceptor to achieve minimal overlap of the emission spectra of the donor and the acceptor; (7) a high extinction coefficient of the donor to be able to study the folding in very dilute solutions (where aggregation problems are minimized) and in particular in the case of single molecule detection experiments. Ideal probes are rare and therefore some of the spectral characteristics are compromised. Under such circumstances, control experiments should accompany the FRET measurements in order to reduce experimental uncertainties.

17.5.2
Strategies for Site-specific Double Labeling of Proteins

Site-specific labeling can be achieved by six basic strategies:

1. Use of natural probes, the tyrosine or tryptophan residues and a few prosthetic groups. Site-directed mutagenesis can be applied for the preparation of single tryptophan mutants and generation of sites for attachment of the acceptor by insertion of cysteine or other residues [33, 40, 41].
2. Nonselective chemical modification of reactive side chains and chain terminal reactive groups (amine, carboxyl, and mainly sulfhydryl groups) combined with high resolution or affinity chromatography separation methods.
3. Solid phase synthesis of protein fragments and conformationally assisted ligation of two synthetic peptide fragments, or of a genetically truncated protein fragment and a synthetic complementing fragment can be applied for efficient site-specific labeling of proteins. Total synthesis of labeled protein molecules is applicable for medium-size proteins (e.g., the total synthesis of chymotrypsin inhibitor 2 (CI2) [42]).
4. Production of edited mutants with no more than one pair of exposed reactive side chains, preferably cysteine residues, at the desired sites, followed by chemical modification and chromatographic separation.

5. Cell-free protein synthesis and the use of edited tRNAs and edited genetic code for the incorporation of nonnatural amino acids into selected sites in the chain [43].
6. Enzyme-catalyzed insertion of fluorescent substrates taking advantage of the specificity and mild conditions of the enzymatic reaction [44].

Strategy 1 avoids the potential artifacts caused by labeling procedures and uncertainties such as the labeling ratio. On the other hand, the main limitations of using the natural probes are the limited arsenal of available probes; their less than ideal spectral characteristics; the limited range of R_o values that can be achieved using them, and the fact that most proteins contain several Trp and Tyr residues. Applications of strategy 2 are limited to small proteins or proteins where a limited number of the reactive side chains can be found [45]. Nitration of tyrosine residues was also successfully applied in few cases [49, 50]. Strategy 5 is the ultimate one, which achieves full freedom for the designer of protein derivatives best suited for testing well-defined working hypotheses. This strategy will most probably become the method of choice in the field of spectroscopic research of protein folding [43, 51]. The limited scope of application of this approach at the present time is mainly due to the very sophisticated and multistep operations involved and the low yields. These yields are limiting only in the case of measurements of folding kinetics, where many samples are employed (e.g., stopped-flow double kinetics experiments). In most FRET folding experiments, the sensitivity of the fluorometric detection systems makes this method satisfactory and promising. Strategy 6 is not of general use due to the limited number of known suitable enzymatic reactions and the multiplicity of the reactive side chains. The well-documented catalysis of peptide bond synthesis by proteolytic enzymes was applied (for C-terminal labeling of RNase A [44]) but it is not easy to apply as a general method. Transglutaminase catalysis of insertion of primary amino groups attached to fluorescent probes was also attempted [46–48] but the development of strategy 4 offers a simpler route.

Strategy 4 is currently the most practical approach for several reasons: first, exposed cysteine residues are rare, and in many proteins these residues can be replaced by Ser or Ala residues without major perturbations of structure and function. Thus, most proteins can be engineered to have two exposed sulfhydryl side chains. Second, the arsenal of sulfhydryl reactive pairs of probes covering a wide range of spectroscopic characteristics and R_o values is abundant, and the experimental design can be optimized to meet a wide variety of distances and labeling conditions. Third, these reagents can be selectively targeted at the cysteine residues and huge amounts of-labeled mutants suitable for kinetic experiments can be produced [40, 52–57].

17.5.3
Preparation of Double-labeled Mutants Using Engineered Cysteine Residues (strategy 4)

Although reagents for selective alkylation of sulfhydryl residues are available, the challenge here is to develop methods for selective modification of each one of the

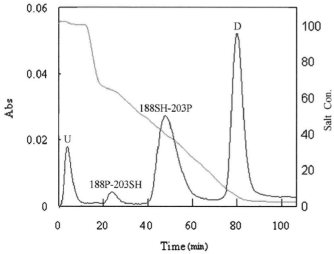

Fig. 17.6. The first step in preparation of a set of labeled AK mutants for trFRET experiments (labeling strategy 4). An engineered AK mutant with two engineered cysteine residues at sites 188 and 203 was prepared and reacted with 1-iodo-acetamido-methyl-pyrene. The reaction was allowed to proceed for 24 hours. The excess reagent was removed by dialysis and the solution was equilibrated with 0.5 M $(NH_4)_2SO_4$ in 20 mM Tris pH 7.5. The reaction mixture was then applied to a phenyl-sepharose column (1 cm diameter, height 7 cm) equilibrated with the same salt solution. The column was eluted with a gradient of decreasing salt concentration. The four fractions that were well resolved were: unlabeled protein (U), the two single labeled proteins (188P-203SH where only cysteine 188 was labeled, 188SH-203P where only cysteine 203 was labeled) and a double-labeled fraction (D). The main product, 188SH-203P, was later on reacted with an acceptor reagent to produce the double-labeled protein. The differential reactivity of the two engineered cysteine residues is demonstrated.

two reactive sulfhydryl groups in the same protein molecule (Figures 17.6 and 17.7).

Submilligram amounts of labeled proteins can suffice for a full set of FRET experiments under equilibrium conditions. Such minute amounts of site-specific double-labeled protein samples can be prepared, based on high-resolution chromatography, i.e. chromatography of a mixture of the possible products, following a nonselective reaction of two-cysteine mutants with a fluorescent reagent. Homogeneous double-labeled products can be obtained in a second cycle of reaction and separation. Hundred-fold larger preparations of labeled proteins are needed for stopped-flow double-kinetic experiments. In this case, it is desirable to develop methods in which the selectivity is achieved by differential reactivity of selected sites on the protein molecule. A general systematic procedure was developed for high-yield selective labeling of each one of two cysteine residues in mutant protein molecules [57, 58]. The determination of individual reaction rate constants of the engineered SH groups and the optimization of reaction conditions enabled high yields (70–90%) of the preparation of pure, site-specific double-labeled AK mutant.

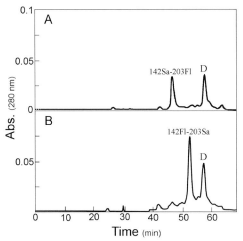

Fig. 17.7. Preparation of site-specific double-labeled AK with two engineered cysteine residues (142 and 203) in a single reaction step (strategy 4). The 10-fold reactivity of cysteine 203 relative to the reactivity of cysteine 142 enabled the preferred labeling of residue 203. The reaction mixture was separated on a Mono-Q ion exchange column (0.5 × 5 cm). Two preparations are shown: A) the acceptor (Fl) was reacted first (equimolar ratio of dye to protein, 2.5 h) and then a 10-fold excess of the donor reagent (Sa) and concentrated GdmCl (to a final concentration of 2 M) were added. After 2.5 h at room temperature. The reaction was stopped by addition of 50-fold excess DTT and dialyzed into 20 mM Tris pH 8.0 before the chromatography. B) The same procedure was repeated, except that the probes reagents were added in the reversed order. Single separation step gave the pure double-labeled products in both cases.

These high yields make some otherwise difficult experiments feasible, such as the double kinetics time-resolved FRET measurements [59, 60].

Jacob et al. [58] improved the prediction of the rates of reactions of engineered cysteine residues at selected sites in a protein. They based this on the observed correlation between reactivity of engineered sulfhydryl groups and the effect of neighboring charged groups (due to dependence of its pK_a), and the freedom of rotation of the side chain. The selectivity of the labeling reaction can be further enhanced by local increase or reduction of reactivity of selected side chains by means of conformational effects. Solvent composition (organic solvents, salt components, and pH) and temperature are the most common variables that can differentially affect the reaction rate constants of selected side chains. Drastic effects can be obtained by ligand binding (substrates, inhibitors, and protein–protein interactions). In this work, the ratio of the reaction rate constants was examined only under a range of temperatures and salt concentrations.

The level of homogeneity required for spectroscopic measurements makes the chromatographic separation steps indispensable. Separation of the reaction products can be enhanced when one of the labels is either charged or very hydrophobic. Yet, either hydrophobicity or extra electrostatic charges can affect the folding pathway of proteins and can be a disadvantage when the modifications are made for structural studies. Affinity chromatography was also used for separating multiple-labeled protein derivatives [45, 61]. It is also possible to use a mixed disulfide resin for separation of fractions with free cysteine residues in the process of labeling two-cysteine mutants. The single-labeled mutants that are used for the reference measurements can be prepared by straightforward labeling of single cysteine mutants. Caution must be taken since in several cases it was shown that the spectroscopic properties of a probe attached at one site on a protein are significantly modified when the second cysteine was inserted at the second site for the double labeling. In this case, it is desirable to use the corresponding single-labeled protein sample of the two-cysteine mutant in which the second cysteine residue is blocked by a nonfluorescent alkylation reagent such as iodoacetamide.

17.5.4
Possible Pitfalls Associated with the Preparation of Labeled Protein Samples for FRET Folding Experiments

Important issues associated with the sample preparations include: (a) the specificity and the extent of labeling; (b) the purity of the labeled samples; and (c) possible effects of the labeling procedures and modifications on the stability, activity, solubility, and folding mechanisms.

(a) Incomplete specificity of the labeling might prohibit meaningful interpretation of trFRET data, since the reference measurements might not be faithful to the double-labeled samples. Consequently, distributions of labeled products might be interpreted as distributions of distances. In principle, under conditions where the extent of distribution of labeling sites is known, it should be possible to adjust the model in the analysis algorithm and resolve subpopulations of distance distributions. This, however, might be possible only under very limited circumstances, and since it requires the addition of parameters in the analysis, the uncertainties might become too high.

(b) Two types of impurities are of concern here: the first is the presence of fluorescent impurities and background emission, and the second is incomplete labeling. As long as the background emission is not too high (ca. $< 20\%$ of the signal), and the emission can be determined independently (control experiments), reliable distance determination is possible by careful subtraction of the background emission. Incomplete labeling of the single-labeled samples should not interfere with analysis of trFRET data. Analysis of data collected with double-labeled samples that contain fractions of single-labeled molecules can be achieved as long as an independent analytical procedure is employed to determine the size of such fractions and that these fractions do not exceed about 25% of the molecules. Consider the results of a FRET measurement using a preparation where a donor–acceptor labeled protein sample contains 20% molecules labeled with donor only, without acceptor. A long lifetime component would be observed in the decay of the donor fluorescence, and this can erroneously be interpreted as a fraction of molecules with long intramolecular distance ("unfolded subpopulation"). When the size of such a fraction is known this decay component can be considered and the correct distance distribution can be recovered. Less strict requirements apply in preparations made for single molecule FRET measurements. Data collected from molecules which lack one of the probes in a double-labeled preparation can be easily recognized and discarded without bias of the computed distributions.

(c) By definition, chemical or mutational modifications of any protein molecule are a perturbation, which should be minimized. Naturally, the modified proteins are different from the native ones, which is not a matter of concern unless the experiments are aimed at studying specific aspects of the folding of a specific protein. In many folding studies, the more important concern regarding the modification is that in a series of modified mutants of any model protein the perturbations should be similar in all mutants. Such common perturbation enables reliable multiprobe test of the cooperativity and steps of the folding mechanism where each pair probes the folding of the same model protein at a selected section of the chain (e.g., secondary structure elements, loops, and nonlocal interactions). Structural perturbations can be minimized by the proper choice of labeling sites and probes. Rigorous control experiments for determination of the extent of perturbation due to the modifications are mandatory. Control experiments can include the determination of biological activities; the detection of structural properties such as CD, IR, and other spectral characteristics and rates of refolding.

17.6
Experimental Aspects of Folding Studies by Distance Determination Based on FRET Measurements

17.6.1
Steady State Determination of Transfer Efficiency

17.6.1.1 Donor Emission

The most commonly used method for steady state determination of transfer efficiency is based on the determination of the extent of quenching of donor emission in the double-labeled samples (Eqs (12)–(14)). Emission of the donor is recorded at a wavelength range free of acceptor emission and then $E = 1 - I_{DA}/I_D$ where I_{DA} and I_D are the donor emission intensities in the presence and absence of an acceptor respectively. Both intensities should be normalized to the same donor concentration. It may be difficult to determine the concentration of the donor in the presence of an acceptor. This difficulty (which is eliminated in the time-resolved experiment) was solved by enzymatic digestion of each sample after determination of I_{DA}. A second emission intensity measurement immediately after the digestion thus provides I_{DA} and an internal reference [62]. The FRET interaction does not affect the shape of the spectra of the probes. Therefore, nonlinear least squares fitting of the absorption and emission spectra of the double- and single-labeled samples can be used to overcome the concentration normalization problem [63]. The nonlinear least squares fitting of the entire emission spectral ranges can improved the accuracy of transfer efficiency calculations when the donor and acceptor emissions overlap [34, 64].

17.6.1.2 Acceptor Excitation Spectroscopy

The most direct and error-free determination of FRET efficiency by steady state experiments is obtained when the excitation spectra of the acceptor recorded for three samples are normalized (at the acceptor excitation range) and superimposed (Eq. (15)). The three samples (and corresponding spectra) are (a) the acceptor single-labeled mutant (I_A^{ex}), (b) the double-labeled mutant (I_{DA}^{ex}), and (c) a special double-labeled reference of well documented transfer efficiency, E_{ref} (preferably a compound in which the two probes are very close, $E_{ref} = 1.0$) (I_{ref}^{ex}). The integrated areas under the three curves in the donor excitation wavelength range provides the transfer efficiency free of instrumental artifacts and concentration normalization:

$$E = \frac{I_{DA}^{ex} - I_A^{ex}}{E_{ref} I_{ref}^{ex} - I_A^{ex}} \tag{31}$$

17.6.2
Time-resolved Measurements

Equations (17)–(19) are used for the determination of the average FRET efficiency by time-resolved measurements. The main technical advantage of this measurement is that there is no need to know the concentrations of the proteins and the accuracy of the lifetime measurements surpasses that of the steady state intensity measurements. Intramolecular distances can be obtained from measurements of the average transfer efficiency only for very uniform structures.

The shape of fluorescence decay curves of the probes in time-resolved FRET experiments contain detailed information on heterogeneity of the ensembles of protein molecules and are therefore much more effective in the folding research. The scheme shown in Figure 17.8 demonstrates the application of trFRET in folding research. The fluorescence decay of the donor (and the acceptor as well) contain the information on the distribution of the distance between the two labeled sites and that can be used for characterization of the conformational state of the labeled protein molecule at equilibrium and during the folding/unfolding transitions.

The technology of time-resolved fluorescence measurements reached a level of maturity during the last quarter of the twentieth century. During this period, the main technological developments that promoted the art of time-resolved measurements to a stage of stable and routine analytical method were (a) the implementation of tunable picosecond laser sources (at present mainly the titanium sapphire solid state lasers); (b) the development of very fast and sophisticated single photon counting devices and the phase modulation instruments; and (c) the development of efficient algorithms for global data analysis. Fluorescence decay can be detected either by the single-photon counting method or by the phase modulation method, and in principle, both methods are equivalent. While phase experiments are rela-

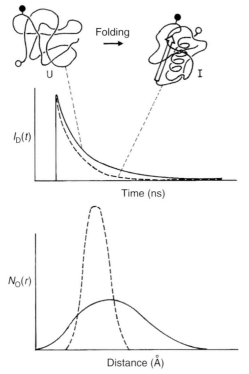

Fig. 17.8. A scheme of the application of the time-resolved energy transfer experiments for determination of conformational changes in globular protein. The scheme describes a hypothetical two states folding transition in a globular protein. The corresponding time domain signals, the decay curves of the donor in the two states, are shown in the center of the scheme. The two corresponding intramolecular distance distributions derived from the time domain data, are shown in the lower section. This scheme is a simulation of the detection of a formation of a nonlocal interaction, indicated by a transition from a broad to a narrow distribution with a smaller mean distance.

tively easy to perform, single-photon counting techniques offer more straightforward methods for incorporation of acceptor decay curves in the analysis and easier control of background subtraction and other corrections.

A standard current state of the art setup for time-correlated single-photon counting measurements includes:

1. An excitation source: a picosecond mode-locked Ti:sapphire laser pumped by a high-power diode laser equipped with broadband optics and a pulse selector which is used to reduce the basic 76 or 80 MHz pulse rate to 1–10 MHz (according to the lifetime of the probes). The laser output is frequency-tripled by a flexible second and third harmonics generator. Such systems can be tuned over ranges from 230 nm to 330 nm, 400 nm to about 500 nm, and 720 nm to above 1000 nm and produce trains of 1.6 ps pulses (before doubling) with minimal drift.

2. A spectrometer which includes a thermostated cell holder equipped with stirring device; a double Fresnel rhomb at the excitation side; a polarizer at the emission side; a double 1/8 meter subtractive monochromator and cooled microchannel plate PMT detector biased at -3200 V.
3. A single-photon counting board (e.g., SPC 630, Becker & Hickel GmbH) fed via an preamplifier and triggered by a photodiode. The response of such system yields a pulse width of 20–30 ps.

17.7
Data Analysis

The basic procedure used for analysis of experiments follows the common practice of analysis of time-resolved experiments. The basic procedure is based on a search for a set of free parameters. These are the parameters of the distribution function; the diffusion coefficient; the fluorescence lifetimes and their associated preexponents. A calculated donor emission impulse response, $i_c^D(t)$ is obtained via numerical solution of Eqs (25)–(28) or (30). The acceptor impulse response, $i_c^A(t)$, is obtained by the convolution of, $N_A(t)$, the calculated population of molecules. The acceptor probes are excited at time t either via the FRET mechanism or via direct excitation, with the acceptor fluorescence lifetime (in the absence of FRET), $i_A(t)$,

$$i_c^A(t) = N_A(t) \otimes i_A(t) \tag{32}$$

A calculated fluorescence decay of the donor, $F_D(t)$, is obtained by the convolution of the donor impulse response with the system response,

$$F_D(t) = k_D(\lambda_{em})[I(t, \lambda_{ex}, \lambda_{em}) \otimes i_c^D(t)] \tag{33}$$

where $k_D(\lambda_{em})$ represents the emission spectral contour of the donor relative to the acceptor and $I(t, \lambda_{ex}, \lambda_{em})$ represents the experimentally measured instrument response function at this particular optical setup (excitation and emission wavelengths setting, monochromator setting, detector spectral response, etc.). A calculated acceptor decay curve, $F_A(t)$, is similarly obtained by:

$$F_A(t) = k_A(\lambda_{em})[i_c^A(t) \otimes I(t, \lambda_{ex}, \lambda_{em})] \tag{34}$$

The final calculated fluorescence decay profile is therefore,

$$F_{calc}(t, \lambda_{ex}, \lambda_{em}) = F_A(t) + F_D(t) \tag{35}$$

Several computational methods were developed for the search of the set of free parameters of $N_o(r)$ and D. This yields the theoretical impulse response, which when convoluted with the system response, yields $F_{calc}(t, \lambda_{ex}, \lambda_{em})$ best fitted to the experimentally recorded fluorescence decay curves $F_{exp}(t, \lambda_{ex}, \lambda_{em})$ [65, 66]. This procedure is inherently difficult since the analysis of any multiexponential decay curve is a mathematically "ill defined" problem [67, 68]. The quality of fit of the experimental and calculated curves is judged by the minimization to the χ^2 value of the fit and the randomness of the deviations [68, 69].

The differential Eq. (25) does not have a simple general analytical solution (except for specific cases). Consequently, a numerical solution based on the finite difference (FA) method [70, 71] was applied. Therefore the currently most common practice is the use of model distribution functions for $N_o(r)$ with a small number of free parameters. The model used in the experiments described below is a Gaussian distribution ($N_o(r) = C 4\pi r^2 \exp -a(r-\mu)^2$ where a and μ are the free parameters which determine the width and the mean of the distribution respectively and C is a normalization factor). Correlations between the free parameters, mainly a, μ and D, searched in the curve-fitting procedure leave a range of uncertainties in these values. This uncertainty is being routinely reduced by global analysis of multiple independent measurements. In particular, coupled analysis of the fluorescence decay of both the donor and the acceptor probes contributes overdetermination of the free parameters. This over-determination is sufficient to allow simultaneous determination of both the parameters of the equilibrium intramolecular distance distribution and the diffusion coefficients, despite the strong correlation between those parameters [38].

The information content of the decay curves of the two probes is not redundant. Due to the r^{-6} dependence of the transfer probability, the fluorescence decay of the donor is insensitive to contributions from the conformers with intramolecular distance $r \ll R_o$ (fast decay). The contribution of these conformers is well detected in the fluorescence decay of the acceptor. On the other hand, the contributions of the fractions of the distance distributions characterized by intramolecular distance $r > R_o$, are well detected by the donor decay and are less weighted in the acceptor fluorescence decay. Both simulations and experiments have shown the following: with the present level of experimental noise of the time-resolved experiments a ratio of fluorescence lifetimes of the donors and the acceptors, $\tau_D/\tau_A \geq 3$ is a necessary condition for reliable simultaneous determination of the parameters of the intramolecular distribution functions and the associated intramolecular diffusion coefficients. Several curve-fitting procedures are available, the most widely used are based on the well-known Marquardt-Levenberg algorithm [38].

Maximum entropy method (MEM) is also applied to the analysis of time-resolved fluorescence [72] and FRET experiments [73]. This analysis is independent of any physical model or mathematical equation describing the distribution of lifetimes. Whatever methods are used for data analysis, the best fit must be tested by the statistical tests. Fine details of the distance distributions can be extracted from each experiment depending on the level of systematic and random noise of the system.

The routine global analysis procedure includes joint analysis of at least one set of four fluorescence decay curves of the probes attached to the protein. Two measurements are conducted on a double-labeled molecule, containing one donor and one acceptor: (a) The "DD-experiment," determination of τ_D the fluorescence decay of the donor in the presence of FRET (excitation at the donor excitation wavelength

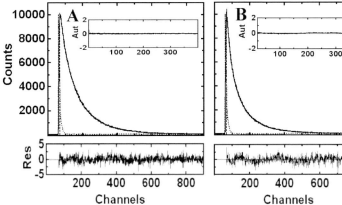

Fig. 17.9. Representative pair of donor fluorescence decay curves used for determination of intramolecular distance distributions in a labeled RNase A mutant. A) Fluorescence decay curves for the tryptophan residue in W^{76}-RNase A (the "D"-experiment) and B) the fluorescence decay of the tryptophan residue in (76–124)-RNase A in the reduced under folding conditions (R_N state) in 40 mM phosphate buffer, pH 7.0, 20 mM DTT (the "DD" experiment) at room temperature. Excitation wavelength 297 nm and emission at 360 nm (bandwidth for emission was 2 nm). (…) the system response to excitation pulse, (---) the experimental trace of the tryptophan fluorescence pulse, and (—) the best fit theoretical curves obtained by the global analysis. The "DD" experiment was fit to a theoretical curve based on the distribution of distances [31]. The residuals of the fits (Res) are shown in the lower blocks, and the autocorrelation functions of the residuals [$C(t)$] are presented for each curve in the upper right inset. The χ^2 obtained in this experiment was 2.02 (reproduced with permission from ref. 40).

and detection at the donor emission wavelength); and (b) The "DA-experiment," determination of τ_A, the acceptor emission in the presence of FRET, with excitation at the donor excitation range, and detection of the acceptor emission at the long wavelength range (Figure 17.9). Two internal reference measurements for determination of the probes lifetime when attached to the same corresponding sites as in the double-labeled mutant but in the absence of FRET. (c) The "D-experiment" is a measurement of τ_D^o, the emission of the donor in the absence of an acceptor, using protein molecule labeled by the donor alone; and (d) the "A-experiment" is a measurement of τ_A^o, the decay of the acceptor emission in the double-labeled protein, excited directly at a wavelength where the donor absorption is negligible or using a protein preparation labeled by the acceptor only.

When collected under the same laboratory conditions (optical geometry, state of the instrument, time calibration, linearity of the response, and solution parameters) used for the first two experiments, the D- and A-experiments serve as "internal standards" for the time calibration of this experiment and the dependence of the spectroscopic characteristics of the probes applicable for the specific labeled sites and experimental conditions of the FRET experiments. These on-site obtained data contribute to the elimination of most possible sources of experimental uncertainty due to variations between independent measurements and conditions.

Why monitor the FRET effect twice, both at the donor and at the acceptor end? Brand and his students [66] showed that this over-determination significantly reduces the uncertainty in the values of the free parameters, searched by curve-fitting procedures. The Brand group introduced the concept of global analysis in the field of time-resolved fluorescence measurements and showed that every additional independent data set that is globally analyzed, further reduces the statistical uncertainty in the most probable values of the searched parameters. Beechem and Haas [38] showed that the global analysis of sets of donor and acceptor decay curves can reduce the uncertainty in the values of the parameters of the distance distributions and their associated diffusion coefficients. Both a shift of the mean and reduction of the width of a distance distribution as well as an enhanced rate of the intramolecular diffusion can cause similar reduction of the donor fluorescence lifetime. The fine differences between the two mechanisms appear in the shape of the decay curves and hence the combined analysis of the donor and the acceptor decay curves can resolve the two mechanisms.

17.7.1
Rigorous Error Analysis

The statistical significance of the value of any one of the free parameters that is determined by any curve-fitting method can be evaluated by a rigorous analysis procedure. In this procedure the iterative curve-fitting procedure is repeated while the select parameter is fixed and the other free parameters adopt values that yield the best possible fit under the limitation of that fixed value of the selected parameter. This procedure can be repeated for a series of values of the selected parameter, and the limits of acceptable values can be determined by the statistical tests and the selected thresholds, in particular those of the χ^2-values (Figure 17.10). This procedure, if repeated for every free parameter, actually generates projections of the χ^2 hypersurface on specific axes in the parameter space. Error limits at any required significance level can then be determined using an F-test [38].

This procedure considers all correlations between the parameters. The evaluation of each analysis and of the significance of the parameters is based on four indicators: (1) The global χ^2-values; (2) the distributions of the residuals; (3) the autocorrelation of the residuals; and (4) the error intervals of the calculated parameters obtained by the rigorous analysis procedure. In addition, each "DD-experiment" is analyzed by multiexponential (three or four exponentials) decay model. The χ^2-values obtained in these analyses serve for evaluation of the significance of the χ^2-values obtained for each set of experiments in the global analysis, based on the solution of Eqs (25)–(28) and (32)–(35).

17.7.2
Elimination of Systematic Errors

The linearity of the systems and possible effects of systematic errors are routinely checked by measurements of the fluorescence decay of well-defined standard ref-

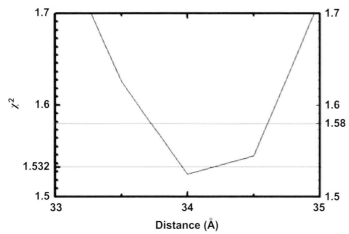

Fig. 17.10. χ^2 surface obtained in rigorous analysis of a typical trFRET experiment for determination of intramolecular distance distribution in double-labeled adenylate kinase mutant labeled at residues 154 and 203. Global analysis was repeated for several fixed values of the intramolecular mean distance and all other parameters were allowed to change. The ranges of one and two standard deviations at χ^2 values of 1.53 and 1.58 respectively are shown and can be used for determination of the uncertainty range for the mean of the distance distribution.

erence materials (e.g., anthracene in cyclohexane). The fluorescence parameters of the reference materials (monoexponential decay of the fluorescence, lifetimes, quantum yields, and spectra) are well known and can be used for testing and fine tuning of the response of the instruments (mainly linearity, time calibration, and dynamic range). Possible effects of other potential systematic errors, introduced by non-FRET effects of conformational changes on the spectral parameters of the probes (e.g., environmental sensitivity, spectral shifts, or quenching) are eliminated by inclusion of the two reference protein derivatives in each experiment. In practice, the error ranges obtained for the mean of the distance distributions determined by trFRET experiments were minimal ($\Delta r < \pm 1$ Å), and the highest uncertainty was encountered in the values of the diffusion parameters.

17.8
Applications of trFRET for Characterization of Unfolded and Partially Folded Conformations of Globular Proteins under Equilibrium Conditions

17.8.1
Bovine Pancreatic Trypsin Inhibitor

Bovine pancreatic trypsin inhibitor (BPTI) (Figure 17.11) is a small, stable globular protein of 58 amino-acyl residues, including four lysyl residues and three disulfide bonds [74, 75]. Four double-labeled BPTI samples were prepared for measure-

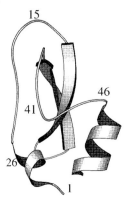

Fig. 17.11. The crystal structure of BPTI, a stable, single folding unit protein. The N-terminus was the site of attachment of the donor probe, 2-methoxy-naphthyl-methylenyl (MNA), and the acceptor, 7-dimethyl-amino-coumarin-4-carboxymethyl (DA-coum), was attached to the lysine residues (15, 26, 41, and 46 in four different preparations), thus labeling a series of monotonically increasing chain length.

ments of a series of increasing chain segmental lengths between the N-terminus (attachment of the donor) and residues 15, 26, 41, and 46 (sites of acceptor attachment) [45].

Relatively narrow distributions with mean distances close to those expected, based on the crystal structure, were obtained for all four-labeled BPTI samples in the native state. This result confirmed the validity of the distance calibration and the orientational averaging of the probes.

17.8.2
The Loop Hypothesis

Reduced BPTI in 6 M GdmCl is fully denatured [75, 76], but under these conditions the mean of the four intramolecular distance distributions did not show monotonous increase as a function of labeled chain length [77]. The mean distances were reduced by raising the temperature from 4 °C to 60 °C. These results are an indication of conformational bias of the distribution even under drastic denaturing conditions (i.e., the protein was partially unfolded). The strong temperature dependence of the mean distances is cold unfolding transition [78] and an indication that the structures that exist in the denatured state are probably stabilized by hydrophobic interactions. Moreover, at least two, one native-like and one unfolded, subpopulations were distinguished in the distributions of end-to-end distances of each segment. The sizes of the two subpopulations were temperature- and denaturant concentration-dependent [79] (i.e., transition to more folding conditions increased the native like subpopulation and reduced the unfolded one).

These observations led to *the loop hypothesis*, which states that loop formation by specific nonlocal interactions (NLIs) prior to the formation of secondary struc-

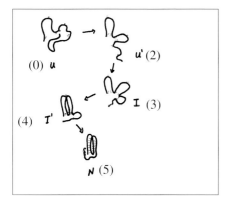

Fig. 17.12. The loop hypothesis. A scheme of a protein molecule with several groups of residues that form nonlocal interactions is shown. A hypothetical pathway is shown where after formation of four nonlocal interactions the chain fold is native-like and the last step of packing all the local interactions completes the folding transition.

tures can be a key factor both in the solution of the Levinthal paradox and in determining the steps which direct the folding pathway [79] (Figure 17.12). The loops, generated by loose cross-links between specific clusters of residues separated by long chain stretches, may bias the intramolecular distance distributions and reduce the chain entropy and the volume of the searched conformational space. The observation of the cold unfolding of the intramolecular loops [77] led to the suggestion that the main interaction between these clusters should be hydrophobic. These long loops define the basic folding unit and thus BPTI is composed of two folding units [79]. A similar mechanism was recently suggested by Klein-Seetharaman et al. [80] based on the NMR study of the denatured state of lysozyme. This specificity of the interactions is in agreement with Lattman and Rose [81], who suggested that a specific stereochemical code directs the folding. FRET experiments can be designed to locate those interactions, and site-directed mutagenesis can be used to test the hypotheses of the role of specific residues in the formation of the NLIs. Berezovsky et al. [82, 83] analyzed the crystal structures of 302 proteins in search for native loop structures. They showed that protein structure can be viewed as a compact linear array of closed loops. Furthermore, they proposed that protein folding is a consecutive looping of the chain with the loops ending primarily at hydrophobic nuclei.

17.8.3
RNase A

Navon et al. [40] applied this same experimental approach to the study of the unfolded and partially folded states of reduced ribonuclease A (RNase A). RNase A represents a higher level of complexity in the folding problem. The chain length is twice that of BPTI (124 residues). Buckler et al. labeled the C-terminus of RNase

A with a donor and attached the acceptor to residues 104, 61, or 1, one at a time. This set of pairs of labeling sites spans the full length, one half of the length of the chain and the C-terminal segment [44, 84]. Navon prepared 11 double-labeled RNase A mutants to probe the folding transitions in the C-terminal subdomain of the protein.

The pattern of the dependence of the means of the end-to-end distance distributions of the labeled segments in the RNase A mutants on the number of residues in each segment resembled the pattern of the segment length dependence of the corresponding Cα-Cα distances in the crystal structure of the molecule. But the width of the distributions was large, an indication for very weak bias by nonlocal interactions.

The distance distribution between the C-terminal residue (residue 124) and residue 76 (a 49-residue chain segment) was another example whereby two subpopulations were resolved. The distributions represent an equilibrium between native-like and unfolded conformers (Figure 17.13). The same equilibrium was found for the chain segment between residues 76 and 115.

The intramolecular segmental diffusion coefficients obtained by Buckler et al. [84] correlated well with the conformational states of the chain. In the native state, only the distance between the two chain termini showed rapid nanosecond fluctuations; in the presence of 6 M GdmCl, the shorter segments also underwent rapid intramolecular distance fluctuations which were further enhanced by the reduction of the disulfide bonds. This demonstrates that the segmental intramolecular diffusion coefficients can be determined and used as another parameter for monitoring specific changes of the states of folding intermediates.

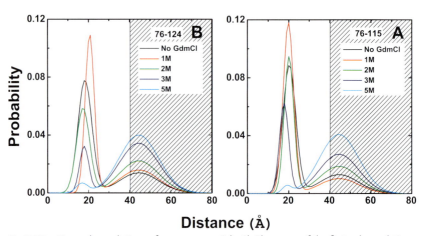

Fig. 17.13. Two subpopulations of intramolecular distance distribution between residues 76 (tryptophan residue, the donor) and 124 (engineered cysteine residue labeled by an aminocoumarin derivative, the acceptor) in the C-terminal subdomain of reduced RNase A were resolved by the trFRET experiment (in 40 mM HEPES buffer pH 7.0, 20 mM DTT, and 5 mM EDTA) with increasing concentrations of GdmCl. The mean of the first subpopulation was close to that of the native state; the mean of second subpopulation was much larger, but could not be determined with a high degree of accuracy due to the smaller R_o. The ratio of the unfolded subpopulations increased from 30% in HEPES buffer to 95% at 6 M GdmCl (Reproduced with permission from ref. 40).

17.8.4
Staphylococcal Nuclease

Brand and coworkers [85, 86] used a similar approach in a search for structural subpopulations and chain dynamics in the compact thermally denatured state of staphylococcal nuclease mutant. Two subpopulations were found in the distribution of distances between residues 78 and 140 in the heat-denatured state. At the highest temperatures measured, both the apo- and holoenzyme showed compact and heterogeneous populations in which the nonnative subpopulation was dominant. The asymmetry found in the distribution of distances between Trp140 and a labeled Cys45 in the destabilized mutant K45C at room temperature (native state) was attributed to static disorder. Partial cold unfolding or chemical denaturation reduced the asymmetry of the distribution and its width. This effect was analyzed in terms of the nanosecond dynamic flexibility of the protein molecule.

17.9
Unfolding Transition via Continuum of Native-like Forms

Lakshmikanth et al. [73] and their collaborators employed the strength of the multiple pairs trFRET experiments to reveal multiple transitions in the unfolding of a small protein. Lakshmikanth et al. [73] used a maximum entropy method for determination of the distributions of intramolecular distances between Trp53 and the trinitrobenzoic acid moiety attached to Cys82 in the bacterial RNase inhibitor barstar. They found that the equilibrium unfolding transition of native barstar occurs through a continuum of native-like conformers. They also concluded that the swelling of the molecule as a function of denaturant concentration to form a molten globule state is separated from the unfolded state by a free energy barrier. This is another demonstration of the effectiveness of FRET-detected folding experiments in the multiple probe test of the cooperativity of folding/unfolding transitions.

17.10
The Third Folding Problem: Domain Motions and Conformational Fluctuations of Enzyme Molecules

Lakowicz and coworkers [41, 89–93] applied frequency domain FRET methods in experiments that focused on the third folding problem, the functional refolding of various proteins. Residues Met25 and Cys98 in rabbit skeletal troponin C located at the N- and C-terminal domains respectively were labeled by a donor and acceptor. At pH 7.5 and in the presence of Mg^{2+}, the mean distance was close to 15 Å and the FWHM was 15 Å, a flexible structure. The addition of calcium caused an increase of the average distance to 22 Å and a decrease of the width by 4 Å, a transition to an extended and more uniform ensemble of conformers. At pH 5.2, the conformation of troponin C is further extended with an average distance of 32 Å, and a narrower FWHM of only 7 Å. Similar results were obtained in the complex of troponin C and troponin I. This finding is a good example of the application of the FRET methods in studies of conformational changes involved in functional refolding of proteins.

Vogel et al. [94, 95] developed applications of FRET measurements in studying conformational transitions of membrane peptides and proteins.

The unique feature of proteins is that they have probably evolved to couple between conformational fluctuations and the specific tasks they perform [96, 97]. A prominent example of this principle is the enzymes that can be viewed as machines designed to accelerate chemical transformations by dynamic adaptation of specific binding sites to metastable transition states [98–100]. This can be achieved by a combination of structural flexibility (fluctuations) and a design that allows a limited number of modes of motion [101]. This is manifested in enzymes called transferases which catalyze the transfer of charged groups. Upon substrate binding, these molecules undergo structural changes to screen the active center from water to avoid abortive hydrolysis [102, 103] and catalyze the transfer reaction.

X-ray analysis of enzyme crystals revealed large-scale structural changes associated with substrate binding. These and diffuse X-ray scattering data suggest that domain displacement induced by substrate binding represents a unique mechanism for the formation of the enzyme active center [102, 104].

The dynamic trFRET and the single molecule FRET experiments go beyond existing methods. These experiments are usually able to describe the full conformational space for a specific mode of motion. Moreover, the FRET experiments provide dynamic information which is essential for understanding how the intramolecular interactions (which are programmed in the sequence information of the protein) control the modes of motion that compose the directed conformational change.

Two phosphoryl transfer enzymes were studied as model systems: yeast phosphoglycerate kinase (PGK) [105], which catalyzes the phosphoryl transfer from (1,3)-diphosphoglyceric acid to an ADP molecule, and *E. coli* adenylate kinase (AK) which catalyzes the phosphoryl transfer reaction (MgATP + AMP $\leftrightarrow$ MgADP + ADP) [106].

The PGK molecule is made of two distinct domains and a domain closure was postulated as the simple mechanism that brings the two substrates into contact (closure of a 10 Å distance) [107]. A mutant PGK was prepared in which one probe was attached to residue 135 (the N-terminal domain) and the second to residue 290 (C-terminal domain). A very wide distribution was obtained from trFRET measurements of the apo form of the labeled PGK molecule. This is an indication of the multiplicity of conformations. Substrates binding induced an increase in the mean of the distributions (increased interdomain distance) and as expected, reduced the FWHM.

The crystal structure of *E. coli* AK shows three distinct movable domains. A comparison of the crystal structures (Figure 17.14) representing the enzyme in different ligand forms revealed large structural changes [108–110]. Very wide distance distributions were obtained for the distances between both minor domains and

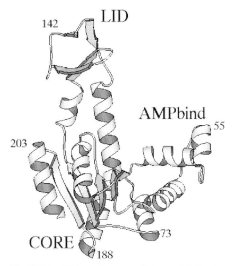

Fig. 17.14. Ribbon diagram of the crystal structure of apo-AK from *E. coli* (1ake) (generated with the program MOLSCRIPT). The sites of attachment of the probes are marked.

the CORE domain in the apo-AK. This an indication for multiple conformers separated by small energy barriers. Furthermore, the crystal structure is only very poorly populated in the apo-AK. Moreover, the wide distributions also include the partial population of the closed conformers, where the interdomain distance is already similar to that of the holoenzyme. This is in support of a select-fit mechanism of the coupling of the domain closure conformational change and the binding of either substrates. Binding of each one of the substrates is associated with substantial reduction of the mean and width of the distributions, as expected for the domain closure mechanism. The interdomain distance distributions observed for the complexes of the two enzymes with their substrates were narrower, i.e., the flexibility is restricted towards a more uniform structure in solution. The folding of the holoenzyme differs from that of the apoenzyme due to the additional interactions between the pairs of the domains and substrate. Electrostatic interactions (see chapter 7) probably contribute a major part of the extra stabilization of the structures of holo-phosphoryl transferases studied here.

17.11
Single Molecule FRET-detected Folding Experiments

The long-range goal of protein folding research is the elucidation of the principles of the solution of the chain entropy problem, i.e., the guided construction of the native structure. Imagine a dream experiment that would create a movie that follows a polypeptide chain during its transition from an unordered state to the native state. Single molecule FRET spectroscopy, based on site-specific labeling and the recent advances in laser technology and pinhole optics, makes this dream a reality. It is already possible to continuously monitor trajectories of changes of intramolecular distances in single protein molecules over periods of few seconds. This ultra high sensitivity helps us arrive as closely as possible to the ultimate analytical resolution. Trajectories collected for series of double-labeled protein mutants that monitor various distances in the protein molecule can report the order of formation of the various structures in the protein and their transitions. This type of experiment can add important insights relevant to the central folding questions, but the folding transition is inherently a stochastic process. Therefore, we cannot deduce any general principle from observations of single molecules at any resolution. Consequently, the principles of the mechanism of the folding transition must be investigated in the behavior of the ensembles, not in the individual molecules. So how do the amazing modern achievements of single molecule FRET spectroscopy serve the goals of protein folding research beyond the knowledge gained by the ensemble studies?

Individual trajectories of single molecule FRET spectroscopy can be grouped to generate both distributions of intramolecular distances and rates of their transitions. So why bother with the tedious, noisy single molecule detection, when trFRET can produce the ensemble distributions with much higher signal-to-noise ratio? The extra value of the distributions constructed from individual trajectories of single molecules can be described as follows: (a) these model-free distributions represent the true statistics of the sample trajectories; (b) subpopulations of protein molecules which may differ in the extent of the conformation space that they populate can be resolved; and (c) slow (longer than the span of a lifetime of a donor-excited state) conformational transitions and fluctuations within an equilibrium conformational space can be recorded and analyzed. This subject is of much interest for studies of conformational dynamics of unfolded and partially folded protein molecules.

The dynamic aspects of single molecule FRET really distinguish it from the ensemble trFRET experiments. So, why not quit the ensemble trFRET experiments and concentrate efforts on the promotion of single molecule FRET folding experiments? In general, the folding problem is multidimensional and therefore, a combination of complementary methods is essential. At present, the main limitations of single molecule FRET folding experiments are: (a) limited time resolution; (b) difficulties in performing perturbation initiated fast kinetics experiments; (c) the contribution of shot noise which affects the observed width of the distributions; (d) the problem of immobilization; and (e) quantitative analyses. Quantitative analyses of single molecule FRET experiments are inherently difficult due to the comparative nature of the measurements. Reference trajectories for the determination of variables that are essential input for the analyses of FRET experiments (e.g., determination of I_D, Φ_D^0, or τ_D^o) and control experiments for the non-FRET contributions, such as molecular orientations or non-FRET dye quenching, are difficult to match with individual FRET measurements. Therefore, it seems that a combination of ensemble and single molecule experiments are and will be the appropriate way to proceed.

17.12
Principles of Applications of Single Molecule FRET Spectroscopy in Folding Studies

In principle, reversible folding–unfolding transitions of single protein molecules can be studied at equilibrium. Through the selection of experimental equilibrium conditions, whereby protein molecules can reversibly cross the barriers between folded and unfolded conformations, single molecule FRET detection allows direct observation of the type of the transitions.

Several recent single molecule FRET studies which employed this strategy were reported. Weiss and coworkers [42] studied the folding of a double-labeled mutant of the two-state folder chymotrypsin inhibitor 2 (CI2) under several partially unfolded conditions (GdmCl concentrations). Subpopulations of folded and unfolded molecules were clearly resolved in this experiment, and their relative contributions under varied denaturant concentrations matched the ensemble denaturation curve. Eaton and coworkers [111, 112] obtained distributions of the folded and unfolded subpopulations of another two-state fast folder, the cold shock protein of *Thermotoga maritime* (CspTm).

Haran and coworkers [113] developed an elegant confinement technique. This technique enabled them to record relatively long (few seconds) trajectories of the time dependence of an intramolecular distance (between residues 73 and 203) in the CORE domain of *E. coli* adenylate kinase under partially folding conditions. Folding and unfolding transitions appeared in experimental time traces as correlated steps in donor and acceptor fluorescence intensities. The size of the spontaneous fluctuations in FRET efficiency shows a very broad distribution. This distribution which peaks at a relatively low value indicates a preference for small-step motion on the energy landscape. The time scale of the transitions is also distributed, and although many transitions are too fast to be time-resolved here, the slowest transitions may take >1 s to complete. These extremely slow changes during the folding of single molecules highlight the possible importance of correlated, non-Markovian conformational dynamics (Figure 17.15).

17.12.1
Design and Analysis of Single Molecule FRET Experiments

17.12.1.1 How is Single Molecule FRET Efficiency Determined?
In a typical single molecule FRET experiment, the single molecule emission intensities at the donor detector and acceptor detector are measured simultaneously (I_D and I_A). These raw intensities are integrated in time according to the desired time resolution. The problem of obtaining time trajectories of single molecules was solved by implementing two basic strategies: (a) free diffusing molecules in solution in which all the photons emitted in a burst (while the molecule is diffusing

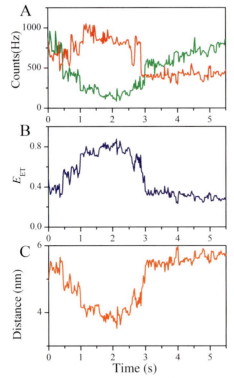

Fig. 17.15. Time-dependent signals from single AK molecules labeled at residues 73 and 203 showing slow folding or unfolding transitions. A) Signals showing a slow folding transition starting at ~0.5 s and ending at ~2 s. The same signals display a fast unfolding transition as well (at ~3 s). The acceptor signal is shown in red, and the donor is shown in green. B) FRET efficiency trajectory calculated from the signals in A. C) The interprobe distance trajectory showing that the slow transition involves a chain compaction by only 20%. The distance was computed from the data presented in trace B using a Förster distance $R_o = 49$ Å.

through the excitation volume) are summed to obtain I_D and I_A; (b) immobilized [114], or confined single molecules [113] in which the photon flux is continuously recorded at time resolution determined by the bining intervals, and $I_D(t)$ $I_A(t)$ are recorded. Following various corrections and filtering techniques (distinguish real molecular signal from noise, correction for leakage of signals due to overlap of spectra) the transfer efficiency, E, is calculated by

$$E(t) = I_A/(I_A + \beta I_D) \tag{36}$$

where β is a correction factor that depends on the quantum yields of the donor and the acceptor and the detection efficiencies of the two channels.

17.12.1.2 The Challenge of Extending the Length of the Time Trajectories

An ideal single molecule FRET folding experiment depends on the ability to follow the temporal trajectory of a single functional biological molecule for a long time under biologically relevant conditions, unperturbed by a trapping method. The free diffusion method which relies on random transit of refolding molecules in the small observation volume (sometimes termed "burst spectroscopy"), avoid any perturbation at the expense of the limitation of the length of the time trajectories to milliseconds. While this technique is very powerful in separating subpopulations and fast conformational dynamics, the free diffusion approach is not suitable for recording the temporal trajectories of the conformational transitions of individual molecules over extended (seconds) periods. Extended observation periods of individual molecules are important for the identification of slow or rare dynamic events. To enable observation periods for timespans much longer than allowed by diffusion, it is necessary to isolate and immobilize molecules. Several immobilization methods have been reported. Methods whereby the protein molecules were directly attached to a solid surface suffer from the influence of the surface interactions. Entrapment in the pores of poly(acrylamide) [115] or agarose [115] gels was also found to be useful for single molecule immobilization with reduced surface interactions. Haran and coworkers [113, 116] trapped single protein molecules in surface-tethered unilamellar lipid vesicles (Figure 17.16). The vesicles are large enough (∼100 nm in diameter) to allow encapsulated protein molecules to diffuse freely within them. Yet because the vesicles are immobilized to a solid

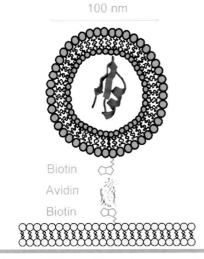

Fig. 17.16. Scheme of the methodology used for surface tethering of single molecule-containing liposomes. Large unilamellar lipid vesicles, each encapsulating a single protein molecule, were attached to a glass-supported lipid bilayer, using biotin-avidin chemistry.

surface, the proteins remain localized within the illuminating laser beam. Therefore, it is possible to record single molecule conformational transitions trajectories whose duration are limited by the photostability of the probes.

17.12.2
Distance and Time Resolution of the Single Molecule FRET Folding Experiments

The distance resolution of the single molecule folding experiments is determined by the characteristics of the pair of dyes and the signal-to-noise ratio, which depends also on the hardware. The scale of the distance determination is defined by the Förster critical distance, R_o, uncertainties in the value of R_o are contributed by the dependence of the parameters of Eq. (6). Uncertainties in the absolute distance calibration are of secondary importance in this case, the main interest in these experiments is in the fast fluctuations of the intramolecular distances due to folding/unfolding transitions. The overlap integral, the dipole orientations, and the donor quantum yield are frequently environmental sensitive, and hence are affected by the time dependent conformational changes. It is impossible to measure trajectories of those parameters simultaneously with the measurements of the FRET trajectories and hence fluctuations of R_o values limit the resolution of the determination of conformational transitions. In addition, in the unfolded or partially unfolded states, most protein molecules undergo conformational fluctuations on time scales shorter than the photon integration time (whether "bining" or burst duration). This can bias the apparent distances towards shorter end. Equation (9) shows that maximal distance resolution is obtained when the interprobe distance, r, equals R_o. An enhanced signal-to-noise ratio of the photon detection, which is essential for reduced uncertainty of the distance determination, calls for the use of dyes with high quantum yields and high extinction coefficients. These in turn contribute to high R_o values, and reduced resolution at the folded states of the proteins, where the distances are reduced. Most of the pairs used to date have R_o values of almost 50 Å. Thus care must be taken in the interpretation of intensity fluctuations results when distances less than 30 Å are involved. A detailed discussion of the signal-to-noise versus distance resolution was presented by Deniz et al. [117].

The time resolution of the single molecule FRET experiments is determined by the photon emission rate. With the current sensitive detection systems the main limiting factor is the extent of the triplet state population which is dependent on the excitation beam intensity and oxygen and triplet quencher agent concentrations [118, 119]. With present-day probes the practical time resolution in determination of trajectories of conformational transitions is in the millisecond range.

At present, the technology of single molecule FRET experiments is in its infancy. Therefore, we have good reason to expect significant further increases of distance and time resolution. The current reported experiments are very promising.

17.13
Folding Kinetics

17.13.1
Steady State and trFRET-detected Folding Kinetics Experiments

A dream folding kinetics experiment (see Chapters 12–15) would be one that would produce a time-dependent series of 3D structures. These structures constitute the transition of an ensemble of molecules from an unfolded state under folding conditions to the folded state. The experiment would probably initiate at very low-resolution blurred 3D images and gradually the profiles would become more resolved and the structures become visible. Would this change to more resolved structures be uniform for the whole molecule or would different folding nuclei or focal points become visible at different time points? FRET-detected rapid kinetic experiments were designed in a search for folding intermediates, the order of formation of structural elements, and the interactions that stabilize them. An analysis of such a dream experiment would hopefully reveal the order of structural transitions and formation of structural elements along very broad and gradually narrowing pathways to the native state. These might hopefully enable inference of the basic principles of the master design of the folding transitions. If such a dream experiment could be combined with site-directed perturbation mutagenesis, it might also be possible to search for the "sequence signals," i.e., inter-residue interactions that stabilize and lock structural elements, either in parallel, simultaneously, or sequentially.

Steady state and trFRET-detected kinetics experiments are far from such dream experiments, but these FRET experiments display some unique qualities that justify the preparative efforts of production of site-specifically labeled protein samples. The goal of these experiments is production of time series of distributions of selected key intramolecular distances (e.g., distances between structural elements). Such series can enable characterization of the transient structures of folding intermediates. The range of detected distances; the time resolution; the direct structural interpretation of the spectroscopic signals; the real time detection and the ability to determine transient distributions of distances add to the kinetic FRET experiment's unique strength.

Both steady state and time-resolved detection were applied together with any method of fast initiation of the folding or unfolding transition, e.g., stopped flow, continuous flow, and T-jump.

17.13.2
Steady State Detection

The ensemble mean transfer efficiency can be detected continuously by rapid recording of the donor emission, or the acceptor emission, or both. As in the case of equilibrium experiments, at least two traces should be recorded: (a) the donor emission ($I_{DD}(t)$) or the acceptor emission ($I_{DA}(t)$) of the double-labeled protein; and (b) the donor or the acceptor emission in the absence of FRET, ($I_D(t)$) or

($I_A(t)$) respectively. Then,

$$\langle E(t) \rangle = 1 - \frac{I_{DD}(t)}{I_D(t)}$$

quantitative determination of $E(t)$ through detection of the acceptor emission can be performed using Eq. (15), provided that the reference of constant transfer efficiency is available. Elegant examples of steady state FRET monitoring of the kinetics of folding were reported, and the text below describes some representative examples (see Chapter 15).

Eaton and Hofrichter [120, 121] used ultrarapid mixing continuous-flow method to study the submillisecond folding of the chemically denatured cytochrome c. Fluorescence quenching due to excitation energy transfer from the tryptophan to the heme was used to monitor the distance between these groups. The biphasic microsecond kinetics were interpreted as a barrier-free, partial collapse to a new equilibrium unfolded state at the lower denaturant concentration. This was followed by a slower crossing of a free energy barrier, separating the unfolded and folded states.

A similar experiment was reported by Roeder and coworkers [122, 123] (see Chapter 15) who studied the early conformational events during refolding of acyl-CoA binding protein (ACBP), an 86-residue α-helical protein. The continuous-flow mixing apparatus was used to measure rapid changes in tryptophan-dansyl FRET. Although the folding of ACBP was initially described as a concerted two-state process, the FRET signal revealed the formation of an ensemble of states on the 100 μs time scale. The kinetic data are fully accounted for by three-state mechanisms with either on- or off-pathway intermediates. The intermediates accumulate to a maximum population of 40%, and their stability depended only on weakly denaturant concentrations, which is consistent with a marginally stable ensemble of partially collapsed states. These experiments demonstrate the strength of fast kinetics FRET measurements, and can reveal and characterize transient accumulation of intermediate states in the folding of a protein considered to have apparent two-state folding mechanisms.

Beechem and coworkers [124, 125] studied the complex unfolding transition of the two-domain protein yeast phosphoglycerate kinase (PGK) using the multiple distances determination approach. Real-time determination of multiple intramolecular distances in the PGK molecule during the unfolding transition was achieved by means of steady state FRET-detected stopped-flow experiments. A series of six site specifically double-labeled PGK mutants was prepared using engineered tryptophan and cysteine residues ("strategy 1"). The unfolding of PGK was found to be a sequential multistep process (native → I_1 → I_2 → unfolded) with rate constants of 0.30, 0.16, and 0.052 s^{-1}, respectively (from native to unfolded). Six intramolecular distance vectors were resolved for both the I_1 and I_2 states. The transition from the native to I_1 state could be modeled as a large hinge-bending motion, in which both domains "swing away" from each other by almost 15 Å. As the domains move apart, the C-terminal domain rotates almost 90° around the hinge region connecting the two domains, while the N-terminal domain remains intact during the native to I_1 transition. This elegant set of measurements demonstrates the potential structural details that can be obtained by FRET determination of multiple intramolecular distances during the unfolding/refolding transitions of protein molecules.

Udgaonkar and Krishnamoorthy and their coworkers used multisite trFRET in a study of the unfolding intermediates in the unfolding transition of barstar [87, 88]. Four different single cysteine-containing mutants of barstar with cysteine residues at positions 25, 40, 62, and 82 were studied. Four different intramolecular distances were measured by steady state detection of the donor emission intensity and multiple, coexisting conformers could be detected on the basis of the time dependence of the apparent mean distances. The authors concluded that during unfolding the protein surface expands faster than, and independently of, water intrusion into the core. Like the works reported previously for the larger proteins, the multisite FRET study of Barstar showed that noncooperative folding, which is not observed by routinely used spectroscopic methods, can be found in unfolding of small proteins as well.

Another application of a multiple-distance probe study of folding of a small two-state protein was recently reported by Magg and Schmid [126]. *Bacillus subtilis* cold shock protein (Bc-CspB) folds very rapidly in a simple two-state mechanism. Magg and Schmid measured the shortening of six intramolecular distances during stopped-flow-initiated refolding by means of steady state FRET. Six mutants labeled by the pair Trp-1,5-AEDANS were prepared by means of labeling strategy 1. The calculated $R_o = 22$ Å fit well with the native dimensions of this small protein. Two pairs of sites were found to have the same inter-residue distance in both the native and the unfolded states. For four donor/acceptor pairs, the probed apparent mean intramolecular distances shorten with almost identical very rapid rates and thus, the two-state folding of this protein was confirmed. At the same time, more than 50% of the total increase in energy transfer upon folding occurred prior to the rate-limiting step. This finding reveals a very rapid collapse before the fast two-state folding reaction of Bc-Csp, and suggests that almost half of the shortening of the intramolecular distances upon folding of Bc-Csp has occurred before the rate-limiting step.

Fink and coworkers [126] used a similar labeling strategy and studied both the equilibrium and transient stopped-flow refolding intermediates of *Staphylococcus* nuclease by means of steady state FRET. The results indicate that there is an initial collapse of the protein in the deadtime of the stopped-flow instrument (regain 60% of the native FRET signal) which precedes the formation of the substantial secondary structure. The distance determination shows similar structures in the equilibrium and transient intermediates.

17.13.3
Time-resolved FRET Detection of Rapid Folding Kinetics: the "Double Kinetics" Experiment

The transient transfer efficiencies determined by steady state detection of the fluorescence intensities of the donor or the acceptor probes report rapid changes of distances. But since the conformations found in ensembles of partially folded protein molecules are inherently heterogeneous, the mean transfer efficiency (Eq. (12)) cannot be used for the determination of any meaningful mean distances. The mean and width of the distributions of distances in these rapid changing ensembles of partially folded protein molecules can be determined by rapid recording of time-resolved fluorescence decay curves of the probes. This was achieved by the "double kinetics" experiment.

The "double kinetics" [59, 127, 129] folding/unfolding experiments combine fast initiation of folding/unfolding transitions by rapid change of the solvent synchronized with the very rapid determination of fluorescence decay curves [129]. Single pulse detection enables the determination of fluorescence decay curves in less than a microsecond per curve, with a satisfactory signal-to-noise ratio. This enables us to determine time-dependent transient intramolecular distance distributions, IDD(t). Two time regimes are involved in this experimental approach: first is the duration of the conformational transition, the *"chemical time regime"* (t_c), (microseconds to seconds) and the second the *"spectroscopic time regime"*, (t_s), the nanosecond fluorescence decay of the probes. This experimental approach is designed to reveal the course of the development of IDD(t)s with millisecond or sub-millisecond time resolution (depending on the time resolution of the folding initiation technique). Combining this instrumental approach with the production of a series of protein samples, site-specifically labeled with donor and acceptor pairs, enables the characterization of the backbone fold and flexibility in transient intermediate states, during the protein folding transitions.

The challenge here is twofold: first, to collect fluorescence decay curves with a sufficiently high signal-to-noise ratio to enable determination of statistically significant parameters of the IDD(t); and second, to synchronize the refolding initiation device with the probe pulsed laser source. A stopped-flow double kinetics device based on a single photon counting method was developed by Beechem and coworkers [127] and applied for the determination of transient anisotropy decays. This approach utilized only one photon in several pulses and hence required both a very large number of stopped-flow experiments, and an amount of labeled protein. A system developed by Ratner et al. [59] based on low-frequency (10 MHz) laser pulses and fast digitizer oscilloscope, overcomes this difficulty (Figure 17.17). The current time resolution of the spectroscopic time scale, t_s, in this mode of double kinetics experiment is 250 ps. Up to 20 fluorescence decay curves can be measured with an acceptable signal-to-noise ratio within a single stopped-flow run. The single pulse detection of fluorescence decay curves can be synchronized with several methods of rapid initiation of refolding or unfolding, e.g., stopped-flow and laser-induced temperature jump.

17.13.4
Multiple Probes Analysis of the Folding Transition

The multiple probe test, in which a variety of probes of secondary and tertiary structures are applied to monitor the unfolding/refolding kinetics (see Chapters 2

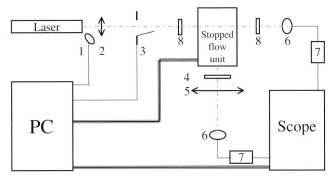

Fig. 17.17. Scheme of double kinetics device in the stopped flow configuration. The laser is a Quata-Ray high finesse OPO YAG Laser (Spectra-Physics, CA, USA), the stopped-flow unit is DX17MV (Applied photophysics, UK), the scope is Infinium 1.5 GHz oscilloscope (Hewlett-Packard, CA, USA), 1 and 5: lenses; 2: fast photodiode (Hamamatsu, Japan); 3: shutter; 4: bandpass filter; 6: biplanar phototubes (R1193U-03 Hamamatsu, Japan); 7: preamplifiers C 5594 (Hamamatsu, Japan); 8: attenuating filters.

and 12), has proved to be useful in distinguishing between parallel and sequential mechanisms. The determination of distributions of selected intramolecular distances by means of a series of double-labeled mutants of the model protein enables the determination of the time dependence of the formation of secondary structure elements; the formation of loops; changes of subdomain structures, and overall compaction of the chain at high time resolution. Changes of the mean and the width of distributions of intramolecular distances in protein molecules during the initial phases of the folding transitions can reveal weak energy bias (see Chapter 6). This favors conformational trends that affect the direction of the folding process. Very few and very weak interactions between chain elements may affect the shape of the distributions at the earliest phases of the folding. From such findings we can ask: Are such interactions effective during the initial collapse making it a biased change of conformation, rather than a random compaction driven only by solvent exclusion mechanism? Are those interactions local or nonlocal?

Ratner et al. [128] applied the double kinetics experiment in studying the early folding transitions of the CORE domain of *E. coli* adenylate kinase (AK) (Figure 17.14). Using the double-mixing stopped-flow mode of the double kinetics experiment the intramolecular distance distributions of a series of double-labeled AK mutants were determined in the unfolded state (in 1.8 M GdmCl). This occurred immediately after the change to refolding conditions (2–5 ms after dilution of denaturant), and upon completion of the refolding transition. This series of experiments was motivated by two questions. First, what is the extent of structure formation in different parts of the chain at the earliest detected intermediate state? And second, do secondary structure elements appear during folding prior to the formation of tertiary folds of the chain? What is the role of local and nonlocal interactions in the earliest phases of folding? This work is in progress, but the results obtained to date with several double-labeled mutants indicate the following: 5 ms after being transferred from unfolded to refolded states in a double-mixing unfolding/refolding cycle, the distributions of distances between two pairs of sites which measure the full length of secondary structure elements in the protein (pairs 169–188, an α-helix and pair 188–203, a β-sheet) maintain the same shape as found in the unfolded state (Figure 17.18). Native intramolecular distance distributions of these two segments are found only after 3 s. This finding shows that at least these secondary structure elements are formed only very late in the folding transition. Do tertiary structure elements appear in the CORE domain at an earlier time? Preliminary experiments in which the distances between residues 18 and 203, indicate that this may be the case, and this question remains under investigation.

Data obtained through multiple FRET experiments can reflect the complexity of the folding transition, but simultaneously help resolve some common principles of the resolution of this complexity. Those are the principles that enable the transition of ensembles of protein molecules from states characterized by multiple conformations to the native state characterized by narrow distributions of conformations.

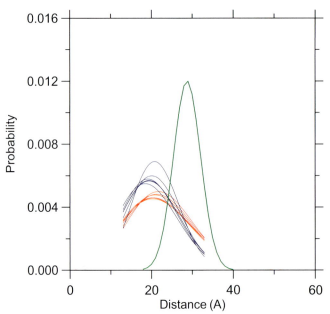

Fig. 17.18. Sets of transient distance distribution between residues 169 and 188 in AK in the denatured state (blue traces), in the 5 ms refolding intermediate state (red traces) and in the native state (after 3 s, green trace). The protein was denatured in 1.8 M GdmCl for 100 s and then the denaturant was diluted to final concentration of 0.3 M. Repeated traces are shown in order to demonstrate the range of experimental uncertainty in the parameters of the distributions. This experiment shows that in the early folding steps the helical segment between residues 169 and 188 (Figure 17.15) is unfolded, very similar to its state in the high denaturant concentration.

17.14
Concluding Remarks

The problem of folding and dynamics of globular proteins is a multidimensional problem. The structures of refolding protein molecules should be characterized by multiple distances and time constants. No single method can provide the full description of the structure and dynamics of a protein molecule in solution. X-ray crystallography and multidimensional NMR measurements are the most informative methods for determination of the average equilibrium conformations of protein molecules. It seems that deciphering the mechanism that overcomes the three folding problems depends on studies of the processes rather than the structures alone. That is where spectroscopy is indispensable.

The elementary steps in all the transitions of the protein molecules are the random fluctuations. Therefore the proteins evolved to have multiple conformations, distributions of the active structures, and combinations of multiple modes of motions. Modulation of these distributions of conformations and modes of motions is the basis for generation of vectorial process from Brownian elementary steps and control of the processes. Therefore it is essential to study the molecular processes at the population level as well as at the single molecule level.

The proposed loop hypothesis raises the next question: where is the information for the nonlocal interactions that form the loops? Our working hypothesis is that pairs of small clusters of mainly hydrophobic residues located on widely separated chain segments, carry the genetic message for the non-local interactions. Therefore an important future direction in FRET applications in folding research is the preparation of series of double-labeled mutants combined with "perturbation mutations," mutations that perturb putative nonlocal interactions. The effect of such mutations on the shape of the time- or denaturant-dependent intramolecular distance distributions is expected to be an adequate test of this hypothesis. It seems that a complete understanding of the folding and action of globular proteins will always require a combination of approaches. Ensemble and single molecule FRET measurements can be a major component in the arsenal of proposed methods.

The modern laser and detector technologies opened a wide window of time and wavelength regimes for probing the processes involved in the folding transition and changes of conformations of proteins in action. The development of protein engineering and advanced separation methods has enabled rational design of protein mutants suitable for experiments designed to answer specific questions. It is the unique combination of these recent advances in three different disciplines as well as new computational techniques that may lead to quantitative analysis of the basic processes. The development of the double kinetics method is an example of combination with a synchronization experiment. Additional technologies such as pressure or temperature jumps will enable determination of detailed correlations between dynamics (both local and global) and the folding transition.

Future development of the applications of FRET methods in folding research will probably be based on the current methods with few new directions. We can expect further improvement of signal-to-noise ratio, distance and time resolution, computational methods of analysis of decay curves, determination of single molecule fluorescence decay rates, methods for preparations of series of multiple distances and the sensitivity of the detection at the level of single molecule level.

Future developments should strive to advance fours aspects of the FRET measurements: (a) the sensitivity of the detection method, (b) the modes of measurements, (c) the biochemical technology of site-specific labeling, and (d) development of new fluorescent probes.

(a) Future directions of the spectroscopic methods can further improve the signal-to-noise ratio for detection of distributions of distances; subpopulations and improved time resolution. Enhancement of the time resolution and dynamic range of single pulse detection will enhance the applications of the double kinetics approach. It is conceivable that faster collection of multiple single molecule trajectories will be achieved. Development of new confinement methods and new methods for ultrafast changes of the solution conditions at the microscopic level for single molecule detection experiments will hopefully enable determination of transient distributions of selected distances during the folding transitions of model proteins.

(b) Future modes of measurements will probably include further development of the double kinetics approach, use of multiparametric measurements including anisotropy measurements, time-resolved spectra and further development of pressure-jump, temperature-jump, pH-jump, and other perturbation modes.

(c) A major future progress in accomplishment of the promising potential of FRET methods in folding research can be gained by investing research efforts in the development of the biochemical methodology of high-yield site-specific labeling methods. The current specific aims of these developmental efforts should be: new methods for selection of nonperturbing modifications of protein molecules, new methods for efficient purification of labeled products and development of new selective reagents. It is expected that the cell-free proteins synthesis methods will become much more widely used.

(d) The main limitation of single molecule FRET spectroscopy in folding research is the photobleaching of the probes. Major developmental efforts for synthesis of photostable probes or methods for reduction of the photo-oxygenation reactions are needed. Parallel efforts for development of probes of smaller volume and pairs of wider ranges of R_o-values are also essential for enhancement of the future applications of FRET spectroscopy in folding research. The current state of the art is very advanced and that is the basis of optimism that these developments are feasible.

Acknowledgments

The author was privileged to work for a long time with a group of devoted and enthusiastic scientists, collaborators and students, who contributed to many aspects of the projects cited in this chapter. Their contributions are deeply acknowledged. The approach described here was influenced by the initiative which was started as early as 1968 by E. Katzir-Katchalsky, Izhak Steinberg, and Meir Wilchek who realized at that time the potential of using this approach in the study of protein conformational dynamics. The author had the privilege of working under their guidance in the early 1970s. This work was supported by the grants from NIH Institute of General Medical Sciences, United States-Israel Binational Science Foundation (BSF), Equipment Grants and Research Grants from the Fund for Basic Research of the Israeli Academy of Sciences (ISF) and the German-Israel foundation (GIF).

References

1 FÖRSTER, T. H. (1948). Zwischen Molekulare Energie Wanderung und Fluoreszenz. *Ann Phys (Leipzig)* 2, 55–75.
2 FÖRSTER, T. H. (1959). Transfer mechanisms of electronic excitation. *Discuss Faraday Soc* 27, 7–17.
3 FÖRSTER, T. H. (1965). Delocalized excitation and excitation transfer. In *Modern Quantum Chemistry, Istanbul lectures Part III: Action of Light and Organic Crystals* (SINAONGLU, O., ed.), pp. 93–137. Academic Press, New York.
4 STEINBERG, I. Z. (1971). Long-range nonradiative transfer of electronic excitation energy in proteins and polypeptides. *Annu Rev Biochem* 40, 83–114.
5 STRYER, L., THOMAS, D. D. & MEARES, C. F. (1982). Diffusion-enhanced fluorescence energy transfer. *Annu Rev Biophys Bioeng* 11, 203–222.
6 VAN DER MEER, W. B., COKER, G. III, & CHEN, S. Y. S. (1994). *Resonance Energy Transfer Theory and Data*. VCH Publishers, New York.
7 MAKHATADZE, G. I. & PRIVALOV, P. L. (1996). On the entropy of protein folding. *Protein Sci* 5, 507–510.
8 PLOTKIN, S. S. & ONUCHIC, J. N. (2002). Understanding protein folding with energy landscape theory. Part II: Quantitative aspects. *Q Rev Biophys* 35, 205–286.
9 FERGUSON, N. & FERSHT, A. R. (2003). Early events in protein folding. *Curr Opin Struct Biol* 13, 75–81.
10 COOPER, A. (1984). Protein fluctuations and the thermodynamic uncertainty principle. *Prog Biophys Mol Biol* 44, 181–214.
11 PRIVALOV, P. L. & POTEKHIN, S. A. (1986). Scanning microcalorimetry in studying temperature-induced changes in proteins. *Methods Enzymol* 131, 4–51.
12 CREIGHTON, T. E. (1993). *Proteins: Structure and Molecular Properties*, 2nd edn. W.H. Freeman, New York.
13 CARERI, G., FASELLA, P. & GRATTON, E. (1979). Enzyme dynamics: the statistical physics approach. *Annu Rev Biophys Bioeng* 8, 69–97.
14 FERSHT, A. R. (1985). *Enzyme Structure and Mechanism*, 2nd edn. W.H. Freeman, New York.
15 SLATER, J. C. & FRANK, N. H. (1947). *Electromagnetism*. McGraw, New York.
16 PERRIN, J. (1927). Fluorescence et induction moleculaire par sesonance. *C R Hebd Seances Acad Sci* 184, 1097–1100.
17 STEINBERG, I. Z., HAAS, E. & KATCHALSKY-KATZIR, E. (1983). Long-range non-radiative transfer of electronic excitation energy. In *Time Resolved Spectroscopy in Biochemistry* (CUNDALL, R. B. & DALE, R. E., eds), pp. 411–451. Plenum Publishing, New York.
18 WEBER, G. & TEAL, F. J. W. (1959). *Discuss Faraday Soc* 27, 134–151.
19 LATT, S. A., CHEUNG, H. T. & BLOUT, E. R. (1965). Energy transfer. A system with relatively fixed donor-acceptor separation. *J Am Chem Soc* 87, 995–1003.

20. Stryer, L. & Haugland, R. P. (1967). Energy transfer: a spectroscopic ruler. *Proc Natl Acad Sci USA* 58, 719–726.
21. Eisinger, J. & Dale, R. E. (1974). Letter: interpretation of intramolecular energy transfer experiments. *J Mol Biol* 84, 643–647.
22. Dale, R. E. & Eisinger, J. (1976). Intramolecular energy transfer and molecular conformation. *Proc Natl Acad Sci USA* 73, 271–273.
23. Dale, R. E., Eisinger, J. & Blumberg, W. E. (1979). The orientational freedom of molecular probes. The orientation factor in intramolecular energy transfer. *Biophys J* 26, 161–193.
24. Selvin, P. R. & Hearst, J. E. (1994). Luminescence energy transfer using a terbium chelate: improvements on fluorescence energy transfer. *Proc Natl Acad Sci USA* 91, 10024–10028.
25. Mersol, J. V., Wang, H., Gafni, A. & Steel, D. G. (1992). Consideration of dipole orientation angles yields accurate rate equations for energy transfer in the rapid diffusion limit. *Biophys J* 61, 1647–1655.
26. Haas, E., Katchalski-Katzir, E. & Steinberg, I. Z. (1978). Effect of the orientation of donor and acceptor on the probability of energy transfer involving electronic transitions of mixed polarization. *Biochemistry* 17, 5064–5070.
27. Albrecht, A. C. (1960). Forbidden characterizing allowed electronic transitions. *J Chem Phys* 33, 156.
28. Eisinger, J., Feuer, B. & Lamola, A. A. (1969). Intramolecular singlet excitation transfer. Applications to polypeptides. *Biochemistry* 8, 3908–3915.
29. Cortijo, M., Steinberg, I. Z. & Shaltiel, S. (1971). Fluorescence of glycogen phosphorylase b. Structural transitions and energy transfer. *J Biol Chem* 246, 933–938.
30. Valeur, B. (2002). *Molecular Fluorescence Principles and Applications.* Wiley-VCH, Weinheim.
31. Lakowicz, J. R. (1999). *Principles of Fluorescence Spectroscopy,* 2nd edn. Kluwer Academic/Plenum, New York.
32. Stryer, L., Thomas, D. D. & Carlsen, W. F. (1982). Fluorescence energy transfer measurements of distances in rhodopsin and the purple membrane protein. *Methods Enzymol* 81, 668–678.
33. Wu, P. & Brand, L. (1994). Resonance energy transfer: methods and applications. *Anal Biochem* 218, 1–13.
34. Clegg, R. M. (1992). Fluorescence resonance energy transfer and nucleic acids. *Methods Enzymol* 211, 353–388.
35. Clegg, R. M. (1996). Fluorescence resonance energy transfer. In *Fluorescence Imaging Spectroscopy and Microscopy* (Wang, X. F. & Herman, B., eds), Chemical Analysis Series Vol. 137, pp. 179–252. John Wiley & Sons, New York.
36. Cantor, C. R. & Pechukas, P. (1971). Determination of distance distribution functions by singlet-singlet energy transfer. *Proc Natl Acad Sci USA* 68, 2099–2101.
37. Haas, E., Katchalsky-Katzir, E. & Steinberg, I. Z. (1978). Brownian motion of the ends of oligopeptide chains in solution as estimated by energy transfer between the chain ends. *Biopolymers* 17, 11–31.
38. Beechem, J. M. & Haas, E. (1989). Simultaneous determination of intramolecular distance distributions and conformational dynamics by global analysis of energy transfer measurements. *Biophys J* 55, 1225–1236.
39. Chandrasekhar, S. (1943). Stochastic problems in physics and astronomy. *Rev Mod Phys* 15, 1–89.
40. Navon, A., Ittah, V., Landsman, P., Scheraga, H. A. & Haas, E. (2001). Distributions of intramolecular distances in the reduced and denatured states of bovine pancreatic ribonuclease A. Folding initiation structures in the C-terminal portions of the reduced protein. *Biochemistry* 40, 105–118.
41. Lakowicz, J. R., Kusba, J., Szmacinski, H. et al. (1991). Resolution of end-to-end diffusion coefficients and distance distributions of flexible molecules using fluorescent donor-acceptor and donor-quencher pairs. *Biopolymers* 31, 1363–1378.

42 DENIZ, A. A., LAURENCE, T. A., BELIGERE, G. S. et al. (2000). Single-molecule protein folding: diffusion fluorescence resonance energy transfer studies of the denaturation of chymotrypsin inhibitor 2. *Proc Natl Acad Sci USA* 97, 5179–5184.

43 LIU, D. R., MAGLIERY, T. J., PASTRNAK, M. & SCHULTZ, P. G. (1997). Engineering a tRNA and aminoacyl-tRNA synthetase for the site-specific incorporation of unnatural amino acids into proteins in vivo. *Proc Natl Acad Sci USA* 94, 10092–10097.

44 BUCKLER, D. R., HAAS, E. & SCHERAGA, H. A. (1993). C-terminal labeling of ribonuclease A with an extrinsic fluorescent probe by carboxypeptidase Y-catalyzed transpeptidation in the presence of urea. *Anal Biochem* 209, 20–31.

45 AMIR, D. & HAAS, E. (1987). Estimation of intramolecular distance distributions in bovine pancreatic trypsin inhibitor by site-specific labeling and nonradiative excitation energy-transfer measurements. *Biochemistry* 26, 2162–2175.

46 TAKASHI, R. (1988). A novel actin label: a fluorescent probe at glutamine-41 and its consequences. *Biochemistry* 27, 938–943.

47 FOLK, J. E. & CHUNG, S. I. (1985). Transglutaminases. *Methods Enzymol* 113, 358–375.

48 FINK, M. L., CHUNG, S. I. & FOLK, J. E. (1980). Gamma-glutamylamine cyclotransferase: specificity toward epsilon-(L-gamma-glutamyl)-L-lysine and related compounds. *Proc Natl Acad Sci USA* 77, 4564–4568.

49 TCHERKASSKAYA, O., PTITSYN, O. B. & KNUTSON, J. R. (2000). Nanosecond dynamics of tryptophans in different conformational states of apomyoglobin proteins. *Biochemistry* 39, 1879–1889.

50 RISCHEL, C., THYBERG, P., RIGLER, F. & POULSEN, F. M. (1996). Time-resolved fluorescence studies of the molten globule state of apomyoglobin. *J Mol Biol* 257, 877–885.

51 SANTORO, S. W., ANDERSON, J. C., LAKSHMAN, V. & SCHULTZ, P. G. (2003). An archaebacteria-derived glutamyl-tRNA synthetase and tRNA pair for unnatural amino acid mutagenesis of proteins in Escherichia coli. *Nucleic Acids Res* 31, 6700–6709.

52 HARAN, G., HAAS, E., SZPIKOWSKA, B. K. & MAS, M. T. (1992). Domain motions in phosphoglycerate kinase: determination of interdomain distance distributions by site-specific labeling and time-resolved fluorescence energy transfer. *Proc Natl Acad Sci USA* 89, 11764–11768.

53 PENNINGTON, M. W. (1994). Site-specific chemical modification procedures. *Methods Mol Biol* 35, 171–185.

54 SINEV, M. A., SINEVA, E. V., ITTAH, V. & HAAS, E. (1996). Domain closure in adenylate kinase. *Biochemistry* 35, 6425–6437.

55 SINEV, M., LANDSMANN, P., SINEVA, E., ITTAH, V. & HAAS, E. (2000). Design consideration and probes for fluorescence resonance energy transfer studies. *Bioconjug Chem* 11, 352–362.

56 ORTIZ, J. O. & BUBIS, J. (2001). Effects of differential sulfhydryl group-specific labeling on the rhodopsin and guanine nucleotide binding activities of transducin. *Arch Biochem Biophys* 387, 233–242.

57 RATNER, V., KAHANA, E., EICHLER, M. & HAAS, E. (2002). A general strategy for site-specific double labeling of globular proteins for kinetic FRET studies. *Bioconjug Chem* 13, 1163–11670.

58 JACOB, M. H., AMIR, D., RATNER, V. & HAAS, E. Straightforward selective protein modification by predictions and utilizing reactivity differences of surface cysteine residues. (in preparation)

59 RATNER, V. & HAAS, E. (1998). An instrument for time resolved monitoring of fast chemical transitions: application to the kinetics of refolding of a globular protein. *Rev Sci Instrum* 69, 2147–2154.

60 RATNER, V., SINEV, M. & HAAS, E. (2000). Determination of intramolecular distance distribution during protein folding on the millisecond timescale. *J Mol Biol* 299, 1363–1371.

61 AMIR, D., LEVY, D. P., LEVIN, Y. & HAAS, E. (1986). Selective fluorescent labeling of amino groups of bovine pancreatic trypsin inhibitor by reductive alkylation. *Biopolymers* 25, 1645–1658.

62 EPE, B., WOOLLEY, P., STEINHAUSER, K. G. & LITTLECHILD, J. (1982). Distance measurement by energy transfer: the 3′ end of 16-S RNA and proteins S4 and S17 of the ribosome of *Escherichia coli*. *Eur J Biochem* 129, 211–219.

63 FLAMION, P. J., CACHIA, C. & SCHREIBER, J. P. (1992). Non-linear least-squares methods applied to the analysis of fluorescence energy transfer measurements. *J Biochem Biophys Methods* 24, 1–13.

64 BRAND, L. (1992). *Methods in Enzymology* (BRAND, L. & JOHNSON, L. M., eds), Vol. 210.

65 BEECHEM, J. M. & BRAND, L. (1985). Time-resolved fluorescence of proteins. *Annu Rev Biochem* 54, 43–71.

66 AMELOOT, M., BEECHEM, J. M. & BRAND, L. (1986). Simultaneous analysis of multiple fluorescence decay curves by Laplace transforms. Deconvolution with reference or excitation profiles. *Biophys Chem* 23, 155–171.

67 LANCZOS, C. (1956). *Applied Analysis*, pp. 272–304. Prentice Hall, Engelwood Cliffs.

68 BEVINGTON, P. R. (1969). *Data Reduction and Error Analysis for the Physical Sciences*. McGraw-Hill, New York.

69 GRINVALD, A. & STEINBERG, I. Z. (1974). On the analysis of fluorescence decay kinetics by the method of least-squares. *Anal Biochem* 59, 583–598.

70 AMES, W. F. (1977). *Ames, William F. Numerical Methods for Partial Differential Equations*, 2nd edn. Academic Press, New York.

71 PRESS, W. H., FLANNERY, B. P., TEUKOLSKI, S. A. & VETTERLING, W. F. (1989). *Numerical Recipes: The Art of Scientific Computing*. Cambridge University Press, Cambridge.

72 BROCHON, J. C. (1994). Maximum entropy method of data analysis in time-resolved spectroscopy. *Methods Enzymol* 240, 262–311.

73 LAKSHMIKANTH, G. S., SRIDEVI, K., KRISHNAMOORTHY, G. & UDGAONKAR, J. B. (2001). Structure is lost incrementally during the unfolding of barstar. *Nat Struct Biol* 8, 799–804.

74 DEISENHOFER, J. & STEIGMANN, W. (1975). Crystallographic refinement of the structure of bovine pancreatic trypsin inhibitor at 1.5 A resolution. *Acta Crystallogr* B31, 238.

75 CREIGHTON, T. E. (1978). Experimental studies of protein folding and unfolding. *Prog Biophys Mol Biol* 33, 231–297.

76 GUSSAKOVSKY, E. E. & HAAS, E. (1992). The compact state of reduced bovine pancreatic trypsin inhibitor is not the compact molten globule. *FEBS Lett* 308, 146–148.

77 GOTTFRIED, D. S. & HAAS, E. (1992). Nonlocal interactions stabilize compact folding intermediates in reduced unfolded bovine pancreatic trypsin inhibitor. *Biochemistry* 31, 12353–12362.

78 GRIKO, Y. V., PRIVALOV, P. L., STURTEVANT, J. M. & VENYAMINOV, S. (1988). Cold denaturation of staphylococcal nuclease. *Proc Natl Acad Sci USA* 85, 3343–3347.

79 ITTAH, V. & HAAS, E. (1995). Nonlocal interactions stabilize long range loops in the initial folding intermediates of reduced bovine pancreatic trypsin inhibitor. *Biochemistry* 34, 4493–4506.

80 KLEIN-SEETHARAMAN, J., OIKAWA, M., GRIMSHAW, S. B. et al. (2002). Long-range interactions within a nonnative protein. *Science* 295, 1719–1722.

81 LATTMAN, E. E. & ROSE, G. D. (1993). Protein folding – what's the question? *Proc Natl Acad Sci USA* 90, 439–441.

82 BEREZOVSKY, I. N., GROSBERG, A. Y. & TRIFONOV, E. N. (2000). Closed loops of nearly standard size: common basic element of protein structure. *FEBS Lett* 466, 283–286.

83 BEREZOVSKY, I. N., KIRZHNER, V. M., KIRZHNER, A. & TRIFONOV, E. N. (2001). Protein folding: looping from hydrophobic nuclei. *Proteins* 45, 346–350.

84. Buckler, D. R., Haas, E. & Scheraga, H. A. (1995). Analysis of the structure of ribonuclease A in native and partially denatured states by time-resolved noradiative dynamic excitation energy transfer between site-specific extrinsic probes. *Biochemistry* 34, 15965–15978.

85. Wu, P. G., James, E. & Brand, L. (1993). Compact thermally-denatured state of a staphylococcal nuclease mutant from resonance energy transfer measurements. *Biophys Chem* 48, 123–133.

86. Wu, P. & Brand, L. (1994). Conformational flexibility in a staphylococcal nuclease mutant K45C from time-resolved resonance energy transfer measurements. *Biochemistry* 33, 10457–10462.

87. Sridevi, K. & Udgaonkar, J. B. (2003). Surface expansion is independent of and occurs faster than core solvation during the unfolding of barstar. *Biochemistry* 42, 1551–1563.

88. Sridevi, K., Lakshmikanth, G. S., Krishnamoorthy, G. & Udgaonkar, J. B. (2004). Increasing stability reduces conformational heterogeneity in a protein folding intermediate ensemble. *J Mol Biol* 337, 699–711.

89. Cheung, H. C., Wang, C. K., Gryczynski, I. et al. (1991). Distance distributions and anisotropy decays of troponin C and its complex with troponin I. *Biochemistry* 30, 5238–5247.

90. Cheung, H. C., Gryczynski, I., Malak, H., Wiczk, W., Johnson, M. L. & Lakowicz, J. R. (1991). Conformational flexibility of the Cys 697-Cys 707 segment of myosin subfragment-1. Distance distributions by frequency-domain fluorometry. *Biophys Chem* 40, 1–17.

91. Lakowicz, J. R., Wiczk, W., Gryczynski, I., Szmacinski, H. & Johnson, M. L. (1990). Influence of end-to-end diffusion on intramolecular energy transfer as observed by frequency-domain fluorometry. *Biophys Chem* 38, 99–109.

92. Lakowicz, J. R., Kusba, J., Wiczk, W., Gryczynski, I., Szmacinski, H. & Johnson, M. L. (1991). Resolution of the conformational distribution and dynamics of a flexible molecule using frequency-domain fluorometry. *Biophys Chem* 39, 79–84.

93. Eis, P. S. & Lakowicz, J. R. (1993). Time-resolved energy transfer measurements of donor-acceptor distance distributions and intramolecular flexibility of a CCHH zinc finger peptide. *Biochemistry* 32, 7981–7993.

94. Vogel, H., Nilsson, L., Rigler, R., Voges, K. P. & Jung, G. (1988). Structural fluctuations of a helical polypeptide traversing a lipid bilayer. *Proc Natl Acad Sci USA* 85, 5067–5071.

95. Vogel, H., Nilsson, L., Rigler, R. et al. (1993). Structural fluctuations between two conformational states of a transmembrane helical peptide are related to its channel-forming properties in planar lipid membranes. *Eur J Biochem* 212, 305–313.

96. Bennett, W. S. & Huber, R. (1984). Structural and functional aspects of domain motions in proteins. *CRC Crit Rev Biochem* 15, 291–384.

97. Huber, R. & Bennett, W. S., Jr. (1983). Functional significance of flexibility in proteins. *Biopolymers* 22, 261–279.

98. Warshel, A. & Aqvist, J. (1991). Electrostatic energy and macromolecular function. *Annu Rev Biophys Biophys Chem* 20, 267–298.

99. Jencks, W. P. (1975). Binding energy, specificity, and enzymic catalysis: the circe effect. *Adv Enzymol Relat Areas Mol Biol* 43, 219–410.

100. Fersht, A. R. (1998). *Structure and Mechanism in Protein Science: A Guide to Enzyme Catalysis and Protein Folding.* W. H. Freeman, New York.

101. Bialek, W. & Onuchic, J. N. (1988). Protein dynamics and reaction rates: mode-specific chemistry in large molecules? *Proc Natl Acad Sci USA* 85, 5908–5912.

102. Schulz, G. E. (1991). Mechanisms of enzyme catalysis from crystal structure analyses. *Ciba Found Symp* 161, 8–22; discussion 22–27.

103. Anderson, C. M., Zucker, F. H. & Steitz, T. A. (1979). Space-filling

models of kinase clefts and conformation changes. *Science* 204, 375–380.

104 PAVLOV, M., SINEV, M. A., TIMCHENKO, A. A. & PTITSYN, O. B. (1986). A study of apo- and holo-forms of horse liver alcohol dehydrogenase in solution by diffuse x-ray scattering. *Biopolymers* 25, 1385–1397.

105 WATSON, H. C., WALKER, N. P., SHAW, P. J. et al. (1982). Sequence and structure of yeast phosphoglycerate kinase. *EMBO J* 1, 1635–1640.

106 NODA, L. (1973). Adenylate kinase. In *The Enzymes* (BOYER, P. D., ed.), Vol. 8, pp. 279–305. Academic Press, New York.

107 BANKS, R. D., BLAKE, C. C., EVANS, P. R. et al. (1979). Sequence, structure and activity of phosphoglycerate kinase: a possible hinge-bending enzyme. *Nature* 279, 773–777.

108 MULLER, C. W., SCHLAUDERER, G. J., REINSTEIN, J. & SCHULZ, G. E. (1996). Adenylate kinase motions during catalysis: an energetic counterweight balancing substrate binding. *Structure* 4, 147–156.

109 MULLER, Y. A., SCHUMACHER, G., RUDOLPH, R. & SCHULZ, G. E. (1994). The refined structures of a stabilized mutant and of wild-type pyruvate oxidase from *Lactobacillus plantarum*. *J Mol Biol* 237, 315–335.

110 MULLER, C. W. & SCHULZ, G. E. (1992). Structure of the complex between adenylate kinase from *Escherichia coli* and the inhibitor Ap5A refined at 1.9 A resolution. A model for a catalytic transition state. *J Mol Biol* 224, 159–177.

111 SCHULER, B., LIPMAN, E. A. & EATON, W. A. (2002). Probing the free-energy surface for protein folding with single-molecule fluorescence spectroscopy. *Nature* 419, 743–747.

112 LIPMAN, E. A., SCHULER, B., BAKAJIN, O. & EATON, W. A. (2003). Single-molecule measurement of protein folding kinetics. *Science* 301, 1233–1235.

113 RHOADES, E., GUSSAKOVSKY, E. & HARAN, G. (2003). Watching proteins fold one molecule at a time. *Proc Natl Acad Sci USA* 100, 3197–3202.

114 TALAGA, D. S., LAU, W. L., RODER, H. et al. (2000). Dynamics and folding of single two-stranded coiled-coil peptides studied by fluorescent energy transfer confocal microscopy. *Proc Natl Acad Sci USA* 97, 13021–13026.

115 DICKSON, R. M., CUBITT, A. B., TSIEN, R. Y. & MOERNER, W. E. (1997). On/off blinking and switching behaviour of single molecules of green fluorescent protein. *Nature* 388, 355–358.

116 BOUKOBZA, R. S. & HARAN, G. (2001). Immobilization in surface-tethered lipid vesicles as a new tool for single biomolecule spectroscopy. *J Phys Chem B* 105, 12165–12170.

117 DENIZ, A. A., DAHAN, M., GRUNWELL, J. R. et al. (1999). Single-pair fluorescence resonance energy transfer on freely diffusing molecules: observation of Forster distance dependence and subpopulations. *Proc Natl Acad Sci USA* 96, 3670–3675.

118 HA, T., ENDERLE, T., CHEMLA, D. S., SELVIN, P. R. & WEISS, S. (1997). Quantum jumps of single molecules at room temperature. *Chem Phys Lett* 271, 1–5.

119 VEERMAN, J. A., GARCIA-PARAJO, M. F., KUIPERS, L. & VAN HULST, N. F. (1999). Time-varying triplet state lifetimes of single molecules. *Phys Rev Lett* 83, 2155–2158.

120 CHAN, C. K., HU, Y., TAKAHASHI, S., ROUSSEAU, D. L., EATON, W. A. & HOFRICHTER, J. (1997). Submillisecond protein folding kinetics studied by ultrarapid mixing. *Proc Natl Acad Sci USA* 94, 1779–1784.

121 EATON, W. A., MUNOZ, V., THOMPSON, P. A., CHAN, C. K. & HOFRICHTER, J. (1997). Submillisecond kinetics of protein folding. *Curr Opin Struct Biol* 7, 10–14.

122 TEILUM, K., KRAGELUND, B. B. & POULSEN, F. M. (2002). Transient structure formation in unfolded acyl-coenzyme A-binding protein observed by site-directed spin labelling. *J Mol Biol* 324, 349–357.

123 TEILUM, K., MAKI, K., KRAGELUND, B. B., POULSEN, F. M. & RODER, H. (2002). Early kinetic intermediate in

the folding of acyl-CoA binding protein detected by fluorescence labeling and ultrarapid mixing. *Proc Natl Acad Sci USA* 99, 9807–9812.

124 LILLO, M. P., SZPIKOWSKA, B. K., MAS, M. T., SUTIN, J. D. & BEECHEM, J. M. (1997). Real-time measurement of multiple intramolecular distances during protein folding reactions: a multisite stopped-flow fluorescence energy-transfer study of yeast phosphoglycerate kinase. *Biochemistry* 36, 11273–11281.

125 LILLO, M. P., BEECHEM, J. M., SZPIKOWSKA, B. K., SHERMAN, M. A. & MAS, M. T. (1997). Design and characterization of a multisite fluorescence energy-transfer system for protein folding studies: a steady-state and time-resolved study of yeast phosphoglycerate kinase. *Biochemistry* 36, 11261–11272.

126 MAGG, C. & SCHMID, F. X. (2004). Rapid collapse precedes the fast two-state folding of the cold shock protein. *J Mol Biol* 335, 1309–1323.

127 JONES, B. E., BEECHEM, J. M. & MATTHEWS, C. R. (1995). Local and global dynamics during the folding of *Escherichia coli* dihydrofolate reductase by time-resolved fluorescence spectroscopy. *Biochemistry* 34, 1867–1877.

128 RATNER, V., KAHANA, E. & HAAS, E. (2002). The natively helical chain segment 169–188 of *Escherichia coli* adenylate kinase is formed in the latest phase of the refolding transition. *J Mol Biol* 320, 1135–1145.

129 BALLEW, R. M., SABELKO, J. & GRUEBELE, M. (1996). Direct observation of fast protein folding: the initial collapse of apomyoglobin *Proc Natl Acad Sci USA* 93, 5759–5764.

18
Application of Hydrogen Exchange Kinetics to Studies of Protein Folding

Kaare Teilum, Birthe B. Kragelund, and Flemming M. Poulsen

18.1
Introduction

Amide hydrogen exchange is one of the most useful methods for obtaining site-specific information about hydrogen bond formation and breaking in the folding protein. In particular, the method allows study of the individual amides in the peptide chain, and for this reason the method has very wide applications in studies of the protein folding mechanism.

Hydrogen exchange is a chemical exchange reaction between labile hydrogen atoms chemically bound to either nitrogen, oxygen, or sulfur atoms. The exchange reaction occurs most commonly between these labile hydrogen atoms in molecules dissolved in a solvent that itself contains labile hydrogen atoms like water. Obviously this type of exchange reaction can take place for many atoms in a protein molecule dissolved in water, where several types of labile hydrogen can exchange with the water hydrogen atoms. The hydrogen atoms that potentially engage in hydrogen exchange typically also have the potential for hydrogen bond formation. The chemistry of the hydrogen exchange reaction is a proton transfer reaction, which requires the release of the hydrogen atom. For this reason the hydrogen exchange reaction can be used to monitor hydrogen bond formation and breaking. In proteins hydrogen bonds involving the peptide backbone amides and carbonyls are the most important structural element in secondary structure formation. Therefore by studying the hydrogen exchange reaction of the individual amides in the folding peptide information becomes available regarding the hydrogen bond formation of the individual segments in the protein folding process. Similarly information about the stability and the kinetics of the hydrogen bonding segments in the peptide chain can be obtained by measuring the hydrogen exchange reaction in native conditions.

The kinetics of hydrogen exchange in proteins can be studied directly by nuclear magnetic resonance (NMR) spectroscopy using solvent saturation transfer or the magnetic relaxation rate to study very fast reactions. Slower reactions can be studied by letting the exchange reaction replace a hydrogen atom in the dissolved molecule by deuterium from deuterium oxide solvent. This replacement can be measured directly by ^{1}H NMR spectroscopy because the NMR active ^{1}H is replaced by the ^{2}H, which is not detectable at the proton NMR frequency. Also the exchange can be measured by mass spectrometry in combination with enzymatic degradation because the molecular mass increases by introduction of the deuteron.

The amide hydrogen exchange reaction has been widely used ever since it was first described as a method for obtaining structural information about the peptide backbone in proteins [1]. Although the methods for measuring the hydrogen exchange were developed at a time where no three-dimensional structures were known, it was being used to describe the content of secondary structures in proteins [2, 3]. However, very early in the history of the method it became clear that the method also carried information about the dynamics of protein structures [4]. In particular after the first three-dimensional structures of proteins had been determined by X-ray crystallography it was realized that amide hydrogen exchange carried the message of dynamics. The compact and firmly packed globular protein structures contained interior amides in the secondary structure, which had no surface contact [5]. Nevertheless, these buried amides were found to be subject to exchange with solvent hydrogen, indicating that these amides do get in contact with solvent and suggesting that dynamic processes had to bring the interior in contact with an exterior water molecule.

Amide hydrogen exchange kinetics was originally measured by methods that were only able to record bulk exchange using either infrared spectroscopy or tritium hydrogen exchange. This meant that the measured exchange was a result of the contribution from all exchanging amides in the protein. The method therefore did not allow measurements of exchange rates at specific sites in the protein. Protein NMR spectroscopy changed this dramatically [6, 7]. With assigned NMR signals from essentially all amide groups in the peptide backbone and access to the three-dimensional structure of the protein molecule in question, hydrogen exchange kinetics could be rationalized and understood in a molecular context that was not available previously. The study of hydrogen exchange using NMR spectroscopy is the key to making the exchange kinetics of protein structures useful.

In the days before NMR spectroscopy was being applied in hydrogen exchange several mechanisms were proposed to explain the observation that amides in the centers of molecules were exchanged [8], however, many NMR studies of the thermodynamics and kinetics of individual hydrogen exchange reactions in many proteins have led to the widely agreed view that individual amides in globular proteins are exposed to exchange with solvent hydrogen atoms by a number of mechanisms. Some amides exchange by local openings, others by subglobular openings and finally the most interior amides exchange by cooperative global protein unfolding.

The latter group of amides exchanging by global unfolding has received considerable attention because it has been proposed that the study of these may provide information about the protein folding and unfolding events in native conditions [9–13]. Considering that the folded state and the unfolded state are in chemical equilibrium, this equilibrium will for most proteins be strongly shifted towards the folded form in native conditions. However, since both states have to be populated in all conditions, even in native conditions a small fraction of molecules will be of the unfolded form. If the mechanism of hydrogen exchange for a specific amide group is by global unfolding, the kinetics of the exchange reaction will carry information about the rate constants of folding and unfolding in native conditions. This consideration has stimulated much interest in the use of hydrogen exchange in native conditions to exploit the nature of the protein folding/unfolding equilibrium.

The group around Linderstøm-Lang suggested a simple, two-step mechanism for hydrogen exchange [1, 4]. The first step is the equilibrium between the closed state and the open state and the second step is the chemical exchange reaction occurring from the open state. Therefore, when the mechanism of exchange for a given amide group is by protein folding/unfolding, the rate constants for opening and for closing would be equivalent to those of folding and unfolding, provided the folding is a simple two-state process. Hence there would be an obvious interest in determining these rate constants. A paragraph in this chapter will address some of the methods used to extract the rate constants of the opening and closing reactions in particular conditions favoring such measurements. The simplest condition is to measure the amide hydrogen exchange kinetics when the rate-determining step in the reaction mechanism is the opening rate, in which case the amide hydrogen exchange rate is simply the unfolding rate [13, 14]. Another method applies the measurement of the hydrogen exchange rates at a set of different pH values at which the chemical exchange rate from the open form is known. The measurement of the amide hydrogen exchange kinetics in an array of pH values provide a sample of data that subjected to nonlinear fitting can provide the two rate constants for the opening and closing reactions [10, 15]. A paragraph in this chapter will describe examples of the application of this type of analysis to several model proteins and present the results of these analyses.

Another approach which is being used to bridge the gap between amide hydrogen exchange in native-like conditions and in denaturing conditions is to study the rate of exchange as a function of a denaturing agent [9, 15–17]. As the concentration of denaturant increases, the unfolding equilibrium constant will typically increase, bringing a larger and larger proportion of the molecules of the unfolded form, resulting in faster amide hydrogen exchange. The study of individual amides has for several proteins shown that the kinetics of the exchange of groups of amides has typical patterns that reflect the mechanism of exchange and allow the distinction between exchange by local, subglobal, or cooperative global exchange mechanisms.

Amide hydrogen exchange has been a very useful tool for measuring the formation of hydrogen bonds during protein folding. The quenched-flow and pulse-labeling techniques, which combine rapid mixing techniques with analysis by NMR spectroscopy or mass spectrometry, have had a great impact on outlining the routes of protein folding [18–26]. These techniques allow determination of the kinetics of the formation of a protected state for individual amides throughout the peptide chain. These methods are extremely powerful because the information they provide has both a structural and kinetic implication.

For an understanding of the folding of proteins, the unfolded state of the peptide chain has been subject to several studies because many studies have indicated the existence of structure in the otherwise unfolded peptide chain. Such structures are of interest because they may be labile precursors of the formation of native structures. Again, amide hydrogen exchange is one method of probing the presence of residual structure in the unfolded state, structures in the molten globule and other folding intermediates by comparison of the exchange kinetics with the exchange rates predicted for the random coil conformation of the peptide [9, 12, 15, 27–43].

The study of protein folding is focused on understanding the thermodynamic and structural routes of a peptide chain transforming from the unfolded state to the folded state, identifying secondary and tertiary interactions in the transition states and intermediates. Amide hydrogen exchange in combination with NMR spectroscopy has played a considerable role in monitoring and describing these states of protein folding. It should be kept in mind, however, that the method of amide hydrogen exchange is an indirect way of detecting the progress of protection of an amide hydrogen atom towards exchange with solvent. The method has no way of providing information regarding the chemical and structural nature of the protection. Therefore, careful evaluation is recommended when structural interpretations of transition states and intermediates based on the results of amide exchange are presented in their own right. The literature has several examples of opposite interpretations of amide hydrogen exchange results. For instance it has been proposed that the slow exchanging core of amides in the interior of a protein is the folding nucleus of proteins [44, 45]. This view has been opposed on the grounds of thermodynamic considerations and by examples of experimental evidence of proteins for which the folding nucleus and the core of slowly exchanging amides did not coincide [46, 47]. It has been suggested that pathways of protein folding may be determined by detection of on-pathway intermediates [48]. Many of the intermediates of protein folding have been identified by ultra-fast kinetic measurements or indirectly detected in the dead time of rapid mixing experiments and recorded by spectroscopic techniques, which do not provide information about the structure of the intermediate. The coincidence of the measures of thermodynamic parameters describing the intermediate and those describing native state hydrogen exchange kinetics have been used to couple such observations [12]. However, even the presence or absence of an on-pathway intermediate in barnase, as determined by amide hydrogen exchange studies in combination with a large number of very carefully conducted folding studies, has turned out to be controversial [11, 49–52].

Nevertheless, amide hydrogen exchange is a powerful tool in studies of protein folding, and used with care the combined results of kinetic and native state hydrogen exchange measurements with other spectroscopic measurements can provide important information about protein folding.

The field of amide hydrogen exchange in protein folding has been reviewed many times, most recently in Refs [53–57].

18.2
The Hydrogen Exchange Reaction

The exchange of amide hydrogen with hydrogen atoms from water is a catalyzed reaction [58]. The catalyst can be the basic hydroxide ion, the neutral water molecule, or the acid hydroxonium ion. The initial step in the reaction is the formation of a transient complex between the reactive group and the catalyst through a hydrogen bond. Subsequently the labile proton is transferred to the catalyst in the complex and the complex is dissociated. The replacement of the hydrogen atom with a hydrogen atom from the solvent molecule represents the hydrogen solvent exchange reactions. This implies that amide hydrogen exchange is pH dependent (Figure 18.1).

18.2.1
Calculating the Intrinsic Hydrogen Exchange Rate Constant, k_{int}

In an unstructured polypeptide, the rate of exchange of an amide hydrogen with solvent depends on the pH, temperature, ionic strength, the isotopes (H or D), and the chemical nature of the neighboring amino acid side chains. Taking these factors into account, intrinsic hydrogen exchange rates of amide hydrogen in pro-

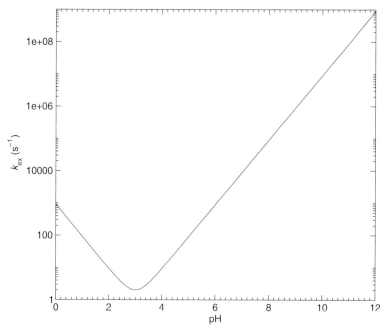

Fig. 18.1. Variation with pH of the amide hydrogen exchange rate, k_{ex}, of poly-D,L-alanine. At pH below 3 acid catalysis dominates. Above pH 3 base catalysis dominates.

Tab. 18.1. Exchange rates of poly-D,L-alanine under various conditions.

Isotopes	Additive	Temp (K)	log k_A	log k_B	log k_W	Ref.
H in D_2O	0.5 M KCl	278	1.19	9.90	−2.5	59
H in D_2O	No salt	293	1.62	10.05	−1.5	136
H in D_2O	0.1 M NaCl	293	1.50	9.66	−1.28	137
D in H_2O	No salt	293	1.40	9.87	−1.6	136
D in 91% H_2O	0.55 M GdmCl	293		10.00		33
H in D_2O	2.0 M GdmCl/0.1 M NaCl	293	1.64	10.01	−1.69	137
H in D_2O	4.0 M GdmCl/0.1 M NaCl	293	1.43	10.05	−1.84	137
H in D_2O	6.0 M GdmCl/0.1 M NaCl	293	1.22	9.78	−1.72	137
D in H_2O	8.0 M urea/0.1 M NaCl	293	2.43	8.52	−1.77	137

The rate constants are in units of $M^{-1}\, min^{-1}$.

teins, k_{int}, can be calculated according to:

$$k_{int} = k_A A_L A_R [H^+] + k_B B_L B_R [OH^-] + k_W B_L B_R \qquad (1)$$

where k_A, k_B, and k_W are second-order rate constants for acid, base, and water catalyzed hydrogen exchange in poly-D,L-alanine (PDLA). These rate constants have been measured under different conditions and are listed in Table 18.1. When calculating k_{int}, a PDLA reference close to the actual experimental conditions should be chosen. In Eq. (1), A_L, A_R, B_L, and B_R are factors correcting the PDLA rates for effects of neighboring side chains other than Ala (Table 18.2) [59]. Different factors for acid and base catalysis should be used. Thus, A_L and A_R are used to correct k_A, whereas B_L and B_R are used to correct k_B and k_W. The L and R subscripts correspond to whether the amide is to the left or to the right of the side chain (see Figure 18.2 for definition of left and right in this situation). Note that the correction factors for left and right in Table 18.2 are different. For calculation of $[OH^-]$ from a pH meter reading, the pK_W values of Covington et al. [60] in H_2O and D_2O at 278 K and 293 K are listed in Table 18.3.

The temperature difference between the temperature under the reference conditions, T_{ref}, and the actual temperature of the experiment, T, is corrected from the Arrhenius equation for each of the reference rate constants k_A, k_B, and k_W according to:

$$k(T) = k(T_{ref}) \exp(-E_a (T^{-1} - T_{ref}^{-1}) R^{-1}) \qquad (2)$$

where E_a is the activation energy and R is the gas constant. E_a is 14, 17, and 19 kcal mol^{-1} for k_A, k_B, and k_W respectively. These E_a values correct the second-order rate constants k_A, k_B, and k_W and the autoprotolysis constants of H_2O or D_2O.

For example, calculate k_{int} in H_2O for the amide hydrogen in the sequence -Phe-NH-Gly- at low salt, pH 7, and 298 K.

From Table 18.1 the PDLA reference for D-exchange in H_2O with no salt at 293 K is chosen. Thus, $k_A = 10^{1.40}$ M^{-1} min^{-1}, $k_B = 10^{9.87}$ M^{-1} min^{-1}, and $k_W = 10^{-1.6}$

Tab. 18.2. Effect of amino acid side chains on hydrogen exchange rates of neighboring amide hydrogens.

Side chain	Acid catalysis		Base catalysis	
	log A_L	log A_R	log B_L	log B_R
Ala	0.00	0.00	0.00	0.00
Arg	−0.59	−0.32	0.08	0.22
Asn	−0.58	−0.13	0.49	0.32
Asp(COO⁻)	(0.9)[1]	0.58	−0.30	−0.18
Asp(COOH)	(−0.9)	−0.12	0.69	(0.6)
Cys (red)	−0.54	−0.46	0.62	0.55
Cys (ox)	−0.74	−0.58	0.55	0.46
Gly	−0.22	0.22	0.27	0.17
Gln	−0.47	−0.27	0.06	0.20
Glu(COO⁻)	(−0.9)	0.31	−0.51	−0.15
Glu(COOH)	(−0.6)	−0.27	0.24	0.29
His			−0.10	0.14
His⁺	(−0.8)	−0.51	(0.8)	0.83
Ile	−0.91	−0.59	−0.73	−0.23
Leu	−0.57	−0.13	−0.58	−0.21
Lys	−0.56	−0.29	−0.04	0.12
Met	−0.64	−0.28	−0.01	0.11
Phe	−0.52	−0.43	−0.24	0.06
Pro (trans)		−0.19		−0.24
Pro (cis)		−0.85		0.60
Ser	−0.44	−0.39	0.37	0.30
Thr	−0.79	−0.47	−0.07	0.20
Trp	−0.40	−0.44	−0.41	−0.11
Tyr	−0.41	−0.37	−0.27	0.05
Val	−0.74	−0.30	−0.70	−0.14
N-term (NH$_3^+$)		−1.32		1.62
C-term (COO⁻)	0.96		(−1.8)	
C-term (COOH)	(0.05)			

[a] Numbers in brackets are determined with low accuracy [59].
Adapted from Bai et al. [59].

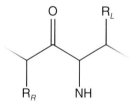

Fig. 18.2. Left and right in Table 18.2 are defined relative to an amino acid side chain. To correct the hydrogen exchange rate of an amide to the right of side chain R_R and to the left of side chain R_L the PDLA reference rate has to be corrected with the "R" value of side chain R_R and the "L" value of side chain R_L.

Tab. 18.3. Autoprotolysis constants (pK_W) of H_2O and D_2O.

Temp (K)	278	293
H_2O	14.734	14.169
D_2O	15.653	15.049

From Covington et al. [60].

M^{-1} min^{-1}. The amide is to the right of Phe, so from Table 18.2 $A_R = 10^{-0.43}$ and $B_R = 10^{0.06}$. The amide is to the left of Gly, so $A_L = 10^{-0.22}$ and $B_L = 10^{0.27}$. From Table 18.3 we get $pOH = pK_W - pH = 15.049 - 7 = 8.049$. Calculating the acid, base, and water terms in Eq. (1) separately gives:

$$k_{acid} = 10^{(1.40-0.22-0.43-7)} \text{ min}^{-1} = 5.6 \times 10^{-7} \text{ min}^{-1}$$

$$k_{base} = 10^{(9.87+0.27+0.06-8.049)} \text{ min}^{-1} = 141.6 \text{ min}^{-1}$$

$$k_{water} = 10^{(-1.6+0.27+0.06)} \text{ min}^{-1} = 53.7 \times 10^{-3} \text{ min}^{-1}$$

Each term must be corrected for the temperature difference between the reference conditions and the experimental conditions according to Eq. (2) and added to get k_{int}. However, k_{acid} and k_{water} are negligible compared with k_B and only the base term need be considered:

$$k_{int} = 141.6 \text{ min}^{-1} \exp\left(\frac{-17 \text{ kcal} \cdot \text{mol}^{-1}((298 \text{ K})^{-1} - (293 \text{ K})^{-1})}{1.987 \times 10^{-3} \text{ kcal} \cdot \text{mol}^{-1} \text{ K}^{-1}}\right)$$

$$= 231.1 \text{ min}^{-1}$$

Hydrogen exchange rates may also be calculated using the web-service "Sphere" from Heinrich Roders laboratory at: http://www.fccc.edu/research/labs/roder/sphere.

Protection factors, P, are defined as the ratio of the calculated intrinsic exchange rate of an amide in the unstructured environment, k_{int}, to the observed rate constant of exchange, k_{obs}, from the folded, intermediate, or unfolded state. Protection factors are calculated individually for each backbone amide and are regarded as a tool to predict the degree of structural protection against exchange, $P = k_{int}/k_{obs}$ (see also Section 18.4.3).

18.3
Protein Dynamics by Hydrogen Exchange in Native and Denaturing Conditions

The hydrogen exchange reactions of amides in native state globular proteins are a result of dynamic processes. From simple measurements of direct amide exchange rates and the subsequent calculation of protection factors (see Section 18.1), the stability and hydrogen bond structure of stable, populated, equilibrium protein folding intermediates of several proteins have been described. Typically, the amides of a protein in its stable native state are protected against exchange and the rates are reduced by factors in the order of 10^6 to 10^{10}, whereas the intermediate states have protection factors in the order of 10^1 to 10^3 [43, 61–64]. For a few proteins studied at high concentrations of chemical denaturants, for example in 8 M urea, protection factors of less than five have been determined [38, 65].

At equilibrium under native conditions, a fraction of the protein population will occupy high-energy intermediate and unfolded states, and this provides an excellent opportunity to map the structure of protein folding intermediates in native conditions, even when sparsely populated. The amide hydrogen exchange process is a suitable process, since any perturbation of the native state stability occurring from a change in any equilibrium in the direction of populating folding/unfolding intermediates or the unfolded state will result in a change in the observed exchange rate. The perturbation of the equilibrium can occur by changing the pH of the sample, by addition of small amounts of chemical denaturants, or by changing the temperature. Information about global and local folding rates and populations of intermediates in the protein folding reactions are all available from equilibrium hydrogen exchange measurements. In the following studies, it is of importance that the native state is always the most populated state.

18.3.1
Mechanisms of Exchange

Amides are protected against exchange either because they are involved in a hydrogen bond or because they are placed in a hydrophobic environment [66]. Exchange of hydrogen-bonded amides requires opening of the hydrogen bond and separation of donor and acceptor by at least 5 Å [67]. The breakage of any given hydrogen bond can occur by three different mechanisms, each of which can be identified by the use of the "Native State Hydrogen Exchange" strategy [9] described below. Either the hydrogen bond is broken as a consequence of a cooperative global unfolding process or by cooperative local unfolding reactions exposing sets of hydrogen bonds [8, 53, 68]. The third possibility for exposure to solvent is by noncooperative (limited) local fluctuations, which only expose one amide proton at a time for exchange [67, 69]. In a globular, native protein the amide hydrogen exchange rate monitored for a particular amide proton will be a result of the sum of exchange by all three mechanisms.

18.3.2
Local Opening and Closing Rates from Hydrogen Exchange Kinetics

Independent of the three mechanisms for exchange described above, and neglecting any populations of intermediates, the amide exchange reaction can be described by a two-step reaction (Figure 18.3) [1, 70]. The first step opens the amide

$$C_{NH} \underset{k_{cl}}{\overset{k_{op}}{\rightleftarrows}} O_{NH} \xrightarrow{k_{int}} O_{ND} \underset{k_{op}}{\overset{k_{cl}}{\rightleftarrows}} C_{ND}$$

Fig. 18.3. The Linderstrøm-Lang model for hydrogen exchange in a protein dissolved in a deuterated medium. An amide may either be found in hydrogen protected state C, or in a hydrogen exchange competent state O. Conversion between C and O is a reversible first-order process with the rate constants k_{op} and k_{cl} for the opening and closing reactions, respectively. From the open state amide hydrogen may exchange with solvent deuterium in a pseudo first-order process with the rate constant k_{int} (see also Section 18.2.1).

from a closed hydrogen exchange protected form, C_{NH}, to an exchange competent form, O_{NH}, described by the first-order rate constants k_{op} and k_{cl}. The second step is a pseudo first-order reaction where the amide hydrogen exchanges with solvent deuterium with the rate constant k_{int} (see also Section 18.2.1). In a hydrogen exchange experiment little or no solvent hydrogen is present and rehydrogenation can be neglected.

As the amide exchange process is strongly pH dependent, the rates of segmental opening and closing of a giving amide, k_{op} and k_{cl}, can be determined quantitatively from measurements of the observed hydrogen exchange rate, k_{obs}, as a function of pH. For several proteins it has been shown that the most slowly exchanging amide protons have opening and closing rates corresponding to the global folding and unfolding rates determined by standard stopped-flow kinetic techniques [71, 72], and this fact opens a route to determine protein folding kinetics by equilibrium hydrogen exchange measurements.

18.3.2.1 The General Amide Exchange Rate Expression – the Linderstrøm-Lang Equation

Without making any assumptions to the relative measures of the three rate constants k_{op}, k_{cl}, and k_{int}, a general expression for the observed exchange rate constant for a given amide hydrogen in a protein can be derived from the kinetic equations (Eqs (3) and (4)) describing the reaction scheme above as follows:

$$\frac{d[C_{NH}]}{dt} = -k_{op}[C_{NH}] + k_{cl}[O_{NH}] \quad (3)$$

$$\frac{d[O_{NH}]}{dt} = k_{op}[C_{NH}] - (k_{int} + k_{cl})[O_{NH}] \quad (4)$$

with the biphasic solutions:

$$[C_{NH}](t) = A_{C0} + A_{C1}e^{-\lambda_1 t} + A_{C2}e^{-\lambda_2 t} \quad (5)$$

$$[O_{NH}](t) = A_{O0} + A_{O1}e^{-\lambda_1 t} + A_{O2}e^{-\lambda_2 t} \quad (6)$$

where A_{x0} is the equilibrium concentration of species x, and A_{x1} and A_{x2} are the amplitudes of the two phases. The rate constants for these phases are given by:

$$\lambda_1 = \frac{1}{2}(k_{op} + k_{cl} + k_{int} - ((k_{op} + k_{cl} + k_{int})^2 - 4k_{op}k_{int})^{1/2}) \tag{7}$$

and

$$\lambda_2 = \frac{1}{2}(k_{op} + k_{cl} + k_{int} + ((k_{op} + k_{cl} + k_{int})^2 - 4k_{op}k_{int})^{1/2}) \tag{8}$$

In a hydrogen exchange experiment λ_2 is generally larger than 10 s^{-1} and the fast phase described by this rate constant will be over in the dead time of the experiment, which typically will be measured in minutes. Thus, only λ_1 will be observed and $k_{obs} = \lambda_1$. This general description of the amide exchange rate was first described by Linderstrøm-Lang [1] and is applicable under all conditions.

18.3.2.2 Limits to the General Rate Expression – EX1 and EX2

In a stable protein under conditions favoring the native state $k_{cl} \gg k_{op}$, Eq. (7) may be reduced as follows using that $(1-x)^{1/2} \approx 1 - x/2$ for $x \ll 1$:

$$\begin{aligned} k_{obs} &\approx \frac{1}{2}\left(k_{cl} + k_{int} - (k_{cl} + k_{int})\left(1 - \frac{4k_{op}k_{int}}{(k_{cl} + k_{int})^2}\right)^{1/2}\right) \\ &\approx \frac{1}{2}\left(k_{cl} + k_{int} - (k_{cl} + k_{int})\left(1 - \frac{2k_{op}k_{int}}{(k_{cl} + k_{int})^2}\right)\right) = \frac{k_{op}k_{int}}{k_{cl} + k_{int}} \end{aligned} \tag{9}$$

Two limiting cases of Eq. (9) may be considered. If $k_{int} \gg k_{cl}$, which is typical under alkaline conditions, Eq. (9) reduces to:

$$k_{obs} \approx k_{op} \tag{10}$$

This is the EX1 limit [4, 70, 73, 74]. In this limit the amide will exchange in every opening event as the rate-limiting step will be the conformational opening from the closed form. In the EX1 limit, the observed rate of exchange, k_{obs}, directly gives k_{op}.

If $k_{cl} \gg k_{int}$, Eq. (9) reduces to:

$$k_{obs} \approx k_{int}k_{op}/k_{cl} = k_{int}K_{op} \tag{11}$$

This is the EX2 limit [4, 73, 74]. In the EX2 limit the amide will not exchange in every opening event as closing competes with exchange. From the observed rate of exchange, k_{obs}, and the intrinsic exchange rate, k_{int}, the equilibrium constant for opening, K_{op}, may be obtained.

From Eqs (10) and (11) it is seen that if k_{op} and k_{cl} are both pH independent, k_{obs} for amides under EX1 conditions is also independent of pH. k_{obs} for amides under EX2 conditions will be pH dependent through the pH dependence of k_{int}. Taking the logarithm of Eqs (10) and (11) yields:

$$\log k_{obs} = \log k_{op} \quad (12)$$

for EX1, and

$$\log k_{obs} = \log k_{int} + \log K_{op} = pH - pK_W + \log B_L + \log B_R + \log k_B + \log K_{op} \quad (13)$$

for EX2. Thus, in a plot of $\log k_{obs}$ versus pH a slope of zero will reflect an amide with exchange in the EX1 limit and a slope of one will reflect an amide with exchange in the EX2 limit (Figure 18.4).

For a given protein, the residue-dependent variations of k_{op}, k_{cl}, and k_{int} will result in observed exchange rates varying over several orders of magnitudes. For a given amide in a fixed amino acid sequence, a set of simulations of the observed exchange rate, k_{obs}, with different opening and closing rates is shown in Figure

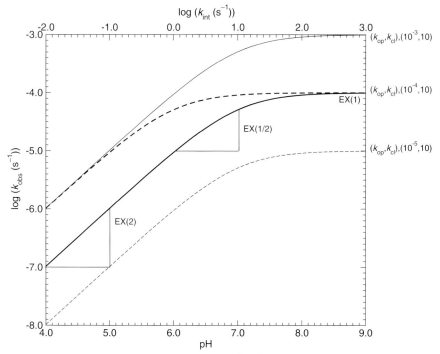

Fig. 18.4. Simulated pH dependence of k_{obs} for amide exchange according to Eq. (7). Data are simulated in the pH range from pH 4.0 to pH 10.0 for the situation that only the base-catalyzed reaction is active and the rate constants k_{op} and k_{cl} are not pH dependent. The chemical exchange rate $k_{int} = 10^8$ s^{-1} M^{-1} × 10$^{(pH-14)}$ M expected for the amide in the sequence Ala-NH-Ala is on the top axis. The three situations of exchange are illustrated in the figure: EX2 (pH dependence = 1), EX(1/2) (0 < pH dependence < 1), and EX1 (no pH dependence).

18.4. Four lines are shown representing simulated pH-dependent exchange in four different situations with different values of k_{op} and k_{cl}. The variation in k_{op} will shift the value for the observed rate constant, log k_{obs}, by a simple transposition where increasing k_{op} (and/or k_{int}) shift log k_{obs} to higher values (bold solid line to thin solid line). Variation in k_{cl} will change the profile of the line and increasing k_{cl} will shift log k_{obs} to the left and lead to exchange in the EX1 limit at lower pH (bold solid line to bold dashed line).

EX1 limit: $k_{obs} = k_{op}$ ($k_{cl} \ll k_{int}$)
EX2 limit: $k_{obs} = K_{op} k_{int}$ ($k_{cl} \gg k_{int}$)

18.3.2.3 The Range between the EX1 and EX2 Limits

In the range between the two limits, the simplifications of the general rate expression as in the EX1 and EX2 limits are not applicable. This range, which has often been ignored, is referred to here as the EX(1/2) range [75]. The change from EX2 to EX1 exchange appears in the alkaline pH range, where k_{int} is large. For amides with exchange kinetics in the EX(1/2) range there is a unique possibility to determine the rate constants k_{op} and k_{cl} simply by nonlinear least squares fitting to the pH dependence of the exchange kinetics using the general expression (Eq. (7)) for k_{obs}.

A nonlinear fitting analysis of Eq. (7) can unambiguously determine k_{op} without any assumptions other than the reaction scheme of Figure 18.3, whereas k_{cl} and k_{int} are too strongly correlated to be determined. If it is assumed that a site-specific random coil rate constant [59] can be applied as k_{int}, the experimental data for a given residue can initially determine k_{op} and subsequently k_{op} and k_{int} are used to determine k_{cl}, or k_{op} and k_{cl} can be determined directly in a two-parameter fit where it is assumed that k_{int} is known [10, 14, 76]. It has been argued that the intrinsic rate constant does not apply to exchange by local fluctuations or local unfolding mechanisms, as these open structures does not resemble the random coil peptides [77]. In these cases only the opening rate, k_{op}, can be determined unambiguously.

18.3.2.4 Identification of Exchange Limit

When the rate constants k_{op} and k_{cl} are themselves pH dependent, the method of determining these from the pH dependence of the amide hydrogen exchange kinetics is clearly not valid. It is therefore important to distinguish between hydrogen bond amides for which the two rate constants are not affected by pH from those for which they are. For an amide with no change in stability with pH, the pH dependence of the exchange kinetics, $\alpha_{pH} = \Delta \log k_{obs}/\Delta pH$ reflect directly the exchange mechanism, such that for an EX2 amide which experiences no pH-induced change in stability, $\alpha_{pH} = 1$, for an EX1 amide the exchange is independent of pH ($\alpha_{pH} = 0$) and for an amide with exchange in the (EX1/2) limit, $0 < \alpha_{pH} < 1$. If the stability is increasing with pH, log k_{obs} will for an EX2 amide increase by less than 1 per unit of pH ($\alpha_{pH} < 1$) and for an EX1 amide α_{pH} will be negative. If the stability is decreasing with pH, the pH dependence will be larger than one for an EX2 amide ($\alpha_{pH} > 1$) and larger than zero for an EX1 amide ($\alpha_{pH} > 0$). An amide that has $\alpha_{pH} < 1$ is therefore either an EX2 amide whose environment is stabilizing with increasing pH, an EX1 amide that is destabilizing with increasing pH, or an amide that exchanges in the EX(1/2) range. When determining the segmental opening and closing rate constants, k_{op} and k_{cl}, these issues must be considered. One way is to assure that exchange takes place by EX1 mechanism at higher pH and by EX2 mechanism at lower pH. Optical techniques may also be invoked to determine the global stability with pH.

pH dependence of k_{obs} and resulting exchange limits When the rate constants k_{op} and k_{cl} are not themselves pH dependent, amides exchanging in the EX(1/2) range can easily be distinguished from amides in the EX1 and the EX2 range by the slope $\alpha_{pH} = \Delta \log k_{obs}/\Delta pH$ of the pH dependence of the hydrogen exchange kinetics.

EX2 amides $\alpha_{pH} = 1$
EX1 amides $\alpha_{pH} = 0$
EX(1/2) amides α_{pH} is between 0 and 1

18.3.2.5 Global Opening and Closing Rates and Protein Folding

Analysis of protein amide exchange by use of the general Linderstrøm-Lang equation together with in-depth analysis of individual segmental opening and closing rates, k_{op} and k_{cl}, have been reported for only six proteins: lysozyme [76], acyl-coenzyme A binding protein (ACBP) [75], turkey ovomucoid third domain (OMTKY3) [10], hisactophilin [78], Csp A [79], and ubiquitin [13]. Either the general rate expression has been applied in the analysis to determine k_{op} and k_{cl} directly, or a combined analysis of exchange under both EX2 or EX1 conditions have provided measures of K_{op} ($= k_{op}/k_{cl}$) and k_{op}, respectively [80].

Global unfolding reactions of the proteins are always more rare than local unfolding reactions. Rates of amide exchange by global unfolding correspond commonly to rates of global unfolding measured by optical techniques. Regardless of the method used, either fitting to the general expression Eq. (7), to the simplified expression for stable proteins Eq. (9), or by combining measurements obtained under both EX1 and EX2 conditions, the opening rate constant, k_{op}, can be determined with high precision.

Determination of k_{cl} is highly dependent on the precision of k_{int} as these are greatly correlated, and when fitting to the reduced expression it is important that the amide do exchange in the EX2 limit. Thus, any local structure will tend to cause k_{int} to be overestimated, k_{cl} to be underestimated, and any pH-dependent change in exchange mechanism will be accumulated as errors in the determination of k_{cl}. The local closings are much faster than the global closing, and k_{cl} are larger for the slowly exchanging amides, resulting from the globally open form being more long lived than the locally open forms. No correlations of either opening or closing rates with overall protein stability have been noticed.

18.3.3
The "Native State Hydrogen Exchange" Strategy

The structure and stability of exchange-competent states, which also includes the structure and stability of low populated, partially unfolded states of a protein, can be deduced and investigated using the "Native State Hydrogen Exchange" strategy [9, 69, 81]. In this strategy, the denaturant dependence of hydrogen exchange rates is measured under EX2 conditions such that the denaturant concentration is varied only in the range favoring the native state. Through the perturbations of the equilibrium between the folded state and any unfolded state by chemical denaturants or by temperature, the dominant exchange process for a given amide can be mapped and cooperative units in protein unfolding, the so-called native-like partially unfolded forms, PUFs (see below) [9] can be located. Importantly, no information regarding the order of the folding reaction can be provided by this strategy, as kinetic mechanisms cannot be derived from an equilibrium analysis [47, 82]. It has been of some debate whether the slow-exchanging amides identify the protein folding nucleus. There seems only to be a correlation for a subset of proteins [44, 83] and certainly not for others [54, 84]. It must be stressed that the order of events cannot be determined from equilibrium measurements. Only when independent and complementary kinetic folding experiments are provided can distinctions between on- and off-pathway intermediates be made.

In the EX2 regime of exchange, the transient equilibrium constant for the opening reaction, K_{op}, can be used to calculate the apparent free energy of exchange (or opening), ΔG_{HX}, (or ΔG_{op}) as

$$\Delta G_{HX} = -RT \ln K_{op} \qquad (14)$$

Earlier studies have demonstrated that the free energy of unfolding is linearly dependent on the concentrations of denaturant [85–87], which is expressed as

$$\Delta G_{unf} = \Delta G°_{unf} - m[\text{denaturant}] \qquad (15)$$

where m is a measure of the change in exposed surface area upon unfolding and $\Delta G°_{unf}$ is the unfolding free energy at 0 M denaturant (see Chapters 12.1, 12.2 and 13). The exchange can be regarded as a sum of two different processes [8, 68]. First, exchange can occur as a result of local fluctuations that do not expose additional surface area upon exchange and that are independent of the concentration of denaturant. Secondly, exchange can occur as a result of an unfolding reaction, which can be either global or local, depends on the concentration of denaturant, and exposes additional surface area. These two processes can be separately defined and the free energy of opening described as the sum of the processes.

$$\Delta G_{HX} = -RT \ln(K_{unf} + K_{fluc}) \qquad (16)$$

The equilibrium constant for unfolding can be rewritten as

$$K_{\text{unf}} = \exp\left(\frac{-\Delta G°_{\text{unf}} + m[\text{denaturant}]}{RT}\right) \quad (17)$$

Combining Eqs (16) and (17), an expression for the free energy of exchange that accounts for both fluctuation and unfolding mechanisms for exchange is obtained as

$$\Delta G_{\text{HX}} = -RT \ln\left(\exp\left(\frac{-\Delta G_{\text{fluc}}}{RT}\right) + \exp\left(\frac{-\Delta G°_{\text{unf}} + m[\text{denaturant}]}{RT}\right)\right) \quad (18)$$

A set of simulations of the different mechanisms of exchange according to the expression above is shown in Figure 18.5. Four different amide-exchange profiles with increasing concentrations of denaturant converted to exchange free energies are illustrated. The four different profiles have the following characteristics:

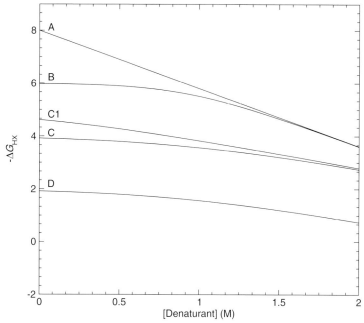

Fig. 18.5. Simulation of different hydrogen exchange mechanisms according to Eq. (18). For amide A the slope of the line (m) equals the m-value from equilibrium unfolding studies monitoring global unfolding (for example fluorescence and CD spectroscopy). For amide A, hydrogen exchange takes place entirely by global unfolding mechanism, and this has the highest free energy determined for the protein. Amide B has an m-value of zero at low concentration of denaturant and thus exchange is due to local fluctuations. The exchange is governed by global unfolding at higher denaturant concentrations. The curve coincide with the curve of amide A. Amides C and C1 shift to local unfolding with increasing denaturant concentration and merge to a common isoenergetic unfolding reaction and is part of the same cluster in the structure. Amide D is an example where global unfolding is never the dominating mechanism of exchange (Adapted from Ref. [135].)

- *Amide A: Global unfolding.* Amide hydrogen exchange takes place by global unfolding and represents the highest free energy at 0 M guanidinium chloride (GdmCl) determined in the protein. The slope of the line equals the m-value in Eq. (18) and corresponds to the m-value obtained from classical equilibrium unfolding studies using, for example, circular dichroism (CD) or fluorescence spectroscopy.
- *Amide B: Local fluctuations to global unfolding.* At low concentrations of denaturant, this amide exchange by local fluctuations. This is seen by the m-value of zero. At higher concentrations of denaturant, there is a shift for amide B to exchange by global unfolding, with an m-value equal to the m-value determined from CD or fluorescence.
- *Amide C and C1: Local fluctuations $\rightarrow$ local unfolding.* At low denaturant concentration amides C and C1 exchange by local fluctuations. At intermediate denaturant concentrations amides C and C1 exchange through local unfolding events with similar change in free energy (see below (PUF)). At higher concentrations of denaturant (>2 M) it is expected that both C and C1 will exchange by global unfolding and their lines will merge with the lines for A and B.
- *Amide D: Local unfolding.* For this amide global unfolding is never the dominating mechanism of exchange in the shown region of denaturant concentrations. This amide exchanges predominantly by local unfolding.

When fitting to Eq. (18), three parameters are determined: the m-value (slope), $\Delta G°_{unf}$, which is the free energy of exchange at 0 M denaturant (extrapolated intercept y-axis, dotted line), and ΔG_{fluc}, which is the free energy arising from fluctuations. The m-value provides direct information on the amide exchange mechanism: Fluctuation mechanisms are characterized by m-values of zero or very low m-values, local unfolding mechanisms have intermediate m-values lower than the m-values describing global unfolding, and finally global unfolding mechanisms give m-values equal to the m-values determined by global unfolding monitored by spectroscopic measures.

18.3.3.1 Localization of Partially Unfolded States, PUFs

Perturbation of the structural equilibrium of a protein by chemical denaturants or by temperature as described above allows mapping of the hydrogen exchange processes for the determination of a structural localization of independent unfolding units. This is done through identification of amides for which the exchange free energies merge to a common isoenergetic line different to the global unfolding line, for example the amides C and C1 of Figure 18.5. A set of such amides, whose exchange free energies, ΔG_{HX}, merge to a common isoenergetic line, originate from the same structural unfolding unit and have been termed a partially unfolded form (PUF) [9]. These unfolded forms of a protein identified from hydrogen exchange measurements have the slowest exchanging amides and generally have a strong dependence on the concentration of chemical denaturant, whereas the fastest exchanging amides are much less dependent on the concentration of denaturant, and require less unfolding of the protein to exchange. For many amides the predominant exchange mechanism at 0 M denaturant is by local fluctuations [67, 88] and since these forms of the exchange competent amides are still highly structured ($m \approx 0$), caution must be taken in interpretation of the results. Again, exchange must occur by the EX2 mechanism in the entire range of denaturant concentrations.

The native state hydrogen exchange strategy has identified both kinetic folding intermediates [12] and partially unfolded forms for several proteins of various structures and with different folding characteristics. PUFs have been identified in proteins known to form kinetic transient intermediates as cytochrome c [9, 89], ribonuclease H [29, 64], apomyoglobin [90], and T4 lysozyme [91] and for other proteins as well such as ribonuclease A [69, 92], cytochrome b_{562} [93], barstar [94], and staphylococcus nuclease [95]. The PUFs observed are in some cases similar to the kinetic folding intermediates observed by optical techniques [9, 64, 89] and also cooperative unfolding units with regions of individual ΔG_{HX} and m-values which represent segments of structure with different intrinsic stabilities, have been localized. For some proteins the absence of any common isoenergetic lines [15, 47, 96, 97] has raised the question of whether the intermediate forms need necessarily to be discrete states (U and I) or if they rather exist as a continuum of states ($U_1 \ldots U_N$) [98, 99]. Some of these proteins fold and unfold in simple two-state reactions without any intermediate forms when investigated with optical techniques. But for barnase and thioredoxin, where kinetic pulse-labeling techniques (see Section 18.4) have identified intermediates, the lack of PUFs are puzzling. It may be due to a change in exchange mechanism from EX2 to EX1 [47] or to destabilization of the intermediates at higher denaturant concentrations.

In many cases, the highest possible energy of opening, ΔG_{HX}, typically derived from the slowest exchanging amides, is found to be equal to the global unfolding energy, ΔG_U. In some cases, however, ΔG_{HX} has been found to be larger than ΔG_U, [77, 97, 100], and this discrepancy has been termed "super protection". This super protection of the most protected amides has been ascribed to several origins: overestimation of k_{int} from residual structure in the open state [101], interference with EX1 behavior [47, 102], cis–trans proline isomerization [77], increased stability in D_2O [103], or a nonlinear dependence of the stability with denaturant concentration leading to increased stability at 0 M denaturant [104–107]. The discrepancy is still under debate.

18.4
Hydrogen Exchange as a Structural Probe in Kinetic Folding Experiments

Following the protein folding reaction by hydrogen exchange in combination with NMR and mass spectrometry is a great supplement to the more general optical detection techniques such as fluorescence and CD. Fluorescence only measures the environment of a single or a few tryptophans, and far-UV CD measures an average of the conformation of all peptide bonds. In contrast to this, hydrogen exchange reports on all amide protons involved in hydrogen bonds or amide protons protected by a hydrophobic environment and in combination with 2D NMR, resolution at the residue level is achievable. In combination with mass spectrometry, hydrogen exchange may report on the heterogeneity of the sample and to some extent sequence-specific information can also be provided by this detection method.

At present, two methods for measuring hydrogen bond formation during the course of the protein folding process are employed; folding/hydrogen exchange competition and pulse labeling hydrogen exchange.

18.4.1
Protein Folding/Hydrogen Exchange Competition

The folding/exchange competition method was developed in Baldwin's laboratory and used to measure bulk hydrogen–tritium exchange [18, 108, 109]. Subsequently, the method was refined by Roder and Wüthrich to using hydrogen–deuterium exchange and NMR to detect exchange of amide hydrogen atoms of individual amides [110]. First, a deuterated unfolded protein is mixed with a hydrogenated refolding buffer under conditions where hydrogen exchange is fast enough to compete with the protein folding reaction. After a period of time, t_{pulse}, exchange is quenched by mixing with a buffer that lowers pH and thus slows hydrogen exchange, and the protein is allowed to completely fold to the native state before analysis (see Figure 18.6 for an outline of the method). If a hydrogen bond is formed during t_{pulse}, deuterium will become trapped in the structure. The amount of deuterium at a specific amide position depends on the rate by which the hydrogen bond is formed at this site, on the stability of this hydrogen bond and on the intrinsic exchange rate, k_{int}, of the amide in the particular conditions of the experiment.

The degree of hydrogen bond formation is determined from the amount of trapped deuterium. This can be measured by NMR spectroscopy by measuring the amount of incorporated proton at a given amide position from the peak volume of the H^N,N cross-peak in a HSQC spectrum or of the H^N,H^α cross-peak in a COSY spectrum. In order to compare the measured proton incorporation the volume, V, of a peak from an amide that becomes protected must be normalized to the volume, V_{ref}, of one or more peaks that do not become protected. This value is then normalized to the corresponding intensities (V_c and $V_{c,ref}$) measured in a uniformly labeled control sample and corrected for the fraction of H_2O, f_{H_2O}, in the labeling buffer. The resulting proton occupancy I can then be expressed:

$$I = \frac{V/V_{ref}}{V_c/V_{c,ref}} f_{H_2O}^{-1} \qquad (19)$$

In the pH interval of an exchange/folding competition experiment the hydrogen exchange is dominated by base catalysis. By varying the pH of the refolding/exchange buffer, the intrinsic exchange rate of the individual amides, k_{int}, may be changed according to:

18.4 Hydrogen Exchange as a Structural Probe in Kinetic Folding Experiments

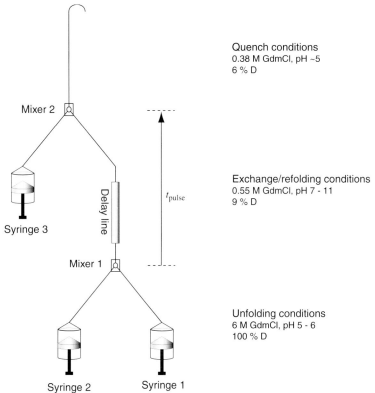

Fig. 18.6. Outline of the folding/exchange competition experiment. The experiment is carried out in a three-syringe quenched flow apparatus. Syringe 1 contains fully deuterated protein in 100% D$_2$O under conditions favoring the unfolded state. In mixer 1 the protein is diluted with 10 volumes of hydrogenated folding and exchange buffer from syringe 2. A series of experiments where pH of the refolding buffer is varied in the range from ~7 to ~11 is performed. Hydrogen exchange and protein folding competes while the protein flows from mixer 1 to mixer 2. After a period of time, t_{pulse} the sample reaches mixer 2 where quench buffer is added from syringe 3. The pH is now lowered to 5 or less and hydrogen exchange becomes slow. The sample is collected and desalted prior to analysis by NMR.

$$k_{int} = k_B B_L B_R [OH^-] \qquad (20)$$

In the following, exchange occurs in a hydrogenated medium and it is assumed that all labile proton sites in the protein initially are deuterated. The equations can readily be modified for the reverse situation. As pH (and thus [OH$^-$]) does not change in a hydrogen exchange experiment, the hydrogen exchange process will be pseudo first-order and the change in proton occupancy, I, can be described by a single exponential decay described by Eq. (21). I is a function of both the length of the pH pulse, t_{pulse}, and of pH (through k_{int}):

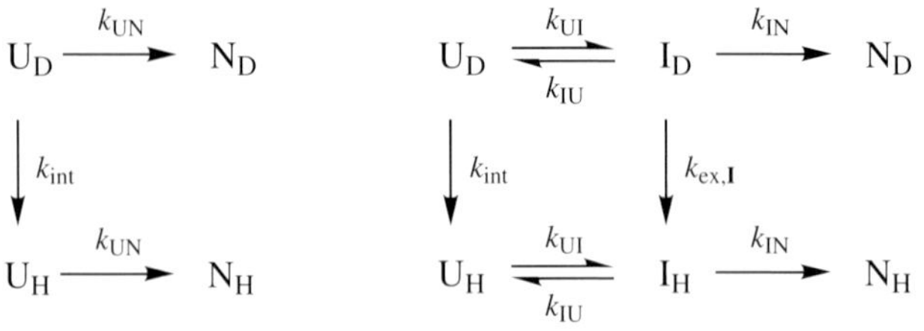

Fig. 18.7. Models of folding and hydrogen exchange of proteins following two- and three-state folding mechanisms. **U** is the unfolded state, **I** an intermediate state, and **N** the native state. Hydrogen exchange may occur directly from the **U**-state, which is supposed to be unstructured, with a rate k_{int}. Exchange from the **I**-state may occur due to local fluctuations with a rate constants $k_{ex,I}$. It is assumed that no hydrogen exchange takes place from an amide in the native state **N**. Back exchange from H to D is not considered due to high excess of H in the exchange medium. It is also assumed that there is no flux away from **N** (i.e., $k_{NU} \approx 0$ and $k_{NI} \approx 0$) as the experiments are conducted under native conditions.

$$I = 1 - \exp(-k_{int} t_{pulse}) \qquad (21)$$

If a hydrogen bond is formed to an amide in a two-state process (Figure 18.7) the proton occupancy may be found from the following rate equations. Remember that U_H is allowed to fold to N_H before measuring the proton occupancy so we are interested in $[U_H] + [N_H]$.

$$\frac{d[U_D]}{dt} = -(k_{int} + k_{UN})[U_D] \qquad (22)$$

$$\frac{dI}{dt} = \frac{d([U_H] + [N_H])}{dt} = k_{UN}[U_D] \qquad (23)$$

Solving Eqs (22) and (23) with the assumption that all protein is found as U_D for $t_{pulse} = 0$ yields:

$$I = \frac{k_{int}}{k_{UN} + k_{int}} (1 - \exp(-(k_{UN} + k_{int}) t_{pulse})) \qquad (24)$$

The proton occupancy versus pH curve of an amide that gets protected from exchange will thus be shifted to the right of the curve calculated for an unstructured amide using Eq. (21) (Figure 18.8). Selecting $t_{pulse} \gg 1/(k_{UN} + k_{int})$ Eq. (24) reduces to:

$$I = \frac{k_{int}}{k_{UN} + k_{int}} \qquad (25)$$

By fitting the measured proton occupancies to Eqs (24) or (25) the rate of structure (hydrogen bond) formation, k_{UN}, can be obtained.

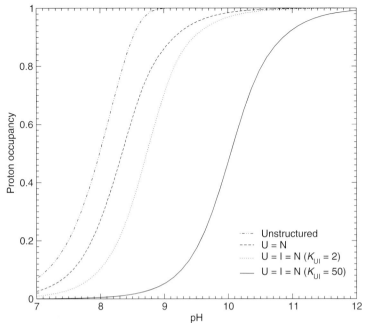

Fig. 18.8. Simulated proton occupancy profiles from a protein folding/hydrogen exchange competition experiment. $(\cdot - \cdot \cdot -)$ exchange from an unprotected amide with $k_{\text{int}} = 1.67 \times 10^8 \text{ s}^{-1} \text{ M}^{-1} [\text{OH}^-]$ according to Eq. (21); $(- - -)$ exchange from an amide in a two-state process with $k_{\text{UN}} = 50 \text{ s}^{-1}$, $k_{\text{int}} = 1.67 \times 10^8 \text{ s}^{-1} \text{ M}^{-1} [\text{OH}^-]$ according to Eq. (24); $(\cdots)$ exchange from an amide in a three-state process with $K_{\text{UI}} = 2$, $k_{\text{IN}} = 50 \text{ s}^{-1}$, $k_{\text{int}} = 1.67 \times 10^8 \text{ s}^{-1} \text{ M}^{-1} [\text{OH}^-]$, and $k_{\text{ex,I}} = 0$ according to Eq. (29); (———) exchange from an amide in a three-state process with $K_{\text{UI}} = 50$, $k_{\text{IN}} = 50 \text{ s}^{-1}$, $k_{\text{int}} = 1.67 \times 10^8 \text{ s}^{-1} \text{ M}^{-1} [\text{OH}^-]$, and $k_{\text{ex,I}} = 0$ according to Eq. (29).

For a protein that folds through an intermediate (Figure 18.7), the folding/competition protocol can be used to estimate the stability of the hydrogen bonds formed in the intermediate. If $k_{\text{UI}}, k_{\text{IU}} \gg k_{\text{int}}, k_{\text{IN}}$, **U** and **I** can be approximated to be in equilibrium at all times. Thus $[\mathbf{U}] = 1/(1 + K_{\text{UI}})([\mathbf{U}] + [\mathbf{I}])$ and $[\mathbf{I}] = K_{\text{UI}}/(1 + K_{\text{UI}})([\mathbf{U}] + [\mathbf{I}])$. The rate equations become:

$$\frac{d[\mathbf{N}_D]}{dt} = k_{\text{IN}}[\mathbf{I}_D] = \frac{k_{\text{IN}} K_{\text{UI}}}{1 + K_{\text{UI}}}([\mathbf{U}_D] + [\mathbf{I}_D]) \tag{26}$$

$$\frac{d([\mathbf{U}_D] + [\mathbf{I}_D])}{dt} = -k_{\text{int}}[\mathbf{U}_D] - (k_{\text{IN}} + k_{\text{ex,I}})[\mathbf{I}_D]$$

$$= -\frac{k_{\text{int}} + (k_{\text{IN}} + k_{\text{ex,I}}) K_{\text{UI}}}{1 + K_{\text{UI}}}([\mathbf{U}_D] + [\mathbf{I}_D]) \tag{27}$$

$$\frac{d[\mathbf{X}_H]}{dt} = k_{\text{int}}[\mathbf{U}_D] + k_{\text{ex,I}}[\mathbf{I}_D] = \frac{k_{\text{int}} + K_{\text{UI}} k_{\text{ex,I}}}{1 + K_{\text{UI}}}([\mathbf{U}_D] + [\mathbf{I}_D]) \tag{28}$$

where X_H is any hydrogenated state ($[X_H] = [U_H] + [I_H] + [N_H]$). During measurement of the proton occupancy I only the N state is observed and essentially all hydrogenated species will be found in this state. Thus $[X_H]$ is equivalent to I. Solving these equations yields the following expression for I:

$$I = \frac{k_{int} + K_{UI}k_{ex,I}}{k_{int} + K_{UI}(k_{IN} + k_{ex,I})} \left(1 - \exp\left(-\frac{K_{UI}(k_{IN} + k_{ex,I}) + k_{int}}{K_{UI} + 1} t_{pulse}\right)\right) \quad (29)$$

I is a function of both pH (through the pH dependence of k_{int}) and t_{pulse}. If $k_{ex,I} \gg k_{IN}$ or $k_{int} \gg K_{UI}k_{IN}$ insignificant amounts of protein will reach the native state N during the exchange period, t_{pulse}, and Eq. (29) is often simplified to:

$$I = 1 - \exp(-k_{app,I} t_{pulse}) \quad (30)$$

where $k_{app,I} = k_{int}/P_{app}$ with the apparent protection factor, $P_{app} = \frac{K_{UI} + 1}{K_{UI}k_{ex,I}/k_{int} + 1}$. If $k_{ex,I} \ll k_{int}$ only little exchange occurs directly from I and $P_{app} \approx 1 + K_{UI}$. If no protection from hydrogen exchange is present in I, $k_{ex,I} = k_{int}$ and $P_{app} = 1$. Great care should be taken when using the simplified Eq. (30). Indeed, Gladwin and Evans [33] and Bieri and Kiefhaber [111] have stressed the importance of accounting for later folding events (i.e. the flux to N) in folding competition experiments. In a study of early folding intermediates in lysozyme and ubiquitin Gladwin and Evans simplified Eq. (29) by considering the U and I states as a single denatured state D in which some protecting structure may be formed. A two-state reaction scheme as in Figure 18.7 was used to describe the situation and an expression analogous to Eq. (24) can be obtained by setting $k_{ex} = (k_{int} + K_{UI}k_{ex,I})/(K_{UI} + 1)$ and $k_f = K_{UI}k_{IN}/(K_{UI} + 1)$ in Eq. (29):

$$I = \frac{k_{ex}}{k_f + k_{ex}}(1 - \exp(-(k_f + k_{ex})t_{pulse})) \quad (31)$$

where k_f is the apparent folding rate and k_{ex} is the apparent exchange rate from D. In order to describe the protective structure in D, a protection factor or inhibition factor, k_{int}/k_{ex}, was reported [33].

An important prerequisite for good estimates of the stability of folding intermediates from the folding/exchange competition protocol is that the stability and kinetics of the folding process have low pH dependence. If the stability of an intermediate changes significantly with pH analysis of the data is not straightforward [111].

18.4.2
Hydrogen Exchange Pulse Labeling

Whereas the folding/exchange competition experiment has proven very useful for characterizing early folding intermediates, the protocol does not have the time resolution required for characterizing the time course of a protein folding process with multiple kinetic phases. The hydrogen exchange pulse labeling experiment is the choice for studying the time course of hydrogen bond formation. The hydrogen exchange pulse labeling protocol was developed on RNase A by Udgaonkar and Baldwin [112] and on cytochrome c by Roder et al. [19]. The great advantage of this technique is its capacity to take a "snap shot" on the millisecond timescale of the hydrogen bonds as they evolve during folding. The pulse labeling experiment consists of three steps: folding, the exchange pulse, and the quench as outlined in Figure 18.9. The similarity with the folding/competition experiment is obvious. The only differences are the initial extra mixing stage, the folding period prior to the labeling pulse and the choice of pH in the various steps.

In a two-state folding experiment the following will occur: pH is low (5–6) during the folding step and $k_{UN} \gg k_{int}$. Therefore, exchange will be negligible for short folding times. During the exchange pulse, pH is raised as much as possible considering the stability of the protein (typically pH 10–11) and $k_{int} \gg k_{UN}$. Now, all deuterated amides that are not forming a protective structure will exchange and become protonated. In the last step pH is lowered to 5 or less whereby $k_{UN} \gg k_{int}$, and exchange is again negligible compared with folding. Folding is now allowed to go to completion. This mixing sequence will result in trapped deuterium at amides sites that have formed hydrogen bonds in the folding step, and will be observed as decreased signals of these amides in proton NMR spectra. The proton occupancy (and thus the NMR signal) will decrease as a function of the folding time, t_{fold}, with a rate equal to k_{UN}. If intermediates are present, additional phases will be observed just as with any other probe.

For an intermediate formed within the dead time of a hydrogen exchange pulse labeling experiment, a folding/exchange competition experiment can be conducted as described above to determine the stability of the structures formed [33]. If the intermediate forms later during folding the stability can be estimated by varying the pH of the exchange pulse in the pulse labeling experiment keeping the folding time, t_{fold}, constant. In this way the intermediate is allowed to populate before the folding/exchange competition is initiated [113].

The advantage of both hydrogen exchange pulse labeling and hydrogen exchange/folding competition in combination with NMR over direct spectroscopic techniques is that the kinetics of structure formation can be resolved at the residue level. Two points should, however, be kept in mind when interpreting the results of hydrogen exchange pulse labeling experiments. First, only amides that are highly protected in the native state are suitable probes. Hydrogens, which are only marginally protected or not protected at all, will exchange with the solvent during sample handling and analysis. Secondly, the experiment cannot distinguish between the formation of native and nonnative hydrogen bonds. If hydrogen bond formation is detected in an intermediate state, information from several experiments must be considered in order to characterize the hydrogen bonds as native or nonnative.

18.4.3
Protection Factors in Folding Intermediates

In order to characterize hydrogen bonds identified in intermediates by hydrogen exchange labeling techniques, protection factors are often reported. For the native

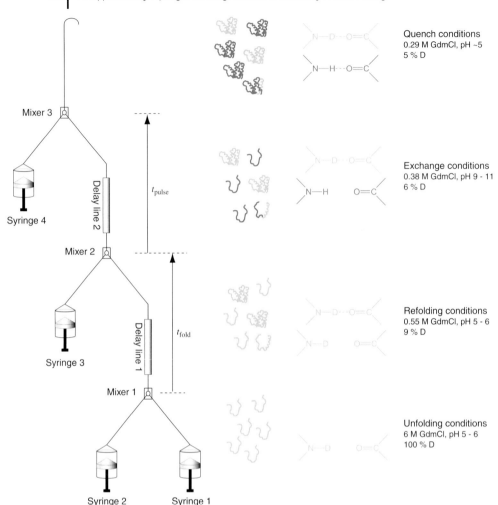

Fig. 18.9. Outline of the quenched flow hydrogen exchange pulse labeling experiment. The experiment is carried out in a four-syringe quenched flow apparatus. Syringe 1 contains fully deuterated protein in deuterated buffer under conditions favoring the unfolded state. In mixer 1 the protein is diluted with 10 volumes of hydrogenated refolding buffer. The protein is allowed to fold while it flows from mixer 1 to mixer 2. The folding time will thus be dependent on the volume between mixers 1 and 2 and the flow rate. At mixer 2 five volumes of hydrogen exchange buffer with high pH is added from syringe 3. At high pH hydrogen exchange is fast and deuterium at amides not engaged in hydrogen bonds will exchange with hydrogen from the solvent. After a period of time, t_{pulse} the sample reaches mixer 3 where quench buffer is added from syringe 4. The pH is now lowered to 5 or less and hydrogen exchange becomes slow. The result is a snapshot of the hydrogen bonds formed at mixer 2 as amides engaged in hydrogen bonds at this point will be labeled with deuterium. The sample is collected and desalted prior to analysis by NMR.

state, the protection factor is defined as $P = k_{int}/k_{obs}$ (Section 18.2.1). In analogy to this, a structural protection factor $P_{struc} = k_{int}/k_{ex,I}$ can be defined, where $k_{ex,I}$ is the apparent rate of hydrogen exchange directly from **I** [114]. P_{struc} thus reports on the stability and local fluctuations of structure in **I**-like protection factors for the native state. In P_{struc}, exchange through preceding unfolding of **I** to **U** is not considered. In order to get a reliable measure of P_{struc} it is important to be able to separate the exchange from the **I**-state from unfolding to and exchange from the **U**-state. This may be achieved if the **I**-state only converts slowly to the **U**-state compared with the exchange directly from **I**, or if previous knowledge of the folding kinetics allows for correction of the contribution to exchange from other states. An example of measurement of P_{struc} for two intermediates in the refolding of RNase A has been reported by Houry and Scheraga, who used an ingenious double jump hydrogen exchange protocol to selectively populate two different intermediates and measure the hydrogen exchange from these [114, 115].

Another way of defining a protection factor for early folding intermediates is as an apparent protection factor, $P_{app} = k_{int}/k_{app,I}$ [114], which was introduced in the discussion of Eq. (30). In this definition, the apparent exchange rate k_{app}, which includes both direct exchange and exchange through unfolding processes, is considered. P_{app} is thus closely related to the overall stability of a hydrogen bond formed in an intermediate state including both the direct exchange and unfolding events. Often P_{app} is more easily measured than P_{struc}, as the contributions from the structure in **I** and the folding events do not have to be dissected.

18.4.4
Kinetic Intermediate Structures Characterized by Hydrogen Exchange

The hydrogen exchange labeling techniques described above have been used to characterize kinetic folding intermediates in numerous proteins. Both transient intermediates formed very early in the folding process and more stable intermediates formed later during folding have been observed by hydrogen exchange labeling.

One of the most well-studied systems is cytochrome c (cyt c) [116]. Three kinetic phases of the folding process have been characterized by tryptophan fluorescence [19, 117]. The fastest phase occurring within the dead time of conventional stopped-flow techniques was characterized by rapid mixing continuous-flow kinetic experiments and has a time constant of 50 µs [117]. The structures formed in this initial phase have been characterized by folding/hydrogen exchange competition. Sauder and Roder [26] found that several amides from three segments of the polypeptide that form α-helices in the folded state acquire significant protection from hydrogen exchange with the solvent within 2 ms of folding in 0.3 M GdmCl. Apparent protection factors in the range from 1.5 to 8 relative to exchange in cyt c unfolded in 2.5 M GdmCl were observed. Larger protection factors of 5–8 were observed in the helical regions forming side-chain interactions between the N- and C-terminal helices in the native structure. These protection factors agree well with the protection factors of 4–5 expected from stopped-flow experiments [26]. The lower protection factors of 1.5–5 were observed at the helical ends, suggesting exchange through local fluctuations in these regions in the intermediate. The following phase in the folding of cyt c has a time constant of 20 ms, during which the structure in the N- and C-terminal helices consolidates. Tight native-like packing of these two helix fragments in the intermediate probably stabilizes the hydrogen bonds. The major fraction of the protein population reaches the native state from the late intermediate by organizing the structure around the heme with a time constant of 370 ms [19].

Lysozyme has also been extensively studied. Folding experiments at pH 5.2 showed a dead-time burst followed by two exponential phases using a variety of detection techniques with time constants of 25 ms and 340 ms respectively [118]. By folding/hydrogen exchange competition, protection factors of 2–15 in the burst phase were observed for amides constituting the α-domain in the native structure [33]. Hydrogen exchange pulse labeling resolved the kinetics of the hydrogen bond formation and showed that the amides in the α-domain become significantly protected in a kinetic phase with varying time constants of 7 ms to 60 ms for the different amides. The β-domain acquires protection in a later phase with a time constant of around 400 ms [22]. The range of time constants observed for the formation of protective structure in the α-domain apparently disagrees with kinetics measured by optical detection. However, simulation of the hydrogen exchange shows, when taking the folding kinetics of lysozyme during the exchange pulse at pH 9.5 into account, that the experimental data are in agreement with the proposed triangular folding model for lysozyme [111, 119]. The results on lysozyme stress the importance of thorough comparison of data from different techniques.

In the case of apomyoglobin, hydrogen exchange pulse labeling was used to identify the order of helix formation [41]. It was found that helices A, G, and H and a part of helix B form very fast within the dead time of the experiment. The rest of helix B forms stable structure at a later stage followed by helix C, the CD loop, and helix E, which dock as a late event in the folding process [41]. Leghemoglobin, which is a distant homolog of myoglobin (13% sequence identity), has also been studied by hydrogen exchange pulse labeling [120]. Like apomyoglobin, apoleghemoglobin forms a substantial amount of helical structure within the dead time of an ordinary stopped-flow experiment. Interestingly, the hydrogen exchange pulse labeling experiment showed that helices E, G, and H are structured in the burst phase intermediate whereas the A and B helices are not. It has thus been demonstrated that though apomyoglobin and apoleghemoglobin both fold through a compact intermediate with helical structure, the order in which the helices form and dock are different [120]. These results have contributed to the discussion about evolutionary conservation in protein folding [121–124] and show that the details of the folding process need not be evolutionary conserved.

On β-lactoglobulin, hydrogen exchange pulse labeling experiments have been performed using both conventional quench-flow mixing and ultra-rapid mixing. By this approach it was possible to resolve the kinetics of the hydrogen bond formation from 240 µs to 1 s [125]. It has been found that during folding of β-lactoglobulin some nonnative helical structure is formed [125, 126]. By hydrogen exchange labeling the location of this transient structure was mapped to the N-terminal part of the protein, which forms a β-strand in the native structure. It was suggested that this nonnative structure protects the N-terminal part of the protein from aggregation during folding until most of the sheet, of which this strand is a part, has formed.

For several other proteins, both well-defined (RNase A [112, 115, 127], barnase [128], and RNase H from *E. coli* [129] and *Thermus thermophilus* [130]) and transient low-stability intermediates (protein G [131] and ACBP [132]) have been observed by hydrogen exchange pulse labeling techniques.

18.5
Experimental Protocols

18.5.1
How to Determine Hydrogen Exchange Kinetics at Equilibrium

The following experiment can be used to measure directly the amide exchange kinetics from any protein state at equilibrium. Typically exchange of hydrogen is measured with solvent deuterium, so any buffers used need to be deuterated before use.

18.5.1.1 Equilibrium Hydrogen Exchange Experiments

For each condition, samples of 0.5–2 mM protein in H_2O are adjusted to the desired pH and freeze dried repeatedly. After the final freeze drying, the protein is dissolved in the appropriate volume (300 or 600 μL) of 99.99% D_2O. For native state hydrogen exchange the desired concentration of deuterated denaturant in deuterated buffer is also included. The concentration of denaturant is always measured by refractive index [133, 134] and all pH measurements in D_2O are normally reported as direct meter readings. The sample is readjusted, if necessary, to the desired pH and transferred to a cold NMR tube to slow exchange. This is immediately placed in the spectrometer, which in advance has been tuned and calibrated using a similar sample. Acquisition of the first experiment is initiated when the desired sample temperature is obtained. Series of two-dimensional $^{15}N-^{1}H$ HSQC spectra, or COSY spectra if ^{15}N-labeled protein is unavailable, are recorded with good resolution for each sample. Peak integrals are measured and decay curves of peak integrals with time are fitted to an exponential three-parameter decay of the form

$$I(t) = I(\infty) + I(0) \exp(-k_{obs}t) \tag{32}$$

where $I(t)$ is the cross-peak intensity at time t after addition of D_2O to the lyophilized protein, $I(0)$ is the cross peak intensity at time zero, $I(\infty)$ is the final cross-peak intensity, and k_{obs} is the observed exchange rate constant. The time, t, is taken as the midpoint of the NMR data acquisition for each data point. For amides with incomplete exchange in the time course of the experiment, $I(\infty)$ is omitted from the fit. Values and standard deviations of $I(\infty)$, $I(0)$, and k_{obs} are obtained from nonlinear fitting procedures, and the uncertainties obtained on peak integrals are conveniently included.

18.5.1.2 Determination of Segmental Opening and Closing Rates, k_{op} and k_{cl}

Determination of the site-specific rate constants k_{op} and k_{cl} of the pre-exchange equilibrium can be determined from the pH dependence of the experimentally determined k_{obs}. Two different approaches can be applied. In the first approach, exchange is measured under both EX2 (to provide measures of k_{op}/k_{cl}) and EX1 (measures k_{op}) conditions and the result are combined to get k_{cl}. This of course requires that the protein is stable in high alkaline conditions and that k_{int} can be applied directly. For amides with exchange in the EX(1/2) regime, nonlinear fitting of experimental data to Eq. (7) or (9) are performed as described.

18.5.1.3 Determination of ΔG_{fluc}, m, and $\Delta G°_{unf}$

In exchange experiments measured as a function of denaturant, values for ΔG_{fluc}, m, and $\Delta G°_{unf}$ are obtained in a nonlinear fitting procedure to ΔG_{HX} as a function of denaturant concentration as described by Eq. (18). The only assumptions are that the unfolding free energies are linearly dependent on the concentration of denaturant and that exchange is within the EX2 limit.

18.5.2
Planning a Hydrogen Exchange Folding Experiment

The following is a general protocol for a folding/competition hydrogen exchange and a hydrogen exchange pulse labeling experiment. The protocol describes a deuterated protein that exchanges with solvent hydrogen. The reverse experiment can also be done. Modifications may be necessary to meet potential problems of protein stability and solubility. If ^{15}N-labeled protein is available it is desirable to use due to the increased sensitivity and simplicity and a better water suppression of a HSQC spectrum compared with a COSY spectrum.

18.5.2.1 Determine a Combination of t_{pulse} and pH_{pulse}

For the pulse labeling experiment the first step is to choose a suitable pH for the labeling pulse and a suitable labeling time. pH should be as high as possible but still well within the stability range of the protein. t_{pulse} should be long enough to ensure full (>99%) exchange of the amide with the lowest intrinsic exchange rate, k_{int}, calculated as described in Section 18.2.1. Typically, pH is in the range from 9 to 10.5 and t_{pulse} in the range of 5 ms to 20 ms.

18.5.2.2 Setup Quench Flow Apparatus

A rapid mixing apparatus (Bio-Logic SFM-400 or similar) has to be configured. For folding/exchange competition experiments the apparatus has to be configured in a two-mixer mode whereas the pulse labeling experiment require three mixers. The easiest way of making proper mixing sequences is to use the manufactures software, which calculates the aging times in the delay lines at the chosen flow rates. In the pulse labeling experiment the apparatus must be operated in either of two modes. In a continuous flow mode the folding time depends on the flow rate and the volume of the first delay line. In interrupted mode the flow is stopped when the first delay line has been filled with mixed protein/refolding buffer. In this mode varying the length of the time the flow is stopped varies the folding time.

In a folding/exchange competition experiment the volume of the solutions in each step will typically be 1:10:5 (protein:refolding/exchange buffer:quench buffer). In the pulse labeling hydrogen exchange experiment the mixing ratio will be 1:10:5:5 (protein:refolding buffer:exchange buffer:quench buffer).

18.5.2.3 Prepare Deuterated Protein and Chemicals

Deuterated protein is prepared by several rounds of incubation in D_2O and freeze drying. If the protein is sensitive to freeze drying hydrogen may be exchanged with deuterium by thorough dialysis. Amide hydrogens that are very stable towards exchange may exchange when the protein is unfolded in deuterated GdmCl. It is a good idea to check the completeness of the exchange by NMR. Deuterated GdmCl is prepared simply by several rounds of incubation in D_2O and freeze drying.

18.5.2.4 Prepare Buffers and Unfolded Protein

The buffers mentioned below are suggestions, which will give proper pH in the various mixing steps; other buffers and pH values can, however, be chosen. The important issue to consider for the choice of buffers and experimental conditions is that the desired pH value is obtained in each mixing step. Prior to the hydrogen exchange step pH should be low (around 5) to ensure minimal hydrogen exchange. In the actual hydrogen exchange step pH is varied in the folding/exchange competition experiments. It should be as high as the stability of the protein allows in the hydrogen exchange pulse labeling experiment to ensure complete exchange of nonhydrogen bonded amides during the short exchange pulse. In the last mixing step, hydrogen exchange has to efficiently quenched. This is achieved by low pH (again around 5).

For a folding/exchange competition experiment the following solutions are suggested:

- Unfolded deuterated protein in 20 mM Na-acetate, 6 M deuterated GdmCl, pH 5.3 (in D_2O)
- Refolding/exchange buffer at varying pH from 7.5 to 11 (for example 20 mM HEPES (pH 6.8–8.2), 20 mM BICINE (pH 7.6–9.0), 20 mM CHES (pH 8.6–10.0), and 20 mM CAPS (pH 9.7–11.1))
- Quench buffer: 0.5 M Na-acetate, pH 4.8.

For a pulse labeling experiment the following solutions are suggested:

- Unfolded deuterated protein in 20 mM Na-acetate, 6 M deuterated GdmCl, pH 5.3 (in D_2O)
- Refolding buffer: 20 mM Na-acetate, pH 5.3
- Exchange buffer: 200 mM Na-borate, pH 11.0
- Quench buffer: 300 mM Na-acetate, pH 3.3.

18.5.2.5 Check pH in the Mixing Steps

It is important to check the pH at each step in the mixing sequence by performing manual mixing of the solutions in the right ratios on the bench. The check should be performed for each new batch of solutions as small variations in the buffer solutions may shift pH. If the desired pH is not obtained adjust pH of the appropriate buffer.

18.5.2.6 Sample Mixing and Preparation

The mixing apparatus is filled with the solutions. Let the solutions temperature equilibrate before the mixing is carried out. For each sample, enough protein for a proper NMR spectrum to be recorded in about 2 hours should be collected. Some loss of protein during the subsequent steps must be expected, but around 5 mg protein per sample (depending on the protein) is usually sufficient for a $^1H,^{15}N$ HSQC. If ^{15}N-labeled protein is unavailable more protein is needed to get the same signal-to-noise ratio from a COSY spectrum.

The samples should be put on ice immediately, and all subsequent steps should be carried out on ice or in the cold to minimize further hydrogen exchange. The samples are concentrated to approximately 1 mL and diluted with H_2O at pH 5 to 10 mL. This is repeated three times to dilute buffer and denaturant components. Finally, the sample is concentrated to 540 μL, 60 μL D_2O is added, and pH is adjusted. It is important to choose a low pH value (around 5) for the NMR samples in order to minimize hydrogen exchange during acquisition of the NMR spectrum.

The sample should be analyzed by NMR as soon as possible. It is our experience that samples kept on ice for up to 3 hours give good results. It is thus possible to prepare two samples in parallel. Longer storage of samples is undesirable. Record a spectrum with good resolution and signal-to-noise in 2–3 hours.

18.5.3
Data Analysis

Transform the spectra and integrate each amide cross-peak. Choose a set of well-resolved cross-peaks that show no protection against hydrogen exchange as reference peaks. Calculate the proton occupancy I according to Eq. (19). In a folding/exchange competition experiment plot I as a function of pH and fit the data to

Eq. (30) or (31) depending on which assumptions apply to the system under investigation. In a hydrogen exchange pulse labeling experiment plot the data against t_{fold} and fit the data to one or more exponential decays.

Acknowledgments

We would like to thank Flemming Hofmann Larsen, Wolfgang Fieber, and Jens Kaalby Thomsen for suggestions and for critically reading the manuscript.

References

1. LINDERSTRØM-LANG, K. (1955). Deuterium exchange between peptides and water. *Chem. Soc. Spec. Publ.* **2**, 1–20.
2. HVIDT, A. (1955). Deuterium exchange between ribonuclease and water. *Biochim. Biophys. Acta* **18**, 306–308.
3. BLOUT, E. R., DELOZE, C. & ASADOURIAN, A. (1961). Deuterium exchange of water-soluble polypeptides and proteins as measured by infrared spectroscopy. *J. Am. Chem. Soc.* **83**, 1895.
4. HVIDT, A. (1973). Isotope hydrogen exchange in solutions of biological macromolecules. In *Dynamic aspects of conformation changes in biological macromolecules* (SADRON, C., ed), pp. 103–115, Reidel and Dordrect, Holland.
5. RICHARDS, F. M. (1979). Packing defects, cavities, volume fluctuations, and access to the interior of proteins – including some general-comments on surface-area and protein-structure. *Carlsberg Res. Commun.* **44**, 47–63.
6. MASSON, A. & WUTHRICH, K. (1973). Proton magnetic resonance investigation of the conformational properties of the basic pancreatic trypsin inhibitor. *FEBS Lett.* **31**, 114–118.
7. CAMPBELL, I. D., DOBSON, C. M., JEMINET, G. & WILLIAMS, R. J. (1974). Pulsed NMR methods for the observation and assignment of exchangeable hydrogens: application to bacitracin. *FEBS Lett.* **49**, 115–119.
8. WOODWARD, C. K. & HILTON, B. D. (1979). Hydrogen exchange kinetics and internal motions in proteins and nucleic acids. *Annu. Rev. Biophys. Bioeng.* **8**, 99–127.
9. BAI, Y., SOSNICK, T. R., MAYNE, L. & ENGLANDER, S. W. (1995). Protein folding intermediates: native-state hydrogen exchange. *Science* **269**, 192–197.
10. ARRINGTON, C. B. & ROBERTSON, A. D. (1997). Microsecond protein folding kinetics from native-state hydrogen exchange. *Biochemistry* **36**, 8686–8691.
11. CHU, R. A., TAKEI, J., BARCHI, J. J. & BAI, Y. (1999). Relationship between the native-state hydrogen exchange and the folding pathways of barnase. *Biochemistry* **38**, 14119–14124.
12. PARKER, M. J. & MARQUSEE, S. (2001). A kinetic folding intermediate probed by native state hydrogen exchange. *J. Mol. Biol.* **305**, 593–602.
13. SIVARAMAN, T., ARRINGTON, C. B. & ROBERTSON, A. D. (2001). Kinetics of unfolding and folding from amide hydrogen exchange in native ubiquitin. *Nat. Struct. Biol.* **8**, 331–333.
14. RODER, H., WAGNER, G. & WUTHRICH, K. (1985). Amide proton exchange in proteins by EX1 kinetics: studies of the basic pancreatic trypsin inhibitor at variable p2H and

temperature. *Biochemistry* **24**, 7396–7407.
15 Itzhaki, L. S., Neira, J. L. & Fersht, A. R. (1997). Hydrogen exchange in chymotrypsin inhibitor 2 probed by denaturants and temperature. *J. Mol. Biol.* **270**, 89–98.
16 Xu, Y., Mayne, L. & Englander, S. W. (1998). Evidence for an unfolding and refolding pathway in cytochrome c. *Nat. Struct. Biol.* **5**, 774–778.
17 Viguera, A. R. & Serrano, L. (2003). Hydrogen-exchange stability analysis of Bergerac-Src homology 3 variants allows the characterization of a folding intermediate in equilibrium. *Proc. Natl. Acad. Sci. USA* **100**, 5730–5735.
18 Kim, P. S. & Baldwin, R. L. (1980). Structural intermediates trapped during the folding of ribonuclease A by amide proton exchange. *Biochemistry* **19**, 6124–6129.
19 Roder, H., Elove, G. A. & Englander, S. W. (1988). Structural characterization of folding intermediates in cytochrome c by H-exchange labelling and proton NMR. *Nature* **335**, 700–704.
20 Miranker, A., Radford, S. E., Karplus, M. & Dobson, C. M. (1991). Demonstration by NMR of folding domains in lysozyme. *Nature* **349**, 633–636.
21 Miranker, A., Robinson, C. V., Radford, S. E., Aplin, R. T. & Dobson, C. M. (1993). Detection of transient protein folding populations by mass spectrometry. *Science* **262**, 896–900.
22 Radford, S. E., Dobson, C. M. & Evans, P. A. (1992). The folding of hen lysozyme involves partially structured intermediates and multible pathways. *Nature* **358**, 302–307.
23 Lu, J. & Dahlquist, F. W. (1992). Detection and characterization of an early folding intermediate of T4 lysozyme using pulsed hydrogen exchange and two-dimensional NMR. *Biochemistry* **31**, 4749–4756.
24 Briggs, M. S. & Roder, H. (1992). Early hydrogen-bonding events in the folding reaction of ubiquitin. *Proc. Natl. Acad. Sci. USA* **89**, 2017–2021.
25 Baldwin, R. L. (1993). Pulsed H/D-exchange studies of folding intermediates. *Curr. Opin. Struct. Biol.* **3**, 84–91.
26 Sauder, J. M. & Roder, H. (1998). Amide protection in an early folding intermediate of cytochrome c. *Fold. Des.* **3**, 293–301.
27 Fong, S., Bycroft, M., Clarke, J. & Freund, S. M. (1998). Characterisation of urea-denatured states of an immunoglobulin superfamily domain by heteronuclear NMR. *J. Mol. Biol.* **278**, 417–429.
28 Neira, J. L., Itzhaki, L. S., Otzen, D. E., Davis, B. & Fersht, A. R. (1997). Hydrogen exchange in chymotrypsin inhibitor 2 probed by mutagenesis. *J. Mol. Biol.* **270**, 99–110.
29 Chamberlain, A. K. & Marqusee, S. (1998). Molten globule unfolding monitored by hydrogen exchange in urea. *Biochemistry* **37**, 1736–1742.
30 Chamberlain, A. K. & Marqusee, S. (1997). Touring the landscapes: partially folded proteins examined by hydrogen exchange. *Structure* **5**, 859–863.
31 Hosszu, L. L., Craven, C. J., Parker, M. J., Lorch, M., Spencer, J., Clarke, A. R. & Waltho, J. P. (1997). Structure of a kinetic protein folding intermediate by equilibrium amide exchange. *Nat. Struct. Biol.* **4**, 801–804.
32 Roder, H. & Colon, W. (1997). Kinetic role of early intermediates in protein folding. *Curr. Opin. Struct. Biol.* **7**, 15–28.
33 Gladwin, S. T. & Evans, P. A. (1996). Structure of very early protein folding intermediates: new insights through a variant of hydrogen exchange labelling. *Fold. Des.* **1**, 407–417.
34 Schulman, B. A., Redfield, C., Peng, Z. Y., Dobson, C. M. & Kim, P. S. (1995). Different subdomains are most protected from hydrogen

exchange in the molten globule and native states of human alpha-lactalbumin. *J. Mol. Biol.* **253**, 651–657.
35 UDGAONKAR, J. B. & BALDWIN, R. L. (1995). Nature of the early folding intermediate of ribonuclease A. *Biochemistry* **34**, 4088–4096.
36 JONES, B. E. & MATTHEWS, C. R. (1995). Early intermediates in the folding of dihydrofolate reductase from Escherichia coli detected by hydrogen exchange and NMR. *Protein Sci.* **4**, 167–177.
37 KOTIK, M., RADFORD, S. E. & DOBSON, C. M. (1995). Comparison of the refolding of hen lysozyme from dimethyl sulfoxide and guanidinium chloride. *Biochemistry* **34**, 1714–1724.
38 BUCK, M., RADFORD, S. E. & DOBSON, C. M. (1994). Amide hydrogen exchange in a highly denatured state. Hen egg-white lysozyme in urea. *J. Mol. Biol.* **237**, 247–254.
39 ALEXANDRESCU, A. T., NG, Y. L. & DOBSON, C. M. (1994). Characterization of a trifluoroethanol-induced partially folded state of alpha-lactalbumin. *J. Mol. Biol.* **235**, 587–599.
40 ALEXANDRESCU, A. T., EVANS, P. A., PITKEATHLY, M., BAUM, J. & DOBSON, C. M. (1993). Structure and dynamics of the acid-denatured molten globule state of alpha-lactalbumin: a two-dimensional NMR study. *Biochemistry* **32**, 1707–1718.
41 JENNINGS, P. A. & WRIGHT, P. E. (1993). Formation of a molten globule intermediate early in the kinetic folding pathway of apomyoglobin. *Science* **262**, 892–896.
42 RADFORD, S. E., BUCK, M., TOPPING, K. D., DOBSON, C. M. & EVANS, P. A. (1992). Hydrogen exchange in native and denatured states of hen egg-white lysozyme. *Proteins* **14**, 237–248.
43 JENG, M. F. & ENGLANDER, S. W. (1991). Stable submolecular folding units in a non-compact form of cytochrome c. *J. Mol. Biol.* **221**, 1045–1061.
44 WOODWARD, C. K. (1993). Is the slow exchange core the protein folding core? *Trends Biochem. Sci.* **18**, 359–360.
45 WOODWARD, C. K. (1994). Hydrogen exchange rates and protein folding. *Curr. Opin. Struct. Biol.* **4**, 112–116.
46 BAI, Y. (1999). Equilibrium amide hydrogen exchange and protein folding kinetics. *J. Biomol. NMR* **15**, 65–70.
47 CLARKE, J. & FERSHT, A. R. (1996). An evaluation of the use of hydrogen exchange at equilibrium to probe intermediates on the protein folding pathway. *Fold. Des.* **1**, 243–254.
48 BALDWIN, R. L. (1996). On-pathway versus off-pathway folding intermediates. *Fold. Des.* **1**, R1–R8.
49 CLARKE, J., HOUNSLOW, A. M., BYCROFT, M. & FERSHT, A. R. (1993). Local breathing and global unfolding in hydrogen exchange of barnase and its relationship to protein folding pathways. *Proc. Natl. Acad. Sci. USA* **90**, 9837–9841.
50 MATOUSCHEK, A., KELLIS, J. T., JR., SERRANO, L., BYCROFT, M. & FERSHT, A. R. (1990). Transient folding intermediates characterized by protein engineering. *Nature* **346**, 440–445.
51 FERSHT, A. R. (2000). A kinetically significant intermediate in the folding of barnase. *Proc. Natl. Acad. Sci. USA* **97**, 14121–14126.
52 TAKEI, J., CHU, R. A. & BAI, Y. (2000). Absence of stable intermediates on the folding pathway of barnase. *Proc. Natl. Acad. Sci. USA* **97**, 10796–10801.
53 ENGLANDER, S. W. (2000). Protein folding intermediates and pathways studied by hydrogen exchange. *Annu. Rev. Biophys. Biomol. Struct.* **29**, 213–238.
54 CLARKE, J. & ITZHAKI, L. S. (1998). Hydrogen exchange and protein folding. *Curr. Opin. Struct. Biol.* **8**, 112–118.
55 ENGLANDER, S. W., MAYNE, L., BAI, Y. & SOSNICK, T. R. (1997). Hydrogen exchange: the modern legacy of Linderstrøm-Lang. *Protein Sci.* **6**, 1101–1109.

56 Dyson, H. J. & Wright, P. E. (1996). Insights into protein folding from NMR. *Annu. Rev. Phys. Chem.* **47**, 369–395.
57 Raschke, T. M. & Marqusee, S. (1998). Hydrogen exchange studies of protein structure. *Curr. Opin. Biotechnol.* **9**, 80–86.
58 Eigen, M. (1964). Proton transfer acid-base catalysis + enzymatic hydrolysis. I. Elementary processes. *Angew. Chem. Int. Ed. Engl.* **3**, 1.
59 Bai, Y., Milne, J. S., Mayne, L. & Englander, S. W. (1993). Primary structure effects on peptide group hydrogen exchange. *Proteins* **17**, 75–86.
60 Covington, A. K., Robinson, R. A. & Bates, R. G. (1966). The Ionization Constant of Deuterium Oxide from 5 to 50°. *J. Phys. Chem.* **70**, 3820–3824.
61 Alexandrescu, A. T., Dames, S. A. & Wiltscheck, R. (1996). A fragment of staphylococcal nuclease with an OB-fold structure shows hydrogen-exchange protection factors in the range reported for "molten globules". *Protein Sci.* **5**, 1942–1946.
62 Guijarro, J. I., Jackson, M., Chaffotte, A. F., Delepierre, M., Mantsch, H. H. & Goldberg, M. E. (1995). Protein folding intermediates with rapidly exchangeable amide protons contain authentic hydrogen-bonded secondary structures. *Biochemistry* **34**, 2998–3008.
63 Yamasaki, K., Yamasaki, T., Kanaya, S. & Oobatake, M. (2003). Acid-induced denaturation of Escherichia coli ribonuclease HI analyzed by CD and NMR spectroscopies. *Biopolymers* **69**, 176–188.
64 Chamberlain, A. K., Handel, T. M. & Marqusee, S. (1996). Detection of rare partially folded molecules in equilibrium with the native conformation of RNaseH. *Nat. Struct. Biol.* **3**, 782–787.
65 Mori, S., van, Z. & Shortle, D. (1997). Measurement of water-amide proton exchange rates in the denatured state of staphylococcal nuclease by a magnetization transfer technique. *Proteins* **28**, 325–332.
66 Spyracopoulos, L. & O'Neil, J. D. J. (1994). Effect of a Hydrophobic Environment on the Hydrogen Exchange Kinetics of Model Amides Determined by 1H-NMR Spectroscopy. *J. Am. Chem. Soc.* **116**, 1395–1402.
67 Milne, J. S., Mayne, L., Roder, H., Wand, A. J. & Englander, S. W. (1998). Determinants of protein hydrogen exchange studied in equine cytochrome c. *Protein Sci.* **7**, 739–745.
68 Woodward, C. K. & Hilton, B. D. (1980). Hydrogen isotope exchange kinetics of single protons in bovine pancreatic trypsin inhibitor. *Biophys. J.* **32**, 561–575.
69 Mayo, S. L. & Baldwin, R. L. (1993). Guanidinium chloride induction of partial unfolding in amide proton exchange in RNase A. *Science* **262**, 873–876.
70 Hvidt, A. & Nielsen, S. O. (1966). Hydrogen exchange in proteins. *Adv. Prot. Chem.* **21**, 287–386.
71 Orban, J., Alexander, P., Bryan, P. & Khare, D. (1995). Assessment of stability differences in the protein G B1 and B2 domains from hydrogen-deuterium exchange: comparison with calorimetric data. *Biochemistry* **34**, 15291–15300.
72 Huyghues-Despointes, B. M., Scholtz, J. M. & Pace, C. N. (1999). Protein conformational stabilities can be determined from hydrogen exchange rates. *Nat. Struct. Biol.* **6**, 910–912.
73 Frost, A. A. & Pearson, R. G. (1953). *Kinetics and mechanisms* John Wiley, New York.
74 Hvidt, A. (1963). A discussion of the pH dependence of the hydrogen-deuterium exchange of proteins. *C. R. Trav. Lab. Carlsberg* **34**, 299–317.
75 Kragelund, B. B., Heinemann, B., Knudsen, J. & Poulsen, F. M. (1998). Mapping the lifetimes of local opening events in a native state protein. *Protein Sci.* **7**, 2237–2248.
76 Pedersen, T. G., Thomsen, N. K.,

Andersen, K. V., Madsen, J. C. & Poulsen, F. M. (1993). Determination of the rate constants k1 and k2 of the Linderstrøm-Lang model for protein amide hydrogen exchange. A study of the individual amides in hen egg-white lysozyme. *J. Mol. Biol.* **230**, 651–660.

77. Bai, Y., Milne, J. S., Mayne, L. & Englander, S. W. (1994). Protein stability parameters measured by hydrogen exchange. *Proteins* **20**, 4–14.

78. Houliston, R. S., Liu, C., Singh, L. M. & Meiering, E. M. (2002). pH and urea dependence of amide hydrogen-deuterium exchange rates in the beta-trefoil protein hisactophilin. *Biochemistry* **41**, 1182–1194.

79. Rodriguez, H. M., Robertson, A. D. & Gregoret, L. M. (2002). Native state EX2 and EX1 hydrogen exchange of Escherichia coli CspA, a small beta-sheet protein. *Biochemistry* **41**, 2140–2148.

80. Arrington, C. B. & Robertson, A. D. (2000). Microsecond to minute dynamics revealed by EX1-type hydrogen exchange at nearly every backbone hydrogen bond in a native protein. *J. Mol. Biol.* **296**, 1307–1317.

81. Englander, S. W. (1998). Native-state HX. *Trends Biochem. Sci.* **23**, 378.

82. Clarke, J., Itzhaki, L. S. & Fersht, A. R. (1997). Hydrogen exchange at equilibrium: a short cut for analysing protein-folding pathways? *Trends Biochem. Sci.* **22**, 284–287.

83. Li, R. & Woodward, C. (1999). The hydrogen exchange core and protein folding. *Protein Sci.* **8**, 1571–1590.

84. Chi, Y. H., Kumar, T. K., Kathir, K. M., Lin, D. H., Zhu, G., Chiu, I. M. & Yu, C. (2002). Investigation of the structural stability of the human acidic fibroblast growth factor by hydrogen-deuterium exchange. *Biochemistry* **41**, 15350–15359.

85. Shortle, D. (1995). Staphylococcal nuclease – a showcase of *m*-value effects. *Adv. Prot. Chem.* 217–247.

86. Tanford, C. (1970). Protein denaturation. C. Theoretical models for the mechanism of denaturation. *Adv. Prot. Chem.* **24**, 1–95.

87. Schellman, J. A. (1987). The thermodynamic stability of proteins. *Annu. Rev. Biophys. Biophys. Chem.* **16**, 115–137.

88. Huyghues-Despointes, B. M., Pace, C. N., Englander, S. W. & Scholtz, J. M. (2001). Measuring the conformational stability of a protein by hydrogen exchange. *Methods Mol. Biol.* **168**, 69–92.

89. Bai, Y. & Englander, S. W. (1996). Future directions in folding: the multi-state nature of protein structure. *Proteins* **24**, 145–151.

90. Feng, Z., Butler, M. C., Alam, S. L. & Loh, S. N. (2001). On the nature of conformational openings: native and unfolded-state hydrogen and thiol-disulfide exchange studies of ferric aquomyoglobin. *J. Mol. Biol.* **314**, 153–166.

91. Llinas, M., Gillespie, B., Dahlquist, F. W. & Marqusee, S. (1999). The energetics of T4 lysozyme reveal a hierarchy of conformations. *Nat. Struct. Biol.* **6**, 1072–1078.

92. Juneja, J. & Udgaonkar, J. B. (2002). Characterization of the unfolding of ribonuclease A by a pulsed hydrogen exchange study: evidence for competing pathways for unfolding. *Biochemistry* **41**, 2641–2654.

93. Fuentes, E. J. & Wand, A. J. (1998). Local dynamics and stability of apocytochrome b562 examined by hydrogen exchange. *Biochemistry* **37**, 3687–3698.

94. Bhuyan, A. K. & Udgaonkar, J. B. (1998). Two structural subdomains of barstar detected by rapid mixing NMR measurement of amide hydrogen exchange. *Proteins* **30**, 295–308.

95. Loh, S. N., Prehoda, K. E., Wang, J. & Markley, J. L. (1993). Hydrogen exchange in unligated and ligated staphylococcal nuclease. *Biochemistry* **32**, 11022–11028.

96. Yi, Q., Scalley, M. L., Simons, K. T.,

GLADWIN, S. T. & BAKER, D. (1997). Characterization of the free energy spectrum of peptostreptococcal protein L. *Fold. Des.* **2**, 271–280.

97 BHUTANI, N. & UDGAONKAR, J. B. (2003). Folding subdomains of thioredoxin characterized by native-state hydrogen exchange. *Protein Sci.* **12**, 1719–1731.

98 PARKER, M. J. & MARQUSEE, S. (1999). The Cooperativity of Burst Phase Reactions Explored. *J. Mol. Biol.* **293**, 1195–1210.

99 PARKER, M. J. & MARQUSEE, S. (2000). A statistical appraisal of native state hydrogen exchange data: evidence for a burst phase continuum? *J. Mol. Biol.* **300**, 1361–1375.

100 SWINT-KRUSE, L. & ROBERTSON, A. D. (1995). Hydrogen bonds and the pH dependence of ovomucoid third domain stability. *Biochemistry* **34**, 4724–4732.

101 NEIRA, J. L. & FERSHT, A. R. (1999). Exploring the folding funnel of a polypeptide chain by biophysical studies on protein fragments. *J. Mol. Biol.* **285**, 1309–1333.

102 PERRETT, S., CLARKE, J., HOUNSLOW, A. M. & FERSHT, A. R. (1995). Relationship between equilibrium amide proton exchange behavior and the folding pathway of barnase. *Biochemistry* **34**, 9288–9298.

103 PARKER, M. J. & CLARKE, A. R. (1997). Amide backbone and water-related H/D isotope effects on the dynamics of a protein folding reaction. *Biochemistry* **36**, 5786–5794.

104 MAKHATADZE, G. I. & PRIVALOV, P. L. (1992). Protein interactions with urea and guanidinium chloride. A calorimetric study. *J. Mol. Biol.* **226**, 491–505.

105 JOHNSON, C. M. & FERSHT, A. R. (1995). Protein stability as a function of denaturant concentration: ther thermal stability of barnase in the presence of urea. *Biochemistry* **34**, 6795–6804.

106 SHORTLE, D., MEEKER, A. K. & GERRING, S. L. (1989). Effects of denaturants at low concentrations on the reversible denaturation of staphylococcal nuclease. *Arch. Biochem. Biophys.* **272**, 103–113.

107 NOZAKI, Y. & TANFORD, C. (1970). The solubility of amino acids, diglycine, and triglycine in aqueous guanidine hydrochloride solutions. *J. Biol. Chem.* **245**, 1648–1652.

108 BREMS, D. N. & BALDWIN, R. L. (1985). Protection of amide protons in folding intermediates of ribonuclease A measured by pH-pulse exchange curves. *Biochemistry* **24**, 1689–1693.

109 SCHMID, F. X. & BALDWIN, R. L. (1979). Detection of an early intermediate in the folding of ribonuclease A by protection of amide protons against exchange. *J. Mol. Biol.* **135**, 199–215.

110 RODER, H. & WÜTHRICH, K. (1986). Protein folding kinetics by combined use of rapid mixing techniques and NMR observation of individual amide protons. *Proteins* **1**, 34–42.

111 BIERI, O. & KIEFHABER, T. (2001). Origin of apparent fast and non-exponential kinetics of lysozyme folding measured in pulsed hydrogen exchange experiments. *J. Mol. Biol.* **310**, 919–935.

112 UDGAONKAR, J. B. & BALDWIN, R. L. (1988). NMR evidence for an early framework intermediate on the folding pathway of ribonuclease A. *Nature* **335**, 694–699.

113 ELÖVE, G. A. & RODER, H. (1991). Structure and Stability of Cytochrome c Folding Intermediates. In *Protein Refolding* (GEORGIOU, G. & DE BERNARDEZ-CLARK, E., eds), pp. 50–63, American Chemical Society, Washington.

114 HOURY, W. A., SAUDER, J. M., RODER, H. & SCHERAGA, H. A. (1998). Definition of amide protection factors for early kinetic intermediates in protein folding. *Proc. Natl. Acad. Sci. USA* **95**, 4299–4302.

115 HOURY, W. A. & SCHERAGA, H. A. (1996). Sturcture of a Hydrophobically Collapsed Intermediate in the Conformational Folding Pathway of Ribonuclease A Probed by Hydrogen-

Deuterium Exchange. *Biochemistry* **35**, 11734–11746.
116. SHASTRY, M. C. R., SAUDER, J. M. & RODER, H. (1998). Kinetic and Structural Analysis of Submillisecond Folding Events in Cytochrome *c*. *Acc. Chem. Res.* **31**, 725.
117. SHASTRY, M. C. & RODER, H. (1998). Evidence for barrier-limited protein folding kinetics on the microsecond time scale. *Nat. Struct. Biol.* **5**, 385–392.
118. ITZHAKI, L. S., EVANS, P. A., DOBSON, C. M. & RADFORD, S. E. (1994). Tertiary interactions in the folding pathway of hen lysozyme: kinetic studies using fluorescent probes. *Biochemistry* **33**, 5212–5220.
119. KIEFHABER, T. (1995). Kinetic traps in lysozyme folding. *Proc. Natl. Acad. Sci. USA* **92**, 9029–9033.
120. NISHIMURA, C., PRYTULLA, S., JANE, D. H. & WRIGHT, P. E. (2000). Conservation of folding pathways in evolutionarily distant globin sequences. *Nat. Struct. Biol.* **7**, 679–686.
121. LARSON, S. M., RUCZINSKI, I., DAVIDSON, A. R., BAKER, D. & PLAXCO, K. W. (2002). Residues participating in the protein folding nucleus do not exhibit preferential evolutionary conservation. *J. Mol. Biol.* **316**, 225–233.
122. PLAXCO, K. W., LARSON, S., RUCZINSKI, I., RIDDLE, D. S., THAYER, E. C., BUCHWITZ, B., DAVIDSON, A. R. & BAKER, D. (2000). Evolutionary conservation in protein folding kinetics. *J. Mol. Biol.* **298**, 303–312.
123. MIRNY, L. A. & SHAKHNOVICH, E. I. (1999). Universally conserved positions in protein folds: reading evolutionary signals about stability, folding kinetics and function. *J. Mol. Biol.* **291**, 177–196.
124. KRAGELUND, B. B., POULSEN, K., ANDERSEN, K. V., BALDURSSON, T., KRØLL, J. B., NEERGÅRD, T. B., JEPSEN, J., ROEPSTORFF, P., KRISTIANSEN, K., POULSEN, F. M. & KNUDSEN, J. (1999). Conserved residues and their role in the structure, function, and stability of acyl-coenzyme A binding protein. *Biochemistry* **38**, 2386–2394.
125. KUWATA, K., SHASTRY, R., CHENG, H., HOSHINO, M., BATT, C. A., GOTO, Y. & RODER, H. (2001). Structural and kinetic characterization of early folding events in beta-lactoglobulin. *Nat. Struct. Biol.* **8**, 151–155.
126. HAMADA, D., SEGAWA, S. & GOTO, Y. (1996). Non-native alpha-helical intermediate in the refolding of beta-lactoglobulin, a predominantly beta-sheet protein. *Nat. Struct. Biol.* **3**, 868–873.
127. UDGAONKAR, J. B. & BALDWIN, R. L. (1990). Early folding intermediate of ribonuclease A. *Proc. Natl. Acad. Sci. USA* **87**, 8197–8201.
128. BYCROFT, M., MATOUSCHEK, A., KELLIS, J. T., SERRANO, L. & FERSHT, A. R. (1990). Detection and characterization of a folding intermediate in barnase by NMR. *Nature* **346**, 488–490.
129. RASCHKE, T. M. & MARQUSEE, S. (1997). The kinetic folding intermediate of ribonuclease H resembles the acid molten globule and partially unfolded molecules detected under native conditions. *Nat. Struct. Biol.* **4**, 298–304.
130. HOLLIEN, J. & MARQUSEE, S. (2002). Comparison of the folding processes of T. thermophilus and E. coli ribonucleases H. *J. Mol. Biol.* **316**, 327–340.
131. KUSZEWSKI, J., CLORE, G. M. & GRONENBORN, A. M. (1994). Fast folding of a prototypic polypeptide: the immunoglobulin binding domain of streptococcal protein G. *Protein Sci.* **3**, 1945–1952.
132. TEILUM, K., KRAGELUND, B. B., KNUDSEN, J. & POULSEN, F. M. (2000). Formation of hydrogen bonds precedes the rate-limiting formation of persistent structure in the folding of ACBP. *J. Mol. Biol.* **301**, 1307–1314.
133. NOZAKI, Y. (1972). The Preparation of Guanidine Hydrochloride. *Methods Enzymol.* **26**, 43–50.

134 PACE, C. N. (1986). Determination and analysis of urea and guanidine hydrochloride denaturation curves. *Methods Enzymol.* **131**, 266–280.

135 THOMSEN, J. K. (2001). Ph.D. Thesis, University of Copenhagen.

136 CONNELLY, G. P., BAI, Y., JENG, M.-F. & ENGLANDER, S. W. (1993). Isotope Effects in Peptide Group Hydrogen Exchange. *Proteins* **17**, 87–92.

137 LOFTUS, D., GBENLE, G. O., KIM, P. S. & BALDWIN, R. L. (1986). Effects of denaturants on amide proton exchange rates: a test for structure in protein fragments and folding intermediates. *Biochemistry* **25**, 1428–1436.

19
Studying Protein Folding and Aggregation by Laser Light Scattering

Klaus Gast and Andreas J. Modler

19.1
Introduction

The term laser light scattering is used here for recent developments in both static light scattering (SLS) and dynamic light scattering (DLS). In principle, DLS can be performed with conventional light sources [1], however, laser light is mandatory in practice, and recent progress in laser technology has greatly improved the capability of both DLS and SLS.

Static or "classical" light scattering, which attains molecular parameters from the angular and concentration dependence of the time-averaged scattering intensity, was one of the main tools for the determination of molecular mass in the middle of the last century. However by the 1970s the method had somewhat lost its significance when it became possible to measure the molecular mass of proteins and other small biological macromolecules with high precision by electrospray ionization (ESI) or matrix-assisted laser desorption/ionization (MALDI) mass spectrometry. Furthermore these alternative methods required less material and less careful preparation of the samples. SLS is now restoring its popularity due mostly to significant technical improvements. Well-calibrated instruments equipped with lasers and sensitive solid-state detectors allow precise measurements over a wide range of particle masses (10^3–10^8 kDa) using sample volumes of the order of 10 μL. The coupling with separation devices (e.g., size exclusion or other chromatographic techniques and field-flow fractionation) has provided the biomolecular community with a powerful molecular analyzer, particularly for polydispersed materials.

The advent of DLS in the early 1970s led to great expectations concerning the study of dynamic processes in solution. A readily measurable quantity by this procedure is the translational diffusion coefficient of macromolecules from which the hydrodynamic Stokes radius R_S can be calculated. Up to the 1990s, however, the method appeared to be experimentally much more difficult than SLS. Laboratory-built or commercially available DLS photometers consisted of gas lasers, expensive, huge correlation electronics and subsequent data evaluation was tedious because of severe requirements on computation: powerful online computers were not gener-

Protein Folding Handbook. Part I. Edited by J. Buchner and T. Kiefhaber
Copyright © 2005 WILEY-VCH Verlag GmbH & Co. KGaA, Weinheim
ISBN: 3-527-30784-2

ally available. The first "compact" devices (DLS-based particle sizers) were only applicable to strongly scattering systems. This situation has fundamentally changed during the last 10 years. A sophisticated single-board correlator, which can be easily plugged into a desktop computer, is essentially the only additional element needed to upgrade an SLS to a DLS instrument.

As a consequence of these developments, laser light scattering can now be considered as a standard laboratory method. Several compact instruments are on the market, which are valuable tools for protein analysis. Nevertheless, for special applications, including also particular studies of protein folding and aggregation, a more flexible laboratory-built set-up assembled of commercially available components is preferred.

We would like to stress that laser light scattering investigations are entirely non-invasive and can be done under any solvent conditions. However, the measurements must be done with solutions that are free of undesired large particles and bubbles. Particular care has to be taken with proteins that absorb light in the visible region. It is worth mentioning that SLS measurements can also be performed using conventional fluorescence spectrometers including stopped-flow fluorescence devices in the case of kinetic experiments.

Important information in the case of protein folding studies can be obtained especially from DLS, which yields the molecular dimensions of proteins in terms of the Stokes radius. However, the combined application of SLS and DLS is very useful for distinguishing changes in the dimensions of individual protein molecules due to molecular folding or association.

In this review we will first deal with the measurement of hydrodynamic dimensions, intermolecular interactions, and the state of association of proteins in differently folded states. The application of kinetic laser light scattering to the study of folding kinetics is also discussed. A special section is dedicated to the study of protein aggregation accompanying misfolding. Useful protocols for some standard applications of light scattering to protein folding studies are listed in Section 19.6.

19.2
Basic Principles of Laser Light Scattering

19.2.1
Light Scattering by Macromolecular Solutions

Much of the experimental methodology, skilful practice, and basic theoretical approaches for static light scattering had already been developed in the first half of the last century. The birth of dynamic light scattering can be dated to the middle of the 1960s, and the next decade yielded much of the fundamental groundwork on the principles and practice of DLS from researchers in the fields of physics, macromolecular biophysics, biochemistry, and biology. This short introduction to both methods is an attempt to supply fellow researchers who are not familiar with these techniques with an intuitive understanding of the basic principles and the applicability to protein folding and aggregation problems. We will therefore only recall some of the more fundamental equations. A more detailed understanding can be achieved on the basis of numerous excellent textbooks (see, for example, Refs [2–7]).

The theory of light scattering can be approached from the so-called single-particle analysis or density fluctuation viewpoint. The single-particle analysis approach, which will be used here, is simpler to visualize and is adequate for studying the structure and dynamics of macromolecules in diluted solutions. The fluctuation approach is appropriate for studying light scattering in liquids. Furthermore, it is useful to distinguish light scattering from small particles and molecules (of $d < \lambda/20$) and large particles, whose maximum dimension, d, are comparable to or larger than the incident wavelength λ (the wavelength of blue-green light is about 500 nm). The former case is easier to handle, but much information about the structure is not accessible. Proteins used for folding studies entirely fulfill the conditions of the first case, provided we exclude large protein complexes or aggregates. The concepts of light scattering from small particles in vacuum can be applied to the scattering of macromolecules in solution by considering the excess of different quantities (e.g., light scattering, polarizability, refractive index) of macromolecules over that of the solvent.

The strength of the scattering effect primarily depends on the polarizibility α of a small scattering element, which might be the molecule itself if it is sufficiently small. Larger molecules are considered as consisting of several scattering elements. The oscillating electric field vector $\vec{E}_0(t, \vec{r}) = \vec{E}_0 \cdot e^{-i(\omega t - \vec{k}_0 \cdot \vec{r})}$ of the incident light beam induces a small oscillating dipole with the dipole moment $\vec{p}(t) = \alpha \cdot \vec{E}(t)$. ω is the angular frequency and $\vec{k}_0$ is the wave vector with the magnitude $|\vec{k}_0| = 2\pi \cdot n_0/\lambda$. This oscillating dipole re-emits electromagnetic radiation, which has the same wavelength in the case of elastic scattering. The intensity of scattered light at distance r is

$$I_S = \frac{4\pi^4 n_0^4}{\lambda^4} \cdot \frac{\alpha^2}{r^2} \cdot I_0 \tag{1}$$

where λ is the wavelength in vacuum, n_0 is the refractive index of the surrounding medium and $I_0 = |\vec{E}_0|^2$ is the intensity of the incident vertically polarized laser beam. Equation (1) differs only by a constant factor for the case of unpolarized light. In general, the scattered light is detected at an angle θ with respect to the incident beam in the plane perpendicular to the polarization of the beam (Figure 19.1). The wave vector pointing in this direction is $\vec{k}_S$, where $\vec{k}_0$ and $\vec{k}_S$ have the same magnitude.

Of particular importance is the vector difference $\vec{q} = \vec{k}_0 - \vec{k}_S$, the so-called scattering vector, which determines the spatial distribution of the phases $\phi_i = \vec{q} \cdot \vec{r}_i$ of the scattered light wave emitted by individual scattering elements i (Figure 19.1). The magnitude of $\vec{q}$ is $q = (4\pi \cdot n/\lambda) \sin(\theta/2)$. The phases play an essential role in the total instantaneous intensity, which results from the superposition of light waves emitted by all scattering elements within the scattering volume v defined by

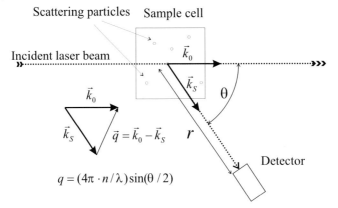

Fig. 19.1. Schematic diagram of the optical part of an light scattering instrument.

the primary beam and the aperture of the detector. The instantaneous intensity fluctuates in time for nonfixed particles, like macromolecules in solution, due to phase fluctuations $\phi_i(t) = \vec{q} \cdot \vec{r}_i(t)$ caused by changes in their location $\vec{r}_i(t)$.

SLS measures the time average of the intensity, thus the term "time-averaged light scattering" would be more appropriate, although "static light scattering" appears the accepted convention now. SLS becomes q-dependent for large particles when light waves emitted from scattering elements within a individual particle have noticeable phase differences.

DLS analyses the above-mentioned temporal fluctuations of the instantaneous intensity of scattered light. Accordingly, DLS can measure several dynamic processes in solution. It is evident that only those changes in the location of scattering elements lead to intensity fluctuations that produce a sufficiently large phase shift. This is always the case for translational diffusion of macromolecules. The motion of segments of chain molecules and rotational motion can be studied for large structures. Rotational motion of monomeric proteins and chain dynamics of unfolded proteins are practically not accessible by DLS.

19.2.2
Molecular Parameters Obtained from Static Light Scattering (SLS)

The expression for the light scattering intensity from a macromolecular solution can be derived from Eq. (1) by summation over the contributions of all macromolecules in the scattering volume v and substituting the polarizability α by related physical parameters. The latter can be done by applying the Clausius-Mosotti equation to macromolecular solutions, which leads to $n^2 - n_0^2 = 4\pi \cdot N' \cdot \alpha$, where n and n_0 are the refractive indices of the solution and the solvent, respectively. N' is the number of particles (molecules) per unit volume and can be expressed by $N' = N_A \cdot c/M$. N_A is Avogadro's number, c and M are the weight concentration and the molecular mass (molar mass) of the macromolecules, respectively. Using the approximation $n + n_0 \sim 2n_0$ we get $n^2 - n_0^2 \sim 2n_0 \cdot (\partial n/\partial c) \cdot c$. $\partial n/\partial c$ is the specific refractive index increment of the macromolecules in the particular solvent. The excess scattering of the solution over that of the solvent $I_{ex} = I_{solution} - I_{solvent}$ of noninteracting small molecules is then

$$I_{ex} = \frac{4\pi^2 n_0^2 (\partial n/\partial c)^2}{\lambda^4 N_A} \cdot \frac{v}{r^2} \cdot c \cdot M \cdot I_0 = H \cdot \frac{v}{r^2} \cdot c \cdot M \cdot I_0 \tag{2}$$

The optical constant $H = 4\pi^2 n_0^2 (\partial n/\partial c)^2/\lambda^4 N_A$ depends only on experimental parameters and the scattering properties of the molecules in the particular solvent, which is reflected by $\partial n/\partial c$. The exact knowledge of $\partial n/\partial c$, the dependence of n on protein concentration in the present case, is very important for absolute measurements of the molecular mass. For proteins in aqueous solvents of low ionic strength it is about 0.19 cm^3 g^{-1} and practically does not depend significantly on the amino acid sequence. However, it is markedly different in solvents containing high concentrations of denaturants.

The instrument parameters in Eq. (2) can be eliminated by using the Rayleigh (excess) ratio $R_q = (I_{ex}/I_0)(r^2/v)$. In practice, R_q of an unknown sample is calculated from the scattering intensities of the sample and that of a reference sample of known Rayleigh ratio R_{ref} by

$$R_q = R_{ref} \cdot f_{corr} \cdot (I_{ex}/I_{ref}) \tag{3}$$

where f_{corr} is an experimental correction factor accounting for differences in the refractive indices of sample and reference sample [5]. In the more general case including intermolecular interactions and large molecules, which have a refractive index n_p and satisfy the condition $4\pi(n_p - n_0)d/\lambda \ll 1$ of the Rayleigh-Debye approximation, static light scattering data are conveniently presented by the relation

$$\frac{H \cdot c}{R_q} = \frac{1}{M \cdot P(q)} + 2A_2 \cdot c \tag{4}$$

$P(q)$ is the particle scattering function, which is mainly expressed in terms of the product $q \cdot \bar{R}_G$, where $\bar{R}_G = \langle R^2 \rangle^{1/2}$ is the root mean square radius of gyration of the particles. Analytical expressions for $P(q)$ are known for different particle shapes, whereas instead of R_G also other characteristic size-dependent parameters are used (e.g., length L for rods or cylinders). In the limit $q \cdot \bar{R}_G \ll 1$, the approximation $P(q)^{-1} = 1 + (q \cdot \bar{R}_G)^2/3$ can be used to estimate $\bar{R}_G$ from the angular dependence of R_q. A perceptible angular dependence of the scattering intensity can only be expected for particles with $R_G > 10$ nm. Thus, light scattering is not an appropriate method to monitor changes of R_G during unfolding and refolding of small monomeric proteins. A substantial angular dependence is observed for large protein aggregates, however.

The concentration dependence of the right hand side of Eq. (4) yields the second virial coefficient A_2, which reflect the strength and the type of intermolecular interactions. The usefulness of measuring A_2 will be discussed below. A_2 is positive for predominant repulsive (covolume and electrolyte effects) and negative for predominant attractive intermolecular interactions.

In general, both extrapolation to zero concentration and zero scattering angle ($q = 0$) are done in a single diagram (Zimm plot) for calculations of M from Eq. (4). The primary mass "moment" or "average" obtained is the weight-average molar mass $M = \sum_i c_i M_i / \sum_i c_i$ in the case of polydisperse systems. This has to be taken into consideration for proteins when monomers and oligomers are present. Molar masses of proteins used in folding studies (i.e., $M < 50\,000$ g mol^{-1}) can be determined at 90° scattering angle because the angular dependence of R_q is negligible. Measurements at different concentration are mandatory, however, because remarkable electrostatic and hydrophobic interactions may exist under particular environmental conditions. In the following, the values of parameters measured at finite concentration are termed apparent values, e.g., M_{app}.

19.2.3
Molecular Parameters Obtained from Dynamic Light Scattering (DLS)

Information about dynamic processes in solution are primarily contained in the temporal fluctuations of scattered electric field $\vec{E}_s(t)$. The time characteristics of these fluctuations can be described by the first-order time autocorrelation function $g^{(1)}(\tau) = \langle \vec{E}_s(t) \cdot \vec{E}_s(t+\tau) \rangle$. The brackets denote an average over many products of $\vec{E}_s(t)$ with its value after a delay time τ. $g^{(1)}(\tau)$ is only accessible in the heterodyne detection mode, where the scattered light is mixed with a small portion of the incident beam on the optical detector. This experimentally complicated detection method must be used if particle motions relative to the laboratory frame, e.g., the electrophoretic mobility in an external electric field, are measured. The less complicated homodyne mode, where only the scattered light intensity is detected, is normally the preferred optical scheme. This has the consequence that only the second-order intensity correlation function $g^{(2)}(\tau) = \langle I(t) \cdot I(t+\tau) \rangle$ is directly available. Under particular conditions, which are met in the case of light scattering from dilute solutions of macromolecules, the Siegert relation $g^{(2)}(\tau) = 1 + |g^{(1)}(\tau)|^2$ can be used to obtain the normalized first-order correlation function $g^{(1)}(\tau)$ from the measured $g^{(2)}(\tau)$. Analytical forms of $g^{(1)}(\tau)$ have been derived for different dynamic processes in solution. As we have already indicated above only translational diffusional motion essentially contributes to the fluctuations of the scattered light in the case of monomeric proteins and we can reasonably neglect rotational effects. $g^{(1)}(\tau)$ for identical particles with a translational diffusion coefficient D has the form of an exponential

$$g^{(1)}(\tau) = e^{-q^2 \cdot D \cdot \tau} \tag{5}$$

D is related to the hydrodynamic Stokes radius R_S by the Stokes-Einstein equation

$$R_S = \frac{k \cdot T}{6\pi \cdot \eta \cdot D} \tag{6}$$

where k is Boltzmann's constant, T is the temperature in K, and η is the solvent viscosity. $g^{(1)}(\tau)$ for a polydisperse solution containing L different macromolecular species (or aggregates) with masses M_i, diffusion coefficients D_i and weight concentrations c_i is

$$g^{(1)}(\tau) = \left(\sum_{i=1}^{L} a_i \cdot e^{-q^2 \cdot D_i \cdot \tau}\right) \Big/ S \tag{7a}$$

where $S = \sum_{i=1}^{L} a_i$ is the normalization factor. The weights $a_i = c_i \cdot M_i = n_i \cdot M_i^2$ reflect the $c \times M$ dependence of the scattered intensity (see Eq. (2)). n_i is the number concentration (molar concentration) of the macromolecular species. Accordingly, even small amounts of large particles are considerably represented in the measured $g^{(1)}(\tau)$. The general case of an arbitrary size distribution, which results in a distribution of D, can be treated by an integral

$$g(\tau) = \left(\int a(D) \cdot e^{-q^2 \cdot D \cdot \tau} \, dD\right) \Big/ S \tag{7b}$$

Equation (7b) has the mathematical form of a Laplace transformation of the distribution function $a(D)$. Thus, an inverse Laplace transformation is needed to reconstruct $a(D)$ or the related distribution functions $c(D)$ and $n(D)$. This is an ill-conditioned problem from the mathematical point of view because of the experimental noise in the measured correlation function. However, there exist numerical procedures termed "regularization", which allow a researcher to obtain stabilized, "smoothed" solutions. A widely used program package for this purpose is "CONTIN" [8]. Nevertheless, the distributions obtained can depend sensitively on the experimental noise and parameters used during data evaluation procedure in special cases. Thus, it might be more appropriate to use simpler but more stable data evaluation schemes like the method of cumulants [9], which yields the z-averaged diffusion coefficient $\bar{D}$ and higher moments reflecting the width and asymmetry of the distribution. $\bar{D}$ can be obtained simply from the limiting slope of the logarithm of $g^{(1)}(\tau)$, viz. $\bar{D} = -q^{-2} \cdot \frac{d}{d\tau}(\ln|g^{(1)}(\tau)|)_{\tau \to 0}$. This approach is very useful for rather narrow distributions.

D, like M, exhibits a concentration dependence, which is usually written in the form

$$D(c) = D_0(1 + k_D \cdot c) \tag{8}$$

where k_D is the diffusive concentration dependence coefficient, which can be used to characterize intermolecular interactions. However, k_D differs from A_2. The concentration dependence of D and other macromolecular parameters was discussed in more detail by Harding and Johnson [10]. Extrapolation to zero protein concentration yielding D_0 is essential in order to calculate the hydrodynamic dimensions in terms of R_S for individual protein molecules.

19.2.4
Advantages of Combined SLS and DLS Experiments

Many of the modern laser light scattering instruments allow both SLS and DLS to be measured in one and the same experiment. This experimental procedure is more complicated, since the optimum optical schemes for SLS and DLS are different. Briefly, DLS needs focused laser beams and only a much smaller detection aperture can be employed because of the required spatial coherence of the scattered light. This reduces the SLS signal and demands higher beam stability.

Combined SLS/DLS methods have two main advances for protein folding studies. The first one concerns the reliability of measurements of the hydrodynamic dimensions during folding or unfolding. The observed changes of R_S could partly or entirely result from an accompanying aggregation reaction, which can be excluded if the molecular mass remains essentially constant.

The second advantage is the capacity to measure molecular masses of proteins in imperfectly clarified solutions that contain protein monomers and an unavoidable amount of aggregates. This situation is often met, for example during a folding reaction or when the amount of protein is too small for an appropriate purification procedure. In this case, SLS alone would measure a meaningless weight average mass. The Stokes radii of monomers and aggregates are mostly well separated, allowing the researcher to estimate the weighting a_i for monomers and aggregates fitting the measured $g^{(1)}(\tau)$ by Eq. (7a). Knowledge of the weighting allows the contributions of monomers and aggregates to the total scattering intensity to be distinguished and, therefore, the proper molecular mass of the protein to be obtained.

19.3
Laser Light Scattering of Proteins in Different Conformational States – Equilibrium Folding/Unfolding Transitions

19.3.1
General Considerations, Hydrodynamic Dimensions in the Natively Folded State

The molecular parameter of interest in folding studies is primarily the hydrodynamic Stokes radius R_S. Additional direct estimations of the molecular mass from SLS data are recommended to check whether the protein is and remains in the monomeric state during folding or unfolding. Formation of a dimer of two globular subunits could easily be misinterpreted as a swollen monomer since its Stokes radius is only larger by a factor of 1.39 than that of the monomer [11]. Hydrodynamic radii in the native state have been measured for many proteins. For globular proteins, a good correlation with the mass or the number of amino acids was found [12] (see also below). This has encouraged some researchers and manufacturers of commercial DLS devices to estimate M from R_S. This procedure has to be applied with great care and clearly contradicts the application of DLS to folding studies.

For correct estimations of molecular dimensions in terms of R_S before and after a folding transition, extrapolation of D to zero protein concentration should be done before applying Eq. (6). The concentration dependence of D indicated by k_D possibly changes during unfolding/refolding transitions as it is shown in Figure 19.2 for thermal unfolding of RNase T1. Strong changes in k_D are often observed

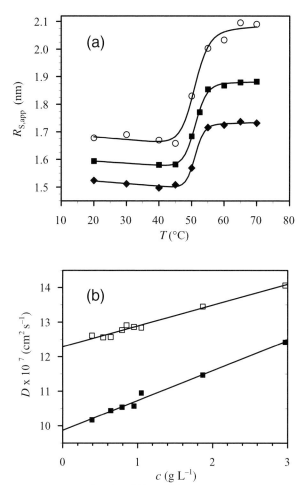

Fig. 19.2. Heat-induced unfolding of ribonuclease T1 in 10 mM sodium cacodylate, pH 7, 1 mM EDTA. a) Temperature dependence of the apparent Stokes radius at concentrations of 0.8 mg mL^{-1} (open circles), 1.9 mg mL^{-1} (filled squares), and 3.9 mg mL^{-1} (filled diamonds). The continuous lines were obtained by nonlinear least squares fits according to a two-state transition yielding $T_m = 51.0 \pm 0.5\ °C$ and $\Delta H_m = 497 \pm 120$ kJ mol^{-1} (average over all three fits). b) Concentration dependence of the diffusion coefficient D at 20 °C (open square) and 60 °C (filled square). Linear fits to the data yield the diffusive concentration dependence coefficient k_D (see text).

in the case of pH-induced denaturation. Apparent Stokes radii, $R_{S,app}$, are measured if extrapolation is not possible due to certain experimental limitations.

19.3.2
Changes in the Hydrodynamic Dimensions during Heat-induced Unfolding

The first DLS studies of the thermal denaturation of proteins were reported by Nicoli and Benedek [13]. These authors investigated heat-induced unfolding of lysozyme in the pH range between 1.2 and 2.3 and different ionic strength. The size transition coincided with the unfolding curve recorded by optical spectroscopic probes. The average radius increased by 18% from 1.85 to 2.18 nm. The same size increase was also found for ribonuclease A (RNase A) at high ionic strength. The proteins had intact disulfide bonds.

A good candidate for thermal unfolding/refolding studies is RNase T1, since the transition is completely reversible even according to light scattering criteria [14]. The unfolding transition curves measured at three different protein concentrations at pH 7 are shown in Figure 19.2a. The concentration dependence of D in the native and unfolded states is shown in Figure 19.2b. The increase of the diffusive concentration dependence coefficient from 49 to 87 mL g^{-1} on the transition to the unfolded state indicates a strengthening of the repulsive intermolecular interactions. The Stokes radii obtained after extrapolation to zero protein concentration are 1.74 nm and 2.16 nm for the native and unfolded states, respectively. The corresponding increase in R_S of 24% is somewhat larger than those for lysozyme and RNase A. This might be due to the fact that RNase T1 differs in the number and position of the disulfide bonds as compared with lysozyme and RNase A.

The thermal unfolding behavior of the wild-type λ Cro repressor (Cro-WT) is more complex, but it allows us to demonstrate the advantage of combined SLS/DLS experiments. The native Cro-WT is active in the dimeric form. Spectroscopic and calorimetric studies [15] (and references cited therein) revealed that thermal unfolding proceeds via an intermediate state. The corresponding changes of the apparent Stokes radius and the relative scattering intensity, which reflects changes in the average mass, are quite different at low and high concentrations (Figure 19.3). At low concentration, Cro-WT first unfolds partly, but remains in the dimeric form between 40 °C and 55 °C according to the increase in R_S between 40 °C and 55 °C and the nearly constant scattering intensity. Above 55 °C, both R_S and the relative scattering intensity decrease due to the dissociation of the dimer. At high concentration, the increase of R_S during the first unfolding step is much stronger and is accompanied by an increase in the scattering intensity. This is due to the formation of dimers of partly unfolded dimers. At temperatures above 55 °C, dissociation into monomers is indicated by the decrease of both R_S and scattering intensity. The transient population of a tetramer is a peculiarity of thermal unfolding of Cro-WT.

Unfortunately, many proteins aggregate upon heat denaturation, thus preventing reliable measurements of the dimensions.

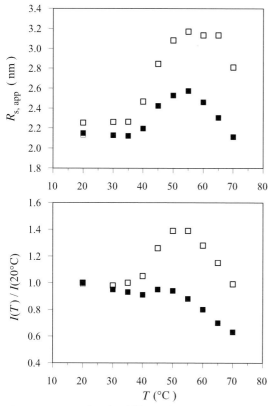

Fig. 19.3. Heat-induced unfolding and association/dissociation of Cro repressor wild-type at 1.8 mg mL^{-1} (filled squares) and 5.8 mg mL^{-1} (open squares) in 10 mM sodium cacodylate, pH 5.5.

19.3.3
Changes in the Hydrodynamic Dimensions upon Cold Denaturation

Since the stability curves for proteins have a maximum at a characteristic temperature, unfolding in the cold is a general phenomenon [16]. However, easily attainable unfolding conditions in the cold exist only for a few proteins. Cold denaturation under destabilizing conditions was reported for phosphoglycerate kinase from yeast (PGK) [17, 18] and barstar [19, 20]. The increases in R_S upon cold denaturation for PGK [18] and barstar (unpublished results) are shown in Figure 19.4. The size increase is a three-state transition in the case of PGK, because of the independent unfolding of the two domains in the cold. Both proteins aggregate on heating preventing to compare the results with those for heat-induced unfolding.

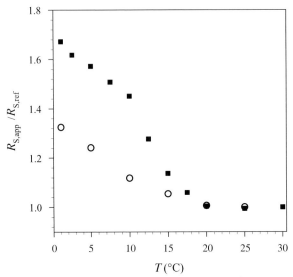

Fig. 19.4. Relative expansion, $R_{S,app}/R_{S,ref}$, upon cold denaturation for barstar (pseudo wild-type), 2.4 mg mL^{-1}, in 50 mM Tris/HCl, pH 8, 0.1 M KCl, 2.2 M urea (open circles) and phosphoglycerate kinase from yeast, 0.97 mg mL^{-1}, in 20 mM sodium phosphate, pH 6.5, 10 mM EDTA, 1 mM DTT, 0.7 M GdmCl (filled squares). $R_{S,ref}$ is the Stokes radius in the reference state at the temperature of maximum stability under slightly destabilizing conditions.

19.3.4
Denaturant-induced Changes of the Hydrodynamic Dimensions

Before the advent of DLS, most data relating to the hydrodynamic dimensions of proteins under strongly denaturing conditions had been obtained by measuring intrinsic viscosities [21]. In a pioneering DLS experiment on protein denaturation, Dubin, Feher, and Benedek [22] studied the unfolding transition of lysozyme in guanidinium chloride (GdmCl). These authors meticulously analyzed the transition at 31 GdmCl concentrations between 0 M and 6.4 M in 100 mM acetate buffer, pH 4.2, at a protein concentration of 10 mg mL^{-1}. It was found that $R_{S,app}$ increases during unfolding by 45% and 86% in the case of intact and reduced disulfide bonds, respectively. Meanwhile, the hydrodynamic dimensions at high concentrations of GdmCl have been measured for many proteins lacking disulfide bonds by using DLS. The results will be discussed below in connection with scaling (i.e., power) laws, which can be derived from this data. Here we consider briefly the influence of disulfide bonds and temperature on the dimensions of proteins in the highly unfolded state. Such experiments were done with RNase A [23]. Figure 19.5 shows the temperature dependence of the relative dimensions ($R_{S,app}/R_{S,native}$) for RNase A with intact and with reduced disulfide bonds. The slight compaction with increasing temperature is typical for proteins in highly unfolded states: for example, similar results are obtained with RNase T1 (Figure 19.5). Surprisingly, RNase

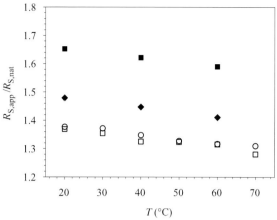

Fig. 19.5. Temperature dependence of the relative expansion $R_{S,app}/R_{S,nat}$ of RNase A and RNase T1 in the unfolded state at high concentrations of GdmCl. RNase T1, 2.2 mg mL^{-1}, in 10 mM sodium cacodylate, pH 7, 1 mM EDTA, 5.3 M GdmCl (open circles), RNase A, 2.9 mg mL^{-1}, in 50 mM MES, pH 5.7, 1 mM EDTA, 6 M GdmCl (open squares), and RNase A (broken disulfide bonds), 2.9 mg mL^{-1}, in 50 mM MES, pH 6.5, 1 mM EDTA, 6 M GdmCl (filled squares). RNase A, 2.5 mg mL^{-1}, with broken disulfide bonds is highly unfolded even in the absence of GdmCl (filled diamonds).

A without disulfide bonds is already unfolded in the absence of GdmCl and has larger dimensions than RNase A with intact disulfide bonds in 6 M GdmCl.

19.3.5
Acid-induced Changes of the Hydrodynamic Dimensions

Many proteins can be denatured by extremes of pH, and most of the studies on this have focused on using acid as denaturant. Proteins respond very differently to acidic pH. For example, lysozyme remains native-like, some other protein adopt the molten globule conformation with hydrodynamic dimensions about 10% larger than in the native state, and many proteins unfold into an expanded conformation. Fink et al. [24] have introduced a classification scheme for the unfolding behavior under acidic conditions. The molecular mechanisms of acid denaturation have been studied in the case of apomyoglobin by Baldwin and co-workers [25]. The dimensions of selected acid-denatured proteins are listed below in Table 19.1. Some proteins are more expanded in the acid-denatured than in the chemical-denatured state, e.g., PGK ($R_{S,unf}/R_{S,native} = 2.49$) and apomyoglobin ($R_{S,unf}/R_{S,native} = 2.05$). Furthermore, those proteins exhibit strong repulsive intermolecular interactions, as reflected by large values of k_D (e.g., 550 mL g^{-1} for PGK at pH 2). This means, that D_{app} is twice as large as D_0 at a concentration of about 2 mg mL^{-1}. This underlines the importance of extrapolation of the measured quantity to zero protein concentration.

Tab. 19.1. Stokes radii R_S, relative compactness $R_S/R_{S,native}$, and diffusive concentration dependence coefficients k_D of selected proteins in differently folded states determined by DLS.

Protein/state	R_S (nm)	$R_S/R_{S,native}$	k_D (mL g^{-1})	References
Bovine α-lactalbumin				
Native state (holoprotein)	1.88		−7	29
Unfolded state (5 M GdmCl)	2.46	1.31	−9	29
A-state (pH 2, 50 mM KCl)	2.08	1.11	15	29
Molten globule (neutral pH)	2.04	1.09	62	29
Molten globule (15% v/v TFE)	2.11	1.12	−100	33
TFE-state (40% v/v TFE)	2.25	1.20	−10	33
Kinetic molten globule	1.99	1.06	−60	33
PGK				
Native state	2.97		0.8	44
Unfolded state (2 M GdmCl)	5.66	1.91	–	44
Cold denatured state (1 °C, 0.7 M GdmCl)	5.10	1.72	–	44
Acid denatured state (pH 2)	7.42	2.50	552	34
TFE-state (50% v/v TFE, pH 2)	7.76	2.61	1030	34
Apomyoglobin				
Native state	2.09		26	30
Acid denatured (U-form, pH 2)	4.29	2.05	520	30
Molten globule (I-form, pH 4)	2.53	1.21	104	30
RNase T1				
Native state	1.74		49	14
Unfolded state (5.3 M GdmCl)	2.40	1.38	–	14
Heat denatured (60 °C)	2.16	1.24	87	14
RNase A				
Native state	1.90		≈0	56
Unfolded state (6 M GdmCl, nonreduced SS)	2.60	1.37	–	56
Unfolded state (6 M GdmCl, reduced SS)	3.14	1.65	–	56
TFE-state (70% v/v TFE)	2.28	1.20	15	33
Barstar				
Native state	1.72		5	87
Unfolded state (6 M urea)	2.83	1.65	50	87
Cold-denatured state (1 °C, 2.2 M urea)	2.50	1.45	−3	87

Experimental errors were omitted for brevity. Errors in R_S are typically less than ±2%. Errors in k_D are of the order of ±10% except for k_D values close to zero.
For exact solvent conditions we refer to the original literature.

19.3.6
Dimensions in Partially Folded States – Molten Globules and Fluoroalcohol-induced States

From the wealth of partly folded states existing for many proteins under different destabilizing conditions, DLS data obtained for some particular types will be considered here. Molten globule states [26–28] have been most extensively studied (see Chapter 23). DLS data on the hydrodynamic dimensions were published for the molten globule states of two widely studied proteins, α-lactalbumin [29] and apomyoglobin [30]. The results including data for the kinetic molten globule of α-lactalbumin are shown in Table 19.1. Data also exist on the geometric dimensions in terms of the radius of gyration, R_G, for both α-lactalbumin [31] and apomyoglobin [30, 32] in different conformational states.

Only a few DLS studies on the hydrodynamic dimension of proteins in aqueous solvents containing structure-promoting substances like trifluoroethanol (TFE) and hexafluoroisopropanol (HFIP) have been reported so far. The Stokes radii at high fluoroalcohol content, at which the structure stabilizing effect saturates, are very different for various proteins. An increase in R_S of only about 20% was found for α-lactalbumin and RNase A [33] with intact disulfide bonds. By contrast, PGK at pH 2 and 50% (v/v) TFE has a Stokes radius $R_S = 7.76$ nm exceeding that of the native state by a factor of 2.6 [34]. According to the high helical content estimated from CD data and the high charge according to the large value of k_D it is conceivable that the entire PGK molecule consists of a long flexible helix. Fluoroalcohols may have a different effect at low volume fractions. In the case of α-lactalbumin, a molten globule-like state was found at 15% (v/v) TFE [33]. This state revealed a native-like secondary structure and a Stokes radius 10% larger than that of native protein. Furthermore, strong attractive intermolecular interactions were indicated by a large negative diffusive concentration dependence coefficient, $k_D = -100$ mL g^{-1}, which result presumably from the exposure of hydrophobic patches in this state. k_D was close to zero both in the absence and at high percentage of TFE.

19.3.7
Comparison of the Dimensions of Proteins in Different Conformational States

The hydrodynamic dimensions in different equilibrium states of five proteins frequently used in folding studies are summarized in Table 19.1. Some general rules become evident from these data, e.g., heat-denatured proteins are relatively compact as compared with those denatured by high concentration of denaturants or cold-denatured proteins. In this connection, a particular class of proteins should be mentioned, namely the so-called intrinsically unstructured proteins [35] (see Chapter 8 in Part II). These proteins are essentially unfolded according to their CD spectra and have much larger Stokes radii under physiological conditions than $R_{S,\text{glob}}$ expected for a globular proteins of the same molecular mass. Typical examples are prothymosin α and α-synuclein, with relative dimensions R_S/R_{glob} of 1.77 [36] and 1.52 (unpublished results), respectively.

19.3.8
Scaling Laws for the Native and Highly Unfolded States, Hydrodynamic Modeling

A systematic dependence of the hydrodynamic dimensions on the number of amino acids N_{aa} or molecular mass M can be established for the dimension of globular proteins in the native state and also for proteins lacking disulfide bonds in the unfolded state induced by high concentrations of GdmCl or urea. Scaling (i.e., "power") laws for native globular proteins have been published by several authors. Damaschun et al. [37] obtained the relation $R_S[nm] = 0.362 \cdot N_{aa}^{1/3}$ on the basis of the Stokes radii calculated for more than 50 globular proteins deposited in the protein databank. Uversky [38] found the dependence $R_S[nm] = 0.0557 \cdot M^{0.369}$ using experimental data from the literature and Stokes radii measured with a carefully calibrated size exclusion column (this is consistent with $R_S[nm] = 0.325 \cdot N_{aa}^{0.369}$ for a mean residue weight of 119.4 g mol^{-1}). An advantage of DLS is that no calibration of the methods is needed. The relation between R_S and N_{aa} determined by Wilkins et al. [39] is $R_S[nm] = 0.475 \cdot N_{aa}^{0.29}$. The scaling laws for the native state are useful for the classification of proteins for which the high-resolution structure is not known or cannot be obtained. Deviation from the scaling law indicates that the protein has no globular shape or does not adopt a compactly folded structure under the given environmental conditions.

In contrast to the native state, proteins in unfolded states populate a large ensemble of configurations, which are restricted only by the allowed ϕ, ψ angles and nonlocal interactions of the polypeptide chain. In the absence of (or at balanced) nonlocal interactions (i.e., in the unperturbed or theta state), polypeptide chains behave as random chains and their conformational properties can be described by the rotational isomeric state theory [40]. Whether or not unfolded proteins are indeed true random chains is a matter of debate [41]. The dimension of a denatured protein is an important determinant of the folding thermodynamics and kinetics [42]. Thus, the chain length dependence of the hydrodynamic dimension in the form of scaling laws is an important experimental basis for the characterization of unfolded polypeptide chains.

A first systematic analysis of the chain length dependence of the hydrodynamic dimensions was done by Tanford et al. [43] by measuring the intrinsic viscosities of unfolded proteins. The Stokes radii measured by DLS [44] for 12 proteins denatured by high concentrations of GdmCl are shown in Figure 19.6. From the data the scaling law $R_S[nm] = 0.28 \cdot N_{aa}^{0.5}$ was obtained. Scaling laws were also published by other authors. Uversky [38] derived $R_S[nm] = 0.0286 \cdot M^{0.502}$ (corresponding to $R_S[nm] = 0.315 \cdot N_{aa}^{0.502}$) from Stokes radii measured by size exclusion chromatography. Wilkins et al. [39] found $R_S[nm] = 0.221 \cdot N_{aa}^{0.57}$ using Stokes radii mostly measured by pulse field gradient NMR. By analysing published Stokes radii measured by different methods Zhou [42] obtained the relation $R_S[nm] = 0.2518 \cdot N_{aa}^{0.522}$.

Two aspects that are closely related to measurements of the hydrodynamic dimensions of proteins are hydrodynamic modeling [45, 46] and the hydration problem [47, 48]. Particularly for globular proteins, both are important for the link between high-resolution data from X-ray crystallography or NMR and hydrodynamic data. Protein hydration appears as an adjustable parameter in recent hydrodynamic modeling procedures [49]. The average values over different proteins are 0.3 g water per g protein or in terms of the hydration shell 0.12 nm. Model calculations are also in progress for unfolded proteins [42].

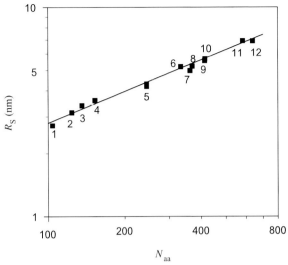

Fig. 19.6. Stokes radii R_S of 12 proteins unfolded by 6 M GdmCl in dependence on the number of amino acid residues N_{aa}. The linear fit yields the scaling law $R_S[nm] = 0.28 \cdot N_{aa}^{0.5}$. 1: apocytchrome c, 2: staphylokinase, 3: RNase A, 4: apomyoglobin, 5: chymotrypsinogen A, 6: glyceraldehydes-3-phosphate dehydrogenase, 7: aldolase, 8: pepsinogen, 9: streptokinase, 10: phosphoglycerate kinase, 11: bovine serum albumin, 12: DnaK. Disulfide bonds, if existing in native proteins, were reduced.

19.4
Studying Folding Kinetics by Laser Light Scattering

19.4.1
General Considerations, Attainable Time Regions

An important characteristic of the folding process is the relation between compaction and structure formation. Therefore, it is intriguing to monitor compactness during folding by measuring the Stokes radius using DLS, which is the fastest among the hydrodynamic methods. Furthermore, R_S and also the radius of gyration, R_G, are direct measures of the overall molecular compactness. The first studies of this kind were measurements of the compaction during lysozyme refolding by continuous-flow DLS [50]. In general, the basic experimental techniques of continuous- and stopped-flow experiments can also be applied for laser light scattering. There are no restrictions from general physical principles concerning the accessible time range for SLS. The data acquisition time T_A, which is needed to obtain a sufficiently high signal-to-noise ratio (S/N), depends solely on the photon flux and can be minimized by increasing the light level. However, there are two fundamental processes that limit the accessible time scales in the case of DLS. These problems are discussed in detail by Gast et al. [51] and will be outlined only briefly in this chapter.

First, unlike other methods including SLS, the enhancement of the signal is not sufficient to increase the (S/N), since the measured signal is itself a statistical quantity. At light levels above the shot-noise limit the signal-to-noise ratio of the measured time correlation function depends on the ratio of T_A to the correlation time τ_C of the diffusion process by $(S/N) = (T_A/\tau_C)^{1/2}$. τ_C is of the order of tens of microseconds for proteins, leading to a total acquisition time $T_A \sim 5$–10 s in order to achieve an acceptable (S/N). This time can be shortened by splitting it into N_S "shots" in a stopped-flow experiment. $N_S = 100$ is an acceptable value leading to acquisition times of 50–100 ms during one "shot." This corresponds to the time resolution in the case of the stopped-flow technique.

The second limiting factor becomes evident when the continuous-flow method is used. Excellent time resolutions have been achieved with this method in the case of fluorescence measurements [52] and small-angle X-ray scattering [53, 54]. Continuous-flow experiments are especially useful for slow methods, since the time resolution depends only on the aging time between mixer and detector and not on the speed of data acquisition. Increasing the flow speed in order to get short aging times also reduces the resting time of the protein molecules within the scattering volume. However, to avoid distortions of the correlation functions, this time must be longer than the correlation time corresponding to the diffusional motion of the molecules. This limits the flow speed to about 15 cm s^{-1} under typical experimental conditions [51], resulting in aging times of about 100 ms. The benefit of continuous-flow measurements over stopped-flow measurements does not exist for DLS. Thus, DLS is only suitable for studying late stages of protein folding in the time range > 100 ms. However, this time range is important for folding and oligomerization or misfolding and aggregation (see Section 19.5). Two examples for the compaction of monomeric protein molecules during folding will be now be considered.

19.4.2
Hydrodynamic Dimensions of the Kinetic Molten Globule of Bovine α-Lactalbumin

The hydrodynamic dimensions of equilibrium molten globule states are well characterized (Table 19.1). It is interesting to compare the dimensions in these states with that of the kinetic counterpart. Stopped-flow DLS investigations [29] were done with the Ca^{2+}-free apoprotein, which folds much slower than the holoprotein. The time course of the changes of the apparent Stokes radius at a protein concentration of 1.35 mg mL^{-1} is shown in Figure 19.7a. Measurements at different protein concentrations (Figure 19.7b) revealed the "sticky" nature of the kinetic molten globule as compared to the completely unfolded and refolded states. The kinetic molten globule appears to be slightly more compact than the equilibrium molten globules. The dimensions of the natively folded apo- and the holoprotein are practically the same. The changes of the dimensions of holo α-lactalbumin were measured later by time-resolved X-ray scattering [55].

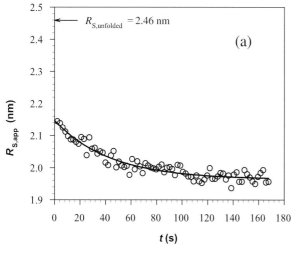

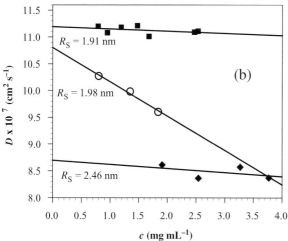

Fig. 19.7. a) Kinetics of compaction of bovine α-lactalbumin after a GdmCl concentration jump from 5 M to 0.5 M, final protein concentration 1.35 mg mL^{-1}, 50 mM sodium cacodylate, pH 7, 50 mM NaCl, 2 mM EGTA, $T = 20\,°C$. The data can be fitted by a single exponential with a decay time $\tau = 43 \pm 6$ s. b) Concentration dependence of the diffusion coefficient D for the apoform of bovine α-lactalbumin under native conditions (filled squares), in the presence of 5 M GdmCl (filled diamonds), and for the kinetic molten globule (open circles). The buffer was same as in (a). The Stokes radii obtained from D extrapolated to zero concentration are shown.

19.4.3
RNase A is Only Weakly Collapsed During the Burst Phase of Folding

In the previous section, it was demonstrated that α-lactalbumin collapses very rapidly into the molten globule with dimensions close to that of the native state. A dif-

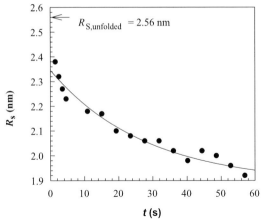

Fig. 19.8. Kinetics of compaction of RNase A after a GdmCl concentration jump from 6 M to 0.67 M, final protein concentration 1.9 mg mL^{-1}, 50 mM sodium cacodylate, pH 6, $T = 10\,°C$. The single exponential fit yields a decay time $\tau = 28 \pm 5$ s.

ferent folding behavior was observed for RNase A [56]. The Stokes radius of RNase is 2.56 nm in the unfolded state at 6 M GdmCl and decreases only to 2.34 nm during the burst phase (Figure 19.8). Most of the compaction towards the native state occurs at later stages in parallel with the final arrangement of secondary structure and the formation of tertiary structure.

19.5
Misfolding and Aggregation Studied by Laser Light Scattering

19.5.1
Overview: Some Typical Light Scattering Studies of Protein Aggregation

Misfolding and aggregation of proteins have been regarded for a long time as undesirable side reactions and the importance of these processes became evident only recently in connection with recombinant protein technology and the observation that the development of particular diseases correlates with protein misfolding and misassembly [57]. The basic principles of protein aggregation have been considered for example by De Young et al. [58].

Light scattering is the most sensitive method in detecting even small amounts of aggregates in macromolecular solutions. From the numerous applications of light scattering to aggregation phenomena only a few selected applications to proteins will be discussed briefly before turning to the more special phenomenon of the combined processes of misfolding and aggregation.

Light scattering studies of aggregation are mostly kinetic experiments. In contrast to the kinetic schemes discussed in Section 19.4, measurements of the intensity, which indicate the increase in the particle mass, are preferred in this case. Typical experiments of this kind have been reported by Zettlmeissl et al. [59], who studied the aggregation kinetics of lactic dehydrogenase using a stopped-flow laser light scattering apparatus. These authors thoroughly investigated the effect of enzyme concentration on the kinetics of aggregation and obtained an apparent reaction order of 2.5 from the concentration dependence of the initial slopes of the light scattering signal. Fast aggregation events in the millisecond time range during refolding of interleukin 1β were measured by Finke et al. [60] using a stopped-flow fluorescence device for the kinetic light scattering experiments.

Many light scattering investigations have been done in connection with heat-induced aggregation of proteins. Jøssang et al. [61] studied the aggregation kinetics of human immunoglobulins by measuring the increase in the relative mass and the relative Stokes radius. From the changes of both quantities the authors obtained quite detailed information about the aggregation process, which could be consistently described by Smoluchowski's coagulation theory [62]. The heat aggregation of the β-lactoglobulin protein was very intensively investigated by light scattering [63–65]. These studies are examples of very careful light scattering investigations of protein aggregation. In many of these studies size-exclusion chromatography has been used in addition to light scattering. This is very important in order to follow the consumption of the initial species (mostly monomers) during aggregation (see below). Monomer consumption cannot be easily derived from light scattering data, since the measured weight averaged mass is dominated by the growing large particles.

If the protein aggregates or polymerization products become large enough, further structural information about the growing species can be obtained from the angular dependence of the light scattering intensity. A careful analysis of fibrin formation using stopped-flow multiangle light scattering was reported by Bernocco et al. [66].

An interesting question is: how can protein aggregation be influenced? Suppression of citrate synthase aggregation and facilitation of correct folding has been demonstrated by Buchner et al. [67] using a commercial fluorometer as a light scattering device.

19.5.2
Studying Misfolding and Amyloid Formation by Laser Light Scattering

19.5.2.1 Overview: Initial States, Critical Oligomers, Protofibrils, Fibrils
Conformational conversion of proteins into misfolded structures with an accompanying assembly into large particles, mostly of fibrillar structure called amyloid, is presently an expanding field of research. In this section of the review we will try to sketch the potentials and limits of laser light scattering for elucidating these processes. It is clear beforehand that SLS and DLS are best applied for this purpose mostly in conjunction with other methods, particularly with those sensitive to

changes in secondary structure (CD, FTIR, NMR) and methods giving evidence of the morphology of the growing species such as electron microscopy (EM) and/or atomic force microscopy (AFM). Nevertheless, we will concentrate mainly upon the contribution from the light scattering studies in the following.

Amyloid formation comprises different stages of particle growth. The information that can be obtained from light scattering experiments strongly depends on the quality of the initial state [68]. A starting solution containing essentially the conversion competent monomeric species is a prerequisite for studies of early stages involving the formation of defined oligomers, later called critical oligomers, and protofibrillar structures. These structures became interesting recently because of their possible toxic role. The size distributions and the kinetics of formation of these rather small particles can be characterized very well by light scattering. A careful experimental design (e.g., multi-angle light scattering or model calculations regarding the angular dependence of the scattered light intensity) is needed for the characterization of late products such as large protofibrils or long "mature" fibrils. The problems involved in light scattering from long fibrillar structures have been discussed in a review article by Lomakin et al. [69]. Quantitative light scattering studies of fibril formation have been reported for a few protein systems. Some of them will be discussed now.

19.5.2.2 Aggregation Kinetics of Aβ Peptides

Aβ peptide is the major protein component found in amyloid deposits of Alzheimer's patients and comprises 39–43 amino acids. The first light scattering investigations of Aβ aggregation kinetics were reported by Tomski and Murphy [70] for synthetic Aβ(1–40). According to their SLS and DLS data, these authors modeled the assembly kinetics as formation of rods with a diameter of 5 nm built up from spontaneously formed small cylinders consisting of eight Aβ monomers. The kinetic aggregation model was based on Smoluchowski's theory in the diffusion-limited case. Later, the same group [71] studied the concentration dependence of the aggregation process. They found that a high molecular weight species is formed rapidly when Aβ(1–40) is diluted from 8 M urea to physiological conditions. The size of this species was largest and constant at the lowest concentration (70 μM), while it was smaller and was growing with time at higher concentrations. Furthermore, dissociation of Aβ monomers from preformed fibrils was observed.

Detailed light scattering investigations of Aβ(1–40) fibrillogenesis in 0.1 M HCl have been reported by Lomakin et al. [72–74]. Aβ(1–40) fibrillization was found to be highly reproducible at low pH. In a review article [69], these authors thoroughly analyzed the requirements to obtain good quantitative data of the fibril length from the measured hydrodynamic radius. Adequate measurements can be done if the fibril length does not exceed about 150 nm. Rate constants for fibril nucleation and elongation were determined from the measured time evolution of the fibril length distribution. The initial fibril elongation rate varied linearly with the initial peptide concentration c_0 below a critical concentration $c^* = 0.1$ mM and was constant above c^*. From these observations particular mechanisms for

Aβ fibril nucleation were derived. Homogeneous nucleation within small ($d \sim 14$ nm) Aβ micelles was proposed for $c_0 > c^*$ and heterogeneous nucleation on seeds for $c < c_0$. From the temperature dependence of the elongation rate an activation energy of 23 kcal mol^{-1} was estimated for the proposed monomer addition to fibrils.

Finally, it should be mentioned that reproducibility of Aβ fibrillogenesis in vitro is a serious problem in general due to possible variations in the starting conformation and the assembly state of the Aβ preparations [68].

19.5.2.3 Kinetics of Oligomer and Fibril Formation of PGK and Recombinant Hamster Prion Protein

Presently more than 60 proteins and peptides are known that form fibrillar structures under appropriate environmental conditions. Many of these proteins are not related to any disease. Particularly interesting are those which have served as model proteins in folding studies and could also be good candidates for studying basic principles of structure conversion and amyloid formation. The potentials of laser light scattering for this purpose will be demonstrated here for PGK from yeast and recombinant prion protein from Syrian hamster, ShaPrP$^{90-232}$. A well-defined conversion competent state free of any "seeds" can be achieved for these proteins under specific environmental conditions thus allowing studies of early stages of misfolding and aggregation.

Conversion of PGK starts from a partially folded state at pH 2, at and above room temperature and in the presence of defined amounts of salt [34]. The time dependence of the relative increases in the average mass and the average Stokes radius after adding Na-TCA are shown in Figure 19.9a. Two growth steps are clearly visible. The mass and size transition curves scale linearly with concentration [75]. Thus, the growth process is a second-order reaction. Both growth steps can be well described within the framework of Smoluchowski's coagulation theory. During the first stage, clustering towards oligomers consisting of about 8–10 PGK molecules occurs. These oligomers, termed "critical oligomers", assemble into protofibrillar structures during the second stage. Such a behavior could be inferred from the relation between mass and size (Figure 19.9c), which yields an idea of the dimensionality of the growth process. Electron microscopy at selected growth stages have confirmed this idea [75]. The mutual processes of growth and secondary structure conversion were analyzed by relating the increase in mass to the changes in the far-UV CD [75].

Recombinant PrP^C can be converted into a β-rich aggregated structure at low pH and under destabilizing conditions [76–78]. The relative changes in the mass and size for ShaPrP$^{90-232}$ in sodium phosphate, pH 4.2, after a GdmCl concentration jump from 0 M to 1 M are shown in Figure 19.9b. The aggregation process resembles that of PGK with respect to the occurrence of a well-populated oligomeric state. The transient size maximum during oligomer formation is an indication of a concentration dependent overshoot phenomenon. Extrapolation of this size maximum to zero protein concentration yielded an octamer as the smallest oligomer

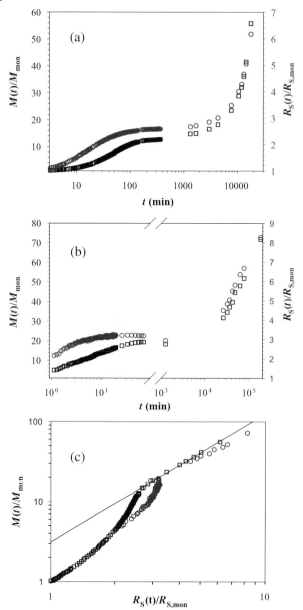

Fig. 19.9. Amyloid assembly of PGK and recombinant Syrian hamster PrP[90–232]. a) Relative increases in mass (squares) and Stokes radius (circles) during assembly of PGK, $c = 2.7$ mg mL^{-1}, into fibrillar structures in 10 mM HCl, pH 2, 9.1 mM sodium trichloroacetate, $T = 20\,°C$. Structural conversion was initiated by adding sodium trichloroacetate. M_{mon} and $R_{S,mon}$ are the molecular mass and the Stokes radius of the monomer, respectively. b) Relative increases in mass (squares) and Stokes radius (circles) during assembly of recombinant Syrian hamster PrP[90–232], $c = 0.59$ mg mL^{-1}, in 20 mM sodium acetate, pH 4.2, 50 mM NaCl, 1 M GdmCl, 20 °C. Structural conversion was initiated by a GdmCl concentration jump from 0 to 1 M. c) Dimensionality plots, $M(t)/M_{mon}$ versus $R_S(t)/R_{S,mon}$ for the data shown in (a) and (b) for PGK (squares) and Syrian hamster PrP[90–232] (circles), respectively. The fit for the data of PGK corresponds to a dimensionality of 1.6.

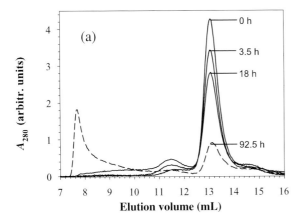

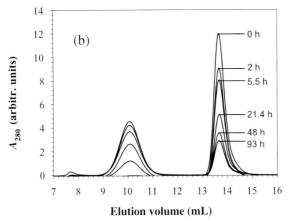

Fig. 19.10. Amyloid assembly monitored by size-exclusion chromatography. a) PGK ($c = 0.14$ mg mL^{-1}, 10 mM HCl, pH 2, 190 mM NaCl) and b) recombinant Syrian hamster PrP$^{90-232}$ ($c = 0.08$ mg mL^{-1}, in 20 mM sodium acetate, pH 4.2, 50 mM NaCl, 1 M GdmCl) at room temperature. The conversion process was triggered either by addition of salt (PGK) or GdmCl (prion protein), respectively.

[79]. The subsequent slow increases in size and mass result from the formation of protofibrillar structures [79]. The change in particle morphology is again indicated by a kink in the dimensionality plot (Figure 19.9c).

The concentration dependence of the growth process revealed remarkable kinetic differences as compared to PGK according to the estimated apparent reaction order of 3. The differences in the formation of the critical oligomers are well documented by size exclusion chromatography (Figure 19.10). Different oligomeric states can be detected during aggregation of PGK, while only monomers and the critical oligomer (octamer) are populated in the case of ShaPrP$^{90-232}$. Accordingly, the octamer is the smallest stable oligomer. Complementary FTIR and CD studies [79] have led to the conclusion that the misfolded secondary structure is stabilized within the octamer. The sole appearance of monomers and octamers and/or higher oligomers tells us that the dimensionality of the first growth process in the case of ShaPrP$^{90-232}$ (Figure 19.9c) is due to changes in the population of two discrete species.

Kinks in the dimensionality plot are also observed when various growth stages are not clearly separated in plots of M and R_S versus time (Figure 19.9a,b) because of comparable growth rates. This is a clear advantage of this data representation. However, one must be careful in relating the (apparent) dimensionality derived from these plots to simple geometrical models.

19.5.2.4 Mechanisms of Misfolding and Misassembly, Some General Remarks

Viewed from the perspective of chemical kinetics amyloid formation is a complex polymerization reaction, since assembly into amyloid fibrils is accompanied by secondary structure rearrangement into a predominant β-sheet structure. β-sheet strands are needed for association at the edges of monomeric subunits to build up the fibrillar end products [80]. This is in marked contrast to polymerization reactions of low-molecular-weight compounds, where the monomers possess already the functional groups for polymerization at initiation of the reaction [81] and to polymerization of proteins like actin, microtubulin [82], and sickle cell hemoglobin [83]. In the latter cases the proteins possess the assembly-competent structure immediately after the start of the reaction. These processes are usually interpreted in the framework of a nucleation polymerization mechanism [82, 83]. This interpretation has been overtaken to rationalize the kinetics of amyloid formation [84] and only little attention has been paid to the relation between secondary and quaternary structure formation in models of amyloid fibril assembly. The latest model proposals have put their focus especially on this point and have emphasized the population of intermediate long-lived oligomeric assemblies on the pathway to amyloid fibrils [75, 85].

19.6
Experimental Protocols

19.6.1
Laser Light Scattering Instrumentation

19.6.1.1 Basic Experimental Set-up, General Requirements

The experimental set-up for light scattering measurements in macromolecular solutions is relatively simple (Figure 19.1). A beam of a continuous wave (cw) laser is focused into a temperature-stabilized cuvette. A temperature stability better than $\pm 0.1\ °C$ is required for DLS experiments because of the temperature dependence of the solvent viscosity. The scattered light intensity is detected at one or different scattering angles using either photomultiplier tubes or avalanche photodiode detectors (APD). Though photomultiplier tubes have still some advantages, an important feature of APDs is the much higher quantum yield (>50%) within the wavelength range of laser light between 400 and 900 nm.

The expense of a light scattering system depends essentially on the type of the laser and the complexity of the detection system. For example, simple systems detect the light only at 90° or in the nearly backward scattering direction, while sophisticated devices are able to monitor the scattered light simultaneously at many selectable scattering angles. An important feature is the construction of the sample cell holder, which should allow the use of different cell types. As discussed earlier, for folding studies and molecular weight determinations of proteins detection at one scattering angle is sufficient. Multi-angle detection is recommended for studies of protein assembly, when the size of the formed particles exceeds about 50 nm.

A criterion for choosing the appropriate cell type is the decision between batch and flow experiments and the available amount of protein. Batch experiments require the lowest amount of substance, since standard microfluorescence cells with volumes of the order of 10 µL can be used. These cells have the further advantage that the protein concentration can be measured in the same cell. Flow-through cells, which are essential for stopped-flow experiments and on-line coupling to particle separation devices, like HPLC, FPLC, and field-flow fractionation, are also useful for batch experiments in the following respect. A direct connection of these cell to a filter unit are the best way to obtain perfectly cleaned samples.

The cells must be cleaned first by flushing a large amount of water and solvent through the filter and the cell. The protein sample is then applied without removing the filter unit. Several cell types can be used including standard fluorescence flow-through cells, sample cells used in stopped-flow systems, or special flow-through cells provided by manufacturers of light scattering equipments. Most of these flow cells have the disadvantage of a restricted angular range. An exception is the cell used in the Wyatt multiangle laser light scattering (MALLS) device. Cylindrical cells in connection with an index matching bath have to be used for precise measurements at different, particularly small scattering angles.

In the case of SLS experiments, measurements of the photocurrent with subsequent analog-to-digital conversion as well as photon-counting techniques are used. The latter technique is preferred in DLS experiments. An appropriate correlation electronics is no longer a problem for modern dynamic light scattering devices since practically ideal digital correlators can be built up.

The minimum protein concentration needed in a DLS experiment depends on the molecular mass and the optical quality of the instrument, particularly on the available laser power. But, even at sufficiently high laser power of about 0.5 W the experiments become impractical below a certain concentration. This happens when the excess scattering of the protein solution falls below the scattering of pure water, e.g., for a protein with $M = 10$ kDa at concentrations below 0.25 mg mL^{-1}. Thus, DLS experiments can be done without special efforts at concentrations satisfying the condition $c \times M \geq 5$ kDa mg mL^{-1}. For SLS experiments the value of $c \times M$ can be about one order of magnitude lower.

Removal of undesired contaminations, such as dust, bubbles, or other large particles, is a crucial step in preparing samples for light scattering experiments. Large particles can readily removed by membrane filters having pore sizes of about 0.1 μm or by centrifugation at 10 000 g or higher, while size-exclusion chromatography or similar separation techniques have to be applied when a protein solution contains oligomers or small aggregates only slightly larger than the monomeric protein. Fluids used to clean the scattering cell must also be free of all particulates.

In an SLS experiment, the time-averaged scattering intensities of solvent, solution, and reference sample (scattering standard) have to be measured. Data evaluation has to be done according to Eqs (3) and (4). Commercial instruments are in general equipped with software packages that generate Zimm plots according to Eq. (4). Additionally, the software contains the Rayleigh ratios for scattering standards at the wavelength of the instrument. A widely used scattering standard is toluene. The Rayleigh ratios of toluene and other scattering standards at different wavelength are given in Ref. [5]. Instrument calibration can also be done using an appropriate protein solution. However, the protein must be absolutely monomeric under the chosen solvent conditions.

DLS measurements demand slightly more experience concerning experiment duration and data evaluation. Calibration is not needed. The basis for obtaining stable and reliable results is an adequately measured correlation function. A visual inspection of this primary data is recommended before any data evaluation is done. The signal-to-noise ratio that is needed depends on the complexity of the system under study. For example, short measurement times are acceptable only in the case of unimodal narrow size distributions, when a noisy correlation function can be well fitted by a single exponential. In practise, two data evaluation schemes are mainly used, the method of cumulants [9] yielding an average Stokes radius and the second and third moments of the size distribution and/or the approximate reconstruction of the size distribution by an inverse Laplace transformation, mostly using the program CONTIN [8]. The corresponding software packages are available from the author or are provided by the manufacturers of the instruments. CONTIN allows the reconstruction of a distribution function in terms of scattering power, weight concentration, or number (molar) concentration. Though the latter two types are more instructive, they should only be calculated when a reasonable particle shape can be anticipated within the considered size range.

19.6.1.2 Supplementary Measurements and Useful Options

For the estimation of the molecular mass from SLS, and the diffusion coefficient and the Stokes radius from DLS, additional physical quantities are required. For the calculation of the optical constant in Eq. (2) the refractive index of the solvent and the refractive index increment of the protein in the particular solvent must be known. The former can be measured easily by an Abbe refractometer, whereas an differential refractometer is needed for precise measurements of $\partial n/\partial c$. Measurements of $\partial n/\partial c$ are often more expensive than the light scattering experiment itself. A comprehensive collection of $\partial n/\partial c$ values can be found in [86]. $\partial n/\partial c = 0.19$ mL g^{-1} is a good approximation for proteins in aqueous solution at moderate (<200 mM) salt concentrations.

For calculations of R_S from DLS data via Eq. (6) the dynamic viscosity η of the solvent must be known. η can be obtained from the kinematic viscosity v (e.g., measured by an Ubbelohde type viscometer) and the density ρ (measured by a digital densitometer, e.g., DMA 58, Anton Paar, Austria) by $\eta = v \times \rho$.

A very useful option for light scattering instruments is the on-line coupling with FPLC, HPLC, or field-flow fractionation (FFF) and a concentration detector, which might be a UV absorption monitor or a refractive index detector. This allows direct measurements of M during the flow. In the case of known $\partial n/\partial c$ for the particular solvent conditions, the molecular mass can be obtained from the output signals of the SLS and refractive index detectors by $M = k_e \times (\partial n/\partial c)^{-1} \times$ (output)$_{SLS}$/(output)$_{RI}$, where k_e is the instrument calibration constant. A parallel estimation of R_S is possible, when the scattering is strong enough allowing a sufficiently precise DLS experiment during a few minutes.

A further option is the coupling to a mixer allowing kinetic light scattering experiments. Two schemes can be used. Either a mixing device is coupled to a light scattering apparatus equipped with a flow cell [66] or the light scattering apparatus is built-up around a stopped-flow device originally constructed for fluorescence detection [56].

19.6.1.3 Commercially Available Light Scattering Instrumentation

There are several companies that produce light scattering instruments and optional units. Many of them have taken into consideration the special demands for studies on protein solutions. Accordingly, most of the systems can handle very small sample volumes, have the sensitivity to study low concentrations of monomeric proteins, and can work in batch and flow mode. For those who are interested in applying SLS and DLS to folding and aggregation studies we have listed several companies and the corresponding websites in alphabetic order.

- ALV-Laservertriebsgesellschaft m.b.H., Germany: http://www.alvgmbh.de
- Brookhaven Instruments Corporation, USA: http://www.bic.com
- Malvern Instruments Ltd., UK: http://www.malvern.co.uk
- Precision Detectors, USA: http://www.lightscatter.com
- Proterion Corporation, USA (formerly Protein Solutions): http://www.proterion.com
- Wyatt Technology Corporation, USA: http://www.wyatt.com, visit also http://www.wyatt.de

19.6.2 Experimental Protocols for the Determination of Molecular Mass and Stokes Radius of a Protein in a Particular Conformational State

The purpose of the protocols is to demonstrate critical steps of combined SLS/DLS measurements. If a commercial instrument with sophisticated data evaluation software is used, the individual steps may be involved, but slightly differently arranged in the software protocol. The protocols are based on the assumption that the instrument provides an average over time T of the scattered intensity $\overline{I_T}$ and a set of N normalized correlation functions $g^{(2)}(\tau, T_A)$ calculated during an accumulation time T_A and is equipped with the software to reconstruct the size distribution (CONTIN or similar packages).

Protocol 1

Determination of the apparent molecular mass M_{app}, diffusion coefficient D_{app}, and the corresponding Stokes radius R_{app} of a protein at a given concentration c_P at 20 °C.

Equipment and reagents

- SLS/DLS instrument with thermostatted cell holder
- UV spectrophotometer, preferentially single-beam instrument
- Abbe-type refractometer
- Flow-through cell for fluorometry, light path 1.5 mm ($V = 25$ μL, e.g., Hellma 176.152) and ultra-microfluorescence cell, light path 1.5 mm ($V = 12$ μL, e.g., Hellma 105.252).
- Whatman filter Anotop and Anodisc, $d = 10$ mm, pore size 0.1 μm with appropriate filter holders, syringes 1–2 mL (laboratory centrifuge and pipettes for method B).
- Suitable buffer, e.g., 50 mM Tris/HCl, pH 8, 100 mM KCl (the viscosity $\eta = 1.018$ cP at 20 °C is not very different from that of water, $\eta = 1.002$ cP).

(A) Measurements with flow-through cells

1. Prepare the protein solution of the required concentration (suitable range 1–5 mg mL^{-1}).
2. Wet the Anotop filter with distilled water and connect it to the flow-through cell (do not disconnect until the end of the experiment).
3. Flush at least 5 mL distilled water through filter and cell.
4. Insert the cell into the light scattering instrument and check for purity (the scattering intensity must be very low and "spikes" in intensity should be totally absent. Otherwise, repeat step 3.
5. Fill a syringe with 1–2 mL buffer and flush it through filter and cell (note that the actual dead volume of filter and cell is larger than the specified volume of the cell).
6. Insert the cell into the light scattering instrument and measure the scattering intensity of the buffer, $\overline{I}_{T,buffer}$. In some cases, it could be useful to run a short DLS experiment in order to check that the baseline is flat.
7. Measure the scattering intensity of the reference sample (toluene) in a separate cuvette of similar geometry.

8. Insert the flow-through cell with buffer into the UV spectrometer and measure the blank spectrum (note that the center height of the cell fits both the SLS/DLS and UV spectrophotometer).
9. Measure the refractive index of the buffer with the Abbe-type refractometer.
10. Flush about 1 mL protein solution slowly through filter and cell in order to replace the buffer by the protein solution (the protein solution that has passed filter and cell can be used to prepare a sample of lower concentration).
11. Measure the absorption spectrum of the protein solution in the flow-through cell, subtract the blank spectrum, and calculate the actual protein concentration in the cell on the basis of known extinction coefficients.
12. Insert the cell into the SLS/DLS instrument and start data acquisition. It is recommended to record the correlation function not in a single long run, but rather by superposition of N short runs (of about 10 s). This enables one to get a better overview about the stability and quality of the sample. Furthermore, good acquisition software allows to discard distorted runs. Though a correlation function is clearly built up within 10 s, it is mostly too noisy for adequate data evaluation.
13. Store all data and prepare data evaluation by entering additional parameters like temperature, refraction index, and viscosity.
14. Data evaluation, especially estimations of size distribution, should be done immediately giving the chance to continue the experiment. This could be useful if the resulting size distribution is still very unstable against variation in data evaluation parameters.
15. The size distribution should consist of one dominating peak in the present case. Otherwise, the protein is strongly prone to aggregation. The influence of stable aggregates of clearly distinct size can be eliminated in many cases as discussed at the end of Section 19.2.
16. Calculate M using Eqs (2), (3), and (4), the Rayleigh ratio of the reference sample (the value depends on the wavelength used in the instrument), and $\partial n/\partial c = 0.19$ mL g^{-1}.

(B) Measurements with ultra-microfluorescence cells

Although Scheme A is superior from the light scattering point of view, a rather large amount of protein is needed. This can be circumvented by using ultra-micro cells. Here, we repeat only those steps that concern the preparation of buffer and protein solutions. Two approaches are applicable.

(B1) Preparation using micro-filter units

Several manufacturers (e.g. Proterion Corporation or Wyatt Technology Corporation) deliver micro-filter holder for Anodisc filter membranes with dead volumes less than 10 µL. This allows sample volumes as small as 25 µL to be filtered. The following steps have to be done.

1. Insert a dry filter membrane into the filter holder and wet the filter with buffer and filter about 100 µL buffer.
2. Flush the fluorescence cell with a strong nitrogen stream in order to remove any dust.
3. Filter about 50 µL buffer into the cell and check for purity in the light scattering instrument. Do the same measurements with the buffer and reference sample as indicated under Scheme A. (Empty the cell, wash is with distilled water, dry it with ethanol (Uvasol) and nitrogen and repeat steps 2 and 3 if dust or bubbles can be detected.)
4. Empty the cell carefully (washing and drying is not necessary) and filter about 25 µL protein solution into the cell.
5. Close the cell, determine the concentration and measure SLS and DLS as under Scheme A.
6. This procedure is nearly as good as filtration according to Scheme A, but the presence of dust and bubbles cannot be totally excluded.

(B2) Preparation by centrifugation

This procedure has to be applied when only about 15 µL of protein solution are available and differs from Scheme B1 only by step 4. The protein solution centrifugated about 15 min at 10 000 g or more has to be transferred into the scattering cell. An alternative is careful centrifugation within the scattering cell (note that ultra-micro cells are rather expensive!).

Protocol 2

Determination of the molecular mass, diffusion coefficient D, Stokes radius R_S and virial coefficients of a protein in a particular state.

This experimental procedure is a repetition of the sequences given in Protocol 1 for at least four protein concentrations. Scheme A is recommended, whereas a stock solution of 1 mL with the highest concentration is needed.

Additional data evaluation concerns extrapolation of $(H \times c)/R_q$ (Eq. (4)) with $P(q) \approx 1$ and D to zero protein concentration.

Acknowledgments

The authors are grateful to Stephen Harding, University Nottingham, for valuable comments and suggestions.

References

1 JAKEMAN, E., PUSEY, P. N. & VAUGHAN, J. M. (1976). Intensity fluctuation light-scattering spectroscopy using a conventional source. *Optics Commun.* **17**, 305–308.
2 HUGLIN, M. (1972). *Light Scattering*

from Polymer Solutions, Academic Press, New York.
3 KRATOCHVIL, P. (1987). *Classical Light Scattering from Polymer Solutions*, Elsevier, Amsterdam.
4 SCHMITZ, K. S. (1990). *An Introduction to Dynamic Light Scattering by Macromolecules.* Academic Press, New York.
5 CHU, B. (1991). *Laser Light Scattering.* Academic Press, New York.
6 BROWN, W. (1993). *Dynamic Light Scattering.* Claredon Press, Oxford.
7 BERNE, B. J. & PECORA, R. (2000). *Dynamic Light Scattering with Applications to Chemistry, Biology, and Physics.* Dover Publications, Mineola, New York.
8 PROVENCHER, S. W. (1982). CONTIN – a general-purpose constrained regularization program for inverting noisy linear algebraic and integralequations. *Com. Phys. Commun.* **27**, 229–242.
9 KOPPEL, D. E. (1972). Analysis of macromolecular polydispersity in intensity correlation spectroscopy: the method of cumulants. *J. Chem. Phys.* **57**, 4814–4820.
10 HARDING, S. E. & JOHNSON, P. (1985). The concentration dependence of macromolecular parameters. *Biochem. J.* **231**, 543–547.
11 GARCIA BERNAL, J. M. & DE LA TORRE, J. G. (1981). Transport-properties of oligomeric subunit structures. *Biopolymers* **20**, 129–139.
12 CLAES, P., DUNFORD, M., KENNEY, A. & PENNY, V. (1992). An on-line dynamic light scattering instrument for macromolecular characterization. In *Laser Light Scattering in Biochemistry* (HARDING, S. E., SATTELLE, D. B. & BLOOMFIELD, V. A., eds), pp. 66–76. Royal Society of Chemistry, Cambridge.
13 NICOLI, D. F. & BENEDEK, G. B. (1976). Study of the thermal denaturation of lysozyme and other globular proteins by light-scattering spectroscopy. *Biopolymers* **15**, 2421–2437.
14 GAST, K., ZIRWER, D., DAMASCHUN, H., HAHN, U., MÜLLER-FROHNE, M., WIRTH, M. & DAMASCHUN, G. (1997). Ribonuclease T1 has different dimensions in the thermally and chemically denatured states: a dynamic light scattering study. *FEBS Lett.* **403**, 245–248.
15 FABIAN, H., FÄLBER, K., GAST, K., REINSTADLER, D., ROGOV, V. V., NAUMANN, D., ZAMYATKIN, D. F. & FILIMONOV, V. V. (1999). Secondary structure and oligomerization behavior of equilibrium unfolding intermediates of the lambda Cro repressor. *Biochemistry* **38**, 5633–5642.
16 PRIVALOV, P. L. (1990). Cold denaturation of proteins. *Crit. Rev. Biochem. Mol. Biol.* **25**, 281–305.
17 GRIKO, Y. V., VENYAMINOV, S. Y. & PRIVALOV, P. L. (1989). Heat and cold denaturation of phosphoglycerate kinase (interaction of domains). *FEBS Lett.* **244**, 276–278.
18 DAMASCHUN, G., DAMASCHUN, H., GAST, K., MISSELWITZ, R., MÜLLER, J. J., PFEIL, W. & ZIRWER, D. (1993). Cold denaturation-induced conformational changes in phosphoglycerate kinase from yeast. *Biochemistry* **32**, 7739–7746.
19 AGASHE, V. R., UDGAONKAR, J. B. (1995). Thermodynamics of denaturation of barstar: evidence for cold denaturation and evaluation of the interaction with guanidine hydrochloride. *Biochemistry* **34**, 3286–3299.
20 WONG, K. B., FREUND, S. M. V. & FERSHT, A. R. (1996). Cold denaturation of barstar: H-1, N-15 and C-13 NMR assignment and characterisation of residual structure. *J. Mol. Biol.* **259**, 805–818.
21 TANFORD, C. (1968). Protein denaturation. *Adv. Protein Chem.* **23**, 121–282.
22 DUBIN, S. B., FEHER, G. & BENEDEK, G. B. (1973). Study of the chemical denaturation of lysozyme by optical mixing spectroscopy. *Biochemistry* **12**, 714–719.
23 NÖPPERT, A., GAST, K., MÜLLER-FROHNE, M., ZIRWER, D. & DAMASCHUN, G. (1996). Reduceddenatured ribonuclease A is not in a compact state. *FEBS Lett.* **380**, 179–182.

24 FINK, A. L., CALCIANO, L. J., GOTO, Y., KUROTSU, T. & PALLEROS, D. R. (1994). Classification of acid denaturation of proteins: intermediates and unfolded states. *Biochemistry* **33**, 12504–12511.

25 BARRICK, D., HUGHSON, F. M. & BALDWIN, R. L. (1994). Molecular mechanism of acid denaturation – the role of histidine-residues in the partial unfolding of apomyoglobin. *J. Mol. Biol.* **237**, 588–601.

26 PTITSYN, O. B. (1995). Molten globule and protein folding. *Adv. Protein Chem.* **47**, 83–229.

27 ARAI, M. & KUWAJIMA, K. (2000). Role of the molten globule state in protein folding. *Adv. Protein Chem.* **53**, 209–282.

28 KUWAJIMA, K. & ARAI, M. (2000). The molten globule state: the physical picture and biological significance. In *Mechanisms of Protein Folding* (PAIN, R. H., ed.), pp. 138–174. Oxford University Press, Oxford.

29 GAST, K., ZIRWER, D., MÜLLER-FROHNE, M. & DAMASCHUN, G. (1998). Compactness of the kinetic molten globule of bovine alpha-lactalbumin: a dynamic light scattering study. *Protein Sci.* **7**, 2004–2011.

30 GAST, K., DAMASCHUN, H., MISSELWITZ, R., MÜLLER-FROHNE, M., ZIRWER, D. & DAMASCHUN, G. (1994). Compactness of protein molten globules: temperature-induced structural changes of the apomyoglobin folding intermediate. *Eur. Biophys. J.* **23**, 297–305.

31 KATAOKA, M., KUWAJIMA, K., TOKUNAGA, F. & GOTO, Y. (1997). Structural characterization of the molten globule of alpha-lactalbumin by solution x-ray scattering. *Protein Sci.* **6**, 422–430.

32 KATAOKA, M., NISHII, I., FUJISAWA, T., UEKI, T., TOKUNAGA, F. & GOTO, Y. (1995). Structural characterization of the molten globule and native states of apomyoglobin by solution X-ray scattering. *J. Mol. Biol.* **249**, 215–228.

33 GAST, K., ZIRWER, D., MÜLLER-FROHNE, M. & DAMASCHUN, G. (1999). Trifluoroethanol-induced conformational transitions of proteins: insights gained from the differences between alpha-lactalbumin and ribonuclease A. *Protein Sci.* **8**, 625–634.

34 DAMASCHUN, G., DAMASCHUN, H., GAST, K. & ZIRWER, D. (1999). Proteins can adopt totally different folded conformations. *J. Mol. Biol.* **291**, 715–725.

35 WRIGHT, P. E. & DYSON, H. J. (1999). Intrinsically unstructured proteins: Re-assessing the protein structure-function paradigm. *J. Mol. Biol.* **293**, 321–331.

36 GAST, K., DAMASCHUN, H., ECKERT, K., SCHULZE-FORSTER, K., MAURER, H. R., MULLER, F. M., ZIRWER, D., CZARNECKI, J. & DAMASCHUN, G. (1995). Prothymosin alpha: A biologically active protein with random coil conformation. *Biochemistry* **34**, 13211–13218.

37 DAMASCHUN, G., DAMASCHUN, H., GAST, K., MISSELWITZ, R., ZIRWER, D., GÜHRS, K. H., HARTMANN, M., SCHLOTT, B., TRIEBEL, H. & BEHNKE, D. (1993). Physical and conformational properties of staphylokinase in solution. *Biochim. Biophys. Acta* **1161**, 244–248.

38 UVERSKY, V. N. (1993). Use of fast protein size-exclusion liquid-chromatography to study the unfolding of proteins which denature through the molten globule. *Biochemistry* **32**, 13288–13298.

39 WILKINS, D. K., GRIMSHAW, S. B., RECEVEUR, V., DOBSON, C. M., JONES, J. A. & SMITH, L. J. (1999). Hydrodynamic radii of native and denatured proteins measured by pulse field gradient NMR techniques. *Biochemistry* **38**, 16424–16431.

40 FLORY, P. J. (1969). *Statistical Mechanics of Chain Molecules*. John Wiley & Sons, New York.

41 BALDWIN, R. L. & ZIMM, B. H. (2000). Are denatured proteins ever random coils? *Proc. Natl Acad. Sci. USA* **97**, 12391–12392.

42 ZHOU, H. X. (2002). Dimensions of denatured protein chains from

hydrodynamic data. *J. Phys. Chem.* **106**, 5769–5775.

43 TANFORD, C., KAWAHARA, K. & LAPANJE, S. (1966). Proteins in 6 M guanidine hydrochloride. Demonstration of random coil behavior. *J. Biol. Chem.* **241**, 1921–1923.

44 DAMASCHUN, G., DAMASCHUN, H., GAST, K. & ZIRWER, D. (1998). Denatured states of yeast phosphoglycerate kinase. *Biochemistry (Moscow)* **63**, 259–275.

45 BYRON, O. (2000). Hydrodynamic bead modeling of biological macromolecules. *Methods Enzymol.* **321**, 278–304.

46 GARCIA DE LA TORRE, J., HUERTAS, M. L. & CARRASCO, B. (2000). Calculation of hydrodynamic properties of globular proteins from their atomic-level structure. *Biophys. J.* **78**, 719–730.

47 HARDING, S. E. (2001). The hydration problem in solution biophysics: an introduction. *Biophys. Chem.* **93**, 8/–91.

48 HALLE, B. & DAVIDOVIC, M. (2003). Biomolecular hydration: From water dynamics to hydrodynamics. *Proc. Natl Acad. Sci. USA* **100**, 12135–12140.

49 GARCIA DE LA TORRE, J. (2001). Hydration from hydrodynamics. General considerations and applications of bead modelling to globular proteins. *Biophys. Chem.* **93**, 159–170.

50 FENG, H. P. & WIDOM, J. (1994). Kinetics of compaction during lysozyme refolding studied by continuous-flow quasielastic light scattering. *Biochemistry* **33**, 13382–13390.

51 GAST, K., ZIRWER, D. & DAMASCHUN, G. (2000). Time-resolved dynamic light scattering as a method to monitor compaction during protein folding. In *Data Evaluation in Light Scattering of Polymers* (HELMSTEDT, M. & GAST, K., eds), pp. 205–220. Wiley-VCH, Weinheim.

52 SHASTRY, M. C. R., LUCK, S. D. & RODER, H. (1998). A continuous-flow capillary mixing method to monitor reactions on the microsecond time scale. *Biophys. J.* **74**, 2714–2721.

53 SEGEL, D. J., BACHMANN, A., HOFRICHTER, J., HODGSON, K. O., DONIACH, S. & KIEFHABER, T. (1999). Characterization of transient intermediates in lysozyme folding with time-resolved small-angle X-ray scattering. *J. Mol. Biol.* **288**, 489–499.

54 POLLACK, L., TATE, M. W., DARNTON, N. C., KNIGHT, J. B., GRUNER, S. M., EATON, W. A. & AUSTIN, R. H. (1999). Compactness of the denatured state of a fast-folding protein measured by submillisecond small-angle x-ray scattering. *Proc. Natl Acad. Sci. USA* **96**, 10115–10117.

55 ARAI, M., ITO, K., INOBE, T., NAKAO, M., MAKI, K., KAMAGATA, K., KIHARA, H., AMEMIYA, Y. & KUWAJIMA, K. (2002). Fast compaction of alpha-lactalbumin during folding studied by stopped-flow X-ray scattering. *J. Mol. Biol.* **321**, 121–132.

56 NÖPPERT, A., GAST, K., ZIRWER, D. & DAMASCHUN, G. (1998). Initial hydrophobic collapse is not necessary for folding RNase A. *Folding Des.* **3**, 213–221.

57 JAENICKE, R. & SECKLER, R. (1997). Protein misassembly in vitro. *Adv. Protein Chem.* **50**, 1–59.

58 DE YOUNG, L. R., DILL, K. A. & FINK, A. L. (1993). Aggregation of globular proteins. *Acc. Chem. Res.* **26**, 614–620.

59 ZETTLMEISSL, G., RUDOLPH, R. & JAENICKE, R. (1979). Reconstitution of lactic dehydrogenase. Noncovalent aggregation vs. reactivation. 1. Physical properties and kinetics of aggregation. *Biochemistry* **18**, 5567–5571.

60 FINKE, J. M., ROY, M., ZIMM, B. H. & JENNINGS, P. A. (2000). Aggregation events occur prior to stable intermediate formation during refolding of interleukin 1 beta. *Biochemistry* **39**, 575–583.

61 JOSSANG, T., FEDER, J. & ROSENQVIST, E. (1985). Heat aggregation kinetics of human IgG. *J. Chem. Phys.* **82**, 574–589.

62 SMOLUCHOWSKI, M. V. (1917). Versuch einer mathematischen Theorie der Koagulationskinetik kolloider Lösungen. *Z. Phys. Chem.* **92**, 129–168.

63 AYMARD, P., NICOLAI, T., DURAND, D. & CLARK, A. (1999). Static and dynamic scattering of beta-lactoglobulin aggregates formed after heat-induced denaturation at pH 2. *Macromolecules* **32**, 2542–2552.

64 LE BON, C., NICOLAI, T. & DURAND, D. (1999). Kinetics of aggregation and gelation of globular proteins after heat-induced denaturation. *Macromolecules* **32**, 6120–6127.

65 BAUER, R., CARROTTA, R., RISCHEL, C. & OGENDAL, L. (2000). Characterization and isolation of intermediates in beta-lactoglobulin heat aggregation at high pH. *Biophys. J.* **79**, 1030–1038.

66 BERNOCCO, S., FERRI, F., PROFUMO, A., CUNIBERTI, C. & ROCCO, M. (2000). Polymerization of rod-like macromolecular monomers studied by stopped-flow, multiangle light scattering: Set-up, data processing, and application to fibrin formation. *Biophys. J.* **79**, 561–583.

67 BUCHNER, J., SCHMIDT, M., FUCHS, M., JAENICKE, R., RUDOLPH, R., SCHMID, F. X. & KIEFHABER, T. (1991). GroE facilitates refolding of citrate synthase by suppressing aggregation. *Biochemistry* **30**, 1586–1591.

68 FEZOUI, Y., HARTLEY, D. M., HARPER, J. D., KHURANA, R., WALSH, D. M., CONDRON, M. M., SELKOE, D. J., LANSBURY, P. T., FINK, A. L. & TEPLOW, D. B. (2000). An improved method of preparing the amyloid beta-protein for fibrillogenesis and neurotoxicity experiments. *Amyloid* **7**, 166–178.

69 LOMAKIN, A., BENEDEK, G. B. & TEPLOW, D. B. (1999). Monitoring protein assembly using quasielastic light scattering spectroscopy. *Methods Enzymol.* **309**, 429–459.

70 TOMSKI, S. J. & MURPHY, R. M. (1992). Kinetics of aggregation of synthetic beta-amyloid peptide. *Arch. Biochem. Biophys.* **294**, 630–638.

71 MURPHY, R. M. & PALLITTO, M. R. (2000). Probing the kinetics of beta-amyloid self-association. *J. Struct. Biol.* **130**, 109–122.

72 LOMAKIN, A., CHUNG, D. S., BENEDEK, G. B., KIRSCHNER, D. A. & TEPLOW, D. B. (1996). On the nucleation and growth of amyloid beta-protein fibrils: Detection of nuclei and quantitation of rate constants. *Proc. Natl Acad. Sci. USA* **93**, 1125–1129.

73 LOMAKIN, A., TEPLOW, D. B., KIRSCHNER, D. A. & BENEDEK, G. B. (1997). Kinetic theory of fibrillogenesis of amyloid beta-protein. *Proc. Natl Acad. Sci. USA* **94**, 7942–7947.

74 KUSUMOTO, Y., LOMAKIN, A., TEPLOW, D. B. & BENEDEK, G. B. (1998). Temperature dependence of amyloid beta-protein fibrillization. *Proc. Natl Acad. Sci. USA* **95**, 12277–12282.

75 MODLER, A. J., GAST, K., LUTSCH, G. & DAMASCHUN, G. (2003). Assembly of amyloid protofibrils via critical oligomers – a novel pathway of amyloid formation. *J. Mol. Biol.* **325**, 135–148.

76 HORNEMANN, S. & GLOCKSHUBER, R. (1998). A scrapie-like unfolding intermediate of the prion protein domain PrP(121–231) induced by acidic pH. *Proc. Natl Acad. Sci. USA* **95**, 6010–6014.

77 MORILLAS, M., VANIK, D. L. & SUREWICZ, W. K. (2001). On the mechanism of alpha-helix to beta-sheet transition in the recombinant prion protein. *Biochemistry* **40**, 6982–6987.

78 BASKAKOV, I. V., LEGNAME, G., BALDWIN, M. A., PRUSINER, S. B. & COHEN, F. E. (2002). Pathway complexity of prion protein assembly into amyloid. *J. Biol. Chem.* **277**, 21140–21148.

79 SOKOLOWSKI, F., MODLER, A. J., MASUCH, R., ZIRWER, D., BAIER, M., LUTSCH, G., MOSS, D. A., GAST, K. & NAUMANN, D. (2003). Formation of critical oligomers is a key event during conformational transition of recombinant Syrian hamster prion protein. *J. Biol. Chem.* **278**, 40481–40492.

80 RICHARDSON, J. S. & RICHARDSON, D. C. (2002). Natural beta-sheet proteins use negative design to avoid edge-to-edge aggregation. *Proc. Natl Acad. Sci. USA* **99**, 2754–2759.

81 FLORY, P. J. (1953). *Principles of Polymer Chemistry*. Cornell University Press, Ithaca.

82 OOSAWA, F. & ASAKURA, S. (1975). *Thermodynamics of the Polymerization of Protein*. Academic Press, London.

83 EATON, W. A. & HOFRICHTER, J. (1990). Sickle cell hemoglobin polymerization. *Adv. Protein Chem.* **40**, 63–279.

84 JARRETT, J. T. & LANSBURY, P. T., JR. (1993). Seeding "one-dimensional crystallization" of amyloid: a pathogenic mechanism in Alzheimer's disease and scrapie? *Cell* **73**, 1055–1058.

85 SERIO, T. R., CASHIKAR, A. G., KOWAL, A. S., SAWICKI, G. J., MOSLEHI, J. J., SERPELL, L., ARNSDORF, M. F. & LINDQUIST, S. L. (2000). Nucleated conformational conversion and the replication of conformational information by a prion determinant. *Science* **289**, 1317–1321.

86 THEISEN, A., JOHANN, C., DEACON, M. P. & HARDING, S. E. (2000). *Refractive Increment Data-book*. Nottingham University Press, Nottingham.

87 GAST, K., MODLER, A. J., DAMASCHUN, H., KRÖBER, R., LUTSCH, G., ZIRWER, D., GOLBIK, R. & DAMASCHUN, G. (2003). Effect of environmental conditions on aggregation and fibril formation of barstar. *Eur. Biophys. J.* **32**, 710–723.

20
Conformational Properties of Unfolded Proteins

Patrick J. Fleming and George D. Rose

20.1
Introduction

The protein folding reaction, U(nfolded) ⇌ N(ative), is a reversible disorder ⇌ order transition. Proteins are disordered (U) at high temperature, high pressure, extremes of pH, or in the presence of denaturing solvents; but they fold to uniquely ordered, biologically relevant conformers (N) under physiological conditions. This folding transition is highly cooperative such that individual molecules within the population are predominantly fully folded or fully unfolded; partially folded chains are transitory and rare. Notably, no covalent bonds are made or broken during folding/unfolding; in effect, the transition is simply a re-equilibration in response to changes in temperature, pressure, pH, or solvent conditions. Currently, there are more than 20 000 examples of native proteins in the protein databank. In contrast, the unfolded population, by its very nature, resists ready structural characterization. In this sense, the folding reaction might be more appropriately denoted as ? ⇌ N.

This chapter traces thinking about the unfolded state from Pauling's and Wu's early suggestions in the 1930s, through the work of Tanford and Flory in the 1960s to the present moment. Early work gave rise to the random coil model for the unfolded state, as described below. Confirmatory findings established this model as the conceptual anchor point for thinking about unfolded proteins – until recently, perhaps. In the past few years, results from both theory and experiment indicate the existence of conformational bias in the unfolded state, a condition that is not addressed by the random coil model. If unfolded conformers are biased toward their native conformation sufficiently, then the random coil model is likely to be superseded by newer, more specific models. Though controversial, such a conceptual shift appears to be underway, as we attempt to describe.

20.1.1
Unfolded vs. Denatured Proteins

The term unfolded protein is generic and inclusive, and it can range from protein solutions in harsh denaturants to protein subdomains that undergo transitory ex-

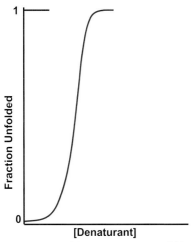

Fig. 20.1. The folding transition. The folding reaction of a typical, small biophysical protein is a highly cooperative, all-or-none transition. At the transition midpoint, half the ensemble is folded and half is unfolded; the population of partially folded/unfolded molecules is negligible. In this idealized plot of an actual experiment [37], the population is followed by a conformational probe (e.g., circular dichroism or fluorescence) as a function of denaturant concentration. Upon addition of sufficient denaturant, the probe signal reaches a plateau, indicating that the transition is complete.

cursions from their native format via spontaneous fluctuations. While conceptually complete, this range is too diverse to be practically useful, and it requires further specification. Accordingly, the field has focused more specifically on denatured proteins, the population of unfolded conformers that can be studied at equilibrium under high concentrations of denaturing solvents, high temperature, high pressure, high/low pH, etc. Early experiments of Ginsburg and Carroll [37] and Tanford [93] demonstrate that such conditions can give rise to a defined equilibrium population in which the unfolding transition is complete (Figure 20.1). In this chapter, we use both terms and rely on the context for specificity.

20.2
Early History

The fact that protein molecules can undergo a reversible disorder ⇌ order transition originated early in the last century, in ideas proposed by Wu [29, 101] and Mirsky and Pauling [59]. Both papers propose that a theory of protein structure is tantamount to a theory of protein denaturation. In particular, these authors recognized that many disparate physical and chemical properties of proteins are abolished coordinately upon heating. This was unlikely to be mere coincidence. Both Wu and Mirsky and Pauling hypothesized that such properties are a consequence of the protein's structure and are abolished when that structure is melted. Their hypothesis was later confirmed by Kauzmann and Simpson [86], at which point the need for an apt characterization of the melt became clear, and protein denaturation emerged as a research discipline.

A widely accepted view assumes that unfolded polypeptide chains can explore conformational space freely, with constraints arising only from short-range local restrictions and longer range excluded volume effects. To a good first approximation, short-range local restrictions refer to repulsive van der Waals interactions between sequentially adjacent residues (i.e., steric effects) captured by the well-known Ramachandran map for a dipeptide [70]. Longer range excluded volume effects also refer to repulsive van der Waals interactions, in this case those between nonbonded atoms that are distant in sequence but juxtaposed in space as the chain wanders at random along a Brownian walk in three dimensions [32, 91]. This random coil model has conditioned most of the thinking in the field.

It is important to realize that the random coil model need not imply an absence of residual structure in the unfolded population. Kauzmann's famous review raised the central question about structure in the unfolded state explicitly [45]:

> ■ *For instance, one would like to know the types of structures actually present in the native and denatured proteins.... The denatured protein in a good solvent such as urea is probably somewhat like a randomly coiled polymer, though the large optical rotation of denatured proteins in urea indicates that much local rigidity must be present in the chain (p. 4).*

20.3
The Random Coil

A chain molecule is a freely jointed random coil if it traces a random walk in three-dimensional space, in incremental steps of fixed length. The random coil model has enjoyed a long and successful history in describing unfolded proteins. By definition, a random coil polymer has no strongly preferred backbone conformations because energy differences among its sterically accessible backbone conformations are of order $\sim kT$. Accordingly, the energy landscape for such a polymer can be visualized as an "egg crate" of high dimensionality, and a Boltzmann-weighted ensemble of such polymers populates this landscape uniformly.

More than others, this elegant theory was developed by Flory [33], pp. 30–31; [17], pp. 991–996) and advanced by Tanford [91–93], who demonstrated that proteins denatured in 6 M guanidinum chloride (a strong denaturant) appear to be structureless, random chains. Tanford's pioneering studies established a compelling framework for interpreting experimental protein denaturation that would survive largely unchallenged for the next 30 years.

Often, the term random coil is used synonymously with the freely jointed chain model (described below), in which there is no correlation between the orientation of two chain monomers at any length scale. That is, configurational properties of a freely jointed chain, such as its end-to-end distance, are Gaussian distributed at all chain lengths. In practice, no actual polymer chain is freely jointed. More realistic models, such as Flory's rotational isomeric-state model, have Gaussian-distributed chain configurations only in the infinite chain limit ([33], pp. 30–31; [17], pp. 991–996). These distinctions notwithstanding, the main characteristic of the random coil holds in all cases, both ideal and real: the unfolded state is structurally featureless because the number of available conformers is large and the energy differences among them are small.

20.3.1
The Random Coil – Theory

Statistical descriptions are the natural way to characterize a large heterogeneous population, such as an ensemble of unfolded proteins. A few key ideas are mentioned here, but they are no substitute for the many excellent treatments of this subject [17, 18, 24, 32, 33].

The fundamental model is the freely jointed chain (or freely jointed random coil or random flight), a linear polymer of n adjoining links, each of fixed length, with complete freedom of rotation at every junction (Figure 20.2). From this definition, it follows that the angles at link junctions (i.e., bond angles) are completely uncorrelated. This model is completely general because it neglects chemical constraints, and therefore its scope is not restricted to any particular type of polymer chain. However, additional constraints such as chain thickness or hindered bond rotation can be added as appropriate, resulting in more specific models. What can be said about a polymer chain that is devoid of chemistry?

The freely jointed chain is equivalent to Brownian motion with a mean free path of fixed length, as described by Einstein-Smoluchowski theory [30]. The basic relationship governing both freely jointed chains and Brownian particles is:

$$\sqrt{\langle r^2 \rangle} = l\sqrt{n} \tag{1}$$

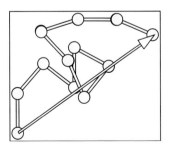

Fig. 20.2. A freely jointed chain. The chain is comprised of links, each of fixed length, l, with freedom of rotation at every junction. For a chain of n links, the vector from the beginning to the end, $\vec{r}_n$, (shown as the long arrow) is given by summing the links, $\vec{l}_i$: $\vec{r}_n = \sum_{i=1}^{n} \vec{l}_i$ and $|\vec{l}_i| = l$.

where $\sqrt{\langle r^2 \rangle}$ is the root-mean-square end-to-end distance (see Figure 20.2), l is the link length, and n is the number of links in the polymer. In other words, the distance between termini increases as the square root of the number of chain links: doubling the distance requires four times as many links.

The end-to-end distance measures the size of a polymer coil. Another such measure is the radius of gyration, R_G, the root-mean-square distance of link termini from their common center of gravity:

$$R_G^2 = \left[\frac{1}{n+1}\right] \sum_{i=0}^{n} R_{Gi}^2 \qquad (2)$$

where R_{Gi} is the distance of link i from the center of gravity and n is the number of links in the polymer. According to a theorem of Lagrange in 1783, R_G can be rewritten in terms of the individual link vectors, r_{ij}, without explicit reference to the center of gravity ([33], appendix A).

$$R_G^2 = \frac{1}{2n^2} \sum_{i=1}^{N} \sum_{j=1}^{N} r_{ij}^2 \qquad (3)$$

The two measures are related:

$$\langle R_G^2 \rangle = \frac{\langle r^2 \rangle}{6} \quad \text{as } n \to \infty \qquad (4)$$

For a freely jointed chain, the values of such configurational measures are Gaussian distributed.

Of course, no real chain is freely jointed. The chemical bonds in real chains restrict motion; bond rotations are never random. Also, each link of a real chain occupies a finite volume, thereby reducing the free volume accessible to remaining links. Accordingly, ideal chains descriptions must be modified if they are to accommodate such real-world constraints.

A strategy for accommodating restricted bond motion is to depart from physical chain links and instead re-represent the chain as though it were comprised of longer, uncorrelated virtual links. The idea underlying this strategy is as follows: a short chain segment (e.g., a dipeptide) is somewhat rigid [70], but a sufficiently long segment is flexible. Therefore, the chain becomes flexible at some length between the dipeptide and the longer peptide. This leads to the idea of an effective segment, $l_{\text{effective}}$, also called a Kuhn segment, the length scale at which chain segments approach independent behavior and correlated orientations between them dwindle away. A chain of length L contains $L/l_{\text{effective}}$ Kuhn segments and can be approximated as a freely jointed chain of Kuhn segments:

$$\langle r^2 \rangle = l_{\text{effective}}^2 \frac{L}{l_{\text{effective}}} = l_{\text{effective}} L \qquad (5)$$

A closely related idea is defined in terms of the chain's persistence length, the length scale over which correlations between bond angles "persist". In effect, the chain retains a "memory" of its direction at distances less than the persistence length. Stated less anthropomorphically, the energy needed to bend the chain through a 90° angle diminishes to $\sim kT/2$ at its persistence length, and thus ambient-temperature fluctuations can randomize the chain direction beyond this length. The size of a Kuhn segment is approximately two persistence lengths (i.e., directional correlations die away in either direction).

Current models strive to capture the properties of real chains with more detail than idealized, freely jointed chains can provide. For example, no actual chemical bond is a freely swiveling joint. To treat bond restrictions more realistically, Flory devised the rotational isomeric state approximation ([33], p. 55), in which bond angles are restricted to discrete values, chosen to correspond to known potential minima (e.g., gauche$^+$, gauche$^-$, and trans).

A real polymer chain cannot evade itself. Inescapably, the volume occupied by a chain element is excluded from occupancy by other chain elements. Otherwise, a steric clash would ensue: nonbonded atoms cannot occupy the same space at the same time. This excluded volume effect is substantial for proteins and results in a major departure from ideal chain dimensions (Eqs (1)–(4)).

As real polymers fluctuate, contracted coils have more opportunities to experience excluded volume steric clashes than expanded coils, perturbing the chain toward larger dimensions than those expected for ideal polymers.

Chain dimensions are also perturbed by the nature of the solvent. A good solvent promotes chain expansion by favoring chain:solvent interactions over chain:chain interactions. Conversely, a poor solvent promotes chain contraction by favoring chain:chain interactions over chain:solvent interactions. Flory introduced the idea of a Θ-solvent in which, on average, chain:chain interactions exactly counterbalance chain:solvent interactions, leading to unperturbed chain behavior. He pointed out that the notion of a Θ-solvent for a liquid is analogous to the Boyle point for a real gas, the temperature at which a pair of gas molecules follow an ideal isotherm because repulsion arising from volume exclusion is compensated exactly by mutual attraction ([33], p. 34).

Flory provided a simple relationship that relates the coil dimensions to solvent quality [33]. For a random coil polymer with excluded volume, the radius of gyration, R_G, is given by:

$$R_G = R_0 n^{\nu} \qquad (6)$$

where R_0 is a constant that is a function of the chain's persistence length, n is the number of links, and ν is the scaling factor of interest that depends on solvent quality. Values of v range from 0.33 for a collapsed, spherical molecule in poor solvent through 0.5 at the Θ-point (Eq. (1)) to 0.6 in good solvent.

Protein molecules are amphipathic, and their interactions with solvent are complex. However, on balance, denaturing agents such as urea and guanidinum chloride can be considered good solvents. Using Eq. (6), the degree to which unfolded proteins are random coil polymers in denaturing solvents can assessed by measuring v, the main topic of Section 20.3.2.

20.3.1.1 The Random Coil Model Prompts Three Questions

The random coil model set the stage for much of the contemporary theoretical work on unfolded proteins. A key question was brought into sharp focus by Levinthal [51]: if the random coil model holds, how can an unfolded protein discover its native conformation in biological real time? In particular, if unfolded protein molecules wander freely in a vast and featureless energy landscape, with barriers of order $\sim kT$, then three related questions arise:

1. *The kinetic question*: How does a protein discover its native conformation in biological real time? If restricted solely to the two most populated regions for a dipeptide, a 100-residue backbone would have $2^{100} \cong 10^{30}$ conformers. With bond rotations of order 10^{-13} s, the mean waiting time en route to the native conformation would be prohibitive just for the backbone. In actuality, experimentally determined folding times range from milliseconds to seconds.
2. *The thermodynamic question*: How does a protein compensate for the large conformational entropy loss on folding? With $2^{100} \cong 10^{30}$ conformers, the entropic price required to populate a single conformation $\cong 30 \times R \ln(10) \cong 40$ kcal mol^{-1} at room temperature, a conservative estimate.
3. *The dynamic question*: How does a protein avoid meta-stable traps en route to its native conformation? An equivalent way of asking this question is: why do proteins have a unique native conformation instead of a Boltzmann-distributed ensemble of native conformations?

Many investigators have sought to provide answers to these questions. Two notable examples are mentioned here, though only in bare outline.

20.3.1.2 The Folding Funnel

Following earlier work of Frauenfelder et al. [2], who suggested an analogy between proteins and spin glasses, Wolynes and coworkers introduced the notion of a folding funnel [16] to describe the progress of a protein population that traverses its energy landscape en route to the folded state. The favorable-high-entropy, unfavorable-high-energy unfolded state is conceptualized as a wide funnel mouth, while the unfavorable-low-entropy, favorable-low-energy native state corresponds to a narrow funnel spout. According to this conception, sloping funnel walls guide the population downhill toward the folded state from any starting point, answering question 1. During this downhill trajectory, lost entropy is progressively compensated by favorable pairwise interactions, answering question 2. Finally, meta-stable traps can be avoided if the funnel walls are sufficiently smooth [23], answering question 3. As a corollary, it is postulated that evolutionary pressures screen protein sequences, selecting those which can fold successfully in a funnel-like manner [94]. The folding funnel evokes a graphic portrait of folding dynamics and thermodynamics but is not intended to address specific structural questions, such as whether a region of interest will be helix or sheet.

20.3.1.3 Transition State Theory

Fersht and coworkers imported transition state theory from small molecule chemical reactions into protein folding [43]. Akin to the folding funnel, transition state kinetics focus on the entire population, with the transition state species pictured at the top of an energy barrier which separates U from N. But, unlike the folding funnel, only a few key residues comprise the organizational "tipping point", viz., those that participate in the transition state.

Questions 1–3 are not at issue for small molecule chemical reactions: (1) the mean waiting time for a reaction to occur depends upon a bond vibration, (2) after barrier crossing, the process is steeply downhill, and (3) intermediates between reactant and product are unstable because bond making/breaking energies are large. To the degree that the transition state approximation holds for protein folding [10], similar answers will obtain.

Transition state theory, expressed in the Eyring rate equation, transforms time-dependent kinetics into time-independent thermodynamics via an internal ticking clock: the rate of product formation depends upon the frequency of vibration of a critical bond. In contrast, no covalent bonds are made or broken in a folding reaction, and structure accretion is incremental and hierarchic en route from U to N [9, 10]. Not surprisingly then, the application of transition state theory to protein folding is complex [20].

Confidence in the application of the transition state approximation to protein folding comes from its success in describing simplified folding reactions [25] and the thermal unfolding of a β-hairpin [60]. However, recent work also illustrates the complexities. The transition state can be shifted dramatically without changing a protein's amino acid sequence [82, 100]. In simulations, the folding reaction can produce a broad ensemble of transition states instead of a single, well-defined species [50]. This blurring of the lines is further compounded by other work showing that the transition state resembles nearby folding intermediates [46] or is simply a distorted form of the native state [77].

20.3.1.4 Other Examples

The preceding examples illustrate how the random coil model has informed current thinking about unfolded proteins and the folding transition. The search for answers to the three questions has motivated other studies as well. In yet another example that focuses on question 3, Sali et al. [74, 75] analyzed the density of states in lattice simulations of folding and found a large energy gap – the e-gap – separating the native state (i.e., the ground state) from the nearest nonnative state (i.e., the 1st excited state). This finding rationalizes the predominance of the native state.

20.3.1.5 Implicit Assumptions from the Random Coil Model

Unfolded state models utilized in computer simulations often incorporate random coil assumptions implicitly. Four such assumptions are mentioned here.

1. The unfolded landscape is smooth. If the energy differences among sterically accessible backbone conformations are of order ~kT, the landscape will be devoid of kinetic traps and conformational bias. This assumption simplifies strategies for exploring the unfolded state in simulations.
2. The isolated-pair hypothesis is valid. Lattice models provide a way to count conformational alternatives explicitly, and they have been used extensively [18]. Most often, residues are represented as single lattice points (i.e., all residues are sterically equivalent on a lattice). Consequently, residue-specific steric restrictions beyond the dipeptide are either underweighted or ignored. This practice is valid to the degree that local steric repulsion does not extend beyond nearest chain neighbors, an assumption made explicit in Flory's isolated-pair hypothesis [33], which posits that each ϕ, ψ pair is sterically independent.
3. The Go approximation holds. A simplifying idea, introduced by Go [38], computes the energy of a conformation by rewarding native-like contacts while ignoring nonnative contacts, i.e., fortuitous nonnative contacts are not allowed to develop into kinetic traps. For simulations, this is a useful artifice that can be rationalized in a featureless landscape, where nonnative contacts dissolve as easily as they form.
4. Peptide backbone solvation is uniform. In other words, solvent water does not induce conformational bias in the unfolded state. If the interaction with water were energetically favored by some particular backbone conformation(s), then the unfolded landscape would be preferentially populated by these favored conformers, in violation of the featureless, random coil model.

These four assumptions are examined in Section 20.4.

20.3.2
The Random Coil – Experiment

Is a denatured protein aptly described as a random coil? It was Charles Tanford's experimental work that convinced the field. In numerous studies, Tanford demonstrated that proteins denatured in 6 M guanidinum chloride (GdmCl) have coil dimensions that obey simple scaling laws, consistent with random coil behavior. His masterful review of protein denaturation in *Advances in Protein Chemistry* [91, 92] is required reading for anyone interested in this topic.

In essence, the experimental strategy involves measuring coil dimensions for unfolded proteins in solution, fitting them to Eq. (6), and determining whether the scaling exponent, v, is consistent with a random coil polymer with excluded volume in good solvent. The excluded volume can be obtained directly from any practical technique that depends upon the colligative properties of the polymer solution, such as osmotic pressure. Using such techniques, concentration-dependent deviations from ideality arising from solute:solvent interactions are measured. To extract the excluded volume, the chemical potential of the polymer solution is expanded as a power series in solute concentration – the virial expansion. For purely repulsive interactions, the molar excluded volume is given by the second virial coefficient [78]. As mentioned above, excluded volume increases chain dimensions, with v ranging from 0.33 for a collapsed, spherical molecule in poor solvent through 0.5 at the Θ-point to 0.6 in good solvent.

Tanford documented many experimental pitfalls [91]. His investigations emphasized the importance of eliminating any potential residual structure in the unfolded protein by showing that the unfolding transition is complete. In fact, some residual structure is evident in heat-denatured proteins [3], but it can be eliminated in 6 M GdmCl. He also cautioned that the radius of gyration alone is an insufficient criterion for assessing random coil behavior, pointing out that a helical rod and a random polypeptide chain have similar values of R_G at chain lengths approximating those of ribonuclease and lysozyme.

20.3.2.1 Intrinsic Viscosity

In classic studies, Tanford used the intrinsic viscosity to determine coil dimensions. The intrinsic viscosity of a protein solution measures its effective hydrodynamic volume per gram in terms of the specific viscosity ([98], chapter 7). In particular, if η is the viscosity of the solution and η_0 is the viscosity of solvent alone, the specific viscosity, $\eta_{sp} = (\eta - \eta_0)/\eta_0$, is the fractional change in viscosity produced by adding solute. While η_{sp} is the quantity of interest, it is expressed in an experimentally inconvenient volume fraction concentration scale. This is remedied by converting to the intrinsic viscosity, $[\eta]$, which is η_{sp} normalized by the protein concentration, c, at infinite dilution: $[\eta] = \lim_{c \to 0} \left(\frac{\eta_{sp}}{c} \right)$. Specific viscosity is a pure number, so intrinsic viscosity has units of reciprocal concentration, milliliters per gram.

The intrinsic viscosity is not a viscosity per se but a viscosity increment owing to the volume fraction of solution occupied by the protein, like η_{sp}. It measures the hydrated protein volume, which will be much larger for randomly coiled molecules than for compactly folded ones; $[\eta]$ scales with chain length, n. The dependence of $[\eta]$ on n is conformation dependent, and Tanford took advantage of this fact. The relevant equation is:

$$[\eta] = Kn^x \tag{7}$$

where K is a constant that depends upon the molecular weight, but only slightly. Intrinsic viscosity is closely related to R_G, and Eqs (6) and (7) have a similar form. If unfolded proteins retain residual structure, each in their own way, the relation between $[\eta]$ and n is expected to be idiosyncratic. Conversely, a series of proteins that conform to Eq. (7) is indicative of random coil behavior.

In fact, for a series of 15 proteins denatured in GdmCl, a plot of $\log[\eta]$ vs. $\log n$ describes a straight line with slope 0.666, the exponent in Eq. (7) ([91], figure 6). The linearity of this series and the value of the exponent are strong evidence in favor of random coil behavior.

20.3.2.2 SAXS and SANS

Small angle X-ray scattering (SAXS) can be used to measure the radius of gyration, R_G, directly [58]. Molecules in a protein solution scatter radiation like tiny antennae ([99], chapter 7). In idealized situations, particles scatter independently (Rayleigh scattering), but significant interference occurs when intramolecular distances are of the same order as the wavelength of incident radiation, λ. This is the situation that obtains when proteins are irradiated with X-rays, and it is the basis for all experimental scattering techniques that yield R_G. In this case, the quantity of interest is $P(\theta)$, the ratio of measured intensity to the intensity expected for independent scattering by particles much smaller than λ, as a function of the scattering angle, θ. For a solution of scatterers,

$$P(\theta) = \frac{1}{n^2} \sum_{i=1}^{N} \sum_{j=1}^{N} \frac{\sinh r_{ij}}{h r_{ij}} \tag{8}$$

where n is the number of scattering centers, r_{ij} is the distance between any pair of centers i and j, and h is a function of the wavelength and scattering angle:

$$h = \frac{4\pi}{\lambda} \sin \frac{\theta}{2} \tag{9}$$

The double sum over all scattering centers is immediately reminiscent of Eq. (3), in which R_G is rewritten in terms of individual vectors, without explicit reference to their center of gravity. Van Holde et al. ([99], p. 321) show that

$$P(\theta) = \frac{1}{n^2} \sum_{i=1}^{N} \sum_{j=1}^{N} (1) - \frac{h^2}{6n^2} \sum_{i=1}^{N} \sum_{j=1}^{N} r_{ij}^2 \tag{10}$$

where the first term is unity and R_G is directly related to the double sum in the second term, as in Eq. (3).

Millett et al. used SAXS to determine R_G for a series of proteins under both denaturing and native conditions ([58], table I and fig. 4). Disulfide cross-links, if any, were reduced in the denatured species. Their experimentally determined values of R_G were fit to Eq. (6), giving values of $\nu = 0.61 \pm 0.03$ for the denatured proteins and $\nu = 0.38 \pm 0.05$ for their native counterparts (Figure 20.3). These values are remarkably close to those expected from theory, viz., $\nu = 0.6$ for a random coil with excluded volume in good solvent and $\nu = 0.33$ for a collapsed, spherical molecule in poor solvent.

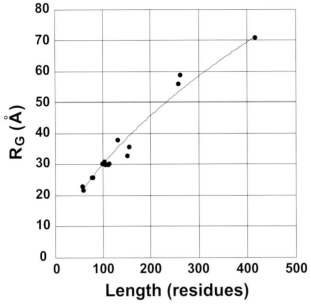

Fig. 20.3. The relationship between chain length and the radius of gyration, R_G, for a series of denatured proteins is well described by Eq. (6). Data points were taken from table 1 in [58], obtained using SAXS. The fitted curve has a value of $v = 0.61 \pm 0.03$, in close agreement with theory. This figure reproduces the one in Millett et al. ([58], fig. 4), but with omission of outliers.

The SAXS data provide the most compelling evidence to date in favor of the random coil model for denatured proteins.

20.4
Questions about the Random Coil Model

The random coil model would seem to be on firm ground at this point. However, recent work from both theory and experiment has raised new questions about the validity of the model – questions that provoke considerable controversy. Are they mere quibbles, or are they the prelude to a deeper understanding of the unfolded state?

Familiarity conditions intuition. At this point, the random coil model has conditioned our expectations for several decades. Should we be surprised that the dimensions of unfolded proteins are well described by a single exponent? Size matters here. As Al Holtzer once remarked, a steel I-beam behaves as a Gaussian coil if you make it long enough. But, at relevant length scales, the fact that proteins and polyvinyl behave similarly is quite unanticipated. After all, proteins adopt a unique folded state, whereas nonbiological polymers do not.

Flory emphasized this difference [33]: "Synthetic analogs of globular proteins are unknown. The capability of adopting a dense globular configuration stabilized by self-interactions and of transforming reversibly to the random coil are characteristics peculiar to the chain molecules of globular proteins alone" (p. 301).

The new questions center around the possibility of conformational bias and/or residual structure in unfolded proteins [11], even those unfolded in strong denaturing solvents like 6 M GdmCl [68]. We turn now to this discussion.

20.4.1
Questions from Theory

Superficially, the question of whether polypeptide chains are true random coils seems amenable to straightforward analysis by computer simulation. In principle, chains of n residues could be constructed, one at a time, using some plausible model (e.g., the Flory rotational isomeric model) to pick backbone dihedral angles. The coil dimensions and other characteristics of interest could then be analyzed by generating a suitable population of such chains. In practice, the excluded volume problem precludes this approach for chains longer than $\sim$20 residues, where the likelihood of encountering a steric clash increases sharply, killing off nascent chains before they can elongate. Naively, one might think that the problem can be solved by randomly adjusting offending residues until the clash is relieved, but this tactic biases the overall outcome. In fact, the only unbiased tactic is to rebuild the chain from scratch, resulting almost invariably in other clashes at new sites for chain lengths of interest. Such problems have thwarted attempts to analyze the unfolded population via simulation and modeling.

20.4.1.1 The Flory Isolated-pair Hypothesis

Nearly all theoretical treatments of the unfolded state assume that local steric repulsion does not extend beyond nearest chain neighbors. This simplifying assumption was made explicit in Flory's isolated-pair hypothesis ([33], p. 252), which posits that each ϕ, ψ pair is sterically independent of its neighbors.

Recently, the isolated-pair hypothesis was tested by exhaustive enumeration of all sterically accessible conformations of short polyalanyl chains [66]. To count, ϕ, ψ space was subdivided into a uniform grid. Every grid square, called a mesostate, encloses a $60° \times 60°$ range of ϕ, ψ values, with 36 mesostates in all. Each such mesostate was sampled extensively and at random to determine whether alanyl dipeptides with ϕ, ψ values in this range are sterically allowed. Only 14 mesostates are populated; the remaining 22 mesostates experience ubiquitous steric clashes throughout their entire range. Reconstruction of allowed ϕ, ψ space from mesostate sampling recreates the dipeptide map [70] and provides an acceptance ratio for each mesostate (Figure 20.4). The acceptance ratio, Λ_i, is the fraction of sterically allowed samples for mesostate i, ranging from 0 to 1.

These Λs were then used to test the isolated-pair hypothesis. Specifically, short polypeptide chains of length $n = 3\ldots7$ were tested by enumerating all possible strings over the 14 allowed mesostates and sampling them extensively. If the iso-

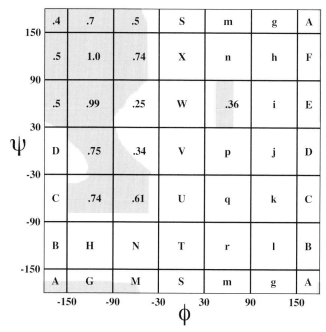

Fig. 20.4. Testing the Flory isolated-pair hypothesis. ϕ, ψ space for a dipeptide was subdivided into 36 alphabetically labeled coarse-grain grid squares, called mesostates. Treating the atoms as hard spheres, a Ramachandran plot (shown in gray) was computed by generating 150 000 randomly chosen ϕ, ψ conformations within each mesostate and testing for steric collisions [65]. Twenty-two meso-states have no allowed population; in each of these cases, every ϕ, ψ value results in a steric collision. In the 14 remaining mesostates, the fraction of sterically allowed samples, $0 < \Lambda_i \leq 1$, was determined, as shown.

lated-pair hypothesis holds, then ϕ, ψ angles in each mesostate are independent, and the fraction of sterically allowed conformers for each string is given by the product of individual acceptance ratios, $\prod_i^n \Lambda_i$. But, if there are steric clashes between nonnearest neighbors in the string, then $\Lambda_{string} < \prod_i^n \Lambda_i$, invalidating the hypothesis.

From this analysis, the isolated-pair hypothesis was found to be valid in the upper left quadrant of ϕ, ψ space but invalid in all other allowed regions. This finding makes sense physically: upon adopting ϕ, ψ values from the upper left quadrant, the chain is extended, like a β-strand, and nonnearest neighbors are separated. However, with ϕ, ψ values from any of the remaining five allowed mesostates (see Figure 20.4), the chain is contracted, like a helix or turn, and nonnearest neighbors are juxtaposed, enhancing opportunities for steric interference.

The failure of the isolated-pair hypothesis for short peptides ($n = 7$) challenges the random coil model, possibly in a major way. Steric restrictions obtain in the folded and unfolded states alike. The failure of the hypothesis for contracted chains implies that such conformers will be reduced selectively in the unfolded state, resulting in a population that is more extended than random coil expectations. Structurally, this shift in the population will result in a more homogeneous ensemble of unfolded conformers, and thermodynamically, it will reduce the entropy loss accompanying folding. But is it significant?

Studies of van Gunsteren et al. [97] and Sosnick and his colleagues [104] concur that the size of conformational space that can be accessed by unfolded molecules is restricted in peptides. However, Ohkubo and Brooks [63] argue that restrictions become rapidly insignificant as chain lengths grow beyond $n \cong 7$, with negligible consequences for the random coil model.

In an inventive approach to the problem, Goldenberg simulated populations of protein-sized chains [39] by adapting a standard software package that generates three-dimensional models from NMR-derived distance constraints. He analyzed the resultant unfolded state population using several measures, including coil dimensions, and found them to be well described as random coils. A note of caution is in order, however, because a substantial fraction of the conformers generated by this method fall within sterically restricted regions of ϕ, ψ space ([39], table 1).

20.4.1.2 Structure vs. Energy Duality

Often, the complex interplay between structure and energy has confounded simulations. Small changes in structure can give rise to large changes in energy, and conversely. From a structural point of view, two conformers are distinguishable when their ϕ, ψ angles differ. From a thermodynamic point of view, two conformers are indistinguishable when they can interconvert via a spontaneous fluctuation.

This structure–energy duality has contributed confusion to the Levinthal paradox (Section 20.3.1.1) and many related size estimates of the unfolded population because a single energy basin can span multiple conformers. For example, most sterically accessible conformers of short polyalanyl chains in good solvent [66] are quite extended, as expected in the absence of stabilizing intramolecular interactions. The ϕ, ψ values for these conformers are densely distributed over a broad region in the upper left quadrant of the ϕ, ψ map, as shown in Figure 20.5. When energy differences among these structures are calculated using a simple soft-sphere potential, the population partitions largely into two distinct energy basins, one that includes β-strands and another that includes polyproline II helices [65]. All conformers within each basin can interconvert spontaneously at room temperature (i.e., $\Delta A_{i,j} \leq RT$ at 300 K, where $\Delta A_{i,j}$ is the Helmholtz free energy difference between any two conformers, i and j, R is the universal gas constant and T the temperature in Kelvin). Thus, apparent structural diversity is reduced to two thermodynamically homogeneous populations.

20.4.1.3 The "Rediscovery" of Polyproline II Conformation

More than three decades ago, Tiffany and Krimm proposed that disordered peptides are comprised of left-handed polyproline II (P_{II}) helical segments inter-

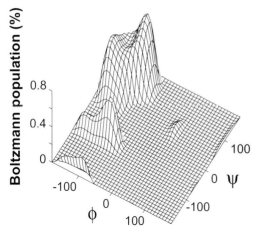

Fig. 20.5. A single energy basin can span multiple conformers. Most sterically accessible conformers of short polyalanyl chains in good solvent are extended. Using a soft-sphere potential, the Boltzmann-weighted population for the alanyl dipeptide is predominantly in the upper left quadrant of ϕ, ψ space and partitions into two distinct energy basins, one that includes polyproline II helices (larger) and another that includes β-strands (smaller) [65]. At 300 K, conformers with each basin interconvert spontaneously.

spersed with bends [95, 96]. They were led to this prescient proposal by the similarity between the optical spectra of P_{II} helices and nonprolyl homopolymers. Even earlier, Schellman and Schellman had already argued that the spectrum of unfolded proteins was unlikely to be that of a true random coil [79]. Following these early studies, the ensuing literature disclosed a noticeable similarity between the spectra of P_{II} and unfolded proteins, but such suggestive hints failed to provoke widespread interest – until recently. See Shi et al. for a thorough review [84].

The designation "polyproline" can be misleading. The circular dichroism (CD) spectrum, characteristic of actual polyproline or collagen peptides, has a pronounced negative band near 200 nm and a positive band near 220 nm. However, similar spectra can be obtained from peptides that are neither "poly" [54] nor proline-containing [96].

The P_{II} conformation is a left-handed helix with three residues per turn ($\phi, \psi \cong -75°, +145°$), resulting in three parallel columns spaced uniformly around the long axis of the helix. This helix has no intrasegment hydrogen bonds, and, in solution, significant fluctuations from the idealized structure are to be expected. The P_{II} conformation is forced by sterics in a polyproline sequence, but it is adopted readily by proline-free sequences as well [21].

Only three repetitive backbone structures are sterically accessible in proteins: α-helix, β-strand and P_{II}-helix [70]. In the folded population, α-helices and β-strands are abundant, whereas P_{II}-helices are rare. More specifically, isolated residues with P_{II} ϕ, ψ values are common in the non-α, non-β regions, accounting for approximately one-third of the remaining residues, but longer runs of consecutive residues with P_{II} ϕ, ψ-values are infrequent [90].

This finding can be rationalized by the fact that P_{II}-helices cannot participate in hydrogen bonds in globular proteins. Hydrogen bonds are eliminated because the spatial orientation of backbone donors and acceptors is incompatible with both intrasegment hydrogen bonding within P_{II}-helices and regular extra-segment hydrogen bonding between P_{II} helices and the three repetitive backbone structures. Upon folding, those backbone polar groups deprived of hydrogen-bonded solvent access can make compensatory hydrogen bonds in α-helices and strands of β-sheet, but not in P_{II}-helices.

Recent work by Creamer and coworkers focused renewed attention on P_{II} [21, 73, 90], raising the question of whether fluctuating P_{II} conformation might contribute substantially to the unfolded population in proteins [96]. Studies performed during the past few years lend support to this idea, as described next.

20.4.1.4 P_{II} in Unfolded Peptides and Proteins

The blocked peptide, N-acetylalanine-N'-methylamide, is a popular backbone model. Many groups have found P_{II} to be an energetically preferred conformation for this peptide in water [1, 26, 40, 41, 44, 69]. Does this finding hold for longer chains?

Again using alanine as a model, Pappu and Rose analyzed the conformational preferences of longer blocked polyalanyl chains, N-acetyl-Ala$_n$-N'-methylamide ($n \leq 7$) [65]. To capture nonspecific solvent effects, they minimized chain:chain interactions, mimicking the chain's expected behavior in good solvent. At physiological temperature, only three energy basins were needed to span $\sim 75\%$ of the population, and within each basin, the population of structures was homogeneous. Notably, the basin corresponding to P_{II} structure was dominant.

Pappu and Rose [65] used soft-sphere repulsion (the repulsive term in a Lennard-Jones potential) to calculate energy. More extensive testing using detailed force fields was performed by Sosnick and coworkers [104].

It is often assumed that the backbone is solvated uniformly in the unfolded state and that the energy of solvent stripping upon folding is not a significant consideration. This assumption follows directly from the random coil model, in which unfolded conformers are readily interconvertible. However, if unfolded state conformers exhibit conformational biases, it becomes important to question this assumption. Is solvation free energy conformation dependent?

A series of papers by Avbelj and Baldwin [4–6] offered a fresh perspective on this issue, motivated by an inconsistency between the measured energy of peptide hydrogen bond formation [81] and the corresponding energy derived from a simple thermodynamic cycle [8]. Specifically, their analysis uncovered a large enthalpy deficit (-7.6 kcal mol^{-1}) upon helix formation that could not be reconciled with data from typical model compounds, such as acetamide derivatives [6]. One or more terms had to be missing.

Avbelj and Baldwin's work prompted a re-examination of peptide solvation in proteins by a number of groups, including themselves [4–6]. Of particular interest are a series of unrelated simulations [4–6, 25, 35, 47, 57, 104], all of which reach a common conclusion: water interacts preferentially with P_{II} peptides, imparting a previously unsuspected conformational bias.

In sum, both peptide:solvent interactions and peptide:peptide interactions [65] favor P_{II} conformers. In the former case, water is simply a better solvent for P_{II} than for other conformers, e.g., β-strands and α-helices. In the latter case, P_{II} affords the chain greater entropic freedom (i.e., more "wiggle room").

20.4.2
Questions from Experiment

Early NMR studies provided evidence for residual structure in the denatured state of both proteins [36, 61] and peptides excised from proteins [28]. However, the structured regions seen in proteins were not extensive. Furthermore, most isolated peptides lacked structure, and the few exceptions did not always retain the conformation adopted in the native protein [27].

Peptide studies tell a similar story. A prime example involves the assessment of autonomous stability in the α-helix. Early evidence indicated that the cooperative unit for stable helix formation is ~ 100 residues [105], a length that exceeds the average protein helix (~ 12 residues) by almost an order of magnitude. Consequently, the prevailing view in the 1970s was that protein-sized helical peptides would be random coils in isolation. This view was reversed in the 1980s, after Bierzynski et al. [12], expanding upon earlier work by Brown and Klee [15], demonstrated helix formation in water at near-physiological temperature for residues 1–13 of ribonuclease, a cyanogen bromide cleavage product. This finding prompted a re-evaluation of helix propensities in peptides [53, 56, 62, 64] and motivated numerous biophysical studies of peptides [80]. Summarizing this large body of work, there is evidence for structure in some short peptides in aqueous solvent at physiological temperature, but it is marginal at best and, more often, undetectable altogether.

20.4.2.1 Residual Structure in Denatured Proteins and Peptides

The limited success of these early attempts to detect residual structure strengthened the conviction that denaturation abolishes structure and reinforced the notion that the unfolded state is a random coil. Consequently, the field was stunned when Shortle and Ackerman [85] demonstrated the persistence of native-like structure in staphylococcal nuclease under strongly denaturing conditions (8 M urea). Shortle and Ackerman's finding was based on evidence from residual dipolar couplings in oriented gels. However, their interpretation that these data provide evidence of global organization was questioned recently by Annila and coworkers [52]. The ultimate conclusions from such work are still unclear, but the perspective has definitely changed and many recent experiments now find evidence for substantial residual structure in the denatured state (e.g., [22, 34, 46, 76, 88, 102]).

In a similar vein, Shi et al. reanalyzed a blocked peptide containing seven consecutive alanine residues for the presence of residual structure [83]. This peptide is too short to form a stable α-helix and should therefore be a random coil. Contrary to this expectation, the peptide is largely in P_{II} conformation, in agreement with predictions from theory [65]. While not all residues are expected to favor the P_{II} conformation [73], this result shows that the unfolded state is predominantly a single conformer, at least in the case of polyalanine.

20.4.3
The Reconciliation Problem

The measured radii of gyration, R_G, of denatured proteins have values [58] that are consistent with those expected for a random coil with excluded volume in good solvent (Section 20.3.2.2). Yet, experimental evidence in both proteins [85] and peptides [83] suggests the presence of residual structure in the unfolded population. How are these seemingly contradictory findings to be reconciled? Millett et al. refer to this as the reconciliation problem ([58], see their discussion, p. 257).

Paradox is often a prelude to perception. Equation (6), in its generality, necessarily neglects the chemical details of any particular polymer type. Accordingly, the resultant chain description is insensitive to short-range order, apart from the proportionality constant R_0, which is a function of the persistence length (Section 20.3.1). Sterically induced local order, encapsulated in R_0, is surely present in unfolded proteins [31, 66], but can it rationalize the apparent contradiction between random coil R_G values and global residual structure [85]? One possible explanation is that multiple regions of local structure dominate the ensemble average to such an extent that they are interpreted as global organization [52, 103].

The coil library may provide a useful clue to the resolution of this puzzle. The coil library is the collection of all nonrepetitive elements in proteins of known structure, that fraction of native structure which remains after α-helix and β-sheet are removed. Given that the library is composed of fragments extracted from solved structures, it is surely not "coil" in the polymer sense. However, the term "coil library" is intended to convey the hypothesis that such fragments do, in fact, represent the full collection of accessible chain conformers in unfolded proteins [87]. Taken to its logical conclusion, this hypothesis posits that the coil library is a collection of structured fragments in folded proteins and, at the same time, a collection of unstructured fragments in unfolded proteins. If so, this library, together with α-helix, β-strand, and P_{II} helix, represents an explicit enumeration of accessible conformers from which the unfolded ensemble might be reconstructed [5].

At this writing, the reconciliation problem remains an ongoing question. Regardless of the eventual outcome, this paradox appears to be moving the field in an informative direction.

20.4.4
Organization in the Unfolded State – the Entropic Conjecture

Are there general principles that lead to organization in the unfolded state? If accessible conformational space is vast and undifferentiated, the entropic cost of populating the native basin exclusively will be large. However, if the unfolded state is largely restricted to a few basins, with nonuniform, sequence-dependent basin preferences, then entropy can function as a chain organizer.

Consider two thermodynamic basins, i and j. The Boltzmann-weighted ratio of their populations, n_i/n_j, is given by $(w_i/w_j)e^{-\beta \Delta U}$, where n_i and n_j designate the number of conformers in i and j, w_i and w_j are the degeneracies of state, β is the Boltzmann factor, and ΔU is the energy difference between the two basins. Both entropy and enthalpy contribute to this ratio. If w_i and w_j are conformational biases (i.e., the number of isoenergetic ways the chain can adopt conformations i and j), and w_i/w_j is the dominant term in the Boltzmann ratio, the entropy difference, $\Delta S_{conf} = R \ln(w_i/w_j)$, would promote organization in the unfolded population.

In particular, if P_{II} is a dominant conformation in polyalanyl peptides, then it is also likely to be favored in unfolded proteins, in which case the unfolded state is not as heterogeneous as previously believed. The usual estimate of about five accessible states per residue in an unfolded protein is based on a familiar argument: the free energy difference between the folded and unfolded populations, ΔG_{conf}, is a small difference between large value of ΔH_{conf} and $T\Delta S_{conf}$ [13, 14]. If $\Delta G_{conf} \cong -10$ kcal mol^{-1} (a typical value) and $\Delta H_{conf} \cong 100$ kcal mol^{-1}, then the counterbalancing $T\Delta S_{conf}$ is also $\sim$100 kcal mol^{-1}. Then $\Delta S_{conf} \cong 3.33$ entropy units per residue for a 100-residue protein at 300 K. Assuming $\Delta S_{conf} = R \ln W$, the number of states per residue, W, is 5.34.

However, instead of a reduction in the number of distinct states, this entropy loss on folding could result from a reduction in the degeneracy of a single state, providing the ϕ, ψ space of occupied regions in the unfolded population is further constricted upon folding. For example, a residue in P_{II} is within a room-temperature fluctuation of any sterically allowed ϕ, ψ value in the upper left quadrant of the dipeptide map ([57], table 1). Consequently, different ϕ, ψ values from these regions would be thermodynamically indistinguishable and therefore not distinct states at all. As a back-of-the-envelope approximation, consider a residue that can visit any allowed region of the upper left quadrant in the unfolded state. Upon folding, let this residue be constrained to lie within $\pm 30°$ of ideal β-sheet ϕ, ψ values. The reduction in ϕ, ψ space would be a factor of 5.58, approximating the value attributed to distinct states. Similar, but less approximate, estimates can be obtained when the unfolded populations are Boltzmann weighted.

What physical factors might underwrite such entropy effects?

20.4.4.1 Steric Restrictions beyond the Dipeptide

It has long been believed that local steric restrictions do not extend beyond the dipeptide boundary [70], but, on re-analysis, this conviction requires revision [11, 66, 89] (see Section 20.4.1.1). In fact, systematic steric restrictions operate over chain regions of several adjacent residues, and they serve to promote organization in unfolded protein chains. Two recent lines of investigation focus on identifying the physical basis for longer range, sterically induced ordering.

In a series of remarkable papers, Banavar, Maritan and their colleagues show that chain thickness alone imposes stringent, previously unrecognized restrictions on conformational space [7, 42, 55]. All the familiar secondary structure motifs emerge automatically when the protein is represented as a self-avoiding tube, coaxial with the main chain, and a single inequality is imposed on all triples of Cα atoms [7, 55]. The further addition of simple hydrogen bond and hydrophobic terms is sufficient to generate the common super-secondary structures [42]. These straightforward geometric considerations demonstrate that sequence-independent steric constraints predispose proteins toward their native repertoire of secondary and super-secondary structural motifs.

Investigating the atomic basis for longer range steric restrictions, Fitzkee and Rose found that a direct transition from an α-helix to a β-strand causes an unavoidable steric collision between backbone atoms [31]. Specifically, a nonnearest neighbor collision occurs between the carbonyl oxygens of an α-residue at position i ($O^\alpha{}_i$) and a β-residue at position i + 3 ($O^\beta{}_{i+3}$). This restriction also holds for the transition from α-helix to P_{II}. These simple steric constraints have pervasive organizational consequences for unfolded proteins because they eliminate all structural hybrids of the form ... $\alpha\alpha\alpha\beta$ ... and ... $\alpha\alpha\alpha P_{II}$ ..., pushing the unfolded population toward pure segments of $\alpha, \beta,$ and P_{II} interconnected by irregular regions such as those found in the coil library.

20.5
Future Directions

The early analysis of steric restrictions in the alanyl dipeptide (more precisely, the compound Cα-CO-NH–CαHR–CO-NH-Cα, which has two degrees of backbone freedom like a dipeptide) by Ramachandran et al. [71] has become one of those rare times in biochemistry where theory is deemed sufficient to validate experiment [49]. The fact that the dipeptide map is based only on "hard sphere" repulsion alone led some to underestimate the generality of this work, but not Richards, who commented [72]:

> *For chemically bonded atoms the distribution is not spherically symmetric nor are the properties of such atoms isotropic. In spite of all this, the use of the hard sphere model has a venerable history and an enviable record in explaining a variety of different observable properties. As applied specifically to proteins, the work of G. N. Ramachandran and his colleagues has provided much of our present thinking about permissible peptide chain conformations.*

The notion that repulsive interactions promote macromolecular organization is not limited to the alanyl dipeptide. Space-filling models [48], which represent each atom literally as a hard sphere, were central to Pauling's successful model of the α-helix [67] and have widespread application throughout chemistry. Much of the theory of liquids is based on the organizing influence of repulsion interactions [19].

Despite such successes, the existence of sterically induced chain organization has had little influence on models of the unfolded state owing to the strongly held conviction that local steric restrictions extend no further than adjacent chain neighbors. Of course, long-range excluded volume effects do affect the population [18, 32], as reflected in the exponent of Eq. (6), but they are not thought to play any role in biasing unfolded proteins toward specific conformations. Given the finding of local steric restrictions beyond the dipeptide (Section 20.4.4.1), it is time to re-analyze the problem.

Re-analysis will involve at least three steps: (1) analysis of local steric restrictions beyond the dipeptide, (2) characterization of elements in the coil library, and (3) combination of the results from these two steps. To the extent that useful insights emerge from this prescription, the folding problem may not be as intractable as previously thought.

Acknowledgments

We thank Buzz Baldwin, Nicholas Fitzkee, Haipeng Gong, Nicholas Panasik, Kevin Plaxco, and Timothy Street for stimulating discussion and The Mathers Foundation for support.

References

1 ANDERSON, A. G., and HERMANS, J. 1988. Microfolding: conformational probability map for the alanine dipeptide in water from molecular dynamics simulations. *Proteins* **3**, 262–265.

2 ANSARI, A., BERENDZEN, J., BOWNE, S. F., FRAUENFELDER, H., IBEN, I. E., SAUKE, T. B., SHYAMSUNDER, E., and YOUNG, R. D. 1985. Protein states and proteinquakes. *Proc. Natl Acad. Sci. USA* **82**, 5000–5004.

3 AUNE, K. C., SALAHUDDIN, A., ZARLENGO, M. H., and TANFORD, C. 1967. *J. Biol. Chem.* **242**, 4486–4489.

4 AVBELJ, F. and BALDWIN, R. L. 2002. Role of backbone solvation in determining thermodynamic beta propensities of the amino acids. *Proc. Natl Acad. Sci. USA* **99**, 1309–1313.

5 AVBELJ, F. and BALDWIN, R. L. 2003. Role of backbone solvation and electrostatics in generating preferred peptide backbone conformations: Distributions of phi. *Proc. Natl Acad. Sci. USA* **100**, 5742–5747.

6 AVBELJ, F., LUO, P., and BALDWIN, R. L. 2000. Energetics of the interaction between water and the helical peptide group and its role in determining helix propensities. *Proc. Natl Acad. Sci. USA* **97**, 10786–10791.

7 BANAVAR, J., MARITAN, A., MICHELETTI, C., and TROVATO, A. 2002. Geometry and physics of proteins. *Proteins* **47**, 315–322.

8 BALDWIN, R. L. 2003. In search of the energetic role of peptide hydrogen bonds. *J. Biol. Chem.* **278**, 17581–17588.

9 BALDWIN, R. L. and ROSE, G. D. 1999a. Is protein folding hierarchic? I. Local structure and peptide folding. *Trends Biochem. Sci.* **24**, 26–33.

10 BALDWIN, R. L. and ROSE, G. D.

1999b. Is protein folding hierarchic? II. Folding intermediates and transition states. *Trends Biochem. Sci.* **24**, 77–83.

11 BALDWIN, R. L. and ZIMM, B. H. 2000. Are denatured proteins ever random coils? *Proc. Natl Acad. Sci. USA* **97**, 12391–12392.

12 BIERZYNSKI, A., KIM, P. S., and BALDWIN, R. L. 1982. A salt-bridge stabilizes the helix formed by isolated C-peptide of RNase A. *Proc. Natl Acad. Sci. USA* **79**, 2470–2474.

13 BRANDTS, J. F. 1964a. The thermodynamics of protein denaturation. I. The denaturation of chymotrypsinogen. *J. Am. Chem. Soc.* **86**, 4291–4301.

14 BRANDTS, J. F. 1964b. The thermodynamics of protein denaturation. II. A model of reversible denaturation and interpretations regarding the stability of chymotrypsinogen. *J. Am. Chem. Soc.* **86**, 4302–4314.

15 BROWN, J. E. and KLEE, W. A. 1971. Helix-coil transition of the isolated amino terminus of ribonuclease. *Biochemistry* **10**, 470–476.

16 BRYNGELSON, J. D., ONUCHIC, J. N., SOCCI, N. D., and WOLYNES, P. G. 1995. Funnels, pathways and the energy landscape of protein folding-a synthesis. *Proteins Struct. Funct. Genet.* **21**, 167–195.

17 CANTOR, C. R. and SCHIMMEL, P. R. 1980. *Biophysical Chemistry*. Part III: *The Behavior of Biological Macromolecules*. Freeman, New York.

18 CHAN, H. S. and DILL, K. A. 1991. Polymer principles in protein structure and stability. *Annu. Rev. Biophys. Chem.* **20**, 447–490.

19 CHANDLER, D., WEEKS, J. D., and ANDERSEN, H. C. 1983. The van der Waals picture of liquids, solids and phase transformations. *Science* **220**, 787–794.

20 CIEPLAK, M. and HOANG, T. X. 2003. Universality classes in folding times of proteins. *Biophys. J.* **84**, 475–488.

21 CREAMER, T. P. 1998. Left-handed polyproline II helix formation is (very) locally driven. *Proteins* **33**, 218–226.

22 DAGGETT, V., LI, A., ITZHAKI, L. S., OTZEN, D. E., and FERSHT, A. R. 1996. Structure of the transition state for folding of a protein derived from experiment and simulation. *J. Mol. Biol.* **257**, 430–440.

23 DILL, K. A. and CHAN, H. S. 1997. From Levinthal to pathways to funnels. *Nat. Struct. Biol.* **4**, 10–19.

24 DILL, K. A. and SHORTLE, D. 1991. Denatured states of proteins. *Annu. Rev. Biochem.* **60**, 795–825.

25 DOYLE, R., SIMONS, K., QIAN, H., and BAKER, D. 1997. *Proteins Struct. Funct. Genet.* **29**, 282–291.

26 DROZDOV, A. N., GROSSFIELD, A., and PAPPU, R. V. 2004. The role of solvent in determining conformational preferences of alanine dipeptide in water. *J. Am. Chem. Soc.* **126**, 2574–2581.

27 DYSON, H. J., RANCE, M., HOUGHTEN, R. A., LERNER, R. A., and WRIGHT, P. E. 1988. Folding of immunogenic peptide fragments of proteins in water solution. I. Sequence requirements for the formation of a reverse turn. *J. Mol. Biol.* **201**, 161–200.

28 DYSON, H. J., SAYRE, J. R., MERUTKA, G., SHIN, H. C., LERNER, R. A., and WRIGHT, P. E. 1992. Folding of peptide fragments comprising the complete sequence of proteins. Models for initiation of protein folding II. Plastocyanin. *J. Mol. Biol.* **226**, 819–835.

29 EDSALL, J. T. 1995. Hsien Wu and the first theory of protein denaturation (1931). In *Advances in Protein Chemistry* (eds D. S. EISENBERG, and F. M. RICHARDS), pp. 1–26. Academic Press, San Diego.

30 EINSTEIN, A. 1956. *Investigations on the Theory of Brownian Movement*. Dover Publications, New York.

31 FITZKEE, N. C. and ROSE, G. D. 2004. Steric restrictions in protein folding: an α-helix cannot be followed by a contiguous β-strand. *Protein Sci.* **13**, 633–639.

32 FLORY, P. J. 1953. *Principles of Polymer Chemistry*. Cornell University Press, New York.

33 FLORY, P. J. 1969. *Statistical Mechanics of Chain Molecules*. Wiley, New York.

34 GARCIA, P., SERRANO, L., DURAND, D., RICO, M., and BRUIX, M. 2001. NMR and SAXS characterization of the denatured state of the chemotactic protein CheY: implications for protein folding initiation. *Protein Science* **10**, 1100–1112.

35 GARCIA, A. E. 2004. Characterization of non-alpha helical conformations in Ala peptides. *Polymer* **45**, 669–676.

36 GARVEY, E. P., SWANK, J., and MATTHEWS, C. R. 1989. A hydrophobic cluster forms early in the folding of dihydrofolate reductase. *Proteins Struct. Funct. Genet.* **6**, 259–266.

37 GINSBURG, A. and CARROLL, W. R. 1965. Some specific ion effects on the conformation and thermal stability of ribonuclease. *Biochemistry* **4**, 2159–2174.

38 GO, N. 1984. The consistency principle in protein structure and pathways of folding. *Adv. Biophys.* **18**, 149–164.

39 GOLDENBERG, D. P. 2003. Computational simulation of the statistidal properties of unfolded proteins. *J. Mol. Biol.* **326**, 1615–1633.

40 GRANT, J. A., WILLIAMS, R. L., and SCHERAGA, H. A. 1990. Ab initio self-consistent field and potential-dependent partial equalization of orbital electronegativity calculations of hydration properties of N-acetyl-N′-methyl-alanineamide. *Biopolymers* **30**, 929–949.

41 HAN, W.-G., JALKANEN, K. J., ELSTNER, M., and SUHAI, S. 1998. Theoretical study of aqueous N-acetyl-L-alanine N′-methylamide: structures and raman, VCD and ROA spectra. *J. Phys. Chem. B* **102**, 2587–2602.

42 HOANG, T. X., TROVATO, A., SENO, F., BANAVAR, J., and MARITAN, A. 2004. What determines the native state folds of proteins? submitted.

43 ITZHAKI, L. S., OTZEN, D. E., and FERSHT, A. R. 1995. The structure of the transition state for folding of chymotrypsin inhibitor 2 analysed by protein engineering methods: evidence for a nucleation-condensation mechanism for protein folding. *J. Mol. Biol.* **254**, 260–288.

44 JALKANEN, K. J. and SUHAI, S. 1996. N-Acetyl-L-Alanine N′-methylamide: a density functional analysis of the vibrational absorption and birational circular dichroism spectra. *Chem. Phys.* **208**, 81–116.

45 KAUZMANN, W. 1959. Some factors in the interpretation of protein denaturation. *Adv. Protein Chem.* **14**, 1–63.

46 KAZMIRSKI, S. L., WONG, K. B., FREUND, S. M. V., TAN, Y. J., FERSHT, A. R., and DAGGETT, V. 2001. Protein folding from a highly disordered denatured state: the folding pathway of chymotrypsin inhibitor 2 at atomic resolution. *Proc. Natl Acad. Sci. USA* **98**, 4349–4354.

47 KENTSIS, A., GINDIN, T., MEZEI, M., and OSMAN, R. 2004. Unfolded state of polyalanine is a segmented polyproline II helix. *Proteins* **55**, 493–501.

48 KOLTUN, W. L. 1965. Precision space-filling atomic models. *Biopolymers* **3**, 665–679.

49 LASKOWSKI, R. A., MACARTHUR, M. W., MOSS, D. S., and THORNTON, J. M. 1993. PROCHECK: a program to check the stereochemical quality of protein structures. *J. Appl. Cryst.* **26**, 283–291.

50 LAZARIDIS, T. and KARPLUS, M. 1997. *Science* **278**, 1928–1931.

51 LEVINTHAL, C. 1969. How to fold graciously. Mossbauer spectroscopy in Biological Systems, Proceedings. *University of Illinois Bull.* **41**, 22–24.

52 LOUHIVUORI, M., PAAKKONEN, K., FREDRIKSSON, K., PERMI, P., LOUNILA, J., and ANNILA, A. 2003. On the origin of residual dipolar couplings from denatured proteins. *J. Am. Chem. Soc.* **125**, 15647–15650.

53 LYU, P. C., LIFF, M. I., MARKY, L. A., and KALLENBACH, N. R. 1990. Side chain contributions to the stability of alpha-helical structure in peptides. *Science* **250**, 669–673.

54 MADISON, V. and SCHELLMAN, J. A. 1970. Diamide model for the optical activity of collagen and polyproline I and II. *Biopolymers* **9**, 65–94.

55 MARITAN, A., MICHELETTI, C., TROVATO, A., and BANAVAR, J. 2000.

Optimal shapes of compact strings. *Nature* **406**, 287–290.

56 MERUTKA, G., LIPTON, W., SHALONGO, W., PARK, S. H., and STELLWAGEN, E. 1990. Effect of central-residue replacements on the helical stability of a monomeric peptide. *Biochemistry* **29**, 7511–7515.

57 MEZEI, M., FLEMING, P. J., SRINIVASAN, R., and ROSE, G. D. 2004. Polyproline II helix is the preferred conformation for unfolded polyalaine in water. *Proteins* **55**, 502–507.

58 MILLETT, I. S., DONIACH, S., and PLAXCO, K. W. 2002. Toward a taxonomy of the denatured state: small angle scattering studies of unfolded proteins. *Adv. Protein Chem.* **62**, 241–262.

59 MIRSKY, A. E. and PAULING, L. 1936. On the structure of native, denatured, and coagulated proteins. *Proc. Natl Acad. Sci. USA* **22**, 439–447.

60 MUNOZ, V., THOMPSON, P. A., HOFRICHTER, J., and EATON, W. A. 1997. *Nature* **390**, 196–199.

61 NERI, D., BILLETER, M., WIDER, G., and WUTHRICHT, K. 1992. NMR determination of residual structure in a urea-denatured protein, the 434-repressor. *Science* **257**, 1559–1563.

62 O'NEIL, K. T. and DEGRADO, W. F. 1990. A thermodynamic scale for the helix-forming tendencies of the commonly occurring amino acids. *Science* **250**, 646–651.

63 OHKUBO, Y. Z. and BROOKS, C. L. 2003. Exploring Flory's isolated-pair hypothesis: statistical mechanics of helix-coil transitions in polyalanine and C-peptide from RNase A. *Proc. Natl Acad. Sci. USA* **100**, 13916–13921.

64 PADMANABHAN, S., MARQUSEE, S., RIDGEWAY, T., LAUE, T. M., and BALDWIN, R. L. 1990. Relative helix-forming tendencies of nonpolar amino acids. *Nature* **344**, 268–270.

65 PAPPU, R. V. and ROSE, G. D. 2002. A simple model for polyproline II structure in unfolded states of alanine-based peptides. *Protein Sci.* **11**, 2437–2455.

66 PAPPU, R. V., SRINIVASAN, R., and ROSE, G. D. 2000. The flory isolated-pair hypothesis is not valid for polypeptide chains: implications for protein folding. *Proc. Natl Acad. Sci. USA* **97**, 12565–12570.

67 PAULING, L., COREY, R. B., and BRANSON, H. R. 1951. The structures of proteins: two hydrogen-bonded helical configurations of the polypeptide chain. *Proc. Natl Acad. Sci. USA* **37**, 205–210.

68 PLAXCO, K. W. and GROSS, M. 2001. Unfolded, yes, but random? Never! *Nat. Struct. Biol.* **8**, 659–660.

69 POON, C.-D. and SAMULSKI, E. T. 2000. Do bridging water molecules dictate the structure of a model dipeptide in aqueous solution? *J. Am. Chem. Soc.* **122**, 5642–5643.

70 RAMACHANDRAN, G. N. and SASISEKHARAN, V. 1968. Conformation of polypeptides and proteins. *Adv. Protein Chem.* **23**, 283–438.

71 RAMACHANDRAN, G. N., RAMAKRISHNAN, C., and SASISEKHARAN, V. 1963. Stereochemistry of polypeptide chain configurations. *J. Mol. Biol.* **7**, 95–99.

72 RICHARDS, F. M. 1977. Areas, volumes, packing, and protein structure. *Annu. Rev. Biophys. Bioeng.* **6**, 151–176.

73 RUCKER, A. L., PAGER, C. T., CAMPBELL, M. N., QUALLS, J. E., and CREAMER, T. P. 2003. Host-guest scale of left-handed polyproline II helix formation. *Proteins* **53**, 68–75.

74 SALI, A., SHAKHNOVICH, E., and KARPLUS, M. 1994a. Kinetics of protein folding – a lattice model study of the requirements for folding to the native state. *J. Mol. Biol.* **5**, 1614–1636.

75 SALI, A., SHAKHNOVICH, E. I., and KARPLUS, M. 1994b. How does a protein fold? *Nature* **477**, 248–251.

76 SANCHEZ, I. E., and KIEFHABER, T. 2003. Origin of Unusual ϕ-values in protein folding: evidence against specific nucleation sites. *J. Mol. Biol.* **334**, 1077–1085.

77 SANCHEZ, I. E. and KIEFHABER, T. 2003. Origin of unusual ϕ-values in protein folding: evidence against specific nucleation sites. *J. Mol. Biol.* **334**, 1077–1085.

78 Schellman, J. A. 2002. Fifty years of solvent denaturation. *Biophys. Chem.* **96**, 91–101.

79 Schellman, J. A. and Schellman, C. G. 1964. In *The Proteins: Composition, Structure and Function*, 2nd edn, Vol 2, pp. 1–37. Academic Press, New York.

80 Scholtz, J. M. and Baldwin, R. L. 1992. The mechanism of alpha-helix formation by peptides. *Annu. Rev. Biophys. Biomol. Struct.* **21**, 95–118.

81 Scholtz, J. M., Marqusee, S., Baldwin, R. L., York, E. J., Stewart, J. M., Santoro, M., and Bolen, D. W. 1991. Calorimetric determination of the enthalpy change for the alpha-helix to coil transition of an alanine peptide in water. *Proc. Natl Acad. Sci. USA* **88**, 2854–2858.

82 Shastry, M. C. and Udgaonkar, J. B. 1995. The folding mechanism of barstar: evidence for multiple pathways and multiple intermediates. *J. Mol. Biol.* **247**, 1013–1027.

83 Shi, Z., Olson, C. A., Rose, G. D., Baldwin, R. L., and Kallenbach, N. R. 2002a. Polyproline II structure in a sequence of seven alanine residues. *Proc. Natl Acad. Sci. USA* **2002**, 9190–9195.

84 Shi, Z., Woody, R. W., and Kallenbach, N. R. 2002b. Is polyproline II a major backbone conformation in unfolded proteins? *Adv. Protein Chem.* **62**, 163–240.

85 Shortle, D. and Ackerman, M. 2001. Persistence of native-like topology in a denatured protein in 8 M urea. *Science* **293**, 487–489.

86 Simpson, R. B. and Kauzmann, W. 1953. The kinetics of protein denaturation. *J. Am. Chem. Soc.* **75**, 5139–5192.

87 Smith, L. J., Bolin, K. A., Schwalbe, H., MacArthur, M. W., Thornton, J. M., and Dobson, C. M. 1996. Analysis of main chain torsion angles in proteins: prediction of NMR coupling constants for native and random coil conformations. *J. Mol. Biol.* **255**, 494–506.

88 Sridevi, K., Lakshmikanth, G. S., Krishnamoorthy, G., and Udgaonkar, J. B. 2004. Increasing stability reduces conformational heterogeneity in a protein folding intermediate ensemble. *J. Mol. Biol.* **337**, 669–711.

89 Srinivasan, R. and Rose, G. D. 1999. A physical basis for protein secondary structure. *Proc. Natl Acad. Sci. USA* **96**, 14258–14263.

90 Stapley, B. J. and Creamer, T. P. 1999. A survey of left-handed polyproline II helices. *Protein Sci.* **8**, 587–595.

91 Tanford, C. 1968. Protein denaturation. *Adv. Protein Chem.* **23**, 121–282.

92 Tanford, C. 1970. Protein denaturation. Part C. Theoretical models for the mechanism of denaturation. *Adv. Protein Chem.* **24**, 1–95.

93 Tanford, C., Pain, R. H., and Otchin, N. S. 1966. Equilibrium and kinetics of the unfolding of lysozyme (muramidase) by guanidine hydrochloride. *J. Mol. Biol.* **15**, 489–504.

94 Tiana, G., Broglia, R. A., and Shakhnovich, E. I. 2000. Hiking in the energy landscape in sequence space: a bumpy road to good folders. *Proteins* **39**, 244–251.

95 Tiffany, M. L. and Krimm, S. 1968a. *Biopolymers* **6**, 1767–1770.

96 Tiffany, M. L. and Krimm, S. 1968b. New chain conformations of poly(glutamic acid) and polylysine. *Biopolymers* **6**, 1379–1382.

97 van Gunsteren, W. F., Bürgi, R., Peter, C., and Daura, X. 2001. The key to solving the protein-folding problem lies in an accurate description of the denatured state. *Angew. Chem. Int. Ed.* **40**, 351–355.

98 van Holde, K. E. 1971. *Physical Biochemistry*. Prentice-Hall, Englewood Cliffs, NJ.

99 van Holde, K. E., Johnson, W. C., and Ho, P. S. 1998. *Physical Biochemistry*. Prentice-Hall, Upper Saddle River, NJ.

100 Viguera, A. R., Serrano, L., and Wilmanns, M. 1996. Different folding transition states may result in the same native structure. *Nat. Struct. Biol.* **3**, 874–880.

101 Wu, H. 1931. Studies on denaturation of proteins. XIII. A theory of denaturation. *Chin. J. Physiol.* **V**, 321–344.

102 Yi, Q., Scalley-Kim, M. L., Alm, E. J., and Baker, D. 2000. NMR characterization of residual structure in the denatured state of protein L. *J. Mol. Biol.* **299**, 1341–1351.

103 Zagrovic, B., Snow, C. Khaliq, S., Shirts, M., and Pande, V. 2002. Native-like mean structure in the unfolded ensemble of small proteins. *J. Mol. Biol.* **323**, 153–164.

104 Zaman, M. H., Shen, M. Y., Berry, R. S., Freed, K. F., and Sosnick, T. R. 2003. Investigations into sequence and conformational dependence of backbone entropy, inter-basin dynamics and the Flory isolated-pair hypothesis for peptides. *J. Mol. Biol.* **331**, 693–711.

105 Zimm, B. H. and Bragg, J. K. 1959. Theory of the phase transition between helix and random coil in polypeptide chains. *J. Chem. Phys.* **31**, 526–535.

21
Conformation and Dynamics of Nonnative States of Proteins studied by NMR Spectroscopy

Julia Wirmer, Christian Schlörb, and Harald Schwalbe

21.1
Introduction

21.1.1
Structural Diversity of Polypeptide Chains

From a structural point of view, protein folding is one of the most fascinating aspects in structural biology. Protein folding proceeds from the disordered random coil polypeptide chain defined by its primary sequence to intermediates with increasing degree of conformational and dynamic order to the final native state of the protein (Figure 21.1). While the starting point of protein folding, probably best described as a statistical coil, but often called the random coil state of a protein, can be defined as a state in which few if any nonlocal interactions exist, the native state can be described by one predominant conformation around which only local fluctuations of low amplitude occur. Order exists and forms on various levels of the protein structure, the primary structure describing the conformational preferences of the amino acid residues differs from secondary structure elements, which are defined by the conformations adopted by consecutive residues to the final arrangement of secondary structure elements in the three-dimensional fold of the protein.

During folding, the polypeptide chain therefore adopts very different conformations and its interaction with the surrounding solvent changes considerably. This variability in conformational space of a polypeptide chain has become even more interesting with the observation of protein misfolding revealing that proteins can also adopt highly ordered, oligomeric states, so-called fibrillic states of proteins, in which residues adopt repetitive conformations very different to the native fold and in which hydrogen bonding is satisfied in an intermolecular fashion. In addition, the conformational equilibria between the various states of a protein are influenced by interactions with chaperones in the cell. The given sequence of a protein therefore can adopt a continuum of different conformational states and degrees of oligomerization, those states interconvert at different rates and therefore vary widely in persistency (Figure 21.1).

Protein Folding Handbook. Part I. Edited by J. Buchner and T. Kiefhaber
Copyright © 2005 WILEY-VCH Verlag GmbH & Co. KGaA, Weinheim
ISBN: 3-527-30784-2

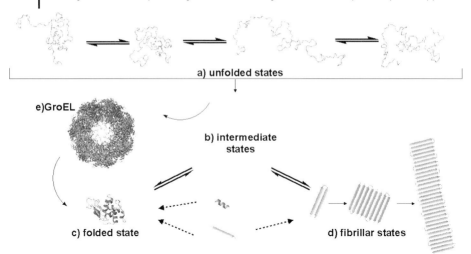

Fig. 21.1. The different conformations adopted by a polypeptide chain can range from the native, often monomeric state (c), in which a single conformation exists and which is build up from secondary structure elements and their specific arrangements, to the ensemble of conformers representing the random coil state of a protein (a). The individual members of this ensemble have widely different compaction, dynamics, local and nonlocal conformations. Protein folding preceeds via formation of folding intermediates (b) whose structure and dynamics may be modulated by protein-protein interactions with molecular chaperones such as GroEL (e) shown in the figure. At the other extreme of conformational states, proteins can aggregate and form oligomeric states called fibrils (d). Liquid and solid state NMR spectroscopy can provide detailed information on structure and dynamics in all of these states.

Spectroscopic techniques to study protein folding are reviewed in Chapter 2. NMR spectroscopy, applied to proteins both in their liquid and solid state, is intrinsically capable of characterizing the variety of conformations and dynamics associated with the polypeptide chain in different states with high precision. In contrast to the native state of proteins, many nonnative states of proteins are ensembles of interconverting conformers. Often, the averaging proceeds locally, e.g., by changes in torsion angles which are rapid (faster than microseconds) on the NMR time scale and lead to sharp NMR signals. In such cases, we will talk in the following of the "random coil state" of a protein. However, averaging can also involve the reorganization of preformed secondary structure elements which is considerably slower. In fact, from an NMR point of view, the observation of line broadening and the absence of detectable NMR signals indicative for a persistent tertiary fold serves as the definition of a "molten globule state" of a protein. NMR is also capable to study the kinetics of protein folding in real time as discussed in Chapter 16.

For nonnative states of proteins, the NMR observables do not directly define conformation and dynamics in these states. Rather, the experimental observables have to be interpreted in the light of models describing conformational averaging, in which the number of participating conformational states and the rate of their interconversion are the two most important parameters. Such models range from theoretical predictions of the properties of the polypeptide chain to extensive molecular dynamics simulations. They are being developed to aid the interpretation of NMR parameters such as chemical shifts δ, spin–spin coupling constants J, homonuclear NOE data (NOEs), residual dipolar couplings (RDCs) and heteronuclear relaxation properties such as relaxation rates (^{15}N-R_1, ^{15}N-R_2) and heteronuclear NOEs ($\{^1H\}$-^{15}N-NOE).

21.1.2
Intrinsically Unstructured and Natively Unfolded Proteins

Nonnative states of proteins have also received attention recently because of the observation that an axiomatic linking of the function of a protein to persistent fold might not be general but a number of proteins have been identified that lack intrinsic globular structure in their normal functional form [1–3]. The expression "intrinsically unstructured" and "natively unfolded" are used synonymously, the latter being coined by Schweers et al. in 1994 in the context of structural studies of the protein tau [4]. Intrinsically unstructured proteins are extremely flexible, noncompact, and reveal little if any secondary structure under physiological conditions. In 2000, the list of natively unfolded protein comprised 100 entries [5] (see Table 21.1b). Natively unfolded proteins are implied in the development of a number of neurodegenerative diseases including Alzheimer's disease (deposition of amyloid-β, tau-protein, a-synuclein), Down's syndrome, and Parkinson disease to name a few (quoted after Ref. [5]). They are predicted to be ubiquituous in the proteome [6, 7] and algorithms available as a web-program (http://dis.embl.de/) have been developed to predict protein disorder [8]. According to the predictions, 35–51% of eukaryotic proteins have at least one long (> 50 residues) disordered region and 11% of proteins in Swiss-Prot and between 6 and 17% of proteins encoded by various genomes are probably fully disordered (quoted from Ref. [6]). Proteins predicted to be intrinsically unstructured show low compositional complexity. These regions sometimes correspond to repetitive structural units in fibrillar proteins. Therefore, it seems likely that the unstructureness of the polypeptide chains in some states of a protein plays an important role in the development of fibrillar states and this further supports the importance for detailed structural and dynamic investigations of nonnative states of proteins. Chapter 8 in Part II is dedicated to the discussion of natively disordered proteins, and Chapter 33 in Part II discusses protein folding diseases.

It has also been noted by Gerstein that the average genomically encoded protein is significantly different in terms of size and amino acid composition from folded proteins in the PDB [9]. This difference would indicate that the structures deposited in the PDB are not random and in turn that they cannot be taken as representative for the entire structural diversity of polypeptide chains.

In this article, we discuss NMR investigations, primarily of nonnative states of proteins based on our investigations in the past using lysozyme and its mutants as well as α-lactalbumin and ubiquitin as model proteins. A number of excellent review articles have been published reporting on the topic of NMR investigations of nonnative states of proteins (see, for example, Refs [10–14]). In order to understand and determine unambiguously those parts of the polypeptide chain in which residual structural elements are present, it is necessary to have a firm understanding of the NMR data to be expected for a random coil state of a protein that lacks any nonlocal structural elements and/or does not reveal sequence-specific differences in dynamics. Therefore, we interpret our data in the framework of a so-called random coil model (Section 21.4). According to this model, the random coil state is assumed to be a state in which there are no nonlocal interactions along the peptide chain, which is built up from the linear sequence of its repetition units, the amino acids. Different amino residues, embedded in their local sequence context, adopt different local conformations and their specific ϕ, ψ distributions have been extracted from residues in coil regions of native proteins to make theoretical predictions of NMR parameters for a random coil. According to this model, the dynamics of the polypeptide chain can be predicted using simple models treating the polypeptide as a branched or unbranched polypeptide chain. Comparisons of experimental NMR parameters with such predictions reveal regions in which residual conformational preferences and dynamic restrictions exist in the random coil state of a protein.

21.2
Prerequisites: NMR Resonance Assignment

For any detailed investigation of the conformation and dynamics of nonnative states of proteins at atomic resolution, the NMR resonances of the amino acids need to be assigned. For a long time, such investigation were considered impossible due to the very low chemical shift dispersion observed in ^{1}H-NMR experiments for nonnative states of proteins different to the native state (Figure 21.2).

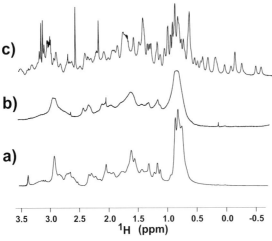

Fig. 21.2. One-dimensional ^{1}H-NMR experiments showing the differences in chemical shift dispersion and linewidth for different states accessible for the protein α-lactalbumin, a protein with 123 amino acid residues and four native disulfide bridges. They range from the random coil state (a) with sharp NMR resonance lines to the molten globule state (b) with extensive line broadening to the native state with well resolved peaks at least for some atomic groups (c).

In addition, while the random coil state of α-lactalbumin obtained by dissolving the protein in high concentrations of denaturant (8 M urea at pH 2) yields very sharp NMR resonances (Figure 21.2a), the molten globule state of α-lactalbumin at pH 2 reveals only an NMR spectra with very broad peaks (Figure 21.2b). The chemical shift dispersion in either of the two states is very limited and the signature of tertiary fold in the native state of the protein, visible in the appearance of upfield shifted methyl groups resonating at ppm values at or below 0.5 ppm is missing (Figure 21.2c).

One approach to overcome the substantial overlap problem observed for nonnative states of proteins was to dissect the protein of interest into smaller peptide fragments (see, for example, Refs [15–17]). These approaches also allowed delineation of those regions of the polypeptide chain that preserves local elements of structure.

Due to introduction of isotope labels and heteronuclear NMR experiments, the low chemical shift dispersion can be overcome in ^{1}H,^{15}N heteronuclear correlation experiments [18] and the modern sequential assignment procedures developed for native proteins by Bax and coworkers [19, 20] can be successfully applied also for the random coil state of proteins [21] (Figure 21.3). The addition of ^{13}C isotope labeling does, however, not provide additional resolution (Figure 21.4).

The surprising observation of relative high chemical shift resolution in the ^{1}H,^{15}N heteronuclear correlation can be predicted based on the analysis of chemi-

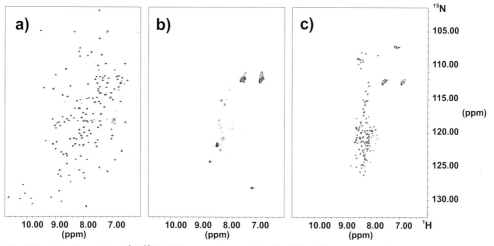

Fig. 21.3. Two-dimensional ^{1}H,^{15}N-NMR correlation experiments showing the differences in chemical shift dispersion for different states accessible for the protein α-lactalbumin, a protein with 123 amino acid residues and four native disulfide bridges. For the random coil state of the protein (c), the resonances cluster in three regions in the spectra, largely reflecting the characteristic ^{15}N chemical shifts of the different types of amino acids when in unfolded conformations. For comparison, the similar spectrum for the native protein is also shown (a). At pH 2, the protein forms a molten globule state and the spectrum shows only very few peaks, while the other peaks are unobservable due to considerable line broadening (b).

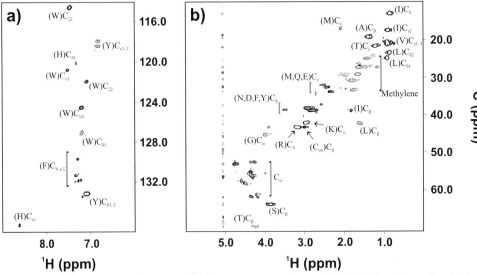

Fig. 21.4. Two-dimensional ^{1}H,^{13}C-NMR correlation experiment of hen egg white lysozyme denatured in 8 M urea at pH 2. The chemical shift resolution in ^{1}H,^{13}C correlation is considerable lower than the ^{1}H,^{15}N-NMR correlation experiment. a) Aromatic carbon region, and b) aliphatic carbon region. Distinct regions of ^{13}C resonances have been labeled with the one-letter amino acid code. The chemical shifts of the ^{13}C resonances are very similar to literature values for linear denatured hexapeptides [37].

Tab. 21.1a. Representative set of proteins that were investigated using heteronuclear NMR spectroscopy in their nonnative state.

Protein	References
α-Lactalbumin	39, 137, 138
Apomyoglobin	99, 139–143
Barnase	144–148
Beta(2)-microglobulin	149, 150
Bovine acyl-coenzyme A binding protein	151, 152
Bovine β-lactoglobulin	153–157
Chemotactic protein CheY	158
drkN SH3 domain	159–175
FK506 binding protein	21, 176
Glutaredoxin 3	177
Hen lysozyme	25, 42, 107, 126, 129, 178–186
HIV-1 protease tethered dimmer	187, 188
OmpX (outer membrane protein X)	189
Protein L	98
Reduced high-potential iron-sulfur protein	190
Ubiquitin	51, 138, 191–195

Tab. 21.1b. Representative set of natively unfolded proteins that were investigated using heteronuclear NMR spectroscopy in their nonnative state.

Protein	References
4E-binding proteins I and II	94
Anti-sigma factor FlgM	95
Antitermination protein N	96
CITED2 (CBP/p300-interacting transactivator with ED-rich tail)	97
Colicin translocation domain	98, 99
Cyclin-dependent kinase inhibitor p21$^{Waf1/Cip1/Sdi1}$	100
Dessication-related protein	101
eIF4G1, functional domain (393–490)	102
Engrailed homeodomain	103
Extracellular domain of beta-dystroglycan	104
Fibronectin-binding domain, D-type, *S. aureus*	105, 106
GCN4, DNA-binding domain	107
Heat shock transcription factor, N-terminal activation domain (*K. lactic* and *S. cerevisiae*)	108
IA3 (aspartic proteinase inhibitor)	109
Negative factor, NEF protein	110
Neutral zinc finger factor 1, two-domain fragment (487–606)	111
Nonhistone chromosomal protein HMG-14	112
Nonhistone chromosomal protein HMG-17	113
Nonhistone chromosomal protein HMG-H6	114
Nonhistone chromosomal protein HMG-T	112
N-terminal domain of p53	115
N-terminal regions of securin and cyclin B	116
Osteocalcin	117
PIR domain of Grb14	118
Prion protein N-terminal part	119, 120
Propeptide of subtilisin	121
Prothymosin α	122
Snc1, cytoplasmic domain	123
Staphylococcal nuclease (SNase), Δ131Δ fragment	58, 124–132
Synaptobrevin cytoplasmic domain	133
TAF$_{II}$-230$_{11–77}$, N-terminal region	134
β-Tubulin, 394–445 fragment	135

Tab. 21.1c. Representative examples of peptides studied in fibrillic states heteronuclear solid state NMR spectroscopy.

Protein	References
Alzheimer's amyloid peptids, β-amyloid	136–140
Transthyretin	141

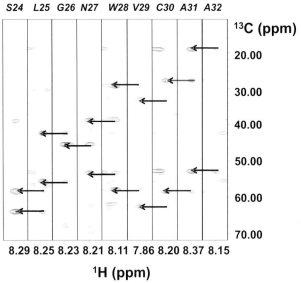

Fig. 21.5. (ω_1, ω_3) strips from the HNCACB experiment of cysteine reduced and methylated lysozyme with the N-terminal additional methionine, a nonnative state of lysozyme in H$_2$O and at pH 2, taken at the ^{1}H^N (ω_3) and ^{15}N (ω_1) frequencies of Ser24, Leu25, Gly26, Asn27, Trp28, Val29, Cys30, Ala31, and Ala32 (the additional methionine residue is labeled M$_{-1}$ to keep the numbering of lysozyme). Intra- and interresidual correlation peaks are marked and clearly visible.

cal shift in protein structures [22] and from quantum chemical calculations [23]. Resonance assignment following this procedure have been reported for a number of different proteins as summarized in Table 21.1a. Table 21.1b gives an overview of natively unfolded proteins for which lack of structure has been identified by NMR spectroscopy. Table 21.1c summarizes solid state NMR investigations of fibrillar states of peptides.

Figure 21.5 shows the parts of three-dimensional spectra for hen egg white lysozyme for which sequential assignment is shown for a stretch of neighboring residues. The resonance assignment for random coil states of proteins with sharp resonance lines is straightforward provided the amino acid sequence is nonrepetitive.

21.3
NMR Parameters

In Section 21.3, the most important NMR parameters summarized in Table 21.2 will be introduced with an emphasis on the relevant aspects in the context of conformational averaging observed in nonnative states of proteins. It is important to stress here that in contrast to native proteins, the dependence of the measurable

Tab. 21.2. Overview of NMR parameters and their conformational dependence.

NMR parameter	Conformational dependence
Chemical shift δ (ppm)	Multiple torsion angles: $\phi, \varphi, \omega, \chi_1$
^nJ couplings (Hz)	Single torsion angles via Karplus equations
Homonuclear NOEs (a.u.)	Distances, dependence on correlation time and motional properties
Heteronuclear relaxation (Hz)	Motional properties, dependence on τ_c, S^2, τ_e
Residual dipolar couplings (RDC) (Hz)	Overall shape, dynamics, S
H/D	exchangeable H^N
Diffusion	Radius of hydration (R_h)
Photo-CIDNP	Accessible Trp, Tyr, His

NMR parameters on conformation and dynamics for nonnative states of proteins is more complicated and requires additional models to predict the random coil NMR spectra. One successful approach to circumvent this problem is to investigate model peptides that should have no preferred conformation and define their properties such as chemical shifts, coupling constants, hydrogen exchange rates (Table 21.2) and such like as the "random coil" value. However, as discussed below, such an approach fails when one predicts NMR observables such as relaxation effects or residual dipolar couplings that show a pronounced dependence on the length of the polypeptide chain. These latter NMR parameters rely on anisotropic interactions and are different for segments of amino acids embedded in a short peptide or in a long polypeptide chain.

A different approach is to predict, using models from polymer theory and distinct torsion angle distributions, for example, the appearance of NMR spectra for the random coil state of a protein. These predictions provide a framework for the interpretation of NMR parameters such as NOEs and J coupling constants [24, 25] or residual dipolar couplings [26] (see Section 21.4).

21.3.1
Chemical Shifts δ

21.3.1.1 Conformational Dependence of Chemical Shifts

Chemical shifts δ depend on the chemical environment of the observed nuclei. Hence, the chemical shifts depend on how and to which degree (parts of) the polypeptide chains are folded. Generally, nonnative states of proteins show a considerably smaller chemical shift dispersion than proteins in their native states (see Figures 21.2a–c and 21.3a–c); NMR active nuclei (^1H, ^{13}C, ^{15}N) resonate close to or at their random coil chemical shift because of conformational averaging. For an ideal random coil polypeptide chain, NMR spectra are similar to the spectra of the mixture of the amino acids [27]. supporting the idea of the absence of any nonlocal interactions in the random coil state. The local amino acid sequence particularly affects the random coil chemical shifts of the $^{15}N^H$, $^1H^N$ and ^{13}CO resonances

[28]. Secondary and tertiary structures contribute to the observed chemical shifts and therefore can be investigated by analysis of the chemical shifts [29]. Deviations from random coil chemical shifts are also called secondary chemical shifts and are indicative for the content of residual structures in unfolded or partially folded proteins. An overall averaging over the deviations from random coil chemical shifts gives an impression of the extent of residual structure in denatured states of proteins.

Proper referencing of the chemical shifts to standard substances like DSS (2,2-dimethyl-2-silapentane-5-sulfonic acid) or TSP (3-(trimethylsilyl) propionate) is required for the comparison of experimental spectra to random coil chemical shifts from the literature [30, 31]. In general, the dispersion of the $^{15}N^H$, $^1H^N$, and ^{13}CO resonances in unfolded proteins is much greater than for the $^{13}C\alpha$, $^{13}C\beta$, $^1H\alpha$, and $^1H\beta$ resonances, reflecting the sensitivity of the former nuclei to the nature of the neighboring amino acids in the primary sequence of the protein [32]. Both $^{13}C\alpha$ and ^{13}CO chemical shifts are shifted downfield when they are in α-helical structures and upfield, when they are located in β-sheets, while $^{13}C\beta$ and $^1H\alpha$ resonances experience upfield shifts in α-helices and downfield shifts in β-structures [30, 31, 33].

So-called chemical shift index (CSI) calculations [33] can routinely be used to detect secondary structure elements in proteins. In this method, chemical shift deviations are normalized to 1, 0, and -1, depending on the extent and direction of the deviation, and then plotted against the residues of the protein sequence. The result of this is a topology plot of the protein from which one easily can spot α-helical structures and β-sheet regions, which differ in the direction of the secondary chemical shifts.

21.3.1.2 Interpretation of Chemical Shifts in the Presence of Conformational Averaging

In nonnative states of proteins we observe conformational averaging (e.g., between different rotameric states). Each rotameric state has a characteristic chemical shift which in a dynamic case will be averaged. However, the NMR signal is influenced by the averaging process both in terms of its resonance position and the observed linewidth (Figure 21.6).

In the case of fast conformational averaging (Figure 21.6c), the observed chemical shift is the population weighted average:

$$\delta^{obs} = \sum_i p_i \delta^i \tag{1}$$

where δ^{obs} is the observed chemical shift, δ^i is the chemical shift in the ith conformation with weights p_i and $\sum_i p_i = 1$. The increase in linewidth at half height $\Delta\Gamma_{1/2}$ due to fast exchange between two site equally populated is given by:

$$\Delta\Gamma_{1/2} = \frac{\pi(\delta v)^2}{2k} \tag{2}$$

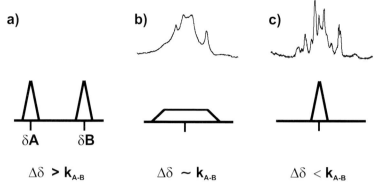

Fig. 21.6. Schematic picture indicating the appearance of NMR spectra for different chemical exchange regimes: a) Slow chemical exchange between conformers, e.g., in the nonnative (A) and native state of a protein (B). The height of the signals indicates the population between the two states, e.g., at a given denaturant concentration. b) Intermediate chemical exchange regime (molten globule state). c) Fast chemical exchange (random coil state).

where $\delta v = v_A - v_B$ and k is the rate of conformational exchange. For fast exchange and small chemical shift differences, a case typically encountered in the random coil conformation of a protein, NMR signals with very narrow linewidths are observed. Under the assumption of two-state fast folding kinetics, in which the unfolded state of a protein is in fast exchange with its folded state, the increase in linewidth (e.g., induced by subsequent addition of increasing amounts of urea starting from the folded state and thereby increasing the population of the unfolded state) can be analyzed to derive microsecond refolding kinetics for two-state folding [34, 35].

However, a detailed description of the random coil state of a protein cannot really be obtained from chemical shift measurement since the prediction of δ^i that is the conformational dependence of chemical shifts is not yet clearly established. It is therefore difficult to predict the chemical shifts for a random coil state of protein from a given distribution of conformations. Rather, reference values for random coil chemical shifts have been determined by measurement of so-called random coil peptides, first of the sequence GGXA [36] and then later of GGYXGG, with Y being either proline or glycine and X any of the 20 amino acids [37, 38].

The NMR spectrum becomes more difficult if the rate of interconversion approaches the differences in chemical shifts δv (Figure 21.6b). In the case of intermediate conformational averaging between two equally populated sites, called intermediate exchange regime, one very broad line is observed. At the point of coalescence, the exchange rate k is given directly by:

$$k = \frac{|\delta v|}{\sqrt{8}} \qquad (3)$$

where $\delta\nu = \nu_A - \nu_B$ and is the rate of conformational exchange. For intermediate exchange, a case associated with the molten globule state of a protein, the dramatic line broadening has made resonance assignment for a molten globule protein impossible so far (see Figures 21.2b and 21.3b). In case of unsymmetric exchange between multiple sites, the prediction of the conformations that participate in the chemical exchange is very difficult and always relies on limiting assumptions, such as fixing one or more of the conformations and the rate of exchange between those sites.

Recently, residues in α-lactalbumin have been identified that are involved in formation of a hydrophobic core undergoing slow conformational exchange and therefore forming a molten globule. Redfield et al. [39] started from the random coil state of the protein, stabilized in high concentrations of denaturants at pH 2, and for decreasing amounts of denaturants observed which residues would disappear first due to intermediate chemical exchange. Interestingly, the pattern of disappearance of resonance can be interpreted assuming a preformed core in the α-domain of the protein.

Under favorable conditions, NMR spectra can be recorded in which the folded and the unfolded state of a protein are in slow exchange; two sets of signals arising from either of the two states can be observed (Figure 21.6a). If exchange takes place on a time scale of hundreds of milliseconds, then exchange peaks between the two states can be observed in NOESY experiments [40], magnetization is transferred from one of the states to the other during the mixing time (of the order of 200–500 ms) of the NOESY experiment [41, 42]. These exchange peaks can be used to assign the spectrum of the unfolded state from known assignments in the folded state and exchange kinetics can be derived.

21.3.2
J Coupling Constants

21.3.2.1 Conformational Dependence of J Coupling Constants

Vicinal (^3J) spin–spin coupling constants can be used for conformational analysis in peptides and proteins. In general, ^3J coupling constants depend on the torsion angles between atoms three bonds apart and the relationship of the dependence of coupling constants on the respective dihedral angles is given by semi-empiric Karplus relations [43] (Table 21.3). Karplus relations are not single-valued, but instead give up to four dihedral angle values for a single coupling constant (Figure 21.7). Karplus parameters have been determined for most of the spin pairs relevant for conformational analysis in the polypeptide backbone and the amino acid side chains (Table 21.3). Coupling constants, especially the coupling constant $^3J(H^N,H_\alpha)$, can be used for the investigation of backbone conformations of partly folded or unfolded states of proteins [36, 44, 45], if the effect of conformational averaging is taken into account.

In addition, ^1J and ^2J coupling constants also contain information about torsion angles, however, their spread in values for different conformations is not as large as for the $^3J(H^N,H_\alpha)$ coupling constant [46–51].

Tab. 21.3. Overview of Karplus parametrization for different ^1J, ^2J, and ^3J coupling constants.

Coupling constant	Karplus parameterization	References
$^3J(H^N,H_\alpha)$	$^3J = 6.51 \cos^2(\phi - 60°) - 1.76 \cos(\phi - 60°) + 1.60$	160
	$^3J = 6.40 \cos^2(\phi - 60°) - 1.40 \cos(\phi - 60°) + 1.90$	161
	$^3J = 6.60 \cos^2(\phi - 60°) - 1.30 \cos(\phi - 60°) + 1.50$	162
	$^3J = 7.90 \cos^2(\phi - 60°) - 1.05 \cos(\phi - 60°) + 0.65$	163
	$^3J = 6.64 \cos^2(\phi - 60°) - 1.43 \cos(\phi - 60°) + 1.86$	164
$^3J(H^N,C')$	$^3J = 4.01 \cos^2(\phi) + 1.09 \cos(\phi) + 0.07$	165
	$^3J = 4.02 \cos^2(\phi) + 1.12 \cos(\phi) + 0.07$	164
$^3J(H^N,C_\beta)$	$^3J = 4.70 \cos^2(\phi + 60°) - 1.50 \cos(\phi + 60°) - 0.20$	166
	$^3J = 2.78 \cos^2(\phi + 60°) - 0.37 \cos(\phi + 60°) + 0.03$	164
$^3J(H_\alpha,C')$	$^3J = 4.50 \cos^2(\phi + 120°) - 1.30 \cos(\phi + 120°) - 1.20$	166
	$^3J = 3.72 \cos^2(\phi + 120°) - 1.71 \cos(\phi + 120°) + 1.07$	167
	$^3J = 3.62 \cos^2(\phi - 60°) - 2.11 \cos(\phi - 60°) + 1.29$	164
$^3J(C',C_\beta)$	$^3J = 1.61 \cos^2(\phi - 120°) - 0.66 \cos(\phi - 120°) + 0.26$	168
	$^3J = 1.28 \cos^2(\phi - 120°) - 1.02 \cos(\phi - 120°) + 0.30$	169
	$^3J = 2.54 \cos^2(\phi - 120°) - 0.55 \cos(\phi - 120°) + 0.37$	167
$^3J(C',C')$	$^3J = 1.33 \cos^2(\phi) - 0.88 \cos(\phi) + 0.62$	170
	$^3J = 1.57 \cos^2(\phi) - 1.07 \cos(\phi) + 0.49$	170
$^1J(N_i,C_{\alpha i})$	$^1J = 9.51 - 0.98 \cos(\psi_i) + 1.70 \cos^2(\psi_i)$	29
	$^1J = 8.65 - 1.21 \cos(\psi_i) + 2.85 \cos^2(\psi_i)$	171
$^2J(N_i,C_{\alpha(i-1)})$	$^2J = 7.82 - 0.17 \cos(\phi_{(i-1)}) - 0.64 \cos^2(\phi_{(i-1)})$ $- 1.39 \cos(\psi_{(i-1)}) - 0.37 \cos^2(\psi_{(i-1)})$	29
	$^2J = 7.85 - 1.52 \cos(\psi_{(i-1)}) - 0.66 \cos^2(\psi_{(i-1)})$	171

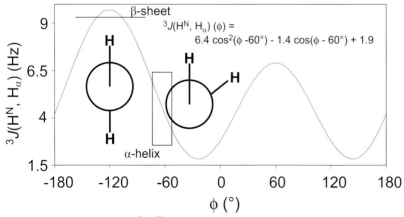

Fig. 21.7. Dependence of the $^3J(H^N,H_\alpha)$ coupling constants on the backbone torsion angle ϕ using the parametrization by Vuister et al. (see Table 21.3).

Scalar couplings through H-bonds ($^{3h}J(N,C')$) provide information on the state of individual hydrogen bonds and therefore on the formation of α-helix [52]. This information is of particular interest for the investigation of folding states of proteins, because the helicity can be observed at atomic level. Thus, $^{3h}J(N,C')$ are valuable complements to chemical shift deviation data.

21.3.2.2 Interpretation of J Coupling Constants in the Presence of Conformational Averaging

In the case of fast conformational averaging, the observed scalar coupling constant is the population weighted average in close analogy to chemical shifts:

$$J^{obs}(\alpha) = \sum_i p_i J^i(\alpha_i) \tag{4}$$

where J^{obs} is the observed scalar coupling constant, J^i is the scalar coupling constant in the ith conformation with torsion angle α_i and weights p_i and $\sum_i p_i = 1$. Two approaches have been developed to interpret the averaged scalar coupling constant. In one approach, random coil scalar coupling constants have been determined for random coil peptides GGXGG in close similarity with the reference chemical shift values [53].

Different to chemical shifts, the dependence of scalar coupling constants on the intervening torsion angle is well established (see Table 21.3) and therefore, coupling constants can be predicted using models for the distribution of torsion angle space in the random coil state of a protein (see Section 21.4).

Table 21.4 shows a comparison of coupling constants predicted from the random coil model and those measured for random coil peptide GGXGG in 6 M GdmCl.

21.3.3
Relaxation: Homonuclear NOEs

21.3.3.1 Distance Dependence of Homonuclear NOEs

NOE interactions between hydrogen atoms form the most important basis for the NMR structure determination (Figure 21.8). Distance information can be obtained through space-mediated transfers in NOESY experiments that reveal crosspeaks between two protons H1 and H2 that are closer than 5 Å together. The cross-relaxation rate between two protons that gives rise to observable cross-peaks in NOESY (ROESY) spectra is defined by the following equation:

$$\sigma^{NOE}_{H1,H2} = \frac{d^2 \tau_c}{5}\left[-1 + \frac{6}{1+4\omega_0^2\tau_c^2}\right]$$
$$\sigma^{ROE}_{H1,H2} = \frac{d^2 \tau_c}{5}\left[2 + \frac{3}{1+\omega_0^2\tau_c^2}\right] \tag{5}$$

where ω_H is the Lamor frequencies of ^1H, $d = \mu_0 h \gamma_H^2/(\sqrt{8} r^3_{H1,H2}\pi^2)$ describes the

Tab. 21.4. Random coil $^3J(H^N,H_\alpha)$ coupling constants predicted on the basis of the random coil model (see Section 21.4) and measured experimentally. Values for $^3J(H^N,H_\alpha)^{pred.}$ are based on the distribution of ϕ angles in the protein database and represent an average over all the different adjacent amino acids, rather than for residues with adjacent glycines. Experimental coupling constants were measured using a GGXGG peptide in 6 M GdmCl at 20 °C, pH 5.0.

	$^3J(H^N,H_\alpha)^{pred.}$ (Hz)	$^3J(H^N,H_\alpha)^{exp.}$ (Hz)
Ala	6.1	6.1
Arg	6.9	7.2
Asn	7.7	7.4
Asp	7.8	7.2
Cys (ox)	7.7	n.d.
Cys (red)	7.3	6.8
Gln	7.1	7.1
Glu	6.7	6.8
His	7.8	7.2
Ile	7.1	7.6
Leu	6.8	7.1
Lys	7	7.1
Met	7.1	7.3
Phe	7.3	7.5
Ser	7	6.7
Thr	7.9	7.6
Trp	7	6.9
Tyr	7.8	7.3
Val	7.2	7.7

constant for dipolar interaction, μ_o is the permeability of the vacuum, h is Planck's constant, γ_H is the gyromagnetic ratio of nucleus i, and $r_{H1,H2}$ is the distance between H1 and H2.

In the following, we wish to discuss some of the assumptions and complications associated with the use of NOE information to derive structural information for nonnative and random coil states of proteins. In the native state of a protein, a

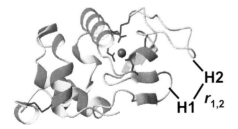

Fig. 21.8. The NOE interaction $\sigma^{NOE}_{H1,H2}$ between two protons H1 and H2 is the most important NMR parameter to determine structures of proteins.

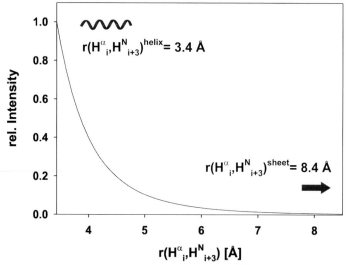

Fig. 21.9. Dependence of the cross-peak intensity $\sigma^{NOE}_{H1,H2}$ in NOESY spectra on the distance between protons H1 and H2. Characteristic secondary structure elements are indicated, the cross-peak for two protons 3.4 Å apart has arbitrarily been scaled to 1.

given cross-peak is interpreted as arising from a single conformation with fixed distance. It is further assumed that the overall rotational tumbling of a molecule is isotropic and can therefore be modeled by a single spectral density function that describes the distribution of motions in a molecule. Furthermore, additional local motions can to good approximation assumed to be absent. These local motions could modulate (decrease) the distance between protons H1 and H2. Under these assumptions, the observed intensity falls off with the inverse sixth power of the distance (Figure 21.9).

Spectral density function The spectral density function $J^q(\omega_q)$ is obtained by evaluating the correlation function of the spherical harmonics

$$J^q(\omega_q) = \int_0^\infty d\tau \overline{F^{(q)}(t)F^{(-q)}(t+\tau)} \exp(-i\omega_q\tau) \tag{6}$$

The bar indicates time average over t. q can assume values of 0, ±1, ±2. The correlation function describes the probability of a given internuclear vector to stay in a fixed orientation in dependence of the time τ. Due to rotational tumbling, the orientation is lost; for proteins of 15 kDa size, a typical time constant τ_c for this process is of the order of 5 ns. However, depending on the model of rotational reorientation (isotropic, axially symmetric, asymmetric), the spectral density function has different forms. Different members of ensemble of conformers in the random coil state of a protein will have very different diffusion tensors, and therefore, the spectral density functions will not be identical for compact conformers compared to extended conformers. In the following, we discuss the isotropic case and the case of axial symmetry.

Spherical top molecules The spectral density function for isotropic rotational diffusion has been derived by Hubbard and coworkers [54–57]:

$$J^q(\omega_q) = \frac{2}{5}\left[\frac{6D}{(6D)^2 + \omega_q^2}\right]$$

$$= \frac{1}{5}\left[\frac{2\tau_c}{1 + (\omega_q \tau_c)^2}\right] \quad (7)$$

where D denotes the diffusion constant, with $1/\tau_c = 6D$. The rotational diffusion constant is inversely proportional to the correlation time τ_c. The correlation time for brownian rotational diffusion can be measured by time-resolved fluorescence spectroscopy, light scattering, or NMR relaxation [58] and can be approximated for isotropic reorientation by the Stoke's-Einstein correlation: $\tau_c = \dfrac{4\pi \eta_w r_H^3}{3 k_B T}$ in which η_w is the viscosity, r_H is the effective hydrodynamic radius of the protien, k_B is the Boltzmann constant, and T the temperature.

Axially symmetric top molecules For the case that $D_{xx} = D_{yy} = D_\perp$, one obtains the spectral density function of a symmetric top rotator ($D_{zz} = D_\parallel$) or of an axially symmetric top molecule whose rotational diffusion is defined by a second rank diffusion tensor with polar coordinates θ and ϕ:

$$J^q(\omega_q) = \frac{1}{20}\{S_0^2 J^{q,0} + S_1^2 J^{q,1} + S_2^2 J^{q,2}\}$$

$$S_0^2 = \langle(3\cos^2\theta - 1)\rangle\langle(3\cos^2\theta - 1)\rangle/4$$

$$S_1^2 = 3\langle(\sin\theta\cos\theta)(\cos\phi)\rangle^2 + \langle(\sin\theta\cos\theta)(\sin\phi)\rangle^2 \quad (8)$$

$$S_2^2 = \frac{3}{4}\langle(\sin^2\theta)(\cos 2\phi)\rangle^2 + \langle(\sin^2\theta)(\sin 2\phi)\rangle^2$$

in which the reduced spectral density functions ($-2 \leq m \leq +2$)

$$J^{q,m} = \frac{2\tau_{c,m}}{1 + (\omega_q \tau_{c,m})^2} \quad (9)$$

have been used. The correlation times $\tau_{c,m}$ can be rewritten as diffusion constants $D_\parallel$ and $D_\perp$ according to

$$1/\tau_{c,m} = 6D_\perp + m^2(D_\parallel - D_\perp) \tag{10}$$

The analysis shows that of cross-peak intensities $\sigma_{H1,H2}^{NOE}$ will therefore be very different depending on the overall compactness of the conformer, ranging from nearly isotropic to highly extended and anisotropic shape, in the ensemble in the random coil state of a protein.

21.3.3.2 Interpretation of Homonuclear NOEs in the Presence of Conformational Averaging

In the ensemble of unfolded states, averaging around intervening torsion angles that connect the two protons of interest will lead to averaging of the distances. Local averaging of torsion angles will lead to modulation of the NOE signal in addition to the global reorientation of the molecule and the effect of local and global motions and the ratio of their characteristic time constants τ_e and τ_c, respectively, will influence the cross-peak intensities $\sigma_{H1,H2}^{NOE}$ in NOESY experiments. The formalism to deal with internal motion has been introduced for the analysis of heteronuclear two-spin systems such as C–H or N–H (see Section 21.3.4) and will be introduced for the treatment of a two-spin H–H system.

Internal motion *Derivation of spectral density functions by explicit calculation of the motion:* Internal motion can be incorporated into the spectral density either by the Lipari and Szabo [59, 60] approach or by explicit calculation of the motion from, for example, motional models of molecular dynamics trajectories. In the above given equations, the spectral densities are Fourier transformations of the rotational diffusive motion of the molecule with respect to the external magnetic field [61, 62].

Internal motions analyzed by Lipari-Szabo approach: The model-free or Lipari-Szabo approach for the analysis of relaxation rate aims at the provision of dynamic parameters describing the dynamic behavior of a two-spin system irrespective of any assumed motional model. This is desirable since the relaxation data does not contain any information on the nature of the motions that cause relaxation. The model-free formalism is the most widely applied one for the dynamic interpretation of relaxation data. In the simplest case (Figure 21.10a), these consist of the generalized order parameter S^2, which is a measure for the spatial restriction of the internal motion, and the effective correlation time τ_e defining the time scale of motion. Assuming isotropic diffusion, the spectral density function can then be expressed as follows:

$$J(\omega) = \frac{S^2 \tau_c}{1 + \omega^2 \tau_c^2} + \frac{(1 - S^2)\tau}{1 + \omega^2 \tau^2} \tag{11}$$

In Eq. (11), τ_c is the overall rotational correlation time of the molecule and $1/\tau = 1/\tau_c + 1/\tau_e$. In the absence of internal motion ($S^2 \sim 1$ and $\tau = \tau_c$), $J(\omega)$ simplifies to:

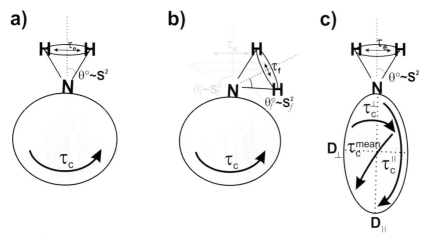

Fig. 21.10. Model-free parameters shown for a N–H spin system within a molecule that tumbles isotropically with the overall rotational correlation time τ_c. A) Simple model with the order parameter S^2 and the effective correlation time τ_e. The overall rotation is assumed to be isotropic. B) Two-time scale extended model with two S^2 parameters, S_s^2 and S_f^2 for slow and fast motions. C) like A but with anistropic overall rotation.

$$J(\omega) = \frac{\tau_c}{1+\omega^2\tau_c^2} \qquad (12)$$

In cases where different fast motions take place, whose time scales differ by at least one order of magnitude, an extended model-free formalism has been developed with a separate S^2 for each motion, S_s^2 and S_f^2 (Figure 21.10b, Eq. (13)).

$$J(\omega) = \frac{S^2\tau_c}{1+\omega^2\tau_c^2} + \frac{(S_f^2 - S^2)\tau}{1+\omega^2\tau^2} \qquad (13)$$

In this two-time scale model it is assumed that the contribution of the faster of the two motions can be neglected, therefore, while it contributes to the overall S^2 since $S^2 = S_s^2 * S_f^2$, the term containing the fast effective correlation time τ_f is left out. The time scale of the slower internal motion τ_s is included in τ ($1/\tau = 1/\tau_c + 1/\tau_s$) similar to τ_e of the single-time scale model (it will therefore not be discriminated from τ_e in the following).

Figure 21.11 shows the dependence of the cross-peak intensity $\sigma_{H1,H2}^{NOE}$ in NOESY experiments on the overall correlation τ_c, the internal correlation time τ_e and the order parameter S^2 as defined in the single-time scale model. It can be seen that for global correlation time of 2 ns, an order parameter S^2 of 0.4, and an internal correlation time τ_e of 150 ps, $\sigma_{H1,H2}^{NOE}$ is scaled down from −0.36 in the absence of any internal motion ($S^2 = 1.0$) to −0.12. If one reduces τ_c to 1 ns, $\sigma_{H1,H2}^{NOE}$ is further reduced to −0.04. It is interesting to note that for longer internal correlation times τ_e, $\sigma_{H1,H2}^{NOE}$ increases again. The partition of τ_c and τ_e is difficult to assess by NMR. In the case, in which local and global motions cannot be entirely uncoupled and in

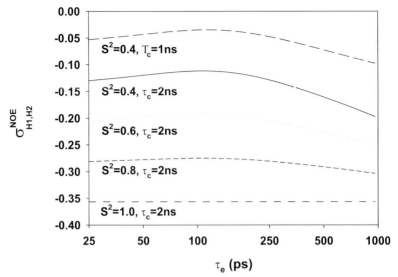

Fig. 21.11. Dependence of the cross-peak intensity $\sigma^{NOE}_{H1,H2}$ in NOESY spectra on the correlation time for local motions τ_e assuming different correlations times τ_2 for the global motion and different order parameters S^2.

which local motions have relative larger amplitudes, interpretation of $\sigma^{NOE}_{H1,H2}$ in terms of arising from an average distance is difficult. In addition, the correlation times for both internal and global motion may be affected in a different manner depending on the specific experimental conditions such as temperature and solvent viscosity; these effects make it difficult to argue that the absence of a specific cross-peak could be taken as evidence for specific conformation sample as is sometimes done in the literature.

Cooperativity in conformational sampling Different to coupling constants J that depend only on a single intervening torsion angle, the distances of atoms further remote is influenced by more than a single torsion and this leads to interesting questions regarding the analysis of cooperativity in sampling different conformations. In this discussion, cooperativity describes the influence of neighbouring residues and their (averaged) conformation on a given torsion angle distribution. Let us assume that the residues of a small peptide can adopt two conformations, α and β, both by 50% (Figure 21.12). Two extreme cases of cooperativity can be distinguished: In the cooperative case of conformational sampling, a residue $i+1$ assumes an $\alpha(\beta)$ conformation if residue i is in $\alpha(\beta)$; in the noncooperative case, a residue $i+1$ assumes either α or β conformation with 50% probability, if residue i is in $\alpha(\beta)$. In the noncooperative case, the eight possible conformations of a tripep-

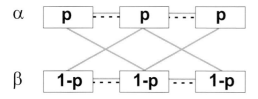

--- cooperative $I(\alpha N)_{i,i+3}^{coop} = 0.5$

— non cooperative $I(\alpha N)_{i,i+3}^{non\ coop} = 0.17$

Fig. 21.12. Model of a consecutive stretch of three amino acid residues. For the sake of the argument, it is assumed that the residues can adopt only two different conformations, α and β. In the case of cooperative sampling of conformational space, which is indicated by the dashed lines, the intensity of the NOE cross-peak $I(\alpha, N)_{i,i+3}^{coop}$ is 50% of the cross-peak observed for a single static conformation. In the case of noncooperative conformational sampling, the intensity $I(\alpha, N)_{i,i+3}^{non\ coop}$ decreases to 17%, for which only two out of the eight possible conformation contribute (see Table 21.5).

tide ($\alpha\alpha\alpha, \beta\alpha\alpha, \alpha\beta\alpha, \beta\beta\alpha, \alpha\alpha\beta, \beta\alpha\beta, \alpha\beta\beta, \beta\beta\beta$) are each sampled by 12.5%. Obviously, the J coupling constant in either of the two cases is the average $(J^\alpha + J^\beta)/2$ and J couplings therefore do not differentiate between different degrees of cooperativity.

This is different for the measurement of an NOE, e.g., between a residue i and $i + 3$. In the cooperative case, there are two populations present ($\alpha\alpha\alpha, \beta\beta\beta$), each by 50%. Let us assume that the $\alpha\alpha\alpha$ conformer represents a helical fragment and the $\beta\beta\beta$ conformer an extended fragment. Then, Figure 21.9 shows that no cross-peak can be observed for $\beta\beta\beta$ conformer. In the cooperative case, in which 50% of all molecules in the ensemble have an $\alpha\alpha\alpha$ conformation, the cross-peak intensity is 50% of the cross-peak intensity predicted for a helix (Table 21.5). In the noncooperative case, in which all eight conformations are equally populated, the cross-peak intensity drops to 0.17 and only two conformers ($\alpha\alpha\alpha, \beta\alpha\alpha$) contribute to the cross-peak intensity.

21.3.4
Heteronuclear Relaxation (^{15}N R$_1$, R$_2$, hetNOE)

21.3.4.1 Correlation Time Dependence of Heteronuclear Relaxation Parameters

Heteronuclear relaxation rates (^{15}N-R$_1$, ^{15}N-R$_2$) and heteronuclear NOEs ({^{1}H}-^{15}N-NOE) are sensitive both to motions on a subnanosecond time scale and to slow conformational exchange in the millisecond time scale [63]. Motions that influence the NMR parameters are the overall rotational tumbling of a molecule and the local fluctuations (e.g., of an **NH** bond vector). The exact formula are given in Eq. (14a–c):

Tab. 21.5. Contribution to NOE cross-peak intensities from different conformations for two models: (i) cooperative sampling of conformational space defined by two different conformations, α and β, or (ii) noncooperative sampling of conformational space (see Figure 21.12).

Conformation of residues $i+1$ to $i+3$	Cooperative sampling		Noncooperative sampling	
	Population	Cross-peak intensity (a.u.)	Population	Cross-peak intensity (a.u.)
$\alpha\alpha\alpha$	50%	0.5	12.5%	0.124
$\beta\beta\beta$	50%	0	12.5%	a
$\alpha\alpha\beta$	–	–	12.5%	a
$\alpha\beta\alpha$	–	–	12.5%	a
$\beta\alpha\alpha$	–	–	12.5%	0.037
$\alpha\beta\beta$	–	–	12.5%	a
$\beta\alpha\beta$	–	–	12.5%	a
$\beta\beta\alpha$	–	–	12.5%	a

[a] All other conformations contribute 0.009 [a.u.] to the cross-peak intensity.

$$R_1 = \frac{d^2}{4}[J(\omega_H - \omega_N) + 3J(\omega_N) + 6J(\omega_H + \omega_N)] + c^2 J(\omega_N)$$

$$R_2 = \frac{d^2}{8}[4J(0) + J(\omega_H - \omega_N) + 3J(\omega_N) + 6J(\omega_H) + 6J(\omega_H + \omega_N)] \quad (14a\text{--}c)$$
$$+ \frac{c^2}{6}[4J(0) + 3J(\omega_N)] + R_{ex}$$

$$\text{NOE} = 1 + (d^2/4R_1)(\gamma_N/\gamma_H)[6J(\omega_H + \omega_N) - J(\omega_H + \omega_N)]$$

In this, ω_H and ω_N are the Lamor frequencies of ^{1}H and ^{15}N, respectively, $J(\omega)$ is the spectral density function at frequency ω; R_{ex} is the parameter accounting for conformational exchange contributions; $d = \mu_0 h \gamma_N \gamma_H/(8 r_{NH}^3 \pi^2)$ and $c = \omega_N \Delta\sigma_N/\sqrt{3}$ describe the dipolar and the chemical shift anisotropy (CSA) interaction, respectively. μ_0 is the permeability of the vacuum; h is Planck's constant; γ_i is the gyromagnetic ratio of nucleus i; r_{NH} is the internuclear distance between N and H; $\Delta\sigma_N$ is the nitrogen CSA.

The dependence of the heteronuclear relaxation parameters on the overall rotational correlation time τ_c in the absence of internal motions and assuming isotropic tumbling is shown in Figure 21.13. While the longitudinal relaxation rate R_1 exhibits a maximum, R_2 increases monotonuously with τ_c. This difference arises from the dependence of R_2 but not of R_1 on a $J(0)$ term (see Eq. (14)), which is negligible at lower τ_c values but becomes more and more relaxation relevant for higher τ_c values. For macromolecules, τ_c usually lies right of the R_1 maximum and consequently small R_1 but large R_2 values are expected.

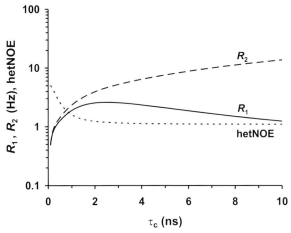

Fig. 21.13. Dependence of R_1, R_2 and the heteronuclear NOE (hetNOE) on τ_c. Calculations are based on the assumption of a rigid molecule tumbling isotropically.

21.3.4.2 Dependence on Internal Motions of Heteronuclear Relaxation Parameters

The relaxation rates are influenced by additional internal motions. In order to show the effect of such internal motions on the observed relaxation, Figure 21.14 shows the simulation of the transverse ^{15}N relaxation rates (^{15}N-R_2) as predicted using the Lipari-Szabo approach and a single-time scale model (Eq. (13)). For a given overall correlation time $\tau_c = 1$ ns, the relaxation rate decreases with increasing internal motion and order parameter. However, the use of the Lipari-Szabo

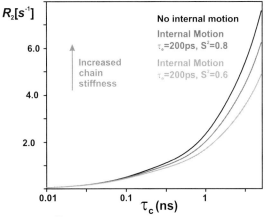

Fig. 21.14. ^{15}N heteronuclear R_2 rates as function of correlation time τ_c for global motions, τ_e for internal motions and the order parameters S^2. The appropriateness of the underlying model-free approach for the description of the dynamics in nonnative states of proteins is discussed in the text.

approach for random coil states of proteins is controversial, for a detailed discussion see, for example, Ref. [64]: the extraction of motional parameters using the Lipari-Szabo approach is based on two (single time scale model) or three (two time scale model) discrete correlation times that can be separated and are uncorrelated to the other motions. In other words, the approach assumes that the time scales of global and local motions can clearly be separated, an assumption questionable in the context of random coil states of proteins.

Alexandrescu and Shortle [65] have used a local model-free approach, according to which one global correlation time is fitted per residue. In the case of extensive averaging over a range of frequency in the nanosecond range, a continuous distribution of correlation times (and subsequently spectral density functions) can be introduced. One of such functions, the Cole-Cole distribution, has been used by Buevich and Baum to describe the relaxation behavior of the unfolded propeptide of subtilisin PPS [66]. An improved treatment of this problem has been proposed by Ochsenbein et al., in which a lorentzian distribution of correlation times (see Eq. (15)) is used to describe the underlying dynamics of the ensemble of conformers.

$$J(\omega) = \int_0^\infty F(\tau) J(\omega, \tau) \, d\tau$$

$$F(\tau) = K \frac{\Delta}{\Delta^2 + (\tau - \tau_0)^2} \quad \text{for } 0 \leq \tau \leq \tau_{max} \tag{15}$$

$$F(\tau) = 0 \quad \text{for } \tau \geq \tau_{max}$$

In yet another approach, we have fitted the sequence dependence of relaxation rates to a simple two-parameter model, which assumes a common intrinsic relaxation rate for a given residue that depends on the viscosity of the solution, and a length dependence of the influence of neighboring residues on the relaxation properties of any member of the chain. Such simple approach fits the random coil behavior of the polypeptide chain but does not provide a deconvolution of the time scales of motion associated with the apparent relaxation behavior. However, the approach reliably predicts the sequence dependence of the relaxation rates in unbranched and disulfide bridge branched polypeptides based on a small number of parameters and is discussed in detail in Section 21.5.3.

21.3.5
Residual Dipolar Couplings

21.3.5.1 Conformational Dependence of Residual Dipolar Couplings
The measurement of residual dipolar couplings (RDCs), although early recognized as having potential [67], has only recently been reintroduced [68, 69] and is now routinely used as a means to obtain information on the orientation of bond vectors relative to an external alignment tensor, that is, NMR information on long-range orientation of individual bond vectors in a macromolecule. The introduction of residual dipolar couplings has considerably changed NMR structure determination

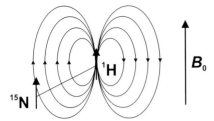

Fig. 21.15. A two-spin system of spin ^{1}H and ^{15}N connected by an internuclear vector r_{NH}. The ^{15}N nucleus exerts a dipolar field, which acts at the side of the ^{1}H nucleus. The magnetic moments of the nuclei are aligned parallel or antiparallel (not shown) relative to the static magnetic field B_0. The dipolar field of the ^{15}N nucleus can increase or decrease the static magnetic field active at the position of the ^{1}H nucleus depending on the orientation of the **NH** vector and the spin state of the proton (parallel or antiparallel to B_0).

for native proteins. A number of excellent review articles have been published [70–73]. Only recently have first measurements using RDCs in the case of nonnative states of proteins [74] been published.

In our description of the conformational dependence of RDCs, we follow a recent review article by Bax [73]: Dipolar couplings between NMR-active nuclei such as a spin N–H are caused by the magnetic dipole field exerted by the N nucleus at the site of nucleus H (Figure 21.14). In the following, we keep only the component of the magnetic dipole field that is parallel to the direction (the z-direction) of the magnetic field B_0. The dipole field exerted by the N nucleus will change the resonance frequency of the H nucleus. The dipolar coupling D^{NH} depends on the internuclear distance r_{NH} and on the orientation θ of the internuclear vector **NH** (bold, italics indicating the vector property) relative to B_0 (Eq. (16)). For a fixed orientation of the vector, the dipolar field exerted by N can either increase or decrease the magnetic field of nucleus H, depending on whether the equally populated spin-states α and β of H are parallel or antiparallel relative to B_0.

$$D^{NH} = D^{NH}_{max}\langle (3\cos^2\theta - 1)/2\rangle \tag{16}$$

where θ is the angle between the internuclear vector **NH** and the magnetic field B_0, the angle brackets denote averaging over the ensemble of molecules, and

$$D^{NH}_{max} = -\mu_0(h/2\pi)\gamma_H\gamma_N/(4\pi^2 r^3_{NH}) \tag{17}$$

is the splitting observed for $\theta = 0$; the additional constants have the same meaning as explained for Eq. (14a–c).

In the case of isotropic rotational tumbling of a molecule in solution (Figure 21.16a), brownian motions average the magnetic dipole effects to zero, while in the solid state spectrum of a molecule, such dipolar couplings are not averaged and every nucleus couples with a large number of other nuclei, resulting in considerable line broadening.

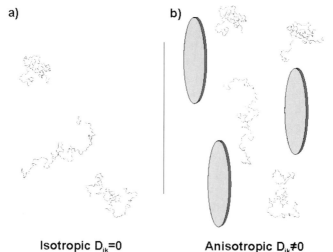

Fig. 21.16. a) Isotropic tumbling of a molecule leads to vanishing dipolar couplings. b) Introduction of non-isotropic medium leads to residual dipolar couplings (RDCs) in solution NMR. RDCs depend on the orientation of a given bond vector relative to an alignment tensor and provide information about the long-range order in high-resolution NMR studies.

However, there are now various methods that impart partial alignment to a dissolved molecule, Many of them are additives to the solution such as liquid crystalline media [75] and partial alignment is induced by steric and/or ionic interaction between the aligned medium and the dissolved macromolecule (Figure 21.16b). In the case of NMR investigations of nonnative states of proteins, the use of polyacrylamide gels that have been compressed in an anisotropic manner have been successful [76, 77].

Yet another different way of inducing alignment of a molecule is to use paramagnetic factors bound to the protein, either by using natural paramagnetic cofactors or metals [78, 79] or specifically designed binding sites for paramagnetic ions [80]. In contrast to external alignment media, alignment is not caused by intermolecular interactions but is rather a property of the molecule itself. This is advantageous in the case of nonnative states of proteins, since different interactions between the members of the ensemble of conformers representing these states and the aligning medium will lead to different extents of alignment for each member of the ensemble.

The aligning of the proteins has the effect that rotational tumbling is not isotropic, and not all orientations are equally likely to occur. The extent of difference in this tumbling can be expressed as an alignment tensor A. Its principal components A_{xx}, A_{yy}, and A_{zz} reflect the probabilities (along the x-, y-, and z-direction) for the diffusional motion relative to the external magnetic field B_0.

The dipolar coupling depends on the polar coordinates of the **NH** vector in the frame of the alignment tensor:

$$D^{NH}(\theta,\varphi) = 3/4 D^{NH}_{max}[(3\cos^2\theta - 1)A_{zz} + \sin^2\theta \cos 2\varphi (A_{xx} - A_{yy})] \tag{18}$$

Equation (18) can be rewritten as:

$$D^{NH}(\theta,\varphi) = D_a[(3\cos^2\theta - 1) + 3/2 R \sin^2\theta \cos 2\varphi] \tag{19}$$

where $D_a = 3/4 D^{NH}_{max} A_{zz}$ is the magnitude of the dipolar coupling tensor and $R = 2/3(A_{xx} - A_{yy})/A_{zz}$ is the rhombicity.

In the context of the analysis of nonnative states based on residual dipolar couplings, it is of importance to discuss the ability to predict the dipolar couplings from a given conformation. In Eq. (19), the magnitude and rhombicity of the alignment tensor relative to the molecular frame need to be known. For a given conformation, this may be predicted on the basis of the molecular shape [81, 82]. However, best results have been obtained so far if the alignment mechanism is steric and due to the large number of members in the ensemble of structures, such approach is particularly difficult for nonnative states of proteins. A second approach is to obtain a tensor that fits the experimental data best in a linear fitting algorithm [83]. This approach, however, cannot be used to predict RDCs from theoretical models for the torsion angle distributions in nonnative states of proteins.

21.3.5.2 Interpretation of Residual Dipolar Couplings in the Presence of Conformational Averaging

Different NMR parameters are influenced by conformational averaging over different time scales. While relaxation data such as homonuclear NOEs or heteronuclear relaxation rates are sensitive to motions that are faster than the overall rotational correlation time, residual dipolar couplings are sensitive to motions on a considerable longer time scale. This observation has only recently be exploited to access the conformational dynamics of proteins in their native state using a combination of NMR experimental dipolar couplings measured in a number of different alignment media and predictions of dynamics from molecular dynamics calculations [84–86].

Louhivuori et al. [26] have calculated the effect of steric obstruction that an external alignment media can exert on a random coil polypeptide chain (i.e., a random flight chain). Such calculation consists of a model for the spatial distribution funcion for an ensemble of conformers and a treatment of the residual dipolar couplings.

The distribution of conformation of the Gaussian chain is different from the isotropic solution in that the aligning medium introduces a barrier to the free diffusion of the chain (Figure 21.17); the distribution function of the random flight chain in the presence of the aligning medium W^{obs} becomes obstructed and deviates from the distribution function W^{free} in isotropic solution. The effect of the aligning medium is also treated as a distribution function W^{bar}, yielding the equation:

$$W^{obs} = W^{free} - W^{bar} \tag{20}$$

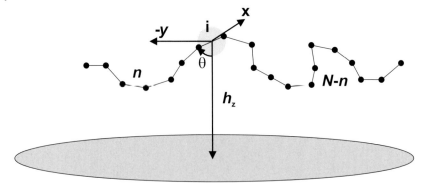

Fig. 21.17. Model for a random flight chain consisting of $N+1$ segment represented by dots and connected by straight lines. The chain tumble at distance h_z from a disk representing the aligning medium. At the locus i, placed at the origin of the molecular frame, the chain is partioned into two half chains containing n and $N-n$ segments. RDCs are functions of the average angle θ of i, the B_o field is parallel to the z-axis.

The random flight chain is composed of $N+1$ segments. For any given element of the chain i, there are n segments before i and $N-n$ segment after i. Therefore, the probability function contains two parts, W_n^{obs}, W_{N-n}^{obs}, and is given by:

$$W(z, h_z, c, n, N) = \frac{1}{N} \{ n W_n^{obs}(z, h_z, c, n) + (N-n) W_n^{obs}(z, h_z, c, N-n) \}$$

$$= \frac{1}{N} \{ n [W_n^{free}(z, c, n) - W_n^{bar}(z, h_z, c, n)]$$

$$+ (N-n)[W_n^{free}(z, c, N-n) - W_n^{bar}(z, h_z, c, N-n)] \}$$

$$= \frac{\sqrt{2n/\pi}}{N} \left\{ \exp\left[-\frac{(2+1/2c)^2}{2n}\right] - \exp\left[-\frac{(2h_z - z + 1/2c)^2}{2n}\right] \right\}$$

$$+ \frac{\sqrt{2(N-n)/\pi}}{N} \left\{ \exp\left[-\frac{(z-1/2c)^2}{2(N-n)}\right] \right.$$

$$\left. - \exp\left[-\frac{(2h_z - z - 1/2c)^2}{2(N-n)}\right] \right\} \quad (21)$$

where $c = \cos\theta$.

For any arbitrary pair of nuclei A and B in the segment i separated by the internuclear vector r_{AB}, the vector r_{AB} will be at an angle α_{AB} with respect to the internal coordinate system of the seqment. The segment itself will be oriented at an angle θ with respect to the magnetic field B_o. The distribution W describing the averaging of the ith segment of the chain is a function of the position of each segment, each segment has its own averaged $\langle P_2 \rangle$ polynomial. As a consequence, each segment has its own alignment tensor A, which is axially symmetric due to the distribution functions for a random flight chain. The dipolar coupling D_{AB} is proportional to the axial component of the alignment tensor:

The average of $\cos^2 \theta$ contained in the expression of the dipolar coupling will be obtained by integration over all angles and space weighted by the distribution and normalized. The following observations can be made: For longer chains, $\langle P_2 \rangle$ will become smaller because the distributions of larger number of uncorrelated segments will become more spherical. Only polypeptide chains up to 100–200 residues will give rise to observable dipolar couplings. These observed residual dipolar couplings for a random flight chain are a consequence of the fact that such chain is restrained by the covalent structure of the chain, each segment is coupled and not free. The RDC originates from the individual loci along the chain, not from the whole chain average; each individual segment exhibits a nonvanishing $\langle P_2 \rangle$, it is time and ensemble averaged, but not averaged over all loci.

The treatment of residual dipolar couplings for random coil states of proteins is yet another example in which the prediction of what to expect for such state is of importance before deviations from random coil behavior can be interpreted with any vigor.

21.3.6
Diffusion

The dimensions of spherical proteins can be determined using diffusion NMR measurements since the hydrodynamic radius (R_h) of a protein is inversely proportional to its diffusion constant. Thus knowing the diffusion constant, R_h can be extracted using the Stokes-Einstein equation which requires the knowledge of the exact diffusion constant, temperature and viscosity of the solution. An easier approach for the determination of diffusion constants is by comparison to an internal probe with known hydrodynamic radius such as dioxane ($R_h = 2.12$ Å) [87]. Measurement of diffusion constants is usually done by PFG (pulse field gradient) NMR experiments performed with a PG-SLED (pulse gradient-simulated echo longitudinal encode-decode) sequence using bipolar gradient pulses for diffusion and varying the strength of the diffusion gradients between the experiments [88], the result of such experiments is shown in Figure 21.18: the signals decay with increasing strength of gradients, the decay rate is faster for the small molecule dioxane (b) than for the protein (a).

Empirical equations that relate the length of the polypeptide chain of isotropic native proteins to the hydrodynamic radius of a protein have been identified [87]. Thus, diffusion measurements are used to identify dimers or higher oligomers in native proteins (references 2, 3, 28 and 29 in Ref. [87]).

The situation is somewhat more complicated for nonnative proteins, since the ensemble of conformers in the denatured state can only be described as a distribution of polypeptides that have different hydrodynamic radii, each radius representing a group of conformations; this distribution cannot be determined experimentally, rather the average of the hydrodynamic radii present in a sample is observed.

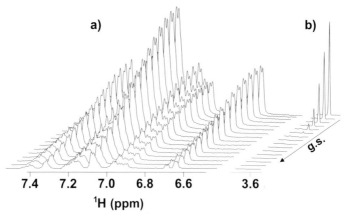

Fig. 21.18. Signal intensity in a ^{1}H 1D NMR spectrum as function of gradient strength (g.s.). a) lysozyme mutant W62G; b) dioxan.

Diffusion measurements have been used to identify urea unfolding of proteins (e.g., of lysozyme [89]) and to characterize different nonnative states of proteins such as BPTI [90] or α-lactalbumin [39]. The R_h of unfolded nondisulfide bridged proteins under highly denaturing conditions (high concentration of urea or GdmCl; extreme pH) can also be correlated to the length of the polypeptide chain [87].

21.3.7
Paramagnetic Spin Labels

Paramagnetic spin labels in proteins can be used to test long-range order or contacts. Spin labels such as nitroxide cause significant enhancement of relaxation rates of resonances that are within a distance of 15 Å away from the label [91, 92]. Spin labels can be attached to a number of residues containing reactive side chains such as Cys [93], His [91], or Lys [92]. Also the N-terminus of the protein can be used to attach a spin label [92]. Solvent-exposed residues especially in folded proteins have to be selected to prevent structure perturbations [94, 95], also paramagnetic metal ions such as Mn^{2+} or Gd^{3+} can be used if specific binding sites exist [96].

In folded proteins, distances can be extracted from the differential line broadening of the diamagnetic protein (reduction of the nitroxide with ascorbate to nitroxylamin) [94] to the paramagnetic protein [91, 92], whereas only a qualitative picture for nonnative states of proteins is gained, since they are composed of fluctuating conformations [97–99]. Using paramagnetic spin labels, long-range interactions were identified (e.g., in highly denatured myglobin) [99], in staphylococcal nuclease [97, 100], and in protein L [98]. Introduction of new binding sites for paramagnetic ions [80] and tagging those to nonnative states of proteins will allow detailed investigations using paramagnetic effects.

21.3.8
H/D Exchange

Hydrogen exchange methods can give insight into protein structure and dynamics by looking at the exchange of amide protons of the protein backbone with solvent protons. In general, protons that are involved in stable hydrogen bonds or are buried within the protein core are protected from exchange with the solvent. If either the buffer contains D_2O or the protein has been deuterated, the exchange of protons for deuterons or vice versa can be monitored by NMR. Hydrogen exchange techniques are valuable for the investigation of protein folding kinetics [101–104], folding intermediates, and unfolded states [105–107], since they can provide information on the degree and dynamics of secondary structure formation, which depends on the formation of stable hydrogen bonds. Hydrogen exchange is slow at low pH and faster at higher pH. In pulse labeling hydrogen exchange experiments [101], the process of folding or refolding could be investigated for some proteins. A typical approach of pulse-labeling hydrogen exchange is to let a deuterated protein initially refold for a short time (ms) (e.g., by diluting into D_2O refolding buffer) and give short pulses of protons at distinct times and high pH. This way, unprotected amide deuterons will exchange for protons and can be detected by 2D-NMR [108]. The reverse approach, namely refolding of nondeuterated protein in H_2O buffer and giving D_2O pulses, is also used [109–111]. The process of protein unfolding has been investigated in pulse-labeling hydrogen exchange NMR studies as well [112]. This topic is further described in Chapter 18.

21.3.9
Photo-CIDNP

Accessible aromatic side chains in proteins can be monitored using the photo-CIDNP (chemically induced dynamic nuclear polarization) NMR technique [113, 114]. The photo-CIDNP NMR technique relies on the temporary interaction of a photoexcited dye – mostly a flavin such as FMN (Figure 21.19) – with an aromatic amino acid side chain (Trp, Tyr, and His). Applying the photo-CIDNP technique to a protein results in signal enhancement as well as a better resolution in the aromatic region of the NMR spectrum (see Figure 21.19). Intensities of photo-CIDNP signals depend on (a) the polarization efficiency of a certain amino acid which is Trp > Tyr ≫ His ≫ Met and (b) the accessibility of the amino acid. Emissive photo-CIDNP signals are observed for Tyr residues, whereas the signals of Trp and His are absorptive – the sign of a signal therefore reports on the type of amino acid.

Using photo-CIDNP to achieve polarization enhancement is mostly seen in proton NMR spectroscopy – nevertheless higher enhancements are observed for ^{15}N nuclei [115]. One particular advantage of photo-CIDNP NMR for unfolded states is the better resolution due to the fact that only few residues are polarized. Using unlabeled protein, it is therefore often possible to assign peaks to the respective amino acid side chain and furthermore following changes in accessibility in differ-

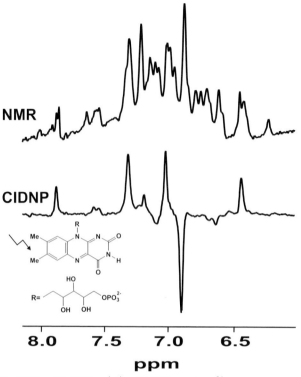

Fig. 21.19. 1D NMR and photo-CIDNP spectra of bovine α-lactalbumin. Photo-CIDNP is induced by irradiation of FMN (structure given at the lower left). Number of scans is 128 for the NMR spectrum and 1 for the CIDNP spectrum.

ent unfolded states [116]. Furthermore, photo-CIDNP NMR is being used in kinetic investigations of protein folding to monitor folding intermediates [117–119] – besides the better resolution since only few amino acids are polarized, better time resolution compared with normal NMR methods can be achieved due to the fact that the interscan delay depends only on the relaxation time of an electron (~10 ms). In addition, CIDNP experiment have been performed that transferred information on accessibilities in nonnative states of proteins onto the native protein using photo-CIDNP by polarizing the protein in the nonnative state, folding it rapidly, and observing polarized spins in the native state [120].

21.4
Model for the Random Coil State of a Protein

In a number of studies investigating nonnative states of lysozyme, α-lactalbumin, and ubiquitin, we have developed a model for the random coil state of a protein. The model assumes that the random coil state of a protein can be described using

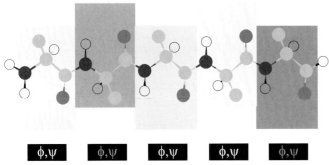

Fig. 21.20. The polypeptide chain consists of a chain of residues. Each residue has an intrinsic conformational preference even in the unfolded chain of protein. For a given distribution, NMR parameters such as J coupling constants and NOE can be predicted.

the notion that a polypeptide chain is a polymer consisting of 20 different amino acids. The conformational properties of the amino acids will depend on the nature of the side chain and therefore, amino acids exhibit distinct preferences in their sampling of conformational space described by the torsion angles ϕ, ψ (for the backbone conformation, depicted in Ramachandran plots) and correlated side-chain angles χ_n. Based on a conformational model for the distribution of torsion angle space, NMR parameters such as J coupling constants and, using some assumptions, NOEs can be predicted. G. Rose et al. further discuss the conformational properties of unfolded proteins in chapter 20.

Figure 21.21 shows the experimental $^3J(H^NH_\alpha)$ coupling constants in reduced and methylated lysozyme recorded in 8 M urea, pH 2. The measured coupling constants all fall in between the values predicted for either α-helical or β-strand secondary structure elements. However, the spread in the mesured values and the uneven distribution also calls for an interpretation of the data. Three models that do not invoke a residue-specific ϕ, ψ distribution could be envisioned to interpret the data: Model I would predict all ϕ-values ($-180° < \phi < 180°$) to be equally likely, the averaged coupling constants would thus be predicted to be 5.3 Hz. Model II assumes an symmetric average between two conformations, namely 50% α-helical, 50% β-strand conformation. Model III assumes a distribution of ϕ-values as derived from the Ramachandran distribution found in all structure of proteins when averaged over all residues. Neither of the three models predicts the observed coupling constants well, especially the spread in couplings calls for a more detailed model including amino acid-specific sampling of the conformational space.

According to this random coil model, the random coil state of a polypeptide chain is presumed to be one in which there are no specific nonlocal interactions between residues [24, 25, 45, 121]. In such a state, individual residues will, however, have local conformational preferences and will sample the low energy ϕ, ψ conformations in the Ramachandran space. In the model, the ϕ, ψ populations for

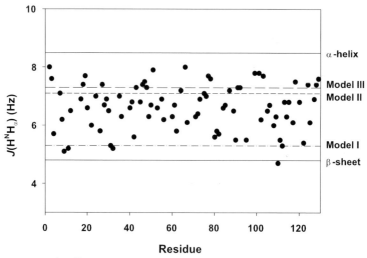

Fig. 21.21. $^3J(H^NH_\alpha)$ coupling constants in unfolded lysozyme (8 M urea, pH 2). Coupling constants expected for an α-helical and a β-sheet conformation are indicated as black lines. Models I–III indicate different averaging models that could be present in a random coil. Model I: all ϕ angles are equally likely (dashed-dotted line); model II: 50% α and 50% β-sheet conformation are populated (small dashed line); model III: populations are equally to the Ramachandran distribution (large dashed line).

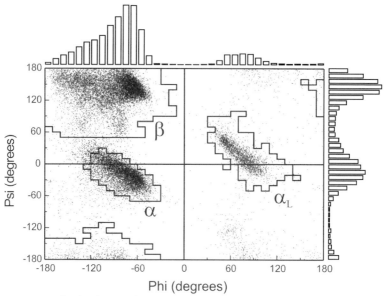

Fig. 21.22. The distribution of ϕ, ψ torsion angles for all amino acids not located in secondary structure elements in a database of 402 high-resolution crystal structures of native folded proteins. The α, β and α_L (combined α_{left} and γ_{left}) regions of ϕ, ψ space are labeled. Similar diagrams can be obtained with sufficient statistic for each amino acid individually.

residues in a random coil are derived from coil regions of 402 proteins from the protein data bank (Figure 21.22). It has been found that the distribution of ϕ, ψ torsion angles in the protein data bank resembles the experimental ϕ, ψ energy surface [24, 45, 122, 123], the effects of specific nonlocal interactions present in individual proteins are thus averaged out over the set of structures.

The amino acid-specific distribution obtained for each amino acid can also be used to calculate residue specific J coupling constants and also to calculate NOE intensities using a Monte-Carlo procedure and a simplified pentapeptide model. The pentapeptide model consists of explicit treatment of the backbone angles ϕ and ψ and a simplistic treatment of the side chains as a pseudoatom. Distributions of interproton distances can then be extracted from the conformational ensembles, and NOE intensities calculated assuming a two-spin approximation and r^{-6} averaging. Predicted values of NOE intensities converged with ensemble sizes of 10^5 conformers.

A different approach is chosen for the interpretation of dynamics in unfolded proteins. Backbone dynamics in proteins can be monitored by ^{15}N amide heteronuclear relaxation measurements. These heteronuclear relaxation measurements are sensitive to motions on a subnanosecond time scale and to slow conformational exchange in the millisecond time scale. Heteronuclear relaxation data unfolded proteins without disulfide bridges show only small variations over the sequence: Values approach a plateau at the middle of the protein while lower values are found at the termini of the sequence (Figure 21.23). This implies that the relaxation properties of a given amide are not influenced by its neighbor, but just by the fact that its part of the polypeptide chain. This relaxation behavior has been found for a number of unfolded proteins [21, 124] and in synthetic polymers [125] (see also Figure 21.35 for random coil state of ubiquitin). It was shown for atactic polystyrene that plateau values of R_1 relaxation rates remain constant for homopolymers between 10 kDa and 860 kDa. Individual segments of the long polymer chain therefore move independently in solutions – the apparent correlation time is therefore largely independent of the overall tumbling rate and only affected by segmental motions.

21.5
Nonnative States of Proteins: Examples from Lysozyme, α-Lactalbumin, and Ubiquitin

After the overview of NMR parameters that can measure aspects of conformation and dynamics of nonnative states of proteins, we wish to show the measurement of these NMR parameters primarily for one particular protein, lysozyme. Many of the experiments have been done together with Chris Dobson and former members of his group.

The native state of lysozyme consists of two domains, the α-domain, which spans from residue 1 to 35 and 85 to 129 and the β-domain that involves residues 36–84. Four disulfides bridges are present in the native structure; two of them in the α-domain of the protein, holding the N- and the C-termini together (C6–C127, C30–C115) one in the β-domain (C64–C80) and one between the domains (C76–C94).

a)

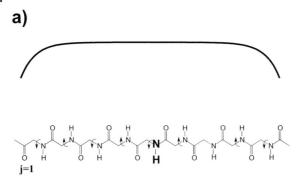

b)

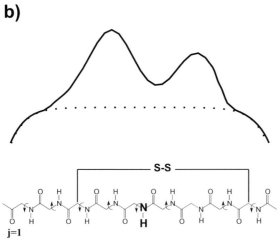

Fig. 21.23. Relaxation dynamics model for the random coil state. The relaxation properties of a given residue is influenced by the number of neighboring residues along the chain. For the interpretation of relaxation rates, one has to distinguish a) the unbranch polymer chain and b) the branched polymer chain, in which disulfide bridges cause additional cross linking of the polypeptide chain and lead to increased relaxation rates.

Lysozyme has been studied under a range of denaturing conditions by NMR techniques [25, 42, 107, 126–129, 183, 253].

21.5.1
Backbone Conformation

21.5.1.1 Interpretation of Chemical Shifts

The measurement of chemical shifts is the most commonly used method to identify secondary structure in nonnative states of proteins. The most easily available chemical shifts are H^N, H^α, and ^{15}N amide chemical shifts as side products of the assignment of the protein backbone using ^{15}N-labeled protein. In addition, ^{13}C chemical shift assignment can be obtained and the interpretation of the data are

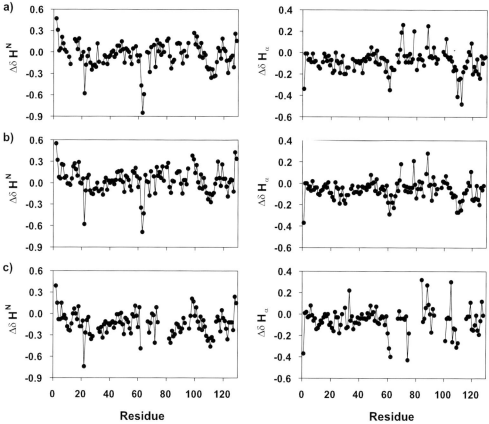

Fig. 21.24. Residual secondary structure in lysozyme as indicated by H^N and H_α chemical shifts pertubations ($\Delta\delta = \delta^{exp} - \delta^{rc}$) from random coil chemical shifts. a) Reduced and methylated hen lysozyme in water, pH 2, b) reduced and methylated hen lysozyme in 8 M urea, pH 2, and c) oxidized hen lysozyme in 8 M urea, pH 2.

shown in Figure 21.25. In the following section, it will be shown how these chemical shifts of random coil proteins can be used to (a) identify residual secondary structure (H^N and H^α) and to (b) investigate mean residue-specific properties of chemical shifts in unfolded states of proteins (^{15}N chemical shift).

H^N and H^α chemical shift perturbations ($\Delta\delta$) in unfolded variants of hen lysozyme from chemical shifts measured in short unstructured peptides are shown in Figure 21.24. The mean chemical shift pertubations are very close to the values measured in small unstructured peptides for all three variants. Figure 21.24a shows perturbations in a lysozyme variant in which all disulfide bridges have been reduced and subsequently methylated. This variant is unfolded at pH 2 in the presence and even in the absence of the denaturant urea. Large downfield deviations are found for Gly22 (H^N), for Val29 and Cys-S^{ME}30 (both H^α) around Trp62/Trp63 (R61 H^α; W62, W63 and C64 H^N), around Trp108/Trp111 (V109, R112 H^α; W111, N113, R114 H^N), and Trp123 indicating residual secondary structure. All these pertubations are found in aromatic patches of the protein, indicating that aromatic amino acids cluster and hereby form secondary structure elements, presumably to prevent contact with water. The cluster around Gly22 consists of the sequence Tyr-X-Gly-Tyr (where X can be any amino acid) and is also found to form a nonrandom cluster in BPTI [130–132]. Comparison of these chemical shift deviations to deviations of unfolded hen lysozyme in 8 M urea (oxidized and reduced) reveal that the formation of residual secunary structure around aromatic residues is a feature of the protein even detectable in 8 M urea.

Single point mutations of residues involved in patches of residual secondary structure in the unfolded state (A9G, W62G, W62Y, W111G, and W123G) do not change the pattern of chemical shift perturbations observed in methylated protein in water at pH 2 (see Figure 21.26) [253].

For an assessment of residue-specific properties of chemical shifts in random coil states of proteins, a data set of three proteins in 8 M urea at pH 2 was compared: human ubiquitin, reduced and methylated hen egg lysozyme and all-Ala α-lactalbumin [138]. As mentioned in Section 21.2, chemical shift dispersion in nonnative states of proteins is considerably smaller than in folded states of proteins, amide ^{15}N chemical shifts, however, display significant chemical shift dispersion and therefore have been analyzed as discussed below. Aromatic amino acids, cysteine, and methionine residues were excluded from the investigations because of the lack of experimental data and due to the fact that aromatic residues have been found to be involved in nonrandom structures in at least one of the three proteins.

Mean observed chemical shifts in native proteins of the BMRB (as of February 1999) are very close to the mean experimental ^{15}N chemical shifts in the three unfolded proteins ($R = 0.98$, see Figure 21.27). However, the spread of chemical shifts in the native state is very large, while there are only small variations in chemical shifts of the unfolded proteins.

21.5.1.2 Interpretation of NOEs

In contrast to the situation in the folded state of a protein, distances that can be observed by NOE measurements in nonnative states of proteins reflect an average of conformations with various distributions of ϕ, ψ, and χ_1 angles. Using the protein database (PDB), ϕ, ψ distributions of all amino acids can be generated. These distributions can be used as a model for a random coil in which no nonlocal interactions are present due to the fact that specific nonlocal interactions that are present in the individual proteins are averaged out over the ensemble of conformers contributing to the spectroscopic properties of the protein (see Section 21.4). Using this data set, interproton distance distributions can be extracted for the nonnative state of a protein. Table 21.6 shows statistics of interproton distance distributions. Effective interproton distances in the random coil state are given as $\langle r^{-6} \rangle^{-1/6}$ using the ϕ, ψ distributions of all proteins in the PDB.

Two features of the random coil state compared to the folded state of a protein are noticeable: (1) In contrast to the folded state of a protein, the distances αN $(i, i+1)$ and NN $(i, i+1)$ are short in every case, which reflects the averaged population of β and α conformations at the same time. (2) The effective αN $(i, i+2)$ dis-

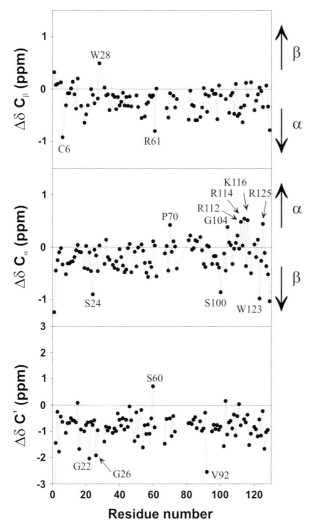

Fig. 21.25. Differences in $^{13}C'$, $^{13}C_\alpha$ and $^{13}C_\beta$ chemical shifts for lysozyme with its four disulfides intact denatured in 8 M urea from empirical values measured in short unstructured peptides ([37] $\Delta\delta = \delta^{exp} - \delta^{emp}$). ^{13}C chemical shifts in the present work are referenced indirectly to the internal standard TSP at pH 2 at 20 °C in 8 M urea. Mean secondary ^{13}C chemical shifts are (with standard deviations given in parentheses): $\langle\Delta\delta C_\alpha\rangle = -0.2$ (0.3) ppm, $\langle\Delta\delta C_\beta\rangle = -0.2$ (0.2) ppm, $\langle\Delta\delta C'\rangle = -0.9$ (0.6) ppm. Residues with $\Delta\delta \geq 0.38$ ppm, $\Delta\delta C_{\alpha,\beta} \leq -0.8$ ppm, and $\Delta\delta C' \leq -1.9$ ppm are labeled by one-letter amino acid code and residue number excluding terminal residues. Appropriate corrections for random coil shifts of Thr69 preceded by Pro70 were employed. Secondary chemical shifts for glutamate and aspartate residues have been omitted as the chemical shifts are affected by the protonation of the amino acid side chains at the low pH value of 2 used in this study. This is reflected by significant average deviations from random coil values for Asp, $\langle\Delta\delta C_\alpha\rangle = -1.6$ ppm, $\langle\Delta\delta C_\beta\rangle = -3.3$ ppm, $\langle\Delta\delta C'\rangle = -1.7$ ppm, and Glu, $\langle\Delta\delta C_\alpha\rangle = -1.3$ ppm, $\langle\Delta\delta C_\beta\rangle = -1.0$ ppm, $\langle\Delta\delta C'\rangle = -1.6$ ppm.

776 *21 Conformation and Dynamics of Nonnative States of Proteins studied by NMR Spectroscopy*

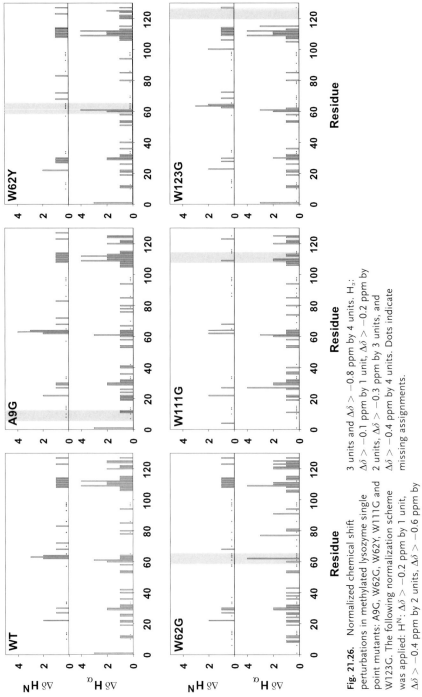

Fig. 21.26. Normalized chemical shift perturbations in methylated lysozyme single point mutants: A9G, W62G, W62Y, W111G and W123G. The following normalization scheme was applied: H^N: $\Delta\delta > -0.2$ ppm by 1 unit, $\Delta\delta > -0.4$ ppm by 2 units, $\Delta\delta > -0.6$ ppm by 3 units and $\Delta\delta > -0.8$ ppm by 4 units. H_α: $\Delta\delta > -0.1$ ppm by 1 unit, $\Delta\delta > -0.2$ ppm by 2 units, $\Delta\delta > -0.3$ ppm by 3 units, and $\Delta\delta > -0.4$ ppm by 4 units. Dots indicate missing assignments.

21.5 Nonnative States of Proteins: Examples from Lysozyme, α-Lactalbumin, and Ubiquitin

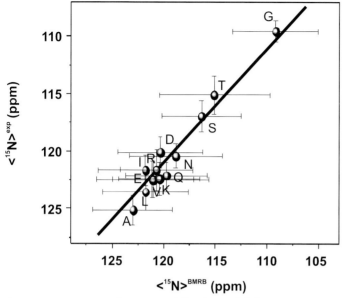

Fig. 21.27. Comparison of experimental ^{15}N chemical shifts in unfolded ubiquitin, hen lysozyme and all-Ala lactalbumin with chemical shifts in native proteins deposited in the databank for biomolecular NMR data (BMRB) (as of February 1999).

tance of 4.2 Å in the random coil state is shorter than in either of the secondary structure types. αN $(i, i+1)$, NN $(i, i+1)$, and αN $(i, i+2)$ as well as NN $(i, i+2)$, αN $(i, i+3)$, and NN $(i, i+3)$ distances have been predicted for hen lysozyme taking into account the amino acid sequence of hen lysozyme and using residue-specific ϕ, ψ distributions extracted from the PDB. The actual observability of NOE cross-peaks in the NOESY-HSQC of hen lysozyme (ox + red) in 8 M urea at pH 2 is dependent on the signal-to-noise ratio in the experiment and the effective correlation time of the protein, which could vary along the protein sequence (for the good signal-to-noise obtainable in the spectra of hen lysozyme, see Figure 21.28) [25].

Tab. 21.6. Comparison of the effective interproton distances in the ensemble of random coil conformers with interproton distances in regions of regular secondary structure. Effective interproton distances $\langle r-6 \rangle^{-1/6}$ are given for an ensemble of random coil conformers generated using the ϕ, ψ torsion angle populations of all residues in the protein database.

Distance	Random coil (Å)	α-helix (Å)	β-strand (Å)
αN $(i, i+1)$	2.5	3.5	2.2
NN $(i, i+1)$	2.7	2.8	4.3
αN $(i, i+2)$	4.2	4.4	6.1
NN $(i, i+2)$	4.6	4.2	6.9

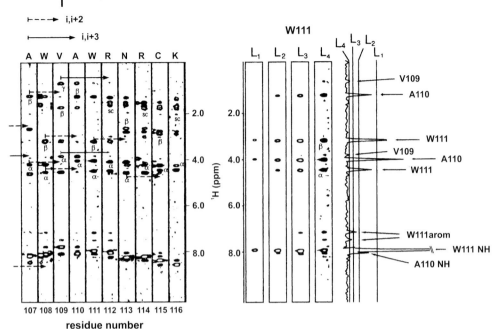

Fig. 21.28. Left: $^1H,^1H$ strips of residues 107–116 from a 3D NOESY spectrum of reduced and denatured lysozyme in 8 M urea at pH 2 (mixing time = 200 ms). Arrows indicated $(i, i+2)$ and $(i, i+3)$ cross-peaks as indicated. Right: $^1H,^1H$ strip of Trp111 is plotted at four different contour levels (= signal-to-noise levels), and 1D trace of Trp111. The different noise levels are indicated at the 1D trace, cross-peaks and the diagonal peak (Trp111 NH) are indicated.

The overall agreement between the predicted and experimental data from NOESY-HSQC spectra is remarkably good (Figure 21.29). αN $(i, i+1)$ and NN $(i, i+1)$ NOEs are predicted and observed throughout the protein sequence, furthermore 56 out of 82 predicted αN $(i, i+2)$ NOEs are seen while two of the αN $(i, i+2)$ NOEs are observed but not predicted. No correlation is observed between predicted and observed NN $(i, i+2)$ NOEs; while only 11 of the 27 predicted NN $(i, i+2)$ NOEs are observed, 10 NN $(i, i+2)$ cross-peaks are observed that are not predicted. A similarly poor correlation is observed for αN $(i, i+3)$ cross-peaks: two out of eight predicted NOEs are observed and seven NOEs are observed but not predicted. However, it is worth noting, that both the αN $(i, i+3)$ and the NN $(i, i+2)$ crosspeaks are predicted to be weak, and that the observability therefore depends very strongly on the signal-to-noise of the spectrum, which can vary locally. In summary, the considerable amount of predicted and observed NOEs in unfolded hen lysozyme, in particular αN $(i, i+1)$, NN $(i, i+1)$, and αN $(i, i+2)$ NOEs indicate that the averaged backbone conformation predicted from all proteins in the PDB describes the local preferences in lysozyme very well. It is important to note that this presence of NOEs is not a sign of persistent nonrandom structure

21.5 *Nonnative States of Proteins: Examples from Lysozyme, α-Lactalbumin, and Ubiquitin* | 779

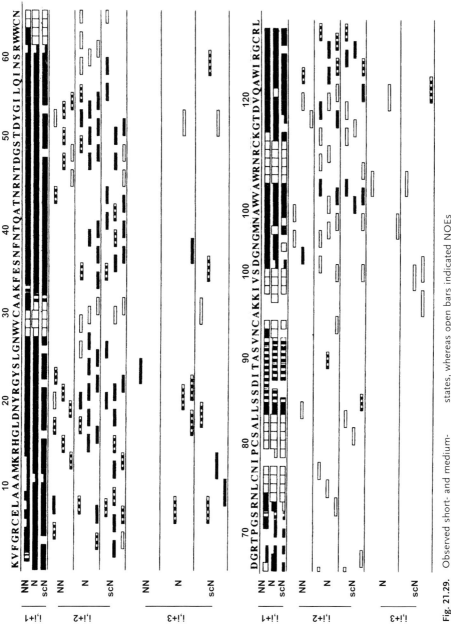

Fig. 21.29. Observed short- and medium-range NOEs in oxidized and reduced lysozyme (8 M urea, pH 2) from 3D NOESY-HSQC spectra (mixing time = 200 ms). Black bars are indicative for NOEs that are present in both states, whereas open bars indicated NOEs observed in the reduced state and dashed bars in the oxidized state only. NOEs to any side chain are summarized as scN NOEs.

in the unfolded state, but arises from statistical averaging of possible conformations in a random coil. This average is dominated by the shorter distances since the NOE is averaged as $\langle r^{-6} \rangle^{-1/6}$.

21.5.1.3 Interpretation of J Coupling Constants

A number of ^3J coupling constants (see Table 21.3) report on the torsion angle ϕ. $^3J(H^NH_\alpha)$, $^3J(C'C_\beta)$ and $^3J(C'C')$ have been measured in unfolded lysozyme (oxidized) in 8 M urea, pH 2 [183]. Moreover $^3J(H^NH_\alpha)$ coupling constants in unfolded ubiquitin and all-Ala lactalbumin (both in 8 M urea, pH 2) are available, to allow residue-specific analysis [138]. The experimental $^3J(H^NH_\alpha)$ coupling constants for all three proteins vary substantially along the protein sequence (see Figure 21.21 for reduced and methylated lysozyme in 8 M urea, pH 2, data for other proteins not shown), the values for a given amino acide type, however, typically cluster in small ranges. Figure 21.30 shows a comparison of experimental $^3J(H^NH_\alpha)$ coupling constants of the three proteins and predictions of $^3J(H^NH_\alpha)$ coupling constants using ϕ, ψ distributions from "coil" regions of 402 protein crystal structures and applying the Karplus parameterization shown in Table 21.3.

A fairly good correlation ($R = 0.85$) is observed for a complete data set including aromatic residues (not shown in Figure 21.30). The correlation improves if Glu and Asp residues that are protonated at pH 2 are excluded ($R = 0.89$). The best correlation, however, is observed if aromatics are left out as well ($R = 0.95$). Deviations of the coupling constants from mean values are similar for the predicted

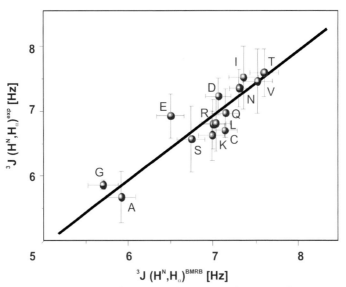

Fig. 21.30. Comparison of experimental $^3J(H^NH_\alpha)$ of unfolded ubiquitin, all-Ala α-lactalbumin, and lysozyme $^3J(H^NH_\alpha)^{exp}$ with $^3J(H^NH_\alpha)$ coupling constants predicted from "coil" regions of 402 PDB structures ($^3J(H^NH_\alpha)^{BMRB}$). Methionine, cysteine, and aromatic residues are left out due to the small amount of available data.

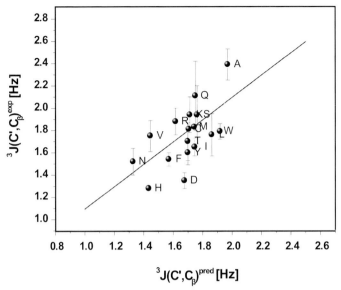

Fig. 21.31. Experimental versus predicted $^3J(C'C\beta)$ coupling constants using the coil model.

and the experimental coupling constants, reflecting the quality of the model used for the predictions. Investigations of the individual data sets (–Glu, –Asp and –aromatics) give rise to similar correlation coefficients ($R = 0.96$ for ubiquitin; $R = 0.94$ for lysozyme and $R = 0.90$ for all-Ala α-lactalbumin). These results indicate the independence of both the primary sequence and the native structure of the protein ubiquitin is composed of mainly β-sheets, while lysozyme and α-lactalbumin contain a significant number of α-helices in the native state.

The variation of $^3J(C'C_\beta)$ and $^3J(C'C')$ coupling constants is smaller than the variation in $^3J(H^NH_\alpha)$ coupling constants (Figure 21.31). The smallest variations are found in the $^3J(C'C')$ values, which range from 0.5 to 1 Hz. Nevertheless clustering of the experimental values is observable: values for eight Gly residues range from 0.5 to 0.7 Hz. A more detailed analysis of residue-specific properties of $^3J(C'C')$ coupling constants is not appropriate due to the small range of measured coupling constants.

Values of $^3J(C'C_\beta)$ in unfolded oxidized lysozyme range from 1.3 to 2.6 Hz. As observed for the $^3J(H^NH_\alpha)$ coupling constants, large variations along the sequence are found. These variations again cluster for different residue types. However a different trend is seen to that for the $^3J(H^NH_\alpha)$ coupling constants: the largest $^3J(C'C')$ are found for alanine residues, which have a high probability of being found in a conformation in the ϕ, ψ Ramachandran space, while the smaller values are found for residues with larger side chains such as histidine, asparagine, and valine residues. Figure 21.31 shows a correlation of experimental coupling constants with predicted coupling constants (Hz) from the coil theory. A correlation coefficient of $R = 0.63$ is observed for the complete data set (axis intercept = 0.07, slope = 1), $R = 0.69$ (axis intercept = 0.0949, slope = 1) if Asp and Glu residues (that are protonated at pH 2 but not in the data sets used for the predictions) are left out.

The discussion thus far has focused on the determination of coupling constants that report on the backbone angle ϕ. Figure 21.32 shows a comparison of the prediction of $^3J(N^H,H_\alpha)$ coupling constants taking (a) all residues into account or (b)

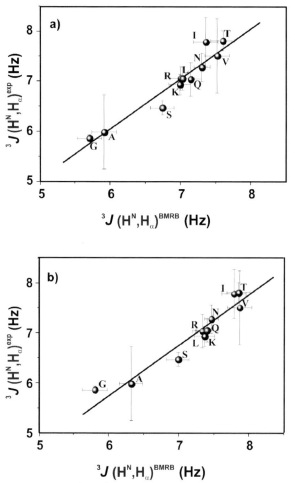

Fig. 21.32. Correlation between experimental $^3J(H^N,H_\alpha)$ averaged for a given residue type, e.g., all alanines in ubiquitin denatured in 8 M urea, pH 2, and $^3J(H^N,H_\alpha)$ coupling constants predicted for amino acids excluding aromatic amino acids and Asp and Glu residues. a) Comparison with predictions for residues that are not located in secondary structure elements in the database of native folded protein structures with ϕ, ψ torsion angles covering all regions of the Ramachandran diagram. b) Comparison with predictions for residues that are not located in secondary structure elements and that have positive ψ torsion angles only.

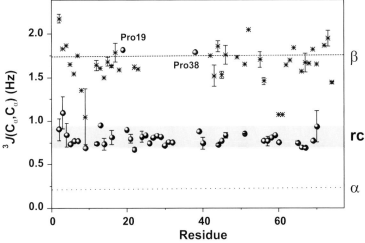

Fig. 21.33. $^3J(C_\alpha,C_\alpha)$ coupling constants in unfolded (dots) and folded (crosses) ubiquitin. Dashed line indicates expected coupling constants in a β-sheet, shaded area indicates expected values in a random coil (rc), and dotted line indicates values of coupling constants in the α-conformation.

only residues with positive ψ angle into account (compare Figure 21.22). A similar correlation ($R = 0.96$ in both cases) is observed showing that $^3J(N^H,H_\alpha)$ coupling constants cannot differentiate between a situation in which only extended conformations are being sampled (positive ψ values) and a situation in which the polypeptide chain samples both α- and β-regions of the Ramachandran space. The presence of experimental NOE contacts H^N_i, H^N_{i+1} and of $H_{\alpha i}$, H^N_{i+1} of similar intensity in nonnative states of proteins indicate that averaging involves sampling of positive and negative ψ torsion angles, which is compatible with predictions of our model. However, a quantitative analysis of populated rotamers from NOE intensities is difficult due to the r^{-6} averaging and complex dynamical properties that are potentially nonuniform along the denatured peptide chain.

$^3J(C_\alpha,C_\alpha)$ coupling constants can probe the conformation around the backbone angle ψ directly and are shown in Figure 21.33 for the random coil state and for the native state of ubiquitin [254]. The dependence of $^3J(C_\alpha,C_\alpha)$ on the angle ψ is unexpected and in contradiction to a simple Karplus-type dependence on the intervening torsion angle ω but was found for data obtained for native ubiquitin. The mean experimental coupling constants $^3J(C_\alpha,C_\alpha)$ for residues in β-sheet regions of native ubiquitin are 1.69 ± 0.1 Hz while coupling constants are too small to be observed in α-helical regions. In the random coil state of ubiquitin, $^3J(C_\alpha,C_\alpha)$ are larger than 0.7 Hz throughout the sequence, including residues 23–34 that are located in the α-helical region in the native state. The mean $^3J(C_\alpha,C_\alpha)$ coupling constant is 0.85 ± 0.2 Hz with a pairwise rmsd of 0.1 Hz for the 41 resonances that are sufficiently resolved to allow coupling constants to be determined. As an exception, large $^3J(C_\alpha,C_\alpha)$ values are observed for Pro19 and Pro38. This might reflect a strong preference of prolines for ϕ, ψ torsion angles in the polyprolyl region of ϕ, ψ space ($\phi \sim -60°$, $\psi \sim +150°$) and potentially restricted ψ sampling.

The correlations found here confirm the previous conclusion based on the $^3J(H^N,H_\alpha)$ coupling constants, that the main-chain torsion angle populations in denatured lysozyme predominantly reflect the intrinsic ϕ, ψ preferences of the amino acids involved, and shows that coupling constant measurements can be used successfully to probe main-chain conformations in the nonnative state of a protein.

21.5.2
Side-chain Conformation

21.5.2.1 Interpretation of J Coupling Constants

According to the random coil model, the side-chain conformation in nonnative states of proteins has to be correlated with the type of amino acid and at least some degree of nonuniform sampling needs to be present between different amino acids. However, the measurement of such torsion angle distribution based on interpretation of NOE data is difficult due to spectral overlap and chemical shift analysis does not provide a detailed picture of the conformational averaging.

For the random coil state lysozyme, however, the measurement of $^3J(C',C_\gamma)$ and $^3J(N,C_\gamma)$ coupling constants has enabled side-chain χ_1 torsion angle populations to be probed in the random coil polypeptide chain [183]. Analysis of the coupling constant data has allowed the relative populations of the three staggered rotamers about χ_1 to be defined for 51 residues. Amino acids can broadly be divided into five classes which show differing side-chain conformational preferences in the random coil state (Figure 21.34).

The coupling constants were interpreted using a three-site jump model to describe qualitatively the χ_1 angle distributions (Pachler analysis). The analysis assumes, as a first approximation, that only the three staggered rotamers around the side-chain angle $\chi_1 = +60°, 180°, -60°$ have significant populations. It is assumed that there is rapid interconversion between these three rotamers so the observed coupling constants are a population weighted mean of the values expected for each of the three staggered rotamers (see Eqs (22a–c)):

$$^3J_{exp}(N,C_\gamma) = p_{180°} \cdot {}^3J_{trans}(N,C_\gamma) + (1 - p_{180°}) \cdot {}^3J_{gauche}(N,C_\gamma)$$

$$^3J_{exp}(C',C_\gamma) = p_{-60°} \cdot {}^3J_{trans}(C',C_\gamma) + (1 - p_{-60°}) \cdot {}^3J_{gauche}(C',C_\gamma) \qquad (22)$$

$$p_{60} = 1 - p_{180} - p_{-60}$$

Using Eqs (22a–c), the relative populations of the χ_1 conformers of 60°, −60°, and 180° have been calculated for each residue from the experimental $^3J(C',C_\gamma)$ and $^3J(N,C_\gamma)$ coupling constants. Analysis of the populations of the three staggered rotamers derived from the $^3J(C',C_\gamma)$ and $^3J(N,C_\gamma)$ coupling constants has enabled

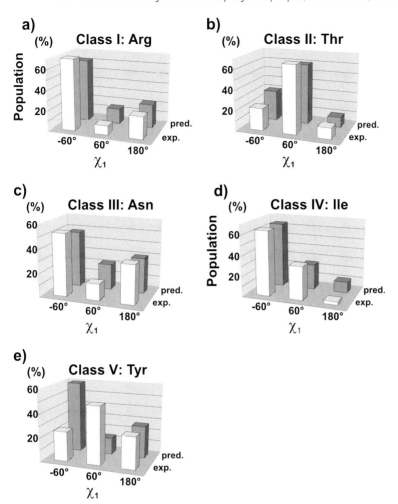

Fig. 21.34. Comparison of the mean percentage populations of the three staggered side chain χ_1 rotamers for different amino acids calculated from the experimental coupling constant data for lysozyme denatured in 8 M urea and predicted from the statistical model for a random coil. The aromatic residues are excluded from this comparison due to their anomalous behavior.

us to recognize several different classes of side-chain behavior (Figure 21.34). For residues with a long aliphatic side chain which is unbranched at C^β (Arg, Gln, Glu, Leu, Met; denoted class I), the $-60°$ χ_1 rotamer is significantly favored (70–79% population) with a very low population (< 10%) of the 60° χ_1 rotamer. In contrast for threonine (class II), it is the $+60°$ χ_1 rotamer that is most highly populated (67%). For aspartate and asparagine (class III) the populations of the $-60°$ χ_1 (53–55%) and 180° χ_1 rotamers (33–37%) are both significant while for isoleucine (class IV) the $-60°$ χ_1 rotamer is most populated (64%) with the lowest population for $180°$ χ_1 (3%). The behavior of the residues with aromatic side chains (class V) varies considerably with a range of populations being seen for a given type of aromatic side chain and different mean populations being observed for the various aromatic residues.

The behavior of the amino acid side chains observed experimentally in the random coil state of lysozyme has been compared with the predictions for a random coil from the statistical model. In particular the χ_1 populations derived from the experimental coupling constants have been compared with the χ_1 distributions in the PDB for residues that are not in recognized regions of secondary structure (COIL parameter set). The choice of this set of residues is important because the steric restraints of regular secondary structure elements can affect the χ_1 distributions significantly [122, 134, 135]. As the Pachler analysis does not consider the population of side-chain conformations other than those of the three staggered rotamers, we compare the relative proportions of χ_1 torsion angles falling within the ranges $\chi_1 = -60° \pm 30°$, $\chi_1 = +60° \pm 30°$ and $\chi_1 = 180° \pm 30°$ in the PDB with the analogous distributions calculated from the coupling constants. In general only a small number of residues in the database have χ_1 torsion angles that do not fall in the three ranges considered (e.g., 87 (5%) of the 1628 aspartate residues in the database). The PDB populations are listed in Table 21.7.

Many of the features of the side-chain populations derived from the experimental coupling constants are also clearly observed in the populations predicted for a random coil from the PDB. Examples of comparisons of the two sets of populations are shown in Figure 21.34. The PDB distributions predict that for all amino acids in a random coil, excluding serine, threonine, valine, isoleucine, cysteine, asparagine, and aspartic acid, the $-60°$ χ_1 rotamer should have the highest population ($> 50\%$) and the $+60°$ χ_1 rotamer the lowest population ($\leq 20\%$). The $-60°$ χ_1 rotamer is most favored as it places the side chain in a position where the steric clash with the main chain is minimized while the $+60°$ χ_1 rotamer is the most sterically restricted and therefore the least populated.

21.5.3
Backbone Dynamics

21.5.3.1 Interpretation of ^{15}N Relaxation Rates

Backbone dynamics in proteins can be monitored by ^{15}N amide heteronuclear relaxation measurements. These heteronuclear relaxation measurements are sensitive to motions on a subnanosecond time scale and to slow conformational exchange in the millisecond time scale. Figure 21.35 shows heteronuclear transverse relaxation rate of the random coil state of ubiquitin in 8 M urea at pH 2. The relaxation rates show only small variations over the sequence: Values approach a plateau at the middle of the protein while lower values are found at the termini of the sequence. This implies that the relaxation properties of a given amide are not influenced by its neighbor, but just by the fact that it is part of the polypeptide chain. The only exception from this behavior is Thr9 with considerable large relax-

Tab. 21.7. Predicted $^3J(C',C_\beta)$ and $^3J(C',C')$ coupling constants and populations of the three staggered side-chain χ_1 rotamers from the statistical model for a random coil. The coupling constant values and were calculated using the residues in all secondary structure elements (ALL) and for residues found neither in α nor β (COIL) in a database of 402 high-resolution crystal structures. The χ_1 populations were calculated using the COIL parameter set considering only the resides whose χ_1 torsion angles fall into 60% ranges centered about the distinct staggered rotamers ($\chi_1 = -60° \pm 30°$; $\chi_1 = 60° \pm 30°$ or $\chi_1 = 180° \pm 30°$). No side-chain populations are given for proline.

	Coupling constants (Hz)				Relative χ_1 population (%)		
	$^3J(C',C_\beta)$ ALL	$^3J(C',C_\beta)$ COIL	$^3J(C',C')$ ALL	$^3J(C',C')$ COIL	$\chi_1 - 60°$ COIL	$\chi_1\ 60°$ COIL	$\chi_1\ 180°$ COIL
Ala	2.1	2.0	0.8	0.8	–	–	–
Asp	1.9	1.7	0.8	0.9	41	27	32
Asn	1.6	1.4	0.9	0.9	48	22	30
Arg	1.9	1.7	0.9	0.9	62	15	23
Cys	1.7	1.6	1.0	1.0	48	25	27
Gln	2.0	1.6	0.8	0.9	65	13	22
Glu	2.0	1.9	0.8	0.8	56	17	26
Gly	–	–	0.9	0.8	–	–	–
His	1.6	1.4	1.0	1.1	57	20	23
Ile	1.8	1.7	0.9	1.0	64	25	11
Leu	2.0	1.8	0.8	0.8	79	4	17
Lys	1.9	1.7	0.8	0.9	65	11	23
Met	1.9	1.7	0.8	0.9	65	11	23
Phe	1.7	1.6	1.0	1.0	62	13	25
Pro	2.4	2.4	0.5	0.5	–	–	–
Ser	1.8	1.7	1.0	1.0	31	46	23
Thr	1.7	1.6	1.0	1.0	30	60	10
Trp	1.8	1.7	0.9	0.9	52	17	31
Tyr	1.6	1.6	1.0	1.0	60	14	27
Val	1.7	1.6	1.0	1.0	26	9	64

ation rates. Besides that, the polypeptide chain of ubiquitin in 8 M urea, pH 2 does not exhibit any amino acid-specific variation of the heteronuclear relaxation rates. This relaxation behavior has also been found for a number of unfolded proteins [21, 124] and in synthetic polymers [125]. It was shown for atactic polystyrene that plateau values of R_1 relaxation rates remain constant for homopolymers between 10 kDa and 860 kDa. Individual segments of the long polymer chain therefore move independently in solutions – the apparent correlation time is therefore largely independent of the overall tumbling rate and only affected by segmental motions.

The relaxation rates for the random coil state of lysozyme are shown in Figure 21.36.

Using a model of segmental motions the relaxation properties can be described, assuming that relaxation contributions due to neighboring residues in a given

Tab. 21.8. Summary of the populations for χ_1 rotamers for denatured lysozyme calculated from the experimental coupling constants for the different classes of amino acids. The experimentally derived populations are mean values for each amino acid, the corresponding standard deviations being given in parentheses. Values for residues where the population of only one of the staggered rotamers could be defined have not been included when calculating the means (except of lysine). For isoleucine the populations listed are a mean of those calculated from coupling constants involving $C_{\gamma 1}$ and $C_{\gamma 2}$.

	Population (%) $\chi_1 = -60°$	Population (%) $\chi_1 = 60°$	Population (%) $\chi_1 = 180°$
Class I: Long aliphatic side chains			
Arg	69.9 (4.1)	8.5 (9.7)	21.6 (11.9)
Gln	70.0 (1.6)	8.1 (1.6)	21.9 (2.7)
Glu	73.3 (0.3)	−1.7 (0.6)	28.5 (1.0)
Lys			35.1 (11.0)
Leu	74.3 (4.3)	−0.5 (2.8)	26.2 (4.7)
Met	78.5	−4.8	26.3
Class II: Side chains with OH at C_β			
Thr	22.1 (8.9)	66.8 (4.5)	11.1 (6.3)
Class III: Asx type residues			
Asn	52.5 (6.4)	14.4 (10.9)	33.1 (8.6)
Asp	55.4 (7.1)	7.4 (11.5)	37.2 (11.5)
Class IV: β-branched side chains			
Ile	63.7 (14.0)	33.2 (6.2)	3.1 (14.6)
Class V: Aromatic side chains			
His	55.5	13.4	31.1
Phe	45.3 (12.6)	23.0 (4.6)	31.7 (10.3)
Trp	24.0 (4.5)	20.9 (16.2)	55.1 (11.7)
Tyr	24.7 (6.5)	48.8 (9.1)	26.5 (15.4)

polypeptide chain decay exponentially as the distance from a given residue increases [25].

$$R_2^{rc}(i) = R_{int} \sum_{j=1}^{N} e^{-(|i-j|/\lambda_0)} \qquad (23)$$

A description of transverse relaxation rates (R_2 and $R_{1\rho}$) in reduced lysozyme (Figure 21.6d) using the segmental motion model (Eq. (23)) is shown in Figure 21.37 [25, 184]. R_2 relaxation rates and $R_{1\rho}$ relaxation rates are very similar, indicating the absence of significant chemical exchange contributions in the reduced and methylated protein. Deviations from relaxation rates predicted by the segmental motion model for a random coil are observed around Trp62/Trp63 and also near Trp108/Trp111 as well as around Trp123. Positive deviations from the segmental motions model indicate additional stiffness of the polypeptide chain, that is induced by interactions of the bulky side chains. A description of these deviations

21.5 Nonnative States of Proteins: Examples from Lysozyme, α-Lactalbumin, and Ubiquitin | 789

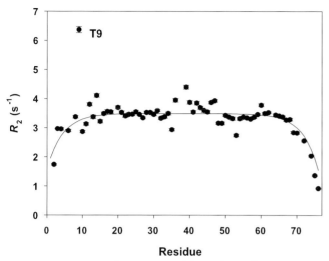

Fig. 21.35. Transverse relaxation rates of the random coil state of ubiquitin (8 M urea, pH 2) at 600 Hz. Fitting parameters for random coil predictions: $R_{int} = 0.4$, $\lambda = 4.35$.

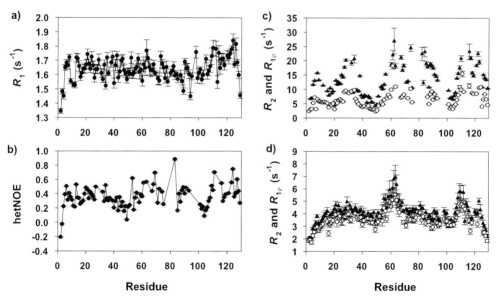

Fig. 21.36. Relaxation data of backbone amides of oxidized and reduced lysozyme in 8 M urea at pH 2. a) R_1 relaxation rates of reduced lysozyme, b) hetNOE of oxidized lysozyme, c) R_2 relaxation rates (filled triangles) and $R_{1\rho}$ relaxation rates (open circles) of oxidized lysozyme, and d) R_2 relaxation rates (filled triangles) and $R_{1\rho}$ relaxation rates (open circles) of reduced lysozyme.

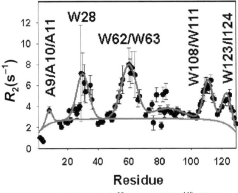
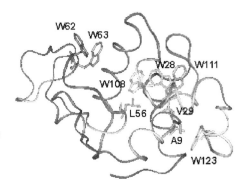

Fig. 21.37. a) ^{15}N R_2 in WT-S^{ME}. The experimental rates are shown as a scatterplot and the rates fitted by a model of segmental motion expected for a random coil with R_2^{rc} ($R_{int} = 0.2$ s^{-1} and $\lambda_0 = 7$) are shown as dashed lines. Translation of the relaxation data on the native structure of lysozyme is shown in b).

from the segmental motions model can be done introducing Gaussian clusters to the segmental motion model:

$$R_2^{exp}(i) = R_{int} \sum_{j=1}^{N} e^{-(|i-j|/\lambda_0)} + \sum_{cluster} R_{cluster} e^{-((i-x_{cluster})/2\lambda_{cluster})^2} \quad (24)$$

The clusters of restricted motions observed by relaxation measurements in reduced and methylated lysozyme correspond to the areas of residual secondary structure identified from chemical shift perturbations (see Figure 21.25). The dynamics of the unfolded peptide chain are also investigated in detail in Chapter 22.

The interpretation of R_2 and $R_{1\rho}$ relaxation rates in oxidized and therefore branched unfolded lysozyme is more complicated: not only residues that are near in sequence influence the relaxation properties of a given amide, but also residues that are near via disulfide bridges. A topological distance matrix, dm_{ij}, which counts the shortest path from residue i to residue j, is used in which a disulfide bond counts the same as a peptide bond in connecting two residues (see Eq. (25)). R_2 and R_1 relaxation rates are clearly different, indicating exchange processes taking place around disulfide bridges. For a complete description of R_2 relaxation rates it is therefore necessary to include a term describing these exchange processes as shown in Eq. (25).

$$R_2^{exp}(i) = R_{int} \sum_{j=1}^{N} e^{-(dm_{ij}/\lambda_0)} + R_{exch} \sum_{k=1}^{N_{cys}} e^{-(|i-Cys_k|/\lambda_2)} \quad (25)$$

Due to the complexity of contributions to relaxation rates in branched systems further investigations concerning the nature of deviations from the segmental motion model have been performed using reduced and methylated variants of lysozyme

21.5 Nonnative States of Proteins: Examples from Lysozyme, α-Lactalbumin, and Ubiquitin

(i.e., unbranched polypeptide chains). Figure 21.37 shows R_2 relaxation rates measured in reduced and methylated lysozyme in water. Six hydrophobic clusters were identified in WT-S^{ME} using the extended segmental motion model (Eq. (24)). These clusters are centered at residues (1) L8, (2) C30, (3) S60, (4) S85, (5) W111, and (6) W123. The most prominent cluster, cluster 3 is found in the middle of the protein sequence which belongs to the β-domain in folded lysozyme. All other clusters (cluster 1, 2, 4, 5, and 6) are found in what is to be the α-domain in the folded state of the protein. While clusters 3, 5, and 6 are also observed in urea, cluster 2 which is near Trp28 is destroyed by the addition of urea. Furthermore the rather small clusters 1 and 4 cannot be identified in the presence of urea. The importance of weak interactions in protein folding is reviewed in Chapter 6.

Stabilization of these clusters can be tested by single point mutations in conjunction with relaxation measurements [184, 253]: Replacement of Trp62 in cluster 3 by glycine essentially abolishes not only its own cluster 3 but also clusters 1 through 4, and diminishes clusters 5 and 6 (as seen in Figure 21.38). The disruption of the clusters far away in sequence clearly indicates the presence of long-range interactions in WT-S^{ME}. Interestingly, clusters also seem to play an important role in refolding: hydrogen exchange measurements have shown that the α-domain of lysozyme folds a lot faster than the β-domain, with exception of residue W63, which becomes exchange-protected as quickly as the α-domain. An association of the region around W62/W63 with the α-domain as shown by relaxation measurements is therefore not only present in the unfolded state, but seems to be important for the folding nucleus of the protein.

In order to test whether this property of Trp62 to stabilize the folding nucleus is unique to tryptophan or may be more generally attributable to aromatic amino acids, the effect of replacing Trp62 by tyrosine is shown in Figure 21.38c. The W62Y mutation has a very minor effect on the relaxation properties of WT-S^{ME} compared with W62G-S^{ME}. This result suggests that aromatic residues play an important role in the stabilization of long-range interactions in unfolded states of lysozyme.

A9G-S^{ME} even more so displays WT-S^{ME} properties. There is not even an effect on its own cluster, cluster 1. The hydrophobic clusters remain of virtually identical size and position, indicating that Ala9 is not important for stabilizing the hydrophobic core.

In contrast, results of the replacement of tryptophan residues at positions 111 and 123 are very significant. When Trp111 (cluster 5) is replaced by glycine, the cluster in which the mutation is located disappears entirely, and cluster 6 also decreases significantly in intensity. Loss is also observed in the intensity of cluster 2 (surrounding Trp28) and a very small decrease in the intensity of cluster 3 (surrounding Trp62 and Trp63). Similarly, when Trp123 at the C-terminus (cluster 6) is changed to a glycine, the cluster around the mutation site essentially disappears, and the intensities of clusters 2, 3, and 5 are lessened. The effect of W123G replacement is less significant than that of W111G, as seen by the complete loss of cluster 5 in W111G-S^{ME} but not in W123G-S^{ME}. The largest changes in R_2 distribution as compared to WT-S^{ME} is observed with W62G-S^{ME}. W62G-S^{ME} on one,

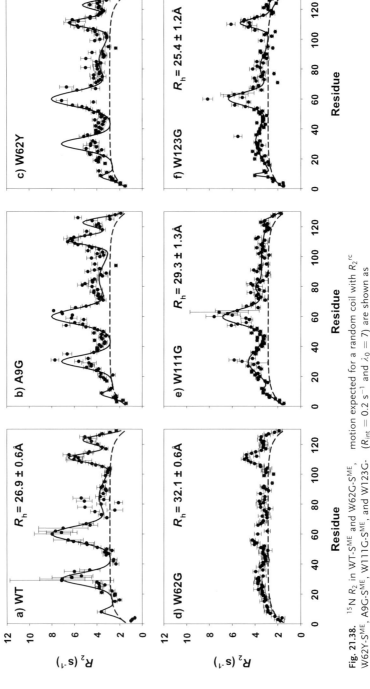

Fig. 21.38. ^{15}N R_2 in WT-S^{ME} and W62C-S^{ME}, W62Y-S^{ME}, A9G-S^{ME}, W111G-S^{ME}, and W123G-S^{ME} are shown in panels a–e, respectively. The experimental rates are shown as a scatter plot and the rates fitted by a model of segmental motion expected for a random coil with R_2^{rc} ($R_{int} = 0.2$ s^{-1} and $\lambda_0 = 7$) are shown as dashed lines; black lines show Gaussian fits for WT-S^{ME} (a,b,c) W111G-S^{ME} (e) and W123G-S^{ME} (f).

W111G-S^{ME} and W123G-S^{ME} on the other side appear complementary to each other: while the W62G mutation causes dramatic disruption of clusters 1–4, and to a lesser extent of clusters 5–6, the W111G and W123G replacements causing more disruption in clusters 2, 5, and 6 and more subtle changes in cluster 3.

21.6 Summary and Outlook

Here we summarize some of the findings from NMR investigations of nonnative states of proteins:

- The investigations show that NMR parameters can be predicted using models for the random coil state of a protein. The applied model treats the polypeptide chain as a (branched or unbranched) homopolymer for the prediction of anisotropic and global parameters such as relaxation rates or RDCs. On the other hand, for the prediction of local parameters, only explicit models such as the random coil model (Section 21.4) with amino acid-specific torsion angle distribution will allow correlation with experimental data such as J-couplings and NOEs, since the experiment is sensitive enough to detect small local differences in the conformational sampling. Second-order effects revealing the more detailed dependence on nearest neighbor effect will need a larger statistical database for their firm interpretation.
- Random coil states of proteins sample the ϕ, ψ and χ_1 conformational space according to distribution functions derived from the analysis of structures prevalent in all protein structures [24, 45, 122, 123]. The assumption that the distribution of ϕ, ψ torsion angles in the PDB resembles the experimental ϕ, ψ energy surface and that the effects of specific nonlocal interactions present in individual proteins are averaged out over the set of structures seems to be valid. More sophisticated approaches including molecular dynamics simulations of nonnative states of proteins will be of importance in the future to go beyond the current conformational and dynamic description of these states.
- On the basis of the delineation of the "random coil" behaviour, significant deviations identifying nonrandom structures in nonnative states can be found for a number of different proteins. These deviations include both nonrandom conformational preferences and differences from rapid dynamics that would interconvert the ensemble of states. The nonrandom structure can be shown to be stabilized both by local and global interactions and same of the global interactions trap the polypeptide chain into conformations not present in the native structure of the protein. In these investigations, combined mutational and NMR spectroscopic investigations have been particularly powerful tools in showing changes in overall compaction in these states. It is tempting to speculate that mutations do alter the energetics of the ensemble of states representing the unfolded states of a protein.

Natively unfolded proteins form a significant part of the proteom and reveal

that the polypeptide chain may not necessarily adopt a globular, rigid fold to exert its function. Bioinformatic analysis suggests these states to be more abundant than initially assumed.
- At the same time, misfolded, yet highly regular fibrillar states of proteins are formed in many protein folding diseases. In light of these considerations, heteronuclear NMR spectroscopy seems to be the most versatile and powerful technique to characterize aspects of conformation and dynamics in all of these states of proteins. In all of these investigations, the clear definition of what to expect for a random coil protein will form a firm baseline to predict deviations in a reliable manner.

Acknowledgments

H.S. would like to acknowledge the fruitful and fascinating collaboration and discussions with C. M. Dobson, K. Fiebig, M. Buck, L. Smith, M. Hennig, W. Peti, S. Grimshaw, C. Redfield, J. Jones, S. Glaser, and C. Griesinger in characterizing nonnative states of proteins. The work in the group of H.S. has been carried out by a number of extremely creative and bright students, who we would like to thank: T. Kühn, E. Collins, J. Klein-Seethamaran, E. Duchardt, and K. Schlepckow. We would like to thank the DFG, the European Commission, the Fonds der Chemischen Industrie, the state of Hesse and the University of Frankfurt for financial Support.

References

1 Dyson, H. J. & Wright, P. E. (2002). Coupling of folding and binding for unstructured proteins. *Curr Opin Struct Biol* 12, 54–60.
2 Wright, P. E. & Dyson, H. J. (1999). Intrinsically unstructured proteins: re-assessing the protein structure-function paradigm. *J Mol Biol* 293, 321–331.
3 Uversky, V. N. (2002). What does it mean to be natively unfolded? *Eur J Biochem* 269, 2–12.
4 Schweers, O., Schonbrunn-Hanebeck, E., Marx, A. & Mandelkow, E. (1994). Structural studies of tau protein and Alzheimer paired helical filaments show no evidence for beta-structure. *J Biol Chem* 269, 24290–24297.
5 Uversky, V. N., Gillespie, J. R. & Fink, A. L. (2000). Why are "natively unfolded" proteins unstructured under physiologic conditions? *Proteins* 41, 415–427.
6 Tompa, P. (2002). Intrinsically unstructured proteins. *Trends Biochem Sci* 27, 527–533.
7 Uversky, V. N. (2002). Natively unfolded proteins: a point where biology waits for physics. *Protein Sci* 11, 739–756.
8 Linding, R., Jensen, L. J., Diella, F., Bork, P., Gibson, T. J. & Russell, R. B. (2003). Protein disorder prediction: implications for structural proteomics. *Structure (Camb)* 11, 1453–1459.
9 Gerstein, M. (1998). How representative are the known structures of the proteins in a complete genome? A comprehensive structural census. *Folding Des* 3, 497–512.
10 Dyson, H. J. & Wright, P. E. (2001).

Nuclear magnetic resonance methods for elucidation of structure and dynamics in disordered states. *Methods Enzymol* 339, 258–270.

11 DOBSON, C. M. & HORE, P. J. (1998). Kinetic studies of protein folding using NMR spectroscopy. *Nat Struct Biol* 5 Suppl, 504–507.

12 DYSON, H. J. & WRIGHT, P. E. (1996). Insights into protein folding from NMR. *Annu Rev Phys Chem* 47, 369–395.

13 FRIEDEN, C., HOELTZLI, S. D. & ROPSON, I. J. (1993). NMR and protein folding: equilibrium and stopped-flow studies. *Protein Sci* 2, 2007–2014.

14 ENGLANDER, S. W. & MAYNE, L. (1992). Protein folding studied using hydrogen-exchange labeling and two-dimensional NMR. *Annu Rev Biophys Biomol Struct* 21, 243–265.

15 DYSON, H. J. & WRIGHT, P. E. (1991). Defining solution conformations of small linear peptides. *Annu Rev Biophys Biophys Chem* 20, 519–538.

16 DYSON, H. J. & WRIGHT, P. E. (1993). Peptide conformation and protein folding. *Curr Opin Struct Biol* 3, 60–65.

17 CASE, D. A., DYSON, H. J. & WRIGHT, P. E. (1994). Use of chemical shifts and coupling constants in nuclear magnetic resonance structural studies on peptides and proteins. *Methods Enzymol* 239, 392–416.

18 NERI, D., BILLETER, M., WIDER, G. & WUTHRICH, K. (1992). NMR determination of residual structure in a urea-denatured protein, the 434-repressor. *Science* 257, 1559–1563.

19 BAX, A. & GRZESIEK, S. (1993). Methodological Advances in Protein NMR. *Acc Chem Res* 26, 131–138.

20 WITTEKIND, M. & MUELLER, L. (1993). HNCACB, a high-sensitivity 3D NMR experiment to correlate amide-proton and nitrogen resonances with the alpha- and beta-carbon resonances in proteins. *J Magn Reson Ser B* 101.

21 LOGAN, T. M., THERIAULT, Y. & FESIK, S. W. (1994). Structural characterization of the FK506 binding protein unfolded in urea and guanidine hydrochloride. *J Mol Biol* 236, 637–648.

22 WISHART, D. S. & CASE, D. A. (2001). Use of chemical shifts in macromolecular structure determination. *Methods Enzymol* 338, 3–34.

23 XU, X.-P. & CASE, D. A. (2001). Automated prediction of 15N, 13Cα, 13Cβ and 13C$'$ chemical shifts in proteins using a density functional database. *J Biomol NMR* 21, 321–333.

24 SMITH, L. J., BOLIN, K. A., SCHWALBE, H., MACARTHUR, M. W., THORNTON, J. M. & DOBSON, C. M. (1996). Analysis of main chain torsion angles in proteins: prediction of NMR coupling constants for native and random coil conformations. *J Mol Biol* 255, 494–506.

25 SCHWALBE, H., FIEBIG, K. M., BUCK, M. et al. (1997). Structural and dynamical properties of a denatured protein. Heteronuclear 3D NMR experiments and theoretical simulations of lysozyme in 8 M urea. *Biochemistry* 36, 8977–8991.

26 LOUHIVUORI, M., PAAKKONEN, K., FREDRIKSSON, K., PERMI, P., LOUNILA, J. & ANNILA, A. (2003). On the origin of residual dipolar couplings from denatured proteins. *J Am Chem Soc* 125, 15647–1550.

27 MCDONALD, C. A. & PHILLIPS, D. C. (1969). Proton magnetic resonance spectra of proteins in random-coil configurations. *J. Am. Chem. Soc.* 91, 1513–1521.

28 SCHWARZINGER, S., KROON, G. J., FOSS, T. R., CHUNG, J., WRIGHT, P. E. & DYSON, H. J. (2001). Sequence-dependent correction of random coil NMR chemical shifts. *J Am Chem Soc* 123, 2970–2978.

29 WISHART, D. S., SYKES, B. D. & RICHARDS, F. M. (1991). Relationship between nuclear magnetic resonance chemical shift and protein secondary structure. *J Mol Biol* 222, 311–333.

30 WISHART, D. S. & SYKES, B. D. (1994). The ^{13}C chemical-shift index: a simple method for the identification of protein secondary structure using ^{13}C chemical-shift data. *J. Biomol. NMR* 4, 171–180.

31 Wishart, D. S. & Sykes, B. D. (1994). Chemical shifts as a tool for structure determination. *Methods Enzymol* 239, 363–392.

32 Yao, J., Dyson, H. J. & Wright, P. E. (1997). Chemical shift dispersion and secondary structure prediction in unfolded and partly folded proteins. *FEBS Lett* 419, 285–289.

33 Wishart, D. S., Sykes, B. D. & Richards, F. M. (1992). The chemical shift index: a fast and simple method for the assignment of protein secondary structure through NMR spectroscopy. *Biochemistry* 31, 1647–1651.

34 Burton, R. E., Huang, G. S., Daugherty, M. A., Fullbright, P. W. & Oas, T. G. (1996). Microsecond protein folding through a compact transition state. *J Mol Biol* 263, 311–322.

35 Sandstrom. (1982). *Dynamic NMR Spectroscopy.* Academic Press, London.

36 Bundi, A. & Wüthrich, K. (1979). ^{1}H NMR parameters of the common amino acid residues measured in aqueous solutions of the linear tetrapeptides H-Gly-Gly-X-L-Ala-OH. *Biopolymers* 18, 285–297.

37 Wishart, D. S., Bigam, C. G., Holm, A., Hodges, R. S. & Sykes, B. D. (1995). ^{1}H, ^{13}C and ^{15}N random coil NMR chemical shifts of the common amino acids. I. Investigations of nearest-neighbor effects. *J Biomol NMR* 5, 67–81.

38 Merutka, G., Dyson, H. J. & Wright, P. E. (1995). 'Random coil' 1H chemical shifts obtained as a function of temperature and trifluoroethanol concentration for the peptide series GGXGG. *J Biomol NMR* 5, 14–24.

39 Redfield, C., Schulman, B. A., Milhollen, M. A., Kim, P. S. & Dobson, C. M. (1999). Alpha-lactalbumin forms a compact molten globule in the absence of disulfide bonds. *Nature Struct Biol* 6, 948–952.

40 Jeener, J., Meier, B. H., Bachmann, P. & Ernst, R. R. (1979). Investigation of exchange processes by two-dimensional NMR spectroscopy. *J Chem Phys* 71, 4546–4553.

41 Dobson, C. M., Evans, P. A. & Williamson, K. L. (1984). Proton NMR studies of denatured lysozyme. *FEBS Lett* 168, 331–334.

42 Evans, P. A., Topping, K. D., Woolfson, D. N. & Dobson, C. M. (1991). Hydrophobic clustering in nonnative states of a protein: Interpretation of chemical shifts in NMR spectra of denatured states of lysozyme. *Proteins Struct Funct Genet* 9, 248–266.

43 Karplus, M. (1959). Contact electron-spin coupling of nuclear magnetic moments. *J Chem Phys* 30, 11–15.

44 Smith, L. J. (1996). Analysis of main chain torsion angles in proteins: prediction of NMR coupling constants for native and random coil conformations. *J Mol Biol* 255, 494–506.

45 Smith, L. J., Fiebig, K. M., Schwalbe, H. & Dobson, C. M. (1996). The concept of a random coil. Residual structure in peptides and denatured proteins. *Folding Des* 1, R95–106.

46 Juranic, N. & Macura, S. (2001). Correlations among 1JNC' and h3JNC' coupling constants in the hydrogen-bonding network of human ubiquitin. *J Am Chem Soc* 123, 4099–4100.

47 Mierke, D. F., Grdadolnik, S. G. & Kessler, H. (1992). Use of one-bond C alpha-H alpha coupling constants a restraints in MD simulations. *J Am Chem Soc* 114, 8283–8284.

48 Edison, A. S., Markley, J. L. & Weinhold, F. (1994). Calculations of one-, two- and three-bond nuclear spin-spin couplings in a model peptide and correlations with experimental data. *J Biomol NMR* 4, 519–542.

49 Edison, A. S., Weinhold, F., Westler, W. M. & Markley, J. L. (1994). Estimates of phi and psi torsion angles in proteins from one-, two- and three-bond nuclear spin-spin couplings: application to staphylococcal nuclease. *J Biomol NMR* 4, 543–551.

50 Delaglio, F., Torchia, D. A. & Bax,

A. (1991). Measurement of ^{15}N-^{13}C J couplings in staphylococcal nuclease. *J Biomol NMR* 1, 439–446.
51 WIRMER, J. & SCHWALBE, H. (2002). Angular dependence of 1J(Ni,Cαi) and 2J(Ni,Cα(i-1)) coupling constants measured in J-modulated HSQCs. *J Biomol NMR* 23, 47–55.
52 JARAVINE, ALEXANDRESCU, A. T. & GRZESIEK, S. (2001). Observation of the closing of individual hydrogen bonds during TFE-induced helix formation in a peptide. *Protein Sci* 10, 943–950.
53 PLAXCO, K. W., MORTON, C. J., GRIMSHAW, S. B. et al. (1997). The effects of guanidine hydrochloride on the 'random coil' conformations and NMR chemical shifts of the peptide series GGXGG. *J Biomol NMR* 10.
54 HUBBARD, P. S. (1958). Nuclear magnetic relaxation of three and four spin molecules in a liquid. *Phys Rev* 109, 1153–1158.
55 HUBBARD, P. S. (1969). Nonexponential relaxation of three-spin systems in nonspherical molecules. *J Chem Phys* 51, 1647–1651.
56 HUBBARD, P. S. (1970). Nonexponential relaxation of rotating three-spin systems in molecules of a liquid. *J Chem Phys* 52, 563–568.
57 KUHLMANN, K. F. & BALDESCHWEILER, J. D. (1965). Analysis of the nuclear Overhauser effect in the difluoro-ethylenes. *J Chem Phys* 43, 572–593.
58 CAVANAGH, J., FAIRBROTHER, W. J., PALMER III, A. G. & SKELTON, N. J. (1996). *Protein NMR Spectroscopy, Principles and Practice*. Academic Press, London.
59 LIPARI, G. & SZABO, A. (1982). Model-free approach to the interpretation of nuclear magnetic-resonance relaxation in macromolecules. 2. Analysis of experimental results. *J Am Chem Soc* 104, 4559–4570.
60 LIPARI, G. & SZABO, A. (1982). Model-free approach to the interpretation of nuclear magnetic-resonance relaxation in macromolecules. 1. Theory and range of validity. *J Am Chem Soc* 104, 4546–4559.

61 BRUTSCHER, B., SKRYNNIKOV, N. R., BREMI, T., BRÜSCHWEILER, R. & ERNST, R. R. (1998). Quantitative investigation of dipole–CSA cross-correlated relaxation by ZQ/DQ spectroscopy. *J Magn Reson* 130, 246–351.
62 BRÜSCHWEILER, R. & CASE, D. A. (1994). Characterization of biomolecular structure and dynamics by NMR cross relaxation. *Prog NMR Spectrosc* 26, 27–58.
63 WAGNER, G. (1993). NMR relaxation and protein mobility. *Curr Opin Struct Biol* 3, 748–754.
64 OCHSENBEIN, F., NEUMANN, J. M., GUITTET, E. & VAN HEIJENOORT, C. (2002). Dynamical characterization of residual and nonnative structures in a partially folded protein by (15)N NMR relaxation using a model based on a distribution of correlation times. *Protein Sci* 11, 957–964.
65 ALEXANDRESCU, A. T. & SHORTLE, D. (1994). Backbone dynamics of a highly disordered 131 residue fragment of staphylococcal nuclease. *J Mol Biol* 242, 527–546.
66 BUEVICH, A. V. & BAUM, J. (1999). Dynamics of unfolded proteins: Incooporation of distributions of correlation times in the model free analysis of NMR relaxation data. *J Am Chem Soc* 121, 8671–8672.
67 GAYATHRI, C., BOTHNERBY, A. A., VAN ZIJL, P. C. M. & MACLEAN, C. (1982). Dipolar magnetic-field effects in NMR-spectra of liquids. *Chem Phys Lett* 87, 192–196.
68 TJANDRA, N. & BAX, A. (1997). Direct measurement of distances and angles in biomolecules by NMR in a dilute liquid crystalline medium. *Science* 278, 1111–1114.
69 TOLMAN, J. R., FLANAGAN, J. M., KENNEDY, M. A. & PRESTEGARD, J. H. (1997). NMR evidence for slow collective motions in cyanometmyoglobin. *Nature Struct Biol* 4, 292–297.
70 PRESTEGARD, J. H. & KISHORE, A. I. (2001). Partial alignment of biomolecules: an aid to NMR characterization. *Curr Opin Chem Biol* 5, 584–590.

71 Tolman, J. R. (2001). Dipolar couplings as a probe of molecular dynamics and structure in solution. *Curr Opin Struct Biol* 11, 532–539.
72 Al-Hashimi, H. M. & Patel, D. J. (2002). Residual dipolar couplings: synergy between NMR and structural genomics. *J Biomol NMR* 22, 1–8.
73 Bax, A. (2003). Weak alignment offers new NMR opportunities to study protein structure and dynamics. *Protein Sci* 12, 1–16.
74 Shortle, D. & Ackerman, M. S. (2001). Persistence of native-like topology in a denatured protein in 8 M urea. *Science* 293, 487–489.
75 Gaemers, S. & Bax, A. (2001). Morphology of three lyotropic liquid crystalline biological NMR media studied by translational diffusion anisotropy. *J Am Chem Soc* 123, 12343–12352.
76 Sass, H. J., Musco, G., Stahl, S. J., Wingfield, P. T. & Grzesiek, S. (2000). Solution NMR of proteins within polyacrylamide gels: diffusional properties and residual alignment by mechanical stress or embedding of oriented purple membranes. *J Biomol NMR* 18, 303–309.
77 Tycko, R., Blanco, F. J. & Ishii, Y. (2000). Alignment of biopolymers in strained gels: A new way to create detectable dipole-dipole coupling in high-resolution biomolecular NMR. *J Am Chem Soc* 122, 9340–9341.
78 Banci, L., Bertini, I., Savellini, G. G. et al. (1997). Pseudocontact shifts as constraints for energy minimization and molecular dynamics calculations on solution structures of paramagnetic metalloproteins. *Proteins* 29, 68–76.
79 Barbieri, R., Bertini, I., Lee, Y. M., Luchinat, C. & Velders, A. H. (2002). Structure-independent cross-validation between residual dipolar couplings originating from internal and external orienting media. *J Biomol NMR* 22, 365–368.
80 Wohnert, J., Franz, K. J., Nitz, M., Imperiali, B. & Schwalbe, H. (2003). Protein alignment by a coexpressed lanthanide-binding tag for the measurement of residual dipolar couplings. *J Am Chem Soc* 125, 13338–13339.
81 Zweckstetter, M. & Bax, A. (2000). Prediction of sterically induced alignment in a dilute liquid crystalline phase: Aid to protein structure determination by NMR. *J Am Chem Soc* 122, 3791–3792.
82 Fernandes, M. X., Bernado, P., Pons, M. & Garcia de la Torre, J. (2001). An analytical solution to the problem of the orientation of rigid particles by planar obstacles. Application to membrane systems and to the calculation of dipolar couplings in protein NMR spectroscopy. *J Am Chem Soc* 123, 12037–12047.
83 Losonczi, J. A., Andrec, M., Fischer, M. W. F. & Prestegard, J. H. (1999). Order Matrix analysis of residual dipolar couplings using singular value decomposition. *J Magn Reson* 138, 334–342.
84 Meiler, J., Prompers, J. J., Peti, W., Griesinger, C. & Bruschweiler, R. (2001). Model-free approach to the dynamic interpretation of residual dipolar couplings in globular proteins. *J Am Chem Soc* 123, 6098–6107.
85 Tolman, J. R. (2002). A novel approach to the retrieval of structural and dynamic information from residual dipolar couplings using several oriented media in biomolecular NMR spectroscopy. *J Am Chem Soc* 124, 12020–12030.
86 Peti, W., Meiler, J., Bruschweiler, R. & Griesinger, C. (2002). Model-free analysis of protein backbone motion from residual dipolar couplings. *J Am Chem Soc* 124, 5822–5833.
87 Wilkins, D. K., Grimshaw, S. B., Receveur, V., Dobson, C. M., Jones, J. A. & Smith, L. J. (1999). Hydrodynamic radii of native and denatured proteins measured by pulse field gradient NMR techniques. *Biochemistry* 38, 16424–16431.
88 Wu, D., Chen, A. & Johnson Jr., C. S. (1995). An improved diffusion-ordered spectroscopy experiment incorporating bipolar-gradient pulses. *J Magn Reson A* 115, 260–264.

89 Jones, J. A., Wilkins, D. K., Smith, L. J. & Dobson, C. M. (1997). Characterisation of protein unfolding by NMR diffusion measurements. *J Biomol NMR* 10, 199–203.

90 Pan, H., Barany, G. & Woodward, C. (1997). Reduced BPTI is collapsed. A pulsed field gradient NMR study of unfolded and partially folded bovine pancreatic trypsin inhibitor. *Protein Sci* 6, 1985–1992.

91 Schmidt, P. G. & Kuntz, I. D. (1984). Distance measurements in spin-labeled lysozyme. *Biochemistry* 23, 4261–4266.

92 Kosen, P. A., Scheek, R. M., Naderi, H. et al. (1986). Two-dimensional 1H NMR of three spin-labeled derivatives of bovine pancreatic trypsin inhibitor. *Biochemistry* 25, 2356–2364.

93 Hubbell, W. L., McHaourab, H. S., Altenbach, C. & Lietzow, M. A. (1996). Watching proteins move using site-directed spin labeling. *Structure* 4, 779–783.

94 Matthews, B. W. (1995). Studies on protein stability with T4 lysozyme. *Adv Protein Chem* 46, 249–278.

95 McHaourab, H. S., Lietzow, M. A., Hideg, K. & Hubbell, W. L. (1996). Motion of spin-labeled side chains in T4 lysozyme. Correlation with protein structure and dynamics. *Biochemistry* 35, 7692–7704.

96 Lee, L. & Sykes, B. D. (1983). Use of lanthanide-induced nuclear magnetic resonance shifts for determination of protein structure in solution: EF calcium binding site of carp parvalbumin. *Biochemistry* 22, 4366–4373.

97 Gillespie, J. R. & Shortle, D. (1997). Characterization of long-range structure in the denatured state of staphylococcal nuclease. II. Distance restraints from paramagnetic relaxation and calculation of an ensemble of structures. *J Mol Biol* 268, 170–184.

98 Yi, Q., Scalley-Kim, M. L., Alm, E. J. & Baker, D. (2000). NMR characterization of residual structure in the denatured state of protein L. *J Mol Biol* 299, 1341–1351.

99 Lietzow, M. A., Jamin, M., Jane Dyson, H. J. & Wright, P. E. (2002). Mapping long-range contacts in a highly unfolded protein. *J Mol Biol* 322, 655–662.

100 Gillespie, J. R. & Shortle, D. (1997). Characterization of long-range structure in the denatured state of staphylococcal nuclease. I. Paramagnetic relaxation enhancement by nitroxide spin labels. *J Mol Biol* 268, 158–169.

101 Roder, H. & Wuthrich, K. (1986). Protein folding kinetics by combined use of rapid mixing techniques and NMR observation of individual amide protons. *Proteins* 1, 34–42.

102 Baldwin, R. (1993). Pulsed H/D-exchange studies of folding intermediates. *Curr Opin Struct Biol* 3, 84–91.

103 Miranker, A., Radford, S. E., Karplus, M. & Dobson, C. M. (1991). Demonstration by NMR of folding domains in lysozyme. *Nature* 349, 633–536.

104 Radford, S. E., Dobson, C. M. & Evans, P. A. (1992). The folding of hen lysozyme involves partially structured intermediates and multiple pathways. *Nature* 358, 302–307.

105 Roder, H., Wagner, G. & Wuthrich, K. (1985). Individual amide proton exchange rates in thermally unfolded basic pancreatic trypsin inhibitor. *Biochemistry* 24, 7407–7411.

106 Robertson, A. D. & Baldwin, R. L. (1991). Hydrogen exchange in thermally denatured ribonuclease A. *Biochemistry* 30, 9907–9914.

107 Buck, M., Radford, S. E. & Dobson, C. M. (1994). Amide hydrogen exchange in a highly denatured state. Hen egg-white lysozyme in urea. *J Mol Biol* 237, 247–254.

108 Jennings, P. A. & Wright, P. E. (1993). Formation of a molten globule intermediate early in the kinetic folding pathway of apomyoglobin. *Science* 262, 892–896.

109 Udgaonkar, J. B. & Baldwin, R. L. (1988). NMR evidence for an early framework intermediate on the

folding pathway of ribonuclease A. *Nature* 335, 694–699.

110 UDGAONKAR, J. B. & BALDWIN, R. L. (1990). Early folding intermediate of ribonuclease A. *Proc Natl Acad Sci USA* 87, 8197–8201.

111 RODER, H., ELOVE, G. A. & ENGLANDER, S. W. (1988). Structural characterization of folding intermediates in cytochrome c by H-exchange labelling and proton NMR. *Nature* 335, 700–704.

112 JUNEJA, J. & UDGAONKAR, J. B. (2002). Characterization of the unfolding of ribonuclease a by a pulsed hydrogen exchange study: evidence for competing pathways for unfolding. *Biochemistry* 41, 2641–2654.

113 KAPTEIN, R. (1982). Photo-CIDNP studies of proteins. *Biol Magn Reson* 4, 145–191.

114 HORE, P. J. & KAPTEIN, R. (1993). Photo-CIDNP of biopolymers. *Prog NMR Spectr* 25, 345–402.

115 LYON, C. E., JONES, J. A., REDFIELD, C., DOBSON, C. M. & HORE, P. J. (1999). Two-dimensional 15N-1H photo-CIDNP as a surface probe of native and partially structured proteins. *J Am Chem Soc* 121, 6505–6506.

116 BROADHURST, R. W., DOBSON, C. M., HORE, P. J., RADFORD, S. E. & REES, M. L. (1991). A photochemically induced dynamic nuclear polarization study of denatured states of lysozyme. *Biochemistry* 30, 405–412.

117 WIRMER, J., KÜHN, T. & SCHWALBE, H. (2001). Millisecond time resolved photo-CIDNP NMR reveals a nonnative folding intermediate on the ion-induced pathway of bovine α-lactalbumin. *Ang Chem Int Ed* 40, 4248–4251.

118 HORE, P. J., WINDER, S. L., ROBERTS, C. H. & DOBSON, C. M. (1997). Stopped-flow photo-CIDNP observation of protein folding. *J Am Chem Soc* 119, 5049–5050.

119 MAEDA, K., LYON, C. E., LOPEZ, J. J., CEMAZAR, M., DOBSON, C. M. & HORE, P. J. (2000). Improved photo-CIDNP methods for studying protein structure and folding. *J Biomol NMR* 16, 235–244.

120 LYON, C. E., SUH, E. S., DOBSON, C. M. & HORE, P. J. (2002). Probing the exposure of tyrosine and tryptophan residues in partially folded proteins and folding intermediates by CIDNP pulse-labeling. *J Am Chem Soc* 124, 13018–13024.

121 FIEBIG, K. M., SCHWALBE, H., BUCK, M., SMITH, L. J. & DOBSON, C. M. (1996). Toward a description of the conformations of denatured states of proteins. Comparison of a random coil model with NMR measurements. *J Phys Chem* 100, 2661–2666.

122 SWINDELLS, M. B., MACARTHUR, M. W. & THORNTON, J. M. (1995). Intrinsic phi, psi propensities of amino acids, derived from the coil regions of known structures. *Nat Struct Biol* 2, 596–603.

123 SERRANO, L. (1995). Comparison between the phi distribution of the amino acids in the protein database and NMR data indicates that amino acids have various phi propensities in the random coil conformation. *J Mol Biol* 254, 322–333.

124 FRANK, M. K., CLORE, G. M. & GRONENBORN, A. M. (1995). Structural and dynamic characterization of the urea denatured state of the immunoglobulin binding domain of streptococcal protein G by multi-dimensional heteronuclear NMR spectroscopy. *Protein Sci* 4, 2605–2615.

125 ALLERHAND, A. & HAILSTONE, R. K. (1972). C-13 Fourier-transform nuclear magnetic-resonance. 10. Effect of molecular-weight on C-13 spin-lattice relaxation-times of polystyrene in solution. *J Chem Phys* 56, 3718.

126 BUCK, M., RADFORD, S. E. & DOBSON, C. M. (1993). A partially folded state of hen egg white lysozyme in trifluoroethanol: structural characterization and implications for protein folding. *Biochemistry* 32, 669–678.

127 BUCK, M., SCHWALBE, H. & DOBSON, C. M. (1995). Characterization of conformational preferences in a partly folded protein by heteronuclear NMR spectroscopy: assignment and secondary structure analysis of hen egg-white lysozyme in trifluoro-

ethanol. *Biochemistry* 34, 13219–13232.
128 BUCK, M., BOYD, J., REDFIELD, C. et al. (1995). Structural determinants of protein dynamics: analysis of 15N NMR relaxation measurements for main-chain and side-chain nuclei of hen egg white lysozyme. *Biochemistry* 34, 4041–4055.
129 BUCK, M., SCHWALBE, H. & DOBSON, C. M. (1996). Main-chain dynamics of a partially folded protein: 15N NMR relaxation measurements of hen egg white lysozyme denatured in trifluoroethanol. *J Mol Biol* 257, 669–683.
130 KEMMINK, J. & CREIGHTON, T. E. (1993). Local conformations of peptides representing the entire sequence of bovine pancreatic trypsin inhibitor and their roles in folding. *J Mol Biol* 234, 861–878.
131 KEMMINK, J. & CREIGHTON, T. E. (1995). The physical properties of local interactions of tyrosine residues in peptides and unfolded proteins. *J Mol Biol* 245, 251–260.
132 LUMB, K. J. & KIM, P. S. (1994). Formation of a hydrophobic cluster in denatured bovine pancreatic trypsin inhibitor. *J Mol Biol* 236, 412–420.
133 HENNIG, M., BERMEL, W., SPENCER, A., DOBSON, C. M., SMITH, L. J. & SCHWALBE, H. (1999). Side-chain conformations in an unfolded protein: chi1 distributions in denatured hen lysozyme determined by heteronuclear 13C, 15N NMR spectroscopy. *J Mol Biol* 288, 705–723.
134 MCGREGOR, M. J., ISLAM, S. A. & STERNBERG, M. J. (1987). Analysis of the relationship between side-chain conformation and secondary structure in globular proteins. *J Mol Biol* 198, 295–310.
135 DUNBRACK, R. L., JR & KARPLUS, M. (1993). Backbone-dependent rotamer library for proteins. Application to side-chain prediction. *J Mol Biol* 230, 543–574.
136 SHORTLE, D. R. (1996). Structural analysis of nonnative states of proteins by NMR methods. *Curr Opin Struct Biol* 6, 24–30.
137 SCHULMAN, B. A., KIM, P. S., DOBSON, C. M. & REDFIELD, C. (1997). A residue-specific NMR view of the non-cooperative unfolding of a molten globule. *Nat Struct Biol* 4, 630–634.
138 PETI, W., SMITH, L. J., REDFIELD, C. & SCHWALBE, H. (2001). Chemical shifts in denatured proteins: resonance assignments for denatured ubiquitin and comparisons with other denatured proteins. *J Biomol NMR* 19, 153–165.
139 ELIEZER, D. & WRIGHT, P. E. (1996). Is apomyoglobin a molten globule? Structural characterization by NMR. *J Mol Biol* 263, 531–538.
140 ELIEZER, D., JENNINGS, P. A., DYSON, H. J. & WRIGHT, P. E. (1997). Populating the equilibrium molten globule state of apomyoglobin under conditions suitable for structural characterization by NMR. *FEBS Lett* 417, 92–96.
141 YAO, J., CHUNG, J., ELIEZER, D., WRIGHT, P. E. & DYSON, H. J. (2001). NMR structural and dynamic characterization of the acid-unfolded state of apomyoglobin provides insights into the early events in protein folding. *Biochemistry* 40, 3561–3571.
142 KITAHARA, R., YAMADA, H., AKASAKA, K. & WRIGHT, P. E. (2002). High pressure NMR reveals that apomyoglobin is an equilibrium mixture from the native to the unfolded. *J Mol Biol* 320, 311–319.
143 SCHWARZINGER, S., WRIGHT, P. E. & DYSON, H. J. (2002). Molecular hinges in protein folding: the urea-denatured state of apomyoglobin. *Biochemistry* 41, 12681–12686.
144 ARCUS, V. L., VUILLEUMIER, S., FREUND, S. M., BYCROFT, M. & FERSHT, A. R. (1994). Toward solving the folding pathway of barnase: the complete backbone 13C, 15N, and 1H NMR assignments of its pH-denatured state. *Proc Natl Acad Sci USA* 91, 9412–9416.
145 ARCUS, V. L., VUILLEUMIER, S., FREUND, S. M., BYCROFT, M. & FERSHT, A. R. (1995). A comparison of the pH, urea, and temperature-denatured states of barnase by

heteronuclear NMR: implications for the initiation of protein folding. *J Mol Biol* 254, 305–321.

146 FREUND, S. M., WONG, K. B. & FERSHT, A. R. (1996). Initiation sites of protein folding by NMR analysis. *Proc Natl Acad Sci USA* 93, 10600–10603.

147 NEIRA, J. L. & FERSHT, A. R. (1996). An NMR study on the beta-hairpin region of barnase. *Folding Des* 1, 231–241.

148 WONG, K. B., CLARKE, J., BOND, C. J. et al. (2000). Towards a complete description of the structural and dynamic properties of the denatured state of barnase and the role of residual structure in folding. *J Mol Biol* 296, 1257–1282.

149 EAKIN, C. M., KNIGHT, J. D., MORGAN, C. J., GELFAND, M. A. & MIRANKER, A. D. (2002). Formation of a copper specific binding site in nonnative states of beta-2-microglobulin. *Biochemistry* 41, 10646–10656.

150 KATOU, H., KANNO, T., HOSHINO, M. et al. (2002). The role of disulfide bond in the amyloidogenic state of beta(2)-microglobulin studied by heteronuclear NMR. *Protein Sci* 11, 2218–2229.

151 THOMSEN, J. K., KRAGELUND, B. B., TEILUM, K., KNUDSEN, J. & POULSEN, F. M. (2002). Transient intermediary states with high and low folding probabilities in the apparent two-state folding equilibrium of ACBP at low pH. *J Mol Biol* 318, 805–814.

152 LINDORFF-LARSEN, K., KRISTJANSDOTTIR, S., TEILUM, K. et al. (2004). Determination of an ensemble of structures representing the denatured state of the bovine acyl-coenzyme a binding protein. *J Am Chem Soc* 126, 3291–3299.

153 RAGONA, L., PUSTERLA, F., ZETTA, L., MONACO, H. L. & MOLINARI, H. (1997). Identification of a conserved hydrophobic cluster in partially folded bovine beta-lactoglobulin at pH 2. *Folding Des* 2, 281–290.

154 UHRINOVA, S., UHRIN, D., DENTON, H., SMITH, M., SAWYER, L. & BARLOW, P. N. (1998). Complete assignment of 1H, 13C and 15N chemical shifts for bovine beta-lactoglobulin: secondary structure and topology of the native state is retained in a partially unfolded form. *J Biomol NMR* 12, 89–107.

155 RAGONA, L., FOGOLARI, F., ROMAGNOLI, S., ZETTA, L., MAUBOIS, J. L. & MOLINARI, H. (1999). Unfolding and refolding of bovine beta-lactoglobulin monitored by hydrogen exchange measurements. *J Mol Biol* 293, 953–969.

156 KATOU, H., HOSHINO, M., KAMIKUBO, H., BATT, C. A. & GOTO, Y. (2001). Native-like beta-hairpin retained in the cold-denatured state of bovine beta-lactoglobulin. *J Mol Biol* 310, 471–484.

157 KUWATA, K., LI, H., YAMADA, H., BATT, C. A., GOTO, Y. & AKASAKA, K. (2001). High pressure NMR reveals a variety of fluctuating conformers in beta-lactoglobulin. *J Mol Biol* 305, 1073–1083.

158 GARCIA, P., SERRANO, L., DURAND, D., RICO, M. & BRUIX, M. (2001). NMR and SAXS characterization of the denatured state of the chemotactic protein CheY: implications for protein folding initiation. *Protein Sci* 10, 1100–1112.

159 ZHANG, O., KAY, L. E., OLIVIER, J. P. & FORMAN-KAY, J. D. (1994). Backbone 1H and 15N resonance assignments of the N-terminal SH3 domain of drk in folded and unfolded states using enhanced-sensitivity pulsed field gradient NMR techniques. *J Biomol NMR* 4, 845–858.

160 ZHANG, O. & FORMAN-KAY, J. D. (1995). Structural characterization of folded and unfolded states of an SH3 domain in equilibrium in aqueous buffer. *Biochemistry* 34, 6784–6794.

161 FARROW, N. A., ZHANG, O., FORMAN-KAY, J. D. & KAY, L. E. (1995). Comparison of the backbone dynamics of a folded and an unfolded SH3 domain existing in equilibrium in aqueous buffer. *Biochemistry* 34, 868–878.

162 FARROW, N. A., ZHANG, O., SZABO, A., TORCHIA, D. A. & KAY, L. E. (1995). Spectral density function mapping

using 15N relaxation data exclusively. *J Biomol NMR* 6, 153–162.
163 Farrow, N. A., Zhang, O., Forman-Kay, J. D. & Kay, L. E. (1997). Characterization of the backbone dynamics of folded and denatured states of an SH3 domain. *Biochemistry* 36, 2390–2402.
164 Yang, D., Mok, Y. K., Forman-Kay, J. D., Farrow, N. A. & Kay, L. E. (1997). Contributions to protein entropy and heat capacity from bond vector motions measured by NMR spin relaxation. *J Mol Biol* 272, 790–804.
165 Zhang, O. & Forman-Kay, J. D. (1997). NMR studies of unfolded states of an SH3 domain in aqueous solution and denaturing conditions. *Biochemistry* 36, 3959–3970.
166 Zhang, O., Forman-Kay, J. D., Shortle, D. & Kay, L. E. (1997). Triple-resonance NOESY-based experiments with improved spectral resolution: applications to structural characterization of unfolded, partially folded and folded proteins. *J Biomol NMR* 9, 181–200.
167 Mok, Y. K., Kay, C. M., Kay, L. E. & Forman-Kay, J. (1999). NOE data demonstrating a compact unfolded state for an SH3 domain under non-denaturing conditions. *J Mol Biol* 289, 619–638.
168 Kortemme, T., Kelly, M. J., Kay, L. E., Forman-Kay, J. & Serrano, L. (2000). Similarities between the spectrin SH3 domain denatured state and its folding transition state. *J Mol Biol* 297, 1217–1229.
169 Choy, W. Y. & Forman-Kay, J. D. (2001). Calculation of ensembles of structures representing the unfolded state of an SH3 domain. *J Mol Biol* 308, 1011–1032.
170 Tollinger, M., Skrynnikov, N. R., Mulder, F. A., Forman-Kay, J. D. & Kay, L. E. (2001). Slow dynamics in folded and unfolded states of an SH3 domain. *J Am Chem Soc* 123, 11341–11352.
171 Mok, Y. K., Elisseeva, E. L., Davidson, A. R. & Forman-Kay, J. D. (2001). Dramatic stabilization of an SH3 domain by a single substitution: roles of the folded and unfolded states. *J Mol Biol* 307, 913–928.
172 Choy, W. Y., Mulder, F. A., Crowhurst, K. A. et al. (2002). Distribution of molecular size within an unfolded state ensemble using small-angle X-ray scattering and pulse field gradient NMR techniques. *J Mol Biol* 316, 101–112.
173 Crowhurst, K. A., Tollinger, M. & Forman-Kay, J. D. (2002). Cooperative interactions and a nonnative buried Trp in the unfolded state of an SH3 domain. *J Mol Biol* 322, 163–178.
174 Crowhurst, K. A. & Forman-Kay, J. D. (2003). Aromatic and methyl NOEs highlight hydrophobic clustering in the unfolded state of an SH3 domain. *Biochemistry* 42, 8687–8695.
175 Tollinger, M., Forman-Kay, J. D. & Kay, L. E. (2002). Measurement of side-chain carboxyl pK(a) values of glutamate and aspartate residues in an unfolded protein by multinuclear NMR spectroscopy. *J Am Chem Soc* 124, 5714–5717.
176 Logan, T. M., Olejniczak, E. T., Xu, R. X. & Fesik, S. W. (1993). A general method for assigning NMR spectra of denatured proteins using 3D HC(CO)NH-TOCSY triple resonance experiments. *J Biomol NMR* 3, 225–231.
177 Nordstrand, K., Ponstingl, H., Holmgren, A. & Otting, G. (1996). Resonance assignment and structural analysis of acid denatured E. coli [U-15N]-glutaredoxin 3: use of 3D 15N-HSQC-(TOCSY-NOESY)-15N-HSQC. *Eur Biophys J* 24, 179–184.
178 Radford, S. E., Woolfson, D. N., Martin, S. R., Lowe, G. & Dobson, C. M. (1991). A three-disulphide derivative of hen lysozyme. Structure, dynamics and stability. *Biochem J* 273(Pt 1), 211–217.
179 Radford, S. E., Buck, M., Topping, K. D., Dobson, C. M. & Evans, P. A. (1992). Hydrogen exchange in native and denatured states of hen egg-white lysozyme. *Proteins* 14, 237–248.
180 Roux, P., Delepierre, M., Goldberg, M. E. & Chaffotte, A. F. (1997). Kinetics of secondary structure

recovery during the refolding of reduced hen egg white lysozyme. *J Biol Chem* 272, 24843–24849.

181 CHUNG, E. W., NETTLETON, E. J., MORGAN, C. J. et al. (1997). Hydrogen exchange properties of proteins in native and denatured states monitored by mass spectrometry and NMR. *Protein Sci* 6, 1316–1324.

182 BUCK, M. (1998). Trifluoroethanol and colleagues: cosolvents come of age. Recent studies with peptides and proteins. *Q Rev Biophys* 31, 297–355.

183 HENNIG, M., BERMEL, W., DOBSON, C. M., SMITH, L. J. & SCHWALBE, H. (1999). Determination of side chain conformations in unfolded proteins: 1 Distribution for lysozyme by heteronuclear 13C,15N NMR spectroscopy. *J Mol Biol* 288, 705–723.

184 KLEIN-SEETHARAMAN, J., OIKAWA, M., GRIMSHAW, S. B. et al. (2002). Long-range interactions within a nonnative protein. *Science* 295, 1719–1722.

185 BALDWIN, R. L. (2002). Making a network of hydrophobic clusters. *Science* 295, 1657–1658.

186 NODA, Y., YOKOTA, A., HORII, D. et al. (2002). NMR structural study of two-disulfide variant of hen lysozyme: 2SS[6–127, 30–115] – a disulfide intermediate with a partly unfolded structure. *Biochemistry* 41, 2130–2139.

187 BHAVESH, N. S., PANCHAL, S. C., MITTAL, R. & HOSUR, R. V. (2001). NMR identification of local structural preferences in HIV-1 protease tethered heterodimer in 6 M guanidine hydrochloride. *FEBS Lett* 509, 218–224.

188 PANCHAL, S. C., BHAVESH, N. S. & HOSUR, R. V. (2001). Improved 3D triple resonance experiments, HNN and HN(C)N, for HN and 15N sequential correlations in (13C, 15N) labeled proteins: application to unfolded proteins. *J Biomol NMR* 20, 135–147.

189 TAFER, H., HILLER, S., HILTY, C., FERNANDEZ, C. & WUTHRICH, K. (2004). Nonrandom structure in the urea-unfolded Escherichia coli outer membrane protein X (OmpX). *Biochemistry* 43, 860–869.

190 BENTROP, D., BERTINI, I., IACOVIELLO, R. et al. (1999). Structural and dynamical properties of a partially unfolded Fe4S4 protein: role of the cofactor in protein folding. *Biochemistry* 38, 4669–4680.

191 HARDING, M. M., WILLIAMS, D. H. & WOOLFSON, D. N. (1991). Characterization of a partially denatured state of a protein by two-dimensional NMR: reduction of the hydrophobic interactions in ubiquitin. *Biochemistry* 30, 3120–3128.

192 STOCKMAN, B. J., EUVRARD, A. & SCAHILL, T. A. (1993). Heteronuclear three-dimensional NMR spectroscopy of a partially denatured protein: the A-state of human ubiquitin. *J Biomol NMR* 3, 285–296.

193 NASH, D. P. & JONAS, J. (1997). Structure of the pressure-assisted cold denatured state of ubiquitin. *Biochem Biophys Res Commun* 238, 289–291.

194 PERMI, P. (2002). Intraresidual HNCA: an experiment for correlating only intraresidual backbone resonances. *J Biomol NMR* 23, 201–209.

195 MIZUSHIMA, T., HIRAO, T., YOSHIDA, Y. et al. (2004). Structural basis of sugar-recognizing ubiquitin ligase. *Nat Struct Mol Biol* 11, 365–370.

196 FLETCHER, C. M., MCGUIRE, A. M., GINGRAS, A. C. et al. (1998). 4E binding proteins inhibit the translation factor eIF4E without folded structure. *Biochemistry* 37, 9–15.

197 DAUGHDRILL, G. W., HANELY, L. J. & DAHLQUIST, F. W. (1998). The C-terminal half of the anti-sigma factor FlgM contains a dynamic equilibrium solution structure favoring helical conformations. *Biochemistry* 37, 1076–1082.

198 MOGRIDGE, J., LEGAULT, P., LI, J., VAN OENE, M. D., KAY, L. E. & GREENBLATT, J. (1998). Independent ligand-induced folding of the RNA-binding domain and two functionally distinct antitermination regions in the phage lambda N protein. *Mol Cell* 1, 265–275.

199 DE GUZMAN, R. N., MARTINEZ-YAMOUT, M. A., DYSON, H. J. & WRIGHT, P. E. (2004). Interaction of

the TAZ1 domain of the CREB-binding protein with the activation domain of CITED2: regulation by competition between intrinsically unstructured ligands for non-identical binding sites. *J Biol Chem* 279, 3042–3049.
200 COLLINS, E. S., WHITTAKER, S. B., TOZAWA, K. et al. (2002). Structural dynamics of the membrane translocation domain of colicin E9 and its interaction with TolB. *J Mol Biol* 318, 787–804.
201 ANDERLUH, G., HONG, Q., BOETZEL, R. et al. (2003). Concerted folding and binding of a flexible colicin domain to its periplasmic receptor TolA. *J Biol Chem* 278, 21860–21868.
202 KRIWACKI, R. W., HENGST, L., TENNANT, L., REED, S. I. & WRIGHT, P. E. (1996). Structural studies of p21Waf1/Cip1/Sdi1 in the free and Cdk2-bound state: conformational disorder mediates binding diversity. *Proc Natl Acad Sci USA* 93, 11504–11509.
203 LISSE, T., BARTELS, D., KALBITZER, H. R. & JAENICKE, R. (1996). The recombinant dehydrin-like desiccation stress protein from the resurrection plant *Craterostigma plantagineum* displays no defined three-dimensional structure in its native state. *Biol Chem* 377, 555–561.
204 HERSHEY, P. E., MCWHIRTER, S. M., GROSS, J. D., WAGNER, G., ALBER, T. & SACHS, A. B. (1999). The Cap-binding protein eIF4E promotes folding of a functional domain of yeast translation initiation factor eIF4G1. *J Biol Chem* 274, 21297–21304.
205 MAYOR, U., GROSSMANN, J. G., FOSTER, N. W., FREUND, S. M. & FERSHT, A. R. (2003). The denatured state of Engrailed Homeodomain under denaturing and native conditions. *J Mol Biol* 333, 977–991.
206 BOZZI, M., BIANCHI, M., SCIANDRA, F. et al. (2003). Structural characterization by NMR of the natively unfolded extracellular domain of beta-dystroglycan: toward the identification of the binding epitope for alpha-dystroglycan. *Biochemistry* 42, 13717–13724.
207 PENKETT, C. J., REDFIELD, C., DODD, I. et al. (1997). NMR analysis of main-chain conformational preferences in an unfolded fibronectin-binding protein. *J Mol Biol* 274, 152–159.
208 UVERSKY, V. N., GILLESPIE, J. R., MILLETT, I. S. et al. (2000). Zn(2+)-mediated structure formation and compaction of the "natively unfolded" human prothymosin alpha. *Biochem Biophys Res Commun* 267, 663–668.
209 WEISS, M. A., ELLENBERGER, T., WOBBE, C. R., LEE, J. P., HARRISON, S. C. & STRUHL, K. (1990). Folding transition in the DNA-binding domain of GCN4 on specific binding to DNA. *Nature* 347, 575–578.
210 CHO, H. S., LIU, C. W., DAMBERGER, F. F., PELTON, J. G., NELSON, H. C. & WEMMER, D. E. (1996). Yeast heat shock transcription factor N-terminal activation domains are unstructured as probed by heteronuclear NMR spectroscopy. *Protein Sci* 5, 262–269.
211 GREEN, T. B., GANESH, O., PERRY, K. et al. (2004). IA3, an aspartic proteinase inhibitor from Saccharomyces cerevisiae, is intrinsically unstructured in solution. *Biochemistry* 43, 4071–4081.
212 GEYER, M., MUNTE, C. E., SCHORR, J., KELLNER, R. & KALBITZER, H. R. (1999). Structure of the anchor-domain of myristoylated and non-myristoylated HIV-1 Nef protein. *J Mol Biol* 289, 123–138.
213 BERKOVITS, H. J. & BERG, J. M. (1999). Metal and DNA binding properties of a two-domain fragment of neural zinc finger factor 1, a CCHC-type zinc binding protein. *Biochemistry* 38, 16826–16830.
214 CARY, P. D., KING, D. S., CRANE-ROBINSON, C. et al. (1980). Structural studies on two high-mobility-group proteins from calf thymus, HMG-14 and HMG-20 (ubiquitin), and their interaction with DNA. *Eur J Biochem* 112, 577–580.
215 ABERCROMBIE, B. D., KNEALE, G. G., CRANE-ROBINSON, C. et al. (1978). Studies on the conformational

215 properties of the high-mobility-group chromosomal protein HMG 17 and its interaction with DNA. *Eur J Biochem* 84, 173–177.

216 CARY, P. D., CRANE-ROBINSON, C., BRADBURY, E. M. & DIXON, G. H. (1981). Structural studies of the non-histone chromosomal proteins HMG-T and H6 from trout testis. *Eur J Biochem* 119, 545–551.

217 ZEEV-BEN-MORDEHAI, T., RYDBERG, E. H., SOLOMON, A. et al. (2003). The intracellular domain of the *Drosophila* cholinesterase-like neural adhesion protein, gliotactin, is natively unfolded. *Proteins* 53, 758–767.

218 COX, C. J., DUTTA, K., PETRI, E. T. et al. (2002). The regions of securin and cyclin B proteins recognized by the ubiquitination machinery are natively unfolded. *FEBS Lett* 527, 303–308.

219 ISBELL, D. T., DU, S., SCHROERING, A. G., COLOMBO, G. & SHELLING, J. G. (1993). Metal ion binding to dog osteocalcin studied by 1H NMR spectroscopy. *Biochemistry* 32, 11352–11362.

220 MONCOQ, K., BROUTIN, I., LARUE, V. et al. (2003). The PIR domain of Grb14 is an intrinsically unstructured protein: implication in insulin signaling. *FEBS Lett* 554, 240–246.

221 DONNE, D. G., VILES, J. H., GROTH, D. et al. (1997). Structure of the recombinant full-length hamster prion protein PrP(29–231): the N terminus is highly flexible. *Proc Natl Acad Sci USA* 94, 13452–13457.

222 LIU, A., RIEK, R., WIDER, G., VON SCHROETTER, C., ZAHN, R. & WUTHRICH, K. (2000). NMR experiments for resonance assignments of 13C, 15N doubly-labeled flexible polypeptides: application to the human prion protein hPrP(23–230). *J Biomol NMR* 16, 127–138.

223 BUEVICH, A. V., SHINDE, U. P., INOUYE, M. & BAUM, J. (2001). Backbone dynamics of the natively unfolded pro-peptide of subtilisin by heteronuclear NMR relaxation studies. *J Biomol NMR* 20, 233–249.

224 GAST, K., DAMASCHUN, H., ECKERT, K. et al. (1995). Prothymosin alpha: a biologically active protein with random coil conformation. *Biochemistry* 34, 13211–13218.

225 FIEBIG, K. M., RICE, L. M., POLLOCK, E. & BRUNGER, A. T. (1999). Folding intermediates of SNARE complex assembly. *Nat Struct Biol* 6, 117–123.

226 ALEXANDRESCU, A. T., ABEYGUNAWARDANA, C. & SHORTLE, D. (1994). Structure and dynamics of a denatured 131-residue fragment of staphylococcal nuclease: a heteronuclear NMR study. *Biochemistry* 33, 1063–1072.

227 ALEXANDRESCU, A. T., JAHNKE, W., WILTSCHECK, R. & BLOMMERS, M. J. (1996). Accretion of structure in staphylococcal nuclease: an 15N NMR relaxation study. *J Mol Biol* 260, 570–587.

228 YE, K. & WANG, J. (2001). Self-association reaction of denatured staphylococcal nuclease fragments characterized by heteronuclear NMR. *J Mol Biol* 307, 309–322.

229 ACKERMAN, M. S. & SHORTLE, D. (2002). Robustness of the long-range structure in denatured staphylococcal nuclease to changes in amino acid sequence. *Biochemistry* 41, 13791–13797.

230 CHOY, W. Y. & KAY, L. E. (2003). Probing residual interactions in unfolded protein states using NMR spin relaxation techniques: an application to delta131delta. *J Am Chem Soc* 125, 11988–11992.

231 CHOY, W. Y., SHORTLE, D. & KAY, L. E. (2003). Side chain dynamics in unfolded protein states: an NMR based 2H spin relaxation study of delta131delta. *J Am Chem Soc* 125, 1748–1758.

232 HAZZARD, J., SUDHOF, T. C. & RIZO, J. (1999). NMR analysis of the structure of synaptobrevin and of its interaction with syntaxin. *J Biomol NMR* 14, 203–207.

233 LIU, D., ISHIMA, R., TONG, K. I. et al. (1998). Solution structure of a TBP-TAF(II)230 complex: protein mimicry of the minor groove surface of the

TATA box unwound by TBP. *Cell* 94, 573–583.

234 JIMENEZ, M. A., EVANGELIO, J. A., ARANDA, C. et al. (1999). Helicity of alpha(404–451) and beta(394–445) tubulin C-terminal recombinant peptides. *Protein Sci* 8, 788–799.

235 SPENCER, R. G., HALVERSON, K. J., AUGER, M., MCDERMOTT, A. E., GRIFFIN, R. G. & LANSBURY, P. T., JR. (1991). An unusual peptide conformation may precipitate amyloid formation in Alzheimer's disease: application of solid-state NMR to the determination of protein secondary structure. *Biochemistry* 30, 10382–10387.

236 BENZINGER, T. L., GREGORY, D. M., BURKOTH, T. S. et al. (2000). Two-dimensional structure of beta-amyloid(10–35) fibrils. *Biochemistry* 39, 3491–3499.

237 ANTZUTKIN, O. N., BALBACH, J. J., LEAPMAN, R. D., RIZZO, N. W., REED, J. & TYCKO, R. (2000). Multiple quantum solid-state NMR indicates a parallel, not antiparallel, organization of beta-sheets in Alzheimer's beta-amyloid fibrils. *Proc Natl Acad Sci USA* 97, 13045–13050.

238 BALBACH, J. J., ISHII, Y., ANTZUTKIN, O. N. et al. (2000). Amyloid fibril formation by A beta 16–22, a seven-residue fragment of the Alzheimer's beta-amyloid peptide, and structural characterization by solid state NMR. *Biochemistry* 39, 13748–13759.

239 PETKOVA, A. T., ISHII, Y., BALBACH, J. J. et al. (2002). A structural model for Alzheimer's beta-amyloid fibrils based on experimental constraints from solid state NMR. *Proc Natl Acad Sci USA* 99, 16742–16747.

240 JARONIEC, C. P., MACPHEE, C. E., ASTROF, N. S., DOBSON, C. M. & GRIFFIN, R. G. (2002). Molecular conformation of a peptide fragment of transthyretin in an amyloid fibril. *Proc Natl Acad Sci USA* 99, 16748–16753.

241 VUISTER, G. W. & BAX, A. (1993). Quantitative J correlation: a new approach for measuring homonuclear three-bond J(HNH.alpha.) coupling constants in 15N-enriched proteins. *J Am Chem Soc* 115, 7772–7777.

242 PARDI, A., BILLETER, M. & WUTHRICH, K. (1984). Calibration of the angular dependence of the amide proton-C alpha proton coupling constants, 3JHN alpha, in a globular protein. Use of 3JHN alpha for identification of helical secondary structure. *J Mol Biol* 180, 741–751.

243 LUDVIGSEN, S., ANDERSEN, K. V. & POULSEN, F. M. (1991). Accurate measurements of coupling constants from two-dimensional nuclear magnetic resonance spectra of proteins and determination of phi-angles. *J Mol Biol* 217, 731–736.

244 SCHMIDT, J. M., BLÜMEL, M., LÖHR, F. & RÜTERJANS, H. (1999). Self-consistent 3J coupling analysis for the joint calibration of Karplus coefficients and evaluation of torsion angles. *J Biomol NMR* 14, 1–12.

245 WANG, A. C. & BAX, A. (1996). Determination of the backbone dihedral angles in human ubiquitin from reparametrized empirical karplus equations. *J Am Chem Soc* 118, 2483–2494.

246 WANG, A. C. & BAX, A. (1995). Reparametrization of the Karplus relation for 3J(Hα-N) and 3J(HN-C') in peptides from uniformly 13C/15N-enriched human ubiquitin. *J Am Chem Soc* 117, 1810–1813.

247 BYSTROV, V. F. (1976). Spin-spin coupling and the conformational states of peptide systems. *Prog NMR Spectrosc* 10, 41–81.

248 LÖHR, F., BLÜMEL, M., SCHMIDT, J. M. & RÜTERJANS, H. (1997). Application of H(N)CA,CO-E.COSY experiments for calibrating the ϕ angular dependences of vicinal couplings J(C'i-1,Hiα), J(C'i-1,Ciβ) and J(C'i-1,C'i) in proteins.

249 HU, J. S. & BAX, A. (1997). Determination of phi and chi1 angles in proteins from 13C–13C three-bond J couplings measured by three-dimensional heteronuclear NMR. How planar is the peptide bond? *J Am Chem Soc* 119.

250 HU, J. S. & BAX, A. (1998).

Measurement of three-bond, 13C′–13C beta *J* couplings in human ubiquitin by a triple resonance, E. COSY-type NMR technique. *J Biomol NMR* 11, 199–203.

251 Hu, J. S. & Bax, A. (1996). Measurement of three-bond 13C–13C *J* couplings between carbonyl and carbonyl/carboxyl carbons in isotopically enriched proteins. *J Am Chem Soc* 118, 8170–8171.

252 Ding, K. & Gronenborn, A. M. (2004). Protein backbone (1)H(N)-(13)C(alpha) and (15)N-(13)C(alpha) residual dipolar and *J* couplings: new constraints for NMR structure determination. *J Am Chem Soc* 126, 6232–6233.

253 Wirmer, J., Schlörb, C., Klein-Scetharaman, J., Hirano, R., Veda, T., Imoto, T. & Schwalbe, H. (2004). Modulation of Compactness and Long-Range Interactions of Unfolded Lysozyme by Single Point Mutations. *Ang Chem Int Ed* 43, in press.

254 Peti, W., Hennig, M., Smith, L. J. & Schwalbe, H. (2002). NMR Spectroscopic Investigation of ψ Torsion Angle Distribution in Unfolded Ubiquitin from Analysis of $^3J(C_\alpha,C_\alpha)$ Coupling Constants and Cross-Correlated $\Gamma^c_{H^N N, C_\alpha H_\alpha}$ Relaxation Rates. *J Am Chem Soc* 122, 12017–12018.

22
Dynamics of Unfolded Polypeptide Chains

Beat Fierz and Thomas Kiefhaber

22.1
Introduction

During protein folding the polypeptide chain has to form specific side-chain and backbone interactions on its way to the native state. An important issue in protein folding is the rate at which a folding polypeptide chain can explore conformational space in its search for energetically favorable interactions. Conformational search within a polypeptide chain is limited by intrachain diffusion processes, i.e., by the rate at which two points along the chain can form an interaction. The knowledge of the rates of intrachain contact formation in polypeptide chains and their dependence on amino acid sequence and chain length is therefore essential for an understanding of the dynamics of the earliest steps in protein folding and for the characterization of the free energy barriers for protein folding reactions. In addition, intrachain diffusion provides an upper limit for the speed at which a protein can reach its native state just like free diffusion provides an upper limit for the rate constant of bimolecular reactions. Free diffusion of particles in solution was treated extensively almost 100 years ago by Einstein and von Smoluchowski [1–3] but until recently, only little was known about the absolute time scales of intrachain diffusion in polypeptides. Numerous theoretical studies have been made to investigate the process of intrachain diffusion in polymers [4–10]. These studies derived scaling laws for intrachain diffusion and made predictions on the kinetic behavior of the diffusion process but they were not able to give absolute numbers. This chapter will briefly introduce some theoretical concepts used to treat chain diffusion and then discuss experimental results on intrachain diffusion from different model systems.

22.2
Equilibrium Properties of Chain Molecules

Since chain dynamics strongly depend on the chain conformation we will briefly present some polymer models for chain conformations. This topic is discussed in

Protein Folding Handbook. Part I. Edited by J. Buchner and T. Kiefhaber
Copyright © 2005 WILEY-VCH Verlag GmbH & Co. KGaA, Weinheim
ISBN: 3-527-30784-2

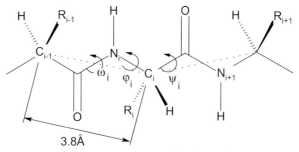

Fig. 22.1. Schematic representation of a polypeptide chain. The R_i denote the different amino acid side chains, the dashed lines denote the virtual chain segment. Adopted from Ref. [12].

more detail in Chapter 20 and in Refs [11–15]. The most simple description on an unfolded state of a protein is the idealized notion of a random coil. In a random coil no specific interactions between residues or more generally chain segments persist and a large conformational space is populated. Polypeptide chains are rather complex polymers as sketched in Figure 22.1. Usually simplified models are applied to calculate general properties of random coils.

22.2.1
The Freely Jointed Chain

The equilibrium properties of an ideal polymer can be described by a three-dimensional random walk. This hypothetical chain, the freely jointed chain, consists of n bonds of length l with equally probable angles at each joint. The chain is described by the $n+1$ position vectors of the joints (chain elements) $\mathbf{R}_i$ or the n bond vectors $\mathbf{r}_i$. The correlation of the bond vectors $\langle \mathbf{r}_m \cdot \mathbf{r}_n \rangle$ is zero for $n \neq m$ because of the independence of the bond vector direction. The average end-to-end distance $\langle r^2 \rangle$ for the freely jointed chain is given by

$$\langle r^2 \rangle = nl^2 \qquad (1)$$

For such an ideal chain the end-to-end vectors are normally distributed. The often-used notion of a Gaussian chain, however, refers to a chain model with Gaussian distributed bond lengths. Thus the end-to-end vector distribution is also Gaussian and Eq. (1) holds. This chain model does not describe the local structure of a chain correctly but the global properties of long chains are modeled in a realistic way.

22.2.2
Chain Stiffness

In a chain with restricted angles between two neighboring segments the correlation between segments n, m is no longer zero for $n \neq m$ but asymptotically approaches zero with increasing distance. The chain is not as flexible as the freely jointed chain because every segment is influenced by its neighbors. This behavior can be described in terms of chain stiffness. The end-to-end distance for such a chain is larger than calculated from Eq. (1). Flory introduced the characteristic ratio (C_n) as a measure for the dimensions of a stiff chain compared to a freely jointed chain [12]

$$\langle r^2 \rangle = C_n n l^2 \qquad (2)$$

For short chains C_n increases with chain length due to preferential chain propagation into one direction (see Figure 22.2). For long chains ($n \to \infty$) C_n reaches a constant limiting value (C_∞).

$$\langle r^2 \rangle = C_\infty n l^2 \qquad (3)$$

In this limit $\langle r^2 \rangle$ grows proportional to n, like an ideal random-walk chain (see Eq. (1)). However, in a real chain $\langle r^2 \rangle$ increases by a factor of C_∞ more per segment compared with an ideal random-walk chain. The value of C_∞ gives the average number of consecutive chain segments that propagate in the same direction ("statistical segment"). Consequently, for an ideal Gaussian chain where there is no correlation between the bond vectors, the limiting characteristic ratio is 1 (see Eq. (1)). For real chains C_∞ is larger than 1 and larger values of C_∞ indicate stiffer chains.

Kuhn showed that for chains with limited flexibility and n bonds of length l an equivalent to the freely jointed chain can be defined introducing an hypothetical bond length b, the Kuhn length ($b > l$) [16, 17]. The Kuhn length is the length of chain segments that can move freely (i.e., without experiencing chain stiffness). The maximal length of the chain (r_{max}) is the same as for a freely jointed chain and thus $r_{max} = nl = n'b$ with $n' < n$. The Kuhn length b is defined as

$$b = \langle r^2 \rangle / r_{max} = C_\infty n l^2 / r_{max} = C_\infty l \qquad (4)$$

Another widely used term is the persistence length l_p. It is a measure for the distance that an infinitely long chain continues in the same direction (i.e., is persistent). The Kuhn and the persistence lengths are connected by the simple relationship

$$b = 2l_p \qquad (5)$$

Both parameters are used as measures for chain stiffness.

22.2.3
Polypeptide Chains

Any real macromolecule like the polypeptide chain has defined segments which are connected by chemical bonds. Bond angles are constrained and the distribution

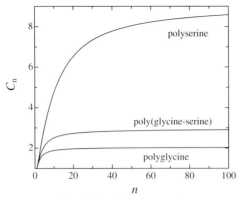

Fig. 22.2. Effect of chain length (n) on the characteristic ratio (C_n) for polyserine, polyglycine and a 50% mixture of serine and glycine. C_n was calculated using average transformation matrices given by Flory [12]. The transformation matrix for alanine was used to calculate the properties of serine.

of torsional angles is given by torsional potentials with several minima. In a peptide chain these torsional potentials are represented in conformational energy plots [18–21] or the Ramachandran map [22]. Flory and coworkers calculated statistical properties of various homopolypeptide chains [12, 18–21] based on the ϕ,ψ-angles from conformational energy plots. An average virtual bond length is defined from $C_{\alpha,i}$ to $C_{\alpha,i+1}$ and taken as 3.8 Å as shown in Figure 22.1 [12]. For polypeptide chains polyglycine represents the most flexible chain with $C_\infty = 2.2$ (see Figure 22.2) [12, 18, 19]. Poly-L-proline gives stiff chains with $C_\infty \approx 100$, due to the formation of a polyproline helix. For all other amino acids, C_∞ is between 8.5 and 9.5 [18]. These values correspond to persistence lengths around 19 Å for chains that do not contain glycine or proline and of 5.7 Å for poly-L-glycine [21]. It should be noted that these values apply for θ conditions, where the attractions between the chain segments (e.g., van der Waals or electrostatic interactions) compensate for the monomer–monomer repulsion due to excluded volume and the chain is apparently unperturbed. For details see Refs [11–15].

22.2.4
Excluded Volume Effects

In real (physical) chains two chain segments cannot occupy the same space. The excluded volume effect was first discussed by Kuhn [16] and leads to a non-ideal behavior of the chain. The chain dimensions increase as a result of the excluded volume, which leads to larger end-to-end distances. Flory obtained an approximate exponent for the chain dimensions including the contributions from the excluded volume effect by simple calculations [11].

$$\sqrt{\langle r^2 \rangle} \propto l \cdot n^\nu, \quad \nu = 0.59 \approx \text{or } 3/5 \tag{6}$$

This shows that the excluded volume effect is especially important for long chains. The excluded volume chain has been a subject of much research and is discussed in detail in the literature [13, 23–25] (see also Chapter 20).

For long real chains the end-to-end distribution function, $p(r)$, depends on the solvent conditions and on the temperature of the system if the intrachain interactions have enthalpic contributions. $p(r)$ for real chains can reasonably well be approximated by a skewed Gaussian function [26]

$$p(r) \propto 4\pi r^2 e^{-((r-r_0)/\sigma)^2} \quad (7)$$

where r is the end-to-end distance, σ is the half-width of the distribution and r_0 indicates the shift of the skewed Gaussian from the origin. Polypeptides in solution are complex molecules which may interact strongly with the solvent and with themselves enabling proteins to fold. The unfolded state of natural proteins was shown to contain both native and nonnative short-range and long-range interactions [27–39] (see Chapter 21). But many of the properties of GdmCl-unfolded proteins may be approximated by statistical chain models (see Section 22.4.2.4), although in some proteins specific intrachain interactions have also been found under strongly denaturing conditions [28, 39]. (see Chapter 21).

22.3
Theory of Polymer Dynamics

One of the major interests in polymer chemistry and biology is to elucidate the dynamic behavior of polymers in solution. Usually the dynamics of such molecules are complex and cannot be easily described by classical concepts. This is especially true for protein folding where the polypeptide chain not only moves freely in solution but also undergoes large cooperative structural transitions involving partially folded and native states. In the following we will give a short introduction to the theoretical concepts of polymer dynamics. More detailed information is given in Refs [8, 13, 23, 24].

22.3.1
The Langevin Equation

The energy surface on which a system moves is often complex with sequential and parallel events during the barrier crossing process (see Chapter 12.2). The motions of the system exhibit a diffusional character. The polymer chain in solution is immersed in solvent molecules which supply the energy for the movement by colliding with the polymer chain and at the same time dissipate this energy by exerting a frictional force on the molecule. The system can be described by the Langevin equation for Brownian motion if one assumes that the relaxation time scale for the solvent fluctuations is extremely short compared with the time scale of polymer motion [40]:

$$\ddot{x} = -M^{-1}\frac{\partial U(x)}{\partial x} - \gamma\dot{x} + M^{-1}F_{\text{fluc}}(t) \tag{8}$$

where M is the particle mass, x is the reaction coordinate, $U(x)$ denotes the energy, and γ is a friction coefficient, which is determined by the interactions of the system with its environment and couples the system to the environment. It can be related to solvent viscosity in real solutions. $F_{\text{fluc}}(t)$ is the random force which represents the thermal motion of the environment. The random force is modeled by Gaussian white noise of zero mean and a δ correlation function

$$\langle F_{\text{fluc}}(t)\rangle = 0$$
$$\langle F_{\text{fluc}}(t)\cdot F_{\text{fluc}}(t')\rangle = 2k_B T\gamma M\delta(t-t') \tag{9}$$

22.3.2
Rouse Model and Zimm Model

For long unfolded polypeptide chains, the simplified models used in polymer theory should be applicable. The system, i.e., the peptide immersed in the solvent, encounters no large barriers and undergoes random conformational changes. The dynamic behavior of Gaussian chains can be described analytically using classical polymer theory. Models of a higher complexity are only accessible, however, by solving the equations numerically in computer simulations. Two models have been initially proposed to describe polymer dynamics in dilute solutions: the Rouse model [5] and the Zimm model [6]. In both models the polymer is treated as a set of beads connected by harmonic springs. The Langevin equation (Eq. (8)) is used to describe the brownian motion of the connected beads in the Rouse model. The frictional and activating force from the solvent independently acts on all beads. The movement of the chain is described as a set of n relaxation modes. These modes can be compared to the vibrational modes of a violin string. Each mode p is coupled to a relaxation time τ_p and involves the motion of a section of the chain with n/p monomers. The mode with the longest relaxation time describes the overall motion (rotational relaxation) of the polymer chain. All other modes represent internal motions.

Yet, the scaling laws obtained for Rouse chains are not consistent with experimental results [24]. When a chain segment moves through a viscous solvent it has to drag surrounding solvent molecules along. This creates a flow field around the particle. Surrounding monomers are additionally affected by this flow field. As a result, the motion of one chain segment influences the motion of all other segments by hydrodynamic interactions.

The resulting chain model including hydrodynamic interactions was put forward by Zimm [6]. This model also gives internal relaxation modes, but the hydrodynamic interactions lead to significant deviations from the Rouse model. The solvent viscosity influences the system in two ways. First, it affects the random force that delivers the energy and dissipates it at the same time (γ, in the Langevin equation). Secondly, viscosity changes the strength of the hydrodynamic interactions. Many studies have been performed to compare the predictions of Rouse model, the Zimm model, and more refined models to actual experiments [41]. The overall chain motions including hydrodynamic interactions agree well with experiment [24]. Measurements of dynamics in DNA molecules using time-resolved optical microscopy directly showed internal relaxation modes [42].

22.3.3
Dynamics of Loop Closure and the Szabo-Schulten-Schulten Theory

One particularly intensively treated dynamic event in polymer chains is the loop closure or end-to-end diffusion reaction. Organic chemists are interested in loop closure probabilities for the synthesis of cyclic compounds. Also in nucleic acids end-to-end diffusion is important during formation of cyclic DNA. The kinetics of such cyclization reactions critically depend on the rate of intrachain contact formation. A similar situation is encountered in protein folding, where residues have to come together and form specific non-covalent interactions. Jacobsen and Stockmayer [4] derived a scaling law for end-to-end ring closure probability as a function of the length of the polymer chain. The calculations were based entirely on entropic contributions and yielded

$$p_{ring} \propto n^{-3/2} \tag{10}$$

where p_{ring} denotes the probability of ring formation and n is the number of monomers in such a ring.

Szabo, Schulten, and Schulten [7] discussed the kinetics that can be expected from a diffusion controlled encounter of two groups attached to the ends of a Gaussian chain (SSS theory). They treated the loop closure reaction as an end-to-end diffusion of a polymer on a potential surface. It was shown that the kinetics of such a process can be approximated by an exponential decay of the probability $\Sigma(t)$ that the system has not reacted (i.e., has not made contact) at time t

$$\Sigma(t) \approx \Sigma_{approx}(t) = e^{(-t/\tau)} \tag{11}$$

This approximation holds if the distribution of chain conformations is stationary throughout the reaction, i.e., the interchange of conformations is fast compared with the rate of reaction, and the probability of reaction is small, i.e., the reactive radius is small compared with the chain length. The time constant τ is the mean first passage time of this reaction and depends on the equilibrium end-to-end distribution of the polymer chain $p(x)$, on the diffusion constant D of the chain ends relative to one another and on the reactive boundary r_a, i.e., the distance which leads to contact between the two ends.

$$\tau = \frac{1}{D} \int_{r_a}^{\infty} p(x)^{-1} \left(\int_{x}^{\infty} p(y)\,dy \right)^2 dx \bigg/ \int_{r_a}^{\infty} p(x)\,dx \tag{12}$$

The mean first passage time of contact formation can be related to the root mean square end-to-end distance, r, by expanding Eq. (12):

$$\tau^{-1} = k_{contact} \propto D(\sqrt{\langle r^2 \rangle})^{-3} \tag{13}$$

With Eq. (1) this gives the well-known relationship [4]:

$$k_{contact} \propto n^{-3/2} \tag{14}$$

for an ideal chain. The effect of viscosity on the rate of reaction is included in the diffusion coefficient D, which is assumed to be inversely proportional to the solvent viscosity. Hydrodynamic interactions, the excluded volume effect and other non-idealities (e.g., a diffusion coefficient dependent on chain conformation) are not considered in these approximations and probably contribute to the dynamics and scaling laws in real systems.

22.4
Experimental Studies on the Dynamics in Unfolded Polypeptide Chains

22.4.1
Experimental Systems for the Study of Intrachain Diffusion

It is now widely accepted that the description of the unfolded state as a complete random coil where the polypeptide chain behaves like an ideal polymer is not accurate. Short stretches along the polypeptide chain behave in a highly non-ideal manner because of chain stiffness, solvent interactions, and excluded volume. In addition, there is evidence for residual short-range and long-range interactions in unfolded proteins in water and even at high concentrations of urea and GdmCl [27, 28, 34, 43, 44]. For the understanding of the dynamic properties of unfolded proteins it is therefore essential to compare scaling laws and absolute rate constants for chain diffusion in homopolymer models with the results from unfolded proteins or protein fragments. Suitable experimental systems to study intrachain diffusion should allow direct measurements of the kinetics of formation of van der Waals contacts between specific groups on a polypeptide chain. To obtain absolute rate constants, the applied method must be faster than chain diffusion and the detection reaction itself should be diffusion controlled (see Appendix).

In the following, we will describe various experimental systems that have been used to study the dynamics of intrachain diffusion. In Section 22.4.2 results from these studies will be discussed in more detail in terms of dynamic properties of unfolded polypeptide chains. A summary of the results is given in Tables 22.1 and 22.2.

22.4.1.1 Early Experimental Studies
Dynamics of synthetic polymers in various organic solvents have been studied for a long time by different means. End-to-end contact formation has been examined

Tab. 22.1. Comparison of end-to-end contact formation rate constants observed in different systems.

k_{app} (s^{-1})	Loop size (n)[a]	Method	Labels[b]	Sequence	Conditions	Reference
$3.3 \cdot 10^8$–$1.1 \cdot 10^8$	4–8, 50[c]	FRET	Dansyl/naphthalene	(heGln)$_n$[d]	Glycerol/TFE	48, 50[c]
$2.8 \cdot 10^4$	60	Bond formation	Heme/Met	Unfolded cytochrome c	5.4 GdmCl, 40 °C	52
$5 \cdot 10^7$–$1.2 \cdot 10^7$	3–9	TTET	Thioxanthone/NAla	(Gly-Ser)$_x$	EtOH	50
$9.1 \cdot 10^6$	10	Triplet quenching	Trp/Cys	(Ala-Gly-Gln)$_3$	H$_2$O, phosphate	59
$1.1 \cdot 10^7$	10	Triplet quenching	Trp/cystine	(Ala-Gly-Gln)$_3$	H$_2$O, phosphate	59
$1.7 \cdot 10^7$	10	Triplet quenching	Trp/lipoate	(Ala-Gly-Gln)$_3$	H$_2$O, phosphate	59
$6.2 \cdot 10^4$	10	Triplet quenching	Trp/lipoate	Pro$_9$	H$_2$O, phosphate	59
$2.4 \cdot 10^7$	10	Triplet quenching	Trp/lipoate	(Ala)$_2$Arg(Ala)$_4$ArgAla	H$_2$O, phosphate	59
$4.1 \cdot 10^7$–$1.1 \cdot 10^7$	1–21	Fluor. quenching	DBO/Trp	(Gly-Ser)$_x$	D$_2$O	65
$3.9 \cdot 10^7$–$1 \cdot 10^5$	7	Fluor. quenching	DBO/Trp	Xaa$_6$[e]	D$_2$O	66
$6.6 \cdot 10^6$	8	Single molecule fluor. quenching	MR121/Trp	Part of human p53	H$_2$O, PBS, 25 °C	67
$8.3 \cdot 10^6$	9	Single molecule fluor. quenching	MR121/Trp	Part of human p53	H$_2$O, PBS, 25 °C	67
$1.8 \cdot 10^8$–$6.5 \cdot 10^6$	3–57	TTET	Xanthone/NAla	(Gly-Ser)$_x$	H$_2$O, 22.5 °C	54
$8 \cdot 10^7$–$3.4 \cdot 10^7$	3–12	TTET	Xanthone/NAla	(Ser)$_x$	H$_2$O, 22.5 °C	54
$2.5 \cdot 10^8$–$2.0 \cdot 10^7$	4	TTET	Xanthone/NAla	Ser-Xaa-Ser[f]	H$_2$O, 22.5 °C	54
$4.0 \cdot 10^6$	15	Triplet quenching	Zn-porphyrin/Ru	Unfolded cytochrome c	5.4 M GdmCl, 22 °C	64
$2.0 \cdot 10^7$	17	TTET	Xanthone/NAla	Carp parvalbumin	H$_2$O, 22.5 °C	55

[a] n is the number of peptide bonds between the reacting groups.
[b] The structures of the labels are shown in Figure 22.3.
[c] Calculated in Ref. [50] from the diffusion coefficients and the donor–acceptor distances given in Ref. [48].
[d] heGln = (N^5-(2-hydroxyethyl)-L-glutamine).
[e] (Xaa)$_6$ = homohexapeptides of 16 different amino acids. The highest (Gly$_6$) and lowest values (Pro$_6$) for the observed rate constants are given.
[f] Xaa = Gly, Ala, Ser, Gly, Arg, His, Ile, Pro. The highest (*cis* Pro) and lowest values (*trans* Pro) for the observed rate constants are given. See Figure 22.9.

Tab. 22.2. Comparison of end-to-end contact formation rates observed in different systems measured or extrapolated to $n \approx 10$.

k_{app} (s^{-1})	Loop size (n)	Method	Labels	Sequence	Conditions[b]	Reference
$1 \cdot 10^6$	10[a]	Bond formation	Heme/Met	Unfolded cytochrome c	5.4 GdmCl, 40 °C	52
$1.2 \cdot 10^7$	9	TTET	Thioxanthone/NAla	(GS)$_4$	EtOH	50
$9.1 \cdot 10^6$	10	Triplet quenching	Trp/Cys	(AGQ)$_3$	H$_2$O, phosphate buffer	59
$1.1 \cdot 10^7$	10	Triplet quenching	Trp/cystine	(AGQ)$_3$	H$_2$O, phosphate buffer	59
$1.7 \cdot 10^7$	10	Triplet quenching	Trp/lipoate	(AGQ)$_3$	H$_2$O, phosphate buffer	59
$3.1 \cdot 10^7$	9	Fluor. quenching	DBO/Trp	(GS)$_4$	D$_2$O	65
$8.3 \cdot 10^6$	9	Single molecule Fluor. quenching	MR121/Trp	Fragment from human p53	H$_2$O, PBS, 25 °C	67
$8.3 \cdot 10^7$	9	TTET	Xanthone/NAla	(GS)$_4$	H$_2$O, 22.5 °C	54
$4.1 \cdot 10^7$	10	TTET	Xanthone/NAla	(Ser)$_9$	H$_2$O, 22.5 °C	54
$1 \cdot 10^7$	10[a]	Triplet quenching	Zn-porphyrin/Ru	Unfolded cytochrome c	5.4 M GdmCl, 22 °C	64

[a] Extrapolated from larger loops to 10 amino acids, as the average loop size in proteins. It should be noted that extrapolations used the Jacobsen Stockmeyer or SSS theory scaling law ($k_c \sim n^{-1.5}$), which gives a slightly smaller length dependence than the experimentally determined scaling laws (cf. Figure 22.8). This would result in larger extrapolated values of k_c.

[b] solvent was H$_2$O unless indicated. The data measured in concentrated GdmCl solutions or EtOH were not corrected for solvent effects. This would lead to a significant increase in k_c (see Figure 22.10).

using phosphorescence quenching of benzophenone in long hydrocarbon polymers [45] and by excimer formation of pyrene groups attached to poly(ethylene glycols) [46]. The dynamics of unfolded polypeptide chains have first been studied using fluorescence resonance energy transfer (FRET) from an energy donor to an energy acceptor group. FRET was shown to be a powerful tool to determine donor–acceptor distances [47], but time-resolved FRET kinetics also contain major contributions from the diffusion of the two FRET probes relative to each other [48, 49]. The number of excited donor molecules ($N^*(r,t)$) with donor–acceptor distance r changes with time in the presence of an acceptor group. This can be described by a second-order partial differential equation for the reduced distance distribution, $\bar{N}(r,t)$, normalized to $N_0(r)$:

$$\frac{\partial \bar{N}(r,t)}{\partial t} = -\frac{1}{\tau}\left\{\left(1+\left(\frac{R_0}{r}\right)^6\right)\bar{N}(r,t)\right\} + \frac{1}{N_0(r)}\frac{\partial}{\partial r}\left\{N_0(r)D\frac{\partial \bar{N}(r,t)}{\partial r}\right\}$$

$$\text{with } \bar{N}(r,t) = \frac{N^*(r,t)}{N_0(r,t)} \tag{15}$$

where R_0 is the distance of 50% FRET efficiency, which is a property of the FRET labels, $N_0(r)$ denotes the equilibrium distribution of distances between donor and acceptor, and τ is the fluorescence lifetime of the donor. The first term (equilibrium term) on the right-hand side describes the change of donor excitation with time through spontaneous decay and nonradiative energy transfer. The second term (diffusion term) describes the replenishment of excited labels through brownian motion. This dynamic component of FRET was exploited to determine diffusion coefficients of segmental motion in short homopolypeptide chains (poly-N^5-(2-hydroxyethyl)-L-glutamine) in glycerol and trifluoroethanol mixtures [48, 49] (see Chapter 17). Naphthyl- and dansyl-groups were used as donor and acceptor, respectively (Figure 22.3A).

These studies allowed the first estimates of the time scale of chain diffusion processes in polypeptide chains. Converting the reported average diffusion constants reported by Haas and coworkers [48] into time constants ($\tau = 1/k$) for contact formation at van der Waals distance gives values between 9 and 3 ns for a donor–acceptor separation of between 4 and 8 peptide bonds, respectively [50]. The dynamic component of FRET is sensitive to motions around the Förster distance (R_0), which is around 22 Å for the dansyl–naphthyl pair used in these studies. It therefore remained unclear whether the same diffusion constants represent the dynamics for formation of van der Waals contact between two groups. In addition, analysis of the FRET data critically depends on the shape of the donor–acceptor distribution function, which had to be introduced in the data analysis and thus did not allow model-free analysis of the data. It was found that this distribution can be described most accurately by the skewed Gaussian function mentioned above [51]. It later turned out that the chain dynamics measured by FRET are significantly faster than rate constants for contact formation (see below), which might indicate a distance-dependent diffusion constant in unfolded polypeptide chains.

820 | *22 Dynamics of Unfolded Polypeptide Chains*

FRET was also used in several proteins to determine donor–acceptor distance distribution functions and diffusion coefficients (see Chapter 17).

Rates of intrachain diffusion have also been estimated from the rate constant for intramolecular bond formation in GdmCl-unfolded cytochrome c [52]. The reaction of a methionine with a heme group separated by 50–60 amino acids along the chain gave time constants around 35–40 µs. Extrapolating these data to shorter distances using SSS or Jacobsen-Stockmeyer theory ($k \sim n^{-1.5}$) gave an estimate of 1 µs for the time constant of contact formation at a distance of 10 peptide bonds, which was predicted to be the distance for the maximum rate of contact formation due to chain stiffness slowing down the diffusion process over shorter distances [53]. A problem with this experimental approach is that the recombination reaction of heme with methionine is not diffusion controlled. Thus, the chain dynamics allow a faster breakage of the heme–methionine contact than formation of the methionine–heme bond. Thus, not every contact was productive and the apparent rate constants obtained in these measurements were slower than the absolute rate constants measured with diffusion-controlled systems (see Experimental Protocols, Section 22.7).

22.4.1.2 Triplet Transfer and Triplet Quenching Studies

The early studies on intrachain diffusion did not yield absolute rate constants for contact formation, but they were important to trigger the development of more direct and model-free methods to study chain diffusion. Recently, several experimental systems using fast photochemical methods to directly monitor contact formation in model peptides and protein fragments have been introduced. Bieri et al. [50] and Krieger et al. [54, 55] used triplet–triplet transfer (TTET) between thioxanthone or xanthone derivatives and naphthalene (see Figure 22.3B,C) to directly measure rate constants for intrachain contact formation in synthetic polypeptide chains. A triplet donor and a triplet acceptor group are introduced at specific points in polypeptide chains. The donor is selectively excited by a laserflash and undergoes fast intersystem crossing into the triplet state (see Figure 22.4). Upon intrachain diffusion the triplet donor and triplet acceptor groups meet and the triplet state is transferred to the acceptor. Since TTET is a two electron exchange process (Dexter mechanism) it has a very strong distance dependence and usually requires

---

Fig. 22.3. Structures of different experimental systems used to measure intrachain contact formation. Results from these systems are summarized in Table 22.1. A) Dansyl and naphthyl-labeled poly-N^5-(2-hydroxyethyl)-L-glutamine used to determine diffusion constants by FRET [48]. B) Thioxanthone and naphthylalanine-labeled poly(Gly-Ser) peptides to measure contact formation by TTET [50]. C) Xanthone and naphthylalanine-labeled peptides to measure contact formation by TTET in various homopolymer chains and natural sequences with up to 57 amino acids between donor and acceptor [54, 55]. D,E) Tryptophan and cysteine or lipoate-labeled peptides to measure contact formation by Trp triplet quenching in various short peptide chains [59, 125, 126]. F) Short peptide chains ($n < 10$) for measuring DBO fluorescence quenching by tryptophan [65]. G) Quenching of Zn-porphyrin triplets by Ru-His complexes in unfolded cytochrome c [64].

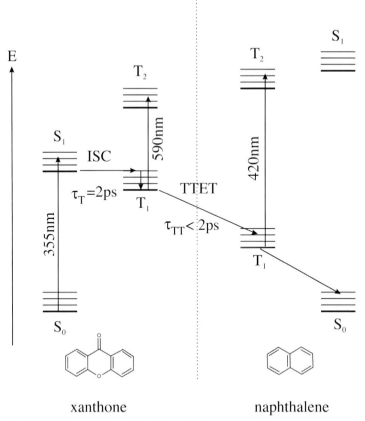

Fig. 22.4. Jablonski diagram for triplet–triplet energy transfer (TTET) from xanthone to naphthalene. Rate constants for triplet formation (k_T) and TTET (k_{TT}) were measured by laserflash photolysis using femtosecond pulses [56].

van der Waals contact to allow for electron transfer. This should be contrasted to FRET, which occurs through dipole–dipole interactions and thus allows energy transfer over larger distances (see above).

The triplet states of the labels have specific absorbance bands, which can easily be monitored to measure the decay of the donor triplet states and the concomitant increase of acceptor triplet states (Figure 22.5). The time constant for formation of xanthone triplet states is around 2 ps [56]. TTET between xanthone and naphthyl acetic acid is faster than 2 ps and has a bimolecular transfer rate constant of $4 \cdot 10^9$ M^{-1} s^{-1} [54, 56], which is the value expected for a diffusion-controlled reaction between small molecules in water (see Experimental Protocols, Section 22.7). Due to this fast photochemistry, TTET between xanthone and naphthalene allows measurements of absolute rate constants for diffusion processes slower than 10–20 ps (Figure 22.6 and Experimental Protocols, Section 22.7). The upper

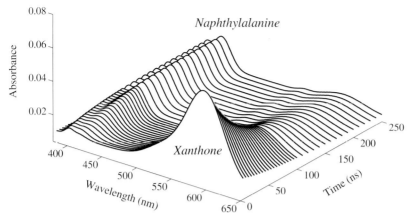

Fig. 22.5. Time-dependent change in the absorbance spectrum of a Xan-(Gly-Ser)$_{14}$-NAla-Ser-Gly peptide after a 4 ns laser flash at 355 nm. The decay in the intensity of the xanthone triplet absorbance band around 590 nm is accompanied by a corresponding increase in the naphthalene triplet absorbance band around 420 nm. Adapted from Ref. [54].

limit of the experimental time window accessible by TTET is set by the intrinsic lifetime of xanthone which is around 20–30 μs in water. Xanthone has a high quantum yield for intersystem crossing ($\phi_{ISC} = 0.99$, $\varepsilon \approx 4000$ M^{-1} cm^{-1}) and the triplet state has a strong absorbance band with a maximum around 590 nm in water ($\varepsilon_{590}^T \approx 10\,000$ M^{-1} cm^{-1}). This allows single-pulse measurements at rather low peptide concentrations (10–50 μM).

The low concentrations applied in the experiments also rule out contributions from intermolecular transfer reactions that would have half-times higher than 50 μs in this concentration range [50, 54]. Also contributions from through-bond transfer processes can be neglected, since this cannot occur over distances beyond eight bonds [57, 58] and even the shortest peptides used in TTET studies had donor and acceptor separated by 11 bonds.

Bieri et al. [50] initially used a derivate of thioxanthone as triplet donor and the nonnatural amino acid naphthylalanine (NAla) as triplet acceptor (see Figure 22.3B). Because of the sensitivity of the triplet energy of the donor on solvent polarity the measurements had to be carried out in ethanol. TTET detected single exponential kinetics in all peptides with time constants of 20 ns for the shortest loops in poly(glycine-serine)-based polypeptide chains [50]. The length dependence of contact formation in these peptides did not exhibit a maximum rate constant for transfer at $n = 10$, in contrast to the predicted behavior [53] (see Section 22.4.2.2). Based on these findings, the minimum time constant for intrachain contact formation was shown to be around 20 ns for short and flexible chains. The thioxanthone used in the initial experiments as triplet donor was later replaced by xanthone [54], which has a higher triplet energy than thioxanthone and thus allowed measurements in water (see Figure 22.3C). These results showed two- to threefold faster rate constants for contact formation compared to the same peptides in ethanol [54], setting the time constant for intrachain diffusion in short flexible chains in water to 5–10 ns. The faster kinetics compared to the thioxanthone/NAla system could be attributed to the effect of ethanol [54] (see below).

A disadvantage of TTET from xanthone to NAla in its application to natural proteins is that Tyr, Trp, and Met interact with the xanthone triplet state either by TTET or by triplet quenching and thus should not be present in the studied polypeptide chains [55].

Lapidus et al. [59] used a related system to measure chain dynamics in short peptides. In this approach contact formation was measured by quenching of tryptophan triplet states by cysteine (see Figure 22.3D). Tryptophan can be selectively excited by a laserflash ($\phi_{ISC} = 0.18$, $\varepsilon_{266} \approx 3500$ M^{-1} cm^{-1}, $\varepsilon_{460}^{T} \approx 5000$ M^{-1} cm^{-1}) and its triplet decay can be monitored by absorbance spectroscopy [60]. The advantage of this system is that donor chromophore and quencher groups are naturally occurring amino acids, which can be introduced at any position in peptides and proteins. As in the case on TTET from xanthone to naphthalene some amino acids interfere with the measurements (e.g., Tyr, Met) since they interact with tryptophan triplets and should thus not be present in the studied polypeptide chains [61].

Major disadvantages of the Trp/Cys system are, (i) that the formation of tryptophan triplets is slow ($\tau = 3$ ns) [62], (ii) that triplet quenching is accompanied by the formation of S·radicals [59], and (iii) that the quenching process is not diffusion controlled [63]. These properties reduce the time window available for the kinetic measurements and do not allow direct and model-free analysis of the quenching time traces (see Experimental Protocols, Section 22.7). In addition, its low quantum yield makes the detection of the triplet states and the data analysis difficult, especially since the kinetics are obscured by radical absorbance bands. In addition to cysteine quenching, two other systems were presented by the same group [59] with cystine or the cyclic disulfide lipoate serving as quencher instead of cysteine (Figure 22.3E). The advantage of using lipoate is that the quenching kinetics are much closer to the diffusion limit. Thus, the rates measured with this system are generally faster. Still, all systems used for quenching of tryptophan triplets gave significantly slower kinetics of intrachain contact formation compared to TTET from xanthone to NAla in the same or in similar sequences (see Tables 22.1 and 22.2). This indicates that tryptophan triplet quenching does not allow measurements of chain dynamics on the absolute time scale [63].

Another recent experimental approach investigated intrachain contact formation in unfolded cytochrome c using electron transfer from a triplet excited Zn-porphyrine group to a Ru complex, which was bound to a specific histidine residue (His33; Figure 22.3G). Since electron transfer is fast and close to the diffusion limit these experiments should also yield absolute rate constants for chain diffusion. Contact formation in the 15-amino-acid loop from cytochrome c was observed with a time constant of 250 ns [64] in the presence of 5.4 M GdmCl. This is significantly faster than the dynamics in unfolded cytochrome c reported by Hagen et al. [52] under similar conditions. However, the dynamics measured by Chang et al. agree well with TTET measurements [50, 54, 55], when the kinetics are extrapolated from 5.4 M GdmCl to water (see below).

22.4.1.3 Fluorescence Quenching

Other approaches to measure contact formation applied quenching of long-lifetime fluorescence probes. Two studies used 2,3-diazabicyclo[2,2,2]oct-2-ene (DBO) as a fluorophore (see Figure 22.3F) [65, 66]. It has a lifetime of up to 1 µs and its excited singlet states can be quenched by tryptophan upon contact. The quenching rate of DBO by tryptophan lies close to the diffusion-controlled limit in water, yet the obtained rates are about a factor of 3 slower than the ones obtained with the xanthone/NAla system in the same peptides [54]. This suggests that the photochemistry of fluorescence quenching may not be faster than breakage of the van der Waals contact between DBO and Trp (see Figure 22.6 and Experimental Protocols, Section 22.7). Another disadvantage of this method is the rather short lifetime of the DBO excited state (< 1 µs), which limits the method to short peptides.

Recently, intrachain diffusion was measured using single molecule fluorescence quenching. Peptides were labeled with a fluorescent dye (MR121) which was quenched by tryptophan [67]. The quenching was reported to only occur efficiently via van der Waals contact. Contact formation between the labels resulted in fluorescence fluctuations, which were recorded by confocal fluorescence microscopy. The rate constants for contact formation were determined using time-correlated single-photon counting. The calculated rates are, however, significantly lower than those determined by more direct methods.

22.4.2 Experimental Results on Dynamic Properties of Unfolded Polypeptide Chains

The different systems presented above were applied to measure intrachain contact formation in a large variety of model peptides. The rates for contact formation have

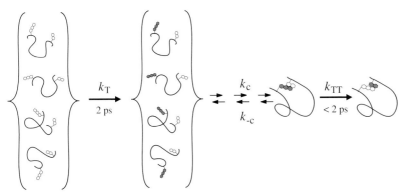

Fig. 22.6. Schematic representation of the triplet–triplet energy transfer (TTET) experiments. Triplet donor and acceptor groups are attached at specific positions on an unstructured polypeptide chain. Triplet states are produced in the donor group by a short laser flash and transferred to the acceptor upon encounter at van der Waals distance in a diffusion-controlled process. The experiments allow determination of the absolute rate constant for contact formation (k_c) in the ensemble of unfolded conformation, if formation of the triplet state (k_T) and the transfer process (k_{TT}) are much faster than chain dynamics ($k_T, k_{TT} \gg k_{-c}, k_c$; see Experimental Protocols, Section 22.7). The rate constants given for k_T and k_{TT} were taken from Ref. [56].

been found to vary depending on the method applied, on the sequence of the peptide and on the peptide length. Results from different groups are compared in Tables 22.1 and 22.2. The results reveal that TTET and electron transfer, which were shown to have fast photochemistry and diffusion-controlled transfer reactions, give virtually identical rate constants and have the fastest contact rates of all applied methods. These methods obviously allow the determination of absolute rate constants for contact formation. In the following, we will discuss results from experiments that used these methods to determine the effect of chain length, amino acid sequence and solvent properties on the rate constants of contact formation.

22.4.2.1 Kinetics of Intrachain Diffusion

All experiments applying TTET [50, 54] and electron transfer [64] to measure dynamics of unfolded polypeptide chains revealed single exponential kinetics on the ns time scale or slower (Figure 22.7). The only exception were short proline-containing peptides, where the kinetics of *cis* and *trans* Xaa-Pro peptide bonds could be resolved [54]. From the observation of single exponential kinetics several conclusions can be drawn. As observed by Szabo, Schulten, and Schulten [7] and pointed out by Zwanzig [68], single exponential kinetics indicate fast interconversion between the different conformations in the ensemble of unfolded states, thus allowing the chain to maintain the equilibrium distribution of the ensemble of conformations that has not made contact. Fast interconversion between individual chain conformations is in agreement with results on conformational relaxation processes in strained peptides, which were shown to occur in the picoseconds time scale [69, 70]. Secondly, the contact radius is small compared to the chain length, which leads to a small probability of contact formation in agreement with the observation that triplet–triplet transfer can only occur when the two labels are in van der Waals contact [7, 68]. A small fraction of chain conformations in equilibrium are predicted to have very short donor–acceptor distances [63] which might allow contact formation on the subnanosecond time scale. From the nanosecond TTET experiments it can be ruled out, however, that a significant fraction ($> 5–10\%$) of chain molecules form donor–acceptor contacts on the subnanosecond time scale, even in the shortest peptides [54]. However, the dynamics of linear peptides in the subnanoseconds time region has not been investigated yet. The fast photochemistry of TTET between xanthone and naphthalene should allow the study of peptide dynamics on the picosecond time scale to detect small fraction of fast transfer processes.

22.4.2.2 Effect of Loop Size on the Dynamics in Flexible Polypeptide Chains

The scaling of the end-to-end diffusion with loop size was measured in TTET experiments by varying the number of amino acids between xanthone and naphthylalanine in two series of homopolypeptides. One series represented flexible polypeptide chains of Xan-(Gly-Ser)$_x$-NAla-Ser-Gly peptides with x varying from 1 to 28. These peptides allowed measurements of contact formation kinetics between two points on the chain separated by between 3 and 57 peptide bonds, which covers the range of side-chain contacts in small proteins. Figure 22.7 displays three repre-

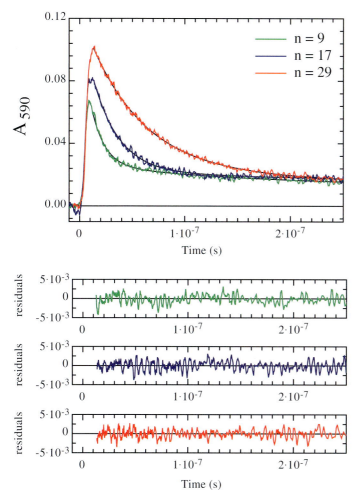

Fig. 22.7. Time course of formation and decay of xanthone triplets in peptides of the form Xan-(Gly-Ser)$_x$-NAla-Ser-Gly after a 4 ns laser flash at $t = 0$ measured by the change in absorbance of the xanthone triplets at 590 nm. Data for different numbers of peptide bonds (n) between donor and acceptor are displayed. Additionally, single exponential fits of the data and the corresponding residuals are shown. The fits gave time constants shown in Figure 22.8. Adapted from Ref. [54].

sentative TTET kinetics for peptides from this series. As mentioned above, single exponential kinetics for contact formation were observed for all peptides.

Figure 22.8 shows the effect of increasing loop size on the rate constant for contact formation. For long loops ($n > 20$) the rate of contact formation decreases with $n^{-1.7 \pm 0.1}$ (n is the number of peptide bonds between donor and acceptor). This indicates a stronger effect of loop size on the rate of contact formation than expected for purely entropy-controlled intrachain diffusion in ideal freely jointed Gaussian

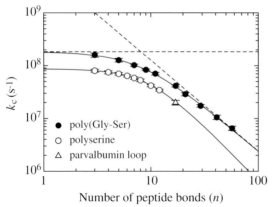

Fig. 22.8. Comparison of the rate constants (k_c) for end-to-end diffusion in poly(glycine-serine) (filled circles), polyserine (open circles) and a parvalbumin loop fragment 85–102 (triangles) measured by TTET between xanthone and naphthylalanine. The data were fitted to Eq. (16) and gave results of $k_a = (1.8 \pm 0.2) \cdot 10^8$ s^{-1}, $k_b = (6.7 \pm 1.6) \cdot 10^9$ s^{-1} and $m = 1.72 \pm 0.08$ for the poly(glycine-serine) series and of $k_a = (8.7 \pm 0.8) \cdot 10^7$ s^{-1}, $k_b = (1.0 \pm 0.8) \cdot 10^{10}$ s^{-1} and $m = 2.1 \pm 0.3$ for polyserine. The dashed lines indicate the limiting cases for dynamics for short chains ($k_c = k_a$) and long chains $k_c = k_b \cdot n^{-m}$). Data were taken from Refs [54, 55].

chains, which should scale with $k \sim n^{-1.5}$ (see above) [4, 7]. However, Flory already pointed out that excluded volume effects should significantly influence the chain dimensions [11]. Accounting for excluded volume effects in the end-to-end diffusion model of Szabo, Schulten, and Schulten [7] gives $k \sim n^{-1.8}$, which is nearly identical to the value found for the long poly(Gly-Ser) chains [54]. This indicates that the dimensions and the dynamics of unfolded polypeptide chains in water are significantly influenced by excluded volume effects, which is in agreement with recent results on conformational properties of polypeptides derived from simplified Ramachandran maps [25]. Additionally, hydrodynamic interactions and the presence of small enthalpic barriers, which were observed in the temperature dependence of intrachain diffusion might contribute to the length dependence (F. Krieger and T. Kiefhaber, unpublished).

The observed simple scaling law breaks down for $n < 20$ and contact formation becomes virtually independent of loop size for very short loops with a limiting value of $k_a = 1.8 \cdot 10^8$ s^{-1}. As pointed out above, the limiting rate constant for contact formation in short loops is not due to limits of TTET (k_{TT} in Figure 22.6), since this process is faster than 2 ps. Obviously, the intrinsic dynamics of polypeptide chains are limited by different processes for motions over short and over long segments. This is in agreement with polymer theory, which suggests that the properties of short chains are strongly influenced by chain stiffness. This leads to a breakdown of theoretically derived scaling laws for ideal chains [12]. To compare the experimental results with predictions from polymer theory the effect of loop size on the rate constants for contact formation (Figure 22.8) can be compared to the length dependence of the characteristic ratio (C_n; Figure 22.2). By comparison with polymer theory the limiting value for k_c for formation of short loops might be related to intrinsic chain stiffness, which causes the increase in C_n with chain length for short chains (see Section 22.2).

The 1:1 mixture of glycine and serine used in these experiments is expected to have a C_∞-value of about 3 and C_∞ should be reached for $n > 10$ (Figure 22.2) [21]. Figure 22.8 shows, however, that the poly(glycine-serine) chains behave like random chains only over distances longer than 20–30 amino acids, where $k_c \sim n^{-1.7}$. This indicates increased chain stiffness compared to the predicted value, which could be due to specific intrachain hydrogen bonds or van der Waals interactions and to excluded volume effects that restrict the number of chain conformations. This model is supported by the measurements of activation energies for the peptides of different length. Formation of long loops shows very low activation enthalpies whereas formation of short loops encounter significant energy barriers (F. Krieger and T. Kiefhaber, unpublished).

The complete effect of loop size on intrachain contact formation can be described by a model where length-dependent diffusional processes, which scale with $k = k_b \cdot n^{-m}$, limit the kinetics of formation of long loops. Contact formation reaches a limiting rate constant (k_a) when these diffusional motions become slower than length-independent short-range motions, which are probably governed by chain stiffness and steric effects. Accordingly, the effect of loop size on contact formation can be described by

$$k_c = \frac{1}{1/k_a + 1/(k_b \cdot n^{-m})} \tag{16}$$

The fit of the experimental results to Eq. (16) is shown in Figure 22.8 and gives a value of $k_a = (1.8 \pm 0.2) \cdot 10^8$ s^{-1}, $k_b = (6.7 \pm 1.6) \cdot 10^9$ s^{-1}, and $m = 1.72 \pm 0.08$ for poly(glycineserine), as discussed above.

22.4.2.3 Effect of Amino Acid Sequence on Chain Dynamics

Both theoretical considerations [18, 19, 21] (see Section 22.2) and experimental results from NMR measurements [71] show that polypeptide chains are especially flexible around glycyl residues. All amino acids except proline ($C_\infty > 100$) and glycine ($C_\infty = 2.2$) are predicted to have C_∞-values around 8.5–9.5 and C_∞ should be reached for intrachain distances longer than about 40–50 amino acids [18, 19, 21] (Figure 22.2). This indicates increased chain stiffness and longer root mean square end-to-end distances ($\sqrt{\langle r^2 \rangle_0}$) compared to the poly(glycine-serine) series (Figure 22.2). The effect of chain stiffness on peptide dynamics was tested in TTET experiments in polyserine chains [54]. Figure 22.8 compares the effect of loop size on intrachain diffusion in a series of Xan-(Ser)$_x$-NAla-Ser-Gly peptides with $x = 2$–11 ($n = 3$–12) to the behavior of poly(Gly-Ser) chains. Single exponential kinetics for contact formation were observed in all polyserine peptides. For short loops ($n < 5$) contact formation is virtually independent of loop size with a limiting value of $k_a = 8.7 \cdot 10^7$ s^{-1} (see Eq. (18)). This indicates that the local dynamics in polyserine are about two- to threefold slower than in the poly(glycine-serine) peptides, which seems to be a small effect compared to the largely different properties expected for the stiffer polyserine chains (see Figure 22.2).

The decreased flexibility and the longer donor-acceptor distances in the polyserine chains are probably compensated by the decreased conformational space available for polyserine compared to poly(glycine-serine) peptides. [54] For longer polyserine chains contact formation slows down with increasing loop size. The effect of increasing loop size on the rates of contact formation seems to be slightly larger in polyserine compared to the poly(glycine-serine) series ($m = 2.1 \pm 0.3$; $k_b = (1.0 \pm 0.8) \cdot 10^{10}$ s^{-1}; see Eq. (16)). However, due to limitations in peptide synthesis it was not possible to obtain longer peptides, which would be required to get a more accurate scaling law for the polyserine peptides.

The kinetics of formation of short loops differ only by a factor of 2 for polyserine compared with poly(glycine-serine), arguing for only little effect of amino acid sequence on local chain dynamics (Figure 22.8). The effect of other amino acids on local chain dynamics was measured by performing TTET experiments on short host–guest peptides of the canonical sequence Xan-Ser-Xaa-Ser-NAla-Ser-Gly using the guest amino acids Xaa = Gly, Ser, Ala, Ile, His, Glu, Arg, and Pro [54]. Figure 22.9 shows that the amino acid side chain indeed has only little effect on the rates of contact formation. All amino acids except proline and glycine show very similar dynamics. Interestingly, there is a small but significant difference in rate between short side chains (Ala, Ser) and amino acids with longer side chains (Ile, Glu, Arg, His). Obviously, chains that extend beyond the C_β-atom slightly decrease the rates

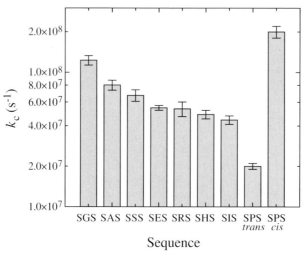

Fig. 22.9. Effect of amino acid sequence on local chain dynamics measured in host–guest peptides with the canonical sequence Xan-Ser-Xaa-Ser-NAla-Ser-Gly. The guest amino acid Xaa was varied and the rate constants for the different guest amino acids are displayed. Data taken from Ref. [54].

of local chain dynamics, whereas charges do not influence the dynamics. Glycyl and prolyl residues show significantly different dynamics compared with all other amino acids, as expected from their largely different conformational properties [19].

As shown before (Figure 22.8) glycine accelerates contact formation about two- to threefold compared with serine. Proline shows slower and more complex kinetics of contact formation with two rate constants of $k_1 = 2.5 \cdot 10^8$ s^{-1} and $k_2 = 2 \cdot 10^7$ s^{-1} and respective amplitudes of $A_1 = 20 \pm 5\%$ and $A_2 = 80 \pm 5\%$. This essentially reflects the cis–trans ratio at the Ser-Pro peptide bond in our host–guest peptide, which has a cis content of $16 \pm 2\%$ as determined by 1D ^{1}H-NMR spectroscopy using the method described by Reimer et al. [72]. Since the rate of cis–trans isomerization is slow ($\tau \sim 20$ s at 22 °C) there is no equilibration between the two isomers on the time scale of the TTET experiments. This allows the measurement of the dynamics of both the trans and the cis form. The results show that the two isomers significantly differ in their dynamic properties of $i, i+4$ contact formation and that the cis-prolyl isomer actually shows the fastest rate of local contact formation of all peptides (Figure 22.9).

22.4.2.4 Effect of the Solvent on Intrachain Diffusion

The dynamics of short poly(glycine-serine) peptides measured with the xanthone–naphthalene TTET pair in water [54] were about 3 times faster than the previously measured rates in the same peptides using the thioxanthone–naphthalene pair in EtOH [50], although both systems were shown to be diffusion-controlled. Figure 22.10 shows that this difference can be attributed to solvent effects [54]. ln k_c linearly decreases with increasing EtOH concentration in a Xan-(Gly-Ser)$_4$-NAla-Ser-Gly peptide (Figure 22.10A). EtOH is a better solvent for polypeptide chains than water and should thus lead to a more extended ensemble of unfolded states. This model was supported by measuring the effect of GdmCl and urea on the contact rates in aqueous solutions. Both denaturants show similar effects on intrachain diffusion as EtOH, with a linear decrease in ln k_c with increasing denaturant concentration. Interestingly, the change in ln k_c with denaturant concentration ($m_c = \partial \ln k_c / \partial$ [Denaturant]) is twofold higher for GdmCl compared with urea, which essentially corresponds to their relative strength in unfolding proteins [73] (see below). Similar m_c-values for EtOH, urea and GdmCl were observed in a Xan-(Ser)$_9$-NAla-Ser-Gly peptides. [54] The observed effect is significantly stronger than expected from the increased solvent viscosity in concentrated GdmCl and urea solutions [50, 74].

The effect of GdmCl and urea on the chain dynamics suggested that these cosolvents may significantly change the chain properties. This was supported by measurements of the effect of GdmCl on the distance dependence of intrachain contact formation. Figure 22.10B compares the effect of loop size on the kinetics of contact formation in the Xan-(Gly-Ser)$_n$-NAla-Ser-Gly series in 8 M GdmCl and water. At high denaturant concentrations the switch from the length-independent dynamics to the length-dependent regime occurs already for formation of shorter loops. This indicates decreased chain stiffness in 8 M GdmCl compared with

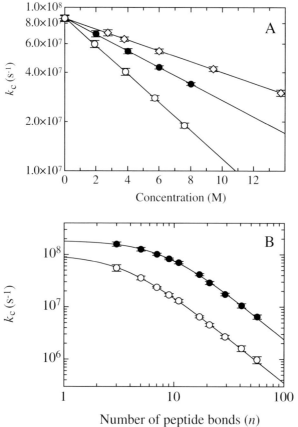

Fig. 22.10. A) Effect of various co-solvents on the dynamics of intrachain contact formation in a Xan-(Gly-Ser)$_4$-NAla-Ser-Gly peptide measured by TTET. Measurements were performed in aqueous solutions in the presence of ethanol (diamonds), urea (filled circles) and GdmCl (open circles). B) Effect of donor–acceptor distance (n) on the rate constant of contact formation in series of poly(Gly-Ser) peptides (cf. Figure 22.8). The rate constants for contact formation in water (filled circles) are compared with the values in 8 M GdmCl (open circles). Data taken from Ref. [54].

water, although the dynamics are significantly slowed down in the presence of GdmCl both for formation of short and long loops (Figure 22.10). For long peptides the effect of chain length on the rates of contact formation is similar in 8 M GdmCl and water with $k_c \sim n^{-1.8 \pm 0.1}$. These results revealed that denaturants like GdmCl and urea slow down local chain dynamics but lead to more flexible chains that behave like ideal polymers already at shorter donor–acceptor distances (Figure 22.10B). These seemingly contradicting findings can be rationalized based on the effect of denaturants on the conformational properties of polypeptide chains. Unlike water, solutions with high concentrations of denaturants represent good solvents for polypeptide chains. This reduces the strength of intramolecular interactions like hydrogen bonds and van der Waals interactions relative to peptide–solvent interactions, which makes unstructured polypeptide chains more flexible and leads to a behavior expected for an unperturbed chain.

In agreement with this interpretation the effect of loop size on the rates of contact formation in 8 M GdmCl is close to the behavior of an unperturbed polypeptide chain predicted by Flory and coworkers [21] (see Figure 22.2). The decreased rate of contact formation at high denaturant concentrations can only in part be explained by an increased solvent viscosity. Additional effects like increased donor–acceptor distance, which is expected in good solvents compared to water and denaturant binding might also contribute to the decreased rate constants [74].

22.4.2.5 Effect of Solvent Viscosity on Intrachain Diffusion

Intrachain contact formation was shown to be strongly viscosity dependent as expected for a diffusional reaction. Viscosity effects on contact formation rates have been measured in TTET experiments on poly(Gly-Ser) peptides [50] (F. Krieger and T. Kiefhaber, unpublished). In all peptides a linear dependency of $\log k_c$ vs. $\log \eta$ was observed and the dependencies could be fitted with the empirical relationship

$$k = \frac{k_0}{\eta^{-\beta}} \tag{17}$$

with $\beta \leq 1$ (Figure 22.11). In the limit for long chains ($n > 15$) the rate of contact formation was found to be inversely proportional to the viscosity ($\beta = 1$) independent of the co-solvent used to alter the viscosity of the solution. In the limit for

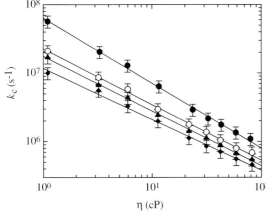

Fig. 22.11. Viscosity dependence of the rate of contact formation for poly(Gly-Ser)$_x$ peptides containing $x = 1$ (filled circles), 2 (open circles), 3 (filled triangles) and 4 (filled diamonds) glycine-serine pairs between thioxanthone and naphthylalanine (cf. Figure 22.3B). Linear fits of the double logarithmic plot of the data give slopes of -0.96 ± 0.05, -0.83 ± 0.05, -0.80 ± 0.05, and -0.81 ± 0.05, respectively. Experiments were carried out in ethanol/glycerol mixtures. Data were taken from Ref. [50].

short chains ($n < 10$), β was smaller than unity (0.8–0.9) [50]. This is not compatible with theoretical consideration which predict $\beta = 1$ as a result from Stokes law relating the solvent viscosity and the diffusion coefficient. β-values < 1 have also been observed for motions of simpler polymers in organic solvents [75–77] and for dynamics of native proteins [78–80].

Several origins have been proposed to explain such behavior like either position-dependent [81] or frequency-dependent [82, 83] friction coefficients. In addition, experiments on myoglobin [80] suggested that deviations from the Stokes law can also be caused by the structure of protein–solvent interfaces, which preferably contain water molecules and have the viscous co-solvent molecules preferentially excluded [84]. This results in a smaller microviscosity around the protein compared with bulk solvent. Additional experiments will be needed to identify the origin of fractional viscosity dependencies of dynamics in short chains.

22.4.2.6 End-to-end Diffusion vs. Intrachain Diffusion

When side-chain contacts are formed within a polypeptide chain during the folding process, the residues are commonly not located near the end of the polypeptide chain. Thus, the end-to-end contact formation experiments represent a rather specific and rare situation during the folding process. Based on the location of the contact sites, three categories of contact formation events in polymers can be distinguished (Figure 22.12). Type I corresponds to end-to-end contact formation. Type II corresponds to end-to-interior contacts and type III contacts are formed between two position remote from the chain ends. Measurements in long hydrocarbon chains yielded trends that cyclization rates are slowed down by additional groups at the ends of the chain [46] without yielding absolute magnitudes of the effects. The xanthone–NAla TTET system has proved to be suitable to perform such experiments in peptides. TTET experiments in type II and III systems have shown that k_c depends on the size of the additional tail (B. Fierz and T. Kiefhaber, unpublished). The strength of the effect is dependent on the nature of the sequence in the observed segment. In long and flexible chains between the labels (i.e., the long chain limit) the dependence on tail length is weaker than for short or stiff chains (the short chain limit).

Such behavior has been predicted by theory [85]. The observed effect should correlate with the surface of the additional amino acids on the ends, if solvent friction limits chain motions as suggested by the viscosity-dependence of the chain diffusion. However the chain movements should only be slowed down to a limiting value. In this limit, the segmental dynamics are independent on further extension of the chain. This limiting rate of innersegmental diffusion corresponds to the maximal rate at which specific side-chain contacts can be formed in the interior of a protein during folding.

22.4.2.7 Chain Diffusion in Natural Protein Sequences

Recently, contact formation rates have been measured in natural protein sequences from cytochrome *c* [64] and carp parvalbumin [55]. The results from these dynam-

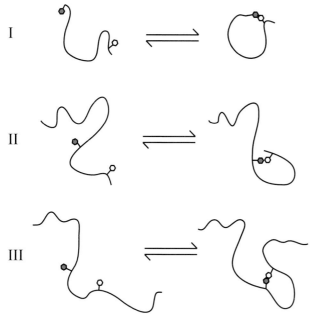

Fig. 22.12. Schematic representation of different types of contact formation in polymer chains. For details see text.

ics are in agreement with dynamics expected from the host–guest studies shown in Figure 22.9. The 18-amino-acid EF-loop of carp parvalbumin connects two α-helices and brings two phenylalanine residues into contact in the native protein, which were replaced by xanthone and naphthylalanine to measure TTET kinetics (Figure 22.13) [55]. The measured kinetics were single exponential (Figure 22.14) and the time constant for contact formation ($\tau = 50$ ns) was comparable to the dynamics for polyserine chains of the same length (Figure 22.8). The EF-loop contains several large amino acids like Ile, Val, and Leu but also four glycyl residues. Obviously, the effects of slower chain dynamics of bulkier side chains and faster dynamics around glycyl residues compensate, which leads to dynamics comparable to polyserine chains. These results indicate that polyserine is a good model to estimate the dynamics of glycine-containing loop sequences in proteins.

The Zn-porphyrine/Ru system was used to measure contact formation in cytochrome c unfolded in 5.4 M GdmCl [64]. The donor and quencher groups were separated by 15 amino acids (Figure 22.13) and contact formation was significantly slower ($\tau = 250$ ns) than for polyserine peptide chains of the same length. However, correcting for the effects of denaturant concentration (see Figure 22.10) and end-extensions (Section 22.4.2.6) on chain dynamics gives rate constants for the cytochrome c sequence which are comparable to the parvalbumin EF-loop in water.

A

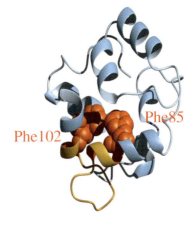

B

Residues 85-102 in carp parvalbumin:

Chemically synthesized fragment with TTET labels:

donor *acceptor*

C

Residues 14-33 in horse heart cytochrome c with triplet quenching labels:

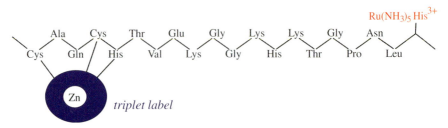

quencher: Ru(NH$_3$)$_5$His^{3+}

triplet label

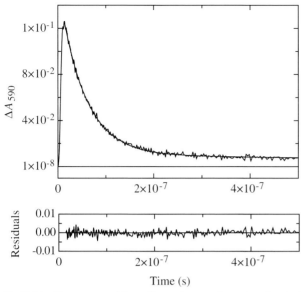

Fig. 22.14. Time course of formation and decay of xanthone triplets in parvalbumin loop fragment 85–102 after a 4 ns laser flash at $t = 0$. The change of xanthone triplet absorbance is measured at 590 nm. The dynamics of contact formation can be described by a single exponential with a time constant of 54 ± 3 ns. Data are taken from Ref. [55].

22.5
Implications for Protein Folding Kinetics

22.5.1
Rate of Contact Formation during the Earliest Steps in Protein Folding

The results on the time scales of intrachain diffusion in various unfolded polypeptide chains show that the amino acid sequence has only little effect on local dynamics of polypeptide chains. All amino acids show very similar rates of end-to-end diffusion with time constants between 12 and 20 ns for the formation of $i, i + 4$

Fig. 22.13. A) Ribbon diagram of the structure of carp muscle β-parvalbumin [127]. Phe85 and Phe102 are shown as space-fill models. The phenylalanine residues have been replaced by the triplet donor and acceptor labels, xanthonic acid and naphthylalanine, respectively. The figure was prepared using the program MolMol [128] and the PDB file 4CPV [127]. B) Sequence of the carp muscle β-parvalbumin EF loop region (residues 85–102) and the synthesized fragment labeled with the two phenylalanine residues at position 85 and 102 replaced by donor (xanthone) and acceptor (naphthylalanine) groups for triplet–triplet energy transfer. C) Sequence from cytochrome c between the Zn-pophyrine group (excited triplet group) and the Ru-His complex used in triplet quenching experiments reported by Chang et al. [64].

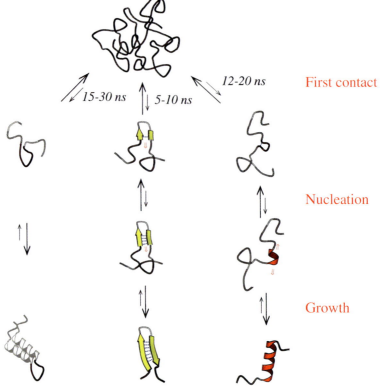

Fig. 22.15. Schematic representation of the time constants for the first steps in formation of loops, β-hairpins and α-helices during protein folding derived from the data measured by TTET in water. Adapted from Ref. [54].

contacts. Polypeptide chains are significantly more flexible around glycine ($\tau = 8$ ns) and stiffer around prolyl residues ($\tau = 50$ ns for the *trans* isomer). Presumably due to the shorter chain dimensions resulting from a *cis* peptide bond the rates of $i, i+4$ contact formation are fastest around the *cis* Ser-Pro bond ($\tau = 4$ ns). These results allow to set an upper limit for the rates of formation of the first productive local contacts during protein folding (Figure 22.15). They can directly be used to estimate the kinetics of loop formation, which are typically of the size of 6–10 amino acids [86]. The results suggest that loops can form with time constants of about 15 ns for glycine-rich loops and 30–40 ns for stiffer loops. β-hairpins, which are the most local structures in proteins, are often rich in glycine and proline [87]. Gly-Ser is actually one of the most frequent sequences found in hairpin loops. The results indicate that the time constants for the first steps in the formation of the tightest turns with $i, i+3$ contacts are around 5 ns for Gly and *cis* Xaa-Pro (Figures 22.8 and 22.9). In glycine- and proline-free turns these rates are slowed down to about 10–20 ns, depending on the amino acid sequence.

Formation of α-helices is most likely initiated by formation of a helical turn, which involves formation of an $i, i + 4$ interaction [88, 89]. Since helices are usually free of glycyl and prolyl residues the initiation cannot occur faster than in about 12–20 ns (Figures 22.8 and 22.9). It should be noted that these values do not represent time constants for nucleation of helices and hairpins (Figure 22.15), which most likely requires formation of more than one specific interaction and most likely encounters additional entropic and enthalpic barriers. The results on the rates of intrachain contact formation rather represent an upper limit for the dynamics of the earliest steps in secondary structure formation (i.e., < for formation of the first contact during helix nucleation). They represent the prefactors (k_0) in the rate equation for secondary structure formation

$$k = k_0 \cdot e^{-\Delta G^{\circ\ddagger}/RT} \tag{18}$$

The measured contact rates thus allow us to calculate the height of the free energy barriers ($\Delta G^{\circ\ddagger}$) when they are compared to measured rate constants (k) for secondary structure formation. However, up to date no direct data on helix or hairpin formation in short model peptides are available. The dynamics of secondary structures were mainly studied by relaxation techniques like dielectric relaxation [90], ultrasonic absorbance [91], or temperature jump [92–94] starting from predominately folded structures (for a review see Ref. [95]). Since helix–coil and hairpin–coil transitions do not represent two-state systems, neither thermodynamically nor kinetically, the time constants for unfolding can not be directly related to rate constants of helix formation [96]. However, the observation of relaxation times around 1 μs for helix unfolding allowed the estimation of time constants for the growth steps of nucleated helices of about 1 to 10 ns, depending on the experimental system [90, 91, 93].

It may be argued that intrachain contacts in a polypeptide chain with a stronger bias towards folded structures can form faster than the observed dynamics in unstructured model polypeptide chains. However, weak interactions like van der Waals contacts between side chains and hydrogen bonds should dominate the early interactions during the folding process and the data on the dynamics of unfolded chains will provide a good model for the earliest events in folding. A stronger energy bias towards the native state will mainly increase the strength or the number of these interactions but not their dynamics of formation.

22.5.2
The Speed Limit of Protein Folding vs. the Pre-exponential Factor

Many small single domain proteins have been found to fold very fast, some of them even on the 10 to 100 μs time scale (for a review see Ref. [97]). These fast folding proteins include α-helical proteins like the monomeric λ-repressor [98, 99] or the engrailed homeodomain [100], β-proteins like cold shock protein of Bacillus subtilis or B. cacodylicus [101] or the WW-domains [102] and α,β-proteins like the single chain arc repressor [103]. There has been effort to design proteins that fold even faster [104, 105]. A folding time constant of 4.1 μs has been reported for a designed Trp-cage.

These very fast folding proteins sparked interest in finding the speed limit for protein folding, which is closely related to the speed limit of the fundamental steps of protein folding and thus also to the dynamics of the unfolded chain. The results on dynamics of contact formation in short peptides showed that no specific interactions between two points on the polypeptide chain can be formed faster than on the 5–10 ns time scale. Considering that dynamics are slowed down if the two groups are located in the interior of a long chain this time constant increases to about 50 ns. Formation of intrachain interactions represents the elementary step in the conformational search on the free energy landscape. Thus, the dynamics of this process set the absolute speed limit for the folding reaction. However, it will not necessarily represent the prefactor (k_0) for the folding process which represents the maximum rate for protein folding in the absence of free energy barriers ($\Delta G^{\circ\ddagger} = 0$, see Eq. (18)). It was shown that even fast apparent two-state folding proceeds through local minima and maxima on the free energy landscape [106, 107] and up to date it is unclear which processes contribute to the rate-limiting steps in protein folding (see Chapters 12.1, 12.2 and 14).

The transition state for folding seems to be native-like in topology but still partly solvated [108–111], Thus, both protein motions and dynamics of protein–solvent interactions in a native-like topology might contribute to the pre-exponential factor and k_0 will probably depend on the protein and on the location of the transition state along the reaction coordinate, which may change upon change in solvent conditions or mutation (see Section 5.3) [44, 106, 107].

22.5.3
Contributions of Chain Dynamics to Rate- and Equilibrium Constants for Protein Folding Reactions

As discussed above, the pre-exponential factor (k_0) for a protein folding reaction should contain major contributions from the rate of intrachain diffusion in an unfolded or partially folded polypeptide chain [112]. Thus, factors that influence intrachain diffusion, like denaturant concentration (Figure 22.10), will also affect the pre-exponential factor for folding. This will influence the observed folding rate constants and thus also the observed equilibrium constants [112]. A method that allows a quantitative description of reaction rates for processes in solution was developed by Kramers [113]. He considered a particle moving on an energy surface with two metastable states (e.g., native and unfolded protein) separated by a barrier (Figure 22.16). The motion of the particle is diffusional and can be described by the Langevin equation (Eq. (8)). The thermal motion of the solvent is modeled as the random force. Solving the diffusion equations, Kramers observed three different regimes which differ in the strength of interaction between system and environment. In the low friction limit the interactions are only weakly coupled. In this limit the rate of the barrier crossing reaction from the educt to the product state increases with increasing friction γ. In the intermediate friction limit the rate is independent on friction and the results from transition state theory and from Kramers' theory approach each other. In the high friction limit, which should cor-

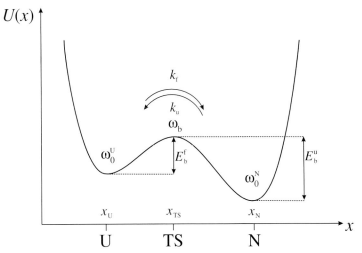

Fig. 22.16. Potential U(x) with two states N and U separated by an energy barriers. Escape from state U to N occurs via the rates k_f and k_u. E_b^f and E_b^u are the activation energies for the forward and back reaction. ω denotes the frequencies of phase point motion at the extremes of the potential. The drawing was adapted from Ref. [114].

respond to most reactions in solution and also to protein folding reactions, the rate of the reaction decreases when the friction is further increased. In this limit the equation for the rate as a function of friction, barrier height and temperature can be written as:

$$k = \frac{\omega_0 \omega_B}{2\pi\gamma} e^{(-E_b/kT)} \tag{19}$$

where ω stands for the frequency of motion of the particle in the starting well (ω_0) and on top of the barrier (ω_B). ω depends on the shape of the barrier and on the nature of the free energy surface [114]. By applying Stokes law this equation can then readily be expressed in terms of thermodynamic parameters and solvent viscosity.

$$k = \frac{C}{\eta(T)} \cdot \exp(-\Delta G^{\circ\ddagger}/RT) = k_0 \cdot \exp(-\Delta G^{\circ\ddagger}/RT) \tag{20}$$

It is not clear whether Kramers' approach in its simple form as described above is directly applicable to protein folding reactions, since it is difficult to measure the viscosity-dependence of folding and unfolding rate constants. The co-solvents used to modify solvent viscosity like glycerol, sucrose, or ethylene glycol generally stabilize the native state [84] and in part also the transition state of folding. This makes a quantification of the effect of solvent viscosity on protein folding rate constants difficult [115, 116].

The results from the viscosity dependence of end-to-end contact formation show that the rate constants in short peptides ($n < 10$) are not proportional to η^{-1} but rather to $\eta^{-0.8}$. This suggests that the presence of barriers in these peptides leads to a more complex behavior than expected from Eq. (20) and indicates that the frictional coefficient from Kramers' theory (γ, see. Eq. (19)) is not only determined by solvent viscosity for protein folding reactions.

Application of Kramers' theory to protein folding reactions shows the problems with quantitatively characterizing the barriers for protein folding. It is very difficult to determine pre-exponential factors for the folding and unfolding reaction, which are most likely different for each protein. Since the pre-exponential factors depend on the shape of the potential and on the dynamics in the individual wells (states), ω_0 will be different for the refolding reaction starting from unfolded protein (ω_0^U) and for the unfolding reaction starting from native protein (ω_0^N) and hence also the prefactors will be different for the forward and backward reactions. However, under all conditions Eq. (21) must be valid [117]:

$$K_{eq} = \frac{k_f}{k_u} \tag{21}$$

Thus, the equilibrium constant (K_{eq}) contains contributions from the pre-exponential factors for folding (k_0^f) and unfolding (k_0^u) and from differences in $\Delta G^{\circ\ddagger}$, according to [112]

$$K_{eq} = \frac{k_0^f \cdot e^{-\Delta G_f^{\circ\ddagger}/RT}}{k_0^u \cdot e^{-\Delta G_u^{\circ\ddagger}/RT}} \tag{22}$$

The contribution of differences in the prefactors for folding and unfolding to $\Delta G^\circ (= -RT \ln K_{eq})$ are consequently [112]

$$\Delta G^\circ = \Delta G_f^{\circ\ddagger} - \Delta G_u^{\circ\ddagger} + \Delta G_{pref}^\circ$$

$$\Delta G_{pref}^\circ = RT \ln \frac{k_0^u}{k_0^f} \tag{23}$$

Differences in k_0 for folding and unfolding can thus significantly contribute to apparent protein folding barriers. A fivefold difference in k_0 would contribute 4 kJ mol^{-1} to the experimentally determined barrier height at 25 °C. Consequently, also changes in the dynamics of the unfolded state, which change the pre-exponential factor for the folding reaction by changing ω_0, will influence K_{eq}, even if the heights of the barriers are not changed. For many proteins significant residual structure in the unfolded state has been observed either directly by NMR [27, 28, 34, 43] or from the analysis of the denaturant dependence [44]. It was shown that this residual structure is sensitive to mutation, which can lead to significant changes in the solvent accessibility [44] and in the dynamics [39] of the unfolded state. This will lead to changes in k_0 for the folding reaction (k_0^f) by changing ω_0^U and will result in a change of the folding rate constant (k_f) and consequently also in the apparent free energy of activation, $\Delta G_f^{\circ\ddagger}$(app) even if $\Delta G_f^{\circ\ddagger}$ is not affected by the mutations [112].

$$\Delta\Delta G_f^{o\ddagger}(\text{app}) = \Delta\Delta G_f^{o\ddagger} + RT \ln \frac{k_0^f(\text{wt})}{k_0^f(\text{mutant})} \qquad (24)$$

As in the example above, a fivefold increase in k_0 of a mutant would apparently decrease $\Delta G_f^{o\ddagger}$ by 4 kJ mol^{-1}. This would have large effects on the interpretation of the results in terms of transition state structure (ϕ-value analysis, see Chapters 12.2 and 13), and might contribute to the large uncertainty in ϕ-value from mutants that lead to small changes in $\Delta G°$ [111].

Mutants that disrupt residual interactions in the unfolded state will most likely have only little effect on k_0 for unfolding (k_0^u), which should be determined by chain motions in the native state. Most native protein structures and transitions state structures were shown to be rather robust against mutations [44]. Thus, mutations which change the dynamics of unfolded proteins will result in an apparent change in $\Delta G°$, in analogy to the effects discussed above (Eq. (24)).

Figure 22.10 shows that denaturants like urea and GdmCl significantly decrease the rate constants for intrachain diffusion (k_c). According to the considerations discussed above, this will also affect the denaturant dependence of the rate constants for protein folding (m_f-values with $m_f = \partial \Delta G_f^{o\ddagger}/\partial[\text{Denaturant}]$). Even if pre-equilibria involving unstable intermediates may dominate the early stages of folding, the observed denaturant dependence of intrachain diffusion will influence these steps by changing the forward rate constants of these equilibria. The m_c-values ($m_c = \partial(-RT \ln k_c)/\partial[\text{Denaturant}]$ for peptide dynamics vary only little with chain length and show values around 0.35 kJ mol^{-1} M^{-1} for urea and 0.50 kJ mol^{-1} M^{-1} for GdmCl [74]. Typical m_f-values for folding of the smallest fast folding proteins with chain length between 40 and 50 amino acids are around 1.0 kJ mol^{-1} M^{-1} for urea, indicating that up to 30% of the measured m_f-values may arise from contributions of chain dynamics. Larger two-state folders consisting of 70–100 amino acids typically show m_f-values around 3 kJ mol^{-1} M^{-1} for urea and 5 kJ mol^{-1} M^{-1} for GdmCl [107, 118], indicating that the denaturant dependence of chain dynamics constitutes up to 10% of the experimental m_f-values.

It is difficult to judge the effect of denaturants on the dynamics of the native state, which will influence the experimental m_u-values. However, it was observed that the effect of denaturants on chain dynamics is mainly based on increased chain dimensions (A. Möglich and T. Kiefhaber, unpublished results). This would argue for only little effects of denaturants on the internal dynamics of the native state and on the transition state structures, which were shown to be structurally robust against changes in denaturant concentration [44, 107, 111]. Thus, the m_u-values should have only little contributions from the effects of denaturants on chain dynamics.

Comparison of kinetic and equilibrium m-values ($m_{eq} = \partial \Delta G°/\partial[\text{Denaturant}]$) is frequently used to characterize protein folding transition states according to the rate-equilibrium free energy relationship [44, 119, 120]:

$$\alpha_D = \frac{m_f}{m_{eq}} \qquad (25)$$

α_D is interpreted as a measure for the solvent accessibility of the transition state. The contributions from chain dynamics to the m_f-values of small proteins lead to apparently higher α_D-values and thus to apparently less solvent exposed transition states. This might explain the commonly observed higher α_D-values compared to α_C-values ($\alpha_C = \dfrac{\Delta C_p^{\circ\ddagger}}{\Delta C_p^{\circ}}$), which are also believed to monitor the solvent accessibility of transition states [44, 120]. Temperature dependence of intrachain diffusion in unstructured peptides showed that this process is not associated with a measurable change in heat capacity (F. Krieger and T. Kiefhaber, unpublished results). This suggests that α_C-values give a more reliable picture of the solvent accessibility of the transition state than α_D-values.

This shows that changes in the dynamics of the unfolded state of a protein lead to changes in the prefactor for the folding reaction which has consequences for both the rate constant and the equilibrium constant of the folding reaction. In general, we cannot expect the pre-exponential factors to be independent of the solvent conditions such as denaturant concentration, temperature, and pressure. This will contribute to the experimentally determined m-values and the activation parameters ($\Delta H^{\circ\ddagger}, \Delta S^{\circ\ddagger}, \Delta V^{\circ\ddagger}$, and $\Delta C_p^{\ddagger}$) and to the equilibrium properties of the protein.

22.6
Conclusions and Outlook

Several experimental systems have been recently developed to measure rate constants for intrachain diffusion processes. Experimental studies on the dynamics of unfolded proteins have revealed single exponential kinetics for formation of specific intrachain interactions on the nanosecond time scale and have been able to elucidate scaling laws and sequence dependence of chain dynamics. It has further been shown that the results derived from homopolypeptide chains and host-guest studies are in agreement with dynamics of loop formation in short natural sequences. Further studies should aim at the investigation of the dynamics in full length unfolded proteins and in folding intermediates to obtain information on the special equilibrium and dynamic properties of free energy landscapes for protein folding reactions.

22.7
Experimental Protocols and Instrumentation

To measure absolute rate constants for contact formation several requirements have to be met by the experimental method and by the equipment for detection. The methods of choice to study contact formation are electron transfer or excited state quenching reactions. The main advantage of energy transfer systems is that they allow both the donor and the acceptor populations to be quantified during the experiment (see Figure 22.5).

22.7.1
Properties of the Electron Transfer Probes and Treatment of the Transfer Kinetics

The lifetime of excited states of the labels used in a specific system determine the time scale on which experiments can be performed. Triplet states are very long lived compared to fluorescence probes and thus allow studies in long chains and in unfolded proteins. Donor groups which undergo very fast intersystem crossing to the triplet state and which have a high quantum yield for intersystem crossing are well-suited. The xanthone derivatives used by Bieri et al. [50] and Krieger et al. [54, 55] form triplet states with a time constant of 2 ps and a quantum yield of 99% [56]. Tryptophan in contrast, which was used by Lapidus et al., has about 18% quantum yield [60] and forms triplet states with a time constant of 3 ns [62], which sets a limit to the fastest processes that can be monitored using tryptophan as a triplet donor. Triplet states usually have strong absorption bands that are easily observable. It is, however, important that the donor has an UV absorption band in a spectral region where no other part of the protein absorbs (i.e., >300 nm). This allows selective excitation of the donor. The requirements for the triplet acceptor are that the molecule should have lower triplet energy than the donor to enable exothermic TTET and it should not absorb light at the excitation wavelength of the donor. Additionally, no radicals should be formed at any time in the reaction, since they may damage the sample and their strong absorbance bands usually interfere with the measurements.

Further requirements of experimental systems which allow to measure the kinetics of contact formation on an absolute time scale are (i) that the process is very strongly distance dependent to allow transfer only at van der Waals distance between donor and acceptor and (ii) that the transfer reaction is faster than dissociation of the complex so that each van der Waals contact between the labels leads to transfer. The coupling of electron transfer to chain dynamics displayed in Figure 22.6 can be kinetically described by a three-state reaction if the excitation of the donor is fast compared to chain dynamics (k_c, k_{-c}) and electron transfer (k_{TT}).

$$O \underset{k_{-c}}{\overset{k_c}{\rightleftharpoons}} C \overset{k_{TT}}{\longrightarrow} C^* \tag{26}$$

O and C represent open chain conformations (no contact) and contact conformations of the chain, respectively. k_c and k_{-c} are apparent rate constant for contact formation and breakage, respectively. If the excited state quenching or transfer (k_{TT}) reaction is on the same time scale or slower than breaking of the contact (k_{-c}), the measured rate constants (λ_i) are functions of all microscopic rate constants.

$$\lambda_{1,2} = \frac{B \pm \sqrt{B^2 - 4C}}{2}$$
$$B = k_c + k_{-c} + k_{TT} \tag{27}$$
$$C = k_c \cdot k_{TT}$$

If the contact states have a much lower probability than the open conformations ($k_{-c} \gg k_c$) the kinetics will be single exponential and the observed rate constant corresponds to the smaller eigenvalue (λ_1) in Eq. (27). Since this is in agreement with experimental results we will only consider this scenario in the following. We can further simplify Eq. (27) if the time scales of electron transfer and chain dynamics are well separated. In the regime of fast electron transfer compared to chain dynamics ($k_{TT} \gg k_c, k_{-c}$) Eq. (27) can be approximated by

$$\lambda_1 = k_c \tag{28}$$

This is the desired case, since the observed kinetics directly reflect the dynamics of contact formation. In the regime of fast formation and breakage of the contact compared to electron transfer ($k_{TT} \ll k_c, k_{-c}$) Eq. (27) can be approximated by

$$\lambda_1 = \frac{k_c}{k_c + k_{-c}} \cdot k_{TT} \cong \frac{k_c}{k_{-c}} \cdot k_{TT} = K_c \cdot k_{TT} \tag{29}$$

where K_c reflects the ratio of contact conformations (C) over open chain conformations (O). In this limit the chain dynamics can not be measured but the fraction of closed conformations can be determined. It should be kept in mind that these simplifications only hold if the formation of excited states is fast compared to the following reactions. If the excitation is on the nanosecond time scale or slower, the solutions of the linear four-state model have to be used to analyze the kinetics [121, 122].

These considerations show that it is crucial to determine the rate constants for the photochemical processes in the applied system to be able to interpret the kinetic data. The photochemistry of the donor excitation and of the electron transfer process are usually characterized by investigating the isolated labels and the kinetics should be studied on the femtosecond to the nanosecond time scale to gain complete information on all photochemical processes in the system [56] (cf. Figure 22.6). To test whether each donor–acceptor contact leads to transfer it should be tested whether the reaction is diffusion controlled. This can be done by studying the bimolecular transfer process from the donor to acceptor groups and determine its rate constants, its temperature-dependence and its viscosity dependence (F. Krieger and T. Kiefhaber, in preparation). These tests will be described in detail in the following. The results showed, that both the xanthone/NAla system [54–56] and the porphyrin/Ru system [64] fulfill these requirements and thus allow measurements of absolute rate constant for contact formation. The Trp/Cys triplet quenching system, in contrast, is not diffusion controlled and gives significantly slower apparent contact rates than the diffusion-controlled systems [59, 63]. Also the DBO/Trp fluorescence quenching system does not seem to be completely diffusion controlled and might have slow electron transfer kinetics, since it gives significantly slower dynamics [65] compared to the xanthone/NAla system in the same peptides [54].

22.7.2
Test for Diffusion-controlled Reactions

22.7.2.1 Determination of Bimolecular Quenching or Transfer Rate Constants

To test whether an energy transfer or energy quenching pair has diffusion-controlled kinetics, the rate constant of energy transfer (k_q) should be measured using the free labels in solution under pseudo first-order conditions. In these experiments the concentration of the quencher or acceptor [Q] should be at least 10 times higher than the concentration of the donor to be in the pseudo first order regime. Since [Q] is approximately constant during the experiment, the apparent first-order rate constant (k) under pseudo first-order conditions is given by [123]

$$k = k_q \cdot [Q] \tag{30}$$

Thus, the bimolecular transfer constant (k_q) can be obtained by varying the quencher concentration [Q] and analyzing the data according to the Stern-Vollmer equation [124] (see Figure 22.17):

$$k = k_0 + k_q[Q] \tag{31}$$

Here, k_0 denotes the rate constant for triplet decay of the donor in absence of quencher (acceptor) and k the time constant in presence of the acceptor or quencher. For diffusion-controlled reactions of small molecules in water k_q is around 4–$6 \cdot 10^9$ s^{-1}.

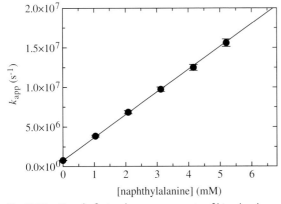

Fig. 22.17. Pseudo first-order measurements of bimolecular TTET from xanthonic acid to naphthylalanine in water. Xanthone concentration was 30 µM. The slope gives a bimolecular transfer constant (k_q) of $3 \cdot 10^9$ M^{-1} s^{-1} (see Eq. (31)). The data were discussed in Ref. [54].

22.7.2.2 Testing the Viscosity Dependence

For a diffusion-controlled reaction, k_q is inversely proportional to solvent viscosity ($k \sim 1/\eta$). The viscosity is usually varied by adding co-solvents like ethylene glycol or glycerol. To determine the final macroscopic viscosity instruments like the falling-ball viscosimeter or the Ubbelohde viscosimeter are used. The molecular size of the co-solvent used in a viscosity study is an important factor. The macroscopic viscosity of a glycerol/water mixture and a solution of polyethylene glycol might be the same, although the microscopic properties of these solutions are very different [80]. In the latter case the dissolved reactant molecules do only feel a fraction of the macroscopic viscosity. Thus it is advisable to use viscous co-solvents of small molecular weight. Glycerol/water mixtures have proven to be most useful because the viscosity can be varied from 1 cP (pure water) over three orders of magnitude by addition of glycerol. It is imperative to control the temperature when working with high concentrations of glycerol because the viscosity is strongly temperature dependent. Note that viscous co-solvents can interfere with the reaction that is studied.

In protein folding studies, the polyols used to vary the viscosity tend to stabilize the native state of proteins [84] making an analysis of the kinetic data difficult. However, for measurements of chain dynamics in unfolded polypeptide chains this effect should not interfere with the measurements as long as the co-solvents do not induce a structure. Specific effects of the polyols on chain conformations/dynamics can be tested by comparing the results from different co-solvents.

22.7.2.3 Determination of Activation Energy

Diffusion-controlled reactions typically have activation energies close to zero. The lack of an activation barrier in a diffusion controlled quenching reaction can be verified by determining the temperature dependence of k_q. The diffusion coefficients of the reactants are, however, temperature dependent, mainly through the effect of temperature on solvent viscosity according to the Stokes-Einstein equation.

$$D = \frac{k_B T}{6\pi r \eta} \tag{32}$$

After correcting for the change in viscosity of the solvent with temperature, the slope in an Arrhenius plot should not be higher than $k_B T$. As any photophysical reaction takes some time, diffusion control is only possible to a certain maximal concentration of quencher molecules. After that the photochemistry of the quenching or transfer process becomes rate limiting. This provides an upper limit of the rate constants that can be measured, even if the system is diffusion controlled at lower quencher concentrations. For the xanthone/naphthalene system TTET has been shown to occur faster than 2 ps and thus this system is suitable to obtain absolute time constants for all processes slower than 10–20 ps [56].

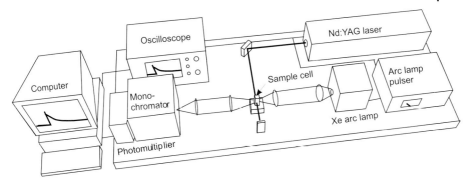

Fig. 22.18. Schematic representation of a laserflash set-up used to measure electron transfer reactions.

22.7.3
Instrumentation

The instrumentation required to perform experiments on contact formation kinetics consists of a high-energy light source to produce excited states and a detection mechanism (Figure 22.18). A pulsed laser is used to produce triplet donor states in TTET or triplet quenching experiments. The duration of the light pulse has to be very short (i.e., shorter than the time scale of the reaction of interest). Additionally, the excitation pulse has to provide enough energy to excite the major portion of the molecules in the sample to order to generate a large signal. A pulsed Nd:YAG laser with a pulse width of <5 ns and a pulse energy of ~100 mJ is well-suited for these purposes. Transient UV absorption is used to detect the triplet states in TTET or triplet quenching experiments. A pulsed flash lamp generates enough light for the absorption measurements. The lamp intensity typically stays at a constant plateau value for several hundred microseconds, which is sufficient for TTET measurements. A monochromator is used to monitor single wavelengths. Transient spectra can be reconstructed from measurements at different wavelengths or by using a CCD camera.

Acknowledgments

We thank Annett Bachmann, Florian Krieger, Robert Kübler and Andreas Möglich for discussion and comments on the manuscript.

References

1 EINSTEIN, A. (1906). Eine neue Bestimmung der Molekül-dimensionen. *Ann. d. Phys.* 19, 289–306.

2 VON SMOLUCHOWSKI, M. (1906). Zur kinetischen Theorie der Brownschen Molekularbewegung und der Suspensionen. *Ann. d. Phys.* 21, 756–780.

3 VON SMOLUCHOWSKI, M. (1916). Drei Vorträge über Diffusion, Brownsche Molekularbewegung und Koagulation von Kolloidteilchen. *Phys. Z.* 17, 557–571.

4 JACOBSEN, H. & STOCKMAYER, W. H. (1950). Intramolecular reaction in polycondensations. I. The theory of linear systems. *J. Phys. Chem.* 18, 1600–1606.

5 ROUSE, P. E. (1953). A theory of the linear viscoelastic properties of dilute solutions of coiling polymers. *J. Chem. Phys.* 21, 1272–1280.

6 ZIMM, B. (1956). Dynamics of polymer molecules in dilute solutions: viscoelasticity, flow bifringence and dielectric loss. *J. Chem. Phys.* 24, 269–278.

7 SZABO, A., SCHULTEN, K. & SCHULTEN, Z. (1980). First passage time approach to diffussion controlled reactions. *J. Chem. Phys.* 72, 4350–4357.

8 DE GENNES, P. G. (1985). Kinetics of collapse for a flexible coil. *J. Phys. Lett.* 46, L639–L642.

9 FIXMAN, M. (1987). Brownian dynamics of chain polymers. *Faraday Discuss. Chem. Soc.* 83, 199–211.

10 THIRUMALAI, D. (1995). From minimal models to real protein: time scales for protein folding kinetics. *J. Phys.* 5, 1457–1467.

11 FLORY, P. J. (1953). *Principles of Polymer Chemistry.* Cornell University Press, Ithaca.

12 FLORY, P. J. (1969). *Statistical Mechanics of Chain Molecules.* Hanser Publishers, Munich.

13 YAMAKAWA, H. (1971). *Modern Theory of Polymer Solutions.* Harper & Row, New York.

14 DOI, M. (1996). *Introduction to Polymer Science.* Oxford University Press, Oxford.

15 RUBINSTEIN, M. & COLBY, R. H. (2003). *Polymer Physics.* Oxford University Press, Oxford.

16 KUHN, W. (1934). Über die Gestalt fadenförmiger Moleküle in Lösungen. *Kolloid-Z* 52, 269.

17 KUHN, W. (1936). Beziehungen zwischen Molekülgrösse, statistischer Molekülgestalt und elastischen Eigenschaften hochpolymerer Stoffe. *Kolloid-Z* 76, 258.

18 BRANT, D. A. & FLORY, P. J. (1965). The configuration of random polypeptide chains. II. Theory. *J. Am. Chem. Soc.* 87, 2791–2800.

19 SCHIMMEL, P. R. & FLORY, P. J. (1967). Conformational energy and configurational statistics of poly-L-proline. *Proc. Natl Acad. Sci. USA* 58, 52–59.

20 BRANT, D. A., MILLER, W. G. & FLORY, P. J. (1967). Conformational energy estimates for statistically coiling polypeptide chains. *J. Mol. Biol.* 23, 47–65.

21 MILLER, W. G., BRANT, D. A. & FLORY, P. J. (1967). Random coil configurations of polypeptide chains. *J. Mol. Biol.* 23, 67–80.

22 RAMACHANDRAN, G. N. & SASISEKHARAN, V. (1968). Conformation of polypeptides and proteins. *Adv. Protein Chem.* 23, 283–437.

23 DE GENNES, P. G. (1979). *Scaling Concepts in Polymer Physics.* Cornell University Press, Ithaca.

24 DOI, M. & EDWARDS, S. F. (1986). *The Theory of Polymer Dynamics.* Oxford University Press, Oxford.

25 PAPPU, R. V., SRINAVASAN, R. & ROSE, G. D. (2000). The Flory isolated-pair hypothesis is not valid for polypeptide chains: Implications for protein folding. *Proc. Natl Acad. Sci. USA* 97, 12565–12570.

26 EDWARDS, S. F. (1965). The statistical mechanics of polymers with excluded volume. *Proc. Phys. Soc.* 85, 613–624.

27 EVANS, P. A., TOPPING, K. D., WOOLFSON, D. N. & DOBSON, C. M. (1991). Hydrophobic clustering in nonnative states of a protein: interpretation of chemical shifts in NMR spectra of denatured states of lysozyme. *Proteins* 9, 248–266.

28 NERI, D., BILLETER, M., WIDER, G. & WÜTHRICH, K. (1992). NMR determination of residual structure in

a urea-denatured protein, the 434-repressor. *Science* 257, 1559–1563.

29 LOGAN, T. M., THERIAULT, Y. & FESIK, S. W. (1994). Structural characterization of the FK506 binding protein unfolded in urea and guanidine hydrochloride. *J. Mol. Biol.* 236, 637–648.

30 SMITH, C. K., BU, Z. M., ANDERSON, K. S., STURTEVANT, J. M., ENGELMAN, D. M. & REGAN, L. (1996). Surface point mutations that significantly alter the structure and stability of a protein's denatured state. *Protein Sci.* 5, 2009–2019.

31 SARI, N., ALEXANDER, P., BRYAN, P. N. & ORBAN, J. (2000). Structure and dynamics of an acid-denatured protein G mutant. *Biochemistry* 39, 965–977.

32 WONG, K. B., CLARKE, J., BOND, C. J. et al. (2000). Towards a complete description of the structural and dynamic properties of the denatured state of barnase and the role of residual structure in folding. *J. Mol. Biol.* 296, 1257–1282.

33 TEILUM, K., KRAGELUND, B. B., KNUDSEN, J. & POULSEN, F. M. (2000). Formation of hydrogen bonds precedes the rate-limiting formation of persistent structure in the folding of ACBP. *J. Mol. Biol.* 301, 1307–1314.

34 KORTEMME, T., KELLY, M. J., KAY, L. E., FORMAN-KAY, J. D. & SERRANO, L. (2000). Similarities between the spectrin SH3 domain denatured state and its folding transition state. *J. Mol. Biol.* 297, 1217–1229.

35 GARCIA, P., SERRANO, L., DURAND, D., RICO, M. & BRUIX, M. (2001). NMR and SAXS characterization of the denatured state of the chemotactic protein CheY: implications for protein folding initiation. *Protein Sci.* 10, 1100–1112.

36 CHOY, W. Y. & FORMAN-KAY, J. D. (2001). Calculation of ensembles of structures representing the unfolded state of an SH3 domain. *J. Mol. Biol.* 308, 1011–1032.

37 KAZMIRSKI, S. L., WONG, K. B., FREUND, S. M., TAN, Y. J., FERSHT, A. R. & DAGGETT, V. (2001). Protein folding from a highly disordered denatured state: the folding pathway of chymotrypsin inhibitor 2 at atomic resolution. *Proc. Natl Acad. Sci. USA* 98, 4349–4354.

38 YI, Q., SCALLEY-KIM, M. L., ALM, E. J. & BAKER, D. (2000). NMR characterization of residual structure in the denatured state of protein L. *J. Mol. Biol.* 299, 1341–1351.

39 KLEIN-SEETHARAMAN, J., OIKAWA, M., GRIMSHAW, S. B., WIRMER, J., DUCHARDT, E., UEDA, T. et al. (2002). Long-range interactions within a nonnative protein. *Science* 295, 1719–1722.

40 GARDINER, C. W. (1985). *Handbook of Stochastic Methods*. Springer Verlag, Berlin, Heidelberg, New York.

41 SAKANISHI, A. (1968). Dynamic viscoelastic properties of dilute polyisobutylene solutions. *J. Chem. Phys.* 48, 3850–3858.

42 PERKINS, T. T., QUAKE, S. R., SMITH, D. E. & CHU, S. (1994). Relaxation of a single DNA molecule observed by optical microscopy. *Science* 264, 822–826.

43 DYSON, H. J. & WRIGHT, P. E. (2002). Insights into the structure and dynamics of unfolded proteins from nuclear magnetic resonance. *Adv. Protein Chem.* 62, 311–340.

44 SÁNCHEZ, I. E. & KIEFHABER, T. (2003). Hammond behavior versus ground state effects in protein folding: evidence for narrow free energy barriers and residual structure in unfolded states. *J. Mol. Biol.* 327, 867–884.

45 MAR, A. & WINNIK, M. A. (1985). End-to-end cyclization of hydrocarbon chains: temperature effects on an intramolecular phosphorescence quenching reaction in solution. *J. Am. Chem. Soc.* 107, 5376–5382.

46 LEE, S. & WINNIK, M. A. (1997). Cyclization rates for two points in the interior of a polymer chain. *Macromolecules* 30, 2633–2641.

47 STRYER, L. & HAUGLAND, R. P. (1967). Energy transfer: a spectroscopic ruler. *Proc. Natl Acad. Sci. USA* 58, 719–726.

48 HAAS, E., KATCHALSKI-KATZIR, E. & STEINBERG, I. Z. (1978). Brownian

motion at the ends of oligopeptid chains as estimated by energy transfer between chain ends. *Biopolymers* 17, 11–31.

49 BEECHEM, J. M. & HAAS, E. (1989). Simultaneous determination of intramolecular distance distributions and conformational dynamics by global analysis of energy transfer measurements. *Biophys J.* 55, 1225–1236.

50 BIERI, O., WIRZ, J., HELLRUNG, B., SCHUTKOWSKI, M., DREWELLO, M. & KIEFHABER, T. (1999). The speed limit for protein folding measured by triplet-triplet energy transfer. *Proc. Natl Acad. Sci. USA* 96, 9597–9601.

51 HAAS, E., WILCHEK, M., KATCHALSKI-KATZIR, E. & STEINBERG, I. Z. (1975). Distribution of end-to-end distances of oligopeptides in solution as estimated by energy-transfer. *Proc. Natl Acad. Sci. USA* 72, 1807–1811.

52 HAGEN, S. J., HOFRICHTER, J., SZABO, A. & EATON, W. A. (1996). Diffusion-limited contact formation in unfolded cytochrome c: Estimating the maximum rate of protein folding. *Proc. Natl Acad. Sci. USA* 93, 11615–11617.

53 CAMACHO, C. J. & THIRUMALAI, D. (1995). Theoretical predictions of folding pathways by using the proximity rule, with applications to bovine pancreatic inhibitor. *Proc. Natl Acad. Sci. USA* 92, 1277–1281.

54 KRIEGER, F., FIERZ, B., BIERI, O., DREWELLO, M. & KIEFHABER, T. (2003). Dynamics of unfolded polypeptide chains as model for the earliest steps in protein folding. *J. Mol. Biol.* 332, 265–274.

55 KRIEGER, F., FIERZ, B., AXTHELM, F., JODER, K., MEYER, D. & KIEFHABER, T. (2004). Intrachain diffusion in a protein loop fragment from carp parvalbumin. *Chem. Phys.*, in press.

56 SATZGER, H., SCHMIDT, B., ROOT, C. et al. (2004). Ultrafast quenching of the xanthone triplet by energy transfer: new insight into the intersystem crossing kinetics in press.

57 CLOSS, G. L., JOHNSON, M. D., MILLER, J. R. & PIOTROWIAK, P. (1989). A connection between intramolecular long-range electron, hole and triplet energy transfer. *J. Am. Chem. Soc.* 111, 3751–3753.

58 WAGNER, P. J. & KLÁN, P. (1999). Intramolecular triplet energy transfer in flexible molecules: electronic, dynamic, and structural aspects. *J. Am. Chem. Soc.* 121, 9626–9635.

59 LAPIDUS, L. J., EATON, W. A. & HOFRICHTER, J. (2000). Measuring the rate of intramolecular contact formation in polypeptides. *Proc. Natl Acad. Sci. USA* 97, 7220–7225.

60 VOLKERT, W. A., KUNTZ, R. R., GHIRON, C. A., EVANS, R. F., SANTUS, R. & BAZIN, M. (1977). Flash photolysis of tryptophan and N-acetyl-L-tryptophanamide; the effect of bromide on transient yields. *Photochem. Photobiol.* 26, 3–9.

61 GONNELLI, M. & STRAMBINI, G. B. (1995). Phosphorescence lifetime of tryptophan in proteins. *Biochemistry* 34, 13847–13857.

62 BENT, D. V. & HAYON, E. (1975). Excited state chemistry of aromatic amino acids and related peptides: III. Tryptophan. *J. Am. Chem. Soc.* 97, 2612–2619.

63 YEH, I. C. & HUMMER, G. (2002). Peptide loop-closure kinetics from microsecond molecular dynamics simulations in explicit solvent. *J. Am. Chem. Soc.* 124, 6563–6568.

64 CHANG, I.-J., LEE, J. C., WINKLER, J. R. & GRAY, H. B. (2003). The protein-folding speedlimit: Intrachain diffusion times set by electron-transfer rates in denatured Ru(NH$_3$)$_5$(His-33)-Zn-cytochrome c. *Proc. Natl Acad. Sci. USA* 100, 3838–3840.

65 HUDGINS, R. R., HUANG, F., GRAMLICH, G. & NAU, W. M. (2002). A fluorescence-based method for direct measurements of submicroscond intramolecular contact formation in biopolymers: an exploratory study with polypeptides. *J. Am. Chem. Soc.* 124, 556–564.

66 HUANG, F. & NAU, W. M. (2003). A conformational flexibility scale for amino acids in peptides. *Angew. Chem. Int. Ed. Engl.* 42, 2269–2272.

67 NEUWEILER, H., SCHULZ, A., BÖHMER, M., ENDERLEIN, J. & SAUER, M. (2003). Measurement of submicrosecond intramolecular contact formation in peptides at the single-molecule level. *J. Am. Chem. Soc.* 125, 5324–5330.

68 ZWANZIG, R. (1997). Two-state models for protein folding. *Proc. Natl Acad. Sci. USA* 94, 148–150.

69 SPÖRLEIN, S., CARSTENS, H., SATZGER, H. et al. (2002). Ultrafast spectroscopy reveals sub-nanosecond peptide conformational dynamics and validates molecular dynamics simulation. *Proc. Natl Acad. Sci. USA* 99, 7998–8002.

70 BREDENBECK, J., HELBING, J., SIEG, A. et al. (2003). Picosecond conformational transition and equilibration of a cyclic peptide. *Proc. Natl Acad. Sci. USA* 100, 6452–6457.

71 SCHWALBE, H., FIEBIG, K. M., BUCK, M. et al. (1997). Structural and dynamical properties of a denatured protein. Heteronuclear 3D NMR experiments and theoretical simulations of lysozyme in 8 M urea. *Biochemistry* 36, 8977–8991.

72 REIMER, U., SCHERER, G., DREWELLO, M., KRUBER, S., SCHUTKOWSKI, M. & FISCHER, G. (1998). Side-chain effects on peptidyl-prolyl cis/trans isomerization. *J. Mol. Biol.* 279, 449–460.

73 PACE, C. N. (1986). Determination and analysis of urea and guanidine hydrochloride denaturation curves. *Methods Enzymol.* 131, 266–280.

74 MÖGLICH, A., KRIEGER, F. & KIEFHABER, T. (2004). Molecular basis of the effect of urea and guanidinium chloride on the dynamics of unfolded proteins. submitted.

75 ZHU, W., GISSER, D. J. & EDIGER, M. D. (1994). C-13 NMR measurements of polybutadiene local dynamics in dilute-solution – further evidence for non-Kramers behavior. *J. Polym. Sci. B Polym. Phys.* 32, 2251–2262.

76 ZHU, W. & EDIGER, M. D. (1997). Viscosity dependence of polystyrene local dynamics in dilute solutions. *Macromolecules* 30, 1205–1210.

77 TYLIANAKIS, E. I., DAIS, P. & HEATLEY, F. (1997). Non-Kramers' behavior of the chain local dynamics of pvc in dilute solution. Carbon-13 NMR relaxation study. *J. Polym. Sci. B* 35, 317–329.

78 BEECE, D., EISENSTEIN, L., FRAUENFELDER, H. et al. (1980). Solvent Viscosity and Protein Dynamics. *Biochemistry* 19, 5147–5157.

79 YEDGAR, S., TETREAU, C., GAVISH, B. & LAVALETTE, D. (1995). Viscosity dependence of O2 escape from respiratory proteins as a function of cosolvent molecular weight. *Biophys. J.* 68, 665–670.

80 KLEINERT, T., DOSTER, W., LEYSER, H., PETRY, W., SCHWARZ, V. & SETTLES, M. (1998). Solvent composition and viscosity effects on the kinetics of CO binding to horse myoglobin. *Biochemistry* 37, 717–733.

81 GAVISH, B. (1980). Position-dependent viscosity effects on rate coefficients. *Phys. Rev. Lett.* 44, 1160–1163.

82 DOSTER, W. (1983). Viscosity scaling and protein dynamics. *Biophys. Chem.* 17, 97–103.

83 SCHLITTER, J. (1988). Viscosity dependence of intramolecular activated processes. *Chem. Phys.* 120, 187–197.

84 TIMASHEFF, S. N. (2002). Protein hydration, thermodynamic binding and preferential hydration. *Biochemistry* 41, 13473–13482.

85 PERICO, A. & BEGGIATO, M. (1990). Intramolecular diffusion-controlled reactions in polymers in the optimized Rouse Zimm approach 1. The effects of chain stiffness, reactive site positions, and site numbers. *Macromolecules* 23, 797–803.

86 LESZCYNSKI, J. F. & ROSE, G. D. (1986). Loops in globular proteins: a novel category of secondary structure. *Science* 234, 849–855.

87 WILMOT, C. M. & THORNTON, J. M. (1988). Analysis and prediction of the different types of β-turns in proteins. *J. Mol. Biol.* 203, 221–232.

88 ZIMM, B. H. & BRAGG, J. K. (1959). Theory of phase transition between helix and random coil in polypeptide chains. *J. Chem. Phys.* 31, 526–535.

89 LIFSON, S. & ROIG, A. (1961). On the

theory of helix-coil transitions in polypeptides. *J. Chem. Phys.* 34, 1963–1974.

90 SCHWARZ, G. & SEELIG, J. (1968). Kinetic properties and the electric field effect of the helix-coil transition of poly(γ-benzyl L-glutamate) determined from dielectric relaxation measurements. *Biopolymers* 6. 1263–1277.

91 GRUENEWALD, B., NICOLA, C. U., LUSTIG, A. & SCHWARZ, G. (1979). Kinetics of the helix-coil transition of a polypeptide with non-ionic side chain groups, derived from ultrasonic relaxation measurements. *Biophys. Chem.* 9, 137–147.

92 WILLIAMS, S., CAUSGROVE, T. P., GILMANSHIN, R. et al. (1996). Fast events in protein folding: helix melting and formation in a small peptide. *Biochemistry* 35, 691–697.

93 THOMPSON, P., EATON, W. & HOFRICHTER, J. (1997). Laser temperature jump study of the helix-coil kinetics of an alanine peptide interpreted with a "kinetic zipper" model. *Biochemistry* 36, 9200–9210.

94 MUNOZ, V., THOMPSON, P., HOFRICHTER, J. & EATON, W. (1997). Folding dynamics and mechanism of β-hairpin formation. *Nature* 390, 196–199.

95 BIERI, O. & KIEFHABER, T. (1999). Elementary steps in protein folding. *Biol. Chem.* 380, 923–929.

96 SCHWARZ, G. (1965). On the kinetics of the helix-coil transition of polypeptides in solution. *J. Mol. Biol.* 11, 64–77.

97 KUBELKA, J., HOFRICHTER, J. & EATON, W. A. (2004). The protein folding speed limit. *Curr. Opin. Struct. Biol.* 14, 76–88.

98 HUANG, G. S. & OAS, T. G. (1995). Submillisecond folding of monomeric lambda repressor. *Proc. Natl Acad. Sci. USA* 92, 6878–6882.

99 BURTON, R. E., HUANG, G. S., DAUGHERTY, M. A., FULLBRIGHT, P. W. & OAS, T. G. (1996). Microsecond protein folding through a compact transition state. *J. Mol. Biol.* 263, 311–322.

100 MAYOR, U., GUYDOSH, N. R., JOHNSON, C. M. et al. (2003). The complete folding pathway of a protein from nanoseconds to microseconds. *Nature* 421, 863–867.

101 PERL, D., WELKER, C., SCHINDLER, T. et al. (1998). Conservation of rapid two-state folding in mesophilic, thermophilic and hyperthermophilic proteins. *Nat. Struct. Biol.* 5, 229–235.

102 FERGUSON, N., JOHNSON, C. M., MACIAS, M., OSCHKINAT, H. & FERSHT, A. R. (2001). Ultrafast folding of WW domains without structured aromatic clusters in the denatured state. *Proc. Natl Acad. Sci. USA* 98, 13002–13007.

103 ROBINSON, C. R. & SAUER, R. T. (1996). Equilibrium stability and submillisecond refolding of a designed single-chain arc repressor. *Biochemistry* 35, 13878–13884.

104 NEIDIGH, J. W., FESINMEYER, R. M. & ANDERSEN, N. H. (2002). Designing a 20-residue protein. *Nat. Struct. Biol.* 9, 425–430.

105 QIU, L., PABIT, S. A., ROITBERG, A. E. & HAGEN, S. J. (2002). Smaller and faster: The 20-residue Trp-cage protein folds in 4 µs. *J. Am. Chem. Soc.* 124, 12952–12953.

106 BACHMANN, A. & KIEFHABER, T. (2001). Apparent two-state tendamistat folding is a sequential process along a defined route. *J. Mol. Biol.* 306, 375–386.

107 SÁNCHEZ, I. E. & KIEFHABER, T. (2003). Evidence for sequential barriers and obligatory intermediates in apparent two-state protein folding. *J. Mol. Biol.* 325, 367–376.

108 KUWAJIMA, K., MITANI, M. & SUGAI, S. (1989). Characterization of the critical state in protein folding. Effects of guanidine hydrochloride and specific Ca^{2+} binding on the folding kinetics of alpha-lactalbumin. *J. Mol. Biol.* 206, 547–561.

109 MAKAROV, D. E., KELLER, C. A., PLAXCO, K. W. & METIU, H. (2002). How the folding rate constant of simple single-domain proteins depends on the number of native contacts. *Proc. Natl Acad. Sci. USA* 99, 3535–3539.

110 MAKAROV, D. E. & PLAXCO, K. W. (2003). The topomer search model: A simple. quanitative theory of two-state protein folding kinetics. *Protein Sci.* 12, 17–26.

111 SÁNCHEZ, I. E. & KIEFHABER, T. (2003). Origin of unusual phi-values in protein folding: Evidence against specific nucleation sites. *J. Mol. Biol.* 334, 1077–1085.

112 BIERI, O. & KIEFHABER, T. (2000). Kinetic models in protein folding. In *Protein Folding: Frontiers in Molecular Biology*, 2nd edn (PAIN, R., ed.), pp. 34–64. Oxford University Press, Oxford.

113 KRAMERS, H. A. (1940). Brownian motion in a field of force and the diffusion model of chemical reactions. *Physica* 4, 284–304.

114 HÄNGGI, P., TALKNER, P. & BORKOVEC, M. (1990). Reaction-rate theory: fifty years after Kramers. *Rev. Mod. Phys.* 62, 251–341.

115 JACOB, M., SCHINDLER, T., BALBACH, J. & SCHMID, F. X. (1997). Diffusion control in an elementary protein folding reaction. *Proc. Natl Acad. Sci. USA* 94, 5622–5627.

116 PLAXCO, K. W. & BAKER, D. (1998). Limited internal friction in the rate-limiting step of a two-state protein folding reaction. *Proc. Natl Acad. Sci. USA* 95, 13591–13596.

117 VAN'T HOFF, J. H. (1884). *Etudes de dynamique*. Muller, Amsterdam.

118 JACKSON, S. E. (1998). How do small single-domain proteins fold? *Folding Des.* 3, R81–R91.

119 TANFORD, C. (1970). Protein denaturation. Part C. Theoretical models for the mechanism of denaturation. *Adv. Protein Chem.* 24, 1–95.

120 SÁNCHEZ, I. E. & KIEFHABER, T. (2003). Non-linear rate-equilibrium free energy relationships and Hammond behavior in protein folding. *Biophys. Chem.* 100, 397–407.

121 SZABO, Z. G. (1969). Kinetic characterization of complex reaction systems. In *Comprehensive Chemical Kinetics* (BAMFORD, C. H. & TIPPER, C. F. H., eds), Vol. 2, pp. 1–80. 7 vols. Elsevier, Amsterdam.

122 KIEFHABER, T., KOHLER, H. H. & SCHMID, F. X. (1992). Kinetic coupling between protein folding and prolyl isomerization. I. Theoretical models. *J. Mol. Biol.* 224, 217–229.

123 MOORE, J. W. & PEARSON, R. G. (1981). *Kinetics and Mechanisms.*, John Wiley & Sons, New York.

124 STERN, O. & VOLMER, M. (1919). Über die Abklingungszeit der Fluoreszenz. *Phys. Z.* 20, 183–188.

125 LAPIDUS, L. J., EATON, W. A. & HOFRICHTER, J. (2001). Dynamics of intramolecular contact fromation in polypeptides: distance dependence of quenching rates in a room-temperature glass. *Phys. Rev. Lett.*, 258101-1–258101-4.

126 LAPIDUS, L. J., STEINBACH, P. J., EATON, W. A., SZABO, A. & HOFRICHTER, J. (2002). Effects of chain stiffness on the dynamics of loop formation in polypeptides. Appendix: Testing a 1-diensional diffusion model for peptide dynamics. *J. Phys. Chem. B* 106, 11628–11640.

127 KUMAR, V. D., LEE, L. & EDWARDS, B. F. (1990). Refined crystal structure of calcium-liganded carp parvalbumin 4.25 at 1.5 Å resolution. *Biochemistry* 1404–1412.

128 KORADI, R., BILLETER, M. & WÜTHRICH, K. (1996). MOLMOL: a program for display and analysis of macromolecular structures. *J. Mol. Graphics* 14, 51–55.

23
Equilibrium and Kinetically Observed Molten Globule States

Kosuke Maki, Kiyoto Kamagata, and Kunihiro Kuwajima

23.1
Introduction

The molten globule state has captured the attention of researchers studying the folding of globular proteins ever since it was proposed that this state is a general intermediate of protein folding [1–4]. Historically, the apparent incompatibility between the efficient folding of a protein into the native structure and the availability of an astronomically large number of different conformations, known as the Levinthal paradox [5], led to the premise that there must be specific pathways of folding, so that by restricting the protein molecule to these pathways, the polypeptide chain can reach its native structure efficiently. According to this view (the "classical" view), the folding occurs in a sequential manner with a series of intermediates populated along the specific pathway of folding (the sequential model of protein folding) [6–8]. Therefore, studies on protein folding were focused on detection and characterization of the specific folding intermediates of globular proteins. The discovery that the intermediates were formed during the kinetic folding of many globular proteins convinced the researchers of the significance of the intermediates in elucidating the mechanism of protein folding. The molten globule is the most typical of these folding intermediates, and has been regarded as a general intermediate of protein folding.

Nevertheless, recent theoretical studies of protein folding have shown that what is required for a protein to reach the unique native structure is not the presence of the specific pathway of folding, but just the presence of a small bias of the free-energy surface toward the native state in the multidimensional conformational hyperspace of the protein [9–11]. Therefore, according to this "new" view of protein folding, the Levinthal paradox is no longer a real paradox. It is obvious that the attempt to detect the folding intermediate to resolve the Levinthal paradox is now irrelevant, casting a fundamental doubt on the role of the molten globule state as a productive intermediate in protein folding [11, 12]. Apparently in support of this new perspective, kinetic folding of small, single-domain proteins with fewer than typically 100 amino acid residues has been shown to occur simply in a two-state manner without accumulation of the folding intermediates [13, 14]. Because the

Protein Folding Handbook. Part I. Edited by J. Buchner and T. Kiefhaber
Copyright © 2005 WILEY-VCH Verlag GmbH & Co. KGaA, Weinheim
ISBN: 3-527-30784-2

folding can take place without the intermediates, they are apparently not prerequisites for successful folding of these proteins. The molten globule state might not be obligatory, but rather produced by misfolding events or kinetic traps at local minima on the free-energy surface of the protein folding [3, 4, 15]. Therefore, whether folding intermediates such as the molten globule state are necessary for directing the folding reactions of globular proteins is still an open question, and there has been intense debate on this issue.

However, it is probably more appropriate to address the following question: Is the sequential model of protein folding truly in conflict with the simple two-state folding of small globular proteins, which has been shown to support the new view of protein folding? In spite of the argument in favor of the new perspective of folding, the experimental studies of the folding intermediates of various globular proteins have clearly demonstrated that the molten globule is a real productive folding intermediate in these proteins [16]. Therefore, there must be a more general model that can accommodate both the sequential and the simple two-state folding reactions of real globular proteins.

We have thus previously proposed a two-stage hierarchical model of protein folding [3, 4], in which the folding process of a protein is, in general, divided into at least two stages: stage I, formation of the molten globule state from the unfolded state; and stage II, formation of the native state from the molten globule, as shown in Scheme 23.1.

$$U \xrightarrow{I} I \xrightarrow{II} N$$

Scheme 23.1

The folding intermediates like the molten globule in this model are not to solve the Levinthal paradox, but they are present because the protein folding takes place in a hierarchical manner [17, 18], reflecting the hierarchy of three-dimensional structure of natural proteins. In this model, the interactions that stabilize the molten globule intermediate should be approximately consistent with the interactions that stabilize the native state. Whether or not the observed folding of a protein is a two-state or non-two-state (multistate) transition may depend on the stability of the intermediate (I) and/or the location of the rate-limiting step of folding. A main purpose of this review is thus to test this hypothesis on the basis of recent studies of various globular proteins.

In the present article, we thus first summarize the structural and thermodynamic properties of the molten globules at equilibrium, and then describe how the molten globule state has been identified as a productive intermediate in kinetic refolding reactions of globular proteins. We also describe our recent statistical analysis of the kinetic refolding data of globular proteins taken from the literature, which clearly demonstrates that the molecular mechanisms behind the non-two-state and two-state protein folding are essentially identical. Finally, practical aspects on the experimental study of the molten globule are presented.

23.2
Equilibrium Molten Globule State

23.2.1
Structural Characteristics of the Molten Globule State

The molten globule state was originally an equilibrium unfolding intermediate assumed by many globular proteins under mildly denaturing conditions [1, 19]. It is structurally characterized by (i) the presence of a substantial amount of secondary structure, (ii) the virtual absence of tertiary structure associated with tight packing of side-chains, (iii) the compact size of the protein molecule with a radius only 10–30% larger than that of the native state, (iv) the presence of a loosely packed hydrophobic core that increases the hydrophobic surface exposed to solvent, and (v) dynamic features of the structure that fluctuates in a time-scale longer than nanoseconds. Thus, in short, the molten globule is a compact globule with a "molten" side chain structure [20].

The mildly denaturing conditions in which we can observe the molten globule state at equilibrium include acid pH [1, 21, 22], moderate concentrations of denaturants [23–26], high temperatures [27, 28], low concentrations of alcohol and fluoroalcohol [29, 30], and high pressure [31–33]. It is, however, dependent on protein species whether the molten globule is stably populated under the conditions; the protein may be still in the native state or more extensively unfolded. Certain proteins extensively unfolded at acid pH are known to refold into the molten globule state in the presence of a stabilizing anion [34]. For ligand-binding proteins, the partial unfolding induced by removal of a tightly bound ligand can bring about the molten globule state [27, 35]. Chemical modification, site-directed mutation and covalent bond cleavage of a natural protein often lead to the molten globule-like partially unfolded state [36–40].

In conventional experimental studies, the molten globule state has been characterized by peptide and aromatic circular dichroism (CD) spectra that detect secondary and tertiary structures, respectively [1, 41], by hydrodynamic techniques that determine the molecular size of the protein [2, 42, 43], and by hydrophobic dye (1-anilinonaphthalene-8-sulfonate (ANS))-binding experiments that detect formation of a loose hydrophobic core accessible to solvent [44].

In addition to these conventional techniques, new experimental techniques, such as hydrogen/deuterium exchange NMR spectroscopy, NMR relaxation and heteronuclear secondary chemical shifts and nuclear Overhauser enhancement (NOE) measurements [45–53], solution small-angle X-ray scattering (SAXS) [54–57], and protein engineering techniques [38–40, 58–63], have been used successfully to further characterize the structure of the molten globule state in many globular proteins during the past decade. Use of these techniques has enabled us to describe the structure of the molten globule states more precisely, to at least the amino acid residue level, and to reveal more detailed characteristics of this state.

Studies using these new techniques have shown the structure of the molten globule state to be more heterogeneous than previously thought, that is, one portion of the structure is more organized and native-like while the other portion is less organized [49, 51, 64–66]. The degree of structural organization and the stability of the molten globule state have both been found to be remarkably dependent on the species of protein conformer, there being a range of species possible within the basic molten globule characteristics of secondary structure, compactness and hydrophobic core. Although the classical concept of little or no persistent tertiary structure still often holds, it is not a universal requirement of the molten globule state.

23.2.2
Typical Examples of the Equilibrium Molten Globule State

Representative examples for showing the diversity of the structure of the molten globule state can be found in α-lactalbumin (α-LA) and Ca^{2+}-binding lysozymes from canine and equine milk [67–70] (Figure 23.1), which are homologous (123–130 amino acid residues). These proteins exhibit the equilibrium molten globule state at a moderate concentration of guanidinium chloride (GdmCl) as well as acidic pH [1, 23–26, 71–73]. The overall structures of the molten globule state of these proteins are similar; the α-helical domain (α-domain) composed of the A-, B-, C-, and D-helices, and the C-terminal 3_{10}-helix is more organized with the β-domain less organized as revealed by hydrogen/deuterium exchange and other techniques [22, 38–40, 47, 74–79]. Despite the similarity in overall structure, the lysozymes are substantially more native-like than α-LA. Significant ellipticity in the near-UV CD spectra and incomplete fluorescence quenching by acrylamide indicate the presence of an immobile tryptophan residue buried inside the molecule [22, 24, 25, 71, 80].

Apomyoglobin (apoMb), which is the apo-form of myoglobin, a 153-residue heme protein with eight helices (A–H) (Figure 23.2) [81], illustrates the diversity in the structure of the molten globule state [34, 82–84]. ApoMb forms a molten globule at pH 4 (the pH 4 intermediate or I_1) while the addition of a stabilizing anion (trichloroacetate (TCA), citrate, or sulfate) refolds the protein to form the second molten globule (I_2) [85, 86]. A similar compact molten globule is formed at acidic pH in the presence of chloride ion [34]. The structure of I_1 has been studied by hydrogen/deuterium exchange NMR as well as by other NMR techniques [45, 50, 66]. It has a compact subdomain consisting of the A-, G-, and H-helical regions as well as part of the B-helix region.

Cytochrome c (cyt c), which is a 104-residue globular protein with a covalently attached heme group surrounded by three α-helices (Figure 23.3) [87], forms a compact intermediate that has the molten globule characteristics [20]. The molten globule state stably populated at acidic pH in the presence of stabilizing anion from salt or acid is more structurally organized and more stable than the molten globule states of α-LA and apoMb [34, 88]. In contrast to the α-LA or apoMb molten globules that correspond to the early folding intermediates, the cyt c molten globule corresponds to a late-folding intermediate that is present after the rate-limiting step in the kinetic refolding [89, 90].

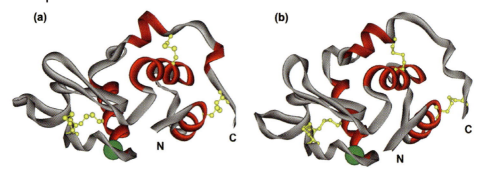

Fig. 23.1. X-ray crystallographic structure of a) human α-lactalbumin (PDB code: 1HML) and b) canine milk lysozyme (PDB code: 1EL1) drawn using DS Modeling. Helices formed in the molten globule state are in red. Four disulfide bonds are shown in yellow in ball-and-stick representation. One Ca^{2+} ion is bound to α-LA and the lysozyme, shown as a green ball.

23.2.3
Thermodynamic Properties of the Molten Globule State

The thermal unfolding of a native globular protein is a highly cooperative, first-order transition. The transition accompanies a large increase in enthalpy, which primarily arises from the disruption of the specific side chain packing interactions, namely, the van der Waals interactions of hydrophobic residues [91]. Because of the disruption of a large fraction of such specific interactions in the molten globule state, thermal unfolding from this state is always less cooperative than the unfold-

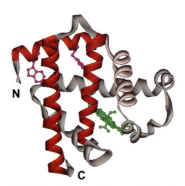

Fig. 23.2. X-ray crystallographic structure of sperm whale myoglobin in the holo form (PDB code: 1MBC) drawn using DS Modeling. The A-, G-, and H-helix regions, which are stably formed in the molten globule state, are in red. The B-helix region, which is partially formed in the structured molten globule state, is in pink. A bound heme and two tryptophan residues (Trp7 and Trp14) are shown in ball-and-stick representation. The heme is in green, and the two tryptophan residues are in purple.

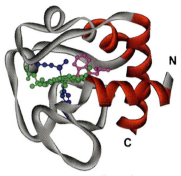

Fig. 23.3. X-ray crystallographic structure of horse cyt c (PDB code: 1HRC) drawn using DS Modeling. The N-terminal and C-terminal helices, and the 60s helix, which are stably formed in the molten globule state, are in red. His18, Trp59, Met80, and a bound heme are shown in ball-and-stick representation. His18 and Met80 are in blue, and Trp59 is in purple. Heme is in green.

ing from the native state. The best-characterized molten globules, such as the molten globule state of α-LA and the pH 4 intermediate of apoMb, do not apparently exhibit cooperative thermal transitions [2, 27, 92–94], so that the absence of the thermal transition was thought to be a characteristic of the molten globule state for some time.

However, it is now clear that the molten globules of certain proteins do exhibit a cooperative thermal unfolding transition with a significant increase in enthalpy [24, 28, 73, 95–97]. Thus, the apparent absence of cooperative thermal unfolding observed for the molten globules of α-LA and apoMb is due to their small enthalpy change. The unfolding transitions of the molten globules are apparently noncooperative for some proteins, but highly cooperative for others.

The most notable example of a cooperative thermal unfolding transition of a molten globule is displayed by equine and canine milk lysozymes [24, 28, 73, 97]. Although the overall structure of these molten globule states is very similar to that of the α-LA molten globule, the unfolding from these molten globules shows a cooperative heat absorption peak. The molten globules of cyt c and staphylococcal nuclease (SNase) also show cooperative thermal transitions [95, 96]. Therefore, it appears that the cooperative thermal unfolding of the molten globule state is more the rule than the exception.

The degree of cooperativity of the molten globule unfolding transition is, however, not only dependent on the protein species, but also, on the experimental conditions or the nature of amino acid replacements for a given protein [59, 90, 98–101]. The chloride-induced molten globule of apoMb at pH 2 shows a small but distinct heat absorption peak on thermal unfolding, and the TCA-induced molten globule is more stable and shows a still more cooperative unfolding transition [85, 99].

An important aspect concerning the variation in molten globule cooperativity is the correlation between the intrinsic stability and the degree of cooperativity of the unfolding transition [101, 102]. The order of stability for the well-characterized molten globules for three proteins (α-LA, apoMb, and cyt c) is the same as the order of the cooperativity of the molten globule unfolding, that is, α-LA < apoMb < cyt c [23, 83, 86, 100, 103]. In apoMb, the molten globule state has been investigated for different mutants in different anion conditions, and there is again a good correlation between stability and cooperativity [101].

The correlation between the intrinsic stability and the degree of cooperativity of the molten globule state may reasonably be interpreted in terms of the two-stage hierarchical model of protein folding (see Scheme 23.1). In this model, the local interactions and nonspecific hydrophobic interactions that stabilize the molten globule state approximate, to a varying degree, the specific tertiary packing interactions that stabilize the native structure. Thus, the presence of specific native interactions further organizes structurally and stabilizes the molten globule state, and makes the unfolding transition more cooperative.

23.3
The Kinetically Observed Molten Globule State

23.3.1
Observation and Identification of the Molten Globule State in Kinetic Refolding

Various experimental techniques, including kinetic CD spectroscopy [41, 104–106], hydrogen/deuterium exchange pulse labeling combined with two-dimensional NMR spectroscopy [46], and stopped-flow SAXS [56, 57, 107], have been used to detect and characterize the kinetic folding intermediates in a number of globular proteins. These studies emphasize the close similarities between the equilibrium molten globule state and kinetic folding intermediates. Kinetic studies have further shown that similar intermediates commonly accumulate during refolding of other proteins whose equilibrium unfolding transitions are represented well by a two-state model. Here we describe how intermediates of protein folding have been identified experimentally as the molten globule states.

Pulsed peptide hydrogen/deuterium exchange combined with two-dimensional NMR and mass spectrometry has been used to characterize transient folding intermediates of several globular proteins, including cyt c, ribonuclease A, lysozyme, apoMb, and ribonuclease HI (RNase HI) [46, 74, 108–113]. The transient folding intermediate and the equilibrium molten globule state have both been characterized by hydrogen/deuterium exchange techniques. The two intermediates have been shown to be equivalent by comparison of their hydrogen/deuterium exchange profile in apoMb and RNase HI [109, 111, 113].

The folding intermediate of apoMb formed within 6 ms of refolding has protected protons in the A-, G-, and H-helices as well as in part of the B-helix and appears to be structurally the same as the pH 4 intermediate at equilibrium [109, 113–116]. The hydrogen/deuterium exchange combined with mass spectrometry indicates that the folding intermediate is obligatory with a significant hydrogen-bonded secondary structure content [112].

The folding intermediate of RNase HI is formed within the dead time (12 ms) of the stopped-flow technique when the refolding starts from the unfolded state, and consists of a structural core region of the protein, namely helices A and D and β-strand 4. This kinetic intermediate resembles both the acid molten globule of the protein and sparsely populated, partially unfolded forms detected under native conditions [111]. Interactions formed in the intermediate persist in the transition and native states, suggesting that RNase HI folds through a hierarchical mechanism [117].

NMR spectroscopy can be applied in real-time to investigating the kinetics of folding and has been used to follow the refolding of bovine α-LA [74, 118, 119]. Although the dead time is approximately 1 s, the kinetics can be monitored directly by a series of one-dimensional NMR spectra and by a single two-dimensional (^{1}H-^{15}N HSQC) experiment, taking advantage of the slow refolding of apo-α-LA. The kinetics monitored by side-chain proton signals in the one-dimensional NMR spectra and by main-chain signals in the two-dmensional experiment are both coincident with those measured by CD and fluorescence [118, 119]. The one-dimensional NMR spectrum observed just after the dead time ($\sim$1 s) was very similar to that of the molten globule state, indicating the folding intermediate of α-LA is identical to the molten globule state [118]. Real-time NMR spectroscopy is a powerful technique, allowing the validity of a specific kinetic scheme to be tested, although it is applicable only for slow-folding proteins that have a rate constant less than $\sim$0.05 s^{-1}. The real-time NMR method is also applied to investigating the unfolding characteristics of interleukin-1β [120].

23.3.2
Kinetics of Formation of the Early Folding Intermediates

The folding intermediates of globular proteins have been observed and characterized as the burst-phase intermediate in most cases because they already accumulated during the dead time of typical stopped-flow instrumentations, which limited the kinetic folding measurements to a time resolution of a few milliseconds. However, recent development of new techniques, which have dead times much shorter than that of conventional stopped-flow instrumentations, allowed us to directly monitor the folding kinetics, including the formation of the molten globule, in submillisecond time region. These techniques include laser photochemically triggering method [121, 122], the temperature-jump method with electrical or laser triggering [123, 124], and the continuous-flow method [125–128].

The folding kinetics of apoMb from the cold-denatured state were monitored by the nanosecond laser temperature-jump method with fluorescence detection, and the subdomain assembly consisting of the A-, G-, and H-helices from the G-H loop and nascent A helix was observed in 5–20 µs after initiation of the folding [129]. CD and SAXS measurements of apoMb using the continuous-flow and stopped-flow techniques have revealed that significant compaction and helix formation occurs within ~300 μs after initiation of the pH-induced refolding, which is followed by the formation of an intermediate similar to the TCA-induced structured molten globule state of this protein [130].

The continuous-flow fluorescence measurements of cyt c revealed the accumulation of a series of intermediates [131]. The earliest intermediate (I_C, see Section 23.3.3) whose Trp fluorescence is fully quenched by the heme is formed with a time constant of ~65 μs, which is much shorter than a typical dead time (a few milliseconds) of the stopped-flow measurement [131]. In fact, this intermediate has been observed as the burst-phase intermediate when the stopped-flow method is used. The continuous-flow CD and SAXS measurements also revealed two intermediates along the sequential folding pathway of cyt c [132, 133]. The first intermediate was, however, observed as the burst-phase intermediate because the dead times of the measurements, 160–390 μs, were much longer than the time constant of formation of the first intermediate. The first intermediate is more compact than the unfolded state with ~20% helicity of the native state, and may correspond to the I_C intermediate. The second intermediate formed with a time constant of ~400 μs is further more compact with a radius of gyration ~30% larger than the native state with ~70% helicity of the native state, and hence its size is close to that of the molten globule state of this protein [132, 133]. The second intermediate may be more similar to the α-LA and apoMb molten globules that are known to be less structurally organized and less stable than the salt-induced cyt c molten globule (see Section 23.2.2).

23.3.3
Late Folding Intermediates and Structural Diversity

Although the molten globule states are formed in less than a few milliseconds as burst-phase intermediates, in many of the proteins shown above, the equilibrium molten globule states occasionally correspond to late-folding intermediates formed after the burst-phase and other early intermediates have accumulated. The late-folding intermediates, which are usually more structured than the burst-phase intermediates, often correspond to the structured molten globules formed by addition of stabilizing anions at equilibrium.

The best-studied example is the late-folding intermediate of cyt c [90, 134, 135]. Cyt c at neutral pH folds by a sequential mechanism, which involves three intermediates (I_C, I_{NC}, N*) as shown in Scheme 23.2 [89, 90, 108, 134].

$$U \rightarrow I_C \rightarrow I_{NC} \rightarrow N^* \rightarrow N$$
Scheme 23.2

I_C is the earliest intermediate characterized by the continuous-flow studies (see Section 23.3.2), and it was observed as the burst-phase intermediate by the stopped-flow method. I_{NC} is a more structured intermediate with tightly interacting N- and C-terminal α-helices, and this was characterized by pulsed hydrogen/

deuterium exchange and mutagenesis techniques [89, 108, 136]. The last intermediate, N*, is known to affect the kinetics of unfolding, but does not accumulate during refolding. It is structurally very close to the native state, but lacks the native Met80 heme ligand. Studies on the KCl-induced kinetic refolding of the protein from the acid-unfolded to the compact molten globule state suggest that the acid compact equilibrium molten globule state may correspond to N* [90].

Similarly, apoMb accumulates at least two intermediates, Ia and Ib, in the kinetic refolding, and the late intermediate, Ib, may correspond to the anion-induced, more structured, molten globule state observed at equilibrium [86]. These intermediates of apoMb may also correspond to the two folding intermediates detected by the continuous-flow experiments (see Section 23.3.2).

Apparently, there is a diversity of molten globule states in terms of where they accumulate along the folding pathway, and this is analogous to the diversity of the equilibrium molten globule state that depends on the protein species and the experimental conditions, as described in Section 23.2. When accumulation takes place at a late stage of refolding, it is more structured than the early intermediate and may correspond to the structured molten globule state observed at equilibrium.

23.3.4
Evidence for the On-pathway Folding Intermediate

The folding kinetics of certain globular proteins have been analyzed very carefully, and hence we can address the question of whether the molten globule intermediate formed during kinetic folding is a productive, on-pathway intermediate, or not. If the intermediate is productive, it must be placed between the fully unfolded and the native states along the folding pathway. Observation of a lag phase during the refolding is known to provide firm evidence that the intermediate is productive. Jennings and her coworkers have studied kinetic refolding of interleukin-1β by pulsed hydrogen/deuterium exchange mass spectrometric analysis, stopped-flow CD, and fluorescence. The transient folding intermediate accumulates with a time constant of 126 ms, and there is a lag phase in the production of the native state of at least 400 ms, clearly indicating that the intermediate is a productive on-pathway intermediate [137]. A similar lag phase has also been reported in SNase and dihydrofolate reductase (DHFR) [138–141], the refolding reactions of which have been characterized by tryptophan fluorescence and methotrexate (an inhibitor of DHFR) absorbance, respectively. In apoMb, two transient intermediates, Ia and Ib, accumulate during refolding from the acid-unfolded state, and the kinetics analyzed by interrupted refolding and interrupted unfolding experiments have been shown to be consistent with a linear folding pathway (U $\rightarrow$ Ia $\rightarrow$ Ib $\rightarrow$ N), in which both are productive on-pathway intermediates [86].

A detailed analysis of Im7 folding and unfolding by monitoring the whole fluorescence signal change using the continuous-flow and stopped-flow methods unambiguously revealed that the kinetic folding intermediate of this protein is an on-pathway, obligatory intermediate [142]. Similar quantitative analysis of the folding and unfolding reactions of several proteins, including the B1 domain of protein G, ubiquitin, cyt *c*, and interleukin-1β, indicates that the experimentally observed folding intermediates are on-pathway and obligatory [16, 143–146].

23.4
Two-stage Hierarchical Folding Funnel

Here, we consider the structural and energetic aspects of protein folding represented by the two-stage hierarchical model, and this together with the funnel representation of protein folding will give us a reasonable picture of the structural diversity of the molten globule state observed experimentally.

Structural changes and their associated energetic properties are different between the two stages in the two-stage hierarchical model of protein folding [3, 4]. In stage I, the protein molecule forms native-like secondary structure, tertiary fold and compact shape, lacking specific side-chain packing, thus acquiring the broad structural architecture of the native molecule. Local interactions that determine preference for secondary structure, and nonspecific hydrophobic interactions that determine the overall backbone topology and the compact shape are important in this stage. On the other hand, in stage II the specific side-chain packing is organized, and specific van der Waals contacts are dominant in this process. Furthermore, the structure formed in stage I may be similar to an expanded version of the structure formed in stage II, and the overall tertiary fold formed in stage I brings the appropriate side chains close in space to enable the specific native packing interaction between them to take place and stabilize the overall native-like tertiary fold.

Taking advantage of the funnel picture of protein folding, it is useful to describe the two-stage hierarchical folding model in terms of the folding funnel. The presence of the two stages indicates the presence of two corresponding folding funnels, I and II (Figure 23.4). The differences in the structural classes and energetics associated with the two stages are reflected in differences in shape between the two folding funnels. The compaction of protein, which occurs in stage I, results in a much larger decrease in the conformational entropy, suggesting that funnel I is much wider at its top than funnel II. The multiple specific tertiary packing interactions occurring in stage II are enthalpic in nature, and, because they are not always native-like, a previously formed interaction must often be broken to form a more stable set of interactions later in stage II. There must, therefore, be a number of enthalpic barriers along funnel II. Because the free-energy barrier is much more entropic in funnel I, the energy landscape must be much more rugged in funnel II.

The two-stage folding funnel provides a nice picture of why the experimentally observed molten globule states are structurally diverse. As described in Section 23.2, some molten globules have no observable tertiary structure, while others often contain native-like side-chain packing in a limited region of the molecule. The degree of structure in the molten globule state depends on where the largest

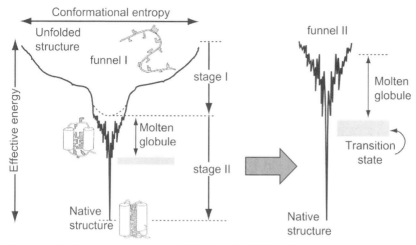

Fig. 23.4. A schematic representation of the hierarchical folding funnel. See the text for details.

energetic barrier is located along funnel II. When it is located at the top of funnel II, there will be no specific side-chain packing in the molten globule, but when close to the bottom, a number of the specific side-chain interactions may already be formed. In cases like apoMb, there may be two large energetic barriers along funnel II, resulting in accumulation of two structurally different molten globules.

23.5
Unification of the Folding Mechanism between Non-two-state and Two-state Proteins

Now we re-address the question raised at the beginning of this chapter, because a number of small globular proteins with fewer than 100 amino acid residues have been found to exhibit two-state folding without any accumulation of the intermediate [14]. These two-state proteins often correspond to a part (domain) of an entire protein molecule, and there are now more than 30 examples of the two-state proteins, including Src homology 3 domain [147–150], E9 colicin-binding immunity domain [151], cold shock protein B [152], N-terminal domain of λ repressor [153], IgG-binding domain of protein L [154], and chymotrypsin inhibitor 2 [155]. Apparently, the accumulation of the folding intermediate is not a prerequisite for the successful folding of these two-state proteins.

Questions thus arise. (i) Is the sequential model of protein folding truly in conflict with the simple two-state folding of small globular proteins? (ii) What is the role of the molten globule intermediate in protein folding if the successful folding takes place for the two-state proteins? (iii) What is a unified view of protein folding? Two possible answers to these questions are: (i) Protein folding is in principle a highly cooperative two-state transition, and any molten globule-like species observed at early stages of folding are not productive, but produced by kinetic trapping and misfolding events; and (ii) non-two-state folding with the molten globule as a productive folding intermediate is more common in globular protein folding, and the two-state folding observed in the small proteins is rather a simplified version of the more common non-two-state folding. Which answer better represents real protein folding is crucial to our understanding of the mechanism of folding.

A recent paper by Kamagata et al. has shed light on this issue, and strongly suggested that the second answer is more appropriate in real globular proteins [156]. They have found that the rate constants for formation of the intermediate and the native state of non-two-state proteins both show essentially the same dependence on the native backbone structure as the rate constant of folding of two-state proteins. This indicates that the folding mechanisms behind the non-two-state and the two-state protein folding are essentially identical.

23.5.1
Statistical Analysis of the Folding Data of Non-two-state and Two-state Proteins

Kamagata et al. have collected the kinetic folding data of globular proteins from the literature, classified the proteins into non-two-state and two-state proteins, and investigated the relationships between the folding kinetics and the native three-dimensional structure of these proteins. Classification of the proteins as two-state folders was based on the criteria: (i) single-exponential refolding kinetics after exclusion of slow isomerization steps such as proline isomerization in the unfolded state, (ii) the absence of rollover behavior in the logarithmic folding rate constant as a function of denaturant concentration, and (iii) the agreement of unfolding parameters between the equilibrium and kinetic experiments. The proteins that did not satisfy these criteria might be non-two-state folders, but they also employed a more rigorous rule to identify non-two-state proteins; namely, the protein that showed the single-exponential refolding kinetics with the rollover behavior was classified as a non-two-state folder only if one of the following criteria was satisfied: (i) the presence of a burst-phase (i.e., missing amplitude) in the folding kinetics, and/or (ii) the accumulation of a well-characterized kinetic folding intermediate. Furthermore, only the proteins for which the kinetic folding mechanism had been determined clearly were used for the analysis, and the proteins with a heme group or disulfide bonds were excluded from the analysis. As a results, there were 16 natural proteins classified as the non-two-state folders; however, if we include proteins that satisfied only the condition of rollover behavior, additional five proteins could be included as well, resulting in 21 non-two-state folders. For 10 proteins among these 21 folders, the folding kinetics were multiphasic, and the rate constant for the formation of the folding intermediate was reported. On the other hand, there were 18 proteins that were classified as two-state folders.

Figure 23.5 shows the relationship between the rate constant (k_I) for formation of the folding intermediate and the parameter (the number of sequence-distant native pairs, Q_D) that represents the backbone topology of the native structure [157], and the relationship between the rate constant (k_N) for formation of the native state and Q_D, in the non-two-state proteins, and these relationships are compared with

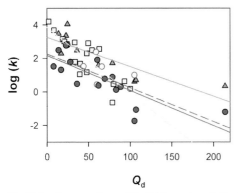

Fig. 23.5. A comparison of the folding rates of non-two-state and two-state folders. The logarithmic rate constants are plotted against the number of sequence-distant native pairs. Filled squares represent the folding of two-state folders. Filled triangles represent the formation of the intermediate of non-two-state folders. Filled circles and open circles represent the formation of the native state for the 16- and 21-protein data set, respectively, of non-two-state folders. The solid lines represent the best linear fit for the formation of the native state for 18 two-state folders, the intermediate for 10 non-two-state folders, and the native state for 16 non-two-state folders, respectively. The dashed line represents the best linear fit for the formation of the native state for 22 non-two-state folders [156].

the relationship between the folding rate constant (k_{UN}) and Q_D in the two-state proteins. It can be seen that all the rate constants (k_I, k_N, and k_{UN}) are significantly correlated with Q_D that is determined by the native backbone topology as given by:

$$Q_D = \sum_{i=2}^{L_p} \sum_{j=1}^{i-1} \Delta_{ij} \tag{1}$$

where i and j are the residues numbers of two contacting residues for which the Cα–Cα distance in space was within 6 Å in the PDB structure, and L_p is the total number of amino acid residues of a protein excluding the disordered terminal regions. $\Delta_{ij} = 1$ if $i - j > 12$, and otherwise $\Delta_{ij} = 0$. Similar correlations of these rate constants were also found with other structure-based parameters, the absolute contact order (ACO) [158] and cliquishness [159], but there were no significant correlations between the rate constants for the non-two-state proteins (k_I and k_N) and the relative contact order (RCO) [160], and this is in contrast with the significant correlation found between k_{UN} and RCO in the two-state proteins. The difference between the non-two-state and the two-state proteins with respect to the correlation with RCO was ascribed to a difference in the chain-length distribution between the two types of proteins. Because RCO is given by ACO/L_p, the correlations of the logarithmic rate constants with ACO and the chain length canceled each other out in the RCO for the non-two-state proteins that had a much wider distribution of L_p.

The significant correlations of both the logarithmic rate constants, log k_I and log k_N, with Q_D (the correlation coefficient $r = -0.74 \sim -0.83$) in the non-two-state proteins indicate that both the processes from the unfolded state (U) to the intermediate (I) and from I to the native state (N) are rate-limited by the process of forming a more native-like backbone topology. Protein molecules thus become progressively more native-like during the folding process from U to N via I, clearly demonstrating that the kinetic intermediate of refolding in the case of non-two-state proteins is a real, productive folding intermediate. Furthermore, both the value of log k_{UN} and its dependence on Q_D in the two-state proteins are very similar to those found in the plots of log k_I and log k_N versus Q_D for the non-two-state proteins (Figure 23.5). This clearly demonstrates that the mechanism of folding does not differ between the two classes of proteins. The non-two-state folding, with the accumulation of a productive folding intermediate, may be a more common mechanism of protein folding, and the two-state folding may be apparently a simplified version of the more common non-two-state folding.

23.5.2
A Unified Mechanism of Protein Folding: Hierarchy

The presence of the molten globule-like intermediate at an early stage of kinetic refolding, together with the productive nature of the intermediate, strongly suggests the two-stage hierarchical model shown earlier in this chapter (Scheme 23.1) as a general mechanism of protein folding. At this point, however, at least two questions arise. First, if the two-stage model is more general, why do certain small globular proteins exhibit two-state (i.e., single-stage) folding? Second, if stage II of Scheme 23.1 corresponds to the process of the specific tertiary packing of side chains from the molten globule to the native state, why does the k_N for non-two-state folders show essentially the same dependence on native backbone topology (i.e., Q_d) as k_I does (see Figure 23.5)?

As regards the first question as it relates to the two-state folders, there are two possible explanations. First, this effect could be due to the movement of the rate-limiting step from stage II in non-two-state folders to stage I in two-state folders. When the size of a protein decreases, it becomes easier to determine the specific conformation of side-chain packing due to the decrease in the number of specific interactions, thereby making the first stage, rather than the second stage, rate-limiting. Thus, this process would lead to the two-state kinetic behavior of the protein. On the other hand, a second explanation would ascribe the two-state behavior to the destabilization of the folding intermediate. When the intermediate is less stable than the unfolded state, a protein must undergo simple two-state folding without any accumulation of the intermediate. If the first explanation is applied to the two-state type of folding, the rate constant k_{UN} of the two-state protein may coincide with the rate constant k_I, at which the intermediate forms in the non-two-state protein. On the other hand, if the second explanation is applied, with the rate-limiting step remaining at stage II, then the rate constant k_{UN} of the two-state protein may coincide with the rate constant k_N, at which the native state for the non-two-state protein is formed. Figure 23.5 shows that the k_{UN} values for the two-state proteins coincide with the k_I values for the non-two-state proteins at Q_d

values of less than 25, suggesting the validity of the first explanation. However, when Q_d is larger than 25, the two-state k_{UN} shows variation in either the non-two-state k_I or k_N, or between the k_I and k_N, and hence both mechanisms given by the two explanations may play a role in the folding of two-state proteins.

However, it still remains unclear why the k_N for non-two-state folders shows essentially the same dependence on Q_d as does the k_I (Figure 23.5). This question may be even more difficult to answer than those discussed above; however, the known structural characteristics of the folding intermediates indicate that stage II includes the process of the specific tertiary packing of side chains. It is possible that this packing process is also dominated by the organization of the native backbone topology. Thus, the folding that occurs in stage II would again be correlated with the native backbone topology, but would take place in a concerted manner together with the tertiary packing of side chains and the organization of the backbone topology.

23.5.3
Hidden Folding Intermediates in Two-state Proteins

Although many small globular proteins fold fast and apparently in a two-state manner without detectable intermediates, it is now becoming increasingly clear that hidden meta-stable folding intermediate may exist along the folding pathway, but behind the rate-limiting transition state. Sanchez and Kiefhaber have shown that nonlinear activation free-energy relationships (i.e., nonlinear relationships of the logarithmic rate constant of folding or unfolding versus denaturant concentration) reported for 23 two-state proteins are caused by sequential folding pathways with consecutive distinct barriers and a few obligatory intermediate that are hidden from direct observation by the high free energies of the intermediates [161]. Bai has also suggested from native-state hydrogen/deuterium exchange data of a number of two-state proteins that partially folded intermediates may exist behind the rate-limiting transition state in these proteins and evade detection by conventional kinetic methods [162]. He proposed two types of the hidden intermediates: type I intermediates, more stable than the unfolded state but hidden behind the rate-limiting transition state located between the unfolded state and the intermediate, and type II intermediates that are less stable than the unfolded state. These are fully consistent with our explanation for the observation of the two-state folding proteins (see Section 23.5.2).

It is thus concluded that there is essentially no difference in mechanism between the non-two-state and the two-state protein folding, and the apparent two-state folding is merely a simplified version of more common non-two-state folding that accumulates obligatory folding intermediates. The major difference between the non-two-state and the two-state folding is the relative stability of the folding intermediates. In fact, non-two-state proteins also exhibit apparent two-state folding and unfolding kinetics under a solution condition where the intermediates are destabilized or by introduction of amino acid replacements that destabilize the intermediates [163].

23.6
Practical Aspects of the Experimental Study of Molten Globules

23.6.1
Observation of the Equilibrium Molten Globule State

As described in Section 23.2, certain proteins show equilibrium unfolding intermediates including the molten globule state, while the other proteins unfold without detectable intermediate. Therefore, when we analyze the experimentally observed unfolding transition of a protein, we must be careful about the possible presence or absence of the equilibrium unfolding intermediates. When there is no intermediate observed, we shall use a two-state model of the unfolding transition, and otherwise a model that involves the intermediates (see Section 23.6.1.2).

The unfolding transition curve of a protein can be obtained by measuring a physical parameter, which reflects the conformational states of the protein, as a function of denaturing perturbations such as denaturant (typically GdmCl or urea) concentration, pH and temperature [1, 23, 28]. The physical parameters often used are optical ones, such as absorption, CD and fluorescence. Here, we will describe a procedure widely used in the analysis of the protein unfolding transitions.

23.6.1.1 Two-state Unfolding Transition
Equilibrium unfolding transitions occurring without detectable intermediate states can be well approximated by the transition between only the two states [164], the native (N) and the fully unfolded (U) states as:

$$N \rightleftarrows U$$

This approximation of the unfolding transition is referred to as the two-state model.

When a denaturant-induced unfolding transition follows the two-state model, the observed value of a given physical parameter (optical absorbance, fluorescence, CD, etc.) $A(c)$ at denaturant concentrations c is given by:

$$A(c) = A_N \times f_N(c) + A_U \times f_U(c) = A_N \times f_N(c) + A_U \times (1 - f_N(c)) \quad (2)$$

where $f_N(c)$ and $f_U(c)$ ($f_N + f_U = 1$) are the fraction native and unfolded at denaturant concentration c, A_N and A_U are the ideal values of the parameter in the native and the unfolded states, respectively, and these are often obtained by linear extrapolations of the dependence of the parameter values on c from the pretransition (native) and the posttransition (unfolded) regions as $A_n = a_1 c + a_2$ and $A_U = a_3 c + a_4$, where a_i is constant. Because an equation essentially the same as Eq. (2) is obtained for the pH- and temperature-induced unfolding transitions, the description of the equilibrium unfolding transition shown below also applies to the pH- and temperature-induced unfolding transitions. Equation (2) is rewritten as:

$$f_N(c) = \frac{A(c) - A_U}{A_N - A_U} \tag{3}$$

This shows that a single observation probe can determine the fraction native and unfolded ($f_U = 1 - f_N$) in the two-state model.

Equation (3) indicates that the fraction native and unfolded calculated from the unfolding transition curves monitored by different probes should be superimposable to each other if the two-state approximation holds. This characteristic of the two-state unfolding transition is often used to distinguish between the two- and multistate unfolding transitions. Figure 23.6A shows an example of the two-state unfolding transition of hen egg white lysozyme [23].

The unfolding transitions approximated by the two-state model exhibit a specific point (for example, at a specific wavelength) in the spectra of the native and the

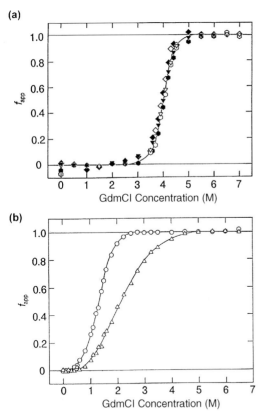

Fig. 23.6. GdmCl-induced equilibrium unfolding transition curves of a) hen egg white lysozyme and (b) α-lactalbumin monitored by CD spectra at a) 222 nm (inverted triangles), 255 nm (hexagons), and 289 nm (diamonds), and b) 222 nm (triangles) and 270 nm (circles) [23].

fully unfolded state, where these two states exhibit the identical value of the parameter used. If there exists such a point(s) in the spectra, all the spectra involved in the unfolding transition at different denaturant concentrations exhibit essentially the same value at that point with the condition that $a_1 \approx a_3 \approx 0$. This is readily evidenced by Eq. (2). Here we take an example of absorbance as an optical probe, and suppose that the absorbance spectra of the native and unfolded states exhibit the same value (A_λ) at a wavelength λ. According to the above assumption ($A_N = A_U = A_\lambda$), the absorbance value ($A(c)$) at λ should be essentially identical (A_λ) at any denaturant concentration [165].

Thus, the existence of these specific points (a single point in many cases), referred to as isosbestic and isodichroic points for absorbance and CD, respectively, lends support to the two-state unfolding transition whereas we have to think about a multistate unfolding transition in the absence of these kinds of points.

23.6.1.2 Multi-state (Three-state) Unfolding Transition

The three-state unfolding transition is a simple extension of the two-state unfolding transition in terms of an additional intermediate state that is stably populated at equilibrium. The unfolding equilibrium of the three-state model is described as:

$$N \rightleftarrows I \rightleftarrows U$$

where I is the equilibrium unfolding intermediate. The observed value of a given physical parameter $A(c)$ at denaturant concentrations c is thus given by

$$A(c) = A_N \times f_N(c) + A_I \times f_I(c) + A_U \times f_U(c) \tag{4}$$

where A_I is the ideal value of the parameter in the intermediate state, $f_I(c)$ is the fraction of the intermediate state at denaturant concentration c, and $f_N + f_I + f_U = 1$. When the intermediate is fully populated at an intermediate concentration of denaturant, we can experimentally determine A_I and extrapolate the values into regions where the intermediate is not stably populated. However, this is not always the case. For the molten globule intermediate, we can often assume that $A_I = A_N$ for the parameters that represent the native secondary structure and that $A_I = A_U$ for the parameters that represent the specific tertiary structure of side chains (see below). Although the two-state unfolding transition in principle requires only a single probe to determine the fraction native or unfolded, the three-state unfolding transition requires two independent probes when the two transitions, $N \rightleftarrows I$ and $I \rightleftarrows U$ are not separated from each other.

As described previously, the molten globule state has a native-like secondary structure with virtually no specific side-chain packing. Taking advantage of these properties of the molten globule state, CD spectroscopy in the far- and near-UV regions, which are sensitive to the backbone secondary structure and the asymmetry of the side chains in aromatic residues, respectively, is extremely useful because they can monitor individually these two properties specific for the molten globule state.

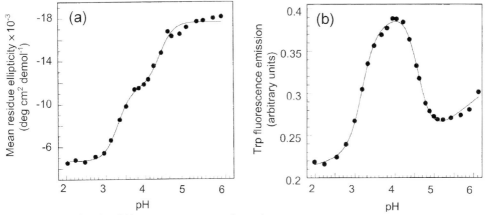

Fig. 23.7. pH-induced unfolding transition curves of apoMb monitored by a) CD at 222 nm and b) tryptophan fluorescence emission. From [84] with permission of Oxford University Press.

The equilibrium unfolding of α-LA monitored using CD spectroscopy in the far- and near-UV regions is an excellent example that clearly demonstrates that the molten globule is stably populated at equilibrium. The transition occurs at a lower denaturant concentration when monitored by the near-UV CD than when monitored by the far-UV CD (Figure 23.6b), and such noncoincidence of the apparent transition curves measured by the different probes is clear evidence that the molten globule intermediate accumulates between the two transition regions [23]. Accumulation of the molten globule state was also observed in the same manner in many other proteins, including canine and equine milk lysozymes [25, 80].

Another example demonstrating the population of the equilibrium unfolding intermediate is the observation of a two-step (or multi-step) unfolding transition measured by a single spectroscopic probe. A good example is the observation of the molten globule intermediate of apoMb at an intermediate pH (pH 4) in the pH-induced unfolding transition. Figure 23.7 shows the pH-induced unfolding transition curve of apoMb monitored by the far-UV CD and the intrinsic tryptophan fluorescence [84]. In contrast to the two-state unfolding transition curves that must have only a single-step transition, apoMb exhibits a two-step transition with a significant enhancement (more intense than the values in the native as well as the unfolded states) in the fluorescence and a plateau region in the far-UV CD at a modest acidic pH region around pH 4. These spectroscopic properties are consistent with population of the molten globule state at the intermediate pH as described in Section 23.2; it contains a portion of the native helical structure with two partially buried tryptophan residues that are located in the A-helix region and give rise to the enhancement of the fluorescence [45, 84].

The apparent difference in the behavior of the unfolding transition curves between α-LA and apoMb is, however, not essential. Because the α-LA molten globule exhibits a substantial amount of the secondary structure with virtual lacking the specific side-chain packing, there is little change in the far- and the near-UV CD intensities on the N ⇌ I and I ⇌ U transitions, respectively, which results in non-coincident single-step unfolding transitions measured by these probes. On the other hand, the apoMb molten globule is sufficiently stable at the intermediate pH values, and exhibits the properties distinct from those in the native and the unfolded states, so that it emphasizes the population of the molten globule state in the unfolding transition curves.

23.6.2
Burst-phase Intermediate Accumulated during the Dead Time of Refolding Kinetics

For many globular proteins, a kinetic folding intermediate is accumulated at an early stage during the refolding from the fully unfolded state under a strongly native condition, and this accumulation of the intermediate often occurs within the dead time of the measurements. Figure 23.8 shows the kinetic trace of α-LA refolding monitored by CD in the far-UV region at 222 nm [106]. In the kinetic trace, a significant fraction of the CD change occurs within the dead time of the measure-

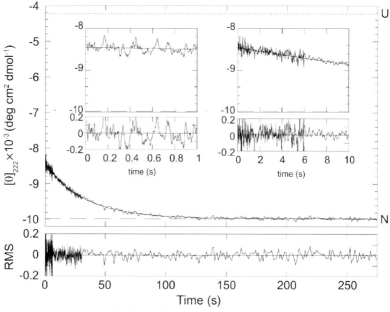

Fig. 23.8. Kinetic trace of refolding of α-LA monitored by CD at 222 nm. The refolding reaction was initiated by a GdmCl concentration jump from 5 M to 1 M. N denotes the ellipticity value of the native state, and U denotes the ellipticity value of the unfolded state at 1 M GdmCl obtained by linear extrapolation of the baseline for the unfolded state. From [106] with permission from Elsevier.

ments, indicating the transient accumulation of an intermediate state that has a substantial amount of the secondary structure.

The unresolved signal change (missing amplitude) occurring within the dead time of measurement in kinetic refolding experiments is referred to as the "burst-phase" change, and the corresponding intermediate is referred to as the "burst-phase intermediate" of refolding. Kinetic traces for the refolding reactions are fitted to a sum of exponential functions. If the significant structural formation occurs within the dead time during refolding, the value obtained by extrapolating the fitted kinetic refolding curve to zero time must be different from the expected value for the unfolded state that can be obtained by extrapolating the values of the unfolded state in the posttransition region of the equilibrium unfolding transition curve to the refolding condition. A more strict definition of the burst-phase change is, thus, a change from the expected value for the unfolded state to the zero-time extrapolated value of the fitted kinetic refolding curve. The burst-phase change results from at least one very fast phase unresolved in the observed kinetic trace, whose rate constant should be much larger than 1/(dead time).

The burst-phase change can be observed by using a variety of conformational probes, including CD, pulsed hydrogen/deuterium exchange, intrinsic and extrinsic fluorescence, and SAXS [57, 80, 104–106, 113, 166]. Among these, SAXS is useful in estimating the size and the overall shape of the burst-phase intermediate. Our previous SAXS measurements of α-LA have directly demonstrated that the burst-phase intermediate, accumulated within $\sim$10 ms after the initiation of the refolding, has a radius of gyration very close to that of the equilibrium molten globule state ($\sim$10% larger than that of the native state) with a globular shape [57], lending support to the idea that the kinetic intermediate of α-LA is identical to the molten globule state.

Whether or not the burst-phase intermediate is observed in the kinetic refolding of a protein depends on the rate of formation of the intermediate and the dead time of measurement. For example, the fluorescence of cyt c is quenched by almost 80% of the total change expected on the refolding reaction within the dead time of stopped-flow measurement (a few milliseconds) while kinetic traces monitored by using continuous-flow fluorescence methods with a dead time of $\sim$50 μs provides the total signal change expected [131].

23.6.3
Testing the Identity of the Molten Globule State with the Burst-Phase Intermediate

Measurements of the kinetic progress curves of refolding at various wavelengths will give us the CD spectrum of the intermediate state. Figure 23.9 shows the far- and near-UV CD spectra of α-LA burst-phase intermediate and they are compared with the equilibrium CD spectra of the protein [104]. The CD spectra of the burst-phase intermediate are similar to the equilibrium CD spectra of the molten globule state. These results further support the idea that the folding intermediate of α-LA is identical to the molten globule.

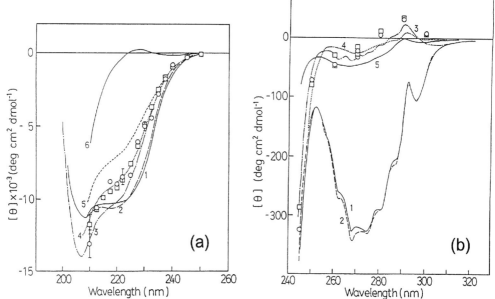

Fig. 23.9. CD spectra of the molten globule state of α-LA at pH 2.0 in the far- and near-UV regions compared with the CD spectra of the native and unfolded states. Open circles and squares show the CD values obtained by extrapolating to zero time of the refolding curves. 1 and 2, the native state of the holo and apo forms, respectively; 3, the molten globule state at pH 2.0; a) 4 and 5, the thermally unfolded states at 41 and 78 °C, respectively; 6, the unfolded state by GdmCl; b) 4, the thermally unfolded state at 62.5 °C; 5, the unfolded state by GdmCl. From [104] with permission from the American Chemical Society.

The formation of secondary structure in the burst-phase of refolding is a very rapid process occurring within a few milliseconds, so that the rapid pre-equilibrium between the unfolded and the intermediate states is established at a very early stage in the refolding reaction. The simplest example of the refolding reactions involving an on-pathway intermediate can be described as follows:

$$U \xrightleftharpoons{\text{very rapid}} I \xrightarrow{\text{slow}} N$$

where U, I, and N represent the unfolded, the intermediate, and the native states, respectively.

The unfolding transition curve of the intermediate state is obtained by measuring the refolding reactions at varying denaturant concentrations and by investigating the dependence of the burst-phase CD spectrum on the denaturant concentration. Comparison of the unfolding transitions of the molten globule state and the burst-phase provide an additional test for the identity of these two states in terms of stability. Figure 23.10 shows the unfolding transition curves of the burst-phase intermediate and the equilibrium molten globule state of α-LA measured by ellip-

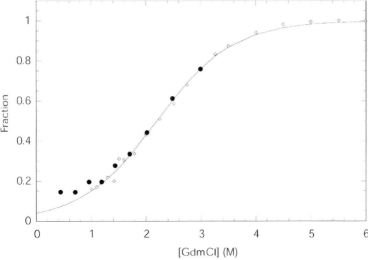

Fig. 23.10. The normalized unfolding transition curves of the burst-phase intermediate of α-LA (filled circles) compared with that of the equilibrium molten globule state at neutral pH (open diamonds). The solid line shows the theoretical curve assuming a two-state transition between the molten globule state and the unfolded state. From [106] with permission from Elsevier.

ticity at 222 nm [106]. The unfolding transition curve of the burst-phase intermediate is superimposable to the unfolding transition of the molten globule state that has been observed at equilibrium, indicating, again, that these two states of α-LA are identical to each other.

Although this method is powerful to compare the stability of the burst-phase intermediate with that of the molten globule, thermodynamic parameters of the burst-phase intermediate may not be able to be reliably obtained. As described in Section 23.2.3, the unfolding transitions of the molten globule state may not occur in a cooperative manner [58], which means that the two-state model may not apply to the transition between the unfolded state and the intermediates. Nevertheless, it is worth measuring these unfolding transitions for the purpose of comparing the molten globule and the burst-phase intermediate.

References

1 K. Kuwajima *Proteins* **1989**, *6*, 87–103.
2 O. B. Ptitsyn *Adv. Protein Chem.* **1995**, *47*, 83–229.
3 M. Arai, K. Kuwajima *Adv. Protein Chem.* **2000**, *53*, 209–282.
4 K. Kuwajima, M. Arai *Mechanisms of Protein Folding*, 2nd edn. Oxford University Press, Oxford, **2000**.
5 C. Levinthal *J. Chim. Phys.* **1968**, *65*, 44–45.
6 P. S. Kim, R. L. Baldwin *Annu. Rev. Biochem.* **1982**, *51*, 459–489.
7 P. S. Kim, R. L. Baldwin *Annu. Rev. Biochem.* **1990**, *59*, 631–660.
8 C. R. Matthews *Annu. Rev. Biochem.* **1993**, *62*, 653–683.

9 J. D. Bryngelson, J. N. Onuchic, N. D. Socci, P. G. Wolynes *Proteins* **1995**, *21*, 167–195.
10 K. A. Dill, S. Bromberg, K. Yue, K. M. Fiebig, D. P. Yee, P. D. Thomas, H. S. Chan *Protein Sci.* **1995**, *4*, 561–602.
11 K. A. Dill, H. S. Chan *Nat. Struct. Biol.* **1997**, *4*, 10–19.
12 A. M. Gutin, V. I. Abkevich, E. I. Shakhnovich *Biochemistry* **1995**, *34*, 3066–3076.
13 A. R. Fersht *Curr. Opin. Struct. Biol.* **1997**, *7*, 3–9.
14 S. E. Jackson *Fold. Des.* **1998**, *3*, R81–R91.
15 D. V. Laurents, R. L. Baldwin *Biophys. J.* **1998**, *75*, 428–434.
16 R. L. Baldwin *Fold. Des.* **1995**, *1*, R1–R8.
17 R. L. Baldwin, G. D. Rose *Trends Biochem. Sci.* **1999**, *24*, 26–33.
18 R. L. Baldwin, G. D. Rose *Trends Biochem. Sci.* **1999**, *24*, 77–83.
19 D. A. Dolgikh, L. V. Abaturov, I. A. Bolotina et al. *Eur. Biophys. J.* **1985**, *13*, 109–121.
20 M. Ohgushi, A. Wada *FEBS Lett.* **1983**, *164*, 21–24.
21 K. Kuwajima *FASEB J.* **1996**, *10*, 102–109.
22 L. A. Morozova-Roche, C. C. Arico-Muendel, D. T. Haynie, V. I. Emelyanenko, H. Van Dael, C. M. Dobson *J. Mol. Biol.* **1997**, *268*, 903–921.
23 M. Ikeguchi, K. Kuwajima, S. Sugai *J. Biochem. (Tokyo)* **1986**, *99*, 1191–1201.
24 H. Van Dael, P. Haezebrouck, L. Morozova, C. Arico-Muendel, C. M. Dobson *Biochemistry* **1993**, *32*, 11886–11894.
25 M. Mizuguchi, M. Arai, Y. Ke, K. Nitta, K. Kuwajima *J. Mol. Biol.* **1998**, *283*, 265–277.
26 M. Kikuchi, K. Kawano, K. Nitta *Protein Sci.* **1998**, *7*, 2150–2155.
27 K. Yutani, K. Ogasahara, K. Kuwajima *J. Mol. Biol.* **1992**, *228*, 347–350.
28 Y. V. Griko, E. Freire, G. Privalov, H. Van Dael, P. L. Privalov *J. Mol. Biol.* **1995**, *252*, 447–459.
29 V. E. Bychkova, A. E. Dujsekina, S. I. Klenin, E. I. Tiktopulo, V. N. Uversky, O. B. Ptitsyn *Biochemistry* **1996**, *35*, 6058–6063.
30 V. N. Uversky, N. V. Narizhneva, S. O. Kirschstein, S. Winter, G. Lober *Fold. Des.* **1997**, *2*, 163–172.
31 G. J. A. Vidugiris, C. A. Royer *Biophys. J.* **1998**, *75*, 463–470.
32 R. Kitahara, II. Yamada, K. Akasaka, P. E. Wright *J. Mol. Biol.* **2002**, *320*, 311–319.
33 M. W. Lassalle, H. Li, H. Yamada, K. Akasaka, C. Redfield *Protein Sci.* **2003**, *12*, 66–72.
34 Y. Goto, N. Takahashi, A. L. Fink *Biochemistry* **1990**, *29*, 3480–3488.
35 V. N. Uversky, V. P. Kutyshenko, N. Y. Protasova, V. V. Rogov, K. S. Vassilenko, A. T. Gudkov *Protein Sci.* **1996**, *5*, 1844–1851.
36 V. N. Uversky, V. V. Leontiev, A. T. Gudkov *Protein Eng.* **1992**, *5*, 781–783.
37 T. E. Creighton, J. J. Ewbank *Biochemistry* **1994**, *33*, 1534–1538.
38 S. J. Demarest, R. Fairman, D. P. Raleigh *J. Mol. Biol.* **1998**, *283*, 279–291.
39 S. J. Demarest, J. A. Boice, R. Fairman, D. P. Raleigh *J. Mol. Biol.* **1999**, *294*, 213–221.
40 P. Polverino de Laureto, D. Vinante, E. Scaramella, E. Frare, A. Fontana *Eur. J. Biochem.* **2001**, *268*, 4324–4333.
41 K. Kuwajima *Circular Dichroism and the Conformational Analysis of Biomolecules*. Plenum, New York, **1996**.
42 V. N. Uversky, O. B. Ptitsyn *Biochemistry* **1994**, *33*, 2782–2791.
43 K. Gast, H. Damaschun, R. Misselwitz, M. Mueller-Frohne, D. Zirwer, G. Damaschun *Eur. Biophys. J.* **1994**, *23*, 297–305.
44 G. V. Semisotnov, N. A. Rodionova, O. I. Razgulyaev, V. N. Uversky, A. F. Gripas, R. I. Gilmanshin *Biopolymers* **1991**, *31*, 119–128.
45 F. M. Hughson, P. E. Wright, R. L. Baldwin *Science* **1990**, *249*, 1544–1548.
46 R. L. Baldwin *Curr. Opin. Struct. Biol.* **1993**, *3*, 84–91.

47 B. A. Schulman, C. Redfield, Z. Y. Peng, C. M. Dobson, P. S. Kim *J. Mol. Biol.* **1995**, *253*, 651–657.
48 J. A. Jones, D. K. Wilkins, L. J. Smith, C. M. Dobson *J. Biomol. NMR* **1997**, *10*, 199–203.
49 B. A. Schulman, P. S. Kim, C. M. Dobson, C. Redfield *Nat. Struct. Biol.* **1997**, *4*, 630–634.
50 D. Eliezer, J. Yao, H. J. Dyson, P. E. Wright *Nat. Struct. Biol.* **1998**, *5*, 148–155.
51 C. Redfield, B. A. Schulman, M. A. Milhollen, P. S. Kim, C. M. Dobson *Nat. Struct. Biol.* **1999**, *6*, 948–952.
52 R. Wijesinha-Bettoni, C. M. Dobson, C. Redfield *J. Mol. Biol.* **2001**, *312*, 261–273.
53 S. Ramboarina, C. Redfield *J. Mol. Biol.* **2003**, *330*, 1177–1188.
54 D. Eliezer, P. A. Jennings, P. E. Wright, S. Doniach, K. O. Hodgson, H. Tsuruta *Science* **1995**, *270*, 487–488.
55 M. Kataoka, Y. Goto *Fold. Des.* **1996**, *1*, R107–R114.
56 M. Arai, T. Ikura, G. V. Semisotnov, H. Kihara, Y. Amemiya, K. Kuwajima *J. Mol. Biol.* **1998**, *275*, 149–162.
57 M. Arai, K. Ito, T. Inobe et al. *J. Mol. Biol.* **2002**, *321*, 121–132.
58 B. A. Schulman, P. S. Kim *Nat. Struct. Biol.* **1996**, *3*, 682–687.
59 J. L. Marmorino, M. Lehti, G. J. Pielak *J. Mol. Biol.* **1998**, *275*, 379–388.
60 L. C. Wu, P. S. Kim *J. Mol. Biol.* **1998**, *280*, 175–182.
61 J. Song, P. Bai, L. Luo, Z. Y. Peng *J. Mol. Biol.* **1998**, *280*, 167–174.
62 P. Bai, L. Luo, Z. Y. Peng *Biochemistry* **2000**, *39*, 372–380.
63 P. Bai, J. Song, L. Luo, Z. Y. Peng *Protein Sci.* **2001**, *10*, 55–62.
64 Z. Peng, P. S. Kim *Biochemistry* **1994**, *33*, 2136–2141.
65 L. C. Wu, Z. Peng, P. S. Kim *Nat. Struct. Biol.* **1995**, *2*, 281–286.
66 D. Eliezer, J. Chung, H. J. Dyson, P. E. Wright *Biochemistry* **2000**, *39*, 2894–2901.
67 H. Tsuge, H. Ago, M. Noma, K. Nitta, S. Sugai, M. Miyano *J. Biochem. (Tokyo)* **1992**, *111*, 141–143.
68 J. Ren, D. I. Stuart, K. R. Acharya *J. Biol. Chem.* **1993**, *268*, 19292–19298.
69 E. D. Chrysina, K. Brew, K. R. Acharya *J. Biol. Chem.* **2000**, *275*, 37021–37029.
70 T. Koshiba, M. Yao, Y. Kobashigawa et al. *Biochemistry* **2000**, *39*, 3248–3257.
71 K. Nitta, H. Tsuge, H. Iwamoto *Int. J. Pept. Protein Res.* **1993**, *41*, 118–123.
72 S. Sugai, M. Ikeguchi *Adv. Biophys.* **1994**, *30*, 37–84.
73 T. Koshiba, T. Hayashi, I. Miwako et al. *Protein Eng.* **1999**, *12*, 429–435.
74 V. Forge, R. T. Wijesinha, J. Balbach et al. *J. Mol. Biol.* **1999**, *288*, 673–688.
75 P. Haezebrouck, K. Noyelle, M. Joniau, H. Van Dael *J. Mol. Biol.* **1999**, *293*, 703–718.
76 M. Mizuguchi, K. Masaki, K. Nitta *J. Mol. Biol.* **1999**, *292*, 1137–1148.
77 M. Mizuguchi, K. Masaki, M. Demura, K. Nitta *J. Mol. Biol.* **2000**, *298*, 985–995.
78 Y. Kobashigawa, M. Demura, T. Koshiba, Y. Kumaki, K. Kuwajima, K. Nitta *Proteins* **2000**, *40*, 579–589.
79 M. Joniau, P. Haezebrouck, K. Noyelle, H. Van Dael *Proteins* **2001**, *44*, 1–11.
80 H. Van Dael, P. Haezebrouck, M. Joniau *Protein Sci.* **2003**, *12*, 609–619.
81 J. Kuriyan, S. Wilz, M. Karplus, G. A. Petsko *J. Mol. Biol.* **1986**, *192*, 133–154.
82 Y. V. Griko, P. L. Privalov, S. Y. Venyaminov, V. P. Kutyshenko *J. Mol. Biol.* **1988**, *202*, 127–138.
83 D. Barrick, R. L. Baldwin *Biochemistry* **1993**, *32*, 3790–3796.
84 P. E. Wright, R. L. Baldwin *Mechanisms of Protein Folding*, 2nd edn. Oxford University Press, Oxford, **2000**.
85 S. N. Loh, M. S. Kay, R. L. Baldwin *Proc. Natl Acad. Sci. USA* **1995**, *92*, 5446–5450.
86 M. Jamin, R. L. Baldwin *J. Mol. Biol.* **1998**, *276*, 491–504.
87 G. W. Bushnell, G. V. Louie, G. D. Brayer *J. Mol. Biol.* **1990**, *214*, 585–595.

88 M. F. Jeng, S. W. Englander, G. A. Elove, A. J. Wand, H. Roder *Biochemistry* **1990**, *29*, 10433–10437.
89 W. Colon, G. A. Eloeve, L. P. Wakem, F. Sherman, H. Roder *Biochemistry* **1996**, *35*, 5538–5549.
90 W. Colon, H. Roder *Nat. Struct. Biol.* **1996**, *3*, 1019–1025.
91 P. L. Privalov, S. J. Gill *Adv. Protein Chem.* **1988**, *39*, 191–234.
92 Y. V. Griko, E. Freire, P. L. Privalov *Biochemistry* **1994**, *33*, 1889–1899.
93 Y. V. Griko, P. L. Privalov *J. Mol. Biol.* **1994**, *235*, 1318–1325.
94 Y. V. Griko *J. Mol. Biol.* **2000**, *297*, 1259–1268.
95 S. Potekhin, W. Pfeil *Biophys. Chem.* **1989**, *34*, 55–62.
96 J. H. Carra, E. A. Anderson, P. L. Privalov *Protein Sci.* **1994**, *3*, 952–959.
97 T. Koshiba, Y. Kobashigawa, M. Demura, K. Nitta *Protein Eng.* **2001**, *14*, 967–974.
98 J. H. Carra, E. A. Anderson, P. L. Privalov *Biochemistry* **1994**, *33*, 10842–10850.
99 I. Nishii, M. Kataoka, Y. Goto *J. Mol. Biol.* **1995**, *250*, 223–238.
100 M. S. Kay, R. L. Baldwin *Nat. Struct. Biol.* **1996**, *3*, 439–445.
101 Y. Z. Luo, M. S. Kay, R. L. Baldwin *Nat. Struct. Biol.* **1997**, *4*, 925–930.
102 M. S. Kay, C. H. Ramos, R. L. Baldwin *Proc. Natl Acad. Sci. USA* **1999**, *96*, 2007–2012.
103 Y. Hagihara, Y. Tan, Y. Goto *J. Mol. Biol.* **1994**, *237*, 336–348.
104 K. Kuwajima, Y. Hiraoka, M. Ikeguchi, S. Sugai *Biochemistry* **1985**, *24*, 874–881.
105 M. Ikeguchi, K. Kuwajima, M. Mitani, S. Sugai *Biochemistry* **1986**, *25*, 6965–6972.
106 M. Arai, K. Kuwajima *Fold. Des.* **1996**, *1*, 275–287.
107 G. V. Semisotnov, H. Kihara, N. V. Kotova et al. *J. Mol. Biol.* **1996**, *262*, 559–574.
108 H. Roder, G. A. Eloeve, S. W. Englander *Nature* **1988**, *335*, 700–704.
109 P. A. Jennings, P. E. Wright *Science* **1993**, *262*, 892–896.
110 A. Miranker, C. V. Robinson, S. E. Radford, R. T. Aplin, C. M. Dobson *Science* **1993**, *262*, 896–900.
111 T. M. Raschke, S. Marqusee *Nat. Struct. Biol.* **1997**, *4*, 298–304.
112 V. Tsui, C. Garcia, S. Cavagnero, G. Siuzdak, H. J. Dyson, P. E. Wright *Protein Sci.* **1999**, *8*, 45–49.
113 C. Nishimura, H. J. Dyson, P. E. Wright *J. Mol. Biol.* **2002**, *322*, 483–489.
114 S. Cavagnero, H. J. Dyson, P. E. Wright *J. Mol. Biol.* **1999**, *285*, 269–282.
115 S. Cavagnero, C. Nishimura, S. Schwarzinger, H. J. Dyson, P. E. Wright *Biochemistry* **2001**, *40*, 14459–14467.
116 C. Nishimura, P. E. Wright, H. J. Dyson *J. Mol. Biol.* **2003**, *334*, 293–307.
117 T. M. Raschke, J. Kho, S. Marqusee *Nat. Struct. Biol.* **1999**, *6*, 825–831.
118 J. Balbach, V. Forge, N. A. van Nuland, S. L. Winder, P. J. Hore, C. M. Dobson *Nat. Struct. Biol.* **1995**, *2*, 865–870.
119 J. Balbach, V. Forge, W. S. Lau, N. A. J. Van Nuland, K. Brew, C. M. Dobson *Science* **1996**, *274*, 1161–1163.
120 M. Roy, P. A. Jennings *J. Mol. Biol.* **2003**, *328*, 693–703.
121 C. M. Jones, E. R. Henry, Y. Hu et al. *Proc. Natl Acad. Sci. USA* **1993**, *90*, 11860–11864.
122 T. Pascher, J. P. Chesick, J. R. Winkler, H. B. Gray *Science* **1996**, *271*, 1558–1560.
123 B. Nolting *Biochem. Biophys. Res. Commun.* **1996**, *227*, 903–908.
124 R. M. Ballew, J. Sabelko, M. Gruebele *Proc. Natl Acad. Sci. USA* **1996**, *93*, 5759–5764.
125 C. K. Chan, Y. Hu, S. Takahashi, D. L. Rousseau, W. A. Eaton, J. Hofrichter *Proc. Natl Acad. Sci. USA* **1997**, *94*, 1779–1784.
126 S. Takahashi, S. R. Yeh, T. K. Das, C. K. Chan, D. S. Gottfried, D. L. Rousseau *Nat. Struct. Biol.* **1997**, *4*, 44–50.
127 M. C. R. Shastry, S. D. Luck, H. Roder *Biophys. J.* **1998**, *74*, 2714–2721.

128 L. Pollack, M. W. Tate, N. C. Darnton et al. *Proc. Natl Acad. Sci. USA* **1999**, *96*, 10115–10117.

129 M. Gruebele *Annu. Rev. Phys. Chem.* **1999**, *50*, 485–516.

130 T. Uzawa, S. Akiyama, T. Kimura et al. *Proc. Natl Acad. Sci. USA* **2004**, *101*, 1171–1176.

131 M. C. R. Shastry, H. Roder *Nat. Struct. Biol.* **1998**, *5*, 385–392.

132 S. Akiyama, S. Takahashi, K. Ishimori, I. Morishima *Nat. Struct. Biol.* **2000**, *7*, 514–520.

133 S. Akiyama, S. Takahashi, T. Kimura et al. *Proc. Natl Acad. Sci. USA* **2002**, *99*, 1329–1334.

134 G. A. Eloeve, A. F. Chaffotte, H. Roder, M. E. Goldberg *Biochemistry* **1992**, *31*, 6876–6883.

135 H. Roder, W. Colon *Curr. Opin. Struct. Biol.* **1997**, *7*, 15–28.

136 J. M. Sauder, H. Roder *Fold. Des.* **1998**, *3*, 293–301.

137 D. K. Heidary, L. A. Gross, M. Roy, P. A. Jennings *Nat. Struct. Biol.* **1997**, *4*, 725–731.

138 W. F. Walkenhorst, S. M. Green, H. Roder *Biochemistry* **1997**, *36*, 5795–5805.

139 K. Maki, T. Ikura, T. Hayano, N. Takahashi, K. Kuwajima *Biochemistry* **1999**, *38*, 2213–2223.

140 K. Kamagata, Y. Sawano, M. Tanokura, K. Kuwajima *J. Mol. Biol.* **2003**, *332*, 1143–1153.

141 D. K. Heidary, J. C. O'Neill, M. Roy, P. A. Jennings *Proc. Natl Acad. Sci. USA* **2000**, *97*, 5866–5870.

142 A. P. Capaldi, M. C. Shastry, C. Kleanthous, H. Roder, S. E. Radford *Nat. Struct. Biol.* **2001**, *8*, 68–72.

143 S. Khorasanizadeh, I. D. Peters, H. Roder *Nat. Struct. Biol.* **1996**, *3*, 193–205.

144 S. H. Park, M. C. R. Shastry, H. Roder *Nat. Struct. Biol.* **1999**, *6*, 943–947.

145 J. M. Finke, P. A. Jennings *Biochemistry* **2002**, *41*, 15056–15067.

146 S. Gianni, C. Travaglini-Allocatelli, F. Cutruzzola, M. Brunori, M. C. Shastry, H. Roder *J. Mol. Biol.* **2003**, *330*, 1145–1152.

147 A. R. Viguera, J. C. Martinez, V. V. Filimonov, P. L. Mateo, L. Serrano *Biochemistry* **1994**, *33*, 2142–2150.

148 V. P. Grantcharova, D. Baker *Biochemistry* **1997**, *36*, 15685–15692.

149 J. I. Guijarro, C. J. Morton, K. W. Plaxco, I. D. Campbell, C. M. Dobson *J. Mol. Biol.* **1998**, *276*, 657–667.

150 K. W. Plaxco, J. I. Guijarro, C. J. Morton, M. Pitkeathly, I. D. Campbell, C. M. Dobson *Biochemistry* **1998**, *37*, 2529–2537.

151 N. Ferguson, A. P. Capaldi, R. James, C. Kleanthous, S. E. Radford *J. Mol. Biol.* **1999**, *286*, 1597–1608.

152 D. Perl, C. Welker, T. Schindler et al. *Nat. Struct. Biol.* **1998**, *5*, 229–235.

153 R. E. Burton, G. S. Huang, M. A. Daugherty, P. W. Fullbright, T. G. Oas *J. Mol. Biol.* **1996**, *263*, 311–322.

154 M. L. Scalley, Q. Yi, H. D. Gu, A. McCormack, I. I. Yates JR, D. Baker *Biochemistry* **1997**, *36*, 3373–3382.

155 S. E. Jackson, A. R. Fersht *Biochemistry* **1991**, *30*, 10428–10435.

156 K. Kamagata, M. Arai, K. Kuwajima *J. Mol. Biol.* **2004**, *339*, 951–965.

157 D. E. Makarov, K. W. Plaxco *Protein Sci.* **2003**, *12*, 17–26.

158 V. Grantcharova, E. J. Alm, D. Baker, A. L. Horwich *Curr. Opin. Struct. Biol.* **2001**, *11*, 70–82.

159 C. Micheletti *Proteins* **2003**, *51*, 74–84.

160 K. W. Plaxco, K. T. Simons, D. Baker *J. Mol. Biol.* **1998**, *277*, 985–994.

161 I. E. Sanchez, T. Kiefhaber *J. Mol. Biol.* **2003**, *325*, 367–376.

162 Y. Bai *Biochem. Biophys. Res. Commun.* **2003**, *305*, 785–788.

163 G. M. Spudich, E. J. Miller, S. Marqusee *J. Mol. Biol.* **2004**, *335*, 609–618.

164 C. Tanford *Adv. Protein Chem.* **1970**, *24*, 1–95.

165 C. D. Snow, H. Nguyen, V. S. Pande, M. Gruebele *Nature* **2002**, *420*, 102–106.

166 M. Engelhard, P. A. Evans *Protein Sci.* **1995**, *4*, 1553–1562.

24
Alcohol- and Salt-induced Partially Folded Intermediates

Daizo Hamada and Yuji Goto

24.1
Introduction

The structures and stabilities of proteins are perturbed by additives such as small organic and inorganic molecules. In this chapter, we discuss the effects of alcohols and salts on inducing partially folded intermediates of proteins. These co-solvents are expected to stabilize otherwise unstable folding intermediates. They can also reveal the conformational preference of the intermediate or unfolded states. This information is valuable for characterizing the structure and stability of the folding intermediates under physiological conditions in the absence of co-solvents. Since this chapter focuses on the effects on the intermediate and unfolded states among various effects on proteins, readers should refer to other reviews on the general effects of alcohols [1, 2] and salts [3–7] on proteins.

In the early 1960s, Tanford and coworkers performed pioneering studies about the effects of organic solvents on the conformational properties of proteins using optical rotatory dispersion (ORD) [3, 4, 8]. For example, the conformational transition of bovine ribonuclease A induced by the addition of 2-chloroethanol into aqueous solvent occurred in two distinct stages involving partial unfolding of the native structure followed by the formation of an α-helical structure. A similar conformational change was also observed for bovine β-lactoglobulin [3, 4] and diisopropylphosphoryl chymotrypsin [9]. Recently, circular dichroism (CD) spectroscopy has become more popular than ORD because of the simplicity in the interpretation of data. However, the results obtained by CD provide essentially the same information as can be obtained by ORD. Further information about the effects of alcohols can be analyzed even to atomic resolution by solution nuclear magnetic resonance (NMR) spectroscopy.

Alcohols are basically categorized as denaturants, as are guanidinium hydrochloride (GdmCl) or urea. Although GdmCl and urea induce random coil structures, alcohols tend to stabilize well-ordered conformations such as α-helical structures. According to the review by Tanford [3], Imahori and Doty first recognized that a high proportion of α-helical proteins form α-helical structures in 2-chloroethanol. Similar effects have been found for many other alcohols. Importantly, Herskovits and Mescanti [10] have reported that α_8-casein, a typical natively unfolded protein [11–14] is also converted into an α-helical structure in the presence of 2-chloroethanol or methanol. Although a significant number of α-helical structures are formed by alcohols, they are certainly not equivalent to the native conformation and are classified as denatured conformations, which lack the native protein's functions.

The properties of alcohols that induce the conformational change in proteins are also widely used for analyzing the intrinsic structural preference of proteins and peptides, particularly to analyze the kinetic intermediates accumulating at the early stages of folding. Although proteins assume significant levels of nonnative α-helical conformations at high concentrations of alcohols, several short peptides form β-turn or β-hairpin conformations according to the structural preference of the amino acid sequence [15–22]. The structures formed in alcohol/water solvent somehow mimic the properties of early intermediates in protein folding, since both structures are stabilized predominantly by local interactions which are formed between residues near each other in amino acid sequence (Figure 24.1) [23, 24]. Moreover, the molten globule state, a compact denatured state with a significant amount of native-like secondary structures but with fluctuating tertiary structures [25–29] (see Chapter 23), is sometimes observed during the alcohol-induced denaturation of several proteins under moderate concentrations of alcohols. Thus, alcohols have the potential to stabilize different types of partially folded structures of polypeptide chains depending on the alcohol species and concentration ranges used (Figure 24.2). It is noted that, in this chapter, our definition of the molten globule state is not as strict as the original proposal and can be also used for the compact denatured state with some native-like secondary structures.

Local interaction

$|i-j| \leq 5$

Nonlocal interaction

$|i-j| > 5$

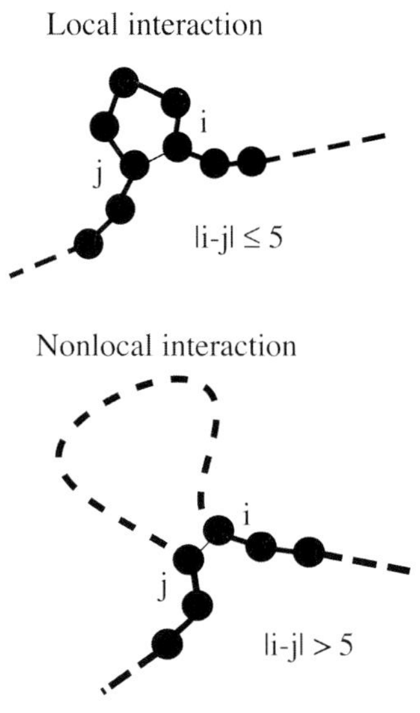

Fig. 24.1. Local (a) and nonlocal (b) interactions [23]. In the figure, circles represent each amino acid. The local interactions are formed between residues near each other in the amino acid sequence, whereas the nonlocal interactions are formed between residues apart each other in the sequence.

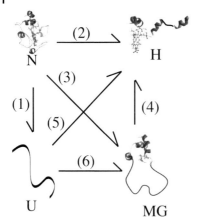

Fig. 24.2. Conformational transitions of proteins. N, native; U, fully unfolded; MG, molten globule; H, alcohol-induced helical states. Transitions by denaturants (GdmCl or urea) or at extreme pH (1), by strong helix inducers such as TFE or HFIP (2), by moderate concentrations of weak helix induces such as methanol or ethanol (3), by high concentrations of weak helix induces (4), by alcohols at extreme pH (5) and by anions at acidic pH (6) are represented schematically.

Salts have been used widely for various purposes in the preparation and characterization of proteins, including salt-induced precipitation (salting out), dissolution of proteins (salting in), and stabilizing or destabilizing protein structures [5–7] (see Chapter 3). Among them, denaturation of proteins by GdmCl is one of the most important applications of salt effects on proteins; we can understand the conformational stability of proteins on the basis of the salt-induced denaturation [1, 3, 4]. However, the findings that horse cytochrome c and several proteins form the molten globule intermediate in the presence of salts, as will be described in detail in this chapter, attracted much attention on the salt effects, suggesting that salts might be useful for specifically stabilizing the folding intermediates. Salt-stabilized molten globule states have been the target of extensive studies addressing the conformation and stability of the folding intermediates. In particular, characterization of the salt-dependent conformational changes of acid-denatured proteins have revealed a more clear view of the acid denaturation and effects of ions in modulating the conformations of the intermediate and unfolded states.

24.2
Alcohol-induced Intermediates of Proteins and Peptides

24.2.1
Formation of Secondary Structures by Alcohols

Alcohols promote the formation of ordered secondary structures in disordered polypeptide chains. The effectiveness of alcohols in inducing the conformational transition varies substantially (see below). Among various alcohols, 2,2,2-trifluoroethanol (TFE) is often used because of its relatively strong effects and low absorbance in the far-UV region, which makes far-UV CD measurements feasible. 1,1,1,3,3,3-hexafluoro-2-propanol (HFIP) has a higher potential than TFE to induce the alcohol-dependent effects, although the formation of alcohol clusters might complicate the interpretation of the alcohol effects (see below).

For many disordered peptides, α-helical structures are induced because they are stabilized by the most local interactions, but various peptides assume β-hairpin or β-turn structures in alcohol/water solvents [15–22]. Importantly, these secondary structures are all stabilized by local hydrogen bonds. The formation of β-sheet structures, which need a network of nonlocal hydrogen bonds, is not usually expected for short peptides unless the peptide molecules are associated into oligomeric structures or aggregates. Nevertheless, there are several examples of longer polypeptides or proteins, including tendamistat and leucocin A [30, 31], showing stabilization of β-sheet structures.

Thus, the exact types of secondary structures stabilized in alcohol/water solvent depend on the amino acid sequence. The intrinsic structural preference of each amino acid sequence is an important factor in determining the types of secondary structures in alcohol/water solvent [15–22, 30–39]. The ability of alcohols to induce the particular secondary structures is, therefore, useful to analyze the intrinsic conformational propensity of a short peptide with no appreciable ordered conformation in aqueous solution. The peptide conformation induced in alcohol/water solvent can be readily analyzed by conventional CD or solution NMR methods. These methods are also useful in detecting the structural properties of early intermediate of protein folding [32, 36, 39–42].

The same experimental system has been successfully applied to elucidate the conformational stabilities and properties of less structured de novo designed proteins in water solvent [19–20]. As is the case for short peptides, de novo proteins often fail to assume ordered conformations in water. In such a case, the addition of alcohol induces the formation of otherwise unstable secondary structures. By analyzing the alcohol-dependent transition, one can estimate the stability of the ordered conformation in water. Such an analysis is helpful to optimize the sequence of de novo proteins in order to create a rigid native-like structure even in the absence of alcohols. Furthermore, the structural properties of membrane proteins can be analyzed in alcohol/water solvent by solution NMR spectroscopy [43–46].

Interestingly, a high correlation was found between the α-helical content expected from secondary structure prediction based on amino acid sequence and the α-helical content estimated by CD in the presence of high concentrations of TFE [47, 48]. This observation supports the idea that the local interactions are dominant in the alcohol-induced protein conformations. Conventional secondary structure predictions consider mainly local interactions. A lesser correlation was found between the α-helical content in the native structure and that in the TFE state, indicating the role of nonlocal interactions for stabilizing the native structures. The α-helical content of the alcohol-induced state tends to be larger than that of the native state. Thus, the TFE-induced structures of polypeptide chains should contain a significant amount of nonnative α-helices.

24.2.2
Alcohol-induced Denaturation of Proteins

As described above, alcohols were originally used as denaturants. However, alcohols are not the first choice reagent for the analysis of protein stability since uncertainty exists in the thermodynamic interpretation of alcohol-induced states. Moreover, alcohols tend to induce nonspecific aggregates of proteins at neutral pH, probably because of the reduced electrostatic repulsion between protein molecules. Therefore, studies have often been performed under acidic conditions where the solubility of proteins in alcohol/water solvent increases.

Shiraki et al. [48] analyzed the effects of TFE on protein conformation using various types of proteins with different secondary structures. The transitions induced by TFE were monitored by far-UV CD. The transition curves obtained by plotting the ellipticity at 222 nm as a function of TFE concentration consisted of two phases. The initial transition was relatively cooperative showing an increase in α-helical content at a certain TFE concentration. At neutral pH, the concentration range of this transition was predominantly determined by the stability of native structure. On the other hand, at acidic pH ~ 2, many proteins were unfolded in the absence of TFE. For most of these unfolded proteins at pH 2, the induction of α-helical structure was observed in the TFE concentrations of 10–20% (v/v). For proteins with a relatively low propensity toward α-helices, the α-helical content just after the first transition was low, but, in the second stage, it gradually increased with further increases in the TFE concentration. When the intrinsic α-helical propensity is high, the posttransitional increase in α-helical content was less obvious because the helical content achieved after the first cooperative transition is already high.

The transition curves observed, for example by the far-UV CD, can be simply analyzed by assuming a two-state mechanism between the native and alcohol-induced states (see chapter by Pace and Scholtz for details of two-state analysis). In most cases, this method works well, providing two parameters, namely $\Delta G(H_2O)$ (a free energy change between the native and TFE states) and the m-value (a measure of cooperativity of transition). The value of $\Delta G(H_2O)$ provided by a such treatment is often consistent with the value obtained from the analysis of GdmCl or urea unfolding curves [2]. This is rather surprising since the TFE denatured state with highly ordered secondary structures is unlikely to be thermodynamically equivalent to the GdmCl unfolded state. Unlike the native state, the denatured state consists of an ensemble of nonnative conformations. Thus, a conformation similar to the alcohol-induced state possibly pre-exists at a low level in such an ensemble of denatured structures even under the native conditions.

The TFE-induced states of several proteins, for example hen lysozyme and bovine β-lactoglobulin, have been well characterized by high-resolution heteronuclear NMR [49, 50]. The chemical shifts of NMR spectra for the TFE state show less dispersed signals compared to the native structures. Beside this, a number of nuclear Overhauser effects (NOEs) consistent with the presence of α-helices can be observed. Other information on the secondary structure, extracted from a chemical shift index or coupling constants, also provides the evidence for the presence of α-helices. On the other hand, the backbone amide hydrogens are much less protected against hydrogen/deuterium exchange even though significant parts of the protein molecule are in an α-helical conformation. Thus, the secondary structures formed in the TFE state are mobile, and parts of polypeptide chain are rapidly interconverted to the extended configurations. In accordance with this, small-angle solution X-ray scattering indicated that the TFE-induced state of hen lysozyme possesses a chain-like character as is the case of fully unfolded structures in high concentrations of urea or GdmCl [51]. On the other hand, the radius of gyration of the TFE state was between the fully unfolded state and the compact native state. Taken together, the TFE state is considered to be an "open helical conformation" in which the α-helical rods are exposed to the solvent because of the absence of strong hydrophobic attraction between them, contrasting with the compact molten globule stabilized by salt. Similar results were obtained with the methanol-induced α-helical states of horse cytochrome c [52] and horse apomyoglobin [53].

24.2.3
Formation of Compact Molten Globule States

Although the alcohol-induced denaturation of proteins is often approximated by a two-state mechanism between the native and TFE-induced α-helical states, accumulation of the molten globule-like intermediate during the transition is sometimes observed in several proteins (Table 24.1) [52–67]. Generally speaking, the alcohols having less potential to induce α-helical structures seem to have a higher potential to induce a three-state transition with a molten globule intermediate. However, even an effective helix inducer such as TFE sometimes stabilizes the molten globule state, as has been shown for α-lactalbumin [60]. Interestingly, many such proteins are also known to form the molten globule states in the pres-

Tab. 24.1. Alcohol or salt-induced molten globule state under equilibrium conditions.

Protein	Conditions	Reference
Apomyoglobin	Salt at acidic pH	103
	Methanol at acidic pH	53
	HFIP at acid pH	67
β-Conglycinin	Ethanol at acidic pH	64
Cytochrome c	Salt at acidic pH	25, 112
	Methanol at acidic pH	54, 52, 57
	Glycerol at acidic pH	66, 59
	Polyol at acidic pH	61
Ervatamin C	Methanol at acidic pH	62
Retinol-binding protein	Methanol at acidic pH	55
β-Lactamase	Salt at acidic pH	102
α-Lactalbumin	TFE at acidic pH	60, 65
Lysozyme	3-Chloro-1,2-propanediol at acidic pH	63

ence of moderate concentrations of urea or GdmCl or by addition of salts in the acid-denatured state (Table 24.1).

The molten globule state is often found in the presence of moderate concentrations of alcohols. Higher concentrations of alcohols tend to stabilize the extended highly α-helical conformation as described above. The mechanism of how alcohols induce such a three-state transition with the molten globule intermediate is still ambiguous. A significant number of the native tertiary contacts between hydrophobic side chains are disrupted in the alcohol state. At the same time, the nonpolar environment introduced by alcohols forces the polypeptide chains to form non-native α-helices. In the presence of moderate concentrations of alcohol, it is likely that such effects of alcohols are less obvious, so that some proteins remain in a compact denatured state with native-like secondary structures.

24.2.4
Example: β-Lactoglobulin

Alcohol can induce α-helical conformations even in proteins consisting of predominantly β-sheets. Most all-β proteins tend to show a less significant increase in α-helical content in alcohol/water solvents. An interesting example revealing a dramatic increase of α-helical content is bovine β-lactoglobulin. The native structure of β-lactoglobulin consists of nine antiparallel β-strands and a long α-helix at the C-terminal region (Figure 24.3A). Shiraki et al. [48] demonstrated that bovine β-lactoglobulin exhibits an extremely high α-helical content upon the addition of TFE (Figure 24.3B). Our systematic analysis of peptide fragments derived from the β-sheet regions of β-lactoglobulin demonstrated that the peptide fragments also have a high α-helical preference [36, 39]. As discussed above, the conformational propensity of peptide fragments in TFE/water solvent is thought to reproduce the situation of early intermediates formed during the refolding processes. Therefore, the intermediate with nonnative α-helical structures is assumed to be accumulated at the early folding stage of β-lactoglobulin. This was confirmed later by the study of folding kinetics using stopped-flow CD in the absence and presence of TFE [68–71].

Bovine β-lactoglobulin assumes a monomeric native state at pH 2, while it is dimeric at neutral pH. In addition, the presence of a free thiol group (Cys121) in addition to two disulfide bonds (Cys66-Cys16 and Cys106-Cys119) (Figure 24.3A) complicates the refolding experiments because of possible thiol/disulfide exchange reactions in the unfolded state [72]. Therefore, the refolding of β-lactoglobulin was initiated by diluting the unfolded protein in a high concentration of GdmCl at pH 2.0 into the refolding buffer at the same pH, and the folding kinetics were monitored by ellipticity measurements at 222 nm. Usually, the signal is expected to change from the unfolded baseline to the value of the native state (Figure 24.4). Contrary to expectation, the ellipticity at 222 nm exceeded the value of the native baseline at the burst phase and subsequently it slowly returned toward the native value. This suggested the transient formation of nonnative α-helices. Similar kinetic traces were monitored at various wavelengths to construct the far-UV CD

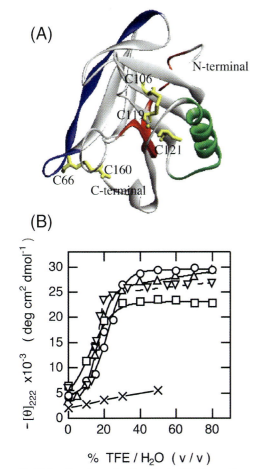

Fig. 24.3. Structure of native β-lactoglobulin (A) and the transitions of its peptide fragments by TFE (B). A) The positions of the fragments are represented by red (fragment 1), blue (fragment 2) and green (fragment 3). The cysteine residues are also shown by yellow. B) The transitions by fragment 1 (circle), fragment 2 (triangle) and fragment 3 (square) are shown. The transitions by intact β-lactoglobulin (reverse triangle) and 14 fragment from hen lysozyme (cross; taken from Ref. [32]) are also shown for comparison. Panel B reproduced from Ref. [36] with permission.

spectrum of the burst phase intermediate (Figure 24.4). The spectrum shows a clear minimum around 222 nm, confirming the presence of an increased amount of α-helical structure. The addition of 9.8% (v/v) TFE in the refolding buffer further increased the α-helical content of the burst phase intermediate, supporting the mechanism that the nonnative α-helices are formed at the early stage of the folding reaction of β-lactoglobulin.

A detailed analysis of the early intermediates of β-lactoglobulin was performed using ultra-rapid mixing techniques in conjunction with fluorescence detection

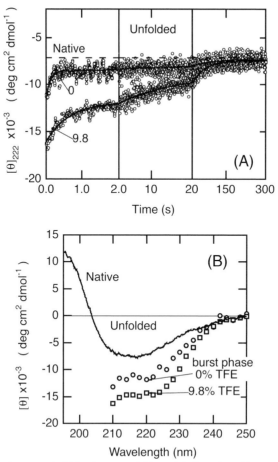

Fig. 24.4. Refolding kinetics of β-lactoglobulin in the presence and absence of TFE detected by stopped-flow CD. A) Time traces for refolding kinetics. The raw data are indicated by circles. The numbers refer to the final concentration of TFE in % (v/v). The ellipticities at 222 nm of the native and unfolded states are indicated by dashed and dotted lines, respectively. B) Far-UV CD spectra of the burst phase intermediate in the absence (circle) and presence (square) of 9.8% TFE. The CD spectra of the native and unfolded states are shown by solid and dotted lines, respectively. Reproduced from Ref. [69] with permission.

and hydrogen/deuterium exchange labeling probed by heteronuclear NMR [70, 71]. It has been shown that the βA-strand region is involved in the formation of nonnative α-helical structures (Figure 24.5). On the other hand, the data also suggested that, in the early intermediate, the native-like structure is also formed around the βG and βH-strands. This region corresponds to the hydrophobic core in the native structure, and both β-strands are linked by disulfide bonds between Cys106 and Cys119. These properties around the βG and βH strands are likely to play an important role in the initial stage of the folding reaction of β-lactoglobulin [73]. On the other hand, the role of nonnative α-helices in the early intermediate state is still unknown. Indeed, the formation of such nonnative secondary structures might be less effective in terms of rapid folding since the structure should be disrupted at the later stage of folding. It is probable, however, that the formation of nonnative α-helical segments is useful for preventing intermolecular aggregation since the polypeptide chain becomes relatively compact. The folding mechanism of β-lactoglobulin is still intriguing to understand the interplay between the local and nonlocal interactions during protein folding, and moreover, to understand the α–β transition suggested from several biologically important processes [74, 75].

24.3
Mechanism of Alcohol-induced Conformational Change

Although alcohols are often used to probe the structural preferences of polypeptide chains, the detailed mechanism of how these co-solvents induce the conformational change of polypeptide chains is still debatable. The chemical properties of alcohols are relatively similar to detergents in a sense that the molecules consist of both hydrophobic (hydrocarbon or halogenated hydrocarbon) and hydrophilic (hydroxyl) groups. Possible mechanisms by which alcohols induce conformational changes in polypeptides, therefore, include (1) destabilizing intramolecular hydrophobic interactions and (2) promoting the formation of local backbone hydrogen bonds [23, 24, 76–79]. In alcohol/water solvents with low polarity or low dielectric constant, hydrophobic interactions stabilizing the native structure are weakened and instead the local hydrogen bonds are strengthened, resulting in denaturation and the simultaneous formation of an "open helical conformation" or "open helical coil", i.e., solvent-exposed helices.

In the case of nonhalogenated alcohols, the efficiency as a helix inducer linearly increases with an increase in the number of carbon atoms in an alcohol molecule [80, 81]. This confirms that the hydrophobic interactions play an important role in the alcohol effects. Although an alcohol is defined by the presence of hydroxyl (OH) group(s), the contribution of the OH group to the alcohol effects is negative. In other words, alcohol effects arise from the dissolved hydrophobic groups, and the OH group is important mainly for dissolving otherwise insoluble hydrocarbon groups into water. In this sense, halogenated alcohols might be expected to be poor alcohols in stabilizing α-helical structures in polypeptide chains, since the halogen atoms have a larger electronegativity than hydrogen atoms. In contrast, halogenated alcohols such as TFE and HFIP have higher potentials to induce the α-helical conformation than nonhalogenated alcohols. Systematic analysis of various halogenated alcohols indicated that the effect of fluoride is the lowest of the various halogens [80, 81]. Nevertheless, the low absorbance of fluoride in the far-UV region makes TFE and HFIP the most useful alcohols to examine the alcohol effects on proteins by CD.

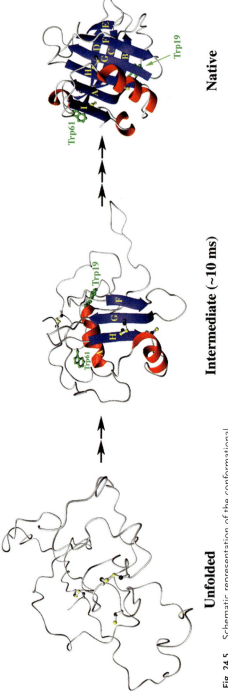

Fig. 24.5. Schematic representation of the conformational states encountered during folding of β-lactoglobulin, including the unfolded ensemble, a partially structured intermediate populated on the millisecond time scale and the native state. Reproduced from Ref. [71] with permission.

24.3 Mechanism of Alcohol-induced Conformational Change

Importantly, studies with solution X-ray scattering provided the evidence that water/alcohol mixtures of some halogenated alcohols such as TFE or HFIP assume micelle-like clusters [82]. This suggests that the halogenated alcohols in fact have higher hydrophobicity than nonhalogenated ones, producing dynamically hydrophobic clusters of alcohols, which can interact effectively with protein molecules, thus enhancing the alcohol effects on proteins. In other words, the direct interactions between proteins and alcohol molecules by hydrophobic interactions play an important role in the alcohol effects on proteins.

From a systematic analysis of the alcohol-induced transitions of melittin, a basic and amphiphilic peptide of 26 amino acid residues present in honey bee venom, Hirota et al. [80, 81] indicated that the effects of each alcohol can be rationalized by the additive contribution from the hydrocarbon (CH), hydroxyl (OH), and halogen groups (F, Cl, and Br), providing the following equations for the m-value, a measure of the cooperativity for the alcohol-induced transition:

$$m = a\ \mathrm{ASA(CH)} + b\ \mathrm{ASA(OH)} + c\ \mathrm{ASA(F)} + d\ \mathrm{ASA(Cl)} + e\ \mathrm{ASA(Br)} \tag{1}$$

$$m = a\ [\mathrm{ASA(CH)}]^2 + b\ [\mathrm{ASA(OH)}]^2 + c\ [\mathrm{ASA(F)}]^2 + d\ [\mathrm{ASA(Cl)}]^2 + e\ [\mathrm{ASA(Br)}]^2 \tag{2}$$

where ASA(X) corresponds to the accessible surface area of each group X and the value of $a, b, c, d,$ and e are the empirically determined proportionality coefficients which are summarized in Table 24.2. Equation (2) provides a better fit than Eq. (1) when the effects of TFE and HFIP are included. It is noted that the coefficient for the OH group is negative while others are positive. A similar relationship was observed for the alcohol-induced denaturation of β-lactoglobulin [83].

These results indicate the importance of the direct interaction between alcohols and polypeptide molecules. Accordingly, the results from NMR spectroscopy and molecular dynamics simulations suggested that TFE molecules preferentially bind to the backbone carbonyl oxygen group and minimize the solvent exposure of amide hydrogen groups, leading to the stabilization of intramolecular hydrogen bonds in a polypeptide chain [84–88].

These results, moreover, suggest that alcohols will be useful for dissolving protein aggregates stabilized by hydrophobic interactions. In fact, some alcohols such

Tab. 24.2. Fitting coefficients of the $m(\mathrm{ts})^a$ and $m(\mathrm{hc})^b$ values for Eqs (1) and (2).

	a	b	c	d	e
Two-state mechanism ($m(\mathrm{TS})$)					
Eq. (1)	0.0251	−0.0197	0.0716	0.0558	0.0998
Eq. (2)	1.06×10^{-4}	-4.54×10^{-5}	3.83×10^{-4}	5.89×10^{-4}	1.34×10^{-3}
Helix/coil mechanism ($m(\mathrm{HC})$)					
Eq. (1)	0.840	0.639	1.92	2.12	3.72
Eq. (2)	3.53×10^{-3}	1.41×10^{-3}	2.05×10^{-2}	1.12×10^{-2}	4.95×10^{-2}

Reproduced from table 3 of Ref. [81].

as TFE or HFIP are useful to dissolve aggregates formed during peptide synthesis. HFIP is often used to dissolve prion and Alzheimer's amyloid β-peptides which tend to form aggregates or amyloid fibrils [89–91]. On the other hand, these alcohols, in particular TFE, are known to induce well-ordered fibrillar architectures similar to the amyloid fibrils [92–95]. These observations are apparently inconsistent with each other and understanding of these contrasting phenomena will be of special importance to clarify the alcohol effects.

24.4
Effects of Alcohols on Folding Kinetics

The studies discussed above are mostly concerned with the behavior of polypeptide chains under equilibrium conditions. During protein folding, many different types of interactions including local hydrogen bonds play important roles. A simple question, therefore, arises: how do proteins behave during kinetic refolding in alcohol/water solvents? Interestingly, Lu et al. [96] first found that the folding kinetics of hen lysozyme is accelerated by the presence of low concentrations of TFE below 5.5% (v/v). In the presence of such a small amount of TFE, the protein is still able to fold into the native structure judging from several spectroscopic data including solution NMR. In addition, pulse-labeling studies suggested that the structural properties of folding intermediates of lysozyme are almost unaffected by the presence of TFE.

An extensive analysis using more than 10 proteins revealed that the acceleration of folding kinetics in the presence of a small amount of TFE is common to all of the proteins examined [97] (Figure 24.6). The logarithm of the refolding rate constant linearly increased with an increase in TFE concentration, but then decreased above a certain concentration of TFE. The concentration range that provides the maximum rate of folding depended on the protein species and other conditions such as pH or temperature. Interestingly, for proteins which fold by a two-state manner without accumulating observable intermediates (two-state proteins), the extent of acceleration by TFE correlates well with the amount of local backbone hydrogen bonds in the native structure (Figure 24.7). This behavior could be rationalized by the fact that TFE stabilizes the local hydrogen bonds in the polypeptide chains, thus forming α-helical or β-turn conformations.

On the other hand, for multi-state proteins which accumulate transient intermediates during the folding process, the acceleration of folding by the presence of TFE was always much less than that anticipated from the correlation found for the two-state proteins (Figure 24.7). The multi-state proteins often assume the molten globule intermediate at the burst phase just after the initiation of folding. According to stopped-flow CD or NMR pulse-labeling analysis, a significant amount of native-like secondary structures is formed in the molten globule intermediates. In such intermediates, the native-like local hydrogen bonds should be already formed. Since the rate-limiting step of folding of the multi-state protein is mainly associated with the rearrangement of partially folded subdomains, the number of local

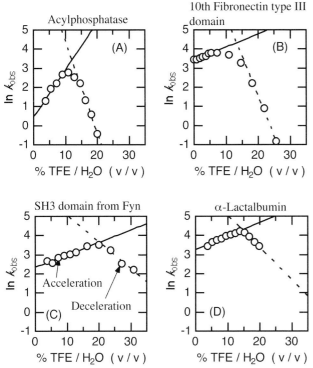

Fig. 24.6. Rate constants of refolding kinetics of proteins in the presence of TFE. Examples are shown for the cases of acylphosphatase (A), 10th type III domain from fibronectin (B), SH3 domain from human Fyn (C), and α-lactalbumin (D). Reproduced from Ref. [97] with permission.

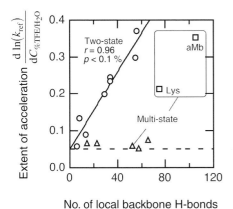

Fig. 24.7. Relationship between the effectiveness of TFE on the acceleration of protein folding and the number of local backbone hydrogen bonds in the native proteins. The data for two-state (circle) and multi-state proteins (triangle and square) are shown. Reproduced from Ref. [97] with permission.

hydrogen bonds which are formed during the rate-limiting step is less than the number in the native structure. The NMR pulse-labeling studies for several proteins have shown that the number of local hydrogen bonds formed in the intermediate states varies depending on proteins. Consistent with this, the acceleration of folding by TFE for multi-state proteins differs significantly even though the number of local hydrogen bonds in the native structure is similar, as is the case for lysozyme and α-lactalbumin.

In the case of acylphosphatase, a two-state protein, the refolding reaction observed by CD indicated that the absolute intensity at 222 nm for the burst phase intermediate increases upon addition of TFE (Figure 24.8), suggesting the formation of partially folded intermediates which do not accumulate in the absence of TFE [98]. The maximum rate of folding was obtained at the TFE concentration where the ellipticity at 222 nm for the burst phase intermediate is the same as that of the native protein. This observation is consistent with the conclusion drawn

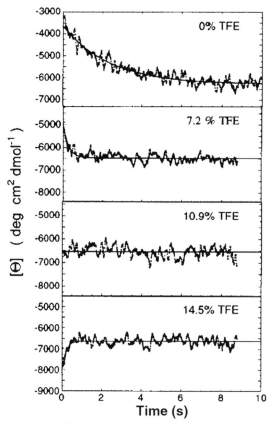

Fig. 24.8. Folding kinetics of acylphosphatase in the presence of TFE monitored by 222 nm far-UV CD. Reproduced from Ref. [98] with permission.

from the analysis of Hammond behavior and nonlinear activation free energy of the folding reaction [99, 100] (see opposing result by Yiu et al. [101]). The analysis indicates that the free energy barriers encountered by a folding polypeptide are generally narrow with robust maxima and that the change in the folding rate by the perturbation of folding landscape, for example, by mutations or changing conditions, is mainly due to the induction of folding intermediates for two-state proteins. Thus, the data suggest that the activation energy is decreased in the presence of TFE because of the decreased stability of the early intermediate relative to the transition state.

24.5
Salt-induced Formation of the Intermediate States

24.5.1
Acid-denatured Proteins

Acid-denatured proteins often assume an intermediate conformation. The most well-known example is the acidic molten globule state of α-lactalbumin [26–29] (see Chapter 23). However, the conformation of acid-denatured proteins depends on the protein species, and the nature of the acid-denatured state was ambiguous in comparison with the unfolded states induced by urea or GdmCl [4, 5] (see Chapter 8). The important results indicating the salt-dependent transition of the acid-denatured state came out with horse cytochrome c [25], followed by *Bacillus subtilis* β-lactamase [102] and horse apomyoglobin [103, 104]. These proteins are substantially unfolded at pH 2 in the absence of salt. The addition of a low concentration of salts induced an intermediate state with notable far-UV CD intensity, while the near-UV CD showed disordering (Figure 24.9). The salt-induced state is similar to the acidic molten globule state of α-lactalbumin [26–29]. In fact, the name of molten globule state was first used for the salt-stabilized acidic state of cytochrome c [25]. Similar observation with β-lactamase [102], apomyoglobin [103, 104], staphylococcal nuclease [105, 106], and other proteins established that the salt-dependent transition is common to various acid-denatured proteins.

The conformation of the acid-denatured state is determined by a balance of various factors stabilizing or destabilizing the folded state. The most important destabilizing force should be the charge repulsion between positive charges because most of the titratable groups are protonated at pH 2. Consistent with this, proteins with a high pI value (for example, cytochrome c) tend to be unfolded substantially. On the other hand, the driving forces for folding would be similar to those at neutral pH except that the electrostatic interactions are affected by pH. The addition of salts somehow shields the unfavorable charge repulsion, thus the folding forces come into play, although the distinct ionizations of the titratable groups prevent the formation of native structure.

The participation of hydrophobic interactions, although less than that of the native state, has been established by a series of studies using calorimetric analysis

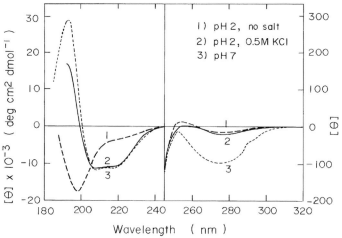

Fig. 24.9. Salt-dependent change of the CD spectrum of the acid-denatured *Bacillus cereus* β-lactamase at 20 °C. 1) The acid-unfolded state at pH 2.0 in the absence of salt; 2) the molten globule state at pH 2.0 in 0.5 M KCl; 3) the native state at pH 7.0. Reproduced from Ref. [102] with permission.

of the salt-stabilized intermediate states [107–111]. Nevertheless, the salt-induced transition itself does not reveal the detailed mechanism of the salt effects.

24.5.2
Acid-induced Unfolding and Refolding Transitions

In an attempt to fully acid-unfold proteins, an intriguing observation was made; increasing the concentration of HCl opposes the unfolding, and rather refolds the protein to the molten globule state [112] (Figure 24.10). The acid-stabilized state was essentially the same intermediate as stabilized by salts. Although the phenomenon was surprising at first, it is straightforward to understand the mechanism once the ionic strength effect is taken into account; decreasing the pH below 2 exponentially increases the concentration of anion as well as of protons by $[Cl^-] = [H^+] = 10^{-pH}$. This suggests that the apparent refolding phenomenon observed by increasing the HCl concentration is caused by the anion effect rather than the pH effects. Consistent with this idea, the acid-induced transition plotted against the concentration of HCl agreed well with that plotted against the concentration of NaCl or KCl [112] (Figure 24.11). By considering the ionic strength effect of HCl, the pH and salt-dependent phase diagram of acid-denatured proteins can be constructed [103, 104, 113] (Figure 24.12). It is important to recognize that decreasing pH increases the chloride concentration, thus producing the prohibited region in the phase. Consequently, decreasing the pH below 2 results in increasing ionic strength, causing the same effects as adding salt at pH 2.

The phase diagram is useful to understand the variation of acid denatured states among proteins. In the cases of horse cytochrome *c* [25, 113] and horse apomyo-

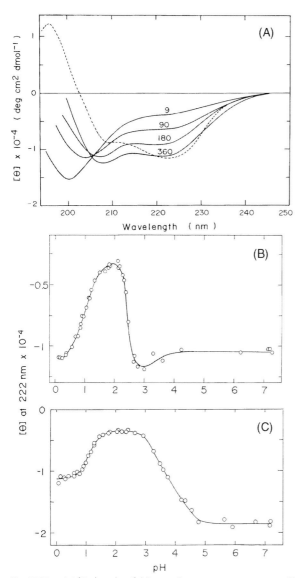

Fig. 24.10. Acid-induced unfolding and refolding transitions of cytochrome c and apomyolobin at 20 °C. A) Far-UV CD spectra of cytochrome c as a function of HCl concentration. The numbers refer to the HCl concentration in millimolar units. The spectrum of the native state (dotted line) is shown for comparison. B,C) Effects of increasing concentration of HCl on the ellipticity at 222 nm of cytochrome c (B) and apomyologobin (C). Upon decreasing pH by addition of HCl, the N (native) → U (acid-unfolded) → MG (molten globule) transition is observed. Reproduced from Ref. [111] with permission.

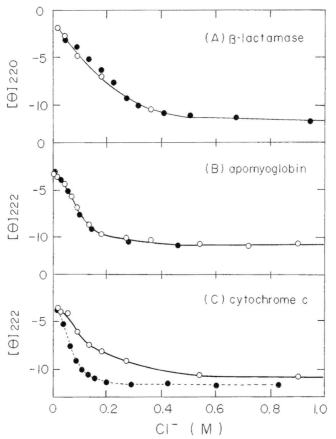

Fig. 24.11. Consistency between the HCl-induced (open circles) and KCl-induced (filled circles) transitions of acid-unfolded proteins as monitored by change in the far-UV CD. A) *Bacillus cereus* β-lactamase, B) horse apomyoglobin, C) horse cytochrome *c*. Reproduced from Ref. [111] with permission.

globin [103, 104], a clear boundary between the fully unfolded state and the molten globule state was observed. For the phase diagram of apomyoglobin, moreover, the unfolding intermediate was observed at moderately acidic pH regions. On the other hand, it is expected for a protein with less charge repulsion that the acid denaturation directly produces the molten globule intermediate without maximally unfolding the protein. This is considered to be the case for bovine α-lactalbumin. Cytochrome *c* species with various degrees of charge repulsion were prepared by acetylating lysine amino groups [107, 114]. The phase diagrams of these variously modifed cytochrome *c* species showed that the boundary between the fully unfolded state and the molten globule state depends on the charge repulsion: decreasing the net charge (that is, decreasing pH) decreases the concentration of salt

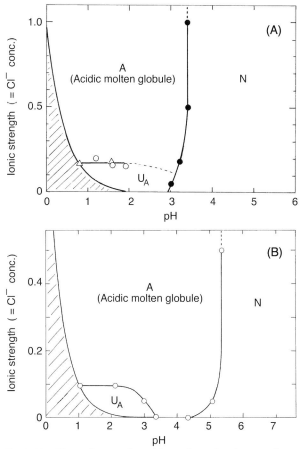

Fig. 24.12. Phase diagrams for acidic conformational states of horse cytochrome c (A) and horse apomyoglobin (B). Phase diagram for apomyoglobin reproduced from Ref. [102] with permission.

required to stabilize the molten globule state [113]. Thus, we can understand consistently the variation of acid denaturation among proteins.

An important implication is that to maximally unfold a protein by acid, pH 2 under the low salt conditions will be best. These conditions would also be optimal for dissolving proteins and peptides taking advantage of the net charge repulsion.

The role of charge repulsion and hydrophobic interactions were characterized by analyzing the salt-dependent transition of horse cytochrome c [107, 114]. Conformational transitions of variously acetylated cytochrome c derivatives, that is, modified cytochrome c species with different pI values, revealed that the net charge repulsion destabilizes the molten globule state by about 400 cal mol^{-1} per charge at 25 °C. In addition, thermal unfolding of these acytylated cytochrome c species verified the contribution of hydrophobic interactions in the stability of the molten globule state, which is about 30–40% of that of the native state [107]. The enthalpy change upon formation of the molten globule state and its temperature dependence were directly determined by isothermal titration calorimetry measurements of the salt-dependent conformational transition, which is consistent with the value determined by DSC measurements [110]. Similar measurements were performed with horse apomyoglobin [111].

24.6
Mechanism of Salt-induced Conformational Change

Although the importance of anions in shielding the charge repulsive forces is evident, anions can work in various ways [115]. (1) Salts can shield the charge repulsion through Debye-Hückel screening effects. (2) Some anions, such as sulfate, are well known to stabilize the native state, which has been interpreted in terms of the effects on water structure (that is, structure maker) consequently strengthening the hydrophobic interactions of proteins. (3) Anions can directly interact with positive charges on proteins to shield the charge repulsion. These three possibilities can be distinguished by examining the effects of various anion species [115].

Debye-Hückel effects are independent of anion species. It is known that the effects of anions in stabilizing protein structure follow the Hofmeister series [5, 6]. The representative series is:

$$\text{sulfate} > \text{phosphate} > \text{fluoride} > \text{chloride} > \text{bromide} > \text{iodide}$$
$$> \text{perchlorate} > \text{thiocyanate} \qquad (3)$$

Although the exact mechanism of stabilization or destabilization by these anions is unknown, the results are consistent with the idea that these salts affect the water structure. On the other hand, the affinity of a particular anion to an anion-exchange resin is called electroselectivity. The selectivity series of various anions depends critically on the structure of the resin and the solution conditions. However, the general trend of selectivity can be seen from the following examples [116, 117]:

$$\text{sulfite} > \text{sulfate} > \text{perchlorate} > \text{thiocyanate} > \text{iodide} > \text{nitrate}$$
$$> \text{bromide} > \text{chloride} > \text{acetate} = \text{fluoride} \qquad (4)$$

$$\text{perchlorate} > \text{iodide} > \text{trichloroacetate} = \text{thiocyante} > \text{nitrate}$$
$$> \text{bromide} > \text{trifluoroacetate} > \text{chloride} > \text{acetate} > \text{fluoride} \qquad (5)$$

The order of the Hofmeister series is distinct from that of electroselectivity. In particular, the orders of monovalent anions are opposite between the two. Moreover, the concentrations of salts to induce the Hofmeister effects are generally higher than those for anion binding (electroselectivity).

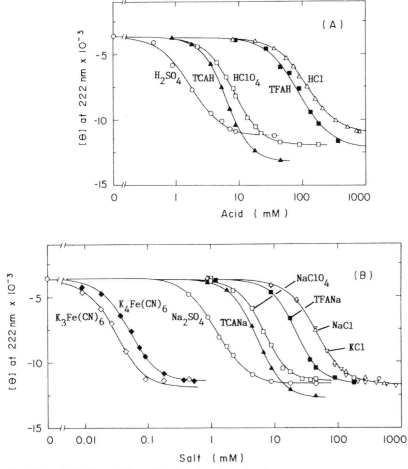

Fig. 24.13. Acid-induced (A) or salt-induced conformational transitions of horse cytochrome c in the presence of 18 mM HCl measured by the ellipticity at 222 nm at 20 °C. Reproduced from Ref. [114] with permission.

The effects of various salts on the acid-unfolded cytochrome c and apomyoglobin were examined [115] (Figure 24.13). Similar experiments with various acids were also performed, producing essentially the same results with respect to the order of anions:

ferricyanide > ferrocyanide > sulfate > trichloroacetate > thiocyanate
> perchlorate > iodide > nitrate > trifluoroacetate
> bromide > chloride   (6)

The results showed convincingly that the effects depend on anion species, and the order of effectiveness was consistent with the electroselectivity series (Eqs (4) and (5)). Anions with multiple charges show stronger potentials. Among the monovalent anions, chaotropic anions such as iodide or bromide show stronger effects while the effects of chloride are the weakest. These results demonstrate that the electrostatic attraction between the positively charged proteins and negatively charged anions causes the direct interactions between them, thus shielding the charge repulsion.

24.7
Generality of the Salt Effects

Although anion binding-induced stabilization of the intermediate state is often observed at acidic pH, it is likely that a similar interaction plays a role in determining the protein conformation under physiological pH when a protein is highly positively charged. The salt-dependent conformational transition has been known for many years for melittin. The conformational transition of melittin has been shown to be dependent on anion binding [118–120]. Moreover, the similar role of anion binding was also shown for another basic and amphiphilic bee venom peptide, mastoparan [121] and a designed amphiphilic peptide [122]. These results imply that the anion binding can control the conformational transition of some of natively unfolded proteins. Recently, a natively unfolded *Bacillus subtilis* ribonuclease P was shown to exhibit anion-dependent folding, indicating that the anion-dependent folding plays a role in the function of some proteins [123]. On the other hand, the role of cations in determining the conformation of proteins is less clear, although it is conceivable that the opposite situation takes place for negatively charged unfolded proteins and peptides.

Finally, it is useful to consider the effects of GdmCl in relation to the anion effects as described in this chapter. GdmCl is a salt made of a guanidinium cation and chloride. Hagihara et al. [124] observed intriguing GdmCl-dependent conformational transitions with the acid-unfolded horse cytochrome *c* and horse apomyoglobin (Figure 24.14). By the addition of low concentrations of GdmCl, the acid-unfolded proteins at first refolded to the molten globule intermediate before unfolding again at high GdmCl concentrations. Although the observation might be surprising, the interpretation is straightforward, as is the case of the acid-induced refolding. The first refolding by GdmCl is caused by the anion effect while subsequent unfolding is driven by the normal chaotropic effects of GdmCl. The similar refolding and unfolding transitions were observed for a designed amphiphilic peptide at neutral pH [122]. Moreover, the stabilizing effects of low concentrations of GdmCl may be common to several proteins under the physiological conditions [125]. Thus, when GdmCl is used, it is important to interpret the data taking into account the salt effects of GdmCl. It is also recommended to compare the results with the denaturation caused by urea, another popular denaturant.

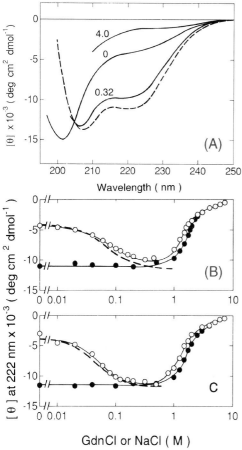

Fig. 24.14. GdmCl-induced refolding and unfolding transitions of acid-unfolded proteins. A) Far-UV CD spectra of horse apomyoglobin as a function of GdmCl concentration in 20 mM HCl (pH 1.8) at 20 °C. The numbers refer to the GdmCl concentration in molar units. The broken line shows the spectrum of the molten globule state stabilized by 0.4 M NaCl in 20 mM HCl. B,C) Conformational transitions of horse apomyoglobin (B) and horse cytochrome c (C) measured by the change in ellipticity at 222 nm. Open and filled circles indicate the GdmCl-induced transitions in the absence and presence of 0.4 M NaCl, respectively. Open squares show the reverse transition, in which proteins were initially unfolded in 4 M GdmCl. For comparison, NaCl-induced refolding transitions in the same buffer are shown by broken lines. Reproduced from Ref. [123] with permission.

24.8
Conclusion

Alcohols and salts affect the conformational stability of partially folded intermediate in various ways. The effects of alcohols are basically explained by destabilization of intramolecular hydrophobic interactions and stabilization of the local hydrogen bonds, often resulting in the formation of the open helical conformation. The molten globule-like compact intermediate is occasionally observed at moderate concentrations of alcohols and, moreover, β-structures are stabilized for some peptides and proteins. Of particular interest is the formation of fibrillar architechtures similar to the amyloid fibrils in alcohol/water solvents. These results indicate that, although major effects of alcohols on protein and peptides are the same, the resultant conformation varies substantially depending on the balance of various forces, the details of which are still unknown. Salts can also stabilize the intermediates in several ways. Major effects of salts are Debye-Hückel screening effects, effects on water structure, resulting in the modulation of hydrophobic interactions, and counter ion binding. Among them, we discussed that the direct interactions of anions with positive charges on proteins play important roles in determining the structure and stability of partially folded intermediates. Understanding kinetic aspects of these alcohol and salt effects will further clarify the conformation and stability of the partially folded intermediates.

References

1 LAPANJE, S. (1978) *Physicochemical Aspects of Protein Denaturation*. John Wiley & Sons, New York.
2 BUCK, M. (1998) Trifluoroethanol and colleagues: co-solvents come of age. Recent studies with peptides and proteins. *Q. Rev. Biophys*. 31, 297–355.
3 TANFORD, C. (1968) Protein denaturation. *Adv. Protein Chem.* 23, 121–282.
4 TANFORD, C. (1970) Protein denaturation. *Adv. Protein Chem.* 24, 1–95.
5 VON HIPPEL, P. H. and SCHLEICH, T. (1969) *Biological Macromolecules* (TIMASHEFF, S. and FASMAN, G., Eds), Vol. II, pp. 417–574. Dekker, New York.
6 COLLINS, K. D. and WASHABAUGH, M. W. (1985) The Hofmeister effect and the behaviour of water at interfaces. *Q. Rev. Biophys.* 18, 323–422.
7 TIMASHEFF, S. and ARAKAWA, T. (1989) Stabilization of protein structure by solvents. In *Protein Structure: A Practical Approach* (CREIGHTON, T. E., Ed.), pp. 331–345. Oxford University Press, Oxford.
8 WEBER, R. E. and TANFORD, C. (1959) The conformation of ribonuclease at low pH in 2-chloroethanol and in 2-chloroethanol-water mixtures. *J. Am. Chem. Soc.* 81, 3255–3260.
9 MARTIN, C. J. and BHATNAGAR, G. M. (1967) Unfolding reactions of proteins. II. Spectral and optical rotatory dispersion studies in urea, guanidinium chloride, and 2-chloroethanol. *Biochemistry* 6, 1638–1650.
10 HERSKOVITS, T. T. and MESCANTI, L. (1965) Conformation of proteins and polypeptides. II. Optical rotatory dispersion and conformation of the milk proteins and other proteins in organic solvents. *J. Biol. Chem.* 240, 639–644.
11 DYSON, H. J. and WRIGHT, P. E. (2002) Coupling of folding and binding for unstructured proteins. *Curr. Opin. Struct. Biol.* 12, 54–60.
12 TOMPA, P. (2002) Intrinsically unstructured proteins. *Trends Biochem. Sci.* 27, 527–533.
13 UVERSKY, V. N. (2002) Natively unfolded proteins: a point where biology waits for physics. *Protein Sci.* 11, 739–756.
14 UVERSKY, V. N. (2002) What does it mean to be natively unfolded? *Eur J. Biochem.* 269, 2–12.
15 BLANCO, F. J., JIMENEZ, M. A.,

Pineda, A., Rico, M., Santoro, J., and Nieto, J. L. (1994) NMR solution structure of the isolated N-terminal fragment of protein-G B1 domain. Evidence of trifluoroethanol induced native-like β-hairpin formation. *Biochemistry* 33, 6004–6014.

16 Cann, J. R., Liu, X., Stewart, J. M., Gera, L., and Kotovych, G. (1994) A CD and an NMR study of multiple bradykinin conformations in aqueous trifluoroethanol solutions. *Biopolymers* 34, 869–878.

17 Vranken, W. F., Budesinsky, M., Fant, F., Boulez, K., and Borremans, F. A. (1995) The complete consensus V3 loop peptide of the envelope protein gp120 of HIV-1 shows pronounced helical character in solution. *FEBS Lett.* 374, 117–121.

18 Wang, J., Hodges, R. S., and Sykes, B. D. (1995) Effect of trifluoroethanol on the solution structure and flexibility of desmopressin: a two-dimensional NMR study. *Int. J. Pept. Protein Res.* 45, 471–481.

19 de Alba, E., Jimenez, M. A., Rico, M., and Nieto, J. L. (1996) Conformational investigation of designed short linear peptides able to fold into β-hairpin structures in aqueous solution. *Folding Des.* 1, 133–144.

20 Reiersen, H., Clarke, A. R., and Rees, A. R. (1998) Short elastin-like peptides exhibit the same temperature-induced structural transitions as elastin polymers: implications for protein engineering. *J. Mol. Biol.* 283, 255–264.

21 Bienkiewicz, E. A., Moon Woody, A., and Woody, R. W. (2000) Conformation of the RNA polymerase II C-terminal domain: circular dichroism of long and short fragments. *J. Mol. Biol.* 297, 119–133.

22 Ramirez-Alvarado, M., Blanco, F. J., and Serrano, L. (2001) Elongation of the BH8 β-hairpin peptide: electrostatic interactions in β-hairpin formation and stability. *Protein Sci.* 10, 1381–1392.

23 Dill, K. A. (1990) Dominant forces in protein folding. *Biochemistry* 29, 7133–7155.

24 Dill, K. A., Bromberg, S., Yue, K., Fiebig, K. M., Yee, D. P., Thomas, P. D., and Chan, H. S. (1995) Principles of protein folding: a perspective from simple exact models. *Protein Sci.* 4, 561–602.

25 Ohgushi, M. and Wada, A. (1983) 'Molten-globule state': a compact form of globular proteins with mobile side-chains. *FEBS Lett.* 164, 21–24.

26 Ptitsyn, O. B. (1991) How does protein synthesis give rise to the 3D-structure? *FEBS Lett.* 285, 176–181.

27 Ptitsyn, O. B. (1995) Molten globule and protein folding. *Adv. Protein Chem.* 47, 83–229.

28 Kuwajima, K. (1989) The molten globule state as a clue for understanding the folding and cooperativity of globular-protein structure. *Proteins* 6, 87–103.

29 Arai, M. and Kuwajima, K. (2000) Role of the molten globule state in protein folding. *Adv. Protein Chem.* 53, 209–282.

30 Fregeau-Gallagher, N. L., Sailer, M., Niemczura, W. P., Nakashima, T. T., Stiles, M. E., and Vederas, J. C. (1997) Three-dimensional structure of leucocin A in trifluoroethanol and dodecylphosphocholine micelles: spatial location of residues critical for biological activity in type IIa bacteriocins from lactic acid bacteria. *Biochemistry* 36, 15062–15072.

31 Schönbrunner, N., Wey, J., Engels, J., Georg, H., and Kiefhaber, T. (1996) Native-like β-structure in a trifluoroethanol-induced partially folded state of the all-beta-sheet protein tendamistat. *J. Mol. Biol.* 260, 432–445.

32 Segawa, S., Fukuno, T., Fujiwara, K., and Noda, Y. (1991) Local structures in unfolded lysozyme and correlation with secondary structures in the native conformation: helix-forming or -breaking propensity of peptide segments. *Biopolymers* 31, 497–509.

33 Sancho, J., Neira, J. L., and Fersht, A. R. (1992) An N-terminal fragment of barnase has residual helical struc-

ture similar to that in a refolding intermediate. *J. Mol. Biol.* 224, 749–758.
34 DYSON, H. J., SAYRE, J. R., MERUTKA, G., SHIN, H. C., LERNER, R. A., and WRIGHT, P. E. (1992) Folding of peptide fragments comprising the complete sequence of proteins. Models for initiation of protein folding. II. Plastocyanin. *J. Mol. Biol.* 226, 819–835.
35 DYSON, H. J., MERUTKA, G., WALTHO, J. P., LERNER, R. A., and WRIGHT, P. E. (1992) Folding of peptide fragments comprising the complete sequence of proteins. Models for initiation of protein folding. I. Myohemerythrin. *J. Mol. Biol.* 226, 795–817.
36 HAMADA, D., KURODA, Y., TANAKA, T., and GOTO, Y. (1995) High helical propensity of the peptide fragments derived from β-lactoglobulin, a predominantly β-sheet protein. *J. Mol. Biol.* 254, 737–746.
37 YANG, J. J., BUCK, M., PITKEATHLY, M., KOTIK, M., HAYNIE, D. T., DOBSON, C. M., and RADFORD, S. E. (1995) Conformational properties of four peptides spanning the sequence of hen lysozyme. *J. Mol. Biol.* 252, 483–491.
38 BOLIN, K. A., PITKEATHLY, M., MIRANKER, A., SMITH, L. J., and DOBSON, C. M. (1996) Insight into a random coil conformation and an isolated helix: structural and dynamical characterisation of the C-helix peptide from hen lysozyme. *J. Mol. Biol.* 261, 443–453.
39 KURODA, Y., HAMADA, D., TANAKA, T., and GOTO, Y. (1996) High helicity of peptide fragments corresponding to β-strand regions of β-lactoglobulin observed by 2D-NMR spectroscopy. *Folding Des.* 1, 255–263.
40 ZITZEWITZ, J. A. and MATTHEWS, C. R. (1991) Molecular dissection of the folding mechanism of the alpha subunit of tryptophan synthase: an amino-terminal autonomous folding unit controls several rate-limiting steps in the folding of a single domain protein. *Biochemistry* 38, 10205–10214.
41 STALEY, J. P. and KIM, P. S. (1994) Formation of a native-like subdomain in a partially folded intermediate of bovine pancreatic trypsin inhibitor. *Protein Sci.* 3, 1822–1832.
42 REYMOND, M. T., MERUTKA, G., DYSON, H. J., and WRIGHT, P. E. (1997) Folding propensities of peptide fragments of myoglobin. *Protein Sci.* 6, 706–716.
43 SALINAS, R. K., SHIDA, C. S., PERTINHEZ, T. A., SPISNI, A., NAKAIE, C. R., PAIVA, A. C., and SCHREIER, S. (2002) Trifluoroethanol and binding to model membranes stabilize a predicted turn in a peptide corresponding to the first extracellular loop of the angiotensin II AT(1A) receptor. *Biopolymers* 65, 21–31.
44 SOULIE, S., NEUMANN, J. M., BERTHOMIEU, C., MOLLER, J. V., LE MAIRE, M., and FORGE, V. (1999) NMR conformational study of the sixth transmembrane segment of sarcoplasmic reticulum Ca^{2+}-ATPase. *Biochemistry* 38, 5813–5821.
45 WRAY, V., KINDER, R., FEDERAU, T., HENKLEIN, P., BECHINGER, B., and SCHUBERT, U. (1999) Solution structure and orientation of the transmembrane anchor domain of the HIV-1-encoded virus protein U by high-resolution and solid-state NMR spectroscopy. *Biochemistry* 38, 5272–5282.
46 TAYLOR, R. M., ZAKHAROV, S. D., BERNARD HEYMANN, J., GIRVIN, M. E., and CRAMER, W. A. (2000) Folded state of the integral membrane colicin E1 immunity protein in solvents of mixed polarity. *Biochemistry* 39, 12131–12139.
47 LUIDENS, M. K., FIGGE, J., BREESE, K., and VAJDA, S. (1996) Predicted and trifluoroethanol-induced α-helicity of polypeptides. *Biopolymers* 39, 367–376.
48 SHIRAKI, K., NISHIKAWA, K., and GOTO, Y. (1995) Trifluoroethanol-induced stabilization of the α-helical structure of β-lactoglobulin: implication for non-hierarchical protein folding. *J. Mol. Biol.* 245, 180–194.
49 BUCK, M., SCHWALBE, H., and DOBSON, C. M. (1995) Characterization of conformational preferences in a partly folded protein by heteronu-

clear NMR spectroscopy: assignment and secondary structure analysis of hen egg-white lysozyme in trifluoroethanol. *Biochemistry* 34, 13219–13232.

50 KUWATA, K., HOSHINO, M., ERA, S., BATT, C. A., and GOTO, Y. (1998) $\alpha \rightarrow \beta$ transition of β-lactoglobulin as evidenced by heteronuclear NMR. *J. Mol. Biol.* 283, 731–739.

51 HOSHINO, M., HAGIHARA, Y., HAMADA, D., KATAOKA, M., and GOTO, Y. (1997) Trifluoroethanol-induced conformational transition of hen egg-white lysozyme studied by small-angle X-ray scattering. *FEBS Lett.* 416, 72–76.

52 KAMATARI, Y. O., KONNO, T., KATAOKA, M., and AKASAKA, K. (1996) The methanol-induced globular and expanded denatured states of cytochrome c: a study by CD, fluorescence, NMR and small-angle X-ray scattering. *J. Mol. Biol.* 259, 512–523.

53 KAMATARI, Y. O., OHJI, S., KONNO, T., SEKI, Y., SODA, K., KATAOKA, M., and AKASAKA, K. (1999) The compact and expanded denatured conformations of apomyoglobin in the methanol-water solvent. *Protein Sci.* 8, 873–882.

54 BYCHKOVA, V. E., DUJSEKINA, A. E., KLENIN, S. I., TIKTOPULO, E. I., UVERSKY, V. N., and PTITSYN, O. B. (1996) Molten globule-like state of cytochrome c under conditions simulating those near the membrane surface. *Biochemistry* 35, 6058–6063.

55 BYCHKOVA, V. E., DUJSEKINA, A. E., FANTUZZI, A., PTITSYN, O. B., and ROSSI, G. L. (1998) Release of retinol and denaturation of its plasma carrier, retinol-binding protein. *Folding Des.* 3, 285–291.

56 UVERSKY, V. N., NARIZHNEVA, N. V., KIRSCHSTEIN, S. O., WINTER, S., and LOBER, G. (1997) Conformational transitions provoked by organic solvents in β-lactoglobulin: can a molten globule like intermediate be induced by the decrease in dielectric constant? *Folding Des.* 2, 163–172.

57 KONNO, T. (1998) Conformational diversity of acid-denatured cytochrome c studied by a matrix analysis of far-UV CD spectra. *Protein Sci.* 7, 975–982.

58 KONNO, T., IWASHITA, J., and NAGAYAMA, K. (2000) Fluorinated alcohol, the third group of co-solvents that stabilize the molten globule state relative to a highly denatured state of cytochrome c. *Protein Sci.* 9, 564–569.

59 SANTUCCI, R., POLIZIO, F., and DESIDERI, A. (1999) Formation of a molten-globule-like state of cytochrome c induced by high concentrations of glycerol. *Biochimie* 81, 745–751.

60 GAST, K., ZIRWER, D., MULLER-FROHNE, M., and DAMASCHUN, G. (1999) Trifluoroethanol-induced conformational transitions of proteins: insights gained from the difference between α-lactalbumin and ribonuclease A. *Protein Sci.* 8, 625–634.

61 KAMIYAMA, T., SADAHIDE, Y., NOGUSA, Y., and GEKKO, K. (1999) Polyol-induced molten globule of cytochrome c: an evidence for stabilization by hydrophobic interaction. *Biochim. Biophys. Acta* 1434, 44–57.

62 SUNDD, M., KUNDU, S., and JAGANNADHAM, M. V. (2000) Alcohol-induced conformational transitions in ervatamin C. An α-helix to β-sheet switchover. *J. Protein Chem.* 19, 169–176.

63 SASIDHAR, Y. U. and PRABHA, C. R. (2000) Conformational features of reduced and disulfide intact forms of hen egg white lysozyme in aqueous solution in presence of 3-chloro-1,2-propanediol and dioxane: implications for protein folding intermediates. *Indian J. Biochem. Biophys.* 37, 97–106.

64 TSUMURA, K., ENATSU, M., KURAMORI, K., MORITA, S., KUGIMIYA, W., KUWADA, M., SHIMURA, Y., and HASUMI, H. (2001) Conformational change in a single molecular species, b3, of β-conglycinin in acidic ethanol solution. *Biosci. Biotechnol. Biochem.* 65, 292–297.

65 POLVERINO DE LAURETO, P., FRARE, E., GOTTARDO, R., and FONTANA, A. (2002) Molten globule of bovine α-lactalbumin at neutral pH induced by heat, trifluoroethanol, and oleic acid: a

comparative analysis by circular dichroism spectroscopy and limited proteolysis. *Proteins* 49, 385–397.

66 BONGIOVANNI, C., SINIBALDI, F., FERRI, T., and SANTUCCI, R. (2002) Glycerol-induced formation of the molten globule from acid-denatured cytochrome *c*: implication for hierarchical folding. *J. Protein Chem.* 21, 35–41.

67 SIRANGELO, I., DAL PIAZ, F., MALMO, C., CASILLO, M., BIROLO, L., PUCCI, P., MARINO, G., and IRACE, G. (2003) Hexafluoroisopropanol and acid destabilized forms of apomyoglobin exhibit structural differences. *Biochemistry* 42, 312–319.

68 KUWAJIMA, K., YAMAYA, H., and SUGAI, S. (1996) The burst-phase intermediate in the refolding of β-lactoglobulin studied by stopped-flow circular dichroism and absorption spectroscopy. *J. Mol. Biol.* 264, 806–822.

69 HAMADA, D., SEGAWA, S., and GOTO, Y. (1996) Non-native α-helical intermediate in the refolding of β-lactoglobulin, a predominantly β-sheet protein. *Nat. Struct. Biol.* 3, 868–873.

70 FORGE, V., HOSHINO, M., KUWATA, K., ARAI, M., KUWAJIMA, K., BATT, C. A., and GOTO, Y. (2000) Is Folding of β-lactoglobulin non-hierarchic? Intermediate with native-like β-sheet and non-native α-helix. *J. Mol. Biol.* 296, 1039–1051.

71 KUWATA, K., SHASTRY, R., CHENG, H., HOSHINO, M., BATT, C. A., GOTO, Y., and RODER, H. (2001) Structural and kinetic characterization of early folding events in β-lactoglobulin. *Nat. Struct. Biol.* 8, 151–155.

72 YAGI, M., SAKURAI, K., KALIDAS, C., BATT, C. A., and GOTO, Y. (2003) Reversible unfolding of bovine β-lactoglobulin mutants without a free thiol group. *J. Biol. Chem.* 278, 47009–47015.

73 KATOU, H., HOSHINO, M., BATT, C. A., and GOTO, Y. (2001) Native-like β-hairpin retained in the cold-denatured state of bovine β-lacto-globulin. *J. Mol. Biol.* 310, 471–484.

74 PRUSINER, S. B. (1997) Prion diseases and the BSE crisis. *Science* 278, 245–251.

75 JACKSON, G. S., HOSSZU, L. L., POWER, A., HILL, A. F., KENNEY, J., SAIBIL, H., CRAVEN, C. J., WALTHO, J. P., CLARKE, A. R., and COLLINGE, J. (1999) Reversible conversion of monomeric human prion protein between native and fibrilogenic conformations. *Science* 283, 1935–1937.

76 THOMAS, P. D. and DILL, K. (1993) Local and nonlocal interactions in globular proteins and mechanisms of alcohol denaturation. *Protein Sci.* 2, 2050–2065.

77 LIU, Y. and BOLEN, D. W. (1995) The peptide backbone plays a dominant role in protein stabilization by naturally occurring osmolytes. *Biochemistry* 34, 12884–12891.

78 CAMMERS-GOODWN, A., ALLEN, T. J., OSLICK, S. L., MCCLURE, K. F., LEE, J. H., and KEMP, D. S. (1996) Mechanism of stabilization of helical conformations of poplypeptides by water containing trifluoroethanol. *J. Am. Chem. Soc.* 118, 3082–3090.

79 LUO, P. and BALDWIN, R. L. (1997) Mechanism of helix induction by trifluoroethanol: a framework for extrapolating the helix-forming properties of peptides from trifluoroethanol/water mixtures back to water. *Biochemistry* 36, 8413–8421.

80 HIROTA, N., MIZUNO, K., and GOTO, Y. (1997) Cooperative α-helix formation of β-lactoglobulin and melittin induced by hexafluoroisopropanol. *Protein Sci.* 6, 416–421.

81 HIROTA, N., MIZUNO, K., and GOTO, Y. (1998) Group additive contributions to the alcohol-induced α-helix formation of melittin: implication for the mechanism of the alcohol effects on proteins. *J. Mol. Biol.* 275, 365–378.

82 HONG, D.-P., KUBOI, R., HOSHINO, M., and GOTO, Y. (1999) Clustering of fluorine-substituted alcohols as a factor responsible for their marked effects on proteins and peptides. *J. Am. Chem. Soc.* 121, 8427–8433.

83 HIROTA-NAKAOKA, N. and GOTO, Y. (1999) Alcohol-induced denaturation of β-lactoglobulin: a close correlation

to the alcohol-induced α-helix formation of melittin. *Bioorg. Med. Chem.* 7, 67–73.

84 LLINAS, M. and KLEIN, M. P. (1975) Charge relay at the peptide bond. A proton magnetic resonance study of solvation effects on the amide electron density distribution. *J. Am. Chem. Soc.* 97, 4731–4737.

85 BROOKS, C. L. and NILSSON, L. (1993) Promotion of helix formation in peptides dissolved in alcohol and water-alcohol mixtures. *J. Am. Chem. Soc.* 115, 11034–11035.

86 GUO, H. and KARPLUS, M. (1994) Solvent influence on the stability of the peptide hydrogen bond: a supramolecular cooperative effect. *J. Phys. Chem.* 98, 7104–7105.

87 IOVINO, M., FALCONI, M., MARCELLINI, A., and DESIDERI, A. (2001) Molecular dynamics simulation of the antimicrobial salivary peptide histatin-5 in water and in trifluoroethanol: a microscopic description of the water destructuring effect. *J. Pept. Res.* 58, 45–55.

88 ROCCATANO, D., COLOMBO, G., FIORONI, M., and MARK, A. E. (2002) Mechanism by which 2,2,2-trifluoroethanol/water mixtures stabilize secondary-structure formation in peptides: a molecular dynamics study. *Proc. Natl Acad. Sci. USA* 99, 12179–12184.

89 BARROW, C. J., YASUDA, A., KENNY, P. T., and ZAGORSKI, M. G. (1992) Solution conformations and aggregational properties of synthetic amyloid β-peptides of Alzheimer's disease. Analysis of circular dichroism spectra. *J. Mol. Biol.* 225, 1075–1093.

90 NILSSON, M. R., NGUYEN, L. L., and RALEIGH, D. P. (2001) Synthesis and purification of amyloidogenic peptides. *Anal. Biochem.* 288, 76–82.

91 HIROTA-NAKAOKA, N., HASEGAWA, K., NAIKI, H., and GOTO, Y. (2003) Dissolution of β2-microglobulin amyloid fibrils by dimethylsulfoxide. *J. Biochem.* 134, 159–164.

92 OHNISHI, S., KOIDE, A., and KOIDE, S. (2000) Solution conformation and amyloid-like fibril formation of a polar peptide derived from a β-hairpin in the OspA single-layer β-sheet. *J. Mol. Biol.* 301, 477–489.

93 GODA, S., TAKANO, K., YAMAGATA, Y., MAKI, S., NAMBA, K., and YUTANI, K. (2002) Elongation in a β-structure promotes amyloid-like fibril formation of human lysozyme. *J. Biochem.* 132, 655–661.

94 ZEROVNIK, E., TURK, V., and WALTHO, J. P. (2002) Amyloid fibril formation by human stefin B: influence of the initial pH-induced intermediate state. *Biochem. Soc. Trans.* 30, 543–547.

95 SRISAILAM, S., KUMAR, T. K., RAJALINGAM, D., KATHIR, K. M., SHEU, H. S., JAN, F. J., CHAO, P. C., and YU, C. (2003) Amyloid-like fibril formation in an all β-barrel protein. Partially structured intermediate state(s) is a precursor for fibril formation. *J. Biol. Chem.* 278, 17701–17709.

96 LU, H., BUCK, M., RADFORD, S. E., and DOBSON, C. M. (1997) Acceleration of the folding of hen lysozyme by trifluoroethanol. *J. Mol. Biol.* 265, 112–117.

97 HAMADA, D., CHITI, F., GUIJARRO, J. I., KATAOKA, M., TADDEI, N., and DOBSON, C. M. (2000) Evidence concerning rate-limiting steps in protein folding from the effects of trifluoroethanol. *Nat. Struct. Biol.* 7, 58–61.

98 CHITI, F., TADDEI, N., WEBSTER, P., HAMADA, D., FIASCHI, T., RAMPONI, G., and DOBSON, C. M. (1999) Acceleration of the folding of acylphosphatase by stabilization of local secondary structure. *Nat. Struct. Biol.* 6, 380–387.

99 SANCHEZ, I. E. and KIEFHABER, T. (2003) Hammond behavior versus ground state effects in protein folding: evidence for narrow free energy barriers and residual structure in unfolded states. *J. Mol. Biol.* 327, 867–884.

100 SANCHEZ, I. E. and KIEFHABER, T. (2003) Non-linear rate-equilibrium free energy relationships and Hammond behavior in protein folding. *Biophys. Chem.* 100, 397–407.

101 YIU, C. P., MATEU, M. G., and FERSHT, A. R. (2000) Protein folding transition

states: elicitation of Hammond effects by 2,2,2-trifluoroethanol. *Chembiochemistry* 1, 49–55.

102 GOTO, Y. and FINK, A. L. (1989) Conformational states of β-lactamase: molten-globule states at acidic and alkaline pH with high salt. *Biochemistry* 28, 945–952.

103 GOTO, Y. and FINK, A. L. (1990) Phase diagram for acidic conformational states of apomyoglobin. *J. Mol. Biol.* 214, 803–805.

104 GOTO, Y. and FINK, A. L. (1994) Acid-induced folding of heme proteins. *Methods Enzymol.* 232, 3–15.

105 FINK, A. L., CALCIANO, L. J., GOTO, Y., NISHIMURA, M., and SWEDBERG, S. A. (1993) Characterization of the stable, acid-induced, molten globule-like state of staphylococcal nuclease. *Protein Sci.* 2, 1155–1160.

106 FINK, A. L., CALCIANO, L. J., GOTO, Y., KUROTSU, T., and PALLEROS, D. R. (1994) Classification of acid denaturation of proteins: Intermediates and unfolded states. *Biochemistry* 33, 12504–12511.

107 HAGIHARA, Y., TAN, Y., and GOTO, Y. (1994) Comparison of the conformational stability of the molten globule and native states of horse cytochrome c: effects of acetylation, heat, urea, and guanidine-hydrochloride. *J. Mol. Biol.* 237, 336–348.

108 NISHII, I., KATAOKA, M., TOKUNAGA, F., and GOTO, Y. (1994) Cold-denaturation of the molten globule states of apomyoglobin and a profile for protein folding. *Biochemistry* 33, 4903–4909.

109 NISHII, I., KATAOKA, M., and GOTO, Y. (1995) Thermodynamic stability of the molten globule states of apomyoglobin. *J. Mol. Biol.* 250, 223–238.

110 HAMADA, D., KIDOKORO, S., FUKADA, H., TAKAHASHI, K., and GOTO, Y. (1994) Salt-induced formation of the molten globule state of cytochrome c studied by isothermal titration calorimetry. *Proc. Natl Acad. Sci. USA* 91, 10325–10329.

111 HAMADA, D., FUKADA, H., TAKAHASHI, K., and GOTO, Y. (1995) Salt-induced formation of the molten globule state of apomyoglobin studied by isothermal titration calorimetry. *Thermochim. Acta* 266, 385–400.

112 GOTO, Y., CALCIANO, L. J., and FINK, A. L. (1990) Acid-induced folding of proteins. *Proc. Natl Acad. Sci. USA* 87, 573–577.

113 GOTO, Y., HAGIHARA, Y., HAMADA, D., HOSHINO, M., and NISHII, I. (1993) Acid-induced unfolding and refolding transitions of cytochrome c: a three-state mechanism in H_2O and D_2O. *Biochemistry* 32, 11878–11885.

114 GOTO, Y. and NISHIKIORI, S. (1991) Role of electrostatic repulsion in the acidic molten globule of cytochrome c. *J. Mol. Biol.* 222, 679–686.

115 GOTO, Y., TAKAHASHI, N., and FINK, A. L. (1990) Mechanism of acid-induced folding of proteins. *Biochemistry* 29, 3480–3488.

116 GJERDE, D. T., SCHMUCHLER, G., and FRITZ, J. S. (1980) Anion chromatography with low-conductivity eluents. II. *J. Choromatogr.* 187, 35–45.

117 GREGOR, H. P., BELLE, J., and MARCUS, R. A. (1955) Studies on ion exchange resins. XIII. Selectivity coefficients of quaternary base anion-exchange resins toward univalent anions. *J. Am. Chem. Soc.* 77, 2713–2719.

118 GOTO, Y. and HAGIHARA, Y. (1992) Mechanism of the conformational transition of melittin. *Biochemistry* 31, 732–738.

119 HAGIHARA, Y., KATAOKA, M., AIMOTO, S., and GOTO, Y. (1992) Charge repulsion in the conformational stability of melittin. *Biochemistry* 31, 11908–11914.

120 HAGIHARA, Y., OOBATAKE, M., and GOTO, Y. (1994) Thermal unfolding of tetrameric melittin: Comparison with the molten globule state of cytochrome c. *Protein Sci.* 3, 1418–1429.

121 HOSHINO, M. and GOTO, Y. (1994) Perchlorate-induced formation of the α-helical structure of mastoparan. *J. Biochem.* 116, 910–915.

122 GOTO, Y. and AIMOTO, S. (1991) Anion and pH-dependent conformational transition of an amphiphilic polypeptide. *J. Mol. Biol.* 218, 387–396.

123 HENKELS, C. H., KURZ, J. C., FIERKE, C. A., and OAS, T. G. (2001) Linked folding and anion binding of the *Bacillus subtilis* ribonuclease P protein. *Biochemistry* 40, 2777–2789.

124 HAGIHARA, Y., AIMOTO, S., FINK, A. L., and GOTO, Y. (1993) Guanidine hydrochloride-induced folding of proteins. *J. Mol. Biol.* 231, 180–184.

125 SANTARO, M. M. and BOLEN, D. W. (1992) A test of the linear extrapolation of unfolding free energy changes over an extended denaturant concentration range. *Biochemistry* 31, 4901–4907.

25
Prolyl Isomerization in Protein Folding

Franz Schmid

25.1
Introduction

Peptide bonds in proteins can occur in two isomeric states: *trans*, when the dihedral angle ω is 180°, and *cis*, when ω is 0°. For the bonds preceding residues other than proline (nonprolyl bonds[1]) the *trans* conformation is strongly favored energetically over *cis*, and therefore *cis* nonprolyl bonds are rare in folded proteins. Peptide bonds preceding proline (prolyl bonds) are frequently in the *cis* conformation. In this case the *trans* form is only slightly favored energetically over *cis*. *Cis* peptide bonds in native proteins complicate the folding process, because the incorrect *trans* forms predominate in the unfolded or nascent protein molecules, and because the *cis* ⇌ *trans* isomerizations are intrinsically slow reactions. Incorrect prolyl isomers do not block the folding of a protein chain right at the beginning, but they strongly decelerate the overall folding process. This is clearly seen for small single-domain proteins. Many of them refold within milliseconds or even less when they contain correct prolyl isomers, but when incorrect isomers are present (typically *trans* isomers of bonds that are *cis* in the native state) folding is decelerated from milliseconds to the time range of several minutes.

Several strategies are useful for identifying proline-limited steps in a folding reaction. They include interrupted unfolding and refolding in double mixing experiments as well as the use of prolyl isomerases, which are enzymes that catalyze prolyl isomerizations (see Chapter 10 in Part II). The prolyl isomerases are ubiquitous proteins. They are found in all organisms, and in bacteria a prolyl isomerase

---

1) To facilitate reading I use the terms *cis* proline and *trans* proline for proline residues that are preceded by a *cis* or a *trans* peptide bond, respectively, in the *folded* protein. "Native-like" and "incorrect, nonnative" denote whether in an unfolded state a particular prolyl peptide bond shows the same conformation as in the native state or not. Further, I use the expression "isomerization of Xaa" for the isomerization of the peptide bond preceding Xaa. Peptide bonds preceding proline are referred to as "prolyl bonds," those preceding residues other than proline as "nonprolyl bonds." The folding reactions that involve Xaa-Pro isomerizations as rate-limiting steps are denoted "proline-limited" reactions.

Protein Folding Handbook. Part I. Edited by J. Buchner and T. Kiefhaber
Copyright © 2005 WILEY-VCH Verlag GmbH & Co. KGaA, Weinheim
ISBN: 3-527-30784-2

is associated with the ribosome. This enzyme, the trigger factor, accelerates proline-limited folding reactions particularly well and might in fact also catalyze prolyl isomerizations in the folding of nascent proteins. Prolyl isomerization in protein folding and its catalysis by prolyl isomerases are covered in a number of review articles [1–10].

25.2
Prolyl Peptide Bonds

In a peptide bond the distance between the carbonyl carbon and the nitrogen is 0.15 Å shorter than expected for a C–N single bond and both the C and the N atom show sp^2 hybridization [11], which indicates that the peptide bond has considerable double bond character. Peptide bonds are thus planar, and the flanking Cα atoms can be either in the *trans* or in the *cis* conformation (equivalent to dihedral angles ω of 180° and 0°, respectively). For peptide bonds preceding residues other than proline the *cis* state is strongly disfavored. Fischer and coworkers found *cis* contents between 0.11 and 0.48% for a number of nonprolyl peptide bonds in oligopeptides. The highest propensity to adopt the *cis* conformation was observed for a Tyr-Ala bond in a dipeptide [12].

In native, folded proteins nonprolyl *cis* peptide bonds are very rare [13–16]. Only 43 *cis* nonprolyl peptide bonds were found in a survey of 571 protein structures [16]. Interestingly carboxypeptidase A contains three of them [17].

For Xaa-Pro peptide bonds ("prolyl bonds") the *cis* and *trans* conformations differ much less in energy because the Cα of Xaa is always arranged in *cis* with a C atom: either with the Cα or the Cδ of the proline (Figure 25.1). The *trans* isomer is thus favored only slightly over the *cis* isomer and in short peptides frequently *cis* contents of 10–30% are observed [18–21]. The *cis*/*trans* ratio depends on the properties of the flanking amino acids and on ionic or van der Waals interactions that are possible in one isomeric state, but not in the other.

In small, well-folded proteins the conformational state of each prolyl bond is usually clearly defined. It is either *cis* or *trans* in every molecule, depending on the structural framework imposed by the folded protein chain. About 5–7% of all prolyl peptide bonds in folded proteins are *cis* [13–16], and 43% of 1435 nonredundant protein structures in the Brookhaven Protein Database contain at least one *cis* proline [20]. Several proteins are heterogeneous and show *cis*/*trans* equilibria at

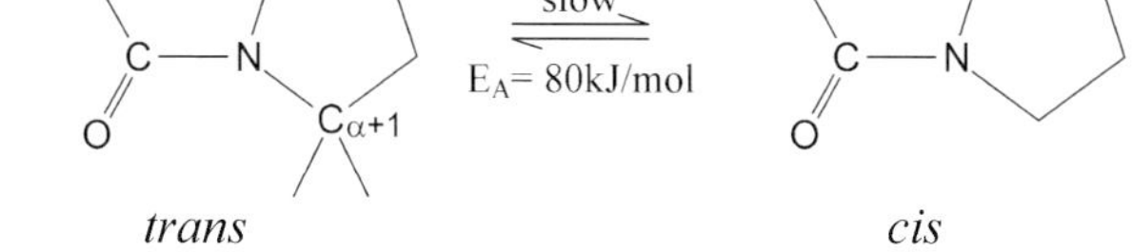

Fig. 25.1. Isomerization between the *cis* and *trans* forms of an Xaa-Pro peptide bond.

one or more prolyl bonds. Most of these heterogeneities were discovered by NMR spectroscopy. They will be discussed later (see Section 25.5).

The energy barrier between the *cis* and the *trans* form is enthalpic in nature. The average activation enthalpy is about 80 kJ mol^{-1} and the activation entropy is close to zero. Because of this high energy barrier prolyl isomerizations are intrinsically slow reactions with time constants that range between 10 and 100 s (at 25 °C). The absence of an activation entropy indicates that the surrounding solvent is not reorganized when the activated state of this intramolecular isomerization reaction is reached.

Secondary amide [22] and prolyl isomerizations are faster in nonpolar solvents than in water. This was traditionally explained by assuming that the transition state for isomerization is less polar than the ground state because the peptide bond is twisted and thus the resonance between the carbonyl group and the nitrogen is lost [23]. Eberhardt et al. [24] found, however, that the rate constant of prolyl isomerization in the dipeptide Ac-Gly-Pro-OCH$_3$ does not correlate with the dielectric constant of the solvent, but with its ability to donate a hydrogen bond to the carbonyl oxygen of the peptide.

The nitrogen of the amide bond can be protonated by very strong acids. This protonation abolishes the resonance between C=O and N of the prolyl bond. Thus, the partial double bond character is lost and the barrier to rotation is diminished. Prolyl isomerization is in fact well catalyzed by a solution of acetic acid in acetic anhydride [25] or by ≥ 7 M HClO$_4$ [26].

In summary, prolyl isomerization is slow because the resonance energy of the CN partial double bond must be overcome. The reaction is decelerated when the resonance is increased, such as in solvents that donate a hydrogen bond or a proton to the carbonyl oxygen. It is accelerated when the resonance is decreased, in particular by N protonation. The mechanism of nonenzymic prolyl isomerization is thoroughly discussed by Stein [23] and Fischer [21].

25.3
Prolyl Isomerizations as Rate-determining Steps of Protein Folding

25.3.1
The Discovery of Fast and Slow Refolding Species

In seminal work Garel and Baldwin [27] discovered that unfolded ribonuclease A (RNase A) consists of a heterogeneous mixture of molecules. They found fast-folding U$_F$ molecules[2] that refolded in less than a second, and slow-folding U$_S$ molecules that required several minutes to fold to completion. Similar U$_F$ and U$_S$

---

2) Abbreviations used: U, I, N, the unfolded, the intermediate, and native state of a protein, respectively; U$_F$, U$_S$, fast- and slow-folding unfolded forms, I$_N$, native-like intermediate; τ and λ, apparent time constant and rate constant, respectively, of a reaction; k, microscopic rate constant.

species were later found in the folding of many other proteins [7, 28, 29]. In 1975, two years after the discovery by Garel and Baldwin, Brandts and coworkers [30] formulated the proline hypothesis and suggested that the U_F and U_S molecules differ in the *cis/trans* isomeric state of one or more Xaa-Pro peptide bonds.

In the native protein (N) usually each prolyl peptide bond is in a defined conformation, either *cis* or *trans*, depending on the constraints dictated by the ordered native structure. After unfolding (N → U_F), however, these bonds become free to isomerize slowly as in short oligopeptides (in the $U_F \rightleftharpoons U_S^i$ reaction), as shown in Scheme 25.1.

$$N \xrightarrow[\text{fast}]{\text{unfolding}} U_F \underset{\text{slow}}{\overset{\text{Pro isomerization}}{\rightleftharpoons}} U_S^i$$

Scheme 25.1. Kinetic model for the coupling betweeen protein unfolding and prolyl isomerization.

The isomerizations thus create a mixture, which consists of a single unfolded species with correct prolyl isomers (U_F) and, depending on the number of prolines, one or more unfolded species with incorrrect prolyl isomers (U_S^i). The U_F molecules with the native-like prolyl isomers can refold directly in a fast reaction to the native conformation. The U_S^i molecules, however, refold slowly, because their refolding involves the re-isomerizations of the incorrect prolyl bonds.

25.3.2
Detection of Proline-limited Folding Processes

Proline-limited reactions in the unfolding and refolding of a protein can be identified and characterized in several ways.

1. *Rates and activation energies of refolding.* Circumstantial evidence for a proline-limited folding reaction is provided when its time constant is in the 10–100 s range (at 25 °C) and when its activation energy is about 80 kJ mol^{-1}. Conformational folding reactions can, however, show very similar kinetic properties, and therefore time constants and activation energies provide weak evidence for a proline-limited process.
2. *Properties of the refolding reactions.* The various U species that are populated at equilibrium in the unfolded protein give rise to parallel direct and proline-limited refolding reactions. The relative amplitudes of these refolding reactions should depend only on the relative populations of these U species. They should be independent of the final folding conditions and independent of the probe that was used to monitor refolding. The amplitudes of refolding should also be independent of the initial unfolding conditions, because the *cis/trans* equilibria at prolyl bonds are largely independent of the denaturant concentration, of temperature, and of the pH (unless the prolines are flanked by ionizable groups).

These criteria are valid only when direct folding is much faster than proline-limited folding and when partially folded intermediates with incorrect prolines do not accumulate during folding. U species with several incorrect prolines can enter alternative refolding routes, and the rank order of the re-isomerizations may change with the refolding conditions. Guidelines for the kinetic analysis of folding reactions that are coupled with prolyl isomerizations are given by Kiefhaber et al. [31, 32].

3. *Slow refolding assays (double jumps)*. Prolyl isomerizations in the unfolded form of a protein (after the N → U_F unfolding step, Scheme 25.1) can be measured by slow refolding assays. The $U_F \rightleftharpoons U_S$ isomerizations usually cannot be monitored directly by a spectroscopic probe. Rather, the amount of U_S formed after different times of unfolding is determined by slow refolding assays in "double jump" experiments [30, 33]. Samples are withdrawn from the unfolding solution at different time intervals after the initiation of unfolding; these samples are transferred to standard refolding conditions, and the amplitudes of the resulting slow refolding reactions, $U_S \rightarrow N$, are determined. These amplitudes are proportional to the concentrations of the U_S species present at the time when the sample was withdrawn for the assay, and their dependence on the time of unfolding reveals the kinetics of the $U_F \rightleftharpoons U_S$ reactions. An experimental protocol for double-jump experiments is given in Section 25.9 (Experimental Protocols). Slow refolding assays work very well when conditions can be found under which the conformational and the proline-limited events are well separated in both the unfolding step and in the subsequent refolding assay. It cannot be used when conformational folding ($U_F \rightarrow N$) in the second step shows a similar rate as the $U_F \rightleftharpoons U_S$ reactions. In this case the prolyl isomerizations couple with the $U_F \rightarrow N$ reaction in the refolding assay, and the amplitudes of refolding no longer reflect the concentration of the U_S molecules, as present at the time of sample transfer.

4. *Catalysis by prolyl isomerases*. Prolyl isomerases are excellent tools to identify proline-limited folding reactions. Since they catalyze only prolyl isomerizations (see Chapter 10 in Part II), the acceleration of a particular folding reaction by a prolyl isomerase provides compelling evidence that it involves a proline-limited step. Partial folding may render prolyl bonds inaccessible for a prolyl isomerase. Therefore a lack of catalysis does not strictly rule out prolyl isomerization as the rate-limiting step of a particular folding reaction. Several families of prolyl isomerases with different substrate specificities are available and can be used as tools in folding studies [3, 7, 8].

5. *Replacement of prolines by other residues*. Site-directed mutagenesis of proline residues is an apparently simple und straightforward means to detect proline-limited folding reactions and to identify the prolines that are involved in a particular folding reaction. The interpretation of the changes in the folding kinetics is usually straightforward when a *trans* prolyl bond of the wild-type protein is changed to a nonprolyl *trans* bond by the mutation, or when a *cis* prolyl bond is replaced by a "normal" nonprolyl *trans* bond. The protein stability and the rate of the direct folding reaction should not be changed strongly by the proline

substitution. For *cis* prolines such conditions are rarely met. Often, mutations of critical *cis* prolines are strongly destabilizing and the *cis* conformation of the respective bond is maintained even after the mutation to a residue other than proline. *Trans* → *cis* isomerizations at nonprolyl peptide bonds are also slow. An unchanged slow folding reaction after the replacement of a particular proline is therefore not sufficient evidence to exclude its isomerization as a possible rate-limiting reaction. It is necessary to analyze the unfolding and refolding kinetics after a proline replacement in detail, in order not to draw incorrect conclusions.

6. *Analysis of the folding kinetics by NMR spectroscopy*. Perhaps the best approach would be to follow the isomerization of a particular proline during folding directly by real-time NMR spectroscopy. The CαH resonance of the residue preceding the proline and the CδH resonance of proline are sensitive to the isomeric state of the prolyl bond and could be used to follow a prolyl isomerization directly in the course of a protein folding reaction. Unfortunately, these resonances are in a very crowded region of the NMR spectrum of a protein, and therefore, to the best of my knowledge, this approach has not yet been employed.

25.3.3
Proline-limited Folding Reactions

Proline-limited steps were found in the folding kinetics of many small single-domain proteins. The direct folding reactions of these proteins are usually fast (often in the time range of milliseconds) and therefore the proline-limited steps can easily be identified in unfolding and in refolding either directly or by double-mixing experiments, as described in Section 25.3.2 and in Section 25.9 (Experimental Protocols).

As expected, proline-limited reactions with large amplitudes are found for proteins with *cis* prolines, such as the immunoglobulin domains [34], thioredoxin [35], barstar [36], staphylococcal nuclease [37, 38], ubiquitin [39], plastocyanin [40], or pseudo-azurin [41]. In most of these cases partially folded intermediates accumulate in refolding before the final slow prolyl *trans* → *cis* isomerization.

Single-domain proteins with only *trans* prolines also show slow refolding reactions. Their amplitudes are usually small, because the *trans* isomer is favored in the unfolded molecules as well. Often, the amplitude of slow folding is smaller than expected from the number of *trans* prolines. This may originate from the fact that in unfolded polypeptides the *trans* form is more favored over *cis* than in short peptides and that some *trans* prolines are nonessential for folding. In addition, native-like intermediates may accumulate prior to re-isomerization, which also decreases the amplitudes of the slow folding reactions. Proteins with only *trans* prolines include cytochrome *c* [42], barnase [43], the chymotrypsin inhibitor CI2 [44], FKBP12 [45, 46], acylphosphatase [47], MerP [48] the immunity proteins Im7 and Im9 [49], the cell cycle protein p13suc1 [50], and the amylase inhibitor Tendamistat [51].

For large oligodomain or oligomeric proteins it is usually difficult to identify prolyl isomerizations in unfolding and refolding, chiefly because direct folding is often slow and not well separated from the proline-limited steps. As a consequence, the two reactions are strongly coupled, and the isomerizations cannot be followed easily by the double-mixing techniques. Nevertheless there are numerous examples for proline-limited steps in the folding of oligodomain and/or oligomeric proteins [52–55].

The immunoglobulins are heterotetramers that are composed of two light chains (with two domains each) and two heavy chains (with four domains each). The individual domains show a related fold (the "immunoglobulin fold") and typically contain a single *cis* proline. Its *trans* → *cis* isomerization determines the folding of the separate domains [34] and of the intact light chains [56]. In the folding of the F_{ab} fragment (which consists of four domain: two from the light chain and two from the heavy chain) proline isomerization can occur late in folding, after association [57]. In the folding of the isolated $C_H 3$ domain, however, prolyl isomerization occurs before this domain associates to a homodimer [58].

The gene-3-protein in the coat of the phage fd is essential for its infectivity. Early in the infection of an *Escherichia coli* cell, the two N-terminal domains of the gene-3-protein (N2 and N1) interact successively with the F pilus and the TolA receptor. To expose the binding site for TolA the domains must disassemble, otherwise infection would be stopped. In refolding, the domains N1 and N2 fold rapidly, but the final domain assembly is unusually slow and shows a time constant of 6200 s (at 25 °C). It is controlled by the *trans* → *cis* isomerization of the Gln212-Pro213 bond in the hinge between the domains. The kinetic block of domain reassembly caused by this very slow isomerization at Pro213 could ensure that after the initial binding of N2 to the F pilus the domains disassemble and that this open state has a lifetime long enough for the N1 domain to approach and interact with TolA. Pro213 isomerization of the gene-3-protein might thus serve as a slow conformational switch during the infection process [59, 60].

Ure2 is a prion-like yeast protein with an unfolded N-terminal prion domain and a compactly folded C-terminal domain, which harbors a *cis* proline at position 166. Unfolded molecules with the native-like *cis* Pro166 fold and dimerize rapidly, whereas molecules with the incorrect *trans* isomer do not reach the native state within several hours [61]. Apparently, Pro166 isomerization decelerates productive folding so strongly that it can no longer compete with side reactions, such as aggregation.

A few important points should be considered when, in the folding of large proteins, the role of many *trans* prolines or of a mixture of *cis* and *trans* prolines is investigated.

1. Not all prolines are necessarily important for the folding kinetics.
2. The *trans* form is usually favored over the *cis*, and therefore *cis* → *trans* isomerizations show small amplitudes and are about 5–10 times faster than *trans* → *cis* isomerizations.

3. The prolyl isomerizations after unfolding are parallel reactions. As a consequence, the rate of U_S formation should increase with the number of prolines.
4. The rate of refolding, however, decreases with the number of incorrect prolines [62], because only the species with all prolines in the native-like isomeric state can complete folding.
5. For large proteins the direct refolding reaction is often rather slow and not well separated in rate from the proline-limited steps. As a consequence, the analysis by slow refolding assays becomes difficult.

Carbonic anhydrase provides a good example. It contains 15 *trans* and 2 *cis* prolines and after long-term denaturation virtually all molecules are in U_S states, i.e., they have nonnative prolines and refold slowly [63]. More than half of the U_S molecules form rapidly after unfolding in a reaction that is about tenfold faster than expected for a single prolyl isomerization [64]. This increase in rate is probably caused by the additive and independent isomerizations of the many *trans* prolines of this protein. The remainder of the slow-folding molecules is then created very slowly in a reaction that is probably caused by the *cis* ⇌ *trans* isomerizations of one or both *cis* prolines. Short-term unfolding of carbonic anhydrase yields U_F molecules with correct prolines, but only a fraction of them can in fact refold rapidly, because molecules with incorrect isomers continue to form early in refolding, in competition with the direct refolding reaction, which is only marginally faster than the multiple *cis* ⇌ *trans* isomerizations. Addition of cyclophilin had a peculiar effect on the refolding of the short-term denatured molecules. At the onset of the experiment this prolyl isomerase catalyzed *cis* ⇌ *trans* isomerization in the unfolded molecules and thus increased the fraction of molecules that fold on slow proline-limited paths. These molecules, however, refolded more rapidly, because cyclophilin accelerated the proline-limited steps in their refolding [64]. A similar behavior was found for the single-chain Fv fragment of an antibody, which contains four *trans* and two *cis* prolines [65].

25.3.4
Interrelation between Prolyl Isomerization and Conformational Folding

When they initially suggested the proline hypothesis, Brandts and coworkers assumed that the nonnative prolyl isomers can block refolding right at its beginning [30], and that the slow $U_S \rightarrow U_F$ isomerization in the unfolded molecules is the first and rate-limiting step of folding. Since then many proline-limited folding reactions have been investigated (see Section 25.4), and it is clear now that conformational folding can start while some prolines are still in their nonnative states, and that partially folded intermediates can tolerate incorrect prolines. The final folding steps, however, require correct prolines and therefore they are limited in rate by the prolyl isomerizations [7, 29, 66, 67].

The extent of conformational folding that can occur prior to prolyl isomerization depends on the location of the nonnative prolyl bonds in the structure, on the stability of partially folded intermediates and on the folding conditions. Generally, incorrect prolyl bonds in exposed or flexible chain regions will not interfere strongly with conformational folding, and solvent conditions that strongly stabilize folded proteins will also stabilize partially folded structure with incorrect isomers. Thus there is a close interdependence between conformational folding steps and prolyl isomerization. On the one hand the presence of incorrect isomers in the chain can decelerate its folding, and on the other hand rapid chain folding can affect the equilibrium and the kinetics of prolyl isomerization. The close interrelationship between structure formation and prolyl peptide bond isomerization is a key feature of slow folding steps and is of central importance for understanding the role of prolyl isomerases in these processes.

A simple $U_S \to U_F \to N$ refolding path is followed only under conditions where the fully folded protein is only marginally stable, such as at the onset of the equilibrium unfolding transition. Here a single incorrect kink in the protein backbone (as introduced by a nonnative prolyl bond) is sufficient to destabilize partially folded conformations.

25.4
Examples of Proline-limited Folding Reactions

25.4.1
Ribonuclease A

Bovine pancreatic RNase A played a central role in the elucidation of proline-limited folding reactions. This protein contains four prolines: Pro42 and Pro117 are *trans* and Pro93 and Pro114 are *cis* in the native protein. Both the proposal of the proline hypothesis and the first experimental evaluations are linked with RNase A. Garel and Baldwin [27] discovered the coexistence of U_F and U_S in unfolded RNase A, and Brandts and coworkers [30] developed the slow refolding assays to measure the $U_F \rightleftharpoons U_S$ equilibration in this protein. The quantitative analyses of the folding kinetics of RNase A in the thermal unfolding transition at pH 3 [68] and in the guanidinium chloride (GdmCl)-induced transition [69] demonstrated that they are in fact well described by a $U_S \rightleftharpoons U_F \rightleftharpoons N$ three-state mechanism. The $U_F \rightleftharpoons U_S$ reaction in unfolded RNase A was found to be independent of denaturants [70] and catalyzed by a strong acid [26], as expected for a proline-limited process.

The first partially folded intermediate with a nonnative prolyl isomer was also discovered in the refolding of RNase A [66, 67]. Under conditions that strongly favor the folded form, the major U_S species (called U_S^{II}) refolds to a native-like form I_N, an intermediate that is folded, as judged by amide circular dichroism and is already enzymatically active. This indicated for the first time that nonnative prolyl bonds do not necessarily block refolding, but that protein molecules with an incorrect prolyl bond could almost reach the native form. The final slow step of folding

($I_N \rightarrow N$) originates from the *trans* → *cis* isomerization of the Tyr92-Pro93 bond [71, 72]. It could be assigned before the advent of site-directed mutagenesis, because it is accompanied by a change in the fluorescence of Tyr92 [73].

The intermediate I_N is less stable and unfolds much faster than native RNase A. The interconversion between I_N and N could thus be measured by unfolding assays which exploited this strong difference between the unfolding rates of I_N and N. In these experiments refolding was interrupted after various times, samples were transferred to standard unfolding conditions and the amplitudes of the fast and slow unfolding reactions were determined as a function of the duration of refolding. These amplitudes reflected the time courses of I_N and N, respectively during folding [74]. Such interrupted folding experiments are excellent tools for discriminating native-like intermediates from the native protein and for following the time courses of these species in complex folding reactions. The method is so sensitive, because the native molecules are often separated from partially folded ones by a high activation barrier, and thus they differ strongly in the rate of unfolding. Section 25.9 gives a protocol for measuring silent slow prolyl isomerizations in largely folded intermediates by such unfolding assays.

The isomeric states of the four prolines in the various U species remained unclear for a long time. The kinetic analyses of the Baldwin, Brandts, and Schmid groups (for reviews see Refs [28, 75]) had indicated already that U_S is heterogeneous and consists of a minor U_S^I species and a major U_S^{II} species, which refolds via the native like intermediate I_N.

The Scheraga group re-evaluated the folding kinetics of RNase A, among others, by making many mutants with alterations at the four prolines (see Ref. [9] for a review). They found that the U_F species is heterogeneous as well [76], being composed of 5% of a very fast folding species U_{vf} and 13% of a fast-folding species U_f. U_{vf} and U_f differ in the isomeric state of Pro114. However, they fold with different rates only at low pH, where the conformational stability of RNase A is low. At neutral pH U_{vf} and U_f fold fast with indistinguishable rates. This is consistent with the original finding of 20% fast-folding U_F molecules [27], and it confirms the suggestion that Pro114 is nonessential for the folding of RNase A [66] under favorable conditions. Under unfavorable conditions, such as at low pH, or when an additional proline is incorrect (such as in the intermediate I_N with incorrect *trans* isomers at both Pro93 and Pro114), a *trans* Pro114 might in fact retard folding. Raines and coworkers replaced Pro114 by 5,5-dimethyl-L-proline, which stays almost exclusively in the native *cis* conformation. This mimic of a *cis* proline slightly accelerated the $U_S^{II} \rightarrow I_N$ reaction. The authors interpreted this to suggest that an incorrect *trans* Pro114 slightly decelerates the folding reaction of the U_S^{II} form of the wild-type protein [77]. Together all this suggests that Pro114 of RNase A is a conditional proline. It is unimportant for folding under favorable conditions, but retards folding slightly under unfavorable conditions.

U_S^{II} contains an incorrect *trans* Pro93. Comparative experiments with homologous RNases from various species [71, 78] and early directed mutagenesis work had already suggested a crucial role of Pro93 for the major folding reaction

$U_S^{II} \rightarrow I_N \rightarrow N$ [79]. The more recent results from the Scheraga group confirm this [9]. They also show that, after the Pro93Ala substitution, Tyr92 and Ala93 form a nonprolyl *cis* peptide bond in the folded form of P93A-RNase A [80].

The minor form U_S^I contains one or more incorrect prolines in addition to the incorrect *trans* Pro93 [71]. This destabilizes partially folded structure further and impairs the formation of intermediates such as I_N. Scheraga et al. have now found that U_S^I contains both an incorrect *trans* Pro93 and an incorrect *cis* Pro117 [9, 80–82]. According to their work, Pro42 is unimportant for folding.

25.4.2
Ribonuclease T1

RNase T1 from *Aspergillus oryzae* [83] is a small single-domain protein of 104 amino acids [83–86], which is not related structurally with RNase A. It contains two disulfide bonds (Cys2-Cys10 and Cys6-Cys103, two *trans* (Trp59-Pro60 and Ser72-Pro73) and two *cis* (Tyr38-Pro39 and Ser54-Pro55) prolyl bonds. RNase T1 is most stable near pH 5 and further stabilized when NaCl is added [87–89]. Importantly, in the absence of the disulfide bonds, RNase T1 can still fold to a native-like conformation when ≥1 M NaCl are present [87, 88, 90].

The equilibrium unfolding transition of RNase T1 is well described by a simple two-state model [91–94], but the unfolding and refolding kinetics are complex. Almost the entire folding of RNase T1 involves prolyl isomerizations [95, 96] both in the presence and in the absence of the two disulfide bonds [97, 98]. All observable phases in the unfolding and the refolding kinetics could be explained by contributions from the two *cis* prolines only [99–101]. RNase T1 is a good substrate for assaying prolyl isomerases [7, 8, 102].

The mechanism of unfolding is well explained by a model in which conformational unfolding ($N \rightarrow U_{55c}^{39c}$) is followed by the two *cis* ⇌ *trans* isomerizations at Pro39 and Pro55 (Scheme 25.2). Both are *cis* in N and in U_{55c}^{39c}, but isomerize

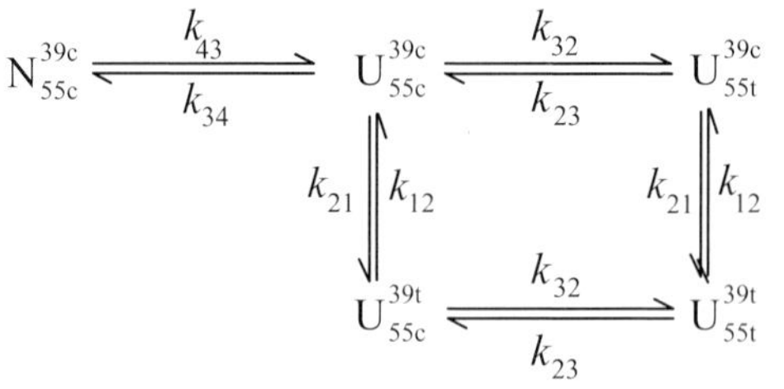

$$N_{55c}^{39c} \underset{k_{34}}{\overset{k_{43}}{\rightleftharpoons}} U_{55c}^{39c} \underset{k_{23}}{\overset{k_{32}}{\rightleftharpoons}} U_{55t}^{39c}$$
$$k_{21} \updownarrow k_{12} \qquad k_{21} \updownarrow k_{12}$$
$$U_{55c}^{39t} \underset{k_{23}}{\overset{k_{32}}{\rightleftharpoons}} U_{55t}^{39t}$$

Scheme 25.2. Kinetic model for the unfolding and isomerization of RNase T1. This model is valid for unfolding only. The superscript and the subscript indicate the isomeric states of prolines 39 and 55, respectively, in the correct, native-like *cis* (c) and in the incorrect, nonnative *trans* (t) isomeric states. As an example, U_{55c}^{39t} is an unfolded species with Pro39 in the incorrect *trans* and Pro55 in the correct *cis* state. In the denatured protein the two isomerizations are independent of each other, therefore the scheme is symmetric with identical rate constants in the horizontal and vertical directions, respectively. At 25 °C and 6.0 M GdmCl, pH 1.6 $k_{43} = 0.49$ s^{-1}, $k_{12} = 2.0 \times 10^{-3}$ s^{-1}, $k_{21} = 22.6 \times 10^{-3}$ s^{-1}, $k_{23} = 8.7 \times 10^{-3}$ s^{-1}, $k_{32} = 50.5 \times 10^{-3}$ s^{-1} [103].

largely to the more favorable *trans* state in the unfolded protein. As a consequence, only 2–4% of all unfolded molecules remain in the U_{55c}^{39c} state with Pro39 and Pro55 in the correct *cis* state. U_{55t}^{39t}, the species with two incorrect isomers, predominates at equilibrium, and the two species with single incorrect isomers (U_{55c}^{39t} and U_{55t}^{39c}) are populated to 10–20% each.

Under strongly native conditions refolding is not a reversal of unfolding and cannot be explained solely with Scheme 25.2. Rather, refolding paths via intermediates originate from the different unfolded species (Scheme 25.3). The two slow-folding species with one incorrect prolyl isomer each (U_{55c}^{39t} and U_{55t}^{39c}) and the species with both prolines in the incorrect isomeric state (U_{55t}^{39t}) can regain rapidly most of their secondary structure in the milliseconds range (the $U_i \rightarrow I_i$ steps, Scheme 25.3) [104]. Subsequently, slow folding steps follow that involve the isomerizations of the incorrect prolyl isomers. A peculiar feature of the folding model in Scheme 25.3 is that the major unfolded species with two incorrect isomers (U_{55t}^{39t}) can enter two alternative folding pathways (the upper or the lower pathway in Scheme 25.3), depending upon which isomerization occurs first. The distribution of refolding molecules on these two pathways is determined by the relative rates of the *trans* $\rightarrow$ *cis* isomerizations of Pro39 and Pro55 at the stage of the intermediate I_{55t}^{39t}. The kinetic models in Schemes 25.2 and 25.3 were supported by experimental data on the slow folding of variants of RNase T1 where the *cis* prolines were replaced by other amino acids [99, 100, 105].

At equilibrium only 2–4% of all unfolded RNase T1 molecules have correct *cis* isomers at Pro39 and Pro55. Their fast direct refolding reaction ($U_{55c}^{39c} \rightarrow N_{55c}^{39c}$) could therefore not be measured initially.

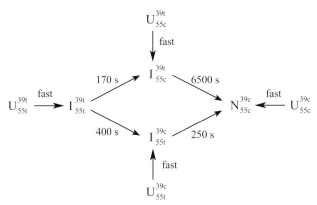

Scheme 25.3. Kinetic model for the slow refolding of RNase T1 under strongly native conditions. U stands for unfolded species, I for intermediates of refolding, and N is the native protein. The superscript and the subscript indicate the isomeric states of prolines 39 and 55, respectively, in the correct *cis* (c) and the incorrect *trans* (t) isomeric states. The rate constants given for the individual steps refer to folding conditions of 0.15 M GdmCl, pH 5.0, 10 °C [106]. The rate constant of the direct $U_{55c}^{39c} \rightarrow N_{55c}^{39c}$ reaction is 5.7 s^{-1} in 1.0 M GdmCl, pH 4.6, 25 °C [103].

Mayr et al. [103] exploited the sequential nature of unfolding (N → U_{55c}^{39c}) and prolyl isomerizations (cf. Scheme 25.2) to characterize the direct U_{55c}^{39c} → N folding reaction. In the first step of a stopped-flow double-mixing experiment they produced U_{55c}^{39c} transiently at a high concentration under conditions, where the N → U_{55c}^{39c} unfolding reaction is much faster than the subsequent prolyl isomerizations and then measured its refolding in the second step. In addition, they varied the duration of the unfolding step and monitored the concentrations of all species of the wild-type protein in Scheme 25.2 as a function of time. Thus the rates of the individual prolyl isomerizations in the unfolded protein (cf. Scheme 25.2) could be determined.

The direct refolding reaction of RNase T1 with correct prolyl isomers shows a time constant of 175 ms (at 25 °C, pH 4.6) [103]. This reaction is almost unaffected by the proline substitutions. It depends nonlinearly on temperature with a maximum near 25 °C, which suggests that the activated state for this reaction resembles the native rather than the unfolded state in heat capacity. The folding of the species with single incorrect prolyl isomers (U_{55c}^{39t} → I_{55c}^{39t} and U_{55t}^{39c} → I_{55t}^{39c}) was only about fivefold slower than direct folding and was also accompanied by a strong decrease in the apparent heat capacity.

Conformational folding and the re-isomerizations of the prolyl peptide bonds are thus tightly interrelated in the folding of RNase T1. As discussed above, rapid partial folding is possible in the presence of nonnative prolyl isomers [104, 106], but the final events of folding are coupled with the prolyl isomerizations and therefore very slow.

The partially folded structure in the intermediates affects the isomerization kinetics. In particular, the *trans* → *cis* isomerization at Pro39 (in the I_{55c}^{39t} → N step) is strongly decelerated by the folded structure in the intermediate I_{55c}^{39t} [106]. In the folding of pancreatic RNase A, however, prolyl isomerization is accelerated in the intermediate I_N [66, 67], and in the folding of dihydrofolate reductase an intramolecular catalysis of a prolyl isomerization was proposed to occur [107].

25.4.3
The Structure of a Folding Intermediate with an Incorrect Prolyl Isomer

In the S54G/P55N variant of RNase T1 [99] the *cis* Ser54-Pro55 prolyl bond is replaced by a *trans* Gly54-Asn55 peptide bond (Hinrichs et al., unpublished results). Thus a *cis* bond is abolished and the branched folding mechanism (Scheme 25.3) is strongly simplified to a sequential mechanism (Scheme 25.4). Only two unfolded species remain: 15% as a fast folding species with the native-like *cis* isomer of Pro39 (U^{39c}), and 85% as a slow folding species with the incorrect *trans* isomer of Pro39 (U^{39t}).

The U^{39t} molecules (Scheme 25.4) refold in two steps. First the intermediate I^{39t} is formed, which then converts to the native protein. This second step is limited in rate by the *trans* → *cis* isomerization of Pro39, which at 10 °C shows a time constant of about 8000 s. The I^{39t} intermediate is thus long-lived enough to obtain highly resolved structure information by kinetic NMR experiments [40, 108–110]

$$U^{39t} \xrightarrow{\text{fast}} I^{39t} \xrightarrow{0.0013 \text{ s}^{-1}} N^{39c} \xleftarrow{2.6 \text{ s}^{-1}} U^{39c}$$
$$\phantom{U^{39t} \xrightarrow{\text{fast}} I^{39t} \xrightarrow{0.0013 \text{ s}^{-1}\;} } 85\% \phantom{N^{39c}\;} 15\%$$

Scheme 25.4. Kinetic model for the slow refolding of S54G/P55N-RNase T1 under strongly native conditions. U stands for unfolded species, I for intermediates of refolding, and N is the native protein. The superscript indicates the isomeric state of proline 39 in the correct *cis* (c) and the incorrect *trans* (t) isomeric states. The rate constants given for the individual steps refer to folding conditions of 1.0 M GdmCl, pH 4.6, 25 °C.

and 2D-NOESY spectroscopy. Balbach et al. determined those protons in I^{39t} that are already surrounded by a native environment and thus show native distances to protons in close vicinity [111]. Surprisingly, amide protons in nonnative environments were found to be located not only close to the incorrect *trans* Tyr38-Pro39 bond in I^{39t}, but were spread throughout the entire protein. The destabilization caused by the incorrect *trans* Pro39 is thus not confined to the local environment of this proline, but involves several regions in the entire molecule.

25.5
Native-state Prolyl Isomerizations

Most prolyl peptide bonds in folded proteins show well defined conformations, being either in the *cis* or in the *trans* conformation. There are, however, a growing number of exceptions to this rule. High-resolution NMR spectroscopy, in particular, identified a series of prolines that are conformationally heterogeneous and exist as a mixture of *cis* and *trans* isomers in the folded state. Examples include staphylococcal nuclease [112], insulin [113], calbindin [114, 115], scorpion venom Lqh-8/6 [116], human interleukin-3 [117] and the TB6 domain of human fibrillin-1 [118, 119]. In folded staphylococcal nuclease *cis/trans* equilibria exist at Pro117 as well as at Pro47 [112, 120, 121], and the two equilibria seem to be independent of each other.

There are two strongly diverging, but not mutually exclusive, interpretations of these findings. The first is that conformationally heterogeneous proline residues occur in locally unfolded regions. They lack strong tertiary interactions, which would be necessary to stabilize one isomer over the other. Alternatively, heterogeneity at prolines might point to a functionally important *cis/trans* isomerization, which might be used as a slow molecular switch, for example, to regulate the function of the corresponding protein. Such switching functions, possibly modulated by prolyl isomerases, were suggested long time ago [122], but the evidence for proline switches remained circumstantial.

Evidence for a proline-dependent conformational switch was obtained for the SH2 domain of the tyrosine kinase Itk [123, 124]. Unlike other proteins with *cis* and *trans* prolyl conformers in the native state [115, 119, 125–127] the SH2 domain of Itk shows a pronounced change in structure upon *cis/trans* isomerization.

The *cis* and the *trans* forms of Itk-SH2 differ slightly in the affinity for the Itk-SH3 domain and for a phosphopeptide, which represent alternative substrates for the SH2 domain [128]. This conformational switch thus seems to modulate substrate recognition and can be regulated by the prolyl isomerase cyclophilin 18. A review of the proline switch in Itk and in other proteins is found in Ref. [10].

The prolyl isomerase, PIN1, catalyzes the isomerization of phosphoserine-proline bonds, and there is compelling evidence that PIN1 integrates phosphorylation-dependent and prolyl-isomerization-dependent switches for a wide range of signaling reactions [129] (see also Chapter 10 in Part II).

25.6
Nonprolyl Isomerizations in Protein Folding

For nonprolyl bonds the *trans* state is strongly favored over *cis*, and the *cis* content as measured for oligopeptides is far below 1% [12, 130]. A large amount of conformational energy is therefore required to lock such a bond in the *cis* state. A shift in the equilibrium from 0.2% to 99% *cis* requires a Gibbs free energy of 15 kJ mol^{-1}. Nevertheless *cis* nonprolyl bonds are found in folded proteins, but they are very rare, and presumably they are important for the function of a protein. In principle, a nonprolyl *cis* peptide bond would be well suited to construct a molecular switch in a protein that can be actuated only under a very strong load. A tight-binding ligand could, for example, supply a sufficient amount of free energy from its own binding to turn such a strong switch. A compilation of proteins with *cis* peptide bonds is found in Ref. [16]. Because the *trans* isomer is so strongly favored over the *cis*, *trans* → *cis* isomerizations at nonprolyl bonds during refolding must occur for such proteins in virtually all molecules.

The first evidence for a nonprolyl isomerization as a rate-determining step in protein folding was obtained for a variant of RNase T1. The replacement of one of the two *cis* prolines, *cis*-Pro39, of the wild-type protein by an alanine residue led to a variant with a *cis* nonprolyl peptide bond between Tyr38 and Ala39 [131]. This *cis* bond is located adjacent to His40, which is a catalytic residue. The Pro39Ala mutation reduced the stability of RNase T1 by about 20 kJ mol^{-1} and caused a major change in the folding mechanism. As shown in Scheme 25.5, the conformational unfolding of P39A-RNase T1 occurs first (in the N → U_{55c}^{39c} reaction) with a time constant of 20 ms. It leads to U_{55c}^{39c}, an unfolded form that has both Ala39 and

$$N_{55c}^{39c} \xrightarrow{\text{unfolding} \atop 54 \text{ s}^{-1}} U_{55c}^{39c} \underset{}{\overset{\text{Ala39 isomerization} \atop 1.46 \text{ s}^{-1}}{\rightleftarrows}} U_{55c}^{39t} \underset{0.0051 \text{ s}^{-1}}{\overset{\text{Pro55 isomerization} \atop 0.057 \text{ s}^{-1}}{\rightleftarrows}} U_{55t}^{39t}$$

Scheme 25.5. Kinetic mechanism for the unfolding and isomerization of P39A-RNase T1. The superscript and the subscript indicate the isomeric states of Ala39 and Pro55, respectively, in the correct, native-like *cis* (c) and in the incorrect, nonnative *trans* (t) isomeric states. The rate constants refer to unfolding in 6.0 M GdmCl (pH 1.6) at 25 °C [96].

Pro55 still in the native-like *cis* conformation. Subsequently, the Tyr38-Ala39 bond isomerizes from *cis* to *trans* (in the $U_{55c}^{39c} \rightleftharpoons U_{55c}^{39t}$ step) in virtually all molecules. This nonprolyl isomerization ($\tau = 730$ ms) is about 60-fold faster than the corresponding prolyl isomerization (of the Tyr38-Pro39 bond) in wild-type RNase T1 [96]. It shows almost the same time constant as the *cis* → *trans* isomerization of the Tyr-Ala bond in the pentapeptide AAYAA ($\tau = 560$ ms [12]). In the unfolding the Tyr-Ala isomerization is then followed by the slow *cis* ⇌ *trans* equilibration at Pro55 ($\tau = 16$ s), which is still present in P39A-RNase T1.

Unfolded molecules in which the Tyr38-Ala39 bond was still in the native-like *cis* conformation (U_{55c}^{39c}) were produced in a stopped-flow double-mixing experiment by a short 100-ms unfolding pulse. They refolded rapidly to the native state with $\tau = 290$ ms (in 1.0 M GdmCl, pH 4.6, 25 °C). After the Ala39 *cis* → *trans* isomerization in the unfolded state, refolding is 1600-fold retarded and shows a time constant of 480 s. This slow refolding is limited in rate by the *trans* → *cis* re-isomerization of the Tyr38-Ala39 bond. Under the same refolding conditions the *trans* → *cis* isomerization of the Tyr38-Pro39 bond in the wild-type protein is only twofold slower ($\tau = 1100$ s) [100, 105]. This comparison with the folding kinetics of wild-type RNase T1 indicates that the kinetics of Tyr38-Pro39 and of Tyr38-Ala39 isomerization differ predominantly in the rate of the *cis* → *trans*, rather than of the *trans* → *cis* reaction. The ratio of the rate constants for the *cis* → *trans* and the *trans* → *cis* reactions at the Tyr38-Ala39 bond gives an equilibrium constant of 0.0015, again in good agreement with the equilibrium constant of 0.0011 in the peptide AAYAA [12].

Very similar results were obtained later for the Pro93Ala mutation in RNase A [80, 132]. As in RNase T1, this mutation converted a *cis* Tyr-Pro bond into a *cis* Tyr-Ala bond and abolished the fast-folding reaction. The *cis* → *trans* isomerization of the Tyr-Ala bond in the unfolded protein occurred with $\tau = 1400$ ms at 15 °C [80], which agrees well with $\tau = 730$ ms, as measured for RNase T1 at 25 °C.

TEM-1 β-lactamase from *E. coli* contains a *cis* peptide bond between Glu166 and Pro167, in a large Ω loop near the active site. It is required for the catalytic activity of β-lactamase [133], and its *trans* → *cis* isomerization is a slow step in refolding [134–137]. Similar to Pro39 of RNase T1 the *cis* character of the 166–167 bond is retained when Pro167 of β-lactamase is replaced by a Thr residue, and the *trans* → *cis* isomerization of the Glu166-Thr167 peptide bond becomes rate limiting for the refolding of the P167T variant [135]. *Cis* peptide bonds between the residues equivalent to Glu166 and Pro167 of TEM-1 β-lactamase of *E. coli* are found in many β-lactamases, but interestingly, these *cis* peptide bonds are not always prolyl bonds. The β-lactamase PC1 from *Staphylococcus aureus* shows a *cis* Glu-Ile bond at this position [138].

In summary, the results obtained for RNase T1, RNase A and the β-lactamases show that the *trans* → *cis* isomerizations of nonprolyl peptide bonds are slow reactions (with time constants of several hundred seconds at 25 °C) and control the folding of proteins with nonprolyl *cis* peptide bonds. The *trans* → *cis* isomerizations of prolyl and nonprolyl bonds show similar rates, the reverse *cis* → *trans* isomerizations are 50–100 fold faster, however, for the nonprolyl peptide bonds.

Because the *trans* conformation is so strongly preferred, only one out of 500 or 1000 nonprolyl peptide bonds is in *cis* in an unfolded protein chain. Furthermore, it re-isomerizes to *trans* within less than a second when the protein folds at 25 °C. It is clear that such rare reactions are not easily identified in experimental folding kinetics.

First evidence for nonprolyl *cis* → *trans* isomerization in the refolding of a protein was obtained for a proline-free mutant of the small protein tendamistat [139]. Ninety-five per cent of all molecules of this protein fold rapidly with $\tau = 50$ ms, but 5% refold more slowly with $\tau = 400$ ms (at 25 °C). The slow-refolding molecules are created by isomerizations in the unfolded protein, most likely by the *cis* ⇌ *trans* isomerizations of nonprolyl peptide bonds. The fraction of molecules with incorrect *cis* nonprolyl isomers is small, because the correct *trans* isomer is favored so strongly over the *cis*.

25.7
Catalysis of Protein Folding by Prolyl Isomerases

Prolyl isomerases are ubiquitous enzymes that catalyze the *cis* ↔ *trans* isomerization of prolyl peptide bonds both in oligopeptides and during the folding of proteins. They are described in Chapter 10 in Part II. The first evidence for a catalysis of protein folding by a prolyl isomerase (porcine cytoplasmic cyclophilin 18, Cyp18) was obtained for the immunoglobulin light chain, porcine ribonuclease (RNase) and the S-protein fragment of bovine RNase A [140]. The slow refolding of RNase A could not be catalyzed, probably because the incorrect *trans* Tyr92-Pro93 is shielded from prolyl isomerases in the native-like intermediate I_N (cf. Section 25.5.1).

25.7.1
Prolyl Isomerases as Tools for Identifying Proline-limited Folding Steps

The prolyl isomerases, in particular Cyp18, are valuable tools to examine whether a slow folding reaction is rate-limited by a prolyl isomerization. A catalysis of proline-limited steps was observed in the folding of many proteins, including barnase [141], carbonic anhydrase [142, 143], β-lactamase (A. Lejeune, unpublished results), chymotrypsin inhibitor CI2 [44], yeast iso-2 cytochrome *c* [144, 145], the immunoglobulin light chain [57, 140], staphylococcal nuclease [146], and *trp* aporepressor [147]. *Cis* → *trans* and *trans* → *cis* isomerizations are catalyzed equally well in the folding of these proteins. In the maturation of the collagen triple helix prolyl and hydroxyprolyl isomerizations are rate-limiting steps, and this folding reaction is also accelerated by Cyp18 [148–150].

Proteins with incorrect prolyl isomers can start to fold before prolyl isomerization, and some of them reach a state that is native-like already. This often hinders prolyl isomerases in their access to the incorrect prolines and impairs catalysis of isomerization. The role of the accessibility of the prolines was studied for iso-2 cytochrome *c*. In aqueous buffer the folding of this protein is barely catalyzed by cyclophilin. When, however, GdmCl is added in increasing, but still nondenaturing concentrations, catalysis is markedly improved, presumably because the denaturant destabilizes folding intermediates and thus improves the accessibility of the prolyl bonds [145].

In native RNase T1 *cis*-Pro55 is solvent exposed in the native protein and presumably also in folding intermediates, and therefore all prolyl isomerases catalyze the isomerization at Pro55 very well. *Cis*-Pro39 is buried in the native protein. Its *trans* → *cis* isomerization is only marginally accelerated, because rapid conformational folding hinders the access of the prolyl isomerases. Cyp18 from *E. coli* at a very high concentration (29 µM) was necessary to obtain a 300-fold acceleration of Pro39 isomerization in the refolding of 2.5 µM RNase T1 [151]. The catalysis at Pro39 is strongly improved when the formation of partially folded intermediates is suppressed by destabilizing mutations, or by breaking of the two disulfide bonds (as in the reduced and carboxymethylated RCM form of RNase T1) [152].

The reduced and carboxymethylated RCM form of the S54G/P55N variant (RCM-(-P55)RNase T1) is a particularly simple model protein for studying catalyzed protein folding, because it involves a single *trans* → *cis* isomerization only (at Pro39; Scheme 25.4) and because the access to this proline is not impaired by premature structure formation, as in the wild-type protein with intact disulfide bonds. The RCM form is unfolded in aqueous buffer but reversible folding to a native-like ordered conformation can be induced by adding 1 M NaCl [152]. Unfolding and refolding in the presence of prolyl isomerases can thus be studied in the absence of denaturants, simply by varying the NaCl concentration [90, 153–155]. This is important because several prolyl isomerases are sensitive to residual concentrations of denaturants, such as guanidinium chloride or urea.

Prolyl isomerases function in protein folding as enzymes. They catalyze *cis* ↔ *trans* isomerization in either direction and carry no information about the isomeric states of the prolyl peptide bonds in the protein substrates. This information is provided by the refolding protein itself. Molecules with native prolyl isomers refold rapidly and are thus no longer substrates for the isomerase. The prolyl isomerases are active in protein *unfolding* as well [152].

25.7.2
Specificity of Prolyl Isomerases

The catalytic function of the prolyl isomerases is restricted to prolyl bonds. They do not catalyze the isomerizations of "normal" nonprolyl peptide bonds. As outlined in Section 25.6, the folded form of the P39A variant of RNase T1 has a *cis* Tyr38-Ala39 bond, and the refolding of virtually all molecules is limited in rate by the very slow *trans* → *cis* re-isomerization of the Tyr38-Ala39 bond. This isomerization is not catalyzed by prolyl isomerases of the cyclophilin, FKBP, and parvulin families. It is also not catalyzed in the peptide Ala-Ala-Tyr-Ala-Ala [156].

Schiene-Fischer et al. discovered that DnaK, a chaperone of the Hsp70 family catalyzes the isomerization of nonprolyl bonds in oligopeptides and in the folding of the P39A variant of RNase T1. This new isomerase is described in Chapter 10 in Part II.

As folding enzymes, the prolyl isomerases can catalyze their own folding. The folding reactions of human cytosolic FKBP12 and of parvulin from *E. coli* are indeed autocatalytic processes [45, 155, 157].

25.7.3
The Trigger Factor

The trigger factor is a ubiquitous bacterial protein, which associates with the large subunit of the ribosome (see Chapter 13 in Part II). Originally it was thought to be involved in the export of secretory proteins [158, 159], but in 1995 Fischer and coworkers [160] discovered that the trigger factor is a prolyl isomerase. This enzymatic activity originates from the central FKBP domain, which encompasses residues 142 to 251. Indeed, a weak sequence homology had been noted between this region of the trigger factor and human FKBP12 [161].

Three domains have been identified in the trigger factor. The central FKBP domain harbors the prolyl isomerase activity of the trigger factor [162, 163] and the N-terminal domain (residues 1–118) mediates the interaction with the ribosome [164–168]. The function of the C-terminal domain is largely unknown. Its presence is required for the high activity of the trigger factor as a catalysts of protein folding [169, 170].

The trigger factor catalyzes the folding of RCM-RNase T1 very efficiently. The addition of as low as 2.5 nM trigger factor led to a doubling of the folding rate of this protein, and in the presence of 20 nM trigger factor this folding reaction is 14-fold accelerated. This remarkable catalytic efficiency of the trigger factor as a folding enzyme is reflected in a specificity constant k_{cat}/K_M of 1.1×10^6 M^{-1} s^{-1}. This is almost 100-fold higher than the respective value for human FKBP12 [170]. The K_M value is 0.7 µM, and the catalytic rate constant k_{cat} is 1.3 s^{-1} [170]. For Cyp18 the catalysis of the *trans* ↔ *cis* prolyl isomerization in a tetrapeptide is characterized by K_M and k_{cat} values of 220 µM and 620 s^{-1}, respectively [64]. This comparison indicates that the high activity of the trigger factor as a folding catalyst does not originate from a high turnover number, but from a high affinity for the protein substrate.

Permanently unfolded proteins are strong, competitive inhibitors of trigger-factor-catalyzed folding. One of those, reduced and carboxymethylated bovine α-lactalbumin (RCM-La) inhibits the trigger factor of *Mycoplasma genitalium* with a K_I value of 50 nM. This binding of inhibitory proteins is independent of proline residues. Unfolded RCM-tendamistat (an α-amylase inhibitor with 74 residues) binds to the trigger factor with equal affinity in the presence and in the absence of its three proline residues [171]. The catalysis of folding of RCM-(-Pro55)-RNase T1 occurs at Pro39. The good inhibition by a nonfolding variant of RNase T1 that lacks Pro39 showed that this proline is dispensable for substrate binding [171]. The affinity of the trigger factor for short peptides is also independent of prolines [172]. The isolated FKBP domain is fully active as a prolyl isomerase towards a short tetrapeptide [162], but in protein folding its activity is about 800-fold reduced, and, moreover, this low residual activity of the FKBP domain alone is not inhibited by unfolded proteins, such as RCM-La [170]. The high enzymatic activity in protein folding thus requires the intact trigger factor [169]. Together, these results suggest that the high-affinity binding site for unfolded proteins is probably distinct from the catalytic site of the trigger factor. It extends over several domains of the intact trigger factor or requires the interaction of these domains.

The good binding of the intact trigger factor to unfolded chain segments probably decelerates the dissociation of the protein substrate from the trigger factor, and the low k_{cat} value of 1.3 s^{-1} may reflect a change in the rate-limiting step from bond rotation (in tetrapeptide substrates) towards product dissociation (in protein substrates). In fact, an unfolded protein dissociates from the trigger factor with a rate constant of 5.8 s^{-1} [173], which is only fourfold higher than the k_{cat} for catalyzed folding. Since high-affinity binding to unfolded proteins is independent of prolines, it is likely that many binding events are nonproductive, because the reactive prolyl peptide bonds are not positioned correctly within the prolyl isomerase site. Indeed, a lowering of both K_M and k_{cat}, as observed for the intact trigger factor, points to nonproductive binding of a substrate to an enzyme [174].

The trigger factor thus seems to have properties of a folding enzyme and of a chaperone. The cooperation of both functions is required for its very high catalytic efficiency in protein folding. Indeed, trigger factor can function as a chaperone *in vitro* [175–177] and presumably also *in vivo* [178, 179]. A combination of the functions as a protein export factor and as a prolyl isomerase seems to be required in the secretion and maturation of a protease of the pathogenic bacterium *Streptococcus pyogenes* [180].

25.7.4
Catalysis of Prolyl Isomerization During de novo Protein Folding

The evidence for a catalysis of proline-limited folding steps during de novo protein folding in the cell remains circumstantial. Early evidence for a role of prolyl isomerization and of prolyl isomerases in cellular folding was provided by studies of collagen folding, which is limited in rate by successive prolyl isomerizations both in vitro and in vivo. Collagen maturation in chicken embryo fibroblasts is retarded by cyclosporin A, possibly by the inhibition of the cyclophilin-catalyzed folding in the endoplasmic reticulum [181]. A similar effect was found for the folding of luciferase in rabbit reticulocyte lysate [182].

Proteins that are targeted to the mitochondrial matrix must unfold outside the mitochondria, cross the two mitochondrial membranes and then refold in the matrix. In a synthetic precursor protein the presequence of subunit 9 of the *Neurospora crassa* F_1F_0-ATPase was linked to mouse cytosolic dihydrofolate reductase (Su9-DHFR) and used to investigate the function of cyclophilins as potential catalysts of protein folding in isolated mitochondria [183–185]. When measured by the resistance against proteinase K, the refolding of DHFR inside the mitochondria showed half-times of about 5 min and folding was about fivefold decelerated when the mitochondria had been preincubated with 2.5–5 µM CsA to inhibit the prolyl isomerase activity of mitochondrial cyclophilin. The refolding of DHFR was similarly retarded in mitochondria that were derived from mutants that lacked a functional mitochondrial cyclophilin. This suggests that the mitochondrial cyclophilins acted as prolyl isomerases and thereby catalyzed protein folding in organello.

25.8
Concluding Remarks

Cis peptide bonds decelerate folding reactions, but nevertheless they occur frequently in folded proteins, mainly before proline and occasionally before other amino acid residues. *Cis* prolines are very well suited to introduce tight turns into proteins, but the structural consequences of *cis* peptide bonds cannot be the sole reason for their widespread occurrence. The coupling with prolyl isomerization is a simple means for increasing the energetic barriers of folding, and thus not only refolding but also unfolding would be decelerated. In such a way a native protein would be protected from sampling the unfolded state too frequently.

Nonprolyl *cis* peptide bonds are strongly destabilizing, because they introduce strain into the protein backbone. They are found preferentially near the active sites of enzymes where strained conformations might be important for the function.

In the overall folding process prolyl isomerizations and conformational steps are linked. Incorrect prolines decrease the stability of partially folded intermediates, and conformational folding can modulate the rate of prolyl isomerization.

Prolyl isomerizations are catalyzed by prolyl isomerases. Their functions reach far beyond the acceleration of de novo protein folding. In addition to the well-understood role in immunosuppression, prolyl isomerases modulate ion channels and transmembrane receptors, participate in hormone receptor complexes, are required for HIV-1 infectivity, and participate in the regulation of mitosis. The PIN1 protein discriminates between phosphorylated and unphosphorylated Ser/Thr-Pro sequences in its target proteins. Two principles of regulation are probably integrated at this point: phosphorylation and prolyl *cis*/*trans* isomerization.

Prolyl isomerization in folded proteins is thus well suited for switching between alternative states of a protein. The positions of the switch (*cis* and *trans*) can be determined by the binding of effectors, the rate of swichting can be modulated by prolyl isomerases. Some isomerases might be effector and isomerase at the same time. For such isomerases the distinction between binding and catalytic function, which has been discussed for a long time, may be inappropriate.

25.9
Experimental Protocols

25.9.1
Slow Refolding Assays ("Double Jumps") to Measure Prolyl Isomerizations in an Unfolded Protein

As shown in Scheme 25.1, the direct N $\rightarrow$ U$_F$ unfolding reaction can be followed by one or more prolyl isomerizations (U$_F$ $\rightleftharpoons$ U$_S$) in the unfolded protein chains.

U_F and U_S are usually equally unfolded species and show similar physical properties. Therefore the kinetics of the $U_F \rightleftharpoons U_S$ equilibration reactions cannot be measured directly. Rather, the amount of U_S formed after different times of unfolding can be determined by slow refolding assays. Samples are withdrawn from the unfolding solution at different time intervals after the initiation of unfolding; these samples are transferred to standard refolding conditions, and the amplitude of the slow refolding reaction, $U_S \rightarrow N$, is determined. Refolding of the U_F species is usually complete within the time of manual mixing. The amplitude of the slow refolding reaction is proportional to the concentration of U_S which was present at the time when the sample was withdrawn for the assay. The dependence of these slow refolding amplitudes on the duration of unfolding yields the kinetics of the $U_F \rightleftharpoons U_S$ reaction. The measured rate constant λ is equal to the sum of the rate constants in the forward (k_{FS}) and the reverse (k_{SF}) directions, $\lambda = k_{FS} + k_{SF}$. When the equilibrium constant $K = [U_S]/[U_F] = k_{FS}/k_{SF}$ is known, the individual rate constants k_{FS} and k_{SF} can be derived.

25.9.1.1 Guidelines for the Design of Double Jump Experiments

Unfolding conditions Any unfolding conditions can be employed; for practical reasons, however, it is of advantage to select conditions (1) where the $N \rightarrow U_F$ unfolding step is rapid and (2) where the $U_F \rightleftharpoons U_S$ equilibration is slow. Under such conditions, $U_F \rightleftharpoons U_S$ can be studied without interference from the preceding unfolding step and manual sampling techniques can be used to perform the slow refolding assays. Usually it is best to perform the unfolding step at a high concentration of a strong denaturant, such as GdmCl, and at low temperature. The rate of conformational unfolding usually increases strongly with the denaturant concentration, and, because of the high activation enthalpy of proline isomerization, the $U_F \rightleftharpoons U_S$ reaction is strongly decelerated by decreasing the temperature. Consequently, the difference in rate between the two processes is highest in concentrated denaturant solutions and at low temperature. Often, the rate of conformational unfolding can be further enhanced by lowering the pH.

Refolding conditions The slow refolding assays are best carried out under conditions where (1) the $U_F \rightarrow N$ refolding reaction is complete within the time of manual mixing, and (2) the slow $U_S \rightarrow N$ reaction occurs in a single phase in a time range that is convenient for manual mixing experiments. This facilitates the accurate determination of the refolding amplitudes. After each refolding assay the actual protein concentration should be determined to correct for variations in the protein concentration in the individual assays. Usually the final absorbance or fluorescence reached after the refolding step is suitable for such a correction. As in the unfolding step, the rate of slow refolding can be fine-tuned by varying the concentration of denaturant, the temperature and the pH.

In cases where more than one U_S species is formed after unfolding, the individual rates of formation of these species can be measured when the refolding assays are carried out under conditions where the refolding of the various U_S species can be separated kinetically.

25.9.1.2 Formation of U_S Species after Unfolding of RNase A

I describe here a protocol that was used to follow the $U_F \rightleftharpoons U_S$ reaction in unfolded RNase A as an example. In these experiments the refolding assays were carried out under conditions where slow refolding is a single, monophasic reaction. Therefore, the refolding assays yielded the kinetics of overall formation of U_S species after unfolding. They did not discriminate between the different U_S species of RNase A (see Section 25.4.1).

Unfolding To initiate unfolding at time zero, 50 μL of native RNase A (1.5 mM in H_2O, 0 °C) is mixed with 250 μL of 6.0 M GdmCl solution in 0.1 M glycine, pH 2.0 at 0 °C. The resulting unfolding conditions are 0.25 mM RNase A in 5.0 M GdmCl, pH 2.0, 0 °C. Under these conditions the $N \rightarrow U_F$ unfolding step of RNase A is very rapid and complete before the onset of the $U_F \rightleftharpoons U_S$ equilibration. Thus the $U_F \rightleftharpoons U_S$ reaction can be measured in a convenient time range without interference from the preceding $N \rightarrow U_F$ unfolding step.

Refolding assays After different time intervals 70-μL samples are withdrawn from the unfolding solution and quickly diluted into 830 μL of the refolding buffer (1.2 M GdmCl in 0.1 M cacodylate, pH 6.2) in the spectrophotometer cell, which is kept at 25 °C. This gives final conditions of 1.5 M GdmCl, pH 6.0 and 25 °C. These conditions are chosen because slow refolding occurs in a single exponential phase in a time range, which is convenient for manual sampling techniques. The slow refolding reaction of U_S is monitored at 287 nm. At the end of each refolding assay the absorbance of the refolded sample is recorded at 277 nm to determine the actual concentration of RNase A in the assay.

Data treatment The refolding kinetics are analyzed. The time constant for refolding in the assay should be independent of the time of sampling, because the final folding conditions are the same in all assays. The amplitudes of refolding are corrected for variations in the protein concentration (if necessary) and plotted as a function of the time of unfolding. The increase in the refolding amplitude with the duration of unfolding yields the kinetics of the formation of the slow-folding species U_S.

25.9.2
Slow Unfolding Assays for Detecting and Measuring Prolyl Isomerizations in Refolding

Often protein molecules with incorrect prolyl isomers can fold to a conformation that appears native-like by spectroscopic properties or even by enzymatic activity. In such cases the re-isomerization of the incorrect prolines in the final step of folding is silent and cannot be followed by spectroscopic probes or by a functional assay.

To follow such silent isomerizations during refolding, a two-step assay is used. It measures the kinetics of the formation of fully folded protein molecules during a refolding reaction [33, 74]. This assay is based on the fact that native protein molecules have passed beyond the highest activation barrier in their refolding and thus are separated from the unfolded state by this energy barrier. They unfold slowly when the conditions are switched to unfolding, because they must cross the high barrier again, now in the reverse direction. Partially folded molecules (such as those with incorrect prolyl isomers) have not yet passed the final transition state and unfold rapidly.

This translates into a double-mixing procedure. First, unfolded protein is mixed with refolding buffer to initiate refolding. Then, after variable time intervals, samples are withdrawn, transferred to standard unfolding conditions, and the amplitude of the subsequent slow unfolding reaction is determined. It is a direct measure for the amount of native molecules with correct prolyl isomers that had been present at the time when refolding was interrupted, and the increase of the amplitude of slow refolding with time gives the kinetics of formation of the native protein.

25.9.2.1 Practical Considerations

The unfolding assays are best carried out under conditions where the N → U unfolding reaction in the assay is monophasic and occurs in a time range that is convenient for manual mixing experiments. Thus the unfolding amplitudes can be determined with a high accuracy. Most proteins unfold in monoexponential reactions under strongly unfolding conditions, i.e., in the presence of a denaturant at high concentration. The rate of unfolding can be varied in a wide range by changes in the pH, the temperature and the denaturant concentration. As in the refolding assays (see above) the actual protein concentration should be determined to correct for variations in the protein concentration in the individual unfolding assays.

25.9.2.2 Kinetics of the Formation of Fully Folded IIHY-G3P* Molecules

Here I describe how this procedure was used to measure the time course of the $trans \to cis$ re-isomerization at Pro213 in the refolding of the gene-3-protein of the phage fd [59, 60]. This isomerization controls the domain docking in the final folding step of this protein. The unfolding assays were performed in 5.0 M GdmCl, pH 7.0 at 25 °C, conditions under which the gene-3-protein unfolds in a monoexponential reaction with a time constant of 25 s.

To follow the formation of native molecules, unfolded protein (50 µM in 5.0 M GdmCl) was first manually 10-fold diluted with buffer (in a test tube) to initiate refolding at 0.5 M GdmCl and then, after times of refolding between 1 and 540 min, samples were withdrawn and unfolded again at 5.0 M GdmCl by a manual 10-fold dilution with a GdmCl solution of 5.5 M in the fluorimeter cell. Unfolding was followed by the change in tyrosine fluorescence at 310 nm after excitation at 280 nm.

References

1 B. T. NALL, Comments Mol. Cell. Biophys. **1985**, 3, 123–143.

2 S. F. GÖTHEL, M. A. MARAHIEL, Cell. Mol. Life Sci. **1999**, 55, 423–436.

3 G. Fischer, *Angew. Chem. Int. Ed.* **1994**, *33*, 1415–1436.
4 G. Fischer, F. X. Schmid, Peptidyl-prolyl cis/trans isomerases. In *Molecular Biology of Chaperones and Folding Catalysts* (B. Bukau, Ed.). Harwood Academic Publishers, New York, 1999, pp. 461–489.
5 F. X. Schmid, Catalysis of protein folding by prolyl isomerases. In *Molecular Chaperones in the Life Cycle of Proteins* (A. L. Fink and Y. Goto, Eds). Marcel Dekker, New York, 1998, pp. 361–389.
6 F. X. Schmid, *Adv. Protein Chem.* **2002**, *59*, 243–282.
7 F. X. Schmid, L. M. Mayr, M. Mücke, E. R. Schönbrunner, *Adv. Protein Chem.* **1993**, *44*, 25–66.
8 J. Balbach, F. X. Schmid, Prolyl isomerization and its catalysis in protein folding. In *Mechanisms of Protein Folding* (R. H. Pain, Ed.). Oxford University Press, Oxford. 2000, pp. 212–237.
9 W. J. Wedemeyer, E. Welker, H. A. Scheraga, *Biochemistry* **2002**, *41*, 14637–14644.
10 A. H. Andreotti, *Biochemistry* **2003**, *42*, 9515–9524.
11 G. E. Schulz, R. E. Schirmer, *Principles of Protein Structure*. Springer Verlag, New York, 1979.
12 G. Scherer, M. L. Kramer, M. Schutkowski, R. U. G. Fischer, *J. Am. Chem. Soc.* **1998**, *120*, 5568–5574.
13 D. E. Stewart, A. Sarkar, J. E. Wampler, *J. Mol. Biol.* **1990**, *214*, 253–260.
14 M. W. Macarthur, J. M. Thornton, *J. Mol. Biol.* **1991**, *218*, 397–412.
15 D. Pal, P. Chakrabarti, *J. Mol. Biol.* **1999**, *294*, 271–288.
16 A. Jabs, M. S. Weiss, R. Hilgenfeld, *J. Mol. Biol.* **1999**, *286*, 291–304.
17 D. C. Rees, M. Lewis, W. N. Lipscomb, *J. Mol. Biol.* **1983**, *168*, 367–387.
18 H. N. Cheng, F. A. Bovey, *Biopolymers* **1977**, *16*, 1465–1472.
19 C. Grathwohl, K. Wüthrich, *Biopolymers* **1981**, *20*, 2623–2633.
20 U. Reimer, G. Scherer, M. Drewello, S. Kruber, M. Schutkowski, G. Fischer, *J. Mol. Biol.* **1998**, *279*, 449–460.
21 G. Fischer, *Chem. Soc. Rev.* **2000**, *29*, 119–127.
22 T. Drakenberg, K.-I. Dahlqvist, S. Forsén, *J. Phys. Chem.* **1972**, *76*, 2178–2183.
23 R. L. Stein, *Adv. Protein Chem.* **1993**, *44*, 1–24.
24 E. S. Eberhardt, S. N. Loh, A. P. Hinck, R. T. Raines, *J. Am. Chem. Soc.* **1992**, *114*, 5437–5439.
25 I. Z. Steinberg, W. F. Harrington, A. Berger, M. Sela, E. Katchalski, *J. Am. Chem. Soc.* **1960**, *82*, 5263–5279.
26 F. X. Schmid, R. L. Baldwin, *Proc. Natl Acad. Sci. USA* **1978**, *75*, 4764–4768.
27 J. R. Garel, R. L. Baldwin, *Proc. Natl Acad. Sci. USA* **1973**, *70*, 3347–3351.
28 P. S. Kim, R. L. Baldwin, *Annu. Rev. Biochem.* **1982**, *51*, 459–489.
29 F. X. Schmid, Kinetics of unfolding and refolding of single-domain proteins. In *Protein Folding* (T. E. Creighton, Ed.). Freeman, New York, 1992, pp. 197–241.
30 J. F. Brandts, H. R. Halvorson, M. Brennan, *Biochemistry* **1975**, *14*, 4953–4963.
31 T. Kiefhaber, F. X. Schmid, *J. Mol. Biol.* **1992**, *224*, 231–240.
32 T. Kiefhaber, H. H. Kohler, F. X. Schmid, *J. Mol. Biol.* **1992**, *224*, 217–229.
33 F. X. Schmid, Fast-folding and slow-folding forms of unfolded proteins. In *Enzyme Structure Part l* (C. H. W. Hirs and S. N. Timasheff, Eds). Academic Press, New York, 1986, pp. 71–82.
34 Y. Goto, K. Hamaguchi, *J. Mol. Biol.* **1982**, *156*, 891–910.
35 R. F. Kelley, F. M. Richards, *Biochemistry* **1987**, *26*, 6765–6774.
36 R. Golbik, G. Fischer, A. R. Fersht, *Protein Sci.* **1999**, *8*, 1505–1514.
37 K. Maki, T. Ikura, T. Hayano, N. Takahashi, K. Kuwajima, *Biochemistry* **1999**, *38*, 2213–2223.
38 W. F. Walkenhorst, S. M. Green, H. Roder, *Biochemistry* **1997**, *36*, 5795–5805.
39 S. Khorasanizadeh, I. D. Peters,

T. R. Butt, H. Roder, *Biochemistry* **1993**, *32*, 7054–7063.

40 S. Koide, H. J. Dyson, P. E. Wright, *Biochemistry* **1993**, *32*, 12299–12310.

41 J. S. Reader, N. A. Van Nuland, G. S. Thompson, S. J. Ferguson, C. M. Dobson, S. E. Radford, *Protein Sci.* **2001**, *10*, 1216–1224.

42 M. M. Pierce, B. T. Nall, *J. Mol. Biol.* **2000**, *298*, 955–969.

43 A. R. Fersht, *FEBS Lett.* **1993**, *325*, 5–16.

44 S. E. Jackson, A. R. Fersht, *Biochemistry* **1991**, *30*, 10436–10443.

45 C. Scholz, T. Zarnt, G. Kern, K. Lang, H. Burtscher, G. Fischer, F. X. Schmid, *J. Biol. Chem.* **1996**, *271*, 12703–12707.

46 A. T. Russo, J. Rosgen, D. W. Bolen, *J. Mol. Biol.* **2003**, *330*, 851–866.

47 N. A. J. Van Nuland, F. Chiti, N. Taddei, G. Raugei, G. Ramponi, C. M. Dobson, *J. Mol. Biol.* **1998**, *283*, 883–891.

48 G. Aronsson, A. C. Brorsson, L. Sahlman, B. H. Jonsson, *FEBS Lett.* **1997**, *411*, 359–364.

49 N. Ferguson, A. P. Capaldi, R. James, C. Kleanthous, S. E. Radford, *J. Mol. Biol.* **1999**, *286*, 1597–1608.

50 F. Rousseau, J. W. Schymkowitz, M. Sanchez del Pino, L. S. Itzhaki, *J. Mol. Biol.* **1998**, *284*, 503–519.

51 G. Pappenberger, A. Bachmann, R. Muller, H. Aygun, J. W. Engels, T. Kiefhaber, *J. Mol. Biol.* **2003**, *326*, 235–246.

52 J. G. Bann, J. Pinkner, S. J. Hultgren, C. Frieden, *Proc. Natl Acad. Sci. USA* **2002**, *99*, 709–714.

53 T. F. Fu, E. S. Boja, M. K. Safo, V. Schirch, *J. Biol. Chem.* **2003**, *278*, 31088–31094.

54 W. J. Satumba, M. C. Mossing, *Biochemistry* **2002**, *41*, 14216–14224.

55 Y. Wu, C. R. Matthews, *J. Mol. Biol.* **2003**, *330*, 1131–1144.

56 M. Tsunenaga, Y. Goto, Y. Kawata, K. Hamaguchi, *Biochemistry* **1987**, *26*, 6044–6051.

57 H. Lilie, R. Rudolph, J. Buchner, *J. Mol. Biol.* **1995**, *248*, 190–201.

58 M. J. W. Thies, J. Mayer, J. G. Augustine, C. A. Frederick, H. Lilie, J. Buchner, *J. Mol. Biol.* **1999**, *293*, 67–79.

59 A. Martin, F. X. Schmid, *J. Mol. Biol.* **2003**, *329*, 599–610.

60 A. Martin, F. X. Schmid, *J. Mol. Biol.* **2003**, *331*, 1131–1140.

61 D. Galani, A. R. Fersht, S. Perrett, *J. Mol. Biol.* **2002**, *315*, 213–227.

62 T. E. Creighton, *J. Mol. Biol.* **1978**, *125*, 401–406.

63 C. Fransson, P. O. Freskgard, H. Herbertsson, A. Johansson, P. Jonasson, L. G. Martensson, M. Svensson, B. H. Jonsson, U. Carlsson, *FEBS Lett.* **1992**, *296*, 90–94.

64 D. Kern, G. Kern, G. Scherer, G. Fischer, T. Drakenberg, *Biochemistry* **1995**, *34*, 13594–13602.

65 M. Jäger, A. Plückthun, *FEBS Lett.* **1997**, *418*, 106–110.

66 F. X. Schmid, H. Blaschek, *Eur. J. Biochem.* **1981**, *114*, 111–117.

67 K. H. Cook, F. X. Schmid, R. L. Baldwin, *Proc. Natl Acad. Sci. USA* **1979**, *76*, 6157–6161.

68 P. J. Hagerman, R. L. Baldwin, *Biochemistry* **1976**, *15*, 1462–1473.

69 B. T. Nall, J.-R. Garel, R. L. Baldwin, *J. Mol. Biol.* **1978**, *118*, 317–330.

70 F. X. Schmid, R. L. Baldwin, *J. Mol. Biol.* **1979**, *133*, 285–287.

71 F. X. Schmid, R. Grafl, A. Wrba, J. J. Beintema, *Proc. Natl Acad. Sci. USA* **1986**, *83*, 872–876.

72 R. A. Sendak, D. M. Rothwarf, W. J. Wedemeyer, W. A. Houry, H. A. Scheraga, *Biochemistry* **1996**, *35*, 12978–12992.

73 A. Rehage, F. X. Schmid, *Biochemistry* **1982**, *21*, 1499–1505.

74 F. X. Schmid, *Biochemistry* **1983**, *22*, 4690–4696.

75 P. S. Kim, R. L. Baldwin, *Annu. Rev. Biochem.* **1990**, *59*, 631–660.

76 W. A. Houry, D. M. Rothwarf, H. A. Scheraga, *Biochemistry* **1994**, *33*, 2516–2530.

77 U. Arnold, M. P. Hinderaker, J. Koditz, R. Golbik, R. Ulbrich-Hofmann, R. T. Raines, *J. Am. Chem. Soc.* **2003**, *125*, 7500–7501.

78 H. KREBS, F. X. SCHMID, R. JAENICKE, J. Mol. Biol. **1983**, *169*, 619–635.
79 D. A. SCHULTZ, F. X. SCHMID, R. L. BALDWIN, Protein Sci. **1992**, *1*, 917–924.
80 R. W. DODGE, H. A. SCHERAGA, Biochemistry **1996**, *35*, 1548–1559.
81 W. A. HOURY, H. A. SCHERAGA, Biochemistry **1996**, *35*, 11719–11733.
82 R. W. DODGE, J. H. LAITY, D. M. ROTHWARF, S. SHIMOTAKAHARA, H. A. SCHERAGA, J. Protein Chem. **1994**, *13*, 409–421.
83 C. N. PACE, U. HEINEMANN, U. HAHN, W. SAENGER, Angew. Chem. Int. Ed. Engl. **1991**, *30*, 343–360.
84 U. HEINEMANN, W. SAENGER, Nature **1982**, *299*, 27–31.
85 U. HEINEMANN, U. HAHN, Structural and functional studies of ribonuclease t1. In Protein–Nucleic Acid Interaction (W. SAENGER and U. HEINEMANN, Eds). Macmillan, London, 1989, pp. 111–141.
86 J. MARTINEZ-OYANEDEL, H.-W. CHOE, U. HEINEMANN, W. SAENGER, J. Mol. Biol. **1991**, *222*, 335–352.
87 M. OOBATAKE, S. TAKAHASHI, T. OOI, J. Biochem. **1979**, *86*, 55–63.
88 C. N. PACE, G. R. GRIMSLEY, J. A. THOMSON, B. J. BARNETT, J. Biol. Chem. **1988**, *263*, 11820–11825.
89 L. M. MAYR, F. X. SCHMID, Biochemistry **1993**, *32*, 7994–7998.
90 M. MÜCKE, F. X. SCHMID, Biochemistry **1994**, *33*, 14608–14619.
91 C. N. PACE, D. V. LAURENTS, Biochemistry **1989**, *28*, 2520–2525.
92 T. KIEFHABER, F. X. SCHMID, M. RENNER, H.-J. HINZ, Biochemistry **1990**, *29*, 8250–8257.
93 C.-Q. HU, J. M. STURTEVANT, J. A. THOMSON, R. E. ERICKSON, C. N. PACE, Biochemistry **1992**, *31*, 4876–4882.
94 Y. YU, G. I. MAKHATADZE, C. N. PACE, P. L. PRIVALOV, Biochemistry **1994**, *33*, 3312–3319.
95 T. KIEFHABER, R. QUAAS, U. HAHN, F. X. SCHMID, Biochemistry **1990**, *29*, 3053–3061.
96 C. ODEFEY, L. M. MAYR, F. X. SCHMID, J. Mol. Biol. **1995**, *245*, 69–78.
97 C. N. PACE, T. E. CREIGHTON, J. Mol. Biol. **1986**, *188*, 477–486.
98 C. FRECH, F. X. SCHMID, J. Mol. Biol. **1995**, *251*, 135–149.
99 T. KIEFHABER, H. P. GRUNERT, U. HAHN, F. X. SCHMID, Biochemistry **1990**, *29*, 6475–6480.
100 L. M. MAYR, O. LANDT, U. HAHN, F. X. SCHMID, J. Mol. Biol. **1993**, *231*, 897–912.
101 T. SCHINDLER, L. M. MAYR, O. LANDT, U. HAHN, F. X. SCHMID, Eur. J. Biochem. **1996**, *241*, 516–524.
102 F. X. SCHMID, Annu. Rev. Biophys. Biomol. Struct. **1993**, *22*, 123–143.
103 L. M. MAYR, C. ODEFEY, M. SCHUTKOWSKI, F. X. SCHMID, Biochemistry **1996**, *35*, 5550–5561.
104 T. KIEFHABER, F. X. SCHMID, K. WILLAERT, Y. ENGELBORGHS, A. CHAFFOTTE, Protein Sci. **1992**, *1*, 1162–1172.
105 L. M. MAYR, F. X. SCHMID, J. Mol. Biol. **1993**, *231*, 913–926.
106 T. KIEFHABER, R. QUAAS, U. HAHN, F. X. SCHMID, Biochemistry **1990**, *29*, 3061–3070.
107 F. L. TEXTER, D. B. SPENCER, R. ROSENSTEIN, C. R. MATTHEWS, Biochemistry **1992**, *31*, 5687–5691.
108 J. BALBACH, V. FORGE, N. A. J. VAN NULAND, S. L. WINDER, P. J. HORE, C. M. DOBSON, Nat. Struct. Biol. **1995**, *2*, 865–870.
109 S. D. HOELTZLI, C. FRIEDEN, Biochemistry **1996**, *35*, 16843–16851.
110 J. BALBACH, V. FORGE, W. S. LAU, N. A. J. VANNULAND, K. BREW, C. M. DOBSON, Science **1996**, *274*, 1161–1163.
111 J. BALBACH, V. FORGE, W. S. LAU, N. A. J. VAN NULAND, K. BREW, C. M. DOBSON, Science **1996**, *274*, 1161–1163.
112 P. A. EVANS, C. M. DOBSON, R. A. KAUTZ, G. HATFULL, R. O. FOX, Nature, **1987**, *329*, 266–268.
113 K. A. HIGGINS, D. J. CRAIK, J. G. HALL, P. R. ANDREWS, Drug Design Delivery **1988**, *3*, 159–170.
114 W. J. CHAZIN, J. KÖRDEL, T. DRAKENBERG, E. THULIN, P. BRODIN, T. GRUNDSTRÖM, S. FORSÉN, Proc.

115 J. Kördel, S. Forsen, T.
Drakenberg, W. J. Chazin,
Biochemistry **1990**, *29*, 4400–4409.
116 E. Adjadj, V. Naudat, E. Quiniou,
D. Wouters, P. Sautiere, C. T.
Craescu, *Eur. J. Biochem.* **1997**, *246*,
218–227.
117 Y. Feng, W. F. Hood, R. W. Forgey,
A. L. Abegg, M. H. Caparon, B. R.
Thiele, R. M. Leimgruber, C. A.
McWherter, *Protein Sci.* **1997**, *6*,
1777–1782.
118 X. Yuan, A. K. Downing, V. Knott,
P. A. Handford, *EMBO J.* **1997**, *16*,
6659–6666.
119 X. Yuan, J. M. Werner, V. Knott,
P. A. Handford, I. D. Campbell, K.
Downing, *Protein Sci.* **1998**, *7*, 2127–
2135.
120 S. N. Loh, C. W. Mcnemar, J. L.
Markley, *Techn. Protein Chem.* **1991**,
II, 275–282.
121 D. M. Truckses, J. R. Somoza, K. E.
Prehoda, S. C. Miller, J. L.
Markley, *Protein Sci.* **1996**, *5*, 1907–
1916.
122 F. X. Schmid, K. Lang, T. Kiefhaber,
S. Mayer, R. Schönbrunner, Prolyl
isomerase. Its role in protein folding
and speculations on its function in the
cell. In *Conformations and Forces in
Protein Folding* (B. T. Nall and K. A.
Dill, Eds). AAAS, Washington, D.C.
1991.
123 R. J. Mallis, K. N. Brazin, D. B.
Fulton, A. H. Andreotti, *Nat. Struct.
Biol.* **2002**, *9*, 900–905.
124 K. N. Brazin, R. J. Mallis, D. B.
Fulton, A. H. Andreotti, *Proc.
Natl Acad. Sci. USA* **2002**, *99*, 1899–
1904.
125 P. A. Evans, C. M. Dobson, R. A.
Kautz, G. Hatfull, R. O. Fox, *Nature*
1987, *329*, 266–268.
126 R. K. Gitti, B. M. Lee, J. Walker,
M. F. Summers, S. Yoo, W. I.
Sundquist, *Science* **1996**, *273*, 231–
235.
127 T. A. Ramelot, L. K. Nicholson,
J. Mol. Biol. **2001**, *307*, 871–884.
128 P. J. Breheny, A. Laederach, D. B.
Fulton, A. H. Andreotti, *J. Am.
Chem. Soc.* **2003**, *125*, 15706–15707.
129 M. B. Yaffe, M. Schutkowski, M. H.
Shen, X. Z. Zhou, P. T. Stukenberg,
J. U. Rahfeld, J. Xu, J. Kuang, M. W.
Kirschner, G. Fischer, L. C.
Cantley, K. P. Lu, *Science* **1997**, *278*,
1957–1960.
130 C. Schiene-Fischer, G. Fischer, *J.
Am. Chem. Soc.* **2001**, *123*, 6227–6231.
131 L. M. Mayr, D. Willbold, P. Rösch,
F. X. Schmid, *J. Mol. Biol.* **1994**, *240*,
288–293.
132 M. A. Pearson, P. A. Karplus, R. W.
Dodge, J. H. Laity, H. A. Scheraga,
Protein Sci. **1998**, *7*, 1255–1258.
133 N. C. Strynadka, H. Adachi, S. E.
Jensen, K. Johns, A. Sielecki, C.
Betzel, K. Sutoh, M. N. James,
Nature **1992**, *359*, 700–705.
134 M. Vanhove, X. Raquet, J.-M. Frere,
Proteins Struct. Funct. Genet. **1995**, *22*,
110–118.
135 M. Vanhove, X. Raquet, T. Palzkill,
R. H. Pain, J. M. Frere, *Proteins
Struct. Funct. Genet.* **1996**, *25*, 104–
111.
136 M. Vanhove, A. Lejeune, R. H. Pain,
Cell Mol. Life Sci. **1998**, *54*, 372–377.
137 M. Vanhove, A. Lejeune, G.
Guillaume, R. Virden, R. H. Pain,
F. X. Schmid, J. M. Frere,
Biochemistry **1998**, *37*, 1941–1950.
138 O. Herzberg, *J. Mol. Biol.* **1991**, *217*,
701–719.
139 G. Pappenberger, H. Aygun, J. W.
Engels, U. Reimer, G. Fischer, T.
Kiefhaber, *Nat. Struct. Biol.* **2001**, *8*,
452–458.
140 K. Lang, F. X. Schmid, G. Fischer,
Nature **1987**, *329*, 268–270.
141 A. Matouschek, J. T. Kellis, L.
Serrano, M. Bycroft, A. R. Fersht,
Nature **1990**, *346*, 440–445.
142 P. O. Freskgård, N. Bergenhem,
B. H. Jonsson, M. Svensson, U.
Carlsson, *Science* **1992**, *258*, 466–468.
143 G. Kern, D. Kern, F. X. Schmid, G.
Fischer, *FEBS Lett.* **1994**, *348*, 145–
148.
144 S. Veeraraghavan, B. T. Nall,
Biochemistry **1994**, *33*, 687–692.
145 S. Veeraraghavan, S. Rodriguez-

Gdiharpour, C. MacKinnon, W. A. McGee, M. M. Pierce, B. T. Nall, *Biochemistry* **1995**, *34*, 12892–12902.
146 S. Veeraraghavan, B. T. Nall, A. L. Fink, *Biochemistry* **1997**, *36*, 15134–15139.
147 C. J. Mann, X. Shao, C. R. Matthews, *Biochemistry* **1995**, *34*, 14573–14580.
148 H. P. Bächinger, P. Bruckner, R. Timpl, D. J. Prockop, J. Engel, *Eur. J. Biochem.* **1980**, *106*, 619–632.
149 H.-P. Bächinger, *J. Biol. Chem.* **1987**, *262*, 17144–17148.
150 J. M. Davis, B. A. Boswell, H. P. Bächinger, *J. Biol. Chem.* **1989**, *264*, 8956–8962.
151 E. R. Schönbrunner, S. Mayer, M. Tropschug, G. Fischer, N. Takahashi, F. X. Schmid, *J. Biol. Chem.* **1991**, *266*, 3630–3635.
152 M. Mücke, F. X. Schmid, *Biochemistry* **1992**, *31*, 7848–7854.
153 M. Mücke, F. X. Schmid, *J. Mol. Biol.* **1994**, *239*, 713–725.
154 J.-U. Rahfeld, A. Schierhorn, K.-H. Mann, G. Fischer, *FEBS Lett.* **1994**b, *343*, 65–69.
155 C. Scholz, J. Rahfeld, G. Fischer, F. X. Schmid, *J. Mol. Biol.* **1997**, *273*, 752–762.
156 C. Scholz, G. Scherer, L. M. Mayr, T. Schindler, G. Fischer, F. X. Schmid, *Biol. Chem.* **1998**, *379*, 361–365.
157 S. Veeraraghavan, T. F. Holzman, B. T. Nall, *Biochemistry* **1996**, *35*, 10601–10607.
158 E. Crooke, W. Wickner, *Proc. Natl Acad. Sci. USA* **1987**, *84*, 5216–5220.
159 R. Lill, E. Crooke, B. Guthrie, W. Wickner, *Cell* **1988**, *54*, 1013–1018.
160 G. Stoller, K. P. Rücknagel, K. Nierhaus, F. X. Schmid, G. Fischer, J.-U. Rahfeld, *EMBO J.* **1995**, *14*, 4939–4948.
161 I. Callebaut, J. P. Mornon, *FEBS Lett.* **1995**, *374*, 211–215.
162 G. Stoller, T. Tradler, J.-U. Rücknagel, G. Fischer, *FEBS Lett.* **1996**, *384*, 117–122.
163 T. Hesterkamp, B. Bukau, *FEBS Lett.* **1996**, *385*, 67–71.
164 T. Hesterkamp, E. Deuerling, B. Bukau, *J. Biol. Chem.* **1997**, *272*, 21865–21871.
165 G. Kramer, T. Rauch, W. Rist, S. Vorderwulbecke, H. Patzelt, A. Schulze-Specking, N. Ban, E. Deuerling, B. Bukau, *Nature* **2002**, *419*, 171–174.
166 R. Maier, B. Eckert, C. Scholz, H. Lilie, F. X. Schmid, *J. Mol. Biol.* **2003**, *326*, 585–592.
167 G. Blaha, D. N. Wilson, G. Stoller, G. Fischer, R. Willumeit, K. H. Nierhaus, *J. Mol. Biol.* **2003**, *326*, 887–897.
168 O. Kristensen, M. Gajhede, *Structure* **2003**, *11*, 1547–1556.
169 T. Zarnt, T. Tradler, G. Stoller, C. Scholz, F. X. Schmid, G. Fischer, *J. Mol. Biol.* **1997**, *271*, 827–837.
170 C. Scholz, G. Stoller, T. Zarnt, G. Fischer, F. X. Schmid, *EMBO J.* **1997**, *16*, 54–58.
171 C. Scholz, M. Mücke, M. Rape, A. Pecht, A. Pahl, H. Bang, F. X. Schmid, *J. Mol. Biol.* **1998**, *277*, 723–732.
172 H. Patzelt, S. Rüdiger, D. Brehmer, G. Kramer, S. Vorderwulbecke, E. Schaffitzel, A. Waitz, T. Hesterkamp, L. Dong, J. Schneider-Mergener, B. Bukau, E. Deuerling, *Proc. Natl Acad. Sci. USA* **2001**, *98*, 14244–14249.
173 R. Maier, C. Scholz, F. X. Schmid, *J. Mol. Biol.* **2001**, *314*, 1181–1190.
174 A. Fersht, *Enzyme Structure and Mechanism*, 2nd edn. W.H. Freeman, New York, 1985.
175 K. Nishihara, M. Kanemori, H. Yanagi, T. Yura, *Appl. Environ. Microbiol.* **2000**, *66*, 884–889.
176 G. C. Huang, Z. Y. Li, J. M. Zhou, G. Fischer, *Protein Sci.* **2000**, *9*, 1254–1261.
177 Z. Y. Li, C. P. Liu, L. Q. Zhu, G. Z. Jing, J. M. Zhou, *FEBS Lett.* **2001**, *506*, 108–112.
178 S. A. Teter, W. A. Houry, D. Ang, T. Tradler, D. Rockabrand, G. Fischer, P. Blum, C. Georgopoulos, F. U. Hartl, *Cell* **1999**, *97*, 755–765.
179 E. Deuerling, A. Schuize-Specking, T. Tomoyasu, A. Mogk, B. Bukau, *Nature* **1999**, *400*, 693–696.

180 W. R. Lyon, M. G. Caparon, *J. Bacteriol.* **2003**, *185*, 3661–3667.
181 B. Steinmann, P. Bruckner, A. Supertifurga, *J. Biol. Chem.* **1991**, *266*, 1299–1303.
182 M. Kruse, M. Brunke, A. Escher, A. A. Szalay, M. Tropschug, R. Zimmermann, *J. Biol. Chem.* **1995**, *270*, 2588–2594.
183 J. Rassow, K. Mohrs, S. Koidl, I. B. Bartiielmess, N. Pfanner, M. Tropschug, *Mol. Cell Biol.* **1995**, *15*, 2654–2662.
184 A. Matouschek, S. Rospert, K. Schmid, B. S. Glick, G. Schatz, *Proc. Natl Acad. Sci. USA* **1995**, *92*, 6319–6323.
185 O. von Ahsen, J. H. Lim, P. Caspers, F. Martin, H. J. Schönfeld, J. Rassow, N. Pfanner, *J. Mol. Biol.* **2000**, *297*, 809–818.

26
Folding and Disulfide Formation

Margherita Ruoppolo, Piero Pucci, and Gennaro Marino

26.1
Chemistry of the Disulfide Bond

The formation of a disulfide bond from two thiol groups is an electron-oxidation reaction that requires an oxidant. Protein thiol oxidation results from a thiol/disulfide exchange process that transfer equivalents from an external disulfide to the protein (Figure 26.1). The reaction occurs via direct attack of a nucleophilic thiolate anion on the most electron-deficient sulfur of the external disulfide bond. The rate constant for the reaction increases as the basicity of the attacking thiolate nucleophile (S_a) increases (pK_a increases) and as the basicity of the leaving thiolate (R_2S^-) decreases (pK_a decreases). The pK_a of the cysteine sulfhydryl group is generally in the range of 8–9; however this may vary considerably from protein to protein due to the effect of the local environment [1].

Protein disulfide formation in a redox buffer (e.g., low-molecular-weight disulfide and its corresponding thiol) is a reversible reaction that involves the formation of a mixed disulfide intermediate followed by an intramolecular attack in which a second cysteine thiol displaces the mixed disulfide to form the protein disulfide bond (Figure 26.1). Glutathione redox buffers are most commonly used in in vitro experiments since glutathione is present at high concentration in the endoplasmic reticulum [2–4]. Dithiothreitol is often used instead of linear reagents because it does not form long-lived mixed disulfides with protein thiols, thereby making oxidative folding studies easier to interpret because of the reduced number of possible intermediates in the experimental system [5]. However, it has to be taken in consideration that dithiothreitol is not physiological. During oxidative folding, the disulfide component of the redox buffer provides oxidizing equivalents for protein disulfide formation. The ability to form a disulfide will depend on the equilibrium and rate constants for the individual steps illustrated in Figure 26.1 and the concentrations and oxidation potential of the redox buffer [6]. A more oxidizing buffer will favor the formation of protein species with more and more disulfide bonds. At the same time, high concentrations of oxidant inhibit folding by formation of nonproductive intermediates which are too oxidized to fold correctly [2, 7].

The relevant intramolecular step for oxidative folding shown in Figure 26.1 heav-

Protein Folding Handbook. Part I. Edited by J. Buchner and T. Kiefhaber
Copyright © 2005 WILEY-VCH Verlag GmbH & Co. KGaA, Weinheim
ISBN: 3-527-30784-2

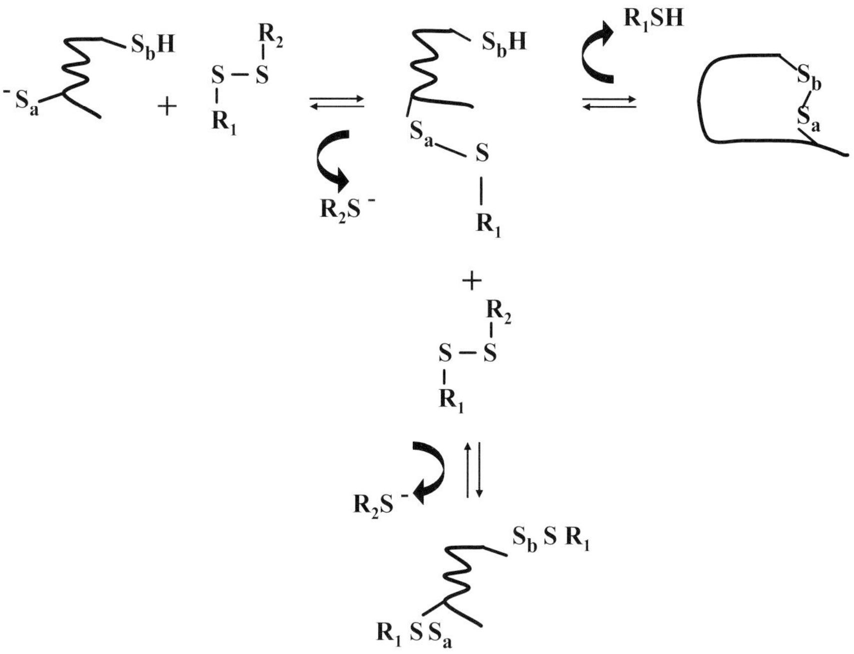

Fig. 26.1. Protein disulfide bond formation.

ily depends on the energetics of the protein conformational changes that bring the two cysteines into proximity to form the disulfide bond [8]. Among intermediates with the same number of intramolecular disulfide bonds, species with the most stable disulfides will have the highest equilibrium concentration regardless the redox buffer composition. The structures of intermediates that result in the formation of the most stable disulfide bond will be present at the highest concentration. Since interconversion among the intermediates with the same number of disulfide bonds constitute a rearrangements that does not involve the net use or production of oxidizing equivalents, the equilibrium distribution of the intermediates with the same number of disulfides will not be affected by the choice of the redox buffer or its composition but it will be an intrinsic property of the protein itself.

26.2
Trapping Protein Disulfides

The strategy of using disulfide bonds as probes to study the folding of disulfide-containing proteins was introduced and developed by Creighton and his group [8]. His pioneering work deserves credit and constitutes a milestone in the field of oxidative folding.

Disulfides present at any instant of time of folding can be trapped in a stable form by simply blocking, rapidly and irreversibly, all thiol groups in the solution. The quenching reaction should be as rapid as possible to ensure that it faithfully traps the species present at the time of addition of the quenching reagent. Trapping with reagents like iodoacetamide or iodoacetic acid has the advantage of being irreversible, but the reaction has to be carried out in highly controlled experimental conditions to be effective. Once all free thiol groups are irreversibly blocked, the trapped species are indefinitely stable and may be then separated and characterized in details provided that disulfide shuffling is prevented.

Acidification also quenches disulfide bond formation, breakage, and rearrangement. Acidification is very rapid and is not prevented by steric accessibility, but it is reversible. There is then a possibility that separation methods used to separate the acidified species will induce disulfide rearrangements. Finally, acid-trapped species have the advantage that they can be isolated and then returned to the folding conditions, to define their kinetic roles.

The conformations that favor the formation of a particular disulfide bond are stabilized to the same extent by the presence of that specific disulfide bond in trapped intermediate. Because of this close relationship between the stability of disulfide bonds and the protein conformations that favored them, the procedures described above trap not only the disulfide bonds but also the conformations of the protein at that time of folding. Some aspects of the conformation of folding intermediates can then be deduced by identifying the cysteine residues that are involved in the disulfides present in the trapped intermediates.

Although the trapping reaction has the advantage to stabilize the folding intermediates, it has the disadvantage of altering the conformation of the intermediates because of the presence of the blocking groups on the cysteine residues not involved in disulfide bonds. Even intermediates trapped with acid have their free cysteine residues fully protonated, whereas the thiol groups should be at least partially ionized during folding process. Some studies were therefore addressed to prepare the analogs of the folding intermediates replacing the free cysteine residues with Ser or Ala by protein engineering methods [9–12]. These methods have the additional advantage that intermediates that do not accumulate to substantial levels and that they can be prepared in quantities sufficient for structural analysis.

26.3
Mass Spectrometric Analysis of Folding Intermediates

Mass spectrometry is used in protein folding studies to characterize the population of species formed on the folding pathway, and the disulfide bonds present at different times of the process [13, 14]. Figure 26.2 outlines the general strategy. Aliquots withdrawn at different times during the folding process are trapped by alkylation of the free thiol and analyzed by electrospray mass spectrometry (ESMS). The alkylation reaction increases the molecular mass of the intermediates by a fixed amount for each reacted free SH group. Intermediates containing different numbers of disulfide bonds can then be separated by mass, and their relative abundance in the sample can be determined. The nature and quantitative distribution

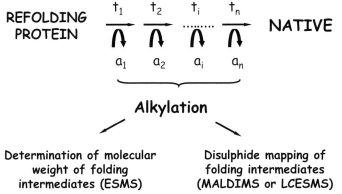

Fig. 26.2. General strategy for mass spectrometry analysis of folding intermediates.

of the disulfide bonded species present at a given time can be used to establish the folding pathway of the protein and to develop a kinetic analysis of the process. The method can also be used to determine the effect of folding catalysts on each step of the folding pathway. Any alteration in the relative distribution of the disulfide bonded species present at a given time due to the catalysts action, in fact, can be identified and quantitated [15].

It should be noted that the intermediates identified by the ESMS analysis are populations of molecular species characterized by the same number of disulfide bonds which may include nearly all possible disulfide bond isomeric species. Once the folding intermediates have been characterized by ESMS in terms of their content of disulfide bonds, a further step would consist in the structural assignment of the various disulfide bonds formed at different times in the entire process by matrix-assisted laser desorption ionization mass spectrometry (MALDIMS) or liquid chromatography ESMS. The experimental approach is based upon determination of the masses of disulfide-linked peptides of unfractionated or partially purified proteolytic digests of folding aliquots in mapping experiments [16, 17]. The folding intermediates should be cleaved at points between the potentially bridged cysteine residues under conditions known to minimize disulfide reduction and reshuffling, using aspecific enzymes or a combination of enzymes, in attempts to isolate any cysteine residue within an individual peptide. The disulfide mapping approach is then used to search for any S–S bonded peptides, which are characterized by their unique masses. The interpretation is then confirmed by performing reduction or EDMAN reactions followed by rerunning the MS spectrum [16].

26.4
Mechanism(s) of Oxidative Folding so Far – Early and Late Folding Steps

During the past three decades, the oxidative folding pathways of various small proteins containing disulfide bonds have been studied by identifying the nature of the intermediates that accumulate during their folding process [18–22]. At present, no single oxidative folding scenario seems to distinctly emerge from these detailed studies. However some common principles have been highlighted. Disulfide folding pathways are not random: they converge to a limited number of intermediates that are more stables than others. The greater thermodynamic stability of some intermediate conformations can have kinetic consequences, as further folding from the stable conformations is favored. This is most clearly illustrated in the case of bovine pancreatic trypsin inhibitor (BPTI) [8], where early folding intermediates appear to be stabilized by strong native-like interactions. A pre-equilibrium occurs very quickly when the protein is placed under folding conditions; the nature of the pre-equilibrium mixture will only depend upon the final folding conditions. Any favorable conformations present in the pre-equilibrium need not to be present at detectable levels in the denatured state. Conversely, nonrandom conformations that might be present in the denatured state may have no relevance for folding. In the case of BPTI, the most stable partially folded conformations in the pre-equilibrium appear to involve interactions between elements of secondary structure, and there is a considerable evidence for the rapid formation of some elements of secondary structure during the folding of many proteins [8].

The rate-limiting step in disulfide folding occurs very late in the process before the acquisition of the native conformation [8, 15, 23, 24]. The highest free energy barrier separates the native conformation from all other intermediates. At late stages in "structured" intermediates, the existing disulfides are sequestered from the thiols of the protein and from the redox reagent, resulting in the slowing of the thiol-disulfide rearrangement reaction by several orders of magnitude (as compared to thiol-disulfide exchange in the "unstructured" intermediate at the early stages). As a result, "structured" species frequently accumulate during the folding process [23].

The rearrangements that take place in the slow steps of oxidative folding may constitute examples of the conformational rearrangements that take place during the rate-limiting steps in folding transitions not involving disulfide bonds. Such conformational rearrangements are the result of the cooperativiity of protein structure and the similarity of the overall folding transition state to the native conformation.

**26.5
Emerging Concepts from Mass Spectrometric Studies**

The study of oxidative folding carried out by our group indicated that in quasi-physiological conditions, the process occurs via reiteration of two sequential steps: (i) formation of a mixed disulfide with glutathione, and (ii) internal attack of a free SH group to form an intramolecular disulfide bond. This sequential pathway seems to be a general mechanism for single-domain disulfide-containing proteins as it has been observed in many different experimental conditions.

According to the above concepts, only a limited number of intermediates were detected in the folding mixture and isomerization between species with the same number of disulfides was found to be only extensive at late stages of the process where slow conformational transitions become significant. Theoretically, the number of all the possible intermediates during the folding process should increase exponentially with the increase in the number of cysteine residues. In contrast, the number of the populations of intermediates increases linearly with the increase in the number of disulfide bonds. Within the frame of this general pathway, differences exist in the amount and accumulation rate of individual intermediates arising from the individual protein under investigation. When the folding pathways of RNase A and toxin α, both containing four disulfide bonds, are compared, some differences can be detected (see below). Species with two disulfide bonds predominate along the RNase A folding [15], while the intermediates containing three disulfides represent the most abundant species in the case of toxin α [15].

Data obtained using different proteins suggests the occurrence of a temporal hierarchy in the formation of disulfide bonds in oxidative folding, in which the sequential order of cysteine couplings greatly affect the rate of the total process. Even the order of formation of native disulfide bonds is in fact important to prevent the accumulation of nonproductive intermediates.

When the folding was carried out in the presence of protein disulfide isomerase (PDI), the folding pathway of reduced proteins was unchanged, but the relative distribution of the various populations of intermediates was altered. All the experiments suggest that PDI catalyzes: (1) formation of mixed disulfides with glutathione, (2) reduction of mixed disulfides, and (3) formation of intramolecular disulfide bonds. These results are not surprising considering the broad range of activities shown by PDI [4].

The oxidative folding pathway of some proteins investigated in our laboratory will be described in detail below.

26.5.1
Three-fingered Toxins

Snake toxins, which are short, all β-proteins, display a complex organization of disulfide bonds. Two S–S bonds connect consecutive cysteine residues (C43-C54, C55-C60) and two bonds intersect when bridging (C3-C24, C17-C41) to form a particular structure classified as "disulfide β-cross" [26] because cysteine residues tend to make a cross symbol when viewed along the length of the β-strands. The general organization of the polypeptide chain in snake toxins generates a trefoil structure termed the "three-fingered fold" (Figure 26.3). "Three-fingered" snake toxins act by blocking ion channels (neurotoxins), or enzymes such as acetylcholinesterase (fasciculin) or Na/K-ATPase and protein kinase C (cardiotoxins), in addition to less-defined targets.

We have shown that three-fingered snake toxins fold according to the general sequential mechanism described above [25, 27]. A single mutation located in an appropriate site of the neurotoxin structure is sufficient to substantially alter the rate of the sequential folding process of the protein and to deeply modify the proportion of intermediates that accumulate during the oxidative folding process. We

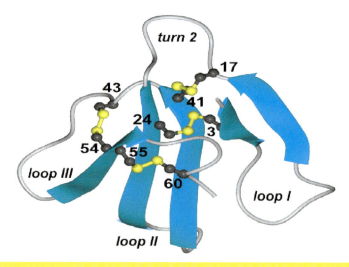

```
α     LECHNQQSSQPPTTKTC-PGETNCYKKVWRDHRGTIIERGCGCPTVKPGIKLNCCTTDKCNN
α₆₂   RICFNHQSSQPQTTKTCSPGESSCYNKQWSDFRGTIIQRGCGCPTVKPGIKLSCCESEVCNN
α₆₀   LECHNQQSSQPPTTKTC-PG-TNCYKKVWRDHRGTIIERGCGCPTVKPGIKLNCCTTDKCNN
       <-------loop I------->      <---------------loop II----------------->   <-------loop III--------->
```

Fig. 26.3. Structure of three-fingered toxin. Sequence alignment of neurotoxins and their variants used in our laboratory.

used three variants for which the length of a large turn, turn 2 (Figure 26.3), was increased from three (variant α60, sequence CPG) to four (toxin α, α61, sequence CPGE) and then to five (variant α62, sequence CSPGE) residues. The increase in the length of loop 2 greatly decreased the rate of the folding process. After 2 hours of incubation, about 90% of the variant α60 was folded, whereas no more than 70% of the 61 residues of toxin α and 20% of the variant α62 had reached the native state. Interestingly, three disulfide-containing intermediates were the predominant species in the folding of all variants but this population was markedly more abundant and persistent for the slowest-folding neurotoxin.

Two intermediates containing three disulfides were found to accumulate to detectable levels during early and late stages of neurotoxin α62 folding. Both intermediates consist of chemically homogeneous species containing three of the four native disulfide bonds and lacking the C43-C54 and the C17-C41 coupling, respectively. The des-[43–54] intermediate is the immediate precursor of the native species. Conversely, the des-[17–41] species is unable to form the fourth disulfide bond and has to rearrange into intermediates that can directly reach the native state. These isomerization reactions provide an explanation for the accumulation of three disulfide intermediates along the pathway, which caused the slow oxidative folding observed for neurotoxin α62.

The particular pathway adopted by neurotoxin α62 shares a number of common features with that followed by BPTI, although the two proteins differ both in the number of disulfides and in their secondary structures. The main similarity is that no scrambled fully oxidized species occur as folding intermediates. In addition, all-but-one-disulfide-containing intermediates (i.e., intermediates containing three and two disulfides) exclusively possess native disulfide bonds [9, 28]. Finally, the native form for both the variant α62 and BPTI results from an intramolecular rearrangement within their respective all-but-one native disulfide populations, with one predominantly productive species, des-[43–54] for the variant α62 and C30-C51 and C5-C55 for BPTI [7]. However, formation of the native form from the productive species seems much faster in BPTI [7] than in neurotoxin α62. It is noteworthy that the folding pathway of variant α62 is much closer to that of BPTI than to those of other all-β disulfide-containing proteins, such as epidermal growth factor (EGF) [29]. In this case disulfide-scrambled isomers accumulate along the pathway and they have to rearrange before forming the native structure.

On more general grounds, the observation that only intermediates with native disulfide bonds accumulate strongly suggests that native disulfide bonds are dominant early in the folding, resulting in funnelling the conformations towards the native state.

26.5.2
RNase A

Bovine pancreatic ribonuclease A has been the model protein for folding studies [30–36] since the landmark discovery by Anfinsen [37] that the amino acid sequence provides all the information required for a protein to fold properly. RNase A contains four disulfide bonds (C26-C84, C40-C95, C65-C72, C58-C110), which are critical both to the function and stability of the native enzyme [38, 39]. The two disulfide bonds C26-C84 and C58-C110 that link a α-helix and a β-sheet in the protein core are the most important to conformational stability. On the other hand, the two disulfide bonds C40-C95 and C65-C72 that link surface loops greatly affect the catalytic activity because of their proximity to active site residues.

The folding of RNase A has been studied in many laboratories, and different mechanisms of folding have been proposed depending on the experimental conditions used. Initial folding studies performed by Creighton and coworkers [40–43] in the presence of reduced and oxidized glutathione suggested that the process proceeds through a single pathway, the rate-determining step being the formation of the C40-C95 disulfide bond.

Parallel studies were performed by Scheraga and coworkers in the presence of reduced and oxidized dithiothreitol [5, 23, 44–46]. These authors suggested that RNase A regenerates through two parallel pathways involving the formation of two native-like species containing three disulfide bonds. The main pathway produces a species lacking C40-C95, in line with former results [40–43], while a species lacking C65-C72 is produced in a minor regeneration pathway.

We showed that the folding of RNase A [15] proceeds throughout the sequential mechanism discussed above. The characterization of the population of one-disulfide intermediates reveals the presence of only 12 disulfide-bonded species out of the expected 28. These results confirmed previous conclusions, that the formation of the S–S bonds during the folding of RNase A proceeds through a nonrandom mechanism [47] and is coupled to the formation of specific stabilizing structures that are either native or nonnative [8, 12, 23].

Mass spectrometry characterization revealed that in the early stages of the process: (1) the native C26-C84 pairing is absent; (2) disulfide bonds containing C26 are underrepresented; and (3) most of the nonnative disulfide bonds are formed by C110 [48].

The nonrandom coupling of cysteine residues is then biased in three respects: (1) towards the formation of short-range linkages; (2) towards the formation of native disulfides; and (3) towards the formation of disulfides involving cysteines located in the C-terminal rather than N-terminal portion of the molecule. While the first two of these trends are relatively easy to rationalize, the last one is more difficult to interpret. It could reflect more complete collapse of the C-terminal portion of the molecule, allowing easy disulfide interchange within a restricted volume or it could reflect formation of some stable structures around Cys26, which inhibit its interaction with other portions of the polypeptide chain.

The characterization of the population of one-disulfide intermediates produced in the presence of PDI reveals that the PDI-catalyzed refolding generates essentially the same disulfide bonds, and hence the same intermediates identified in the noncatalyzed reoxidation [48]. The similar distribution of isomeric species within the population of one-disulfide intermediates in the uncatalyzed and in the PDI-assisted process indicates that at the early stages of the folding, catalysis of the isomerization of disulfide bonds has no significant effect on the folding pathway. The observation that the population of one-disulfide intermediates is hardly affected by the presence of PDI further proves that thermodynamic control operates at this stage of RNase A refolding. Since nonnative disulfide bonds need to isomerize to produce native RNase A, PDI-dependent catalysis of the rearrangements of disulfide bonds is significant for the overall pathway at later stages of the refolding process. At that stage, the intermediates would already have acquired some tertiary structure and the key function of PDI is to catalyze disulfide rearrangements within kinetically trapped, structured folding intermediates as reported for the PDI-assisted refolding of BPTI [49].

The removal of the C65-C72 disulfide bond has no effect on the kinetics of folding of RNase A. The folding of the C65A/C72A mutant occurs in fact on the same time scale as the folding of wild-type protein [12]. Furthermore the individual native and nonnative disulfides appear at the same times during the folding processes of the mutant and the native protein (with the obvious exception of those involving the two absent cysteine residues). These results suggest that the folding processes of the wild-type and the mutant RNase A are driven by similar interatomic interactions without any great influence of the C65-C72 bond. Interestingly, the C65-C72 disulfide bond is the only disulfide bond that is not conserved throughout the RNase superfamily [50].

The C58A/C110A and C26A/C84A mutants fold much more slowly than the wild-type protein. A steady state is established between two- and three-disulfide containing species in the folding processes of these two mutants, thus suggesting that the rate-limiting step might be the formation of the third disulfide bond. The assignments of disulfide bonds during the folding of the C58A/C110A and C26A/C84A mutants show that many nonnative disulfide bonds are still present in solution at late stages of the reaction, indicating that many scrambled molecular species have not yet completed the folding process. The formation of C58-C110 or C26-C84 disulfide bonds can function as a lock on the native structure, hampering the three-disulfide-containing species reshuffling in the folding of wild-type RNase A.

Finally, the identification of disulfide bonds formed during the folding of RNase A mutants showed that C110 is the most actively engaged cysteine residue in the formation of disulfide bonds, as occurred in the folding of the wild-type protein. C110 can then function as an internal catalyst able to promote reshuffling of disulfides in order to accelerate isomerization reactions in the regaining of the native state.

26.5.3
Antibody Fragments

Antibodies are multimeric proteins consisting of different domains characterized by two antiparallel β-sheets linked by an intradomain disulfide bond [51–53]. A characteristic feature of antibodies is that the intradomain disulfide bond, which connects residues far apart in the sequence, is completely buried in the core of the protein. Mechanistic studies on oxidative folding of immunoglobulins were previously carried out on antibody fragments and single antibody domains [54–59].

The analysis of immunoglobulin folding by our group was dissected by investigating the folding process of different portions of the antibody molecule. The elucidation of the folding pathway of the noncovalent homodimer formed by the C-terminal domain C_H3 containing a single intramolecular disulfide bond was the first step. We showed that folding, oxidation, and association of the immunoglobulin domain C_H3 requires a fine-tuned and well-coordinated interplay between structure formation to bring the cysteine residues into proximity and to shield the disulfide bond from the solvent and structural flexibility that is required for redox shuffling and rearrangements of structural elements [59].

Analysis of the folding of the Fc fragment constituted by C_H2 and C_H3 domains showed the presence of a kinetic trap during the process that impairs the formation of the second S–S bond from the species containing one intramolecular disulfide. The results indicated that the two domains present in the Fc fragment fold independently. A hierarchy of events exists in the overall process with the disulfide of the C_H3 domain forming faster than the C_H2 S–S bond. During the early stages of folding, the C_H3 domain attains a structure which is able to bring the two cysteines in the right orientation to form the first disulfide bond and to shield it from the solvent as happens for isolated C_H3 [59]. Conversely, the C_H2 domain seems to have a flexible conformation that prevents the formation of the second S–S bond. The rate-limiting step in the overall Fc folding process is then the coupling of the two cysteines of the C_H2 domain. Consequently, spontaneous folding in the absence of folding helpers only occurs at a very slow rate. The addition of PDI catalyzes the process at various steps, promoting the formation of the C_H2 disulfide.

26.5.4
Human Nerve Growth Factor

Nerve growth factor (NGF), together with brain-derived neurotropic factor (BDNF), neurotrophin-3 (NT-3) and NT-4/5, belongs to the neurotrophin family of protein growth factors. Neurotrophins promote growth, survival, and plasticity of specific neuronal populations during developmental and adult life phases [60, 61]. NGF, BDNF, and other growth factors such as transforming growth factor (TGFβ2), bone morphogenetic protein 2 (BMP-2), and platelet-derived growth factor (PDGF) share characteristic tertiary and quaternary features: the proteins are homodimeric in their native and active conformations and have a typical tertiary fold, the cysteine knot. Our studies showed that the pro-region of recombinant human (rh)-NGF facilitates protein folding. Folding yields and kinetics of rh-pro-NGF were significantly enhanced when compared with the in vitro folding of mature rh-NGF. The characterization of the folding pathway of rh-pro-NGF indicates that structure formation is very efficient since a unique 3S species containing only the native disulfide bonds was detected at early time points of folding. The characterization of intermediates produced in the folding of mature rh-NGF revealed that the process is very slow with the formation of multiple unproductive intermediates containing nonnative cysteine couplings. Two out of the three native disulfide bonds formed after 1 hour, while the formation of the complete set of native disulfide bonds was reached only after 24 hours. Since denatured, mature rh-NGF is known to fold quantitatively as long as the cysteine knot is intact [62], it has been suggested that the pro-sequence assists the oxidative folding by possibly contributing to the formation of the cysteine knot. However, unlike BPTI, the pro-sequence of NGF does not contain a cysteine that could facilitate disulfide bond shuffling during folding of NGF [63].

A possible function of the pro-peptide may be that it acts as a specific scaffold during structure formation of the cysteine knot. Once the native disulfide bonds are formed, the characteristic β-sheet structure is built on and the two monomers can associate to form the rh-pro-NGF dimer. Alternatively, the pro-sequence of NGF could passively confer solubility to aggregation prone folding intermediates by shielding hydrophobic patches and thus enhancing folding yields.

26.6
Unanswered Questions

It is quite clear that many aspects of oxidative folding have been clarified so far. However some questions remain unsolved. An aspect that is still to be clarified concerns how far the disulfide folding mechanism can be extrapolated to protein folding mechanisms in general. Some common aspects have been highlighted throughout the chapter. However oxidative folding shows peculiar features due to the fact that disulfide-containing proteins constitute a unique class of proteins. In our opinion, the main unanswered questions are related to the observation that some disulfide-containing proteins fold with high efficiency into their native structure while others do not. These aspects are intimately related to the role played in vivo by the multiple redox machineries that cooperate to make folding fast and efficient. These aspects are discussed in Chapters 9 and 19 in Part II.

26.7
Concluding Remarks

Practical considerations drawn from the study of oxidative folding are very important in the production of recombinant proteins with disulfide bonds, many of which are therapeutically useful. A systematic exploration of pH, salt concentration, effectors, cofactors, temperature, and protein concentrations has been made in order to define the best experimental conditions to achieve the highest yield of folded protein. However, fast and high-throughtput methods are still required to investigate optimal folding conditions. There is no doubt that mass spectrometry can be a very useful tool in these kinds of studies. Few solid conclusions have been reached that suggest widely applicable methods for folding denatured and reduced proteins [64]. Guidelines for the optimization of folding methods will be discussed in other chapters.

26.8
Experimental Protocols

26.8.1
How to Prepare Folding Solutions

1. Reduction and denaturation buffer: 0.1 M Tris–HCl, 1 mM EDTA containing 6 M guanidinium chloride, pH 8.5, stable at room temperature for up to 1 month.
2. Folding buffer: 0.1 M Tris–HCl, 1 mM EDTA, pH 7.5; stable at room temperature for up to 1 month.
3. Reduced glutathione stock solution: 50 mM reduced glutathione in folding buffer; prepare fresh daily.
4. Oxidized glutathione stock solution: 50 mM oxidized glutathione in folding buffer; prepare fresh daily.
5. Iodoacetamide (IAM) solution: 2.2 M iodoacetamide solution in folding buffer; prepare fresh daily. IAM is freshly dissolved in folding buffer at 65 °C and cooled to room temperature before use. During preparation of the reagents, the solutions should be protected from light to minimize iodine production, which is a very potent oxidizing agent for thiols.

26.8.2
How to Carry Out Folding Reactions

Preparation of reduced samples

1. Reduce protein with the reduction and denaturation buffer by incubation with reduced DTT (DTT mol/S–S mol = 50/1) for 2 h at 37 °C, under nitrogen atmosphere.
2. Add 0.2 vol of 1 M HCl.
3. Desalt the reaction mixture on a gel-filtration column equilibrated and eluted with 0.01 M HCl.
4. Recover the protein fraction, test for the SH content, lyophilize and store at -20 °C.

Folding reactions

1. Dissolve reduced proteins in 0.01 M HCl and then dilute into the folding buffer to the desired final concentration.
2. Add the desired amounts of GSH and GSSG stock solutions to initiate folding. The choice of the concentrations of reduced and oxidized glutathione depends on the protein under investigation and is based on conditions giving the highest yield of native protein at the end of the folding process. Adjust the pH of the solution to 7.5 with Tris–base and incubate at 25 °C under nitrogen atmosphere. In the case of antibody fragments start folding by diluting 100-fold the reduced and denatured protein in 0.1 M Tris–HCl (pH 8.0), 1 mM EDTA, containing 6 mM GSSG. Carry out the process at 4 °C.
3. Monitor the folding processes by removing 50–100 μL samples of the folding mixture at appropriate intervals.
4. Alkylate the protein samples as described below.
5. Purify from the excess of blocking reagent by rapid HPLC desalting.
6. Recover the protein fraction and lyophilize.
7. Alternatively, quench the folding by adding hydrochloric acid to a final concentration of 3% (experimentally determined pH $\sim$ 2).

Use of folding catalysts

1. Dissolve folding catalysts in folding buffer.
2. Preincubate for 10 min at 25 °C. Add this mixture to reduced proteins and continue the folding at 25 °C under nitrogen atmosphere, as described.

Alkylation of the folding aliquots

1. Add the folding aliquots (50–100 μL) to an equal volume of a 2.2 M iodoacetamide solution. Perform the alkylation for 30 s, in the dark, at room temperature, under nitrogen atmosphere as described [65].

2. Add 100 µL of 5% trifluoroacetic acid, vortex the aliquots, and store on ice prior loading on the HPLC for the desalting.

26.8.3
How to Choose the Best Mass Spectrometric Equipment for your Study

The determination of the molecular weight of intermediates containing a different numbers of disulfide bonds has to be carried out with a mass spectrometer equipped with an electrospray ion source. The analyzer can easily be a single quadrupole. You do not need very high mass resolution for this kind of analysis.

The assignment of disulfide bonds has to be carried out on the peptide mixtures by using a MALDI time-of-flight mass spectrometer. The instrument has to be a reflectron TOF. You need very high mass resolution for this kind of analysis.

26.8.4
How to Perform Electrospray (ES)MS Analysis

1. Perform mass-scale calibration by means of multiply charged ions from a separate injection of hen egg white lysozyme (average molecular mass 14 305.99 Da) or of horse heart myoglobin (average molecular mass 16 951.5 Da).
2. Dissolve the protein samples in a mixture of H_2O/CH_3CN (50/50) containing 1% acetic acid.
3. Inject the protein samples (10 µL) in concentrations ranging from 10 to 20 pmol $µL^{-1}$ into the ion source via loop injection at a flow rate of 10 µL min^{-1}.
4. Record the spectra by scanning the quadrupole at 10 s per scan. Data are acquired and analyzed by the MassLynx software. Figure 26.4 shows, as an example, the deconvoluted ES spectra of the mixtures of species sampled at 1 min and 4 h from the folding mixture starting from reduced toxin α. The different populations of disulfide intermediates present were identified on the basis of their molecular mass. Each population of trapped intermediates is characterized by a different number of intramolecular disulfide bonds (indicated as nS), mixed disulfides with the exogenous glutathione (nG) and carboxyamidomethyl (CAM) groups. The number of CAM groups corresponds to the number of free thiols present in the folding intermediates and is therefore indicated as nH.

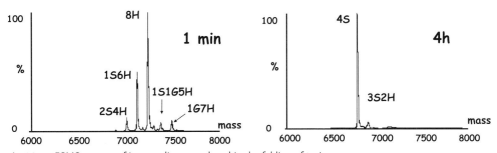

Fig. 26.4. ESMS spectra of intermediates produced in the folding of toxin α.

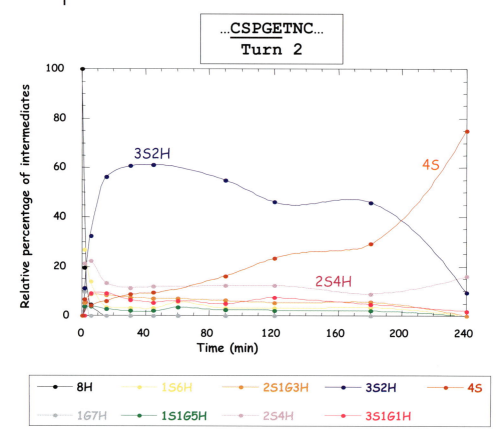

Fig. 26.5. Time course of folding of toxin α62.

Accurately quantify each population of intermediates by measuring the total ion current produced by each species provided that the different components are endowed with comparable ionization capabilities. Each set of folding data should be obtained as the mean of three independent folding experiments. The differences between folding experiments performed completely independently of each other should be about 5%. The time course of folding of toxin α62, plotted in Figure 26.5, shows that intermediates containing three disulfides are predominant from the beginning of the reaction up to about 200 min, when the relative concentration of the species 4S increases.

26.8.5
How to Perform Matrix-assisted Laser Desorption Ionization (MALDI) MS Analysis

1. Hydrolyze the protein samples withdrawn at different incubation times of folding.
2. Dissolve samples in 0.1% trifluoroacetic acid at 10 pmol/μL.
3. Apply 1 μL of sample to a sample slide and allow to air-dry.

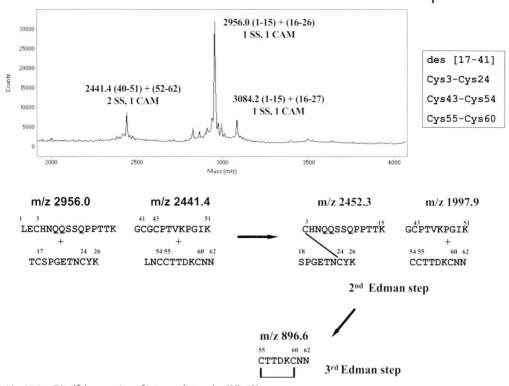

Fig. 26.6. Disulfide mapping of intermediates des-[17–41] accumulating in the folding of toxin α62.

4. Apply 1 µL of bovine insulin to the sample slide and allow to air-dry.
5. Apply 1 µL of α-cyano-4-hydroxycinnamic acid (10 mg mL^{-1}) in ethanol/acetonitrile/0.1% trifluoroacetic acid 1:1:1 (v:v:v) and allow to air-dry.
6. Collect spectra. Mass spectra are generated from the sum of 50 laser shots.
7. Calibrate the mass range using bovine insulin (average molecular mass 5734.6 Da) and a matrix peak (379.1 Da) as internal standards.

Figure 26.6 shows the MALDI mass spectrum of the tryptic mixtures of peptides derived by intermediate des-[17–41] accumulating in the folding of toxin α62. The Edman strategy adopted to assign the disulfide bonds present in the intermediate is also reported.

References

1 H. F. GILBERT (1994). The formation of native disulfide bonds, in *Mechanism of Protein Folding* (R. H. PAIN eds), Oxford University Press, Oxford, pp 104–136.

2 C. HWANG, A. J. SINSKEY and H. F. LODISH (1992). Oxidized redox state of glutathione in the endoplasmic reticulum. *Science* **257**, 1496–1502.

3 R. B. FREEDMAN, H. C. HAWKINS and

S. H. McLaughlin (1995). Protein disulfide isomerase. *Methods Enzymol* **251**, 387–406.

4 R. B. Freedman, P. Klappa and L. W. Ruddock (2002). Protein disulfide isomerases exploit synergy between catalytic and specific binding domains. *EMBO Rep* **3**, 136–140.

5 D. M. Rothwarf and H. A. Scheraga (1993). Regeneration of bovine pancreatic Ribonuclease A. 1. Steady-state distribution. *Biochemistry* **32**, 2671–2680.

6 H. F. Gilbert (1990). Molecular and cellular aspects of thiol-disulfide exchange. *Adv Enzymol* **63**, 69–75. 69–72.

7 J. S. Weissman and P. S. Kim (1991). Re-examination of the folding of BTPI: Predominance of native intermediates. *Science* **253**, 1386–1393.

8 T. E. Creighton (1992). Folding pathways determined using disulfide bonds, in *Protein Folding* (T. E. Creighton ed.), Freeman, New York, pp 301–352.

9 C. P. Van Mierlo, N. J. Darby, D. Neuhaus and T. E. Creighton (1991). Two-dimensional 1H nuclear magnetic resonance study of the (5–55) single-disulfide folding intermediate of bovine pancreatic trypsin inhibitor. *J Mol Biol* **222**, 373–390.

10 J. P. Staley and P. S. Kim (1992). Complete folding of bovine pancreatic trypsin inhibitor with only a single disulfide bond. *Proc Natl Acad Sci USA* **89**, 1519–1523.

11 T. A. Klink, K. J. Woycechowsky, K. M. Taylor and R. T. Raines (2000). Contribution of disulfide bonds to the conformational stability and catalytic activity of ribonuclease A. *Eur J Biochem* **267**, 566–572.

12 M. Ruoppolo, F. Vinci, T. A. Klink, R. T. Raines and G. Marino (2000). Contribution of individual disulfide bonds to the oxidative folding of ribonuclease A. *Biochemistry* **39**, 12033–12042.

13 M. Ruoppolo, R. B. Freedman, P. Pucci and G. Marino (1996). The glutathione dependent pathways of refolding of RNase T1 by oxidation and disulfide isomerization. Catalysis by protein disulfide isomerase. *Biochemistry* **35**, 13636–13646.

14 M. Ruoppolo, C. Torella, F. Kanda, G. Marino, M. Panico, P. Pucci and H. R. Morris (1996). Identification of disulfide bonds in the refolding of bovine pancreatic RNase A. *Folding Des* **1**, 381–390.

15 M. Ruoppolo, J. Lundstrom-Ljung, F. Talamo, P. Pucci and G. Marino (1997). Effect of Glutaredoxin and Protein Disulfide Isomerase on the glutathione-dependent folding of ribonuclease A. *Biochemistry* **36**, 12259–12267.

16 H. R. Morris and P. Pucci (1985). A new method for rapid assignment of S–S bridges in proteins. *Biochem Biophys Res Commun* **126**, 122–128.

17 H. R. Morris, P. Pucci, M. Panico and G. Marino. (1990). Protein folding/refolding analysis by mass spectrometry. *Biochem J* **268**, 803–806.

18 T. E. Creighton, N. J. Darby and J. Kemmink (1996). The roles of partly folded intermediates in protein folding. *FASEB J* **10**, 110–118.

19 N. W. Isaacs (1995). Cystine knots. *Curr Opin Struct Biol* **5**, 391–395.

20 R. S. Norton and P. K. Pallaghy (1998). The cystine knot structure of ion channel toxins and related polypeptides. *Toxicon* **36**, 1573–1583.

21 L. R. De Young, C. H. Schmelzer and L. E. Burton (1999). A common mechanism for recombinant human NGF, BDNF, NT-3, and murine NGF slow unfolding. *Protein Sci* **8**, 2513–2518.

22 R. J. Darling, R. W. Ruddon, F. Perini and E. Bedow (2000). Cystine knot mutations affect the folding of the glycoprotein hormone alpha-subunit. Differential secretion and assembly of partially folded intermediates. *J Biol Chem* **275**, 15413–15421.

23 M. Narayan, E. Welker, C. Wanjalla, G. Xu and H. A. Scheraga (2003). Shifting the competition between the intramolecular Reshuffling reaction and the direct oxidation reaction during the oxidative folding of

kinetically trapped disulfide-insecure intermediates. *Biochemistry* **42**, 10783–10789.

24 W. J. WEDEMEYER, E. WELKER, M. NARAYAN and H. A. SCHERAGA (2000). Disulfide bonds and protein folding. *Biochemistry* **39**, 4207–4216.

25 M. RUOPPOLO, M. MOUTIEZ, M. F. MAZZEO, P. PUCCI, A. MENEZ, G. MARINO and E. QUÉMÉNEUR (1998). The length of a single turn controls the overall folding rate of "three-fingered" snake toxins. *Biochemistry* **37**, 16060–16068.

26 P. M. HARRISON and M. J. E. STERNBERG (1996). The disulfide beta-cross: from cystine geometry and clustering to classification of small disulfide-rich protein folds. *J Mol Biol* **264**, 603–623.

27 M. RUOPPOLO, F. TALAMO, P. PUCCI, M. MOUTIEZ, E. QUÈMÈNEUR, A. MÈNEZ and G. MARINO (2001). Slow folding of three-fingered toxins is associated with the accumulation of native disulfide-bonded intermediates. *Biochemistry* **40**, 15257–15266.

28 C. EIGENBROT, M. RANDAL and A. A. KOSSIAKOFF (1990). Structural effects induced by removal of a disulfide-bridge: the X-ray structure of the C30A/C51A mutant of basic pancreatic trypsin inhibitor at 1.6 A. *Protein Eng* **3**, 591–598.

29 J. Y. CHANG, L. LI and A. BULYCHEV (2000). The underlying mechanism for the diversity of disulfide folding pathways. *J Biol Chem* **275**, 8287–8289.

30 D. B. WETLAUFER and S. RISTOW (1973). Acquisition of three-dimensional structure of proteins. *Annu Rev Biochem* **42**, 135–158.

31 D. B. WETLAUFER (1973). Nucleation, rapid folding and globular intra-chain regions in proteins. *Proc Natl Acad Sci USA* **70**, 697–701.

32 M. KARPLUS and D. C. WEAVER (1976). Protein-folding dynamics *Nature (London)* **260**, 404–406.

33 R. T. RAINES (1998). Ribonuclease A. *Chem Rev* **98**, 1045–1065.

34 J. R. GAREL and R. L. BALDWIN (1973). Both the fast and slow refolding reactions of ribonuclease A yield native enzyme. *Proc Natl Acad Sci USA* **70**, 3347–3351.

35 L. N. LIN and J. F. BRANDTS (1984). Involvement of prolines-114 and -117 in the slow refolding phase of ribonuclease A as determined by isomer-specific proteolysis. *Biochemistry* **23**, 5713–5723.

36 P. S. KIM and R. L. BALDWIN (1982). Specific intermediates in the folding reactions of small proteins and the mechanism of protein folding. *Annu Rev Biochem* **51**, 459–489.

37 C. B. ANFINSEN (1973). Principles that govern the folding of protein chains *Science* **181**, 223–230.

38 S. SHIMOTAKAHARA, C. B. RIOS, J. H. LAITY, D. E. ZIMMERMANN, H. A. SCHERAGA and G. T. MONTELIONE (1997). NMR structural analysis of an analog of an intermediate formed in the rate-determining step of one pathway in the oxidative folding of bovine pancreatic ribonuclease A: automated analysis of 1H, 13C, and 15N resonance assignments for wild-type and [C65S, C72S] mutant forms. *Biochemistry* **36**, 6915–6929.

39 B. R. KELEMEN, L. W. SCHULTZ, R. Y. SWEENEY and R. T. RAINES (2000). Excavating an active site: the nucleobase specificity of ribonuclease A. *Biochemistry* **39**, 14487–14494.

40 T. E. CREIGHTON (1979). Intermediates in the refolding of reduced ribonuclease A. *J Mol Biol* **129**, 411–431.

41 T. E. CREIGHTON (1980). A three-disulfide intermediate in refolding of reduced Ribonuclease A with a folded conformation. *FEBS Lett* **118**, 283–288.

42 T. E. CREIGHTON (1983). An empirical approach to protein conformation stability and flexibility. *Biopolymers* **22**, 49–60.

43 T. E. CREIGHTON (1990). Protein folding. *Biochem J* **270**, 1–10.

44 D. M. ROTHWARF and H. A. SCHERAGA (1993). Regeneration of bovine pancreatic Ribonuclease A. 2. Kinetics of regeneration. *Biochemistry* **32**, 2680–2690.

45 D. M. ROTHWARF and H. A. SCHERAGA

(1993). Regeneration of bovine pancreatic Ribonuclease A. 3. Dependence on the nature of the redox reagent. *Biochemistry* **32**, 2690–2697.

46 D. M. ROTHWARF and H. A. SCHERAGA (1993). Regeneration of bovine pancreatic ribonuclease A. 4. Temperature dependence of the regeneration rate. *Biochemistry* **32**, 2698–2706.

47 X. XU, D. M. ROTHWARF and H. A. SCHERAGA (1996). Non-random distribution of the one-disulfide intermediates in the regeneration of ribonuclease A. *Biochemistry* **35**, 6406.

48 F. VINCI, M. RUOPPOLO, P. PUCCI, R. B. FREEDMAN and G. MARINO (2000). Early intermediates in the PDI-assisted folding of ribonuclease A. *Protein Sci* **9**, 525–535.

49 J. S. WEISSMANN and P. S. KIM (1993). Efficient catalysis of disulfide bond rearrangements by protein disulfide isomerase. *Nature* **365**, 185–188.

50 J. J. BEINTEMA, C. SCHULLER, M. IRIE and A. CARSANA (1988). Molecular evolution of the ribonuclease superfamily. *Progr Biophys Mol Biol* **51**, 165–192.

51 R. HUBER, J. DEISENHOFER, P. M. COLMAN, M. MATSUSHIMA and W. PALM (1976). Crystallographic structure studies of an IgG molecule and an Fc fragment. *Nature* **264**, 415–420.

52 J. DEISENHOFER (1981). Crystallographic refinement and atomic models of a human Fc fragment and its complex with fragment B of protein A from *Staphylococcus aureus* at 2.9 and 2.8 Å resolution. *Biochemistry* **20**, 2361–2370.

53 J. G. AUGUSTINE, A. DE LA CALLE, G. KNARR, J. BUCHNER and C. A. FREDERICK (2001). The crystal structure of the fab fragment of the monoclonal antibody MAK33. Implications for folding and interaction with the chaperone bip. *J Biol Chem* **276**, 3287–3294.

54 Y. GOTO and Y. TAMAGUCHI (1979). The role of the intrachain disulfide bond in the conformation and stability of the constant fragment of the immuno-globulin light chain. *J Biochem* **86**, 1433–1441.

55 Y. GOTO and Y. TAMAGUCHI (1982). Unfolding and refolding of the constant fragment of the immuno-globulin light chain. Kinetic role of the intrachain disulfide bond. *J Mol Biol* **156**, 911–926.

56 J. BUCHNER and R. RUDOLPH (1991). Renaturation, purification and characterization of recombinant Fab-fragments produced in *Escherichia coli*. *Bio/technology* **9**, 157–162.

57 H. LILIE, S. MC LAUGHLIN, R. FREEDMAN and J. BUCHNER (1994). Influence of protein disulfide isomerase (PDI) on antibody folding in vitro. *J Biol Chem* **269**, 14290–14296.

58 M. J. W. THIES, J. MAYER, J. G. AUGUSTINE, C. A. FREDERICK, H. LILIE and J. BUCHNER (1999). Folding and association of the antibody domain C_H3: prolyl isomerization preceeds dimerization. *J Mol Biol* **293**, 67–79.

59 M. J. W. THIES, F. TALAMO, M. MAYER, S. BELL, M. RUOPPOLO, G. MARINO and J. BUCHNER (2002). Folding and oxidation of the antibody domain C_H3. *J Mol Biol* **319**, 1267–1277.

60 Y. A. BARDE (1990). The nerve growth factor family. *Prog Growth Factor Res* **2**, 237–248.

61 H. THOENEN (1995). Neurotrophins and neuronal plasticity. *Science* **270**, 593–598.

62 L. R. DE YOUNG, L. E. BURTON, J. LIU, M. F. POWELL, C. H. SCHMELZER and N. J. SKELTON (1996). RhNGF slow unfolding is not due to proline isomerization: possibility of a cystine knot loop-threading mechanism. *Protein Sci* **5**, 1554–1566.

63 J. S. WEISSMAN and P. S. KIM (1992). The pro region of BPTI facilitates folding. *Cell* **71**, 841–851.

64 D. R. THATCHER and M. HITCHCOCK (1994). Protein folding in biotechnology in *Mechanism of Protein Folding* (R. H. PAIN eds). Oxford University Press, Oxford, pp 229–261.

65 W. R. GRAY (1993). Disulfide structures of highly bridged peptides: a new strategy for analysis. *Protein Sci* **2**, 1732–1748.

27
Concurrent Association and Folding of Small Oligomeric Proteins

Hans Rudolf Bosshard

27.1
Introduction

Our current understanding of protein folding has been mostly obtained through the study of monomeric, globular proteins. However, oligomeric proteins are probably more abundant in nature. There is strong evolutionary pressure for monomeric proteins to associate into oligomers [1]. Many results and conclusions acquired from the study of monomeric proteins can be applied to oligomeric proteins. For example, the stabilities of both monomeric and oligomeric proteins result from a delicate balance between opposing forces, the major contributors to stability being the hydrophobic effect, van der Waals interactions, and hydrogen bonds among polar residues [2–4]. In contrast to monomeric proteins, the stabilizing and destabilizing forces operate not only within single subunits but also between the subunits of an oligomeric protein; that means, both *intra*molecular and *inter*molecular forces contribute to stability. Nevertheless, there are no fundamental differences between the mechanisms of folding of monomeric and oligomeric proteins. For both types of proteins singular as well as multiple folding pathways are observed with no, a few, or many intermediates and sometimes with nonnative, dead-end products.

However, one difference is important: At least one step in the folding of an oligomeric protein must be concentration dependent. This is because the formation of an oligomeric protein includes folding *and* association; and folding and association are concurrent reactions. In some cases, folding and association can be studied as separate and virtually independent reactions, in others they are fully coupled into a single reaction step ("two-state" mechanism or "single-step" mechanism), yet most often they can be distinguished but are not independent of each other. The degree of coupling between folding and association largely depends on the balance between the intra- and intermolecular forces that stabilize the oligomeric structure. If the isolated monomeric subunits are stabilized by strong intramolecular forces and the folded monomeric subunits themselves are thermodynamically stable folded structures, then folding and association are separate steps: formation of the oligomer corresponds to the association of preformed folded monomers. If the isolated monomeric subunits are unstable and stability is gained only by intermolecular forces within the oligomeric structure, then folding and association are intimately dependent on each other and are tightly coupled.

The concurrence of intra- and intermolecular reactions and the dependence of folding on concentration adds an extra level of complexity to folding. This is perhaps the main reason why our knowledge of the stability and folding of multisubunit proteins is still rudimentary and often of only a qualitative nature. Indeed, quantitative thermodynamic and kinetic analyses have been restricted to mainly dimeric proteins. Reliable quantitative thermodynamic and kinetic data on trimeric and higher order oligomeric structures are still scarce.

The present chapter focuses on the detailed, quantitative in vitro analysis of the thermodynamics and kinetics of folding of small dimeric proteins and protein motifs, with only brief and passing reference to trimeric and tetrameric structures. The emphasis is on concepts and methods. We do not present a comprehensive review of the assembly of dimeric proteins but select appropriate examples to illustrate experimental methods, concepts and principles. In vitro and in vivo folding and association of oligomeric and multimeric proteins has been reviewed before (see Refs [5–8]). In vivo folding and assembly of large, higher order multisubunit proteins is covered in Chapter 2 in Part II.

27.2
Experimental Methods Used to Follow the Folding of Oligomeric Proteins

In principle, the methods used in the study of monomeric proteins can be used to follow dissociation and unfolding of oligomeric proteins under equilibrium conditions and to measure the kinetics of folding, unfolding, and refolding. The main difference is in data analysis, which has to consider the association step that is lacking in monomeric proteins.

27.2.1
Equilibrium Methods

Chemical unfolding by urea or guanidinium chloride (GdmCl) is the most commonly used method to measure the fraction of folded and unfolded protein, respectively, and to calculate equilibrium constants. Free energies of unfolding at standard conditions, $\Delta G_U(H_2O)$, are obtained by linear extrapolation to zero denaturant as is done for monomeric proteins [9] (see Appendix: Section A2.2 and Chapter 3 for more details). Extrapolation is performed on any suitable signal change paralleling the denaturant-induced change of molecular state; examples are far-UV circular dichroism (CD) to follow secondary structure changes, or tryptophan fluorescence to monitor molecular packing density and polarity. In applying linear extrapolation one tacitly assumes that the effects of the denaturant on both unfolding and dissociation of the oligomeric protein are linear. This assumption is difficult to check. Extrapolating the same value of $\Delta G_U(H_2O)$ from urea and

GdmCl unfolding is a good indication of a linear denaturant dependence of dissociation and unfolding [10, 11].

Another popular equilibrium method is thermal unfolding. If the equilibrium between unfolded subunits and folded oligomer is reversible, the change with temperature of an appropriate spectroscopic signal such as fluorescence emission or CD can be used to determine the fraction of protein in the folded and unfolded states and, occasionally, also the fraction of intermediate states. The immediate result from the thermal melting curve of a protein is its midpoint temperature of unfolding, T_m. If the reversible unfolding of an oligomeric protein follows a two-state mechanism, half of the protein is unfolded and dissociated and half is folded and associated at T_m. (This does not apply to the midpoint T_m of a melting curve obtained by differential scanning calorimetry [12].) The melting curve provides the fraction of unfolded, monomeric protein, f_M, from which the equilibrium unfolding constant K_U and the free energy of unfolding $\Delta G_U = -RT \ln K_U$ is calculated. Note that in using K_U and ΔG_U, the folded, native state is the reference state.

From the temperature dependence of the equilibrium constants of unfolding, the van't Hoff enthalpy of the dissociation–unfolding transition at T_m can be calculated (see Appendix: Sections A2.3 and A2.4). From differential scanning calorimetry experiments the calorimetric enthalpy can be obtained, which is model independent and may differ from the van't Hoff enthalpy if the unfolding reaction is not two-state. Extrapolation of enthalpies and free energies to standard conditions (for example 25 °C) are based on the formalism of equilibrium thermodynamics (see Appendix: Section A2.5). Such extrapolation can be inaccurate if it extends over a large temperature range and the heat capacity change is not accurately known.

The fraction of protein in the oligomeric and monomeric state may also be obtained by following an appropriate signal after simple dilution. This approach is an added benefit of the concentration dependence of the monomer/oligomer equilibrium and can be used under benign buffer conditions, that is, in the absence of denaturant and at ambient temperature. If dissociation and unfolding are fully coupled, the dilution method yields the overall free energy of unfolding plus dissociation. However, if different free energies of unfolding are obtained from the dilution method and from chemical or thermal unfolding, this indicates that folding and association are not, or only partly, coupled. For example, a lower free energy of unfolding from dilution than from chemical unfolding is an indication for a folding intermediate that is populated under benign conditions but not in the presence of chemical denaturant. In other words, the oligomeric protein dissociates into (partly) folded subunits during dilution under benign conditions, yet it dissociates into unfolded subunits in the presence of a denaturant. The difference between $\Delta G_U(H_2O)$ from dilution and $\Delta G_U(H_2O)$ from chemical unfolding can then be ascribed to the unfolding free energy of the subunits.

Other methods to monitor the fraction of folded and unfolded oligomeric proteins under benign buffer conditions are sedimentation equilibrium analysis, gel-filtration chromatography, light-scattering analysis, and isothermal titration calorimetry. In the last method, the heat evolved or taken up when two (or more) molecules interact with each other is measured and the reaction enthalpy is calculated from the integrated heat change. Isothermal titration calorimetry can only be applied to heterodimeric proteins whose subunits do not form homomeric structures. The procedure yields the change of the enthalpy, the Gibbs free energy and the subunit stoichiometry of the oligomeric structure [12–14].

It should be noted that all the methods based on dissociation under benign buffer conditions work only if the association between the subunits is relatively weak and association/dissociation occurs in the micromolar to nanomolar concentration range. If an oligomeric protein assembles at less than nanomolar concentration, spectroscopic signal changes and calorimetric heat changes may be too weak to monitor by dilution.

27.2.2
Kinetic Methods

The same kinetic methods used for monomeric proteins can be applied to study the time course of folding and association or unfolding and dissociation of oligomeric proteins. Time-resolved methods include conventional stopped-flow and quenched-flow techniques for the time range of milliseconds to seconds, and fast relaxation methods like temperature jump to follow more rapid reactions (see Chapter 14). Variation of the observation signal allows different aspects of the reaction to be investigated. For example, far-UV CD spectroscopy, pulse-labeling NMR and mass spectroscopy are used to follow secondary structure formation and formation of hydrogen bonds; intrinsic and extrinsic fluorescence measurements monitor molecular packing; fluorescence anisotropy and small angle X-ray estimate changes in molecule size; ligand binding reports the gain or loss of biological activity. (See Chapter 2 and Ref. [15] for spectroscopic methods used to study protein folding.)

As for monomeric proteins, unfolding and refolding is often studied by rapid dilution into denaturant and out of denaturant, respectively. This necessitates extrapolation to standard conditions in water. To avoid extrapolation, unfolding under benign buffer conditions may be studied from relaxation after a rapid change of temperature, pH or simply after rapid dilution (see Refs [16] and [17] and Appendix: Section A3). If an oligomeric protein is composed of different subunits, folding can be studied in a very direct way by rapid mixing of the different subunits. This means there is no need to first unfold the heterooligomeric protein and thereafter follow its refolding by returning to folding conditions.

The rates of dissociation of an oligomeric protein can also be measured by subunit exchange. To this end, the protein has to be labeled with a fluorescence tag or another appropriate marker. Folded protein with the marker is rapidly diluted with a large excess of folded protein lacking the marker. The change of the signal of the marker is then proportional to the rate of dissociation of the marked protein provided that reassociation is much more rapid than dissociation, which can be achieved by increasing the protein concentration [17, 18].

27.3 Dimeric Proteins

The simplest folding mechanism of a dimeric protein is a two-state or one-step reaction between two unfolded monomers M_u and the folded dimer D_f as described by Eq. (1).

$$2M_u \underset{k_u}{\overset{k_f}{\rightleftarrows}} D_f$$

$$K_d = \frac{k_u}{k_f} = \frac{[M_u]^2}{[D_f]} \tag{1}$$

Here and elsewhere, we use the notation M for monomer and D for dimer. The subscripts indicate unfolded and folded, respectively. For a monomeric protein both k_f and k_u are monomolecular rate constants (s^{-1}). In the case of a dimeric protein, k_f is a bimolecular rate constant ($M^{-1} s^{-1}$) and only k_u is unimolecular (s^{-1}). Square brackets indicate molar concentrations. Also, in contrast to a monomeric protein, the equilibrium dissociation constant K_d has the unit of a concentration. It is related to the thermodynamic parameters of unfolding by the well-known thermodynamic equation:

$$\Delta G_U = \Delta H_U - T\Delta S_U = -RT \ln K_d \tag{2}$$

Equation (1) says that the folding and association reactions are completely linked. In practice, this means that the association of the monomeric subunits and their folding to the native conformation of the final dimer cannot be experimentally separated. However, this does not exclude the fleeting existence of intermediates. Indeed, there are several instances where folding is thermodynamically two-state but kinetics reveal transitory intermediates (see the examples in Table 27.3). Hence, a more physically meaningful mechanism for the formation of a dimeric protein is described by Eq. (3).

$$2M_u \underset{k_{-1}}{\overset{k_1}{\rightleftarrows}} 2M_i \underset{k_{-2}}{\overset{k_2}{\rightleftarrows}} D_i \underset{k_{-3}}{\overset{k_3}{\rightleftarrows}} D_f$$

$$K_{d1} = \frac{[M_u]}{[M_i]} = \frac{k_{-1}}{k_1} \quad K_{d2} = \frac{[M_i]^2}{[D_i]} = \frac{k_{-2}}{k_2} \quad K_{d3} = \frac{[D_i]}{[D]} = \frac{k_{-3}}{k_3} \tag{3}$$

Here, monomeric and dimeric intermediates (subscript i) are at equilibrium with the unfolded monomer M_u and the folded dimer D_f, respectively. Intermediate M_i can be considered as an association-competent subset of conformational forms of the monomer, perhaps a set of partly folded monomers. Only M_i associates to the dimeric intermediate D_i, which then rearranges to the final folded dimer D_f. The first and third equilibria are unimolecular, the middle equilibrium is bimolecular. If the first equilibrium is far on the side of M_u and the last is far on the side of D_f, Eq. (3) simplifies to the one-step Eq. (1). In this case, only M_u and D_f are detected under equilibrium conditions, but the kinetics of folding and unfolding may exhibit different phases indicative of transitory formation of intermediates.

27.3.1
Two-state Folding of Dimeric Proteins

Several dimeric proteins fold according to Eq. (1), exhibiting two-state transition between two unfolded monomers and a folded dimer. For monomeric proteins it has been shown that a two-state folding model is justified if the protein rapidly equilibrates between many unfolded conformations prior to complete folding [19]. By the same reasoning, two-state folding seems justified if intermediates M_i and D_i of Eq. (3) are ensembles of barely populated conformations at rapid equilibrium with M_u and D_f, respectively. One has to keep in mind that two-state folding is an operational concept: The sole observation of a folded dimer and an unfolded monomer under a given set of experimental conditions does not rule out the existence of folding intermediates. Table 27.1 lists operational criteria that typically apply to two-state folding. At least one of criteria i–iv referring to equilibrium experiments, and one of criteria v–vii referring to kinetic experiments, as well as criterion viii should be met to demonstrate two-state folding. The calorimetric and van't Hoff equality criterion ix is more difficult to test because thermal unfolding has to be over 90% reversible, which is not always the case.

The thermodynamic and kinetic formalism for two-state folding is relatively simple and has often been described (see the Appendix for details). As for monomeric proteins, the equilibrium dissociation constant K_d is obtained from the fraction of monomeric protein f_M, or the fraction of dimeric protein $f_D = 1 - f_M$. For a

Tab. 27.1. Criteria for two-state folding.

(i)	Equilibrium unfolding followed with different physical probes such as light absorption, fluorescence emission, circular dichroism yield superimposable curves
(ii)	Equilibrium measurements are dependent on protein concentration
(iii)	The change of the free energy of unfolding with denaturant concentration is linear
(iv)	The reciprocal midpoint temperature of unfolding ($1/T_m$) is linearly proportional to the logarithm of the total protein concentration
(v)	Refolding kinetics is concentration dependent, unfolding kinetics is concentration independent
(vi)	Kinetics of folding and unfolding exhibit single phases, and these account for the entire amplitude of the signal changes of folding and unfolding, respectively
(vii)	The change of the unfolding and refolding rate constants with denaturant is linear
(viii)	Equilibrium constants determined experimentally are the same within error as those calculated from the ratio of the association and dissociation rate constants
(ix)	The enthalpy change of the dimer to monomer transition obtained by van't Hoff analysis of thermal unfolding curves is the same within error as the calorimetrically measured enthalpy change

monomeric protein the equilibrium constant, defined as $K_d = f_M/(1 - f_M)$, is independent of the protein concentration. In contrast, to determine K_d of a dimeric protein, the protein concentration has to be known. Moreover, the relationship between K_d and concentration is different for homodimeric and heterodimer proteins, as shown by Eqs (4) and (5) (see also Appendix: Section A1).

$$K_d = \frac{2[P_t]f_M^2}{1 - f_M} \qquad (4)$$

$$K_d = \frac{[P_t]f_M^2}{2(1 - f_M)} \qquad (5)$$

In the above equations, $[P_t]$ is the total protein concentration expressed as the total monomer concentration: $[P_t] = [M] + 2[D]$. It is important to note that the protein concentration has to be accounted for in analyzing the thermodynamic and kinetic data. For example, the well-known linear extrapolation of chemical unfolding data to zero denaturant concentration has to consider protein concentration because ΔG_U is not zero at the transition midpoint of chemical unfolding (see Appendix: Section A2.2).

27.3.1.1 Examples of Dimeric Proteins Obeying Two-state Folding

Table 27.2 lists five representative examples of small dimeric proteins for which two-state folding has been demonstrated based on several of the criteria listed in Table 27.1. The Arc repressor of bacteriophage P22 is a small homodimeric protein composed of two 53 residue polypeptide chains [10]. Figure 27.1 shows that the fluorescence and CD changes induced by denaturant are superimposable (criterion i of Table 27.1). The position of the unfolding trace is dependent on protein concentration as predicted for a bimolecular reaction (criterion v of Table 27.1). The change of unfolding free energies with denaturant concentration is linear (criterion iii of Table 27.1) and extrapolation to water yields the same free energy of unfolding for both urea and GdmCl unfolding, supporting of the validity of the linear extrapolation approach for both the unfolding reaction and the dissociation of dimers into monomers.

Figure 27.2 shows the urea dependence of unfolding and refolding rates of the dimeric Arc repressor. The rate of refolding increases linearly when the denaturant concentration is lowered; the rate of unfolding increases linearly when the denaturant concentration increases (criterion vii). Refolding is concentration dependent, unfolding is concentration independent, and the single kinetic phases account for the entire amplitude of refolding and unfolding (criterion vi). Finally, the free energy of unfolding and dissociation extrapolated from equilibrium measurements and calculated from the ratio of the refolding and unfolding rate are very similar, as demanded by criterion viii. In the middle range of denaturant concentrations, both unfolding and refolding reactions have to be taken into account when analyzing the unfolding and refolding traces, respectively. The analysis was performed using Eq. (A29) [20]. Note that, unlike in the case of a monomeric protein, the

Tab. 27.2. Examples of dimeric proteins folding through a simple two-state mechanism: $2M_u \underset{k_u}{\overset{k_f}{\rightleftarrows}} D_f$.

Protein	Equilibrium analysis[a]	Kinetic analysis[a]	Comments
Arc repressor (homodimer)	$\Delta G_U = 40$ kJ mol^{-1}, $\Delta C_P = 6.7$ kJ M^{-1} K^{-1}, $\Delta H_U(54\,°C) = 297$ kJ M^{-1} [10]	$k_f = 9 \times 10^6$ M^{-1} s^{-1}, $k_u = 0.2$ s^{-1}, $\Delta G_U = 41$ kJ mol^{-1} [20]	Many native-like interactions in the dimeric transition state [21–23]
GCN4 leucine zipper (homodimer)	$\Delta G_U(5\,°C) = 39$ kJ mol^{-1} [72]	$k_f = 4 \times 10^5$ M^{-1} s^{-1}, $k_u = 3.3 \times 10^{-3}$ s^{-1}, $\Delta G_U(5\,°C) = 43$ kJ mol^{-1} [45]	Calorimetric analysis indicates two monomolecular pretransitions at 20 °C and 50 °C before the main bimolecular transition [26] Evidence for association-competent, partly helical monomer [48, 49]
Leucine zipper AB (heterodimer)	$\Delta G_U = 44$ kJ mol^{-1}, $\Delta C_P = 3–4$ kJ M^{-1} K^{-1}, $\Delta H_U(25\,°C) = 120$ kJ M^{-1} [27]	$k_f = 2 \times 10^7$ M^{-1} s^{-1}, $k_u = 0.2$ s^{-1}, $\Delta G_U = 46$ kJ mol^{-1} [28]	Folding, but not unfolding, is strongly dependent on ionic strength indicating electrostatically assisted association [28]
Nerve growth factor (homodimer)	$\Delta G_U = 81$ kJ mol^{-1} [11]		
Mannose-binding lectin (heterodimer)	$\Delta G_U = 55$ kJ mol^{-1}, $\Delta C_P = 13$ kJ M^{-1} K^{-1}, $\Delta H_U(66\,°C) = 728$ kJ M^{-1} [73]		ΔC_P-values from experiment and calculated from the buried subunit interface agree well

[a] Unless stated, thermodynamic and kinetic parameters refer to ambient temperature (20 to 25 °C).

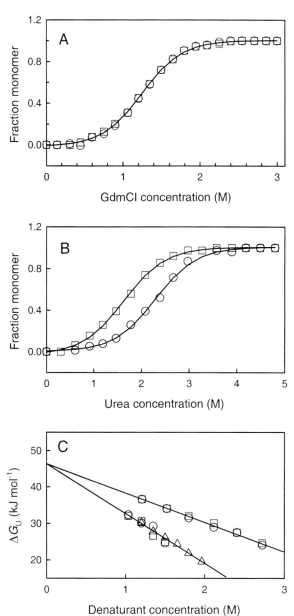

Fig. 27.1. Equilibrium unfolding of dimeric Arc repressor according to a two-state mechanism. A) Guanidinium chloride (GdmCl) denaturation curves monitored by fluorescence (squares) and CD (circles) are superimposable and the solid line is described by Eq. (4) for two-state folding. B) Urea denaturation is concentration dependent as predicted by Eq. (4). Experiments were performed with 1.6 µM (squares) and 16 µM (circles) total protein concentration. C) Free energies of unfolding measured with different total peptide concentrations (different symbols) are linearly dependent on GdmCl concentration (lower data) and urea concentration (upper data). Linear extrapolation of urea and GdmCl according to Eq. (A12) yields the same free energy of unfolding of $\Delta G_U(H_2O) = 46$ kJ mol^{-1}, supporting the validity of the linear extrapolation method. Figure adapted from Ref. [10] with permission.

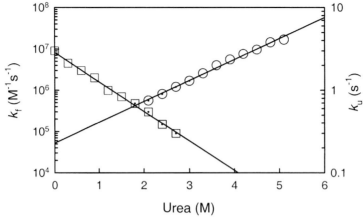

Fig. 27.2. Urea dependence of the rate constants of refolding (squares, left ordinate) and unfolding (circles, right ordinate) of Arc repressor. Empty symbols refer to data analysis neglecting unfolding (Eq. (A22)) or refolding (Eq. (A25)). Dotted symbols refer to data analysis including both unfolding and refolding as described by Eqs (A27) and (A29). Experiments were performed at pH 7.5 and 20 °C. Figure adapted from Ref. [20] with permission.

crossover point of the unfolding and refolding lines in Figure 27.2 do not yield the midpoint concentration of the denaturant at which $[M] = [D]/2$ and $\Delta G_U = 0$. This is because the refolding reaction and hence the midpoint of denaturation depends on the protein concentration.

Extrapolation of the folding rate constant of the Arc repressor to zero denaturant yields k_f of about 9×10^6 M^{-1} s^{-1}, which is about two orders of magnitude below the diffusion limit. Moreover, the folding rate is nearly independent of solvent viscosity. It seems that only a small fraction of collisions of unfolded monomers are productive, otherwise the rate should approach the diffusion limit and depend on solvent viscosity. This is supported by a dimeric transition state of folding exhibiting regions of native-like structure. In particular, there are native-like backbone interactions in the transition state of folding of the Arc repressor [21–23]. Mutating a cluster of buried charged residues forming a complex salt bridge network into hydrophobic residues speeds up folding, makes the folding rate viscosity dependent and brings the structure of the dimeric transition state closer to the unfolded state [24].

The leucine zipper domain of the yeast transcriptional activator protein GCN4 forms a coiled-coil structure in which two α-helices are wound around each other to form a superhelix [25]. This leucine zipper domain of GCN4 folds by a two-state mechanism. However, calorimetric analysis indicates two monomolecular pretransitions at 20 °C and 50 °C before the main concentration-dependent, bimolecular transition takes place [26]. Another coiled coil whose folding has been analyzed in much detail is the heterodimeric leucine zipper AB [27, 28]. This short, designed leucine zipper consists of two different 30 residue peptide chains whose sequences

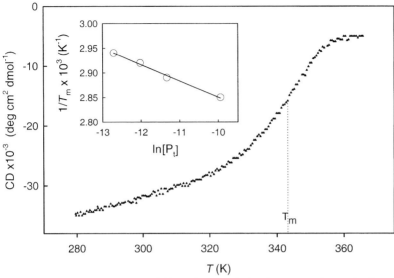

Fig. 27.3. Temperature-induced unfolding of the heterodimeric leucine zipper AB. The unfolding transition was monitored by the change in the CD signal at 222 nm. From $T_m = 343$ K and $[P_t] = 3.9 \times 10^{-5}$ M, the dissociation constant K_d of 1×10^{-5} M at T_m is calculated, corresponding to the free energy of unfolding of $\Delta G_U = 32.9$ kJ mol^{-1}. To calculate K_d at any temperature T, the pre- and post-transition slopes (thin lines) have to be taken into account. Figure adapted from Ref. [27] with permission.

are related to the GCN4 leucine zipper [25]. One chain is acidic (A-chain) and the other is basic (B-chain) so that inter-subunit electrostatic interactions are maximized [27, 29]. Figure 27.3 shows thermal unfolding of leucine zipper AB. The sigmoidal shape is described by Eq. (5) for two-state unfolding of a heterodimeric protein. The midpoint of the thermal transition yields the transition temperature T_m where half of the protein is dimeric and half is monomeric. The enthalpy of unfolding at T_m, $\Delta H_{U,m}$, is obtained from the data in Figure 27.3 by plotting $\ln K$ versus T according to the van't Hoff equation (Eq. (A14)). A further test for two-state unfolding is provided by a plot of $1/T_m$ against the logarithm of the total protein concentration, as shown in the inset of Figure 27.3 (Eq. (A18)).

To obtain the unfolding free energy, $\Delta G_U(T)$, at any temperature T, the data from the thermal unfolding curve of Figure 27.3 need to be extrapolated by the Gibbs-Helmholtz equation describing the change of ΔG_U with temperature.

$$\Delta G_U(T) = \Delta H_{U,m}\left(1 - \frac{T}{T_m}\right) + \Delta C_P\left[T - T_m - T\ln\left(\frac{T}{T_m}\right)\right]$$
$$- RT\ln[K_d(T_m)] \tag{6}$$

The last term of Eq. (6) accounts for the concentration dependence of ΔG_U. This term is missing in the equation used for monomeric proteins. Figure 27.4 shows

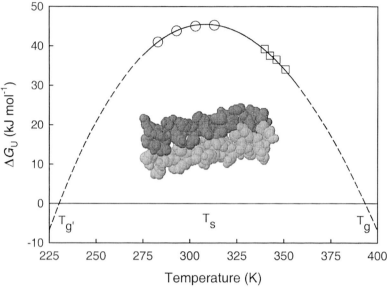

Fig. 27.4. Stability curve of the heterodimeric leucine zipper AB at pH 7.2 and 15 mM ionic strength. The curve was calculated by fitting experimental values of ΔG_U obtained by thermal unfolding (squares) and isothermal titration calorimetry (circles) with the help of Eq. (6). The maximum stability is at $T_s =$ 36 °C, $\Delta G_U = 0$ is at $T_g = 120$ °C and $T_{g'} = -42$ °C. The solid line is a best fit in the range of the experimental data, the dashed line is the extrapolation to T_g and $T_{g'}$. Surface contour model of a leucine zipper is from 2ZTA.pdb. Figure adapted from Ref. [27] with permission.

the thermal stability curve of the heterodimeric leucine zipper AB. The curve was constructed using data sets from thermal unfolding in the range 340–350 K (T_m was varied with protein concentration) and from isothermal titration calorimetry performed between 283 and 313 K. Since neither peptide chain alone associates to a homodimer, the formation of the heterodimeric "AB-zipper" can be followed by titrating one chain to the other by isothermal titration calorimeter. The experiment yields the enthalpy and free energy changes of the coupled association and folding reaction [27]. In this way, the data from thermal unfolding can be complemented with thermodynamic data pertaining to a temperature range in which the folded protein dominates. The stability of the leucine zipper peaks at 309 K (36 °C) and decreases above and below this temperature. This is an example of cold denaturation visible already at ambient temperature. Extrapolation of the stability curve yields the temperatures of heat and cold denaturation, T_g and T'_g, at which $\Delta G_U = 0$.

The shape of the stability curve is governed by the heat capacity change, ΔC_P, of the unfolding reaction. ΔC_P of 4.3 kJ M^{-1} K^{-1} is obtained from a fit of the data in Figure 27.4 using Eq. (6). The positive value of ΔC_P implies that the unfolded monomeric peptide chains can take up more heat per degree K than the folded leucine zipper. The specific ΔC_P normalized per g of protein is 0.6 J g^{-1} K^{-1}. This is of the same magnitude as the average specific ΔC_P for the unfolding of monomeric globular proteins [30] and points to similar types of inter- and intramolecular forces stabilizing dimeric and monomeric proteins. There have been many attempts to calculate ΔC_P from structural data based on the amount of polar and nonpolar surface buried at the interface of a dimeric protein [31, 32].

The kinetics of folding and unfolding of the leucine zipper AB are presented in Figure 27.5. This dimeric leucine zipper is characterized by several interheli-

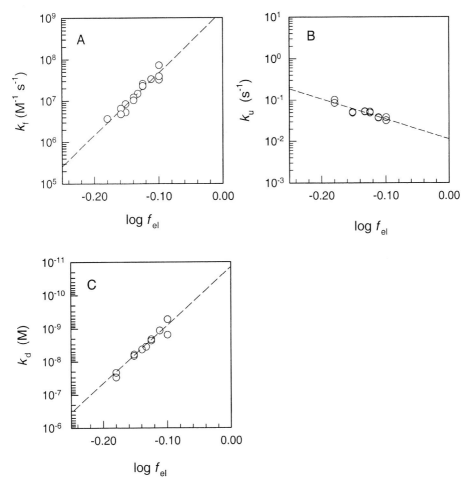

Fig. 27.5. Dependence of the rate constants and equilibrium dissociation constants of folding and association of the leucine zipper AB on ionic strength. A) Rate constant k_f of coupled association and folding. B) Rate constant k_u of coupled dissociation and unfolding. C) Equilibrium dissociation constant calculated as $K_d = k_u/k_f$. The ionic strength is expressed as the logarithm of the mean rational activity coefficient of NaCl [33]. Rate constants were determined under benign buffer conditions by rapid mixing of the unfolded peptide chains. Kinetic traces were analyzed with the help of Eq. (A29). Figure adapted from Ref. [28] with permission.

cal charge–charge interactions (salt bridges). Therefore, the rates of association/folding and dissociation/unfolding depend on ionic strength. The rate of association and folding decreases steeply when the ionic strength is raised (panel A) while dissociation and unfolding is only weakly ionic strength dependent (panel B). Consequently, the equilibrium constant is dominated by the ionic strength effect on the rate of folding (panel C). The strong decrease with ionic strength of the rate of folding could indicate a general effect on the free energy of activation of the folding and association reaction due to long-range electrostatic attraction between the oppositely charged, unfolded polypeptide chains [33]. A small effect of the ionic strength on the rate of unfolding and dissociation has been interpreted by the presence of native-like electrostatic interactions in the transition state of folding [34, 35]. However, in the case of several leucine zippers, interhelical salt bridges contribute negligibly to the stability of the coiled-coil structure, or even destabilize the folded dimer [36–39]. Therefore, the small ionic strength effect on the dissociation rate constant k_u confirms a minor role of electrostatic forces in stabilizing the folded structure [28].

The bimolecular rate constant of association and folding extrapolated to zero ionic strength is very large: $k_f = 9 \times 10^9$ M^{-1} s^{-1}, larger than the rate of about 10^9 M^{-1} s^{-1} estimated for the diffusion-limited association of two short, uncharged polypeptide chains [40–42]. The extremely large value of k_f in the absence of salt points to electrostatic acceleration of the association step as reported for other protein association reactions [33, 43]. Such very rapid association may indicate that the interaction is not much geometrically restricted. In other words, the transition state of folding of the leucine zipper is not well structured, which has been supported by kinetic transition state analysis [44].

The folding and association of leucine zipper AB as well as of other leucine zippers [45] are very tightly coupled and very rapid, yet there seems to be a transitory population of association-competent monomers preceding the actual association-folding reaction [$M_u \rightarrow M_i$ in Eq. (3)]. There is indirect evidenced for a small helix content in the association-competent conformation of the peptide chains forming a leucine zipper [46–49]. The impact of preformed helix on the rate of association and folding of a coiled coil has been manipulated by engineering a metal-binding site into the helix sequence to stabilize the helix conformation [50].

27.3.2
Folding of Dimeric Proteins through Intermediate States

As noted above, two-state folding is an operational concept since the lack of observable intermediates (or ensembles of intermediates) may have experimental reasons. Detection of intermediates depends on their stability as well as on the sensitivity and time resolution of thermodynamic and kinetic detection methods. Table 27.3 lists representative examples of dimeric proteins folding by a multistate mechanism, that means through at least one detectable intermediate. A good indication for a thermodynamic intermediate is the disparity between the free energy of unfolding determined by different methods. For example, the dimeric cAMP receptor

Tab. 27.3. Examples of dimeric proteins folding through intermediate states.

Protein	Mechanism	Comments[a]
cAMP receptor protein (CPR) (homodimer) [51]	$2M_u \rightarrow 2M_i \rightarrow D_f$	Thermodynamically stable monomeric intermediate. $\Delta G_U = 30$ kJ mol^{-1} for $D_f \rightarrow 2M_i$ and 50 kJ mol^{-1} for $2M_i \rightarrow 2M_u$. Equilibrium transition between folded dimer and monomeric intermediate from sedimentation equilibrium, overall unfolding and dissociation from unfolding by denaturant
Disulfoferredoxin (homodimer) [74]	$2M_u \rightarrow 2M_i \rightarrow D_f$	$\Delta G_U = 23$ kJ mol^{-1} for $D_f \rightarrow 2M_i$ and 50 kJ mol^{-1} for $2M_i \rightarrow 2M_u$
LC8, light chain of dynein (homodimer) [75]	$2M_u \rightarrow 2M_i \rightarrow D_f$	$G_U = 35$ kJ mol^{-1} for $D_f \rightarrow 2M_i$ and 62 kJ mol^{-1} for $2M_i \rightarrow 2M_u$
SecA dimeric bacterial ATPase (homodimer) [52]	$2M_u \rightarrow D_i \rightarrow D_f$	Plateau in denaturant unfolding curve indicates intermediate shown to be dimeric by sedimentation equilibrium. $\Delta G_U = 35$ kJ mol^{-1} for $D_f \rightarrow D_i$ and 59 kJ mol^{-1} for $D_i \rightarrow 2M_u$
Ketosteroid isomerase (homodimer) [53]	$2M_u \rightarrow 2M_i \rightarrow D_i \rightarrow D_f$	Two-state folding under equilibrium conditions with $\Delta G_U = 90$ kJ mol^{-1}. Four-state folding in kinetic analysis with $k = 60$ s^{-1} for $M_u \rightarrow M_i$, $k = 5.4 \times 10^4$ M^{-1} s^{-1} for $2M_i \rightarrow D_i$, $k = 0.017$ s^{-1} for $D_i \rightarrow D_f$
Glutathione transferase A1-1 (homodimer) [76, 77]	$2M_u \rightarrow 2M_i \rightarrow D_i \rightarrow D_i'$ $\rightarrow D_f$	Under equilibrium conditions folding is two-state [77]. Folding through a monomeric and two dimeric intermediates; folding pathway complicated by proline *cis* $\rightarrow$ *trans* isomerization in M_u and M_i [76].
Trp repressor protein (TRP) (homodimer) [54, 58]	$2M_u \rightarrow 2M_i \rightarrow D_i \rightarrow D_f$	Rapid transition to two classes of association-competent monomers which associate in a combinatorial way to three different dimeric intermediates, and these rearrange in the rate-limiting step to the folded dimer [58]. Under equilibrium conditions, folding appears two state [54]
Bacterial luciferase (heterodimer) [59]	$M_{\alpha u} \rightarrow M_{\alpha i}$, $M_{\beta u} \rightarrow M_{\beta i}$ $M_{\alpha i} + M_{\beta i} \rightarrow D_i \rightarrow D_f$	Under equilibrium conditions folding is two state. Folding proceeds through monomeric and dimeric intermediates. The β-subunit can irreversibly form monomeric and dimeric dead-end species

[a] Unless stated, thermodynamic and kinetic parameters refer to ambient temperature (20–25 °C).

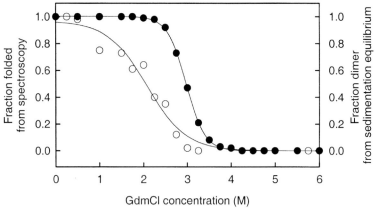

Fig. 27.6. Guanidinium chloride-induced unfolding of the homodimeric cAMP receptor protein (CPR) followed by different spectroscopic methods shows a sharp transition at 3 M GdmCl (filled circles, left ordinate). In contrast, the monomer/dimer ratio followed by sedimentation equilibrium analysis has a transition midpoint at 2 M GdmCl (empty circles, right ordinate). The latter midpoint is assigned to an equilibrium between folded dimer and (partly folded) monomeric intermediate. The midpoint at 3 M GdmCl is assigned to the equilibrium between the fully folded dimer and the fully unfolded monomers. Figure redrawn from Ref. [51] with permission.

protein (CRP) exhibits different unfolding curves depending on whether unfolding is observed by a spectroscopic probe or by sedimentation equilibrium analysis in the ultracentrifuge (Figure 27.6). The midpoint of the denaturant-induced unfolding curve obtained by measuring the monomer/dimer distribution in the ultracentrifuge is at 2 M GdmCl. However, the midpoint shifts to 3 M GdmCl when unfolding and dissociation is followed from the change of the far-UV CD spectrum, from fluorescence emission or from fluorescence anisotropy. The higher midpoint corresponds to unfolding and dissociation, the lower midpoint to dissociation alone, which does not significantly change the CD and fluorescence signals [51]. Thus, CRP folds through a thermodynamically stable monomeric intermediate. The overall free energy of unfolding and dissociation is $\Delta G_U = 80$ kJ mol^{-1}, that for dissociation alone (from ultracentrifugation) is $\Delta G_U = 30$ kJ mol^{-1}. The difference of 50 kJ mol^{-1} is the free energy of unfolding of the stable monomeric intermediate (Table 27.3).

The fourth protein in Table 27.3, dimeric bacterial enzyme SecA, folds through a thermodynamically stable dimeric intermediate, which appears as a plateau in the denaturant unfolding curve [52]. A larger amount of energy is necessary for the unfolding and dissociation reaction $D_i \rightarrow 2M_u$ than for unfolding of D_f into D_i, in accord with a thermodynamically stabilized dimeric intermediate. In the kinetics of folding, the monomolecular transition $D_i \rightarrow D_f$ is rate-determining since even at low protein concentration the bimolecular association $2M_u \rightarrow D_i$ is very rapid [52].

The first four proteins of Table 27.3 fold through thermodynamically stable intermediates. In many cases, however, the equilibrium distribution between folded dimer and unfolded monomer is well described by a two-state mechanism yet transitory intermediates are detectable in the kinetic analysis. Representative examples in Table 27.3 are the ketosteroid isomerase, the glutathione transferase A1-1 and the tryptophan repressor protein (TRP). Equilibrium analysis of all these proteins conforms to two state folding and fulfills at least one of criteria i–iv but none of criteria v–viii listed in Table 27.1. Thus, dimeric ketosteroid isomerase is stabilized by $\Delta G_U = 90$ kJ mol^{-1} under equilibrium conditions [53]. However, the time course of unfolding exhibits three kinetic phases, a slow monomolecular phase $D_f \rightarrow D_i$, a bimolecular phase $D_i \rightarrow 2M_i$, and a rapid monomolecular phase $M_i \rightarrow M_u$. Thus, two additional states are seen in kinetics. One may say that the simple Eq. (1) applies to the enzyme at equilibrium yet the more realistic Eq. (3) to the time-dependent nonequilibrium situation of folding and unfolding.

Folding of TRP has been studied in much detail [54–58]. Matthews and co-workers have formulated the folding scheme shown in Figure 27.7 based on a large

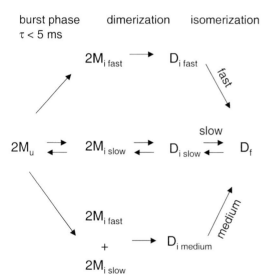

Fig. 27.7. Complex folding scheme for tryptophan repressor protein (TRP) from *Escherichia coli* through three "folding channels." Following the formation of two ensembles of partly folded monomeric intermediates, $M_{i\,fast}$ and $M_{i\,slow}$, folding to the dimer is controlled by three kinetic phases ("channels"). These are due to the association of two molecules of $M_{i\,fast}$ or $M_{i\,slow}$ or of $M_{i\,fast} + M_{i\,slow}$. "Fast," "slow," and "medium" refer to the rate of the final conformational rearrangement from the dimeric intermediate populations $D_{i\,fast}$, $D_{i\,slow}$, and $D_{i\,medium}$ to the unique folded dimer D_f. $D_{i\,slow}$ is a native-like intermediate and $D_{i\,fast}$ and $D_{i\,medium}$ are nonnative intermediates. Only the slow folding reaction through native-like intermediate $D_{i\,slow}$ is reversible. Figure adapted from Ref. [58] with permission.

series of kinetic experiments. In a burst phase following the removal of denaturant, unfolded monomeric TRP rearranges to two different ensembles of partly folded monomers, $M_{i\ fast}$ and $M_{i\ slow}$. These rapidly combine to the three dimeric intermediates $D_{i\ fast}$, $D_{i\ slow}$, and $D_{i\ medium}$. The unique folded dimer is formed in a fast, a medium, and a slow reaction from $D_{i\ fast}$, $D_{i\ medium}$, and $D_{i\ slow}$, respectively. The rate of this final unimolecular reaction exhibits very little dependence on denaturant concentration suggesting an isomerization reaction proceeding with no detectable change in solvent-accessible surface area. The structural basis for the three "folding channels" is the presence of two prefolded monomeric species that randomly associate in different pairwise combinations. The intermediates differ in their compactness. The fast and medium channels access the native structure through the most compact yet least native-like intermediates, $D_{i\ fast}$ and $D_{i\ medium}$, respectively. The slow channel, on the other hand, proceeds through a more "direct" folding route since the dimeric intermediate $D_{i\ slow}$ is native-like. Only this slow route is reversible. The example of TRP shows that the native structure can be accessed through alternative routes, that folding through nonnative states can accelerate folding, and that nonnative states need not lead into kinetic traps [58].

Kinetic traps have been detected in the folding of bacterial luciferase, the last example in Table 27.3. Figure 27.8 shows the complex folding scheme for this heterodimeric enzyme composed of an α- and a β-subunit. The unfolded monomeric subunits $M_{\alpha u}$ and $M_{\beta u}$ are converted to the monomeric intermediates $M_{\alpha i}$ and $M_{\beta i.}$, which associate to the heterodimeric intermediate D_i in a bimolecular reaction followed by the monomolecular rearrangement to the final folded dimer D_f. However, the β-subunit can enter into a dead-end trap as indicated by the irreversible reactions $M_{\beta i} \rightarrow M_{\beta x}$ and $2M_{\beta i} \rightarrow D_{\beta x}$ where $D_{\beta x}$ is a dead-end homodimer of the β-subunit [59].

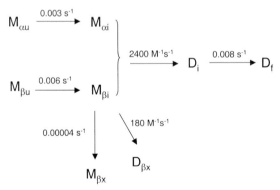

Fig. 27.8. Complex folding scheme for the heterodimeric bacterial luciferase composed of an α-subunit and a β-subunit. Folding proceeds through the monomeric intermediates $M_{\alpha i}$ and $M_{\beta i}$ and the dimeric intermediate D_i. Monomeric ($M_{\beta x}$) and dimeric ($D_{\beta x}$) dead-end species formed from the β-subunit intermediate $M_{\beta i}$ slow the folding and reduce the folding yield. Figure adapted from Ref. [59] with permission.

27.4
Trimeric and Tetrameric Proteins

We only briefly discuss a few examples of small trimeric and tetrameric proteins and refer to Chapter 2 in Part II presenting the folding and assembly in vivo of large, higher order multisubunit proteins. For statistical reasons, any reaction higher than second order is very slow. Therefore, simultaneous association of three polypeptide chains to a trimer is statistically unlikely, and simultaneous association of four chains to a tetramer is impossible. Thus, trimers (Tr) and tetramers (Te) are most likely to form by consecutive bimolecular reactions. For a trimer one can write:

$$3M \underset{k_{-1}}{\overset{k_1}{\rightleftarrows}} D + M \underset{k_{-2}}{\overset{k_2}{\rightleftarrows}} Tr$$

$$K_{d1} = \frac{[M]^2}{[D]} = \frac{k_{-1}}{k_1} \quad K_{d2} = \frac{[M][D]}{[Tr]} = \frac{k_{-2}}{k_2} \quad \text{two-step folding of trimer} \quad (7)$$

Subscripts u, f, and i have been omitted in Eq. (7) to leave open the conformational state of the three species. Of course, there can be monomeric, dimeric, and trimeric intermediates, making the folding mechanism very complex.

Three-state or two-step folding described by Eq. (7) has been reported for the designed homotrimeric coiled-coil protein LZ16A composed of three 29-residue peptide chains [60]. In the range of 1–100 µM total protein concentration, monomer, dimer, and trimer are populated in a concentration-dependent equilibrium [61]. Figure 27.9 shows the decrease of the CD spectral minimum as a function of total protein concentration. The data are best fit by two equilibria with $K_{d1} = 7.7 \times 10^{-7}$ M and $K_{d2} = 2.9 \times 10^{-6}$ M, where K_{d1} refers to the monomer/dimer equilibrium and K_{d2} to the dimer/trimer equilibrium. $K_{d1} < K_{d2}$ means that the dimer is a stable intermediate dominating at lower protein concentration. For example, at 10 µM total protein concentration, there is 50% dimer, 35% trimer, and 15% monomer. From kinetic refolding experiments the four rate constants defined by Eq. (7) have been determined and were found to be in agreement with the equilibrium data (see the legend to Figure 27.9).

So it may seem surprising that a trimeric protein can apparently fold in a single-step reaction, which means by a two-state folding mechanism according to Eq. (8).

$$3M_u \underset{k_u}{\overset{k_f}{\rightleftarrows}} Tr_f, \quad K_d = \frac{[M_u]^3}{[Tr_f]} = \frac{k_u}{k_f} \quad \text{one-step folding of trimer} \quad (8)$$

Here, k_f is a trimolecular association rate constant ($M^{-2} s^{-1}$) and k_u is a monomolecular dissociation rate constant. One-step folding according to Eq. (8) occurs if the dimer (Eq. (7)) is very unstable and scarcely populated so that the first equilibrium of Eq. (7) is far on the side of the unfolded monomer and the second far on the side of the folded trimer. Under these conditions, trimer formation appears

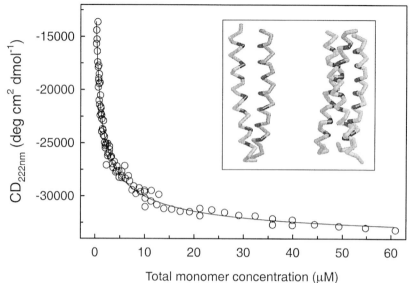

Fig. 27.9. Concentration dependence of the CD signal at 222 nm of the trimeric coiled coil LZ16A. The solid line is a best fit for a monomer–dimer–trimer equilibrium according to Eq. (7) with $K_{d1} = 7.7 \times 10^{-7}$ M and $K_{d2} = 2.9 \times 10^{-6}$ M. These values are in good agreement with the K_1 and K_2 calculated from the rate constants $k_1 = 7.8 \times 10^4$ M^{-1} s^{-1}, $k_{-1} = 1.5 \times 10^{-2}$ s^{-1}, $k_2 = 6.5 \times 10^5$ M^{-1} s^{-1}, and $k_{-2} = 1.1$ s^{-1}. Inset: backbone structure of dimer and trimer; ribbon models are from 2ZTA.pdb (dimer) and 1gcm.pdb (trimer). Figure adapted from Ref. [60] with permission.

two-state. It should be noted that 3M → Tr can appear two-state despite the presence of a dimeric intermediate if 2M → D is spectroscopically silent under the chosen detection conditions so that the entire spectral signal change is caused by M + D → Tr [60].

An example of a homotrimeric protein folding through only a single reaction step is the six-helix bundle at the core of the gp41 envelope protein of the human and simian immunodeficiency viruses [62]. The structure is shown in Figure 27.10 and is composed of three "hairpins" formed by two α-helices linked through a short peptide loop [63]. The trimer folds from three unfolded monomers in a single step according to Eq. (8). Equilibrium and kinetic data are in excellent agreement: $\Delta G_U(H_2O)$ determined by denaturant unfolding is 116 kJ mol^{-1}, and 114 kJ mol^{-1} when calculated from kinetic rate constants. The transition state of folding seems closer to the unfolded than the folded state. Only 20–40% of the surface buried in the folded trimer is already buried in the transition state of folding [62].

Formation of the catalytic trimer of aspartate transcarbamoylase is at the other extreme of possible folding mechanisms. Here, partly folded monomers are rapidly formed initially and thereafter rearrange in a slow step to association-competent monomers. These monomeric intermediates are thermodynamically stable. They

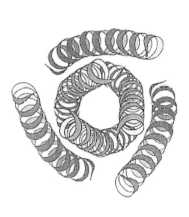
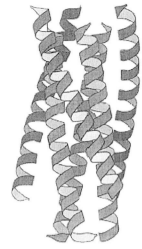

Fig. 27.10. Crystal structure of a six-helix bundle composed of three "hairpins." Each "hairpin" comprises two helices linked by a short loop; the loop is not visible in this crystal structure from Ref. [63]. Left: Axial view looking down the threefold axis of the helix bundle. Right: Lateral view. The trimer folds from unfolded monomers in a single step according to Eq. (8). $\Delta G_U(H_2O)$ determined by denaturant unfolding is 116 kJ mol^{-1}, and 114 kJ mol^{-1} when calculated from kinetic rate constants. The trimolecular rate constant of association is 1.3×10^{15} M^{-2} s^{-1} and the unimolecular rate constant of dissociation is 1.08×10^{-5} s^{-1} [62].

associate through at least two steps to the folded trimeric enzyme [64]. The folding scheme is $3M_u \rightarrow 3M_i \rightarrow \rightarrow Tr_f$.

Finally, we consider two tetrameric proteins: the C-terminal 30-residue domain of the tumor suppressor protein p53 [65] and a small bacterial dihydrofolate reductase composed of four 78-residue polypeptide chains [66]. The C-terminal p53 domain folds into a homotetramer in a thermodynamically fully reversible reaction. Interestingly, under equilibrium conditions, only folded tetramer and unfolded monomer are observed [67]. However, the time course of folding shows two consecutive bimolecular reactions with two kinetic, dimeric intermediates (Figure 27.11) [65]. Folding can be described as an association of dimers, in accord with the crystal structure, which has been likened to a dimer of dimers. The two intermediates shown in Figure 27.11 were deduced from extensive Φ-value analysis [65]. The first bimolecular association reaction is between highly unstructured monomers and can be described by a "nucleation-condensation" mechanism [68]. The first dimeric intermediate $D_{i,1}$ is formed through an early transition state that has little similarity to the native structural organization of the dimer. The second dimeric intermediate $D_{i,2}$ is highly structured and already includes most of the structural features of the native protein. The transition $D_{i,2} \rightarrow Te_f$ is an example of a "framework mechanism" in which stable preformed structural elements "diffuse and collide" [68].

A final example of a tetrameric protein is the bacterial R67 dihydrofolate reduc-

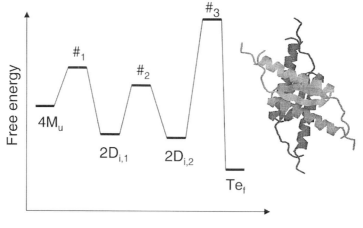

Fig. 27.11. Qualitative reaction coordinate diagram for the folding of the tetrameric protein domain from tumor suppressor p53. Under equilibrium conditions, only a single transition between unfolded monomer (M_u) and folded tetramer (Te_f) is seen [67]. Kinetic refolding and unfolding experiments show a single bimolecular association reaction and a single monomolecular dissociation reaction, which could be assigned as indicated [65]. Ribbon model of tetrameric C-terminal domain of tumor suppressor p53 from sak.pdb. See the text for a more detailed discussion. Figure adapted from Ref. [65] with permission.

tase [66]. Here, two monomeric intermediates (the formation of one being due to slow isomerization of peptidyl-proline bonds) precede dimer formation. Thereafter, two dimers associate to the folded tetramer. The folding reaction can be summarized as $4M_u \rightarrow 4M_{i,1} \rightarrow 4M_{i,2} \rightarrow 2D_i \rightarrow Te$.

27.5
Concluding Remarks

The mechanisms of folding of small oligomeric proteins is highly variable. At one extreme is the fully concerted association of unfolded monomers to folded dimers, trimers and even tetramers. Such folding without a thermodynamically stable intermediate is surprisingly frequent. Even in the case of the tetrameric p53 domain, whose folding proceeds through several kinetic folding intermediates among which the final dimeric intermediate is very native-like, none of the intermediates is sufficiently stable to be noticed at equilibrium. The reason for such scarcity of stable folding intermediates may be due, in part, to destabilizing nonpolar surface exposed in the transitory intermediates. This surface becomes buried only in the final folded oligomer. Indeed, intermolecular or intersubunit forces dominate the stability of many oligomeric proteins. However, the interface between the subunits of the folded oligomeric proteins is highly variable. Shape complementarity between the contacting interfaces is a common feature and a main contributor to the specificity of intersubunit interactions. In a survey of 136 homodimeric proteins, a mixture of hydrophobic and intersubunit contacts was identified [69]. Small hydrophobic patches, polar interactions and water molecules are scattered over the intersubunit contact area in two-thirds of the proteins studied. Only in one-third of the dimers there was a single, large and contiguous hydrophobic core between the subunits. Though the hydrophobic effect plays an important role in subunit interactions, it seems to be less strong than that observed in the interior of monomeric proteins [70]. Only in a few cases, oligomerization conforms to the association of preformed stable subunits. The case of the trimeric coiled-coil LZ16A at equilibrium with a stable dimeric coiled coil is probably an exception.

From studies addressing the nature of the transition states of folding, it follows that the same principles deduced for monomeric proteins [71] apply also to oligomers. Folding of the subunits and their concurrent association to the oligomeric protein occurs by diffusion and collision of preformed structural frameworks ("framework model") as well as by simultaneous nucleation and condensation. This is not surprising since the folded structure is always based on an extensive interplay of secondary and tertiary interactions in monomeric proteins and of secondary, tertiary, and quaternary interactions in oligomeric proteins. Because there is no intrinsic difference between tertiary and quaternary interactions there is no principal difference of folding between monomers and oligomers. The main difference is in the association step. Whereas the folding of a monomeric protein can be described by one or several consecutive or parallel monomolecular (single exponential) reactions, the folding of an oligomeric protein exhibits at least one concentration-dependent reaction step. On the one hand this feature complicates the thermodynamic and kinetic analysis. On the other hand the concentration dependence adds a useful variable to assist in the measurement of folding since the monomer/oligomer equilibrium can be studied by simple dilution under benign buffer conditions. Comparison of the data from dilution and from chemical or heat denaturation can give information about the contribution of intermolecular and intramolecular forces to the stability of the oligomeric protein. Also, determination of thermodynamic parameters such as the free energy of unfolding or the heat capacity change of unfolding is facilitated when the protein concentration can be adduced as an additional variable. Finally, in the case of heterooligomeric proteins, folding can be probed by simple mixing of the different subunits.

Appendix – Concurrent Association and Folding of Small Oligomeric Proteins

In this Appendix we have put together a set of equations for the analysis of thermodynamic and kinetic data of two-state folding. Many equations are equivalent to those for the folding of monomeric proteins except for an additional term accounting for protein concentration. Derivation of equilibrium association constants for two-state folding of oligomeric systems and the relationship between equilibrium constants and the thermodynamic parameters $\Delta G, \Delta H$, and ΔS have been described in more detail by Marky and Breslauer [78]. The folded (native) state is taken as the reference state so that the thermodynamic parameters refer to the unfolding reaction ($\Delta G_U, \Delta H_U$, etc.). A special form of Eq. (A29) describing reversible two-state folding of a homodimer has been published by Milla and Sauer [20]. The method of deducing kinetic rate constants from the shift of a pre-existing equilibrium is briefly summarized (method of Bernasconi [16]).

The kinetic and equilibrium equations presented here are useful for the analysis of simple reaction steps, or steps that are kinetically or thermodynamically well separated. If the reaction steps are linked, numerical integration can be used to analyze kinetic and equilibrium data. There are good computer programs to aid kinetic analysis, some of the programs are freely available on the Web (for example DynaFit [79]). Depending on the scatter of experimental data, numerical fitting of linked reactions may result in large errors and distinguishing between different models can be very difficult.

Numbering of equations Equations appearing only in this Appendix have the prefix A. Other equations are numbered as in the main text.

Abbreviations M, monomer; D, dimer; N, n-mer; [P_t], total protein concentration expressed as total concentration of monomer; f_M, fraction of free monomer defined as $[M]/[P_t]$. Subscripts; u, unfolded; i, intermediate; f, folded; m, midpoint; K_d, equilibrium dissociation constant ($= 1/K$).

A1
Equilibrium Constants for Two-state Folding

Association and dissociation of oligomeric proteins results in concentration-dependent equilibria. Expressions for the equilibrium constants of homooligomeric and heterooligomeric proteins are different.

A1.1 Homooligomeric Protein
For a homooligomeric protein, the equilibrium transition between n identical, unfolded monomeric subunits M and the folded n-meric protein N is described by

$$nM \rightleftarrows N \tag{A1}$$

The equilibrium dissociation constant K_d is defined as

$$K_d = \frac{[M]^n}{[N]} \tag{A2}$$

Defining the fraction of free monomeric subunits as

$$f_M = \frac{[M]}{[P_t]} \quad (0 \le f_M \le 1) \tag{A3}$$

we can write K_d as

$$K_d = \frac{n[P_t]^{n-1} f_M^n}{1 - f_M} \tag{A4}$$

where $[P_t] = [M] + n[N]$ is the total protein concentration expressed as total monomer concentration. At $f_M = 0.5$ we have

$$K_{d, f_M = 0.5} = n([P_t]/2)^{n-1} \tag{A5}$$

A1.2 Heterooligomeric Protein

For a heterooligomeric protein, the equilibrium transition between n different, unfolded monomeric subunits $M_1, M_2 \ldots M_n$ and the folded n-meric protein N is described by

$$M_1 + M_2 + \cdots + M_n \rightleftarrows N \tag{A6}$$

The equilibrium dissociation constant K_d is defined as

$$K_d = \frac{[M_1][M_2]\ldots[M_n]}{[N]} \tag{A7}$$

The corresponding expression in terms of the fraction f_M of free monomeric subunits is

$$K_d = \frac{[P_t]^{n-1} f_M^n}{n^{n-1}(1 - f_M)} \tag{A8}$$

and at $f_M = 0.5$

$$K_{d, f_M = 0.5} = \left(\frac{[P_t]}{2n}\right)^{n-1} \tag{A9}$$

In Eqs (A8) and (A9), $[P_t] = [M_1] + [M_2] + \cdots + [M_n] + n[N]$. The difference between Eqs (A4), (A5) and (A8), (A9) reflects the statistical difference between homooligomeric and heterooligomeric equilibria. Equations (4) and (5) of the main text correspond to Eqs (A4) and (A8), respectively.

A2
Calculation of Thermodynamic Parameters from Equilibrium Constants

A2.1 Basic Thermodynamic Relationships

The changes of free energy, enthalpy, and entropy are related by the well-known equation

$$\Delta G_U = \Delta H_U - T\Delta S_U = -RT \ln K_d \tag{2}$$

The superscript ° indicating standard conditions is omitted for simplicity. The van't Hoff enthalpy of unfolding is

$$\Delta H_U^{vH} = RT^2 \left(\frac{d \ln K_d}{dT} \right) \tag{A10}$$

The heat capacity change at constant pressure, ΔC_P, is defined by

$$\Delta C_P = \left(\frac{\partial \Delta H}{dT} \right)_P \tag{A11}$$

The calculation of thermodynamic parameters from equilibrium constants is based on Eqs (2), (A10), and (A11).

A2.2 Linear Extrapolation of Denaturant Unfolding Curves of Two-state Reaction

Linear extrapolation is performed by the same procedure as applied to denaturant unfolding of monomeric proteins, except that the concentration has to be taken into account. The linear extrapolation method is discussed in Chapter 3. Linear extrapolation of the free energy of two-state unfolding of homooligomeric and heterooligomeric proteins is described by

$$-RT \ln \left(\frac{n[P_t]^{n-1} f_M^n}{1 - f_M} \right) = \Delta G_U(H_2O) - m[\text{denaturant}] \quad \text{homooligomer} \tag{A12}$$

$$-RT \ln \left(\frac{[P_t]^{n-1} f_M^n}{n^{n-1}(1 - f_M)} \right) = \Delta G_U(H_2O) - m[\text{denaturant}] \quad \text{heterooligomer} \tag{A13}$$

The term on the left hand side of Eq. (A12) follows from Eqs (2) and (A4), that on the left hand side of Eq. (A13) from Eqs (2) and (A8). The slope m is the change of ΔG_U with denaturant concentration (J mol^{-1} M^{-1}).

A2.3 Calculation of the van't Hoff Enthalpy Change from Thermal Unfolding Data

From the change of K_d with T, the van't Hoff enthalpy is calculated according to Eq. (A10). In practice, one calculates ΔH_U^{vH} at T_m from a plot of $\ln K_d$ against $1/T$. With Eq. (2) one obtains the Arrhenius equation

$$\ln K_d = -\frac{\Delta H_{U,m}}{RT} + \frac{\Delta S_U}{R} \qquad (A14)$$

where $\Delta H_{U,m}$ is the van't Hoff enthalpy at T_m. Only data points in a narrow interval around T_m should be used to calculate $\Delta H_{U,m}$ from the slope $\Delta H_{U,m}/R$ since Eq. (A14) neglects the change of the enthalpy with T (heat capacity change assumed to be zero).

A2.4 Calculation of the van't Hoff Enthalpy Change from the Concentration-dependence of T_m

Since association of an oligomeric protein is concentration dependent, T_m changes with concentration. Measuring T_m for varying $[P_t]$ is another way to calculate $\Delta H_{U,m}$. From Eqs (A5) and (A9) it follows that at T_m, K_d is described by

$$K_d(T_m) = n([P_t]/2)^{n-1} \quad \text{homooligomer} \qquad (A15)$$

$$K_d(T_m) = ([P_t]/2n)^{n-1} \quad \text{heterooligomer} \qquad (A16)$$

From Eqs (2), (A15), and (A16) one obtains after rearranging [78]:

$$\frac{1}{T_m} = \frac{(n-1)R}{\Delta H_m} \ln[P_t] + \frac{\Delta S - (n-1)R \ln 2 + R \ln n}{\Delta H_m} \quad \text{homooligomer} \qquad (A17)$$

$$\frac{1}{T_m} = \frac{(n-1)R}{\Delta H_m} \ln[P_t] + \frac{\Delta S - (n-1)R \ln 2n}{\Delta H_m} \quad \text{heterooligomer} \qquad (A18)$$

A2.5 Extrapolation of Thermodynamic Parameters to Different Temperatures: Gibbs-Helmholtz Equation

The calculation of the van't Hoff enthalpy described above refers to the transition temperature T_m. To obtain the thermodynamic parameters at any other temperature, the change of ΔH and ΔS with T has to be taken into account according to the Gibbs-Helmholtz equation:

$$\Delta G_U(T) = \Delta H_{U,m}\left(1 - \frac{T}{T_m}\right) + \Delta C_P\left[T - T_m - T \ln\left(\frac{T}{T_m}\right)\right] - RT \ln[K_d(T_m)] \qquad (6)$$

The term $-RT \ln[K_d(T_m)]$ on the right hand side is necessary to account for the concentration dependence of ΔG_U. $K_d(T_m)$ corresponds to K_d at $f_M = 0.5$. Depending on the number of subunits and on whether the protein is a homo- or a heterooligomer, the appropriate form of K_d from Eq. (A5) or (A9) is inserted in Eq. (6). For example, the correction term for a homodimeric protein is $-RT \ln([P_t])$. The concentration-independent form of the Gibbs-Helmholtz equation is obtained by using the reference temperature T_g at which $\Delta G_U = 0$

$$\Delta G_U(T) = \Delta H(T_g)\left(1 - \frac{T}{T_g}\right) + \Delta C_P\left[T - T_g - T \ln\left(\frac{T}{T_g}\right)\right] \qquad (A19)$$

A3
Kinetics of Reversible Two-state Folding and Unfolding: Integrated Rate Equations

The following integrated rate equations for two-state folding of a dimeric protein are used to calculate the folding rate constant k_f and the unfolding rate constant k_u from kinetic traces. Depending on experimental conditions, only folding (A3.1), only unfolding (A3.2), or both (A3.3) have to be taken into account.

A3.1 Two-state Folding of Dimeric Protein

$$2M \xrightarrow{k_f} D \qquad (A20)$$

The time course of folding is given by

$$\frac{d[M]}{dt} = -k_f [M]^2 \qquad (A21)$$

After integration, the time course of disappearance of the monomer fraction becomes

$$f_M(t) = \frac{[M_0]}{[P_t](k_f t + 1)} \qquad (A22)$$

where $[M_0]$ is the initial monomer concentration at $t = 0$. $[M_0] = [P_t]$ if $[D] = 0$ at the beginning of the folding reaction. Equation (A22) is used to obtain k_f from a folding trace expressed as $f_M(t)$ if the unfolding reaction can be neglected (for example at low denaturant concentration).

A3.2 Two-state Unfolding of Dimeric Protein

$$D \xrightarrow{k_u} 2M \qquad (A23)$$

The time course of unfolding is given by

$$\frac{d[D]}{dt} = -k_u [D] \qquad (A24)$$

After integration, the time course of disappearance of the dimer fraction $f_D = 1 - f_M$ becomes

$$f_D(t) = \frac{2[D_0]}{[P_t]} \exp(-k_u t) \qquad (A25)$$

where $[D_0]$ is the initial dimer concentration at $t = 0$. $[D_0] = [P_t]/2$ if $[M] = 0$ at $t = 0$. Equation (A25) is used to obtain k_u from an unfolding trace expressed as

$f_D(t) = 1 - f_M(t)$ if refolding can be neglected (for example at high denaturant concentration).

A3.3 Reversible Two-state Folding and Unfolding

A3.3.1 Homodimeric protein

$$2M \underset{k_u}{\overset{k_f}{\rightleftarrows}} D \tag{1}$$

The time course of disappearance of the monomer can be written as

$$\frac{d[M]}{dt} = -k_f[M]^2 + 2k_u[D] \tag{A26}$$

which can be recast into

$$\frac{d[M]}{a + b[M] + c[M]^2} = dt \tag{A27}$$

where

$$a = k_u[M_0], \quad b = -k_u, \quad c = -k_f \tag{A28}$$

After integration, the time course of disappearance of the monomer fraction is

$$f_M(t) = \frac{(b+s)(Z-1)}{[P_t]2c(1-Z)} \tag{A29}$$

where $Z = \left(\dfrac{2c[M_0] + b - s}{2c[M_0] + b + s}\right) \exp(st)$ and $s = \sqrt{b^2 - 4ac}$

Equation (A29) is used to obtain k_f and k_u from a folding trace expressed as $f_M(t)$ under conditions where both folding and unfolding have to be taken into account (for example at intermediate denaturant concentration).

A3.3.2 Heterodimeric protein

$$M_1 + M_2 \underset{k_u}{\overset{k_f}{\rightleftarrows}} D \tag{A30}$$

The change of $[M_1]$ with time can be written as

$$\frac{d[M_1]}{dt} = -k_f[M_1][M_2] + k_u[D] \tag{A31}$$

The time course of disappearance of the fraction of M_1 or M_2 is again described by

Eq. (A27) except that the constants a, b, and c are different. If $[M_1] = [M_2]$, a, b, and c are

$$a = k_u[M_0]/2, \quad b = -k_u, \quad c = -k_f \tag{A32}$$

If $[M_1] \neq [M_2]$

$$a = k_u[M_{1,0}], \quad b = k_f([M_{1,0}] - [M_{2,0}]) - k_u, \quad c = -k_f \tag{A33}$$

or

$$a = k_u[M_{2,0}], \quad b = k_f([M_{2,0}] - [M_{1,0}]) - k_u, \quad c = -k_f \tag{A34}$$

Equation (A33) is valid for $f_{M1}(t)$ and (A34) for $f_{M2}(t)$; subscript 0 indicates concentration at $t = 0$.

A4
Kinetics of Reversible Two-state Folding: Relaxation after Disturbance of a Pre-existing Equilibrium (Method of Bernasconi)

Rate equations can be simplified if the reaction corresponds to a relaxation after a relatively small disturbance of the equilibrium. The procedure, pioneered by Bernasconi [16], allows rate constants to be obtained in those cases where there is no analytical solution of the differential rate equation. In practise, a pre-existing equi-

Tab. 27.4. Relaxation times for various reversible folding reactions.

Reaction	Equation for relaxation time[a]	
$2M \underset{k_u}{\overset{k_f}{\rightleftarrows}} D$	$\dfrac{1}{\tau} = 4k_f[\bar{M}] + k_u$	(A35)
$nM \underset{k_u}{\overset{k_f}{\rightleftarrows}} N$	$\dfrac{1}{\tau} = n^2 k_f[\bar{M}]^{n-1} + k_u$	(A36)
$M_1 + M_2 \underset{k_u}{\overset{k_f}{\rightleftarrows}} D$	$\dfrac{1}{\tau} = k_f([\bar{M}_1] + [\bar{M}_2]) + k_u$	(A37)
$2M \underset{k_{-1}}{\overset{k_1}{\rightleftarrows}} D_i \underset{k_{-2}}{\overset{k_2}{\rightleftarrows}} D$	$\tau_1 < \tau_2$: bimolecular association is faster than monomolecular rearrangement $\dfrac{1}{\tau_1} = 4k_1([\bar{M}]) + k_{-1}, \quad \dfrac{1}{\tau_2} = \dfrac{4k_1 k_2 [\bar{M}]}{4k_1([M]) + k_{-1}} + k_{-2}$	(A38)
$2M \underset{k_{-1}}{\overset{k_1}{\rightleftarrows}} D_i \underset{k_{-2}}{\overset{k_2}{\rightleftarrows}} D$	$\tau_1 > \tau_2$: bimolecular association is slower than monomolecular rearrangement $\dfrac{1}{\tau_1} = k_1 + k_{-1}, \quad \dfrac{1}{\tau_2} = 4k_1[\bar{M}] + \dfrac{k_{-1} k_{-2}}{k_2 + k_{-2}}$	(A39)

[a] Concentrations with overbars are equilibrium concentrations after relaxation to the new equilibrium.

librium between folded and unfolded protein is rapidly disturbed, for example by rapid dilution or rapid increase of temperature (T-jump), and the subsequent relaxation to the new equilibrium is observed. This relaxation is characterized by the relaxation time $\tau = 1/k_{app}$, where k_{app} (s^{-1}) is the apparent rate constant of the exponential decay to the new equilibrium. The number of relaxation times equals the number of reaction steps. Table 27.4 lists a few equations for different reversible folding reactions. They describe the relationship between the experimentally measured relaxation time τ and the individual rate constants of the reversible folding–unfolding reaction. The main difficulty with this procedure is that the equilibrium concentrations of the reactants at the new equilibrium, that is after relaxation, have to be known (concentrations with overbars in Table 27.4). These concentrations can be calculated if the equilibrium constants are known from independent experiments. Alternatively, a fitting procedure can be used in which initial estimates of the equilibrium constants are iteratively adapted until a best fit is obtained.

Acknowledgments

I thank Ilian Jelesarov and Daniel Marti for fruitful discussions and helpful comments on the manuscript. Work from my own laboratory has been supported in part by the Swiss National Science Foundation.

References

1 D. S. GODSELL and A. J. OLSON, *Trends Biochem. Sci.* **1993**, *18*, 65–68.
2 K. A. DILL, *Biochemistry* **1990**, *29*, 7133–7155.
3 W. KAUZMANN, *Adv. Protein Chem.* **1959**, *14*, 1–63.
4 C. N. PACE, *Biochemistry* **2001**, *40*, 310–313.
5 R. JAENICKE and H. LILIE, *Adv. Protein Chem.* **2000**, 53.
6 K. E. NEET and D. E. TIMM, *Protein Sci.* **1994**, *3*, 2167–2174.
7 R. SECKLER, Assembly of multi-subunit structures. In *Mechanisms of Protein Folding* (R. H. PAIN, ed.), pp. 279–308. Oxford University Press, Oxford, 2000.
8 R. JAENICKE, *Biol. Chem.* **1998**, *379*, 237–43.
9 C. N. PACE, *Methods Enzymol.* **1986**, *131*, 266–280.
10 J. U. BOWIE and R. T. SAUER, *Biochemistry* **1989**, *28*, 7139–7143.
11 D. E. TIMM and K. E. NEET, *Protein Sci.* **1992**, *1*, 236–44.
12 I. JELESAROV and H. R. BOSSHARD, *J. Mol. Recogn.* **1999**, *12*, 13–18.
13 E. FREIRE, O. L. MAYORGA and M. STRAUME, *Anal. Chem.* **1990**, *62*, 950A–959A.
14 T. WISEMAN, S. WILLISTON, J. F. BRANDTS and L.-N. LIN, *Anal. Biochem.* **1989**, *179*, 131–137.
15 K. W. PLAXCO and C. M. DOBSON, *Curr. Opin. Struct. Biol.* **1996**, *6*, 630–636.
16 C. F. BERNASCONI, *Relaxation Kinetics*, 1st edn. Academic Press, New York, 1976.
17 H. WENDT, C. BERGER, A. BAICI, R. M. THOMAS and H. R. BOSSHARD, *Biochemistry* **1995**, *34*, 4097–4107.
18 T. JONSSON, C. D. WALDBURGER and R. T. SAUER, *Biochemistry* **1996**, *35*, 4795–4802.
19 R. ZWANZIG, *Proc. Natl Acad. Sci. USA* **1997**, *94*, 148–150.

20 M. E. MILLA and R. T. SAUER, Biochemistry **1994**, *33*, 1125–1133.
21 T. JONSSON, C. D. WALDBURGER and R. T. SAUER, Biochemistry **1996**, *35*, 4795–4802.
22 A. K. SRIVASTAVA and R. T. SAUER, Biochemistry **2000**, *39*, 8308–8314.
23 M. E. MILLA, B. M. BROWN, C. D. WALDBURGER and R. T. SAUER, Biochemistry **1995**, *34*, 13914–13919.
24 C. D. WALDBURGER, T. JONSSON and R. T. SAUER, Proc. Natl Acad. Sci. USA **1996**, *93*, 2629–2634.
25 I. A. HOPE and K. STRUHL, EMBO J. **1987**, *6*, 2781–2784.
26 A. I. DRAGAN and P. L. PRIVALOV, J. Mol. Biol. **2002**, *321*, 891–908.
27 I. JELESAROV and H. R. BOSSHARD, J. Mol. Biol. **1996**, *263*, 344–358.
28 H. WENDT, L. LEDER, H. HÄRMÄ, I. JELESAROV, A. BAICI and H. R. BOSSHARD, Biochemistry **1997**, *36*, 204–213.
29 E. K. O'SHEA, K. J. LUMB and P. S. KIM, Curr. Biol. **1993**, *3*, 658–667.
30 P. L. PRIVALOV and S. A. POTHEKIN, Methods Enzymol. **1986**, *131*, 4–51.
31 R. S. SPOLAR and M. T. RECORD, JR., Science **1994**, *263*, 777–784.
32 G. I. MAKHATADZE and P. L. PRIVALOV, Adv. Protein Chem. **1995**, *47*, 307–425.
33 G. SCHREIBER and A. R. FERSHT, Nature Struct. Biol. **1996**, *3*, 427–431.
34 H. X. ZHOU, Biopolymers **2001**, *59*, 427–433.
35 H. X. ZHOU, Protein Sci. **2003**, *12*, 2379–2382.
36 H. R. BOSSHARD, D. N. MARTI, and I. JELESAROV, J. Mol. Recogn. **2004**, *17*, 1–16.
37 K. J. LUMB and P. S. KIM, Science **1995**, *268*, 436–439.
38 P. PHELAN, A. A. GORFE, J. JELESAROV, D. N. MARTI, J. WARWICKER and H. R. BOSSHARD, Biochemistry **2002**, *41*, 2998–3008.
39 D. N. MARTI and H. R. BOSSHARD, J. Mol. Biol. **2003**, *330*, 621–637.
40 C. C. MOSER and P. L. DUTTON, Biochemistry **1988**, *27*, 2450–2461.
41 E. DÜRR, I. JELESAROV and H. R. BOSSHARD, Biochemistry **1999**, *38*, 870–880.
42 R. KOREN and G. G. HAMMES, Biochemistry **1976**, *15*, 1165–1171.
43 M. VIJAYAKUMAR, K. Y. WONG, G. SCHREIBER, A. R. FERSHT, A. SZABO and H. X. ZHOU, J. Mol. Biol. **1998**, *278*, 1015–1024.
44 H. R. BOSSHARD, E. DÜRR, T. HITZ and I. JELESAROV, Biochemistry **2001**, *40*, 3544–3552.
45 J. A. ZITZEWITZ, O. BILSEL, J. B. LUO, B. E. JONES and C. R. MATTHEWS, Biochemistry **1995**, *34*, 12812–12819.
46 R. A. KAMMERER, T. SCHULTHESS, R. LANDWEHR et al. Proc. Natl Acad. Sci. USA **1998**, *95*, 13419–13424.
47 J. K. MYERS and T. G. OAS, J. Mol. Biol. **1999**, *289*, 205–209.
48 R. A. KAMMERER, V. A. JARAVINE, S. FRANK et al. J. Biol. Chem. **2001**, *276*, 13685–13688.
49 J. A. ZITZEWITZ, B. IBARRA-MOLERO, D. R. FISHEL, K. L. TERRY and C. R. MATTHEWS, J. Mol. Biol. **2000**, *296*, 1105–1116.
50 B. A. KRANTZ and T. R. SOSNICK, Nature Struct. Biol. **2001**, *8*, 1042–7.
51 X. CHENG, M. L. GONZALEZ and J. C. LEE, Biochemistry **1993**, *32*, 8130–9.
52 S. M. DOYLE, E. H. BRASWELL and C. M. TESCHKE, Biochemistry **2000**, *39*, 11667–76.
53 D. H. KIM, D. S. JANG, G. H. NAM et al. Biochemistry. **2000**, *39*, 13084–92.
54 M. S. GITTELMAN and C. R. MATTHEWS, Biochemistry **1990**, *29*, 7011–7020.
55 C. J. MANN, S. XHAO and C. R. MATTHEWS, Biochemistry **1995**, *34*, 14573–14580.
56 L. M. GLOSS and C. R. MATTHEWS, Biochemistry **1997**, *36*, 5612–5623.
57 L. M. GLOSS and C. R. MATTHEWS, Biochemistry **1998**, *37*, 15990–15999.
58 L. M. GLOSS, B. R. SIMLER and C. R. MATTHEWS, J. Mol. Biol. **2001**, *312*, 1121–34.
59 A. C. CLARK, S. W. RASO, J. F. SINCLAIR, M. M. ZIEGLER, A. F. CHAFFOTTE and T. O. BALDWIN, Biochemistry **1997**, *36*, 1891–9.
60 E. DÜRR and H. R. BOSSHARD, Protein Sci. **2000**, *9*, 1410–1415.
61 R. M. THOMAS, H. WENDT, A. ZAMPIERI and H. R. BOSSHARD, Progr. Colloid. Polym. Sci. **1995**, *99*, 24–30.

62 D. N. Marti, M. Lu, H. R. Bosshard and I. Jelesarov, *J. Mol. Biol.* **2004**, *336*, 1–8.
63 D. C. Chan, D. Fass, J. M. Berger and P. S. Kim, *Cell* **1997**, *89*, 263–273.
64 D. L. Burns and H. K. Schachman, *J. Biol. Chem.* **1982**, *257*, 8648–8654.
65 M. G. Mateu, M. M. S. Del Pino and A. R. Fersht, *Nature Struct. Biol.* **1999**, *6*, 191–198.
66 C. Bodenreider, N. Kellershohn, M. E. Goldberg and A. Mejean, *Biochemistry* **2002**, *41*, 14988–14999.
67 C. R. Johnson, P. E. Morin, C. H. Arrowsmith and E. Freire, *Biochemistry* **1995**, *34*, 5309–5316.
68 A. R. Fersht, *Curr. Opin. Struct. Biol.* **1997**, *7*, 3–9.
69 T. A. Larsen, A. J. Olson and D. S. Goodsell, *Structure* **1998**, *6*, 421–7.
70 C. J. Tsai, S. L. Lin, H. J. Wolfson and R. Nussinov, *Protein Sci.* **1997**, *6*, 53–64.
71 V. Daggett and A. R. Fersht, *Trends Biochem. Sci.* **2003**, *28*, 18–25.
72 K. S. Thompson, C. R. Vinson and E. Freire, *Biochemistry* **1993**, *32*, 5491–5496.
73 K. Bachhawat, M. Kapoor, T. K. Dam and A. Surolia, *Biochemistry* **2001**, *40*, 7291–300.
74 D. Apiyo, K. Jones, J. Guidry and P. Wittung-Stafshede, *Biochemistry* **2001**, *40*, 4940–4948.
75 E. Barbar, B. Kleinman, D. Imhoff, M. Li, T. S. Hays and M. Hare, *Biochemistry* **2001**, *40*, 1596–1605.
76 L. A. Wallace and H. W. Dirr, *Biochemistry* **1999**, *38*, 16686–16694.
77 L. A. Wallace, N. Sluis-Cremer and H. W. Dirr, *Biochemistry* **1998**, *37*, 5320–5328.
78 L. A. Marky and K. J. Breslauer, *Biopolymers* **1987**, *26*, 1601–1620.
79 P. Kuzmic, *Anal. Biochem.* **1996**, *237*, 260–273.

28
Folding of Membrane Proteins

Lukas K. Tamm and Heedeok Hong

28.1
Introduction

The major forces that govern the folding and thermodynamic stability of water-soluble proteins have been studied in significant detail for over half a century [1, 2]. Even though the reliable prediction of protein structure ab initio from amino acid sequence is still an elusive goal, a huge amount of information on the mechanisms by which soluble proteins fold into well-defined three-dimensional structures has been obtained over this long time period [3]. In marked contrast, our current knowledge on thermodynamic and mechanistic aspects of the folding of membrane proteins is several orders of magnitude less advanced than that of soluble proteins, although integral membrane proteins comprise about 30% of open reading frames in prokaryotic and eukaryotic organisms [4]. There are several reasons why the research on membrane protein folding lags far behind similar research on soluble proteins.

First, the first structure of a soluble protein (myoglobin) was solved in 1958, but the first membrane protein structure (photosynthetic reaction center) was only determined 27 years later in 1985. Even today, of the approximately 25 000 protein structures in the Protein Structure Databank (PDB), only about 140 (i.e., less than 0.6%) are structures of integral membrane proteins.

Second, membrane proteins are more difficult to express, purify, and handle in large batches than soluble proteins. They require the presence of nondenaturing detergents or lipid bilayers to maintain their native structure, which adds an additional layer of complexity to these systems compared with soluble protein systems. The requirement for detergents or membranes also explains why it is more difficult to crystallize membrane proteins or subject them to nuclear magnetic resonance (NMR) for high-resolution structural studies.

Third, it is very difficult to completely denature membrane proteins [5]. Solvent denaturation with urea or guanidinium chloride (GdmCl) is generally not applicable to membrane proteins because secondary structures that are buried in the membrane are often not accessible to these denaturants. Membrane proteins are also quite resistant to heat denaturation, and even if they can be (partially) dena-

Protein Folding Handbook. Part I. Edited by J. Buchner and T. Kiefhaber
Copyright © 2005 WILEY-VCH Verlag GmbH & Co. KGaA, Weinheim
ISBN: 3-527-30784-2

tured, they are prone to irreversible aggregation. These properties have not helped to make membrane proteins favorable targets for folding studies. However, as will be shown in this chapter, new and different techniques are being developed to study the folding of membrane proteins. If this is done with sufficient care, membrane proteins can be brought back "into the fold" and very useful and much needed information can be gathered on the molecular forces that determine their thermodynamic stability and mechanisms that lead to their native structures. Even though most helical membrane proteins are inserted into biological membranes by means of the translocon [6], knowledge of the energetics and kinetics of membrane protein structure formation are important from a basic science standpoint and for structure prediction and protein engineering of this still neglected class of proteins, which constitutes the largest fraction of all targets of currently available drugs.

To understand the folding of membrane proteins, we first need to understand the dynamic structure of the lipid bilayer (i.e., the "solvent") into which these proteins fold. Fluid lipid bilayers are highly dynamic, yet well-ordered two-dimensional arrays of two layers of lipids with their polar headgroups exposed towards water and their apolar fatty acyl chains shielded from water and facing the center of the bilayer. This structure [7, 8] is maintained almost entirely by the hydrophobic effect [9]. The hydrocarbon chains in the core of the bilayer are quite well ordered, but their order degrades towards the midplane of the bilayer. This region comprises a width of about 30 Å in a "typical" bilayer of dioleoylphosphatidylcholine. The two interfaces of the bilayer are chemically and structurally quite complex and comprise a width of about 15 Å each. These regions contain the phosphate and choline headgroups, the glycerol backbone and about 25 molecules of bound, but disordered water for each phosphatidylcholine (PC) headgroup. The lipids are free to diffuse laterally (~ 1 μm^2 s^{-1}) and rotationally (~ 1 μs^{-1}), but basically do not flip-flop across the membrane.

The molecular packing of lipids in a fluid lipid bilayer is maintained mechanically by a combination of three types of opposing forces: lateral chain repulsions in the core region, headgroup repulsions, and surface tension at the polar–nonpolar interface [10]. These forces create a lateral pressure profile along the membrane normal that cannot be directly measured, but that has been calculated based on the known magnitudes of the relevant forces [11]. This packing and lateral pressure profile has profound consequences on the lipid hydrocarbon chain dynamics [12] and the folding of membrane proteins as will be summarized in later sections of this chapter. Therefore, the mechanical and dynamical properties as well as the chemical properties of membranes change dramatically as the composition of chemical groupings changes along the bilayer.

To dehydrate and bury a single peptide bond in the lipid bilayer is energetically unfavorable and costs 1.15 kcal mol^{-1} [13]. The energetic gain of folding a polypeptide into an internally hydrogen-bonded secondary structure in membranes is much smaller (i.e., of the order of -0.1 to -0.4 kcal per residue) [14–17]. This shows that partitioning of the bare backbone (e.g., polyglycine) into membranes is energetically unfavorable irrespective of whether the peptide is folded or not. Productive membrane insertion of a polypeptide segment is only promoted if the segment contains a sufficiently large number of apolar side chains, which by virtue of their favorable partitioning can overcome the energy cost of transferring the peptide backbone into the membrane. As a rule of thumb, the equivalent of a minimum of five leucine substitutions are required to offset the unfavorable energy of trying to insert a 20-residue polyalanine helix (see below).

Forming intramolecular hydrogen bonds is moderately favorable at the interface and presumably quite favorable in the apolar core of the bilayer [18]. The energetic cost of opening a hydrogen bond in a membrane likely depends on the concentration of water or the probability of finding another hydrogen-bonding partner in that location. Since water and other hydrogen-bond acceptors are rare in membranes, one generally finds only fully hydrogen-bonded secondary structures in membrane proteins. Therefore, the predominant building blocks of membrane proteins are transmembrane (TM) helices and closed TM β-barrels. Other fully hydrogen-bonded secondary structures such as 3_{10}-helices and π-bulges are rare, but do occasionally occur in membrane proteins.

Since the α-helix and closed β-sheets are the main building blocks and since these two elementary secondary structures are rarely mixed in membrane proteins, we treat α-helical and β-barrel membrane proteins (for examples of two prototype membrane protein structures, see Figure 28.1) in separate sections of this chapter. These sections are preceded by a general section on the thermodynamics of side-chain partitioning, which applies to both classes of membrane proteins and which also forms the basis of membrane protein topology and structure prediction algorithms.

28.2
Thermodyamics of Residue Partitioning into Lipid Bilayers

The partitioning of apolar amino acid side chains into the membrane provides the major driving force for folding of α-helical and β-barrel membrane proteins. Folding of membrane proteins is always coupled to partitioning of the polypeptide from water into the membrane environment. A large number of hydropathy scales have been developed over the years. The different scales have been derived based on very different physical or statistical principles and, therefore, differ for different residues sometimes quite significantly. It is beyond the scope of this chapter to review the different scales that are in use. The only scales that are based on comprehensive partitioning measurements of peptides into membranes and membrane-mimetic environments are the two Wimley & White (WW) scales, namely the WW interface [19] and the WW octanol scale [13, 20].

The WW interface scale refers to partitioning into the interface region of phosphatidylcholine bilayers and the octanol scale refers to partitioning into octanol, which is thought to be a good mimetic of the core of the lipid bilayer. As has been demonstrated for the hydrophobic core of soluble proteins [21], the dielectric constant in the core of a lipid bilayer may be as high as 10 due to water penetration (i.e., considerably larger than 2 in pure hydrocarbon and closer to that of wet octanol). Both scales are whole-residue scales, which means that they take into account the thermodynamic penalty of also partitioning the peptide backbone into the respective membrane locations. Based on these partition data, one may group the amino acid residues into three groups, namely those that do not favor membranes (Ala, Ser, Asn, Gly, Glu$^-$, Asp$^-$, Asp^0, Lys$^+$, Arg$^+$, His$^+$), those that favor the interface (Trp, Tyr, Cys, Thr, Gln, His^0, Glu^0), and those that favor the hydrocarbon core (Phe, Leu, Ile, Met, Val, Pro). Sliding a 19-residue window of summed WW hydropathy values over the amino acid sequences of membrane proteins and using the thermodynamic $\Delta G = 0$ cutoff predicts TM helices with high accuracy [13]. The residues that fall in the first and third group are not surprising. However, the residues that prefer the interface deserve a few comments. Statistical analyses and model studies of membrane proteins have indicated for quite some time that the aromatic residues Trp and Tyr residues have a strong preference for membrane interfaces [22] (see also Figure 28.1). It has been suggested that their bulkiness, aromaticity and slightly polar character are the major factors that determine their preferred location in the interface [23].

Hydrogen bonding is probably not important for this localization. The moderately polar residues in this group are not surprising. However, the uncharged form of glutamic acid interestingly is apolar enough to localize into the interface. The differential partitioning of glutamate depending on its charge state likely has profound consequences on the pK_a value of this residue at membrane surface because the pK_a is now coupled to partitioning. A manifestation of this shifted pK_a is seen, for example, when the binding and membrane penetration of the fusion peptide of influenza hemagglutinin is examined [16]. This peptide contains two Glu residues in its N-terminal portion. A shift of the pH from 7 to 5 is sufficient to bind the peptide more tightly and more deeply (i.e., deep enough to cause membrane fusion at pH 5, but not at pH 7).

28.3
Stability of β-Barrel Proteins

All β-barrel membrane proteins whose structures are known at high resolution are outer membrane proteins of gram-negative bacteria [24]. However, several mitochondrial and chloroplast outer membrane proteins, namely those that are part of the protein translocation machinery of these membranes also appear to be β-barrel membrane proteins [25, 26]. Bacterial outer membrane proteins are synthesized with an N-terminal signal sequence and their translocation through the inner membrane is mediated by the translocon consisting of the SecY/E/G complex with the assistance of the signal recognition particle and the SecA ATPase, which together constitute the prokaryotic protein export machinery [27]. Helical membrane proteins that reside in the inner membrane have contiguous stretches of about 20 hydrophobic residues that form TM helices, which presumably can exit the translocon laterally into the lipid bilayer [6]. The TM strands of outer membrane proteins on the other hand are composed of alternating hydrophilic and hydrophobic residues [28]. Therefore, they lack the signal for lateral exit from the translocon into the inner membrane and become secreted into the periplasmic space, where they are presumably greeted by periplasmic chaperones such as Skp and SurA [29, 30]. Skp binds unfolded outer membrane proteins, prevents them from aggregation, but does not appear to accelerate their insertion into lipid model membranes (D. Rinehart, A. Arora, and L. Tamm, unpublished results).

The outer membrane proteins of gram-negative bacteria may be grouped into six families according to their functions [31]. Structures of members of each of these families have been solved by X-ray crystallography or NMR spectroscopy. The six families with representative solved structures are: (i) the general porins such as OmpF and PhoE [32, 33], (ii) passive sugar transporters such as LamB and ScrY [34, 35], (iii) active transporters of siderophores such as FepA, FecA, and FhuA [36–39] and of vitamin B_{12} such as BtuB [40], (iv) enzymes such as the phospholipase OmpLA [41], (v) defensive proteins such as OmpX [42], and (vi) structural proteins such OmpA [43]. All outer membrane proteins form β-barrels. The β-barrels of these proteins consist of even numbers of β-strands ranging from 8 to 22. The average length of the TM β-strands is 11 amino acid residues in trimeric porins and 13–14 residues in monomeric β-barrels. Since the strands are usually inclined at about 40° from the membrane normal, they span about 27–35 Å of the outer membrane, respectively.

Most outer membrane proteins have long extracellular loops and short periplamsic turns that alternatingly connect the TM β-strands in a meandering pattern. The loops exhibit the largest sequence variability within each outer membrane protein family [31]. The largest and smallest outer membrane proteins are monomeric, but others such as OmpLA are dimers and the porins are trimers of β-barrels.

Two general approaches have been taken to examine the thermodynamic stability of β-barrel membrane proteins: solvent denaturation by urea and guanidinium chloride and differential scanning calorimetry (DSC). In the following, we will discuss the thermodynamic stability of the simple monomeric β-barrel protein OmpA and then extend the discussion to complexities that arise by the presence of the central plug domain of FepA and FhuA and trimer formation in the case of porins. The most extensive thermodynamic stability studies have been carried out with the outer membrane protein A (OmpA) of *Escherichia coli*. OmpA is an abundant structural protein of the outer membrane of gram-negative bacteria. Its main function is to anchor the outer membrane to the peptidoglycan layer in the periplasmic space and thus to maintain the structural integrity of the outer cell envelope [44]. The 325-residue protein is a two-domain protein whose N-terminal 171 residues constitute an eight-stranded β-barrel membrane-anchoring domain and whose C-terminal 154 residues form the periplasmic domain that interacts with the peptidoglycan. The three-dimensional structure of the TM domain has been solved by X-ray crystallography [43] and, more recently, by solution NMR spectroscopy [45]. The structure resembles a reverse micelle with most polar residues facing the interior, where they form numerous hydrogen bonds and salt bridges, and the apolar residues facing the lipid bilayer [43]. Two girdles of aromatic side chains, trypto-

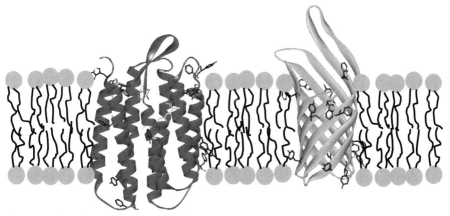

Fig. 28.1. Three-dimensional structures of representative α-helical and β-barrel membrane proteins. Left: Bacteriorhodopsin from *Halobacterium salinarium* (PDB code: 1C3W). Right: Transmembrane domain of OmpA (1QJP) from *Escherichia coli*. The aromatic side chains (Trp, Tyr, Phe) are concentrated at the bilayer–water interfaces as shown in both structures.

phans and tyrosines, delineate the rims of the barrel where they interact with the two membrane interfaces (Figure 28.1).

Extensive mutagenesis studies show that OmpA is quite robust against many mutations especially in the loop, turn, and lipid bilayer-facing regions [46]. OmpA can even be circularly permutated without impairing its assembly and function in outer membranes [47]. Dornmair et al. [48] showed that OmpA could be extracted from the outer membrane and denatured and solubilized in 6–8 M urea and that the protein spontaneously refolded into detergent micelles by rapid dilution of the denaturant. They subsequently showed that urea-unfolded OmpA can also be quantitatively refolded into preformed lipid bilayers by rapid dilution of urea [49].

Recently, we could achieve the complete and reversible refolding of OmpA in lipid bilayers [50]. The reaction was shown to be a coupled two-state membrane partition folding reaction with the unfolded state in urea being completely dissociated from the membrane and the folded state completely integrated into the membrane (Figure 28.2). This represents the first example of an integral membrane protein, for which the thermodynamic stability could be quantitatively accessed. To achieve this goal, a urea-induced equilibrium folding system was developed for OmpA. The system consisted of small unilamellar vesicles composed of 92.5 mol% phosphatidylcholine and 7.5 mol% phosphatidylglycerol in weakly basic (pH 7.0–10) and low ionic strength conditions. These conditions facilitate both the folding and unfolding reactions and dissociation of the unfolded form from the membrane surface. Folding of OmpA is monitored by tryptophan fluorescence or CD spectroscopy, or by sodium dodecylsulfate polyacrylamide gel electrophoresis (SDS-PAGE) (Figure 28.3).

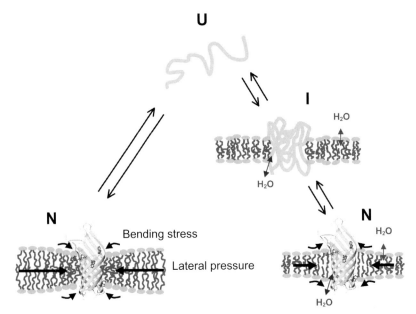

Thick bilayer: $\Delta G°_u$, m : high

Thin bilayer: $\Delta G°_u$, m : low

Fig. 28.2. Cartoon depicting the folding of OmpA into lipid bilayers. Left path: Folding into most bilayers is a thermodynamic two-state process. Right path: Folding into thin bilayers is multi-state (i.e., at least one equilibrium intermediate occurs). Bilayer forces acting on OmpA folding are indicated with arrows. The large black arrows indicate lateral bilayer pressure imparted on the lipid/protein interface in the hydrophobic core of the bilayer. Increasing this pressure increases the thermodynamic stability of the protein. The small black arrows indicate lipid deformation forces caused by hydrophobic mismatch between the protein and unstressed bilayers. These forces decrease the thermodynamic stability of the protein. Water molecules penetrate more easily into the hydrophobic core of thin bilayers (blue arrows) and stabilize equilibrium intermediates until, in very thin bilayers, complete unfolding is no longer observed. The unfolded state in urea is dissociated from the membrane. Adapted from Ref. [50].

Each of these techniques monitors a different aspect of the folding reaction. Trp fluorescence reports on the insertion of tryptophans into the lipid bilayer, far-UV CD on the formation secondary structure, and SDS-PAGE on the completion of the tertiary β-barrel structure. If samples are not boiled prior to loading onto the SDS gels, they run at an apparent molecular mass of 30 kDa if the protein is completely folded, but at 35 kDa if it is unfolded or incompletely folded.

This shift on SDS gels has proven to be a very useful assay for tertiary structure formation of OmpA and other outer membrane proteins [51]. Complete refolding as measured by the SDS-PAGE shift correlates with the reacquisition of the ion channel activity of OmpA [52]. Since unfolding/refolding curves as a function of denaturant concentration measured by three techniques that report on vastly different kinetic processes superimpose, it is very likely that folding/unfolding is a ther-

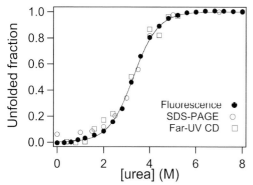

Fig. 28.3. Unfolded fraction of OmpA as a function of urea concentration obtained from the average fluorescence emission wavelength ⟨λ⟩, far-UV CD spectroscopy, and the SDS-PAGE shift assay. These measures of lipid binding, secondary structure, and tertiary structure, respectively, superimpose in equilibrium measurements although they develop at different times in kinetic experiments. Adapted from Ref. [50].

modynamic two-state process that can be analyzed using a two-state equilibrium folding model.

From such an analysis, Hong and Tamm [50] found a free energy of folding $\Delta G°_{H2O} = -3.4$ kcal mol^{-1} and an m-value, which describes the cooperativity parameter of the folding/unfolding reaction, of $m = 1.1$. The rather small $\Delta G°_{H2O}$ of OmpA, which is of the same order of magnitude as that of water-soluble proteins, is perhaps surprising in view of the quite extreme heat resistance of this and other β-barrel membrane proteins. However, if one simply calculates the free energy of transfer of all residues that are transferred into the lipid bilayer with the augmented WW hydrophobicity scale, one finds that the net $\Delta G°_{H2O}$ amounts to only about -1 kcal mol^{-1}. This value may be further decreased by adding a few (negative) kcal mol^{-1} for folding, which would bring the prediction close to the experimentally determined value. The folding contribution may be the sum of contributions for secondary structure formation, estimated to be about 80×-0.2 kcal mol^{-1}, and on the order of 10 kcal mol^{-1} of entropy and other costs for packing polar residues in the core of the protein.

The take-home message from these measurements and theoretical considerations is that the overall stability of membrane proteins is not as large as one might have anticipated, but rather similar in magnitude to that of soluble proteins of similar size. As is true for soluble proteins, the thermodynamic stability of this and perhaps most membrane proteins is determined by the sum of many relatively large thermodynamic contributions that ultimately cancel to yield a relatively small net free energy of folding.

The study by Hong and Tamm [50] also examined the effect of lipid bilayer forces on the thermodynamic stability of OmpA. The effects of elastic lipid deformation due to a mismatch of the hydrophobic length of the protein compared with the bilayer thickness and due to curvature stress imposed by cone-shaped lipids

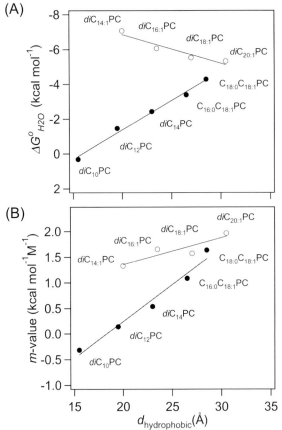

Fig. 28.4. Dependence of A) the free energy of folding $\Delta G°_{H2O}$ and B) the m-value reflecting the cooperativity of folding on the hydrophobic thickness of phosphatidylcholine bilayers. Filled circles: saturated acyl chain series with small lateral bilayer pressure. Open circles: double-unsaturated acyl chain series with increasing lateral bilayer pressure as hydrophobic thickness decreases. Adapted from Ref. [50].

were investigated. Increasing the thickness of the lipid bilayer increases the stability (i.e., decreases $\Delta G°_{H2O}$) of OmpA. It also increases the cooperativity (m-value) of folding (Figure 28.4). The decrease in $\Delta G°_{H2O}$ is -0.34 kcal mol^{-1} per Å of increased bilayer thickness, which converts to -4 cal mol^{-1} per Å^2 of increased hydrophobic contact area. This is only about 20% of the standard value for the hydrophobic effect [9]. We conclude that elastic energy due to lipid deformation counteracts the energy gain that would be expected for a full development of the hydrophobic effect.

Including cone-shaped lipids with a smaller cross-sectional headgroup than acyl chain area increases the lateral pressure within lipid bilayers [53]. Increasing the lateral pressure in membranes by including lipids with increasing relative

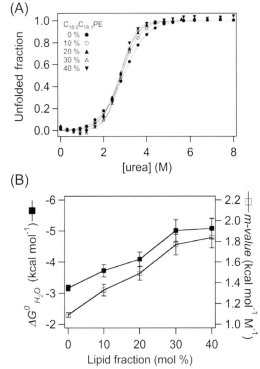

Fig. 28.5. Effect of increasing mol fraction of phosphatidylethanolamine ($C_{16:0}C_{18:1}$ PE) in phosphatidylcholine ($C_{16:0}C_{18:1}$ PC) bilayers on the thermodynamic stability of OmpA. A) Unfolding curves measured by Trp fluorescence. B) Dependence of $\Delta G°_{H2O}$ and m-value of OmpA folding on the cone-shaped and lateral pressure-inducing lipid $C_{16:0}C_{18:1}$ PE. Adapted from Ref. [50].

amounts of *cis* double bonds in the hydrocarbon region increased the stability (i.e., decreased $\Delta G°_{H2O}$) of OmpA from -3.4 to -7 kcal mol^{-1} and more negative values when phosphatidylethanolamines were included as cone-shape lipids (Figure 28.5). The *m*-value exhibited an increase that was correlated with the $\Delta G°_{H2O}$ decrease, in the case of bilayer thickening, but a decrease when the bilayer pressure was increased with *cis* double bonds. The various bilayer forces that act on OmpA and modulate its thermodynamic stability in membranes are illustrated in Figure 28.2.

Gram-negative bacteria possess active transport systems for the uptake of iron–siderophore complexes, vitamin B_{12}, and other essential nutrients. The systems consist of active ATP-requiring transporters in the inner membrane, the TonB linker protein that spans the periplasm, and a family of proteins called TonB-dependent transporters in the outer membrane. The TonB-dependent transporters share common structural features. They are monomeric 22-stranded β-barrels. The lumina of these large-diameter barrels are filled with a N-terminal "cork" or "plug" domain with a globular mixed α-helix/β-structure. The plug domains are tightly inserted into the barrels and make extensive salt bridge and hydrogen bond contacts with the inner barrel wall. The mechanism of substrate transport through these transporters is presently not well understood.

A thermal denaturation study using DSC of FhuA demonstrated that the plug domain and the surrounding β-barrel are autonomous folding units [54]. In the absence of bound substrate (ferrichrome), a reversible transition centered at 65 °C was well separated from an irreversible transition centered at 74 °C. The binding of the substrate increased the lower transition temperature to 71 °C, while the higher transition temperature was not changed. Since the higher temperature transition was accompanied by a significant change of the CD signal at 198 nm and 215 nm, this transition was assigned to the denaturation of the β-barrel. The lower temperature transition was assigned to the denaturation of the plug domain and neighboring loops emerging from the β-barrel because antibodies targeting unfolded outer loops bound to FhuA only above the lower transition temperature of 65 °C. A deletion mutant (Δ21–128) that lacked the plug domain underwent a thermal denaturation at a lower temperature ($T_m = 62$ °C) than the wild-type protein, implying that the presence of the plug stabilizes the barrel structure.

The stability of FepA in Triton-X100 micelles was probed by solvent-induced denaturation and EPR spectroscopy [55, 56]. In these studies, FepA was functionally refolded from the denatured state by dialysing the denaturant in the presence of TX-100 micelles. When GdmCl- and urea-induced denaturation was monitored by site-directed spin-label EPR spectroscopy at an extracellular loop site, unfolding occurred in a sharp transition at 2.0 M GdmCl or 5.5 M urea, respectively. The free energies of unfolding were approximately 6 kcal mol^{-1} with both denaturants. The rate of unfolding of the substrate-bound protein was significantly smaller than that of the free protein. This confirms that the substrate has a stabilizing effect on FepA as has also been observed by DSC with FhuA. When residues pointing towards the center of the barrel were spin-labeled, their EPR spectra indicated completely mobile residues in 4 M GdmCl. However, residues facing the detergent were still quite immobilized at this GdmCl concentration. The authors argued that denatured FepA retained substantial residual hydrophobic interactions with the detergent micelle. This observation is reminiscent of residual hydrophobic interactions that have been observed in soluble proteins at high denaturant concentrations [57]. However, in these studies with FepA, the degree of denaturation was unfortunately not recorded by a global method such as CD spectroscopy and therefore, it is not clear whether the transition is two-state and whether fully denaturing conditions have been reached in this work.

Porins facilitate the general diffusion of polar solutes (< 400 Da) across the outer membranes of gram-negative bacteria. The porins are composed of 16-stranded β-barrels that assemble into rigid homotrimers whose trimer interfaces are formed by close van der Waals packing of nonpolar residues and aromatic ring stacking interactions [32, 33]. The lumen of each monomer is aqueous and surrounded by hydrophilic residues. An interesting structural and functional feature of the porins is that the long loop L3 folds back into the lumen of the channel where it engages in a hydrophobic contact with a few residues inside the barrel. The loop forms a constriction in the channel and its acidic residues together with a cluster of basic residues on the opposite channel wall are thought to exert a transverse electric field across the channel opening. The shorter loop L2 forms a "latch" that reaches over from one subunit to a neighboring subunit and thereby further stabilizes intersubunit interactions. Porins are extremely stable towards heat denaturation, protease digestion, and chemical denaturation with urea and GdmCl [31]. For example, extraction of OmpF from the *E. coli* outer membrane requires the heating of outer membranes in mixtures of isopropanol and 6 M GdmCl at 75 °C for more than 30 min [58]. The trimer structure itself is maintained up to 70 °C in 1% SDS [59].

A mutagenesis study combined with DSC and SDS-PAGE shift assays demonstrates that the inter-subunit salt bridge and hydrogen bonding interactions involving loop L2 contribute significantly to the trimer stability [59]. For example, mutations breaking the salt bridge between Glu71 and Arg100 decrease the trimer–monomer transition temperature from 72 to 47–60 °C and ΔH_{cal} from 430 to 200–350 kcal mol^{-1}. A similar behavior was observed when residues 69–77 of L2 were deleted. The energetic contribution of the nonpolar contact between neighboring barrel walls has not yet been investigated in porins, but would be interesting to examine and compare with lateral associations of interacting TM helices (see below).

28.4
Stability of Helical Membrane Proteins

In 1990, Popot and Engelman proposed a two-stage model for the folding of α-helical membrane proteins [60]. In the framework of this model, TM helices are thought to be independently folded units (stage I) that may laterally associate into TM helix bundles (stage II). The model was inspired by the finding of the same group that the seven-TM-helix protein bacteriorhodopsin (BR) can be split into helical fragments that would still integrate as TM helices and then laterally assemble into a seven-helix protein that was capable of binding to cofactor retinal just as the wild-type protein [61]. Since then, it has been realized that more stages are needed to describe the folding of helical proteins.

The interface is a membrane compartment that may be important in early stages of folding [62]. In addition, cofactors and polypeptide hinges and loops can be additional factors that sometimes play important roles in late stages of helical membrane protein folding [63]. Although the real world of membrane proteins is clearly more complex, the two-stage model has some merits because it is conceptually simple and because it appears to correctly describe the hierarchy of folding of at least a few simple membrane proteins. Therefore, we describe in this section how individual TM helices are established and then proceed in the following section with lateral helix associations and higher order structural principles of membrane protein folding.

As indicated above, open hydrogen bonds are energetically unfavorable in membranes. Sequence segments with a sufficient number of apolar residues are driven into membranes by the hydrophobic effect where they fold into TM or interfacial α-helices. The thermodynamic gain of folding a 20-residue peptide into a TM helix is about -5 kcal mol^{-1}, if we take -0.25 kcal mol^{-1} as the per-residue increment of forming an helix in membranes [16]. Generally and in contrast to soluble protein α-helices, TM helices are quite rich in glycines. Prolines are also not uncommon in TM helices, where they often induce bends, but do not completely break the helix. Finally, serines, threonines, and cysteines are frequently interspersed into otherwise apolar TM sequences. These latter residues often form hydrogen bonds with backbone carbonyls and may therefore have a more apolar character in TM helices than in aqueous environments.

It is generally not possible to obtain interpretable results on the stability of helical integral membrane proteins from thermal denaturation studies because thermal denaturation of these proteins usually leads to irreversibly denatured states [5]. However, the thermodynamic stabilities of a few polytopic helical membrane proteins have been studied by partial denaturation with strong ionic detergents. Early refolding studies by Khorana and coworkers showed that BR could be quantitatively refolded by step-wise transfer from organic solvent to SDS to renaturing detergents or lipids [64, 65]. These protocols were later refined to measure the kinetics of refolding of BR upon transfer from the ionic detergent SDS into mixed lipid/detergent micelles [66, 67]. It should be kept in mind that although tertiary contacts are lost, most secondary structure is retained or perhaps even partially refolded in SDS. Therefore, the reactions observed with these protocols report only on the last stages of membrane protein folding. In 1997, Lau and Bowie developed a system to reversibly unfold native diacylglycerol kinase (DAGK) in n-decyl-β-maltoside (DM) micelles by the addition of SDS [68]. In these carefully controlled studies, the ratios of the denaturing and native structure-supporting detergents were systematically varied, so that reversibility of the partial unfolding reaction could be demonstrated. By analyzing their data with the two-state model of protein folding, these authors determined the stability of the three-helix TM domain of DAGK in DM relative to SDS to be -16 kcal mol^{-1}. This large (negative) value is somewhat surprising and indicates an extremely high thermodynamic stability of DAGK. A similar study has been carried out recently with BR [69].

28.5
Helix and Other Lateral Interactions in Membrane Proteins

The packing density and the hydrophobicity of amino acid residues involved in the intramolecular contact of α-helical membrane proteins are similar to those in the hydrophobic cores of water-soluble proteins. The basic principles of the hydrophobic organization of residues in water-soluble proteins appear to hold true also for membrane proteins. However, the energy gain of the hydrophobic effect has already been expended on inserting individual α-helices into membranes and cannot be a driving force for helix association. From comprehensive mutagenesis work on the TM domain of glycophorin A [70] and later from genome-wide database searches [71], it has become clear that tight van der Waals contacts between complementary helix surfaces provide a strong driving force for lateral helix association in membranes. This force is perhaps best understood if one considers the possibility that lipids even in their fluid liquid-crystalline state do not conform as well to corrugated exposed helix interaction sites as the complementary helix binding partner. Lipids also gain entropy upon release from a helix surface into the bulk liquid-crystalline bilayer ("lipophobic effect"). Since thermodynamic driving forces are always determined by the energetic differences between two states, it is the unfavorable lipid solvation of corrugated helix surfaces that ultimately drives helix association.

Helices of membrane proteins are rarely oriented exactly perpendicular to the plane of the membrane. They are frequently tilted by about 20° from the membrane normal in structurally rugged membrane proteins [72] and by much larger angles in some recently solved structures of ion channels and complex transporters. TM helices interact with each other with left- or right-handed crossing angles [72, 73]. The specificity of helix interactions in membranes is thought to arise from complementary knob-in-the-hole van der Waals interactions. Glycines provide the holes and the β-branched side-chains of valine and isoleucine provide the most frequent knobs on the complementary surfaces. The generality of this interaction is supported by the frequent occurrence of GxxxG motifs in interacting TM segments [71]. The interaction energy of the helix dimer of the glycophorin A TM domain has been measured in C_8E_5 micelles by analytical ultracentrifugation and was found to be -9 kcal mol^{-1} [74]. Although this measurement offers a rare and important glimpse on the thermodynamics of TM helix–helix interactions, the value was determined in detergent micelles and may be different in lipid bilayers. Subsequent alanine scanning experiments indicated that mutation of Gly79 and Gly83 resulted in the largest energy penalty (1.5–3 kcal mol^{-1} relative to wild-type) for dimer formation of glycophorin [75]. The helix interaction "knobs" Leu75 and Ile76 also cost more than 1 kcal mol^{-1} in stabilization energy when replaced by alanines.

Another alanine scanning mutagenesis study examined the effect of all side chains of helix B of BR on the thermodynamic stability of helix interactions in this protein [69]. The most destabilizing mutations were Y57A (3.7 kcal mol^{-1} relative to wt) and T46A (2.2 kcal mol^{-1} relative to wt). These were followed by I45A, F42A, and K41A (1.9–1.6 kcal mol^{-1}). Apparently, very different types of side chains (polar, apolar, aromatic, charged) can cause large effects. These studies show that helix packing interactions may be more complex (and less predictable) than previously thought based on extrapolations from the glycophorin paradigm. A remarkable and important result from the Faham et al. [69] study is that each Å^2 of buried surface area yields about 26 cal mol^{-1} of stabilizing energy. This is about the same number as observed for the burying of residues in soluble proteins, which is usually thought to be driven by the hydrophobic effect. Since the hydrophobic effect is already consumed by inserting side chains into a hydrophobic environment (SDS micelle) and since polar residues essentially yield the same number, the 26 cal mol^{-1} per Å^2, packing energy may be more universal than previously thought and may reflect van der Waals, hydrophobic, and lipophobic interactions. Alternatively, water may access helical surfaces in partially denaturing SDS micelles better than in native structure-supporting DMPC/CHAPSO micelles, in which case the result may still conform to the classical hydrophobic effect.

An additional mechanism of helix interaction is thought to arise from inter-helix hydrogen bonds. Side chain-to-side chain (Asn, Gln, Asp, Glu, His) hydrogen bonds have been introduced into engineered TM helical bundles and were found to stabilize homodimers and homotrimers [76–79]. Ser and Thr side chains were also examined in these host–guest model sequences, but were insufficient to support helix association. It has also been suggested that backbone Cα-H to backbone carbonyl hydrogen bonds might stabilize membrane proteins [80]. However, although such hydrogen bonds are observed quite frequently in crystal structures of membrane (and soluble) proteins, they may merely allow close packing and may not make significant energetic contributions to the association of neighboring helices as has been demonstrated for one such bond in BR [81]. Pore loops are found to intercalate into open angled helical membrane protein structures as exemplified by the KcsA potassium channel structure [82]. The loops are generally also fixed to the helical framework of the protein by side chain-to-side chain hydrogen bonds, namely a hydrogen bond from a loop Tyr to a neighboring helix Trp and Thr in the case of KcsA. Finally, many membrane protein structures contain cofactors and pigments that likely contribute significantly to the stability and interaction specificity of these structures. Intersubunit interactions that determine quaternary structures of membrane proteins generally follow the same helix packing rules as those determining their tertiary structures [83]. In case of BR, van der Waals packing between nonpolar surfaces, hydrogen bonding, and aromatic ring stacking of tyrosines in the bilayer interface all contribute to trimer formation.

28.6
The Membrane Interface as an Important Contributor to Membrane Protein Folding

In this and the following sections we turn to pathways and possible intermediates of membrane protein folding. Jacobs and White [84] proposed a three-step thermodynamic model of membrane protein folding based on structural and partitioning data of small hydrophobic peptides. In this model, the first three steps of membrane protein folding are thought include partitioning into the interface, folding in the interface, and translocation across the membrane to establish a TM helix. The model was later extended to a four-step model, which added helix–helix interactions as the fourth step [62]. The model is rather a thermodynamic concept than an actual pathway of folding of constitutive helical membrane proteins because their TM helices are inserted with the assistance of the translocon in biological membranes. However, the concept is relevant for the insertion of helical toxins into membranes, which occurs without the assistance of other proteins, and for the placement of interfacial helices into membranes. Separation of a partition and folding step in the interface is probably also mostly semantic because in all observable reactions these two processes appear to be coupled. Nevertheless, to think of protein folding as a coupled process of two components is useful for separating the thermodynamic contributions of each component [14–16]. Partition–folding coupling was observed more than 20 years ago for toxic peptides [85], peptide hormones [86], signal peptides [87, 88], and membrane fusion peptides [89]. Beta sheet forming proteins and peptides may also insert and fold in the membrane interface before they become completely inserted into and translocated across the membrane [90–93]. An example of the respective contributions of $\Delta G, \Delta H,$ and ΔS to partitioning and folding of an α-helical fusion peptide in lipid bilayers is shown in Figure 28.6.

28.7
Membrane Toxins as Models for Helical Membrane Protein Insertion

There are several proteins that assume globular structures in solution, but upon interaction with a membrane receptor or at low pH insert into and sometimes translocate across membranes. Among them are several colicins, which are translocated across outer bacterial membranes via porins and subsequently inserted into the inner membrane where they form toxic pores. Other toxins such as diphtheria, tetanus, and botulinum toxins spontaneously insert their translocation domains into mammalian cell membranes and thereby permit the translocation of linked catalytic domains into the cells. Since these proteins form helical structures in solution and in membranes, they are good models to study the refolding of soluble helical proteins into membranes. The channel-forming colicins are perhaps in this regard the best-studied bacterial toxins. The structures of the soluble forms of the channel domains of several colicins are known [94]. They consist of a central hydrophobic helical hairpin that is surrounded by eight amphipathic helices.

Refolding at the membrane interface has been implicated in many models of colicin insertion into lipid bilayers. An early study found a molten globule intermediate at an early stage of colicin A insertion into membranes [95]. The helices were formed, but tertiary contacts were not. Subsequent fluorescence studies indicated that the helices of colicin E were located in the interface, but widely dispersed in this state [96, 97]. The central hydrophobic helical hairpin of colicin A appears to be more deeply inserted, but does not assume a complete TM topology [98]. How the voltage-gated ion channel is opened from this stage is still unclear. Apparently, long stretches of sequence translocate across the membrane upon the application of a membrane potential by a process that is still poorly understood structurally and mechanistically [99, 100].

The process of membrane insertion of the translocation domain of diphtheria toxin likely resembles that of the colicins. The structure of the soluble form features a hydrophobic helical hairpin similar to that of the colicins [101]. It is thought that this hairpin inserts first into the lipid bilayer. The membrane-bound protein can exist in two conformations with an either shallow or deeply inserted

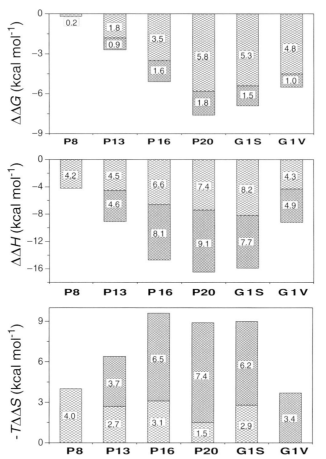

Fig. 28.6. Thermodynamics of partition-folding of the helical fusion peptide from influenza hemagglutinin. The lighter bars represent contributions from partitioning and the darker bars represent contributions from the coil → α-helix transition to ΔG, ΔH, and $T\Delta S$ upon peptide insertion into mixed phosphatidylcholine/phosphatidylglycerol bilayers. P8 through P20 are 8- to 20-residue peptides of the same sequence, and G1S and G1V are single amino acid mutations in position 1 of P20 that cause hemi-fusion or are defective in causing fusion, respectively. The numbers in the bars refer to the respective energies in kcal mol^{-1}. Adapted from Ref. [16].

helical hairpin [102]. Similar to the colicins, diphtheria toxin switches between two very different topologies in membranes and their relative populations depend on the applied membrane potential [103]. Again, the translocation mechanism is only poorly understood at the present time. However, despite many still unexplored details, the bacterial toxins appear to be an interesting class of facultative membrane proteins, from which we expect to learn a great deal about the mechanisms and energetics of refolding helical proteins in lipid bilayers.

28.8
Mechanisms of β-Barrel Membrane Protein Folding

The mechanism of folding and membrane insertion of the eight-stranded β-barrel protein OmpA has been studied by a number of kinetic experiments. The kinetics of folding into lipid bilayers composed of dimyristoylphosphatidylcholine in the fluid phase were found to be rather slow (i.e., on the order of many minutes at 30 °C) [104]. A more detailed kinetic study carried out with dioleoylphosphatidylcholine (DOPC) lipid bilayers in the 2–40 °C temperature range revealed three distinct kinetic phases [105]. The fastest phase detected by tryptophan (Trp) fluorescence changes had a time constant of 6 min and was rather independent of temperature. This phase was attributed to the initial binding of the unfolded protein to the bilayer surface. A second phase was strongly temperature dependent and had time constants in the 15 min to 3 h range (40–2 °C). The activation energy determined from an Arrhenius plot was 11 kcal mol^{-1}. This phase presumably corresponds to a deeper insertion, but not yet complete translocation of the β-strands in the lipid bilayer. The slowest phase was observed by the SDS gel-shift assay, which reports on the completion of the β-barrel. Complete folding in DOPC bilayers was only observed at temperatures greater than 30 °C, had a time constant of about 2 h, and took about 6 h to go to completion at 37 °C.

The translocation process of Trps of OmpA across the lipid bilayer was monitored by time-resolved Trp fluorescence quenching (TDFQ) [106]. In this technique, quenchers of Trp fluorescence are placed at different depths into the membrane and the time course of passage of Trps past these zones of quenchers is followed. When the technique was applied to single Trp mutants that were placed on different membrane-crossing β-hairpins, it was found that all four β-hairpins of OmpA crossed the membrane following the same time course (i.e., using a synchronized concerted mechanism of folding and insertion) [107]. This result is in accordance with the notion that interstrand hydrogen bonds and the barrel itself have to form while the protein translocates across the membrane. Again and very similar to the partition–folding coupling that was discussed in the context of helical membrane protein folding at membrane interfaces, we observe a folding–translocation coupling for this class of membrane proteins.

Thus the mechanism of β-barrel membrane protein folding in lipid bilayers is very different from the two (or more) stage models discussed above for helical membrane protein folding. The requirement that the barrel needs to be completely folded in order to translocate across the membrane probably also explains why the time constants of this process are so slow and the activation energies are so high. Folding and translocation of the barrel requires a large defect to be created in the membrane in order to insert a barrel of the size of OmpA. Moreover, the membrane provides a 100- to 1000-fold more viscous environment than water for folding and inserting membrane proteins.

In another kinetic study, the dependence of the rates of folding and insertion of OmpA on the bilayer thickness was examined [108]. As one might expect from the material properties of lipid bilayers, the folding and insertion rates increased significantly as the bilayer thickness decreased. When the bilayers were sufficiently thin, folding of OmpA into large unilamellar vesicles was observed, whereas in average or thick bilayers complete folding and insertion occurs only in small unilamellar vesicles. Small vesicles are more strained and therefore exhibit more defects than large vesicles, which permits the quantitative refolding of OmpA in small, but not in large unilamellar vesicles if the hydrocarbon thickness of the bilayer is more than 20 Å (i.e., that of diC_{12}-phosphatidylcholine). Interestingly, the hydrophobic thickness of outer bacterial membranes is thought to be thinner than that of the inner membranes [109].

28.9
Experimental Protocols

28.9.1
SDS Gel Shift Assay for Heat-modifiable Membrane Proteins

The heat modifiability can be defined as a property of a protein whose electrophoretic mobility depends on the treatment of heat. It is characteristic for many bacterial outer membrane proteins [110]. For example, detergent-solubilized OmpA from E. coli migrates on SDS-polyacrylamide gels as a 30-kDa form without boiling, but as a 35-kDa form if boiled in SDS. Likewise, the iron transporter FepA shows an apparent molecular mass of 58 kDa if not boiled, but exhibits its actual molecular mass of 81 kDa after boiling [55]. The correlation between conformational states and electrophoretic mobilities on SDS-PAGE has been studied most extensively for unboiled samples OmpA. The 30-kDa form represents the membrane-inserted native state, which also exhibits a characteristic ion channel activity [45, 49, 52]. The 35-kDa form may represent a completely denatured state in the presence or absence of lipids [49–51], a misfolded state in the absence of lipid [111], a surface-adsorbed partially folded state [49, 90], or a kinetic intermediate at a stage before the complete insertion into membranes [105, 108].

28.9.1.1 Reversible Folding and Unfolding Protocol Using OmpA as an Example

1. Dilute OmpA in 8 M urea more than 100-fold into sonicated small unilamellar vesicles (SUV) composed of the desired lipids to give a final OmpA concentration of 12 µM and lipid-to-protein molar ratio of 800.
2. Incubate for 3 h at 37 °C for refolding.
3. Divide the refolded protein-lipid complex into 10 aliquots.
4. Add appropriate amounts of a freshly made 10 M urea and buffer (10 mM glycine, pH 10, 2 mM EDTA) solutions to the aliquots to reach a 2.5-fold dilution of OmpA and urea concentrations in the range of interest.
5. Incubate reactions overnight at 37 °C.
6. Terminate reactions by mixing the samples with an equal volume of 0.125 M Tris buffer, pH 6.8, 4% SDS, 20% glycerol, and 10% 2-mercaptoethanol at room temperature.

7. Without boiling, load equilibrated samples on 12.5% SDS polyacrylamide gels, run electrophoresis, and stain with Coomassie Blue.
8. Measure fractions folded and unfolded, F and U, by densitometry and fit to two-state equilibrium folding equation:

$$[U] = \frac{([U]+[F])\exp(\{m[\text{urea}] - \Delta G°_{H2O}\}/RT)}{1+\exp(\{m[\text{urea}]-\Delta G°_{H2O}\}/RT)} \quad (1)$$

The fit parameters are the free energy of unfolding, $\Delta G°_{H2O}$, and the *m*-value.

An example of a reversible unfolding/refolding reaction using this protocol is shown in Figure 28.7.

(A)

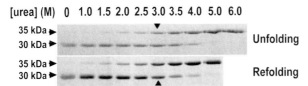

(B)

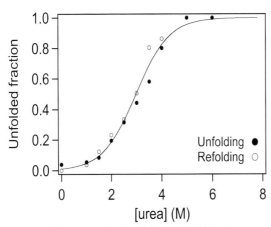

Fig. 28.7. Reversible urea-induced unfolding/refolding of OmpA in lipid bilayers composed of 92.5% phosphatidylcholine ($C_{16:0}C_{18:1}$ PC) and 7.5% phosphatidylglycerol ($C_{16:0}C_{18:1}$ PG) at pH 10 and 37 °C monitored by the SDS-PAGE shift assay. A) Unfolding (upper panel) and refolding (lower panel) of OmpA. The 30-kDa form represents the native state, and the 35-kDa form represents denatured states. B) Unfolded fractions as a function of urea concentration estimated from A. The data are fitted with the two-state equilibrium folding equation Eq. (1).

28.9.2
Tryptophan Fluorescence and Time-resolved Distance Determination by Tryptophan Fluorescence Quenching

Steady-state and time-resolved Trp fluorescence spectroscopy have been widely used to the study folding and stability of membrane proteins. The relatively large spectral shifts and fluorescence intensity changes of Trps as a result of changing local environments have made Trp fluorescence an excellent tool for such studies. Figure 28.8A shows typical fluorescence spectral changes of OmpA–lipid complexes equilibrated in various urea concentrations. The unfolding transition curves are constructed as a function of urea by parametrizing each spectrum to an average emission wavelength:

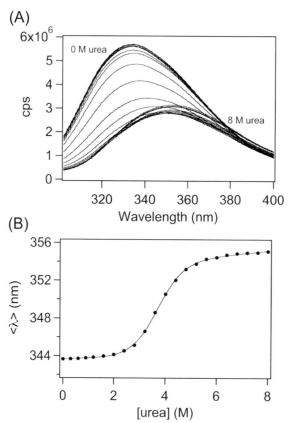

Fig. 28.8. Urea-induced unfolding of OmpA in lipid bilayers composed of 92.5% phosphatidylcholine ($C_{16:0}C_{18:1}$ PC) and 7.5% phosphatidylglycerol ($C_{16:0}C_{18:1}$ PG) at pH 10 and 37 °C monitored by Trp fluorescence. A) Fluorescence emission spectra of OmpA as a function of urea concentration. The protein and lipid concentrations were 1.2 µM and 10 mM, respectively, and the urea concentration ranged from 0 to 8 M (left to right). B) Unfolding transition monitored by the average emission wavelength $\langle \lambda \rangle$ (Eq. (2)) calculated from the spectra in A and fitted with the two-state equilibrium folding equation Eq. (3).

$$\langle \lambda \rangle = \frac{\sum_i (F_i \lambda_i)}{\sum_i (\lambda_i)} \quad (2)$$

and fitting the data to the two-state equilibrium folding equation:

$$\langle \lambda \rangle = \frac{\langle \lambda \rangle_F + \langle \lambda \rangle_U \frac{1}{Q_R} \exp[m([\text{denaturant}] - C_m)/RT]}{1 + \frac{1}{Q_R} \exp[m([\text{denaturant}] - C_m)/RT]} \quad (3)$$

Inclusion of the weighting factor $Q_R =$ (total fluorescence intensity in native state)/(total fluorescence intensity in denatured state) makes $\langle \lambda \rangle$ proportional to species concentrations and therefore permits $\langle \lambda \rangle$ to be taken as a measure of the fraction of folded protein [112]. The free energy of unfolding is calculated from the fitted denaturant concentration at the mid-point of the transition, C_m, and the m-value using $\Delta G°_{H2O} = mC_m$. Figure 28.8B shows an example of equilibrium unfolding of OmpA measured by Trp fluorescence.

The degree of solvent exposure of Trp residues of membrane proteins and thus their degree of immersion in the lipid bilayer is often determined by the effect of polar (e.g., cesium, iodide, bromide, acrylamide) or nonpolar (e.g., molecular oxygen) quenchers on the fluorescence spectra. A more informative method to determine the depth of Trps in membranes involves the use of spin-labeled or brominated lipids as quenchers. Phospholipids selectively labeled with bromines or spin labels in various positions of the acyl chains are commercially available. Depths in membranes of specific Trps are often measured by using a whole set of these lipids and analyzing the resulting spectra with the "parallax" or "distribution analysis" methods [113]. These steady-state methods may be extended to measure the time dependence of Trp positions and more specifically the translocation of Trps across lipid bilayers during the folding of membrane proteins [106]. Reliable distances in membranes are obtained by the time-resolved distance determination by fluorescence quenching (TDFQ) technique because the positions and distributions of the bromines in the bilayer are known from X-ray diffraction studies [114].

28.9.2.1 TDFQ Protocol for Monitoring the Translocation of Tryptophans across Membranes

1. Mix solutions of DOPC and one each of DOPC plus 4,5-, 6,7-, 9,10-, and 11,12-diBrPC, respectively, at molar ratios of 7:3 (DOPC:m,n-diBrPC) in chloroform.
2. Dry the mixtures on the bottom of test tubes under a stream of nitrogen and further in a desiccator under a high vacuum overnight.
3. Disperse dried lipids in 1.0 mL buffer (10 mM glycine, pH 8.5, 1 mM EDTA, 150 mM NaCl) to a lipid concentration of 2 mg mL^{-1}.
4. Prepare SUVs by sonication of the lipid dispersions for 50 min using the microtip of a Branson ultrasonifier at 50% duty cycle in an ice/water bath.

5. Remove titanium dust by low-speed centrifugation (e.g., at 6000 rpm for 10 min on a tabletop centrifuge) and store vesicles at 4 °C overnight for equilibration.
6. Measure blank fluorescence emission spectra for each vesicle solution at 1 mM lipid concentration and at the controlled temperature of interest with instrument settings for Trp fluorescence (e.g., excitation at 290 nm with 2-nm slit; emission from 310 to 370 nm with 5-nm slit).
7. Add small volume of concentrated OmpA (stock is typically 20–50 mg mL^{-1}) in 8 M urea to give a final OmpA concentration of 1.4 µM and mix well.
8. Record fluorescence spectra with above settings every 2–5 min. (The first spectrum can be taken 30 s after mixing.) Repeat measurements for each of the five vesicle solutions.
9. Measure fluorescence intensities of smoothed spectra at 330 nm as a function of time and normalize values for each of the quenching vesicles to the corresponding values of vesicles without lipid quencher.
10. Fit normalized fluorescence intensities to the Gaussian distribution equation:

$$\frac{F(d_Q)}{F_0} = \exp\left\{ -\frac{S}{\sigma\sqrt{2\pi}} \exp\left\{ -\frac{1}{2}\left(\frac{d_Q - d_{Trp}}{\sigma}\right)^2 \right\} \right\} \tag{4}$$

F_0 and $F(d_Q)$ are the fluorescence intensities in the absence and presence of a particular m,n-diBrPC quencher. d_Q are the known distances of the quenchers from the bilayer center, and d_{Trp} is the distance of the fluorophor from the center. The dispersion σ is a measure of the width of the distribution of the fluorophor in the membrane, which in turn depends on the sizes and thermal fluctuations of the fluorophors and quenchers. S is a function of the quenching efficiency and quencher concentration in the membrane. The fits yield d_{Trp}, σ, and S.

11. Plot d_{Trp} and σ as a function of time.

The accuracy of the distribution analysis could be improved if more than four membrane-bound quenchers at different depths in the membrane and/or a quencher attached to the polar lipid headgroup were used. Figure 28.9 shows two examples of Trp movements of OmpA into and across lipid bilayers, respectively, measured by TDFQ using four different bromine quenchers in the acyl chain region.

28.9.3
Circular Dichroism Spectroscopy

Circular dichroism (CD) refers to the differential absorption of the right- and left-handed circular polarized lights by a sample that exhibits molecular asymmetry. Far-UV CD spectroscopy has been used to characterize the secondary structures of

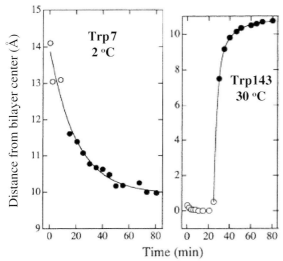

Fig. 28.9. Measurement of the translocation rate of single tryopophans of OmpA into and across lipid bilayers composed of phosphatidylcholine (diC$_{18:1}$PC) by time-resolved fluorescence quenching (TDFQ). A) Time course of the movement of Trp7 of OmpA into lipid bilayer at 2 °C. This Trp stays on the *cis* side and does not cross the bilayer. B) Time course of the movement of Trp143 across the lipid bilayer at 30 °C. This Trp translocates across the bilayer with a time constant of 3.3 min at 30 °C. The solid lines are fits of the data to exponential functions. Adapted from Ref. [107].

proteins including membrane proteins because the differential amide absorptions of α-helices, β-sheets, turns, and random coils in the range between 180 and 250 nm yield characteristic CD signals that can be used to determine their secondary structure compositions. The CD spectra of α-helices exhibit minima at 208 and 222 nm. Beta-sheets show a weaker minimum at 216 nm. Both of these structures exhibit maxima just below 200 nm, but the spectra of random coils are negative in the entire accessible far-UV spectral range with a minimum just below 200 nm.

CD spectroscopy of membrane proteins is difficult because the proteo-liposomes, which are necessary for functional reconstitution of membrane proteins, are strong scatterers of UV light resulting in overwhelming spectral noise. Therefore, only SUVs, which are typically obtained by sonication and which are typically 30 nm in diameter, or protein–detergent mixed micelles are recommended for recording CD spectra of membrane proteins. Even with these samples, CD spectra cannot usually be recorded to below 200 nm, which limits the accuracy of secondary structure determinations. Good results can be obtained with 0.05-mm pathlength cells and protein concentrations of 0.5–1.0 mg mL^{-1} in buffers that do not absorb in the far UV. Relatively crude estimates of secondary structure content can be made from the spectral shapes in the 200–240 nm spectral region and using the following rule of thumb values for 100% secondary structure: $\theta_{222} = -30\,000$ deg cm^2 dmol^{-1} for α-helix and $\theta_{216} = -15\,000$ deg cm^2 dmol^{-1} for β-sheet.

28.9.4
Fourier Transform Infrared Spectroscopy

Fourier transform infrared (FTIR) spectroscopy is a useful technique for determining the conformation and orientation of membrane proteins and lipids in lipid bilayers. The method has no light scattering artifacts and works well with turbid and even solid samples. Since the absorption bands of liquid water overlap with several bands that are of interest in protein and membrane spectroscopy, it is convenient to work in D$_2$O solutions. In conventional transmittance spectroscopy, the strong absorbances of residual H$_2$O and bulk D$_2$O are minimized by using high protein concentrations and measuring cells of short path lengths.

Another very convenient method for suppressing unwanted water signals is to record attenuated total reflection (ATR) FTIR spectra. In this method, the IR beam is reflected within an IR-transparent internal reflection element such as a germanium plate. An evanescent wave of the same frequency as the incoming IR light builds up at the germanium–water interface. The electric field of the evanescent wave decreases exponentially away from the interface with a characteristic decay length, d_p, which is on the order of a few hundred nanometers in typical applications. This wave probes with high sensitivity samples that are adsorbed at the germanium–water interface such as a reconstituted protein–lipid bilayer. Since absorptions from the bathing buffer solution are greatly suppressed in this surface-sensitive technique, small amounts of H$_2$O in D$_2$O are not a problem in ATR-FTIR spectroscopy. An additional advantage is that the orientation of membrane proteins and lipids in the bilayers can be obtained by using polarized light. Microgram quantities of membrane proteins in single supported bilayers are sufficient to record high quality spectra. Finally, buffer conditions can be varied in situ using a perfusable ATR measuring cell. A comprehensive description of ATR-FTIR spectroscopy on supported membranes has appeared [115].

The most widely used conformation-sensitive vibrational mode of proteins is the amide I' band (1600–1700 cm^{-1}). The secondary structures of proteins are correlated with the amide I' frequencies, which enables the assignment of the major secondary structure elements of proteins by spectral decomposition of this band (α-helix: 1648–1660 cm^{-1}; parallel and antiparallel β-sheet: 1625–1640 cm^{-1}; antiparallel β-sheet: 1675–1695 cm^{-1}; aggregated strands: 1610–1628 cm^{-1}; turns: 1660–1685 cm^{-1}; unordered structures: 1652–1660 cm^{-1} in H$_2$O and 1640–1648 cm^{-1} in D$_2$O). The peak corresponding to the α-helix is often not well resolved from nearby peaks of turns and unordered structures. However, β-sheet can usually be reliably assigned due to its distinctive components at about 1630 cm^{-1} and 1690 cm^{-1}.

The following bands of lipid vibrations are well separated from protein amide bands and are frequently analyzed in studies of protein–lipid interactions: symmetric and antisymmetric methylene stretches at 2850 cm^{-1} and 2920 cm^{-1}, respectively, and ester carbonyl stretch at 1730 cm^{-1}.

28.9.4.1 Protocol for Obtaining Conformation and Orientation of Membrane Proteins and Peptides by Polarized ATR-FTIR Spectroscopy

1. Fill liquid ATR cell containing germanium plate with D_2O buffer (e.g., 5 mM HEPES/10 mM MES pH 7.3, 135 mM NaCl) and record FTIR spectra at vertical and horizontal polarizations with a spectral resolution of 2 cm^{-1} to obtain reference spectra.
2. Deposit lipid monolayer on germanium plate by the Langmuir-Blogett technique at 32 mN m^{-1}.
3. Add proteoliposomes (typically about 0.5 mM lipid and 20 µg mL^{-1} protein in same H_2O buffer) to supported monolayer in assembled ATR cell and incubate with gentle shaking for 1 h. The proteoliposomes will fuse with the monolayer during this time and form a complete planar bilayer on the surface of the germanium plate.
4. Flush excess proteoliposomes out of measuring cell with 10 volumes of D_2O buffer. This step also exchanges H_2O for D_2O.
5. Incubate exchanged sample for 1 h at room temperature to equilibrate labile amide protons with deuteriums. This is done in the FTIR spectrometer and the spectrometer compartment is purged with dry air or nitrogen at the same time to remove water vapor.
6. Collect the sample spectra using the same experimental settings as reference spectra at vertical and horizontal polarizations.
7. Ratio sample and reference spectra and calculate absorption spectra.

The following steps describe a recipe to extract secondary structure information from complex (overlapped) amide I' bands:

8. Calculate second-derivative spectra by differentiation with 11-point Savitsky-Golay smoothing. The minima in the second derivatives indicate possible component peaks.
9. Using only prominent and clearly identified component peaks as starting points, carry out spectral decomposition by least-squares fitting with Gaussian component line shapes. Use as few components as possible to reproduce the experimental line shape. Accept only fits that keep the Gaussian peak components centered close to the component frequencies that were identified in the second derivative spectra.
10. Integrate spectral components, assign to secondary structures, and calculate area fractions to indicate respective secondary structure fractions.

The following steps describe how to obtain orientations of secondary structure elements of proteins and orientations (order parameters) of lipid chemical groups:

11. Calculate ATR dichroic ratio for the amide I' bands and symmetric or antisymmetric methylene stretching band intensities according to

$$R_{\text{amideI}}^{\text{ATR}} = \frac{\int A_{//}(\nu)\,d\nu}{\int A_{\perp}(\nu)\,d\nu} \tag{5}$$

and

$$R_{\text{Lipid}}^{\text{ATR}} = \frac{A_{//}(\nu)}{A_{\perp}(\nu)} \tag{6}$$

12. The order parameter is defined as

$$S_\theta = (3\langle \cos^2 \theta \rangle - 1)/2 \tag{7}$$

where θ is the angle between the axis of rotational symmetry of an axially symmetric molecule (long axis of the lipid, α-helix, β-barrel, etc.) and the normal to the germanium plate and the angular brackets denote a time and space average over all angles in a given sample.

13. Calculate the order parameter for an amide I′ band component as:

$$S_{\text{amideI}} = \frac{E_x^2 - R_{\text{amideI}}^{\text{ATR}} E_y^2 + E_z^2}{E_x^2 - R_{\text{amideI}}^{\text{ATR}} E_y^2 - 2E_z^2} \tag{8}$$

where E_x, E_y, and E_z are the electric field amplitudes at the germanium–buffer interface. For a 45° germanium plate with a thin attached membrane in D_2O buffer $E_x^2 = 1.9691$, $E_y^2 = 2.2486$, and $E_z^2 = 1.8917$. For α-helices, where the angle between the amide I′ transition dipole moment and the helix axis is 39°, Eq. (8) transforms to

$$S_H = 2.46 \frac{E_x^2 - R_{\text{amideI}}^{\text{ATR}} E_y^2 + E_z^2}{E_x^2 - R_{\text{amideI}}^{\text{ATR}} E_y^2 - 2E_z^2} \tag{9}$$

S_H is the order parameter that describes the helix orientation relative to the germanium plate and membrane normal. For β-sheets, there is no helical symmetry and Eq. (8) describes the order parameter of the transition dipole moment of the amide I′ band, which roughly coincides with the order parameter of the amide carbonyl bond orientations.

14. Lipid order parameters are obtained from the experimental dichroic ratios of the symmetric and/or antisymmetric methylene stretch vibrations:

$$S_L = -2 \frac{E_x^2 - R_L^{\text{ATR}} E_y^2 + E_z^2}{E_x^2 - R_L^{\text{ATR}} E_y^2 - 2E_z^2} \tag{10}$$

Lipid order parameters should always be recorded along with those of embedded proteins because they provide essential controls of the bilayer quality.

Results on protein orientations are only meaningful if lipid orientations (to the germanium plate) are satisfactory.

Lipid-to-protein ratios in supported bilayers:

15. Use the following form of Beer-Lambert's law modified for polarized ATR spectroscopy to calculate the lipid/protein ratio in supported bilayers [116]:

$$L/P = 0.208(n_{res} - 1)\frac{1 - S_{amideI}}{1 + S_L/2} \frac{\int_{2800}^{2980} A_\perp(v_L)\,dv}{\int_{1600}^{1690} A_\perp(v_{amideI})\,dv} \quad (11)$$

where n_{res} is the number of residues in the polypeptide and v_L and v_{amideI} are the wavenumbers of the lipid methylene stretch and the protein amide I' vibrations, respectively.

Acknowledgments

We thank members of the Tamm laboratory, past and present, for their contributions and many discussions. The work from our laboratory was supported by grants from the National Institutes of Health.

References

1. DILL, K. A. (1990). Dominant forces in protein folding. *Biochemistry* 29, 7133–7155.
2. BALDWIN, R. L. (1999). Protein folding from 1962 to 1982. *Nature Struct. Biol.* 6, 814–817.
3. FERSHT, A. (1999). *Structure and Mechanism in Protein Science*. Freeman, New York.
4. WALLIN, E. & VON HEIJNE, G. (1998). Genome-wide analysis of integral membrane proteins from eubacterial, archaean, and eukaryotic organisms. *Protein Sci.* 7, 1029–1038.
5. HALTIA, T. & FREIRE, E. (1995). Forces and factors that contribute to the structural stability of membrane proteins. *Biochim. Biophys. Acta* 1228, 1–27.
6. VAN DEN BERG, B., CLEMONS, W. M., JR., COLLINSON, I., KODIS, Y., HARTMAN, E., HARRISON, S. C. & RAPOPORT, T. A. (2004). X-ray structure of a protein-conducting channel. *Nature* 427, 37–44.
7. WHITE, S. H. & WIENER, M. C. (1996). The liquid-crystallographic structure of fluid lipid bilayer membranes. In *Biological Membranes: A Molecular Perspective from Computation and Experiment* (MERZ, K. M. & ROUX, B., eds), pp. 127–144. Birkhauser, Boston.
8. NAGLE, J. F. & TRISTRAM-NAGLE, S. (2000). Structure of lipid bilayers. *Biochim. Biophys. Acta* 1469, 159–195.
9. TANFORD, C. (1980). *The Hydrophobic Effect*, 2nd edn. Wiley, New York.
10. MARSH, D. (1996). Lateral pressure in membranes. *Biochim. Biophys. Acta* 1286, 183–223.
11. CANTOR, S. C. (1999). Lipid composition and the lateral pressure

profile in bilayers. *Biophys. J.* 76, 2625–2639.
12 BROWN, M. F., THURMOND, R. L., DODD, S. W., OTTEN, D. & BEYER, K. (2002). Elastic deformation of membrane bilayers probed by deuterium NMR relaxation. *J. Am. Chem. Soc.* 124, 8471–8484.
13 JAYASINGHE, S., HRISTOVA, K. & WHITE, S. II. (2001). Energetics, stability, and prediction of transmembrane helices. *J. Mol. Biol.* 312, 927–934.
14 LADOKHIN, A. S. & WHITE, S. H. (1999). Folding of amphipathic α-helices on membranes: energetics of helix formation by melittin. *J. Mol. Biol.* 285, 1363–1369.
15 WIEPRECHT, T., APOSTOLOV, O., BEYERMANN, M. & SEELIG, J. (1999). Thermodynamics of the α-helix-coil transition of amphipathic peptides in a membrane environment: implications for the peptide-membrane binding equilibrium. *J. Mol. Biol.* 294, 785–794.
16 LI, Y., HAN, X. & TAMM, L. K. (2003). Thermodynamics of fusion peptide-membrane interactions. *Biochemistry* 42, 7245–7251.
17 WIMLEY, W. C., HRISTOVA, K., LADOKHIN, A. S., SILVESTRO, L., AXELSEN, P. H. & WHITE, S. H. (1998). Folding of β-sheet membrane proteins: A hydrophobic hexapeptide model. *J. Mol. Biol.* 277, 1091–1110.
18 BENTAL, N., SITKOFF, D., TOPOL, I. A., YANG, A.-S., BURT, S. K. & HONIG, B. (1997). Free energy of amide hydrogen bond formation in vacuum, in water, and in liquid alkane solution. *J. Phys. Chem. B* 101, 450–457.
19 WIMLEY, W. C. & WHITE, S. H. (1996). Experimentally determined hydrophobicity scale for proteins at membrane interfaces. *Nature Struct. Biol.* 3, 842–848.
20 WIMLEY, W. C., CREAMER, T. P. & WHITE, S. H. (1996). Solvation energies of amino acid side chains and backbone in a family of host-guest pentapeptides. *Biochemistry* 35, 5109–5124.
21 DWYER, J. J., GITTIS, A. G., KARP, D. A., LATTMAN, E. E., SPENCER, D. S., STITES, W. E. & GARCIA-MORENO, E. B. (2000). High apparent dielectric constants in the interior of a protein reflect water penetration. *Biophys. J.* 79, 1610–1620.
22 KILLIAN, J. A. & VON HEIJNE, G. (2000). How protein adapt to a membrane-water interface. *Trends Biochem. Sci.* 25, 429–434.
23 YAU, W.-M., WIMLEY, W. C., GAWRISCH, K. & WHITE, S. H. (1998). The preference of tryptophan for membrane interfaces. *Biochemistry* 37, 14713–14718.
24 SCHULZ, G. E. (2002). The structure of bacterial outer membrane proteins. *Biochim. Biophys. Acta* 1565, 308–317.
25 GABRIEL, K., BUCHANAN, S. K. & LITHGOW, T. (2001). The alpha and the beta: protein translocation across mitochondrial and plastid outer membranes. *Trends Biochem. Sci.* 26, 36–40.
26 PASCHEN, S. A., WAIZENEGGER, T., STAN, T., PREUSS, M., CYRKLAFF, M., HELL, K., RAPAPORT, D. & NEUPERT, W. (2003). Evolutionary conservation of biogenesis of β-barrel membrane proteins. *Nature* 426, 862–866.
27 DANESE, P. N. & SILHAVY, T. J. (1998). Targeting and assembly of perplasmic and outer-membrane proteins in *Escherichia coli*. *Annu. Rev. Genet.* 32, 59–94.
28 WIMLEY, W. C. (2002). Toward genomic identification of β-barrel membrane proteins: Composition and architecture of known structures. *Protein Sci.* 11, 301–312.
29 CHEN, R. & HENNING, U. (1996). A periplasmic protein (Skp) of *Escherichia coli* selectively binds a class of outer membrane proteins. *Mol. Microbiol.* 19, 1287–1294.
30 LAZAR, S. W. & KOLTER, R. (1996). SurA assists the folding of *Escherichia coli* outer membrane proteins. *J. Bacteriol.* 178, 1770–1773.
31 KOEBNIK, R., LOCHER, K. P. & VAN GELDER, P. (2000). Structure and function of bacterial outer membrane proteins: barrels in a nutshell. *Mol. Microbiol.* 37, 239–253.

32. COWAN, S. W., SCHIRMER, T., RUMMEL, G., STEIERT, M., GHOSH, R., PAUPTIT, R. A., JANSONIUS, J. N. & ROSENBUSCH, J. P. (1992). Crystal structures explain functional properties of two E. coli porins. *Nature* 358, 727–733.

33. WEISS, M. S., ABELE, U., WECKESSER, J., WELTE, W., SCHILTZ, E. & SCHULZ, G. E. (1991). Molecular architecture and electrostatic properties of a bacterial porin. *Science* 254, 1627–1630.

34. SCHIRMER, T., KELLER, T. A., WANG, Y. F. & ROSENBUSCH, J. P. (1995). Structural basis for sugar translocation through maltoporin channels in 3.1 A resolution. *Science* 267, 512–514.

35. FORST, D., WELTE, W., WACKER, T. & DIEDERICHS, K. (1998). Structure of the sucrose-specific porin ScrY from *Salmonella typhimurium* and its complex with sucrose. *Nature Struct. Biol.* 5, 37–46.

36. FERGUSON, A. D., HOFMANN, E., COULTON, J. W., DIEDERICHS, K. & WELTE, W. (1998). Siderophore-mediated iron transport: crystal structure of FhuA with bound lipopolysaccharide. *Science* 282, 2215–2220.

37. LOCHER, K. P., REES, B., KOEBNIK, R., MITSCHLER, A., MOULINIER, L. & ROSENBUSCH, J. P. (1998). Transmembrane signaling across the ligand-gated FhuA receptor: crystal structures of free and ferrichrome-bound states reveal allosteric changes. *Cell* 95, 771–778.

38. BUCHANAN, S. K., SMITH, B. S., VENKATRAMANI, L. et al. (1999). Crystal structure of the outer membrane active transporter FepA from *Escherichia coli. Nature Struct. Biol.* 6, 56–63.

39. FERGUSON, A. D., CHAKRABORTY, R., SMITH, B. S., ESSER, L., VAN DER HELM, D. & DEISENHOFER, J. (2002). Structural basis of gating by the outer membrane transporter FecA. *Science* 295, 1658–1659.

40. CHIMENTO, D. P., MOHANTY, A. K., KADNER, R. J. & WIENER, M. C. (2003). Substrate-induced transmembrane signaling in the cobalamin transporter BtuB. *Nature Struct. Biol.* 10, 394–401.

41. SNIJDER, J. J., UBARRETXENA-BALANDIA, I., BLAAUW, M. et al. (1999). Structural evidence for dimerization-regulated activation of an integral membrane protein phospholipase. *Nature* 401, 717–721.

42. VOGT, J. & SCHULZ, G. E. (1999). The structure of the outer membrane protein OmpX from *Escherichia coli* reveals possible mechanisms of virulence. *Structure* 7, 1301–1309.

43. PAUTSCH, A. & SCHULZ, G. E. (1998). Structure of the outer membrane protein A transmembrane domain. *Nature Struct. Biol.* 5, 1013–1017.

44. SONNTAG, I., SCHWARZ, H., HIROTA, Y. & HENNING, U. (1978). Cell envelope and shape of *Escherichia coli*: multiple mutants missing the outer membrane lipoprotein and other major outer membrane proteins. *J. Bacteriol.* 136, 280–285.

45. ARORA, A., ABILDGAARD, F., BUSHWELLER, J. H. & TAMM, L. K. (2001). Structure of outer membrane protein A transmembrane domain by NMR spectroscopy. *Nature Struct. Biol.* 8, 334–338.

46. KOEBNIK, R. (1999). Membrane assembly of the *Escherichia coli* outer membrane protein OmpA: Exploring sequence constraints on transmembrane β-strands. *J. Mol. Biol.* 285, 1801–1810.

47. KOEBNIK, R. & KRÄMER, L. (1995). Membrane assembly of circularly permuted variants of the E. coli outer membrane protein OmpA. *J. Mol. Biol.* 250, 617–626.

48. DORNMAIR, K., KIEFER, H. & JÄHNIG, F. (1990). Refolding of an integral membrane protein. OmpA of *Escherichia coli. J. Biol. Chem.* 265, 18907–18911.

49. SURREY, T. & JÄHNIG, F. (1992). Refolding and oriented insertion of a membrane protein into a lipid bilayer. *Proc. Natl Acad. Sci. USA* 89, 7457–7461.

50. HONG, H. & TAMM, L. K. (2004). Elastic coupling of integral membrane

protein stability to lipid bilayer forces. *Proc. Natl Acad. Sci. USA* 101, 4065–4070.

51 SCHWEIZER, M., HINDENNACH, I., GARTEN, W. & HENNING, U. (1978). Major proteins of the *Escherichia coli* outer cell envelope membrane. Interaction of protein II with lipopolysaccharide. *Eur. J. Biochem.* 82, 211–217.

52 ARORA, A., RINEHART, D., SZABO, G. & TAMM, L. K. (2000). Refolded outer membrane protein A of *Escherichia coli* forms ion channels with two conductance states in planar lipid bilayers. *J. Biol. Chem.* 275, 1594–1600.

53 GRUNER, S. (1985). Intrinsic curvature hypothesis for biomembrane lipid composition: a role for nonbilayer lipids. *Proc. Natl Acad. Sci. USA* 82, 3665–3669.

54 BONHIVERS, M., DESMADRIL, M., MOECK, G. S., BOULANGER, P., COLOMER-PALLAS, A. & LETELIER, L. (2001). Stability studies of FhuA, a two-domain outer membrane protein from *Escherichia coli*. *Biochemistry* 40, 2606–2613.

55 KLUG, C. S., SU, W., LIU, J., KLEBBA, P. E. & FEIX, J. B. (1995). Denaturant unfolding of the ferric enterobactin receptor and ligand-induced stabilization studied by site-directed spin labeling. *Biochemistry* 34, 14230–14236.

56 KLUG, C. S. & FEIX, J. B. (1998). Guanidine hydrochloride unfolding of a transmembrane β-strand in FepA using site-directed spin labeling. *Protein Sci.* 7, 1469–1476.

57 ACKERMAN, M. S. & SHORTLE, D. (2001). Persistence of native-like topology in a denatured protein in 8 M urea. *Science* 293, 487–489.

58 SURREY, T., SCHMID, A. & JÄHNIG, F. (1996). Folding and membrane insertion of the trimeric β-barrel protein OmpF. *Biochemistry* 35, 2283–2288.

59 PHALE, P. S., PHILLIPPSEN, A., KIEFHABER, T., KOEBNIK, R., PHALE, V. P., SCHIRMER, T. & ROSENBUSCH, J. P. (1998). Stability of trimeric OmpF porin: The contributions of the latching loop L2. *Biochemistry* 37, 15663–15670.

60 POPOT, J.-L. & ENGELMAN, D. M. (1990). Membrane protein folding and oligomerization: the two-stage model. *Biochemistry* 29, 4031–4037.

61 POPOT, J. L., GERCHMAN, S. E. & ENGELMAN, D. M. (1987). Refolding of bacteriorhodopsin in lipid bilayers. A thermodynamically controlled two-stage process. *J. Mol. Biol.* 198, 655–676.

62 WHITE, S. H. & WIMLEY, W. C. (1999). Membrane protein folding and stability: Physical principles. *Annu. Rev. Biophys. Biomol. Struct.* 28, 319–365.

63 ENGELMAN, D. M., CHEN, Y., CHIN, C.-N. et al. (2003). Membrane protein folding: beyond the two stage model. *FEBS Lett.* 555, 122–125.

64 HUANG, K. S., BAYLEY, H., LIAO, M. J., LONDON, E. & KHORANA, H. G. (1981). Refolding of an integral membrane protein. Denaturation, renaturation, and reconstitution of intact bacteriorhodopsin and two proteolytic fragments. *J. Biol. Chem.* 256, 3802–3809.

65 LONDON, E. & KHORANA, H. G. (1982). Denaturation and renaturation of bacteriorhodopsin in detergents and lipid-detergent mixtures. *J. Biol. Chem.* 257, 7003–7011.

66 BOOTH, P. J., FLITSCH, S. L., STERN, L. J., GREENHALGH, D. A., KIM, P. S. & KHORANA, H. G. (1995). Intermediates in the folding of the membrane protein bacteriorhodopsin. *Nature Struct. Biol.* 2, 139–143.

67 LU, H. & BOOTH, P. J. (2000). The final stages of folding of the membrane protein bacteriorhodopsin occur by kinetically indistinguishable parallel folding paths that are mediated by pH. *J. Mol. Biol.* 299, 233–243.

68 LAU, F. W. & BOWIE, J. U. (1997). A method for assessing the stability of a membrane protein. *Biochemistry* 36, 5884–5892.

69 FAHAM, S., YANG, D., BARE, E., YOHANNAN, S., WHITELEGGE, J. P. &

Bowie, J. U. (2004). Side-chain contributions to membrane protein structure and stability. *J. Mol. Biol.* 335, 297–305.

70. Lemmon, M. A., Flanagan, J. M., Treutlein, H. R., Zhang, J. & Engelman, D. M. (1992). Sequence specificity in the dimerization of transmembrane alpha helices. *Biochemistry* 31, 12719–12725.

71. Senes, A., Gerstein, M. & Engelman, D. M. (2000). Statistical analysis of amino acid patterns in transmembrane helices: the GxxxG motif occurs frequently and in association with beta-branched residues at neighboring positions. *J. Mol. Biol.* 296, 921–936.

72. Bowie, J. U. (1997). Helix packing in membrane proteins. *J. Mol. Biol.* 272, 780–789.

73. Popot, J. L. & Engelman, D. M. (2000). Helical membrane protein folding, stability, and evolution. *Annu. Rev. Biochem.* 69, 881–922.

74. Fleming, K. G., Ackerman, A. L. & Engelman, D. M. (1997). The effect of point mutations on the free energy of transmembrane alpha-helix dimerization. *J. Mol. Biol.* 272, 266–275.

75. Fleming, K. G. & Engelman, D. M. (2001). Specificity in transmembrane helix-helix interactions can define a hierarchy of stability for sequence variants. *Proc. Natl Acad. Sci. USA* 98, 14340–14344.

76. Zhou, F. X., Cocco, M. J., Russ, W. P., Brünger, A. T. & Engelman, D. M. (2000). Interhelical hydrogen bonding drives strong interactions in membrane proteins. *Nature Struct. Biol.* 7, 154–160.

77. Choma, C., Gratkowski, H., Lear, J. D. & DeGrado, W. F. (2000). Asparagine-mediated self-association of a model transmembrane helix. *Nature Struct. Biol.* 7, 161–166.

78. Gratkowski, H., Lear, J. D. & DeGrado, W. F. (2001). Polar side chains drive the association of model transmembrane peptides. *Proc. Natl Acad. Sci. USA* 98, 880–885.

79. Zhou, F. X., Merianos, H. J., Brünger, A. T. & Engelman, D. M. (2001). Polar residues drive association of polyleucine transmembrane helices. *Proc. Natl Acad. Sci. USA* 98, 2250–2255.

80. Senes, A., Ubarretxena-Belandia, I. & Engelman, D. M. (2001). The Calpha −H ⋯ O hydrogen bond: a determinant of stability and specificity in transmembrane helix interactions. *Proc. Natl Acad. Sci. USA* 98, 9056–9061.

81. Yohannan, S., Faham, S., Yang, D., Grosfeld, D., Chamberlain, A. K. & Bowie, J. U. (2004). A C(alpha)–H ⋯ O hydrogen bond in a membrane protein is not stabilizing. *J. Am. Chem. Soc.* 126, 2284–2285.

82. Doyle, D. A., Morais Cabral, J., Pfuetzner, R. A. et al. (1998). The structure of the potassium channel: molecular basis of K+ conduction and selectivity. *Science* 280, 69–77.

83. Stowell, M. H. B. & Rees, D. C. (1995). Structure and stability of membrane proteins. *Adv. Protein Chem.* 46, 279–311.

84. Jacobs, R. E. & White, S. H. (1989). The nature of the hydrophobic binding of small peptides at the bilayer interface: implications for the insertion of transbilayer helices. *Biochemistry* 28, 3421–3437.

85. Vogel, H. (1981). Incorporation of melittin into phosphatidylcholine bilayers. Study of binding and conformational changes. *FEBS Lett.* 134, 37–42.

86. Kaiser, E. T. & Kezdy, F. J. (1984). Amphiphilic secondary structure: design of peptide hormones. *Science* 223, 249–255.

87. Briggs, M. S., Cornell, D. G., Dluhy, R. A. & Gierasch, L. M. (1986). Conformations of signal peptides induced by lipids suggest initial steps in protein export. *Science* 233, 206–208.

88. Tamm, L. K. (1991). Membrane insertion and lateral mobility of synthetic amphiphilic signal peptides in lipid model membranes. *Biochim. Biophys. Acta* 1071, 123–148.

89. Lear, J. D. & DeGrado, W. F. (1987). Membrane binding and confor-

mational properties of peptides representing the NH$_2$ terminus of influenza HA-2. *J. Biol. Chem.* 262, 6500–6505.
90 RODIONOVA, N. A., TATULIAN, S. A., SURREY, T., JÄHNIG, F. & TAMM, L. K. (1995). Characterization of two membrane-bound forms of OmpA. *Biochemistry* 34, 1921–1929.
91 VECSEY-SEMJEN, B., LESIEUR, C., MOLLBY, R. & VAN DER GOOT, F. G. (1997). Conformational changes due to membrane binding and channel formation by staphylococcal alpha-toxin. *J. Biol. Chem.* 272, 5709–5717.
92 TERZI, E., HÖLZEMANN, G. & SEELIG, J. (1994). Reversible random coil-beta-sheet transition of the Alzheimer beta-amyloid fragment (25–35). *Biochemistry* 33, 1345–1350.
93 BISHOP, C. M., WALKENHORST, W. F. & WIMLEY, W. C. (2001). Folding of beta-sheets in membranes: specificity and promiscuity in peptide model systems. *J. Mol. Biol.* 309, 975–988.
94 ZAKHAROV, S. D. & CRAMER, W. A. (2002). Colicin crystal structures: pathways and mechanisms for colicin insertion into membranes. *Biochim. Biophys. Acta* 1565, 333–346.
95 VAN DER GOOT, F. G., GONZALEZ-MANAS, J. M., LAKEY, J. H. & PATTUS, F. (1991). A "molten-globule" membrane-insertion intermediate of the pore-forming domain of colicin A. *Nature* 354, 408–410.
96 ZAKHAROV, S. D., LINDEBERG, M., GRIKO, Y. et al. (1998). Membrane-bound state of the colicin E1 channel domain as an extended two-dimensional helical array. *Proc. Natl Acad. Sci. USA* 95, 4282–4287.
97 ZAKHAROV, S. D., LINDEBERG, M. & CRAMER, W. A. (1999). Kinetic description of structural changes linked to membrane import of the colicin E1 channel protein. *Biochemistry* 38, 11325–11332.
98 LAKEY, J. H., DUCHE, D., GONZALEZ-MANAS, J. M., BATY, D. & PATTUS, F. (1993). Fluorescence energy transfer distance measurements. The hydrophobic helical hairpin of colicin A in the membrane bound state. *J. Mol. Biol.* 230, 1055–1067.
99 QIU, X. Q., JAKES, K. S., KIENKER, P. K., FINKELSTEIN, A. & SLATIN, S. L. (1996). Major transmembrane movement associated with colicin Ia channel gating. *J. Gen. Physiol.* 107, 313–328.
100 KIENKER, P. K., QIU, X., SLATIN, S. L., FINKELSTEIN, A. & JAKES, K. S. (1997). Transmembrane insertion of the colicin Ia hydrophobic hairpin. *J. Membr. Biol.* 157, 27–37.
101 WEISS, M. S., BLANKE, S. R., COLLIER, R. J. & EISENBERG, D. (1995). Structure of the isolated catalytic domain of diphtheria toxin. *Biochemistry* 34, 773–781.
102 WANG, Y., MALENBAUM, S. E., KACHEL, K., ZHAN, H., COLLIER, R. J. & LONDON, E. (1997). Identification of shallow and deep membrane-penetrating forms of diphtheria toxin T domain that are regulated by protein concentration and bilayer width. *J. Biol. Chem.* 272, 25091–25098.
103 SENZEL, L., GORDON, M., BLAUSTEIN, R. O., OH, K. J., COLLIER, R. J. & FINKELSTEIN, A. (2000). Topography of diphtheria toxin's T domain in the open channel state. *J. Gen. Physiol.* 115, 421–434.
104 SURREY, T. & JÄHNIG, F. (1995). Kinetics of folding and membrane insertion of a β-barrel membrane protein. *J. Biol. Chem.* 270, 28199–28203.
105 KLEINSCHMIDT, J. H. & TAMM, L. K. (1996). Folding intermediates of a beta-barrel membrane protein. Kinetic evidence for a multi-step membrane insertion mechanism. *Biochemistry* 35, 12993–13000.
106 KLEINSCHMIDT, J. H. & TAMM, L. K. (1999). Time-resolved distance determination by tryptophan fluorescence quenching (TDFQ): Probing intermediates in membrane protein folding. *Biochemistry* 38, 4996–5005.
107 KLEINSCHMIDT, J. H., DEN BLAAUWEN, T., DRIESSEN, A. J. M. & TAMM, L. K. (1999). Outer membrane protein A of *Escherichia coli* inserts and folds

into lipid bilayers by a concerted mechanism. *Biochemistry* 38, 5006–5016.

108 KLEINSCHMIDT, J. H. & TAMM, L. K. (2002). Secondary and tertiary structure formation of the beta-barrel membrane protein OmpA is synchronized and depends on membrane thickness. *J. Mol. Biol.* 324, 319–330.

109 RAETZ, C. R. & WHITFIELD, C. (2002). Lipopolysaccharide endotoxins. *Annu. Rev. Biochem.* 71, 635–700.

110 HINDENNACH, I. & HENNING, U. (1975). The major proteins of the *Escherichia coli* outer cell envelope membrane. Preparative isolation of all major membrane proteins. *Eur. J. Biochem.* 59, 207–213.

111 KLEINSCHMIDT, J. H., WIENER, M. C. & TAMM, L. K. (1999). Outer membrane protein A of *E. coli* folds into detergent micelles, but not in the presence of monomeric detergent. *Protein Sci.* 8, 2065–2071.

112 MANN, C. J., ROYER, C. A. & MATTHEWS, C. R. (1993). Tryptophan replacements in the trp aporepressor from *Escherichia coli*: probing the equilibrium and kinetic folding models. *Protein Sci.* 2, 1853–1861.

113 LONDON, E. & LADOKHIN, A. S. (2002). Measuring the depth of amino acid residues in membrane-inserted peptides by fluorescence quenching. *Curr. Topics Membr.* 52, 89–115.

114 MCINTOSH, T. J. & HOLLOWAY, P. W. (1987). Determination of the depth of bromine atoms in bilayers formed from bromolipid probes. *Biochemistry* 26, 1783–1788.

115 TAMM, L. K. & TATULIAN, S. A. (1997). Infrared spectroscopy of proteins and peptides in lipid bilayers. *Q. Rev. Biophys.* 30, 365–429.

116 TAMM, L. K. & TATULIAN, S. A. (1993). Orientation of functional and nonfunctional PTS permease signal sequences in lipid bilayers. A polarized attenuated total reflection infrared study. *Biochemistry* 32, 7720–7726.

29
Protein Folding Catalysis by Pro-domains

Philip N. Bryan

29.1
Introduction

The activation of secreted proteases is tightly regulated by a variety of maturation mechanisms [1]. The zymogen form of some consists of a domain-sized extension attached to the N-terminus of the mature protease. Biosynthesis of these proteases involves the folding of the pro-protease followed by processing of the pro-domain from the folded protease. The biological imperative for tight regulation of protease activation creates a situation in which the protease domain must coevolve with the pro-domain. In many cases the mature protease does not have the capacity to fold independently of the pro-domain even though the mature protease appears to be a stable globular entity without obvious structural characteristics which would distinguish it from facile-folding proteins.

Some pro-domains will catalyze folding of the mature form of the protease even as a separate polypeptide chain. The bimolecular reaction is an intriguing case of complementation. The folded protease is stable in a practical sense. That is, the native conformation persists under favorable solvent conditions for many months. Yet, once denatured and returned to these same favorable conditions, it remains in an unfolded conformation for many months. When the isolated pro-domain is added to unfolded protein, however, folding of the two components into a stable complex occurs spontaneously. Examples of pro-domain mediated folding have been found in all four mechanistic families of proteases: serine proteases [2–7]; Aspartic proteases [8–10]; metalloproteases [11–15], and cysteine proteases [16].

For most well-studied proteins, sufficient information is contained within the protein sequence itself to guide the folding process without the requirement for additional factors. Thus the amino acid sequence of these proteins has evolved to encode both a stable native state and a folding pathway efficient enough to reach the native conformation on a biological time scale. The existence of high kinetic barriers to folding challenges many widely accepted ideas, namely thermodynamic determination of native structure and the sufficiency of thermodynamic stability to determine a pathway. Thus the evolution of pro-proteins as folding units appears to have created folding mechanisms which are outside the paradigm established from data on facile-folding proteins.

Protein Folding Handbook. Part I. Edited by J. Buchner and T. Kiefhaber
Copyright © 2005 WILEY-VCH Verlag GmbH & Co. KGaA, Weinheim
ISBN: 3-527-30784-2

This review will focus on what general inferences about protein folding can be made from pro-domain catalyzed folding. General questions include whether native protein conformation is always thermodynamically determined, what causes high kinetic barriers between unfolded and native states, and what do pro-domains do to reduce these kinetic barriers. The proteases that have provided the most mechanistic information on catalyzed folding are subtilisin (SBT) and α-lytic protease (ALP). Both are serine proteases secreted from bacteria but are not evolutionarily related [17, 18]. Each has eukaryotic homologs. ALP is a structural homolog of chymotrypsin and trypsin and SBT is homologous in structure to prohormone convertases [19–21].

29.2
Bimolecular Folding Mechanisms

With regard to general questions about protein folding mechanisms, analysis of the bimolecular folding reaction has been particularly informative. Unlike unimolecular folding of pro-protease, which results in a metastable product, the product of bimolecular folding is a stable complex between the native protease and the folded pro-domain. The overall goal of many experiments on SBT and ALP has been to obtain a quantitative understanding of all steps in the bimolecular folding reaction. The mechanism of pro-domain-catalyzed folding is as follows:

$$P + U \underset{k_{-1}}{\overset{k_1}{\rightleftharpoons}} P\text{-}I \underset{k_{-2}}{\overset{k_2}{\rightleftharpoons}} P\text{-}N \underset{k_{on}}{\overset{k_{off}}{\rightleftharpoons}} N + P$$

where P is pro-domain, U is unfolded protease, N is native protease, P-I is a collision complex of a partially folded protease and pro-domain, P-N is the complex of native protease and pro-domain. To define mechanistically how the pro-domain participates in folding, it is necessary to measure the rates of all steps in the bimolecular folding reaction. It would also be useful to understand the structure of all populated species in this reaction. The approach has been to characterize the individual steps in the reaction and then use this information to understand the overall reaction.

29.3
Structures of Reactants and Products

29.3.1
Structure of Free SBT

Subtilisin BPN′ is the canonical member of the subtilisin family of proteases [22]. It is a heart-shaped protein consisting of a seven-stranded parallel β-sheet packed between two clusters of α-helices (Figure 29.1). Six α-helices are packed against one face of the central β-sheet and two α-helices are packed against the other [23–

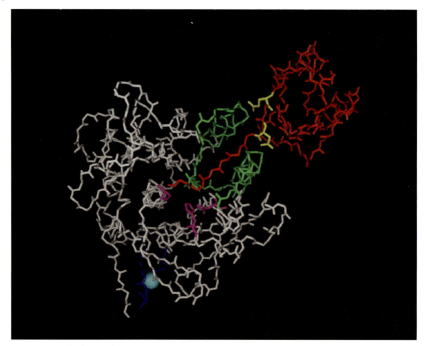

Fig. 29.1. Subtilisin (white) in complex with its pro-domain (red). Active-site triad is shown in violet; α-β-α structure to which pro-domain binds is shown in green; calcium at site A is light blue sphere; calcium-binding loop 75–83 is shown in dark blue. From 1SCJ in the Protein Data Bank [31].

25]. The catalytic triad is distributed as follows: The active-site Ser221 is on the N-terminal end of a long, central helix (helix F); Asp32 is on an interior strand of the β-sheet (β1); His64 is in the N-terminal end of the adjacent helix (helix C). Subtilisin contains a number of notable structural features which are highly conserved across the family. A *cis*-peptide bond occurs between Tyr167 and Pro168 and creates part of the substrate binding pocket. A rare left-handed cross-over occurs between strands β2 and β3. A loop of nine amino acids (75–83) interrupts the last turn of helix C [25]. The loop is part of the left-handed cross-over and forms the central part of a high affinity calcium-binding site (site A). Four carbonyl oxygen ligands to the calcium are provided by the loop (Figure 29.2). The other calcium ligands are a bidentate carboxylate (Asp41) which occurs on an ω-loop (36–45) and the side chain Gln2. Three hydrogen bonds link the N-terminal segment (1–4) to residues 78–82 in parallel-β arrangement. The 75–83 loop also has extensive interactions with a β-hairpin (202–219) so that the calcium ion is buried within the protein structure.

A second ion-binding site (site B) is located 32 Å from site A in a shallow crevice between two segments of polypeptide chain near the surface of the molecule. Evi-

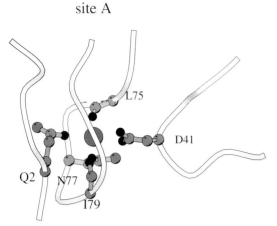

Fig. 29.2. Calcium site A of subtilisin. Ligands to calcium are labeled. From 1SUD in the Protein Data Bank [29].

dence that site B binds calcium comes from determining the occupancy of the site in a series of X-ray structures from crystals grown in 50 mM NaCl with calcium concentrations ranging from 1 to 40 mM [26]. In the absence of excess calcium, this locus was found to bind a sodium ion. The binding of these two ions appears to be mutually exclusive so that as the calcium concentration increases, the sodium ion is displaced, and a water molecule appears in its place directly coordinated to the bound calcium [26]. When calcium is bound, the coordination geometry of this site closely resembles a distorted pentagonal-bipyramid. Three of the ligands are derived from the protein and include the carbonyl oxygen atom of Glu195 and the two side-chain carboxylate oxygens of Asp197. Four water molecules complete the coordination sphere. When sodium is bound, its position is removed by 2.7 Å from that of calcium and the ligation pattern changes. The sodium ligands derived from the protein include a side-chain carboxylate oxygen of Asp197 and the carbonyl oxygen atoms of Gly169, Tyr171, Val174, and Glu195. Two water molecules complete the coordination sphere.

It is not understood how any of these structural features affect folding rate, except for the ion-binding sites. The architecture of site A is the major impediment to folding. Deleting amino acids 75–83 accelerates folding rate by at least 10^7-fold [27–29]. The X-ray structure of the deletion mutant has shown that except for the region of the deleted calcium-binding loop and N-terminal amino acids 1–4, the structure of the mutant and wild-type protein are remarkably similar (Figure 29.3). The structures of SBT with and without the deletion superimpose with an rms difference between 261 Cα positions of 0.17 Å with helix C exhibiting normal helical geometry over its entire length. The deletion also abolishes the calcium binding at site A. As will be discussed below, the formation of calcium site A appears to have a high activation barrier which may contribute to the inability of wild-type SBT to fold.

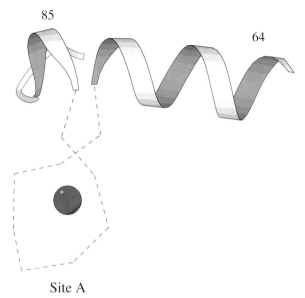

Fig. 29.3. Calcium-binding loop of subtilisin. Region of deletion (75–83) is shown by the dashed line. From 1SUC in the Protein Data Bank [29].

29.3.2
Structure of SBT/Pro-domain Complex

The structure of the native SBT is almost identical in free and complexed states. In the complex, the pro-domain is a single compact domain with an antiparallel four-stranded β-sheet and two three turn α-helices (Figure 29.1) [30, 31]. The β-sheet of the pro-domain packs tightly against the two parallel surface α-helices Asp and Glu (residues 104–116 and 133–144). The pro-domain residues Glu69 and Asp71 form helix caps for the N-termini of the two SBT helices. In another charge-dipole interaction, the carboxylate of Glu112 of SBT accepts H-bonds from the peptide nitrogens of pro residues 42, 43 and 44. The C-terminal residues 72–77 extend out from the central part of the pro-domain and bind in a substrate-like manner along SBT's active-site cleft. Residues Tyr77, Ala76, His75, and Ala74 of the pro-domain occupy subsites S1–S4 of SBT, respectively. If Tyr77 occupies the same position in unprocessed pro-SBT then residues 1–10 of mature SBT would be displaced from their native positions. This hypothetical conformer can be modeled by minor torsional reorganizations involving residues 1–10 with Ala1 and Gln2 of mature SBT occupying the S1′ and S2′ substrate sites. A major consequence for this "active site" conformer, is that the calcium site A cannot fully form until after processing because one of the calcium ligands (Gln2) cannot simultaneously bind in the S2′ subsite and bind to calcium in the A-site.

29.3.3
Structure of Free ALP

The overall fold of ALP is of the chymotrypsin type [17]. The fold is typified by the association of two similar domains each consisting of a six-stranded antiparallel β-barrel [32]. The active-site triad occurs in the interface between the domains with His57 and Asp102 in domain 1 and Ser221 in domain 2 (Figure 29.4). Numbering of ALP residues will be given according to the homologous position in chymotrypsin. Proteases in this family typically contain multiple intradomain disulfide bonds and ALP contains three. One of these (137–159) is not found in the mammalian enzymes. The structural basis for the inability of ALP to fold independently is not apparent from its basic fold. The activation pathways of trypsin and chymotrypsin do not involve pro-domains but rather small N-terminal activation peptides. In fact active trypsin can be produced in *E. coli* from a gene lacking the hexapeptide activation region [33–35]. Hence the basic fold is not intrinsically problematic for folding. Unique structural features found in ALP and bacterial relatives with pro-domains include a high glycine content (16%) and the extended β-hairpin (167–179). Both features have been suggested as possible impediments to folding [36].

An additional feature which should be noted is a buried salt bridge between Arg138 and Asp194 (Figure 29.4). Asp194 is a key residue in zymogen activation of trypsinogen and chymotrypsinogen. The proteolytic event which creates the mature N-terminus (e.g., Ile16 in chymotrypsin) causes amino acids 189–194 to rearrange into the active conformation [1]. The rearrangement results from a salt bridge formed between the new N-terminus (Ile16) and Asp194. Thus a buried salt bridge occurs in both ALP and mammalian relatives but the burial of the salt bridge in trypsin and chymotrysin follows folding of the zymogen.

29.3.4
Structure of the ALP/Pro-domain Complex

The 166-amino-acid ALP pro-region wraps around the C-terminal half of mature ALP [37]. The contact interface is more than 4000 Å^2, almost four-times that of SBT and its pro-domain (Figure 29.4). Most of the contacts to ALP are made with the C-terminal half of the pro-domain (amino acids 65–166) [38], although both N- and C-terminal portions of the pro-domain are required for folding of ALP [39]. Amino acids 160–166 of the pro-domain are inserted like a substrate into the active site, reminiscent of the SBT complex. The edges of a three-stranded β-sheet in the C-terminal part of pro and a β-hairpin (167–179) of ALP abut and form a five-stranded β-sheet in the complex.

As with SBT, the C-terminus of the pro-domain binds tightly in the active site of ALP in the bimolecular complex [37] (Figure 29.4). Thus in the unprocessed molecule, the N-terminus of the active-site conformer would be displaced by 24 Å from

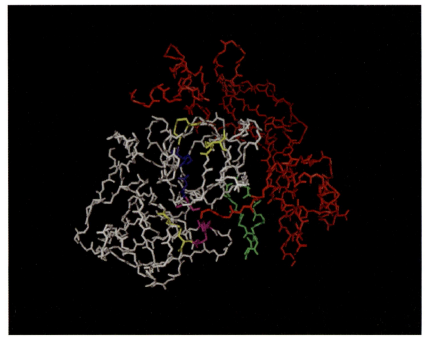

Fig. 29.4. Alpha-lytic protease (white) in complex with its pro-domain (red). Active-site triad is shown in violet; extended β-hairpin 167–179 is shown in green; buried salt bridge Asp194 and Arg138 is shown in dark blue; disulfide bridges are shown in yellow. From 2ALP in the Protein Data Bank [37].

its position in native ALP. Unlike SBT, ALP does not have the complications of metal binding, but it is possible that pro-domain binding organizes Asp194 so that it can form the buried salt bridge with Arg138.

There are some general features of the C-terminal half of the ALP pro-domain that are topologically similar to the SBT pro-domain [38]. Specifically, β-strands 2, 3, 4 and the α-helices connecting strands 1 to 3 and 2 to 4 in the SBT pro-domain have structural counterparts in the ALP pro-domain. Apart from the general similarity of the substrate-binding cleft interactions outlined above, however, there is little similarity in the interactions of pro-SBT and pro-ALP at their binding interfaces with the mature proteases.

ALP pro-domain has no similarity to the small activation peptides of its eukaryotic homologs trypsin and chymotrypsin [1]. The SBT pro-domain on the other hand has high structural homology to the pro-hormone convertase pro-domain, although little sequence similarity [40, 41]. Surprisingly, the SBT pro-domain also has the same basic fold as the pro-domains of eukaryotic carboxypeptidases [42, 43]. This is in spite of the lack of any structural similarity between the mature regions of SBT and the carboxypeptidases.

29.4
Stability of the Mature Protease

29.4.1
Stability of ALP

The most basic folding experiment is determining the free energy of the folding reaction. This is difficult with ALP and SBT because of the problem of establishing equilibrium under conditions in which the ratio of folded and unfolded states can be accurately measured. Denatured ALP diluted into native conditions rapidly collapses into a ensemble of conformations with an average hydrodynamic radius about midway between the native and guanidium chloride (GdmCl)-denatured forms and with almost the same amount of regular secondary structure as native ALP [44]. This collapsed state is stable for weeks at neutral pH without detectable conversion to the native state [44]. The free energy of a two-state reaction can sometimes be determined, however, by (1) measuring a folding rate at a standard state; (2) measuring unfolding rates as a function of denaturant; or (3) extrapolating the unfolding rate to the standard state. From the ratio of the rates of folding and unfolding at the standard state, the free energy of folding is determined. This was the approach taken with ALP [45]. To minimize the influence of autodigestion during the folding reaction, the measurements were made at pH 5.0. As the pH is decreased below the pK_a of the active-site histidine, the proteolytic activity of serine proteases decreases. A pH around 5 is commonly used as a compromise between low activity and preserving some conformational stability [46]. The refolding rate of ALP was determined by incubating denatured protein in native conditions and removing aliquots at intervals. The amount of active ALP was determined by measuring peptidase activity using a synthetic substrate. Over the course of days small but linear increases in activity were detected. Quantitation in the experiment is tricky because the accumulation of active ALP causes digestion of the unfolded protein, but the initial rate of the folding reaction is 6×10^{-12} s^{-1} at 25 °C. This was compared with the unfolding rate which was determined as a function of [GdmCl]. The rate of unfolding without GdmCl was estimated by linear extrapolation to be 1×10^{-7} s^{-1} at 25 °C. Since $\Delta G_{unfolding} = -RT \ln(k_{unfolding}/k_{folding})$ in a two-state system, the simplest interpretation of the unfolding and refolding rates would mean that $\Delta G_{unfolding}$ is about -6 kcal mol^{-1} at 25 °C.

Even given the uncertainties in determining the folding and unfolding rates, it is convincing that the unfolded state of ALP is more stable than the folded state at pH 5.0. The effect of pH 5.0 on the protonation of Asp194 also may bear consideration. If the formation of the buried salt bridge between Asp194 and Arg138 is involved in the transition state for folding, then partial protonation of Asp194 at pH 5.0 could destabilize its interaction with Arg138 and slow folding significantly.

The stability measurements on ALP were made with the three disulfide bonds intact in all forms, U, I and N. An unstrained disulfide cross-link should stabilize a protein by decreasing the entropic cost of folding. The loss of conformational entropy in a polymer due to a cross-link has been estimated by calculating the probability that the ends of a polymer will simultaneously occur in the same volume element. According to statistical mechanics, the three disulfide bonds in ALP should destabilize the unfolded state by $\sim$10 kcal mol^{-1} at 25 °C [47, 48]. Since the lifespan of a protease depends upon the kinetics rather than the thermodynamics of unfolding, however, disulfide bonds are a surprising feature. The effects of cross-links on the stability of the unfolded state would not generally be manifested in the activation energy of the unfolding reaction, because the transition states for unfolding reactions usually are compact, with a slightly larger heat capacity than the native state. In folding experiments on ALP, the disulfide bonds are present in both the I and N states of the protein and thus do not contribute to the thermodynamic measurements [45, 49].

29.4.2
Stability of Subtilisin

Ion binding has a dominant role in stabilizing SBT. The active-site mutant Ala221 of SBT can be unfolded at 25 °C by incubation in 30 mM Tris-HCl, pH 7.5, 0.1 mM EDTA. The rate of unfolding under these conditions is $\sim$0.5 h [1]. (The S221A mutation reduces proteolytic activity of SBT by about 10^6-fold [50] and eliminates the problem of autoproteolysis during the unfolding process.) Addition of micromolar concentrations of free calcium or millimolar concentrations of sodium will reduce the unfolding rate to >days^{-1}. The role of calcium site A in the kinetic stability of SBT is well-documented [51]. SBT with the calcium site A occupied (e.g., 100 mM CaCl$_2$, 100 mM NaCl) unfolds 1000-times slower than the apo-form of SBT (e.g., 1 mM EDTA, 100 mM NaCl) [52] and the activation energy for calcium dissociation from the folded state is 23 kcal mol^{-1}. The binding parameters of calcium at site A have been determined previously by titration calorimetry: $\Delta H_{binding} = -11$ kcal mol^{-1} and $\Delta G_{binding} = -9.3$ kcal mol^{-1} at 25 °C [28]. Thus the binding of calcium is primarily enthalpically driven with only a small net loss in entropy ($\Delta S_{binding} = -6.7$ cal deg^{-1} mol^{-1}). This is surprising since transfer of calcium into water results in a loss of entropy of -60 cal deg^{-1} mol^{-1}. Therefore the freeing of water upon calcium binding to the protein will make a major contribution to the overall ΔS of the process. The gain in solvent entropy upon binding must be compensated for by a loss in entropy of the protein. Loop amino acids 75–83, N-terminal residues 1–5, the ω-loop (36–45) and the β-hairpin (202–219) have increased mobility when calcium is absent from the A-site [27].

Ion site B also influences stability but to a lesser extent. The K_d of site B for sodium is $\sim$1 mM and 15 µM for calcium [52]. By binding at specific sites in the tertiary structure of subtilisin, cations contribute their binding energy to the stability of the native state and should have predictable effects on thermodynamic stability. If cation binding is exclusively to the native state then the contribution of cation binding to the free energy of unfolding would be:

$$\Delta G_{binding} = -RT \ln(1 + K_a \, [\text{cation}])$$

29.4 Stability of the Mature Protease

For example, the free energy of unfolding in 1 mM calcium would be increased by 5.5 kcal mol^{-1} due to the influence of binding at the A-site and an additional 2.5 kcal mol^{-1} due to site B, for a total of 8 kcal mol^{-1} per mM calcium (25 °C).

The distinction between thermodynamic and kinetic effects is blurred, however, by the observations of Eder and Fersht showing that a structured intermediate state of SBT can be induced by high salt (1 M KCl or 0.5 M (NH$_4$)$_2$SO$_4$) and calcium (1 mM) [53]. Ala221 SBT was denatured in 6 M HCl pH 1.8 and then either dialysed or diluted into refolding buffer around neutral pH. Upon return to native conditions in high salt (e.g., 1.0 M KCl) mature SBT slowly folds into a state reminiscent of the I-state of ALP: hydrodynamic radius intermediate between native and unfolded and a high percentage of secondary structure. Remarkably, the SBT intermediate state binds a stoichiometric amount of calcium with an affinity ≤1 μM. No conversion of the intermediate of the native form was detected in the absence of added pro-domain [54]. This experiment has a detection limit of ~1%, which would mean that uncatalyzed folding of mature Ala221 SBT occurs at a rate of <10^{-8} s^{-1}.

If the salt-induced intermediate is part of the relevant folding landscape, this means that the net contribution of cation binding to the thermodynamics of unfolding would be much less than predicted based on exclusive binding to the native state. If high salt is avoided (e.g., ≤0.2 M KCl), however, unfolded SBT remains in a largely unstructured state, indicating the compact state may not be populated under more common folding conditions.

The most straightforward demonstration that mature SBT is thermodynamically stable in high salt was achieved by denaturing and renaturing SBT immobilized on agarose beads to prevent autodigestion [55]. Quantitative renaturation was achieved in 24 h in the presence of 2 M potassium salts, indicating that the folding rate under these conditions must be >10^{-5} s^{-1}. The unfolding rate is not known under these conditions, but the $\Delta G_{unfolding}$ must be greater than 2 kcal mol^{-1} since recovery of native activity approaches 100%. This is surprising since Ala221 SBT in solution could not be renatured in high KCl. It should be noted, however, that SBT might be stabilized by immobilization.

Deleting the calcium A-loop creates a subtilisin that is thermodynamically stable at KCl concentrations above 10 mM. The folding rate of the deletion mutant is strongly ionic dependent [28]. In 30 mM Tris-HCl, pH 7.5 the mutant is not observed to fold. As NaCl or KCl is added, the folding rate increases with the log of the salt concentration. In 0.2 M KCl, pH 7.5, the mutant refolds to the native form in a single exponential process (k_{fold} = 0.0027 s^{-1}) [56] (Figure 29.5).

In summary, one can conclude that Ala221 SBT with the wild-type calcium loop is thermodynamically unstable at low ionic strength (e.g., 30 mM Tris-HCl pH 7.5, 0.1 M EDTA). Thermodynamic stability increases as calcium and monovalent salts are added but it is not certain under what ionic conditions the $\Delta G_{unfolding}$ exceeds zero because the folding and unfolding rates are so slow.

In spite of whether ALP or SBT is thermodynamically stable under a given set of conditions, their slow folding reactions are not due simply to instability of the folded state. If this were true folding could be thermodynamically driven by adding

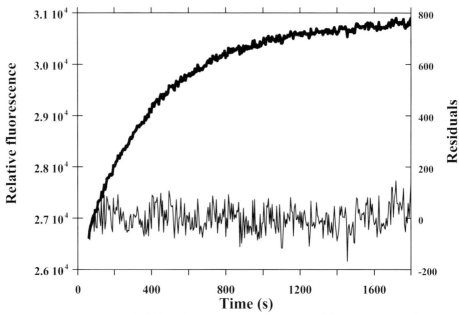

Fig. 29.5. Kinetics of refolding Ala221 Δ75–83 SBT. Refolding in 0.2 M KCl, 30 mM Tris-HCl, pH 7.5, and 25 °C is followed by the increase in tryptophan fluorescence. The curve plotted is the average of three experiments. The residuals after subtracting the data from a single exponential fit ($k = 0.0027$ s^{-1}) are plotted on a five-times expanded scale.

a ligand which binds tightly and preferentially to the native state. For example, the streptomyces SBT inhibitor protein (SSI) has no detectable influence on the folding rate of SBT even though its binds to SBT more tightly than the pro-domain does [56]. Thus pro-domains must be capable of stablizing some structure in the folding pathway whose formation is limiting in its absence.

29.5
Analysis of Pro-domain Binding to the Folded Protease

$$P + N \xrightleftharpoons[k_{off}]{k_{on}} P\text{-}N$$

The SBT pro-domain is a competitive inhibitor of the active enzyme (K_i of 5.4×10^{-7}) [57]. Binding of ALP to its pro-domain is even tighter ($K_i \sim 10^{-10}$) [58]. The high affinity of the pro-domain for the native conformation suggests that its role in folding might be to stabilize native-like structures in the transition state for folding [59]. The details of the complementary interfaces between the respective pro-domains and proteases have been discussed above.

A feature of the isolated pro-domains which influences binding properties is independent stability. The ALP pro-domain is fully folded at 4 °C, independent of binding to ALP [60]. In contrast, the isolated SBT pro-domain is essentially unfolded at all temperatures [56], even though it is a compact globular structured in complex with SBT. This situation suggests that there is thermodynamic linkage between the SBT pro-domain stability and its binding affinity. This linkage can be established by examining the effects of mutations that do not directly affect contacts with the protease but that affect the stability of pro-domain. Although the stability of the isolated pro-domain is low, it has been measured by titration with stabilizing co-solvents and found to have a $\Delta G_{\text{unfolding}} \sim -2$ kcal mol^{-1} at 25 °C [61]. This propensity to fold can be altered by mutations. For example, the mutation I30T in the hydrophobic core of the SBT pro-domain weakens binding to SBT by ~ 100-fold [62].

A number of mutations have been described that increase the independent stability of the SBT pro-domain with concomitant increases in binding affinity to native SBT [63–67]. Sequentially introducing stabilizing mutations into the pro-domain shifted the equilibrium for independent folding from $\sim 97\%$ unfolded to $>98\%$ folded. As the independent stability of the pro-domain increases, the binding affinity for SBT increases concomitantly. The most stable pro-domain mutant bound about ~ 30-times more tightly than wild-type pro-domain [68]. Since these mutations were introduced in regions of the pro-domain that did not directly contact with SBT, their effects on binding to SBT were linked their to their stabilizing effects on the pro-domain. The linked equilibria for pro-domain folding and binding are:

$$P_u \overset{K_f}{\leftrightarrow} P + S \overset{K_P}{\leftrightarrow} P\text{-}S$$

where K_f is the equilibrium constant for folding the pro-domain and K_P is the association constant of folded pro-domain for SBT. The observed binding constant is expected to be: $K_{(P+Pu)} = [K_f/(1+K_f)]K_P$. The data show that as the fraction of folded pro-domain approaches one, the observed association constant approaches its maximum, $\sim 10^{10}$ M^{-1} [63].

The kinetics of the pro-domain binding to SBT also have been determined by measuring the rate of fluorescence change that occurs upon binding. If the reaction is carried out with a 10-fold or greater excess of P, then one observes a pseudo first-order kinetic process with a rate constant equal to $k_{\text{on}}[P] + k_{\text{off}}$. The k_{on} for binding the pro-domain to folded SBT is $\sim 10^6$ M^{-1} s^{-1} [69]. This value is fairly insensitive to mutations that change the stability of the pro-domain.

29.6
Analysis of Folding Steps

$$P + U \underset{k_{-1}}{\overset{k_1}{\rightleftarrows}} P\text{-}I \underset{k_{-2}}{\overset{k_2}{\rightleftarrows}} P\text{-}N$$

In the kinetic mechanism above, P is pro-domain, P-I is an initial complex between pro-domain and partially folded protease, and P-N is the folded complex of pro-domain and native protease. The reaction has been studied for both SBT and ALP as a function of [P]. Under similar conditions the catalyzed folding of SBT is much slower than that of ALP. For example, adding isolated pro-domain (5 μM) to unfolded Ala221 SBT (5 μM) results in only 50% recovery of the native complex after 8 days at 4 °C [53]. Under similar conditions ALP is expect to fold quantitatively within minutes [44]. At first glance this result suggests that the mechanisms of pro-domain catalyzed folding of ALP and SBT differ in some fundamental way. Closer examination reveals, however, that the mechanisms are more similar than they first appear.

Most of the difference in catalyzed folding rate can be explained by examination of the independent stabilities of the two pro-domains. Recall that the SBT pro-domain is largely unfolded at 25 °C ($\Delta G_{unfolding} = -2$ kcal mol^{-1}). In comparison, the ALP pro-domain is about 30–40% unfolded at 25 °C ($\Delta G_{unfolding} \sim 0.3$ kcal mol^{-1}) [60]. Further, the analysis of ALP folding usually is carried out at 4 °C, where its pro-domain is essentially fully folded. The affect of pro-domain stability on catalyzed folding can be seen by examining mutants of the SBT pro-domain which stabilize its independent folding. These mutants were selected based on examination of the X-ray structure of the pro-domain SBT complex and none of the mutations directly contact SBT. Using a fully folded pro-domain mutant (denoted proR9) the kinetics of the conversion of U + P to P-N were determined as a function of [P] ranging from 5 μM to 100 μM [69]. When the concentration of pro-domain is high relative to protease, the kinetics of a single turnover of folding are pseudo first order. The observed rate of folding can be explained by a rapid equilibrium model where k_{-1} is large relative to k_2. In this model, the equilibrium constant K_1 is equal to [P-I]/[U][P]. Since [P-I]/[P$_{total}$] = K_1[P]/(1 + K_1[P]), the observed rate of formation of P-N would be equal to

$$\{k_2 K_1 [P]/(1 + K_1 [P])\} + k_{-2}$$

At micromolar concentrations of pro-domain, the complex P-I reaches saturation and the reaction is limited by the folding of P-I to P-N. The rate of the isomerization step (k_2) is 0.1 min^{-1} for the folding Ala221 SBT and the stabilized pro-domain proR9 (Ruan and Bryan, unpublished). The kinetics of this reaction are unaffected by either calcium ($\leq$ 100 mM) or EDTA ($\leq$ 10 mM) even though these salts greatly affect the stability of the product.

For ALP folding the concentration dependence of the rate also follows a binding hyperbola, with $K_1 = 23$ μM [60]. The maximum rate of the reaction (k_2) is 0.03 s^{-1}. It appears from the plots that k_{-2} is very small compared with k_2.

Analysis of mutations of the ALP pro-domain indicate that interactions at or near the active site are involved in stabilizing the transition state between the initial complex and the native complex [49, 60]. In contrast these interactions have less effect on the stability of the initial complex (P-I). For example, successive deletions of the last four amino acids of the pro-domain decrease k_2 on average

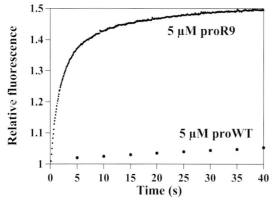

Fig. 29.6. Folding rate of Ala221, Δ75–83 SBT in the presence of the wild-type pro-domain and proR9. Pro-domain (5 μM) and 1 μM denatured SBT mutant were mixed in 5 mM KPO$_4$, 30 mM Tris pH 7.5 at 25 °C. The reaction was followed by the increase in tryptophan fluorescence which occurs upon folding of SBT into the pro-domain SBT complex.

>10-fold for each lost amino acid. Effects of K_1 are not detectable until three or more amino acids are deleted. Recall that these amino acids bind to ALP in a substrate-like manner (Figure 29.4). It is tempting to speculate that the reduction of k_2 is due to the diminished capacity of the deletion mutants to organize the 189–194 loop.

The catalyzed folding reaction of SBT has also been analyzed using an A221 mutant which lacks the calcium binding loop of site A and which folds more rapidly than wild-type SBT. Using wild-type pro-domain as catalyst, the rate of SBT folding increases linearly as a function of [P] ([P] ≤ 100 μM) [56]. The absence of curvature in the plot implies that the formation of the initial complex, P-I, is the limiting step in the reaction up to [P] = 100 μM. These results indicated that initial association of wild-type pro-domain with Δ75–83 SBT is weak. As the pro-domain is stabilized, the folding reaction becomes faster and distinctly biphasic (Figure 29.6). The fast phase of the reaction corresponds to the second-order binding step (1×10^5 M^{-1} s^{-1}) and the slow phase to the isomerization step (0.15 s^{-1}). With the folded pro-domain proR9 the maximum isomerization rates for SBT and Δ75–83 SBT are similar (∼0.1 s^{-1}).

It was possible to demonstrate that proline isomerization is rate limiting in folding Δ75–83 SBT. There are 13 proline residues in SBT. All but one (Pro168) exist as *trans* isomers in the native structure [25]. The peptide bond between proline and its preceding amino acid (Xaa-Pro bonds) exist as a mixture of *cis* and *trans* isomers in solution unless structural constraints, such as in folded proteins, stabilize one of the two isomers. In the absence of ordered structure, the *trans* isomer is favored slightly over the *cis* isomer. The isomerization *trans* ↔ *cis* is an intrinsically slow reaction with rates at 0.1–0.01 s^{-1} at 25 °C. Upon denaturation of SBT, the structural constraints are removed and the *trans* and *cis* isomers of all 13 prolyl peptide bonds gradually come to equilibrium in the unfolded state, resulting in 2^{13} different proline isomer combinations. To study the catalyzed folding rate of SBT in the absence of prolyl peptide bond isomerization, folding kinetics were measured after a short denaturation time [70]. The Δ75–83 SBT mutant used in these studies can be denatured in acid in <1 s. The rapid denaturation time minimizes the amount of prolyl peptide bond isomerization occurring during the time require to unfold. In the absence of proline isomerization, the observed rate of folding increases as concentration of pro-domain increases, and follows a linear relationship with no intermediate detectable in the course of the reaction. This occurs because the formation of P-N from P-I is faster than the formation of initial collision complex P-I from P and U. The mechanism of folding reaction for denatured Δ75–83 SBT with native proline isomers is as follows:

$$P + U \underset{<0.05\ s^{-1}}{\overset{9 \times 10^4\ M^{-1}\ s^{-1}}{\rightleftarrows}} P\text{-}I \underset{<0.01\ s^{-1}}{\overset{>3\ s^{-1}}{\rightleftarrows}} P\text{-}N$$

where U is denatured SBT with all prolyl peptide bond isomers the same as in the native structure. According to the mechanism proposed above, the pseudo first-order rate constant would reach its maximum when [P] is high enough that the value $k_1 \cdot [P] > (k_2 + k_{-2})$. That point has not been reached at $[P] = 20$ μM. Thus, in the presence of an independently folded pro-domain mutant, the rate of SBT folding into a complex becomes typical of in vitro folding rates for many small globular proteins. That is, independent of proline isomerization, folding is largely complete within a few seconds [71].

In comparison, the rate of ALP folding plateaus at 0.03 s^{-1}, which is in the range of proline isomerization. ALP has three *trans* and one *cis* proline. It is not known how proline isomerization is involved in the kinetics of the P-I to P-N transition, however.

29.7
Why are Pro-domains Required for Folding?

The question remains as to why folding is so slow without the pro-domain. An efficient folding pathway implies that productive intermediates are significantly more stable than the surrounding landscape of unfolded conformations. Native state H-D exchange experiments for several proteins have shown that partially folded states exist with significantly lower energy than the globally unfolded state [72–74]. Native state H-D exchange experiments for ALP [75] and SBT (Sari et al., unpublished), show highly cooperative exchange behavior. Amides protons in all elements of secondary structure exchange with solvent deuterons at a rate in the range of year^{-1}. In ALP, there are 103 amide protons with protection factors $> 10^4$ and 31 amide proton with protection factor $> 10^9$. The most protected protons occur throughout both domains of ALP. In comparison, there are 131 protons in SBT with protection factors $\geq 10^5$. The 49 slowest exchangers have protection factors $\geq 10^9$. These strongly protected residues occur throughout the main structural elements of SBT, except for the short N-terminal α-helices, A and B, and

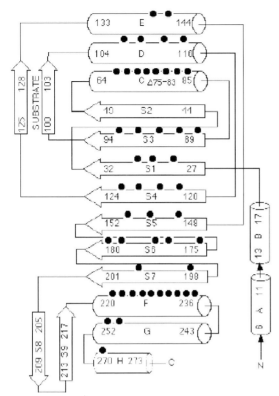

Fig. 29.7. Schematic representation of the subtilisin topology. The positions of the 49 slowest exchanging amide protons in SBT are shown by black dots. The slowest 49 exchanging protons reside throughout the central seven-stranded parallel β-sheet and in all the five major α-helical segments. Adapted from Siezen et al. [85].

strands, β8 and β9 (Figure 29.7). The large cooperative folding core may create a folding problem, because most of the tertiary structure must be acquired before a free energy well is encountered in conformational space. The enormous loss of conformational entropy before that energy well occurs would result in a large transition state barrier to folding [66, 75]. In this model, the transition state of uncatalyzed folding is of high energy because of its low conformational entropy. The prodomain decreases the entropic barrier by pushing the transition state back toward a less folded form of the proteinase. Thus much less conformational entropy is lost in the transition state. Once over the barrier, folding of P-I to P-N is rapid.

29.8
What is the Origin of High Cooperativity?

Formation of the calcium A-site may be responsible for the cooperative folding reaction of SBT. SBT is unstable in the absence of calcium and at low ionic strength.

Addition of calcium increases stability by 8 kcal mol^{-1} per mM but does accelerate the folding reaction to an observable level. The reason that the calcium does not affect folding rate may be that site A is not fully formed until considerable tertiary structure has accrued. The structure elements involved in forming the A-site are the N-terminus, the ω-loop, the 75–83 loop and the 202–219 β-hairpin. Thus calcium binding may not contribute to the stability of intermediately folded states but only to a substantially folded molecule. This situation creates a highly cooperative folding reaction. In addition, the burial of the charged calcium and the requirement for its desolvation prior to burial within the protein creates an additional kinetic barrier to folding. This suggestion is supported by the observation that the Δ75–83 mutant SBT folds $\geq 10^7$-times faster than SBT with the calcium A-site intact.

The origins of high cooperativity in ALP may also be related to a buried ionic interaction: the formation of the buried salt bridge between Arg138 and Asp194. Formation of a structurally homologous ionic interaction is involved in zymogen activation in trypsin and chymotrypsin.

Buried ionic interactions have been implicated previously in raising the activation barrier to folding because of unfavorable protein–solvent rearrangements. For example, replacing a buried salt bridge in ARC repressor dimer with a hydrophobic cluster has little effect on thermodynamic stability but increases the folding rate of the dimer by up to 1000-fold [76, 77]. Desolvating these charged residues in ALP may raise kinetic but not necessarily the thermodynamic barriers to unfolding.

29.9
How Does the Pro-domain Accelerate Folding?

The simplest model of catalyzed folding is one in which the pro-domain accelerates folding by stabilizing a substructure on the folding pathway which does not form frequently in the absence of pro-domain binding. Once the substructure forms, it acts as a folding nucleus with subsequent folding propagating into other regions. The nature of this substructure can be guessed at based on the structures of the folded complexes. In the bimolecular SBT complex, the pro-domain binds to an α–β–α structure (amino acids 100–144) which includes the parallel surface α-helices D and E. It also supplies caps to the N-termini of the two helices (Figure 29.1). In the intermediate P-I, the pro-domain may stabilize the α–β–α structure relative to other unfolded states. Subsequent folding may propagate from this folding nucleus into N- and C-terminal regions of SBT. In the absence of the pro-domain, the α–β–α structure may not have sufficient independent stability to initiate folding very frequently. It should be noted that unimolecular folding of proSBT may circumvent the problem of calcium A-site formation. The covalent attachment of the pro-domain to SBT prevents the final folding of the calcium A-site region, until after the active site is sufficiently formed to cleave the pro-domain from the mature [30, 78].

In ALP, the pro-domain may stabilize the formation of a long β-hairpin (118–

130) in the C-terminal half of ALP [36, 37] but may also be involved in organizing the 189–194 loop through the interaction of the C-terminus of the pro-domain with the active-site Ser195.

29.10
Are High Kinetic Stability and Facile Folding Mutually Exclusive?

Broad specificity proteinases must be resistant to unfolding. Assuming that partially folded intermediates are good substrates for autolysis, the half-life of active protease would be determined by the equilibrium between the proteinase-resistant native state and the lowest energy, proteinase-labile intermediate [28, 36, 75, 79]. Thus to avoid autolysis, partially folded conformations must be of much higher energy than the native state so that excursions between the two states are rare. Since partially folded states with significant independent stability are important intermediates in the folding pathway of many proteins does resistance to partial unfolding cause proteins to be hard to fold? The answer appears to be not necessarily.

It has been possible artificially to evolved the facile-folding $\Delta 75–83$ SBT into a protease which is hyperstable and facile-folding [80, 81]. The initial deletion mutant is unfolds 100-times faster than wild type in 100 mM calcium, but deleting the loop creates the potential to use facile-folding structural solutions for stabilization. Four regions of subtilisin are directly affected by the deletion of the calcium loop: the N-terminal amino acids 1–5; the ω-loop amino acids 36–45; the α-helix amino acids 70–74; the β-ribbon amino acids 202–219. As a result the naturally occurring amino acids in these regions were not optimal for stability in the loop-deleted subtilisin. Residues in these regions were randomized and stabilizing mutations were selected in five iterations [80]. The final result was a mutant which unfolds 1000-times slower than wild-type SBT in 100 mM calcium [81] but, nevertheless, has an uncatalyzed folding rate of $0.02\ \text{s}^{-1}$. This protease can be expressed at high levels through its natural biosynthetic pathway in *Bacillus*, although obviously its long-term evolutionary fitness is unknown. The fact that facile folding and high kinetic stability are not mutually exclusive in the SBT fold suggests that the inability of wild-type SBT and ALP to fold without their pro-domains may be part of the mechanism to regulate the timing of protease activation. It is also possible that when evolution of the folding unit is uncoupled from the sequence of the mature enzyme that the facile folding of the mature enzyme is easily lost.

29.11
Experimental Protocols for Studying SBT Folding

29.11.1
Fermentation and Purification of Active Subtilisin

Subtilisin BPN′ and active mutants were expressed in *B. subtilis* in a 1.5-L New Brunswick fermenter at a level of $\sim 500\ \text{mg L}^{-1}$. After fermentation 16 g of Tris–

base and 22 g of $CaCl_2$ (dihydrate) were added to ~1400 mL of broth. Cells and precipitate were pelleted into 250-mL bottles by centrifugation at 12 000 g for 30 min, 4 °C. Acetone (70% final volume) was added to the supernatant. The 70% acetone mixture was then centrifuged in 500-mL bottles at 12 000 g for 30 min, 4 °C. The pellet was resuspended in ~150 mL 20 mM HEPES, pH 7.0, 1 mM $CaCl_2$. Resuspended material was centrifuged at 12 000 g for 10 min, 4 °C to remove insoluble material. Using a vacuum funnel, the sample was passed over 150 g DE52, equilibrated in 20 mM HEPES, pH 7.0. The DE52 was washed twice with 150 mL 20 mM HEPES, 1 mM $CaCl_2$ buffer and the all washes pooled. Solid NH_4SO_4 was added to the sample to a final concentration of 1.8 M. Final purification was carried out using a 2 × 30 cm Poros HP 20 column on a Biocad Sprint. The sample was loaded and washed in 1.8 M NH_4SO_4, 20 mM HEPES, pH 7.0 and then eluted with a linear gradient (1.8–0 M NH_4SO_4 in 20 mM HEPES, pH 7.0). Subtilisin BPN' eluted at a conductivity of 108 mS. Assays of peptidase activity were performed by monitoring the hydrolysis of sAAPF-pNA [82]. The [subtilisin] was determined using 1 mg mL^{-1} = 1.17 at 280 nm.

29.11.2
Fermentation and Purification of Facile-folding Ala221 Subtilisin from E. coli

Mutagenesis and protein expression of inactive Δ75–83 SBT mutants were performed using a vector called pJ1. Vector pJ1 is identical to pG5 [83] except that the ClaI site of pG5 has been removed. The SBT gene was inserted between NdeI and HindIII sites in the polycloning site of pJ1 and transformed into E. coli production strain BL21(DE3).

The transformed production strain BL21(DE3) was plated out on selection plates. Ten milliliters of 1LB broth (10 g of tryptone, 5 g of yeast extract, 10 g of NaCl L^{-1} media) supplemented with 100 µg mL^{-1} ampicillin was inoculated with 4–6 ampicillin-resistant colonies in a 250-mL baffled flask. The culture was grown at 37 °C, 300 rpm, until mid-log phase. This culture was used to inoculate 1.5 L of L Broth buffered with 2.3 g of KH_2PO_4, 12.5 g of K_2HPO_4 L^{-1} media supplemented with 100 µg mL^{-1} ampicillin. The culture was grown at 37 °C in a BioFlo Model C30 fermenter (New Brunswick Scientific Co., Edison, NJ, USA) until an A_{600} 1–1.2 was attained, upon which 1 mM IPTG was added to induce the production of T7 RNA polymerase that directs synthesis of target DNA message. Three hours after induction, the cells were harversted by centrifugation for 30 min at 4 °C, 10 000 rpm in a J2-21 centrifuge (Beckman Instruments, Fullerton, CA, USA) with a JA-14 rotor.

E. coli paste from a 1.5-L fermentation (~5 g) was resuspended in 50 mL lysis buffer (50 mM Tris-HCl pH 8.0/100 mM NaCl/1 mM EDTA) and PMSF was added to a final concentration of 1 mM and DNase I (in 40 mM Tris-HCl, 1 M $MgCl_2$) to 20 µg mL^{-1}. The cells were lysed by two passes through a French Press and additional DNase I was added to a final concentration of 25 µg mL^{-1}. Inclusion bodies containing mutant SBT were recovered by centrifugation at 10 000 g, 4 °C with a JA-17 rotor for 20 min in a Beckman J2-21 centrifuge. The resulting pellet was washed three times by repeated resuspension in 60 mL of ice-cold 20 mM HEPES, pH 7.0, and centrifugation for 10 min at the same condition as above. The final pellet was then resuspended in 50 mL of ice-cold 20 mM HEPES, pH 7.0, containing 2 M urea, frozen on dry ice and allowed to thaw at room temperature to clarify the solution. If a precipitate formed, the volume was increased and the freeze/thaw cycle was repeated. The solution then was centrifuged for 10 min and the supernatant applied to a Productive DE cartridge (Metachem Technologies Inc., Torrance, CA, USA) equilibrated with 20 mM HEPES, pH 7.0/2 M urea. The cartridge was washed with 40 mL of cold 20 mM HEPES, pH 7.0/2 M urea. Flow through and wash fractions were combined and dialyzed 36–48 h at room temperature against 3.5 L of 100 mM KPi, pH 7.0 to fold the SBT.

An affinity agarose column was prepared by coupling the tetrapeptide A-L-A-L (Sigma, St. Louis, MO, USA) to Affi-Gel 10 (Bio-Rad, Hercules, CA, USA) following the manufacturer's protocol. The DE cartridge-purified protein was split and each half loaded onto a 4.5-mL column of A-L-A-L-agarose equilibrated with 20 mM HEPES, pH 7.0. The column was washed with 20 mL of 20 mM HEPES, pH 7.0/1 M NaCl followed by 10 mL of 20 mM HEPES, pH 7.0 and the protein was eluted with 50 mM of triethylamine, pH 11.1. Peak fractions (1 mL each) of the protein were immediately neutralized by the addition of 0.1 mL of 1 M KPi, pH 7.0 and pooled together, then dialyzed overnight against 3.5 L of 2 mM ammonium carbonate. Peak fractions were checked on SDS-polyacrymide (SDS-PAGE) gels (Novex Experimental Technoolgy, San Diego, CA, USA). The desired fractions were pooled and dialyzed against 2 mM ammonium bicarbonate, pH 7.0 at 4 °C. After lyophilization, the protein sample was stored at −20 °C.

29.11.3
Mutagenesis and Protein Expression of Pro-domain Mutants

The gene fragment encoding the 77-amino-acid pro-domain was cloned into vector pJ1. When superinfected with helper phage M13K07, the M13 origin enables the DNA replication in single-stranded form for mutagenesis and sequencing. Mutagenesis of the cloned pro-domain gene was performed according to the oligonucleotide-directed in vitro mutagenesis system, version 2, Sculptor system (Amersham International plc., Bucks, UK) or Kunkel dUTP incorporation method [84, 85]. *Escherichia coli* TG1 [K12, Δ(lac-pro), supE, thi, hsdD5/F′traD36, proA + B$^+$, lacI^q, lacZΔM15] was obtained from the oligonucleotide-directed in vitro mutagenesis system version 2 (Amersham International plc.). Competent *E. coli* TG1 cells were transformed with mutagenesis product and single-stranded phagemid DNA was then prepared from TG1 cells superinfected with helper phage, while double-stranded phagemid DNA was isolated using Wizard Plus Minipreps DNA Purification System. Phagemid DNA was sequenced as described in Sequenase Protocol (United States Biochemical, Cleveland, OH, USA), using either single-stranded or double-stranded phagemid DNA. Upon confirmation of the mutations, the mutant double-stranded phagemid DNA was used to transform the *E. coli* production strain BL21(DE3)[F$^-$, hsdS, gal] [86], BL21(DE3) cells contain the bacteriophage T7 gene 10 which encodes the bacteriophage T7 RNA polymerase. Gene 10 is under the control of *lac* UV5 promoter, inducible by IPTG.

29.11.4
Purification of Pro-domain

Fermentation of of pro-domain was carried out in *E. coli* as described for SBT. Pro-domain was purified using a freeze-thaw method to release soluble protein from the cells [84]. *E. coli* paste from a 1.5-L fermentation ($\sim$5 g) was suspended in 10 mL of cold phosphate-buffered saline (PBS) and 20 mL water. PMSF was added to a final concentration of 1 mM, DNase I to 10 µg mL^{-1}, and the oxidized form of glutathione (100 mg in 1 mL total volume of 100 mM KPi, pH 7.0) to 1.6 mg mL^{-1}. The resuspension was frozen by submerging in a dry-ice/ethanol bath until it was completely frozen, and then thawed in a room temperature water bath. This cycle was repeated two additional times, with addition of 100 µL glutathione stock solution before each cycle. The final mixture was centrifuged at 4 °C for 10 min at 10 000 rpm in a J2-21 centrifuge (Beckman Instruments) with a JA-14 rotor. The supernatant containing the soluble protein was diluted four-fold with 20 mM HEPES, pH 7.0, and further purified as described [56]. The concentration of proR9 was determined by UV absorbance using 1 mg mL^{-1} = A$_{275}$ of 0.67.

29.11.5
Kinetics of Pro-domain Binding to Native SBT

The rate of binding of the pro-domain to folded SBT was monitored by fluorescence (excitation λ = 300 nm, emission λ, 340 nm cutoff filter) using a KinTek Stopped-Flow Model SF2001. The reaction was followed by the 1.2-fold increase in the tryptophan fluorescence of SBT upon its binding of the pro-domain [56]. Pro-domain solutions of 10–160 µM in 30 mM Tris-HCl, pH, 5 mM KPi, pH 7.5 were mixed with an equal volume of 2 µM subtilisin, 30 mM Tris-HCl, pH, 5 mM KPi, pH 7.5 in a single mixing step. Typically 10–15 kinetics traces were collected for each [P]. Final [SBT] was 1 µM and final [P] were 5–80 µM.

29.11.6
Kinetic Analysis of Pro-domain Facilitated Subtilisin Folding

29.11.6.1 Single Mixing

Refolding of subtilisin is accompanied by a 1.5-fold increase in the tryptophan fluorescence of subtilisin upon folding into its complex with the pro-domain. There are three tryptophan residues in the subtilisin mutants and none in pro-domain. Reaction kinetics were measured using a KinTek Stopped-Flow Model SF2001 (excitation λ = 300 nm, emission: 340 nm cutoff filter) as described [61]. A stock solution of subtilisin at a concentration of 100 µM in 100 mM KPi, pH 7.0 was prepared for refolding studies. Subtilisin was denatured by diluting 20 µL of the stock solution into 1 mL of 50 mM HCl. The samples were neutralized by mixing the subtilisin and HCl solution into an equal volume of 60 mM Tris-base, pH 9.4, KPi and pro-domain in the KinTek Stopped-flow (final buffer concentrations were 30 mM Tris-HCl, pH, 5 mM KPi, pH 7.5). The final concentration of subtilisin was 1 µM, while the pro-domain concentration was varied from 5 to 20 µM. Typically 10–15 kinetics traces were collected for each [P].

29.11.6.2 Double Jump: Renaturation–Denaturation

The kinetics of the formation of the folded complex, PS, were determined by a double-jump renaturation-denaturation experiment. The KinTek Stopped Flow in the three syringe configuration was used to perform the two mixing steps. In the first mixing step, 3 µM subtilisin in 50 mM HCl is mixed with a variable concentration of pro-domain in 60 mM Tris-base and KPi. The resulting solution is 30 mM Tris-HCl, 5 mM KPi, pH 7.5. The final concentration of subtilisin was 1.5 µM, while the pro-domain concentration was varied from 5 to 20 µM after the first mixing step. Subtilisin and pro-domain are allowed to fold under native conditions in the delay line of the stopped-flow instrument for aging times ranging from 0.5 to 60 s. After the prescribed aging time the subtilisin/pro-domain solution is mixed 2:1 with 0.1 M H_3PO_4 to bring the pH to 2.3. Fluorescence data for the denaturation of folded complex was collected after the second mixing step. At each renaturation time point, the denaturation curve was fit to a single exponential decay curve. The amplitude of the decay curve was recorded as a function of refolding time to assess the amount of folded complex which had accumulated at each renaturation time. Typically 10–15 kinetics traces were collected for each renaturation time point.

29.11.6.3 Double Jump: Denaturation–Renaturation

A double jump denaturation–renaturation experiment was used to study the effect of denaturation time on the rates and amplitudes of the folding reaction. The KinTek Stopped Flow in the three syringe configuration was used to perform the two mixing steps. In the first mixing reaction 3 µM of subtilisin and 30 µM proR9 in 0.05 M KPi, pH 7.2 was denatured by mixing with an equal volume of 0.1 M phosphoric acid. The resulting solution has a pH of 2.15. The denaturation reaction was aged for varied lengths of time and then mixed with one-half volume of 0.15 M KPO_4, pH 12.0. The final conditions were 1 µM of sbt15 and 10 µM of proR9 in 0.1 M KPi, pH 7.2, at 25 °C. The folding process is then followed by fluorescence change.

A second double mixing denaturation–renaturation experiment was used to follow folding kinetics after 0.5 s of denaturation. The two mixing steps were performed using a BioLogic SFM-4 Q/S in the stopped flow mode. In the first step, 2 µM SBT in 10 mM KPi, pH 7.2 is mixed with an equal volume of 100 mM HCl. The resulting solution is 1 µM SBT, 50 mM HCl, 5 mM KPi, pH 2.1. After 0.5 s, the denatured SBT solution is mixed with an equal volume of 60 mM Tris-base, 5 mM KPi and variable amounts of proR9. The resulting solution is 0.5 µM SBT,

5–20 µM proR9, 30 mM Tris-HCl, 5 mM KPi, pH 7.5. The renaturation process is followed by fluorescence change.

29.11.6.4 Triple Jump: Denaturation–Renaturation–Denaturation

A triple mixing denaturation–renaturation–denaturation experiment to measure directly the accumulation of native complex after a denaturation time of 0.5 s. The three mixing steps were performed using a BioLogic SFM-4 Q/S in the stopped flow mode. In the first step, 5 µM SBT in 10 mM KPi, pH 7.2 is mixed with an equal volume of 100 mM HCl. The resulting solution is 2.5 µM SBT, 50 mM HCl, 5 mM KPi, pH 2.1. After 0.5 s, the denatured SBT solution is mixed with an equal volume of 60 mM Tris-base, 5 mM KPi and 10 µM proR9. The resulting solution is 1.25 µM SBT, 5 µM proR9, 30 mM Tris-HCl, 5 mM KPi, pH 7.5. SBT and proR9 are allowed to fold under native conditions in the delay line for aging times ranging from 0.5 to 60 s. After the prescribed aging time the SBT-proR9 solution is mixed 3:1 with 0.132 M H_3PO_4 to bring the pH to 2.3 and denature the folding reaction. At each renaturation time point, the denaturation curve was fit to a single exponential decay curve. The amplitude of the decay curve was recorded as a function of refolding time to assess the amount of folded complex which had accumulated at each renaturation time.

References

1 KHAN, A. R. & JAMES, M. N. (1998). Molecular mechanisms for the conversion of zymogens to active proteolytic enzymes. *Protein Sci* 7, 815–36.

2 ZHOU, Y. & LINDBERG, I. (1993). Purification and characterization of the prohormone convertase PC1(PC3). *J Biol Chem* 268, 5615–23.

3 BAIER, K., NICKLISCH, S., MALDENER, I. & LOCKAU, W. (1996). Evidence for propeptide-assisted folding of the calcium-dependent protease of the cyanobacterium Anabaena. *Eur J Biochem* 241, 750–5.

4 FABRE, E., NICAUD, J. M., LOPEZ, M. C. & GAILLARDIN, C. (1991). Role of the proregion in the production and secretion of the *Yarrowia lipolytica* alkaline extracellular protease. *J Biol Chem* 266, 3782–90.

5 FABRE, E., THARAUD, C. & GAILLARDIN, C. (1992). Intracellular transit of a yeast protease is rescued by trans-complementation with its prodomain. *J Biol Chem* 267, 15049–55.

6 CHANG, Y. C., KADOKURA, H., YODA, K. & YAMASAKI, M. (1996). Secretion of active subtilisin YaB by a simultaneous expression of separate pre-pro and pre-mature polypeptides in *Bacillus subtilis*. *Biochem Biophys Res Commun* 219, 463–8.

7 BAARDSNES, J., SIDHU, S., MACLEOD, A., ELLIOTT, J., MORDEN, D., WATSON, J. & BORGFORD, T. (1998). *Streptomyces griseus* protease B: secretion correlates with the length of the propeptide. *J Bacteriol* 180, 3241–4.

8 VAN DEN HAZEL, H. B., KIELLAND-BRANDT, M. C. & WINTHER, J. R. (1993). The propeptide is required for in vivo formation of stable active yeast proteinase A and can function even when not covalently linked to the mature region. *J Biol Chem* 268, 18002–7.

9 CAWLEY, N. X., OLSEN, V., ZHANG, C. F., CHEN, H. C., TAN, M. & LOH, Y. P. (1998). Activation and processing of non-anchored yapsin 1 (Yap3p). *J Biol Chem* 273, 584–91.

10 FUKUDA, R., HORIUCHI, H., OHTA, A. & TAKAGI, M. (1994). The prosequence of Rhizopus niveus aspartic proteinase-I supports correct folding and secretion of its mature part in Saccharomyces cerevisiae. *J Biol Chem* 269, 9556–61.

11 NIRASAWA, S., NAKAJIMA, Y., ZHANG, Z. Z., YOSHIDA, M. & HAYASHI, K. (1999). Intramolecular chaperone and inhibitor activities of a propeptide from a bacterial zinc aminopeptidase. *Biochem J* 341 (Pt 1), 25–31.

12 MARIE-CLAIRE, C., RUFFET, E., BEAUMONT, A. & ROQUES, B. P. (1999). The prosequence of thermolysin acts as an intramolecular chaperone when expressed in trans with the mature sequence in *Escherichia coli*. *J Mol Biol* 285, 1911–15.

13 CAO, J., HYMOWITZ, M., CONNER, C., BAHOU, W. F. & ZUCKER, S. (2000). The propeptide domain of membrane type 1-matrix metalloproteinase acts as an intramolecular chaperone when expressed in trans with the mature sequence in COS-1 cells. *J Biol Chem* 275, 29648–53.

14 VENTURA, S., VILLEGAS, V., STERNER, J., LARSON, J., VENDRELL, J., HERSHBERGER, C. L. & AVILES, F. X. (1999). Mapping the pro-region of carboxypeptidase B by protein engineering. Cloning, overexpression, and mutagenesis of the porcine proenzyme. *J Biol Chem* 274, 19925–33.

15 WETMORE, D. R. & HARDMAN, K. D. (1996). Roles of the propeptide and metal ions in the folding and stability of the catalytic domain of stromelysin (matrix metalloproteinase 3). *Biochemistry* 35, 6549–58.

16 YAMAMOTO, Y., WATABE, S., KAGEYAMA, T. & TAKAHASHI, S. Y. (1999). Proregion of Bombyx mori cysteine proteinase functions as an intramolecular chaperone to promote proper folding of the mature enzyme. *Arch Insect Biochem Physiol* 42, 167–78.

17 JAMES, M. N., DELBAERE, L. T. & BRAYER, G. D. (1978). Amino acid sequence alignment of bacterial and mammalian pancreatic serine proteases based on topological equivalences. *Can J Biochem* 56, 396–402.

18 SIEZEN, R. J. & LEUNISSEN, J. A. (1997). Subtilases: the superfamily of subtilisin-like serine proteases. *Protein Sci* 6, 501–23.

19 STEINER, D., SMEEKENS, S. P., OHAGI, S. & CHAN, S. J. (1992). The new enzymology of precursor processing endoproteases. *J. Mol. Biol.* 267, 23435–8.

20 HOLYOAK, T., WILSON, M. A., FENN, T. D., KETTNER, C. A., PETSKO, G. A., FULLER, R. S. & RINGE, D. (2003). 2.4 A resolution crystal structure of the prototypical hormone-processing protease Kex2 in complex with an Ala-Lys-Arg boronic acid inhibitor. *Biochemistry* 42, 6709–18.

21 HENRICH, S., CAMERON, A., BOURENKOV, G. P., KIEFERSAUER, R., HUBER, R., LINDBERG, I., BODE, W. & THAN, M. E. (2003). The crystal structure of the proprotein processing proteinase furin explains its stringent specificity. *Nat Struct Biol* 10, 520–6.

22 KRAUT, J. (1971). Subtilisin x-ray structure. In *The Enzymes* 3rd edn, Vol. 3, pp. 547–60. Academic, New York.

23 MCPHALEN, C. A., SCHNEBLI, H. P. & JAMES, M. N. (1985). Crystal and molecular structure of the inhibitor eglin from leeches in complex with subtilisin Carlsberg. *FEBS Lett* 188, 55–8.

24 MCPHALEN, C. A. & JAMES, M. N. G. (1987). Crystal and molecular structure of the serine proteinase inhibitor CI-2 from barley seeds. *Biochemistry* 26, 261–9.

25 MCPHALEN, C. A. & JAMES, M. N. G. (1988). Structural comparison of two serine proteinase–protein inhibitor complexes: Eglin-C-Subtilisin Carlsberg and CI-2-Subtilisin novo. *Biochemistry* 27, 6582–98.

26 PANTOLIANO, M. W., WHITLOW, M., WOOD, J. F. et al. (1988). The engineering of binding affinity at metal ion binding sites for the stabilization of proteins: Subtilisin as a test case. *Biochemistry* 27, 8311–17.

27 Gallagher, T. D., Bryan, P. & Gilliland, G. (1993). Calcium-free subtilisin by design. *Proteins Struct Funct Genetics* 16, 205–13.

28 Bryan, P., Alexander, P., Strausberg, S. et al. (1992). Energetics of folding subtilisin BPN'. *Biochemistry* 31, 4937–45.

29 Almog, O., Gallagher, T., Tordova, M., Hoskins, J., Bryan, P. & Gilliland, G. L. (1998). Crystal structure of calcium-independent subtilisin BPN' with restored thermal stability folded without the pro-domain. *Proteins* 31, 21–32.

30 Gallagher, T. D., Gilliland, G., Wang, L. & Bryan, P. (1995). The prosegment-subtilisin BPN' complex: crystal structure of a specific foldase. *Structure* 3, 907–14.

31 Jain, S. C., Shinde, U., Li, Y., Inouye, M. & Berman, H. M. (1998). The crystal structure of an autoprocessed Ser221Cys-subtilisin E-propeptide complex at 2.0 A resolution. *J Mol Biol* 284, 137–44.

32 Lesk, A. M. & Fordham, W. D. (1996). Conservation and variability in the structures of serine proteinases of the chymotrypsin family. *J Mol Biol* 258, 501–37.

33 Higaki, J. N., Evnin, L. B. & Craik, C. S. (1989). Introduction of a cysteine protease active site into trypsin. *Biochemistry* 28, 9256–63.

34 Vasquez, J. R., Evnin, L. B., Higaki, J. N. & Craik, C. S. (1989). An expression system for trypsin. *J Cell Biochem* 39, 265–76.

35 McGrath, M. E., Wilke, M. E., Higaki, J. N., Craik, C. S. & Fletterick, R. J. (1989). Crystal structures of two engineered thiol trypsins. *Biochemistry* 28, 9264–70.

36 Cunningham, E. L., Jaswal, S. S., Sohl, J. L. & Agard, D. A. (1999). Kinetic stability as a mechanism for protease longevity. *Proc Natl Acad Sci USA* 96, 11008–14.

37 Sauter, N. K., Mau, T., Rader, S. D. & Agard, D. A. (1998). Structure of alpha-lytic protease complexed with its pro region. *Nat Struct Biol* 5, 945–50.

38 Baker, D. (1998). Metastable states and folding free energy barriers. *Nat Struct Biol* 5, 1021–4.

39 Cunningham, E. L., Mau, T., Truhlar, S. M. & Agard, D. A. (2002). The Pro region N-terminal domain provides specific interactions required for catalysis of alpha-lytic protease folding. *Biochemistry* 41, 8860–7.

40 Tangrea, M. A., Alexander, P., Bryan, P. N., Eisenstein, E., Toedt, J. & Orban, J. (2001). Stability and global fold of the mouse prohormone convertase 1 pro-domain. *Biochemistry* 40, 5488–95.

41 Tangrea, M. A., Bryan, P. N., Sari, N. & Orban, J. (2002). Solution structure of the Pro-hormone convertase 1 pro-domain from *Mus musculus*. *J Mol Biol* 320, 801–12.

42 Guasch, A., Coll, M., Aviles, F. X. & Huber, R. (1992). Three-dimensional structure of porcine pancreatic procarboxypeptidase A. A comparison of the A and B zymogens and their determinants for inhibition and activation. *J Mol Biol* 224, 141–57.

43 Coll, M., Guasch, A., Aviles, F. X. & Huber, R. (1991). Three-dimensional structure of porcine procarboxypeptidase B: a structural basis of its inactivity. *EMBO J* 10, 1–9.

44 Baker, D., Sohl, J. L. & Agard, D. A. (1992). A protein-folding reaction under kinetic control. *Nature* 356, 263–5.

45 Sohl, J. L., Jaswal, S. S. & Agard, D. A. (1998). Unfolded conformations of alpha-lytic protease are more stable than its native state. *Nature* 395, 817–19.

46 Ottesen, M. & Svendsen, I. (1970). The subtilisins. *Methods Enzymol* 19, 199–215.

47 Poland, D. C. & Scheraga, H. A. (1965). Statistical mechanics of noncovalent bonds in polyamino acids. VIII. Covalent loops in proteins. *Biopolymers* 3, 379–99.

48 Pace, C. N., Grimsley, G. R., Thomson, J. A. & Barnett, B. J. (1988). Conformational stabilities and activity of ribonuclease T1 with zero,

one and two intact disulfide bonds. *J Biol Chem* 263, 11820–5.

49 DERMAN, A. I. & AGARD, D. A. (2000). Two energetically disparate folding pathways of alpha-lytic protease share a single transition state. *Nat Struct Biol* 7, 394–7.

50 CARTER, P. & WELLS, J. A. (1988). Dissecting the catalytic triad of a serine protease. *Nature* 332, 564–8.

51 VOORDOUW, G., MILO, C. & ROCHE, R. S. (1976). Role of bound calcium in thermostable, proteolytic enzymes. Separation of intrinsic and calcium ion contributions to the kinetic thermal stability. *Biochemistry* 15, 3716–24.

52 ALEXANDER, P. A., RUAN, B. & BRYAN, P. N. (2001). Cation-dependent stability of subtilisin. *Biochemistry* 40, 10634–9.

53 EDER, J., RHEINNECKER, M. & FERSHT, A. R. (1993). Folding of subtilisin BPN′: Characterization of a folding intermediate. *Biochemistry* 32, 18–26.

54 EDER, J., RHEINNECKER, M. & FERSHT, A. R. (1993). Folding of subtilisin BPN′: role of the pro-sequence. *J Mol Biol* 233, 293–304.

55 HAYASHI, T., MATSUBARA, M., NOHARA, D., KOJIMA, S., MIURA, K. & SAKAI, T. (1994). Renaturation of the mature subtilisin BPN′ immobilized on agarose beads. *FEBS Lett* 350, 109–12.

56 STRAUSBERG, S., ALEXANDER, P., WANG, L., SCHWARZ, F. & BRYAN, P. (1993). Catalysis of a protein folding reaction: Thermodynamic and kinetic analysis of subtilisin BPN′ interactions with its propeptide fragment. *Biochemistry* 32, 8112–19.

57 OHTA, Y., HOJO, H., AIMOTO, S., KOBAYASHI, T., ZHU, X., JORDAN, F. & INOUYE, M. (1991). Pro-peptide as an intramolecular chaperone: renaturation of denatured subtilisin E with a synthetic pro-peptide [corrected]. *Mol Microbiol* 5, 1507–10.

58 BAKER, D., SILEN, J. L. & AGARD, D. A. (1992). Protease pro region required for folding is a potent inhibitor of the mature enzyme. *Proteins* 12, 339–44.

59 SILEN, J. L. & AGARD, D. A. (1989). The alpha-lytic protease pro-region does not require a physical linkage to activate the protease domain in vivo. *Nature* 341, 462–4.

60 PETERS, R. J., SHIAU, A. K., SOHL, J. L. et al. (1998). Pro region C-terminus:protease active site interactions are critical in catalyzing the folding of alpha-lytic protease. *Biochemistry* 37, 12058–67.

61 BRYAN, P., WANG, L., HOSKINS, J. et al. (1995). Catalysis of a protein folding reaction: Mechanistic implications of the 2.0 Å structure of the subtilisin-prodomain complex. *Biochemistry* 34, 10310–18.

62 LI, Y., HU, Z., JORDAN, F. & INOUYE, M. (1995). Functional analysis of the propeptide of subtilisin E as an intramolecular chaperone for protein folding. Refolding and inhibitory abilities of propeptide mutants. *J Biol Chem* 270, 25127–32.

63 RUVINOV, S., WANG, L., RUAN, B. et al. (1997). Engineering the independent folding of the subtilisin BPN′ prodomain: analysis of two-state folding vs. protein stability. *Biochemistry* 36, 10414–21.

64 KOJIMA, S., MINAGAWA, T. & MIURA, K. (1997). The propeptide of subtilisin BPN′ as a temporary inhibitor and effect of an amino acid replacement on its inhibitory activity. *FEBS Lett* 411, 128–32.

65 KOJIMA, S., MINAGAWA, T. & MIURA, K. (1998). Tertiary structure formation in the propeptide of subtilisin BPN′ by successive amino acid replacements and its close relation to function. *J Mol Biol* 277, 1007–13.

66 RUAN, B., HOSKINS, J. & BRYAN, P. N. (1999). Rapid folding of calcium-free subtilisin by a stabilized pro-domain mutant. *Biochemistry* 38, 8562–71.

67 KOJIMA, S., YANAI, H. & MIURA, K. (2001). Accelerated refolding of subtilisin BPN′ by tertiary-structure-forming mutants of its propeptide. *J Biochem (Tokyo)* 130, 471–4.

68 RUAN, B., HOSKINS, J., WANG, L. & BRYAN, P. N. (1998). Stabilizing the subtilisin BPN′ pro-domain by phage display selection: how restrictive is the

amino acid code for maximum protein stability? *Protein Sci* 7, 2345–53.

69 WANG, L., RUAN, B., RUVINOV, S. & BRYAN, P. N. (1998). Engineering the independent folding of the subtilisin BPN' pro-domain: correlation of pro-domain stability with the rate of subtilisin folding. *Biochemistry* 37, 3165–71.

70 RUAN, B. (1998). Folding of subtilisin: Study of independent folding and pro-domain catalyzed folding. PhD Dissertation, University of Maryland, College Park.

71 BRANDTS, J. F., HALVORSON, H. R. & BRENNAN, M. (1975). Consideration of the Possibility that the slow step in protein denaturation reactions is due to cis-trans isomerism of proline residues. *Biochemistry* 14, 4953–63.

72 BAI, Y., SOSNICK, T. R., MAYNE, L. & ENGLANDER, S. W. (1995). Protein folding intermediates: native-state hydrogen exchange. *Science* 269, 192–7.

73 CHAMBERLAIN, A. K., HANDEL, T. M. & MARQUSEE, S. (1996). Detection of rare partially folded molecules in equilibrium with the native conformation of RNaseH. *Nat Struct Biol* 3, 782–7.

74 CHUNG, E. W., NETTLETON, E. J., MORGAN, C. J. et al. (1997). Hydrogen exchange properties of proteins in native and denatured states monitored by mass spectrometry and NMR. *Protein Sci* 6, 1316–24.

75 JASWAL, S. S., SOHL, J. L., DAVIS, J. H. & AGARD, D. A. (2002). Energetic landscape of alpha-lytic protease optimizes longevity through kinetic stability. *Nature* 415, 343–6.

76 WALDBURGER, C. D., JONSSON, T. & SAUER, R. T. (1996). Barriers to protein folding: formation of buried polar interactions is a slow step in acquisition of structure. *Proc Natl Acad Sci USA* 93, 2629–34.

77 WALDBURGER, C. D., SCHILDBACH, J. F. & SAUER, R. T. (1995). Are buried salt bridges important for protein stability and conformational specificity? *Nat Struct Biol* 2, 122–8.

78 YABUTA, Y., SUBBIAN, E., TAKAGI, H., SHINDE, U. & INOUYE, M. (2002). Folding pathway mediated by an intramolecular chaperone: dissecting conformational changes coincident with autoprocessing and the role of Ca^{2+} in subtilisin maturation. *J Biochem (Tokyo)* 131, 31–7.

79 BAKER, D., SHIAU, A. K. & AGARD, D. A. (1993). The role of pro regions in protein folding. *Curr Opin Cell Biol* 5, 966–70.

80 STRAUSBERG, S., ALEXANDER, P., GALLAGHER, D. T., GILLILAND, G., BARNETT, B. L. & BRYAN, P. (1995). Directed evolution of a subtilisin with calcium-independent stability. *Bio/technology* 13, 669–73.

81 ALEXANDER, P. A., RUAN, B., STRAUSBERG, S. L. & BRYAN, P. N. (2001). Stabilizing mutations and calcium-dependent stability of subtilisin. *Biochemistry* 40, 10640–4.

82 DELMAR, E., LARGMAN, C., BRODRICK, J. & GEOKAS, M. (1979). A sensitive new substrate for chymotrypsin. *Anal Biochem* 99, 316–20.

83 ALEXANDER, P., FAHNESTOCK, S., LEE, T., ORBAN, J. & BRYAN, P. (1992). Thermodynamic analysis of the folding of the streptococcal protein G IgG-binding domains B1 and B2: why small proteins tend to have high denaturation temperatures. *Biochemistry* 31, 3597–603.

84 KUNKEL, T. A. (1985). Rapid and efficient site-specific mutagenesis without phenotypic selection. *Proc Natl Acad Sci USA* 82, 488–92.

85 KUNKEL, T. A., SABATINO, R. D. & BAMBARA, R. A. (1987). Exonucleolytic proofreading by calf thymus DNA polymerase delta. *Proc Natl Acad Sci USA* 84, 4865–9.

86 STUDIER, F. W. & MOFFATT, B. A. (1986). Use of bacteriophage T7 RNA polymerase to direct selective high-level expression of cloned genes. *J Mol Biol* 189, 113–30.

30
The Thermodynamics and Kinetics of Collagen Folding

Hans Peter Bächinger and Jürgen Engel

30.1
Introduction

30.1.1
The Collagen Family

The human collagen family of proteins now consists of 27 types. The individual members are numbered with roman numerals. The family is subdivided in different classes: The fibrillar collagens (types I*, II, III, V*, XI*, XXIV, and XXVII), basement membrane collagens (type IV*), fibril-associated collagens with interrupted triple helices (FACIT collagens, types IX*, XII, XIV, XVI, XIX, XX and XXII), short chain collagens (types VIII* and X), anchoring fibril collagen (type VII), multiplexins (types XV and XVIII), membrane-associated collagens with interrupted triple helices (MACIT collagens, types XIII, XVII, XXIII, and XXV), and collagen type VI*. The types indicated by an asterisk are heterotrimers, consisting of two or three different polypeptide chains. All others consist of three identical chains or the chain composition is unknown. For type IV collagens six different polypeptide chains are known that form at least three distinct molecules and type V collagens contain three polypeptide chains in probably three molecules. The common feature of all collagens is the occurrence of triple helical domains with the repeated sequence -Gly-Xaa-Yaa- in the primary structure and the high content of proline and hydroxyproline residues. This sequence allows for the formation of a tertiary structure consisting of three polypeptide chains in a left-handed polyproline II helix. The three chains form a right-handed supercoiled triple helix, which is stabilized by hydrogen bonds. The glycine residues in every third position are packed tightly in the center of the triple helix. The residues in the Xaa and Yaa positions are exposed to the solvent and are often proline and hydroxyproline, respectively. All collagens are extracellular matrix proteins responsible for the architecture of connective tissues, such as bone, tendon, cartilage, skin, and basement membranes. Besides their structural roles, collagens interact with numerous other molecules and are crucial for development and homeostasis of connective tissue.

30.1.2
Biosynthesis of Collagens

The biosynthesis of procollagens is now fairly well understood at least for the most abundant fibrillar collagens [1, 2]. The current hypothesis of fibrillar procollagen biosynthesis involves the following steps:

1. Translocation of the emerging N-terminal end of the pro-α-chains into the rER.
2. Soon after the synthesis of the N-terminal propeptide, this domain folds and forms intrachain disulfide bonds. HSP47, a collagen-specific heat shock protein, may specifically bind to the N-terminal propeptide, but additionally has binding sites along the helical portion of the molecule.
3. The synthesis continues and some of the proline residues become hydroxylated by prolyl 4-hydroxylase (EC 1.14.11.2) and prolyl 3-hydroxylase (EC 1.14.11.7). During the continued synthesis of the major helical sequences, some interactions with molecular chaperones take place to prevent premature triple helix formation. Some lysine residues are hydroxylated by lysyl hydroxylase (EC 1.14.11.4).
4. After completion of the synthesis of the C-terminal propeptide, this part of the molecule folds, forms intrachain disulfide bonds, and interacts directly or indirectly (with a docking molecule) with the lipid bilayer of the rER.
5. Selection and association of the correct chains occur by diffusion of the C-terminal propeptides attached to the rER membrane.
6. A nucleus for triple helix formation is formed that aligns the chains in the right order. This nucleus initiates triple helix formation and is between propeptides stabilized by interchain disulfide bonds being formed, a reaction probably catalyzed by protein disulfide isomerase (PDI). In mutations of the C-terminal propeptides, an association with GRP78/BiP is found, indicating a potential role for GRP78/BiP in the assembly of procollagen chains.
7. Hydroxylation of proline residues and some lysine residues continues and triple helix formation proceeds from the C-terminal end towards the N-terminal end. The fast propagation of the triple helix formation is followed by a slower folding determined by *cis* peptide bond isomerization at proline residues. These peptide bonds need to be isomerized into *trans* conformation to allow triple helix formation to continue. This step is catalyzed by peptidyl-prolyl *cis-trans* isomerases (PPIases).
8. After completion of the folding of the major helix, the N-terminal propeptides associate and form the small triple helix within this domain. Further modifications occur during the transport through the Golgi stack by cisternal maturation. The long-standing problem of how a 300-nm-long procollagen molecule is transported to the Golgi in 50-nm transport vesicles has recently been investigated [3, 4]. Procollagen traverses the Golgi stack without leaving the lumen of the cisternae, but rather is transported by cisternal maturation. Nothing is known about what proteins direct that process.

The rate of synthesis of procollagen chains was measured for type I and II collagen, and found to be 209 residues per minute [5, 6]. For a fibrillar procollagen molecule, the synthesis time then is about 7 min. The time for secretion was 30 min when analyzed in bone [7] and biphasic secretion kinetics with half-times of 14 and 115 min were found in fibroblasts [8].

30.1.3
The Triple Helical Domain in Collagens and Other Proteins

The length of the triple helix in fibrillar collagens is about 300 nm and consists of 330–340 Gly-Xaa-Yaa tripeptide units. These molecules can be described as semi-rigid rods, well suited for the formation of fibrils. The fibrils have an axial 67-nm periodicity, which results from electrostatic and hydrophobic sidechain interactions [9]. In other collagens, stretches of tripeptide units are interspersed by interruptions, making these molecules more flexible. A large variation in the length of triple helices can be found in nature, the shortest collagen described is 14 nm long (mini-collagen of hydra) and the longest one 2400 nm (cuticle collagen of annelids) [10]. Gly-Xaa-Yaa repeats that form triple helices are also found in a number of other proteins: complement protein C1q, lung surfactant proteins A and D, mannose-binding protein, macrophage scavenger receptors A (types I, II, and III) and MARCO, ectodysplasin-A, scavenger receptor with C-type lectin (SRCL), the ficolins (L, M and H), the asymmetric form of acetylcholinesterase, adiponectin, and hibernation proteins HP-20, 25, and 27.

30.1.4
N- and C-Propeptide, Telopeptides, Flanking Coiled-Coil Domains

All collagens are synthesized with noncollagenous domains. The fibrillar collagens contain an N-terminal propeptide, an N-terminal telopeptide, the major triple helix, a C-terminal telopeptide and a C-terminal propeptide. The propeptides prevent premature fibril formation within the cell and are cleaved by specific proteases within both the N- and C-terminal telopeptides. The C-terminal propeptide is responsible for chain selection and association. The N-terminal end of the C-terminal propeptide contains three to four heptad repeats indicative of a coiled-coil structure that might facilitate its trimerization and the nucleation of the triple helix [11]. After cleavage, the remaining telopeptides play important roles in the formation of fibrils. The telopeptides also contain the lysine residues for the formation of covalent cross-links between molecules. The N-terminal propeptide consists of a globular domain and a minor triple helix. The globular domain potentially interacts with growth factors. Coiled-coil domains are prevalent in nonfibrillar collagens as well. The transmembrane domain-containing molecules with triple helices show a preference for a coiled-coil domain at the N-terminal end of the ectodomain.

30.1.5
Why is the Folding of the Triple Helix of Interest?

Folding studies of collagens are of interest because of the unusual structure of the collagen triple helix, which is an obligatory trimer. The three chains have to combine to the native state and do not have a defined structure if dissociated. Collagen belongs to the class of filamentous proteins and the mechanism of the folding process is distinctly different from folding pathways in globular proteins. Because of the repeating sequence and the uniform structure, wrong alignments of chains are likely and these are prevented by oligomerization domains.

There are also a number of practical aspects, which render folding studies interesting. Inherited and spontaneous mutations decrease triple helix stability and slow down the folding kinetics. Folding studies of wild-type and mutated collagens are needed for an understanding of theses diseases, which cause severe tissue disorders. The present review surveys the current knowledge of collagen folding. An earlier recent review on collagen folding is by Baum and Brodsky [12] and a useful general chapter on collagens was written by Bateman et al. [13].

30.2
Thermodynamics of Collagen Folding

30.2.1
Stability of the Triple Helix

The collagen molecules undergo a highly cooperative transition from a triple helix to an unfolded state upon heating. The temperature at the midpoint of this transition is termed T_m and is specific for different collagens from different species. Interestingly, the T_m of the collagens is around the body temperature of the organism from which it is isolated [14]. A linear correlation between the T_m and the number of imino acid residues was noticed early on [15]. Later, a better correlation was found between the T_m and the 4(R)-hydroxyproline content [16, 17]. Hydroxylation of proline residues is therefore important for the stability of the collagen triple helix and a summary of the posttranslational modifications of collagen molecules is needed. Further stabilization of the collagen molecules is achieved by the formation of supramolecular structures such as fibrils. In this review, we will only deal with individual collagen molecules in solution. The unfolding transition of triple helices occurs in a very narrow temperature interval, indicating a highly cooperative process. The T_m was found to be dependent on the rate of the increase in temperature and a hysteresis or apparent irreversibility is observed for the thermal transition of most extracted collagens. These difficulties in measuring such transition curves have led to the proposal that the collagen molecule is only kinetically stable [18, 19]. True equilibrium transition curves have been measured only recently for such collagens [14, 20]. From these measurements and from transition curves of short fragments of collagen or collagen-like synthetic peptides, it is now established that the collagen triple helix is a thermodynamically stable structure but establishment of equilibrium is slow. Synthetic collagen-like peptides have been extensively used in the characterization of the triple helical structure and its stability.

30.2.2
The Role of Posttranslational Modifications

Collagen molecules undergo extensive posttranslational modifications in the rough endoplasmic reticulum during biosynthesis (for a review see Refs [21, 22]). These enzymatic modifications occur on the nascent chains and require the collagen chains in an unfolded structure. Triple helical molecules are no longer substrates for these enzymes and it is thought that this regulates the stability of the triple helix. Three enzymes are involved in the hydroxylation of proline and lysine residues. Prolyl 4-hydroxylase (EC 1.14.11.2) hydroxylates proline residues in the Yaa position of collagens to form 4(R)-hydroxyproline [22]. Prolyl 3-hydroxylase (EC 1.14.11.7) can hydroxylate proline residues in the Xaa position to 3(S)-hydroxyproline, if the Yaa position is 4(R)-hydroxyproline. This modification occurs much less frequently in vertebrate collagens than the "obligatory" 4-hydroxylation. Certain lysine residues in collagen sequences are hydroxylated by lysine hydroxylase (EC 1.14.11.4) and O-linked oligosaccharides are attached. The carbohydrates were identified as 2-O-α-D-glucosyl-O-β-D-galactosylhydroxylysine and O-β-D-galactosylhydroxylysine. Hydroxylysine residues are also used for cross-linking of different collagen molecules in fibrils.

All these modifications affect the thermal stability of the collagen triple helix. When the stability of procollagen is investigated from cells that are incubated with α,α'-dipyridyl, an inhibitor of hydroxylases, the T_m was found to be about 15 °C lower than in the presence of hydroxylation. Such unhydroxylated procollagens are poorly secreted and tend to accumulate in the rough endoplasmic reticulum of the cell.

30.2.3
Energies Involved in the Stability of the Triple Helix

The enthalpy change of the coil ⇌ triple helix transition for collagens was determined to be $\Delta H° - (15–18)$ kJ mol^{-1} tripeptide units [15]. The enthalpy change increases with increasing imino acid content. On a per residue basis, the change of negative enthalpy for structure formation of triple helices is significantly larger than that of globular proteins. The source of this large change in enthalpy is still controversial.

Another unique feature is the temperature independence of $\Delta H°$. The specific heat of the reaction is zero within the error limits of the measurements [15]. Of the classical energies that determine protein stability, the electrostatic effect is the easiest to deal with in collagen. The stability of the collagen triple helix is only minimally affected by pH and salt concentrations, therefore electrostatic interactions can only play a minor role in the stability of the triple helix. The hydrophobic effect probably plays a role in stabilizing the triple helix, but its effect is not as dominant as in the stability of globular proteins. From the structure of the triple helix, it is evident that an interchain hydrogen bond between the NH group of glycine and the CO group of the amino acid in the Xaa position of a neighboring chain is present. Hydrogen exchange studies indicate that two classes of hydrogen bonds exist. The first class consists of 1.0 ± 0.1 very slowly exchanging hydrogens per tripeptide unit and a second class of 0.7 ± 0.1 slowly exchanging hydrogens per tripeptide unit [166]. It is likely that the first class comprises the above-mentioned hydrogen bond. If the Xaa and Yaa positions are not occupied by proline residues, it was suggested that an additional interchain hydrogen bond can form between the carbonyl group of glycine and the amino group of the Xaa residue, probably involving one or two water molecules.

The particular effectiveness of 4(R)-hydroxyproline in stabilizing the triple helix has led to suggestions that it may form additional hydrogen bonds through an extended water structure [23]. This notion became weaker, when it was shown that 4(R)-fluoroproline formed more stable triple helices than 4(R)-hydroxyproline [24, 25]. The replacement of the hydroxyl group by fluorine has several effects. The inductive effect of the fluorine group influences the pucker of the pyrrolidine ring. This is a stereoelectronic effect as it depends on the configuration of the substituent. 4(R) substituents stabilize the $C\gamma$-exo pucker, while 4(S) substituents stabilize the $C\gamma$-endo pucker. The X-ray structure of $(Pro-Pro-Gly)_{10}$ showed a preference of proline residues in the Xaa position in $C\gamma$-endo pucker, while the Yaa position prolines preferred $C\gamma$-exo puckering [26, 27]. This puckering influences the range of the main chain dihedral angles φ and ψ of proline, which are required for optimal packing in the triple helix. In addition, the trans/cis ratio of the peptide bond is influenced by the substituent [24]. Because all peptide bonds in the triple helix have to be trans, the helix is stabilized, if the amount of trans peptide bonds in the unfolded state is increased. None of these individual effects can fully explain the observed stability of the triple helix and it is difficult to quantitate the contribution of each of them.

A thermodynamic analysis of the triple coil $\rightleftarrows$ triple helix transition of fibrillar collagens shows that the calorimetric enthalpy and the van't Hoff enthalpy are significantly different [15]. The ratio of the van't Hoff enthalpy and the calorimetric enthalpy determines the cooperative length. For fibrillar collagens type I, II, and III a cooperative length of about 80–100 tripeptide units was determined [28]. This corresponds to about one-tenth of the total length of the triple helix. Therefore, the all-or-none model should be a good approximation for the shorter synthetic peptides. The cooperative length is an average parameter that applies to the whole triple helix and does not take into account the sequence variations that occur within a given molecule.

Because the rate of unfolding of the collagen triple helix is very slow in the transition region (see Section 30.3.4.4), great care must be given to the experimental rate of heating. Figures 30.1 and 30.2 show the dependence of the T_m of type III collagen as a function of the rate of heating. The T_m does not vary linearly with the heating rate and true equilibrium curves are only obtained by extrapolation of the heating rate to 0 (Figure 30.2). It should also be noted that most collagens show an irreversible denaturation curve, because the molecules were proteolytically cleaved and lack important domains for folding.

No equilibrium intermediates were observed during the unfolding transition for

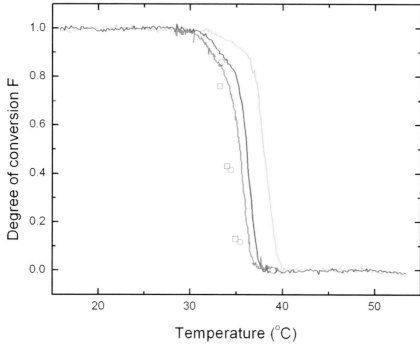

Fig. 30.1. Thermal unfolding of bovine pN type III collagen at different heating rates (thin line = 10 °C h^{-1}, thick line = 2 °C h^{-1}, and dashed line = 0.5 °C h^{-1}). The fraction of triple helix formation was measured by ORD at 365 nm. The endpoints of the unfolding (squares) and refolding (circles) kinetics at the indicated temperature are also shown.

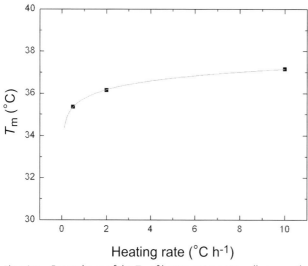

Fig. 30.2. Dependence of the T_m of bovine pN type III collagen on the rate of heating.

the triple helices of type I, II, and III collagen, despite the heterogeneity of the amino acid sequences along the triple helix. Several attempts have been made to calculate differences in stabilities of triple helical regions with the collagen triple helices. Sequence-dependent stabilities were calculated and these can explain the occurrence of unfolding intermediates in the transition of types V and XI collagen [29].

30.2.4
Model Peptides Forming the Collagen Triple Helix

30.2.4.1 Type of Peptides

Synthetic peptides containing Gly-Xaa-Yaa repeats have been used extensively to study the thermal stability and folding of the collagen triple helix. These peptides can be synthesized as either single chains or cross-linked peptides. The first peptides were synthesized by polycondensation of tri- or hexa-peptide units (for a review see Ref. [30]). This procedure resulted in a broad molecular weight distribution that was difficult to fractionate. The introduction of solid-phase methods allowed the synthesis of peptides with a defined chain length. In 1968 Sakakibara synthesized (Pro-Pro-Gly)$_n$ with $n = 5, 10, 15,$ and 20 [31]. Later, the oligomers of (Gly-Pro-Gly)$_n$ $n = 1-8$ [32], (Gly-Pro-Pro)$_n$ $n = 3-7$ [33], (Pro-4(R)Hyp-Gly)$_n$ $n = 5-10$ [34], and (Pro-Pro-Gly)$_n$ $n = 10, 12, 14,$ and 15 [35] were synthesized. Sutoh and Noda introduced the concept of block copolymers, where two blocks of (Pro-Pro-Gly)$_n$ $n = 5, 6,$ or 7 were separated by a block of (Ala-Pro-Gly)$_m$ $m = 5, 3,$ and 1, respectively [35]. This concept was later extended to include all amino acids and called host-guest peptide system. The most studied host-guest system uses the sequence Ac-(Gly-Pro-4(R)Hyp)$_3$-Gly-Xaa-Yaa-(Gly-Pro-4(R)Hyp)$_4$-Gly-Gly-NH$_2$ [36], but other variations were also studied [37, 38].

Because the stability of collagen-like peptides is highly concentration dependent, covalently linked peptides were synthesized by various methods. Heidemann used propane tricarboxylate tris(pentachlorophenyl) and N-tris(6-amino hexanoyl)-lysyl-lysine as covalent bridging molecules for the three chains [39]. Other covalent linkers include Kemp triacid [40], a modified dilysine system [41], the sequence found at the C-terminal end of type III collagen Gly-Pro-Cys-Cys-Gly [42–46], and tris(2-aminoethyl)amine with succinic acid spacers [47]. Peptides with covalent cross-links at both ends were also synthesized [48]. The noncovalent obligatory trimer foldon [46], peptide amphiphiles [49–51], and the iron (II) complex of bipyridine [52] have also been used to study trimeric peptides.

Initially, all peptides were studied as homotrimers, but strategies were developed to synthesize heterotrimers. Fields proposed the dilysine system to synthesize heterotrimeric peptides of type IV collagen [41] and Moroder used a simplified cysteine bridge to synthesize the collagenase cleavage site of type I collagen [53].

30.2.4.2 The All-or-none Transition of Short Model Peptides

We will start with the triple helix $\rightleftarrows$ coil transition of single chains. The simplest mechanism for this transition can be written as follows [54]:

$$3C \overset{\beta s^x}{\rightleftarrows} H_x \overset{s}{\rightleftarrows} \cdots \overset{s}{\rightleftarrows} H_i \overset{s}{\rightleftarrows} \cdots \overset{s}{\rightleftarrows} H_{3n-2} \tag{1}$$

where C is a coiled chain, H_i is a triple helical species with i tripeptide units in a triple helical conformation, β is the cooperativity parameter for the nucleation step, βs^x is the equilibrium constant for the step forming nucleus H_x, c_o is the total concentration of chains and s is the equilibrium constant for the addition of a tripeptide unit after nucleation. Because of the staggering of the chains, only $3n - 2$ tripeptide units can form hydrogen bonds in the triple helix and the complete triple helix is designated as H_{3n-2}. After nucleation, the triple helix is completed in propagation steps, which is the conversion of a coiled tripeptide unit adjacent to a triple helical tripeptide unit into a triple helical tripeptide unit. Since all these steps are essentially identical, the same equilibrium constant is applied to all of them. The process can be simplified if n is relatively small and βc_o^2 is less than 1. Under these conditions, the concentration of H_i becomes small compared with the concentrations of C and H_{3n-2}. The formation of the triple helix can then be approximated by an "all-or-none" reaction and the overall reaction can be described by an apparent equilibrium constant K.

$$K = \frac{[H_{3n-2}]}{[C]^3} = \beta s^{3n-2} = \frac{F}{3c_0^2(1-F)^3} \tag{2}$$

where F is the degree of helicity in terms of the fraction of chains or tripeptide units in the triple helical state and c_o the total molar concentration of chains, normally in mol tripeptide units l^{-1}.

$$F = \frac{3[H_{3n-2}]}{c_0} \tag{3}$$

From

$$\Delta G° = -RT \ln K = \Delta H° - T\Delta S° \tag{4}$$

and from Eq. (2) it follows that at the midpoint of the transition ($F = 0.5$ and $T = T_m$),

$$T_m = \frac{\Delta H°}{\Delta S° + R \ln(0.75 c_0^2)} \tag{5}$$

where $\Delta G°$ is the standard Gibbs free energy, $\Delta H°$ is the standard free enthalpy, and $\Delta S°$ is the standard entropy. It is important to note that the T_m is concentration dependant and stability values should only be compared at identical molar chain concentrations or concentrations which were corrected for concentration differences with Eq. (5).

$\Delta H°$ can be obtained from the slope of the transition curves. A first approximation of $\Delta H°$ can be obtained from the van't Hoff equation

$$\Delta H° = 8RT_m^2 \left(\frac{dF}{dT}\right)_{F=0.5} \tag{6}$$

It is much more accurate to obtain $\Delta H°$ by curve fitting of the entire transition curve with the equation

$$K = \exp\left[\frac{\Delta H°}{RT}\left(\frac{T}{T_m} - 1\right) - \ln 0.75 c_0^2\right] \tag{7}$$

which is obtained by solving Eq. (5) for $\Delta S°$ and substituting for $\Delta S°$ in Eq. (4). In this equation the specific heat c_p is assumed to be zero, as measured for several collagens by Privalov (see Section 30.2.3). In this regard, it is important to calculate F as a function of temperature. In the helical region and especially in the coiled region the circular dichroism (CD) signal shows a linear decrease with increasing temperature.

For a direct fit of F by K and c_o Eq. (2) is rewritten as the cubic equation:

$$F^3 - 3F^2 + bF - 1 = 0 \tag{8}$$

with:

$$b = \frac{9Kc_0^2 + 1}{3Kc_0^2} \tag{9}$$

By the method of Cardano [55] the real solution of this equation was obtained as:

$$F = u + v + 1 \tag{10}$$

with:

$$u = \sqrt[3]{-\frac{q}{2} + \sqrt{\left(\frac{q^2}{4} + \frac{p^3}{27}\right)}} \tag{11}$$

and:

$$v = \sqrt[3]{-\frac{q}{2} - \sqrt{\left(\frac{q^2}{4} + \frac{p^3}{27}\right)}} \tag{12}$$

in which $p = q = \frac{1}{3Kc_0^2}$.

When measuring CD the molar ellipticity is given by

$$[\Theta] = F([\Theta]_h - [\Theta]_c) + [\Theta]_c$$

where $[\Theta]_h$ and $[\Theta]_c$ are the ellipticities for the peptide in either the completely triple helical conformation or in the unfolded conformation. The linear temperature dependencies of $[\Theta]_h$ and $[\Theta]_c$ can be described by

$$[\Theta]_h = [\Theta]_{h,Tm} + p(T - T_m)$$
$$[\Theta]_c = [\Theta]_{c,Tm} + q(T - T_m)$$

Here the ellipticities at the midpoint temperature T_m are arbitrarily used as reference points. For a best fit of the transition curves, initially the T_m is kept constant and $\Delta H°$, p and q are varied. After an initial fit, the T_m is then also varied.

For peptides that contain a cross-link, the triple helix ⇌ coil transition becomes concentration independent and the equations for the all-or-none model simplify to:

$$K = \frac{F}{1 - F}$$

then F becomes

$$F = \frac{K}{1 + K}$$

and

$$T_m = \frac{\Delta H°}{\Delta S°}.$$

The fitting procedure for the determination of the thermodynamic parameters is otherwise the same as for nonlinked chains. In all fitting procedures the temperature dependence of $\Delta H°$ was neglected since the heat capacity c_p was determined to be very small or zero [15].

30.2.4.3 Thermodynamic Parameters for Different Model Systems

The simplest way to study the stability of collagen-like peptides is to measure the thermal transition curve by circular dichroism. The temperature at the midpoint of the transition T_m is normally used as a comparison for the stability between different peptides. However, as shown above this value is dependent on the concentration of the peptide and the rate of heating, and only values either measured or corrected to the same concentration and heating rate should be compared. Heating rates above $10 °C\ h^{-1}$ lead to transition curves that are too steep, therefore an enthalpy change that is too large if evaluated by the van't Hoff equation, and to T_m values that are too high. The enthalpy change of the triple helix coil transition can be calculated by fitting of the transition curve with the van't Hoff equation (see Section 30.2.4.2) or measured directly by differential scanning calorimetry (DSC). Van't Hoff enthalpies are usually less accurate than calorimetric values and both methods suffer from uncertainties in baselines. Uncertainties in the thermody-

Tab. 30.1. Thermodynamic data of the triple helix–coil transition of model peptides in aqueous solution determined by circular dichroism ($\Delta H°_{vH}, \Delta S°$) and differential scanning calorimetry ($\Delta H°_{cal}$). The values are expressed per mol of tripeptide units in a triple helix. The number of tripeptide units in the triple helix was calculated as $3n - 2$ for the peptide (Gly-Xaa-Yaa)$_n$.

Peptide	T_m (°C)	$\Delta H°_{vH}$ (kJ mol^{-1} tripeptide)	$\Delta S°$ (J mol^{-1} tripeptide K^{-1})	$\Delta H°_{cal}$ (kJ mol^{-1} tripeptide)	Reference
(Pro-Pro-Gly)$_{10}$	28	−10.6	−26.5		35
(Pro-Pro-Gly)$_{10}$	24.6	−7.91	−22.4	−7.68	54
(Pro-Pro-Gly)$_{10}$	25	−18.8	−58.6		159
(Pro-Pro-Gly)$_{10}$	34	−18.6	−60.9		112
(Pro-Pro-Gly)$_n$ $n = 10, 15, 20$		−8.2	−19.2		160
(Pro-Pro-Gly)$_n$ $n = 12, 14, 15$		−10.6	−26.5		35
(Pro-Pro-Gly)$_{10}$	32.6			−6.43	59
(Pro-4(R)Hyp-Gly)$_{10}$	57.3	−13.4	−36.4	−13.4	54
(Pro-4(R)Hyp-Gly)$_{10}$, pH 1	60.8	−23.6	−65.6		57
(Pro-4(R)Hyp-Gly)$_{10}$, pH 7	57.8	−23.3	−65.2		57
(Pro-4(R)Hyp-Gly)$_{10}$, pH 13	60.8	−25.1	−70.7		57
(Pro-4(R)Hyp-Gly)$_{10}$	60.0			−13.9	59
(Pro-4(R)Flp-Gly)$_{10}$	80	−17.1			112
(4(S)Flp-Pro-Gly)$_{10}$	58	−12.0			112
Ac-(Gly-4(R)Hyp-Thr)$_{10}$-NH$_2$	18	−27.1	−87.1	−27.5	98
Ac-(Gly-Pro-Thr(βGal))$_{10}$-NH$_2$	38.8	−14.0	−39.6		98
Ac-(Gly-4(R)Hyp-Thr(βGal))$_{10}$-NH$_2$	50.0	−11.1	−29.0		98
EK peptide,[a] pH 1	43.8	−27.4	−81.0		57
EK peptide,[a] pH 7	45.8	−26.3	−77.3		57
EK peptide,[a] pH 13	48.8	−22.6	−65.3		57
PTC((Pro-Ala-Gly)$_{12}$)$_3$[b]	12	−3.81	−13.3		161
PTC((Pro-Ala-Gly)$_{12}$)$_3$[c]	15	−3.63	−12.6		161
((Ala-Gly-Pro)$_9$)$_3$Lys-Lys	16.5	−7.32	−25.5		30
((Ala-Gly-Pro)$_{10}$)$_3$Lys-Lys	17.3	−6.69	−23.0		30
((Ala-Gly-Pro)$_{11}$)$_3$Lys-Lys	19.0	−6.65	−23.0		30
((Ala-Gly-Pro)$_{12}$)$_3$Lys-Lys	19.7	−7.28	−25.6		30
((Ala-Gly-Pro)$_{13}$)$_3$Lys-Lys	20.3	−5.69	−19.2		30
((Ala-Gly-Pro)$_{14}$)$_3$Lys-Lys	23.6	−5.31	−18.0		30
((Ala-Gly-Pro)$_{15}$)$_3$Lys-Lys	26.2	−5.36	−18.0		30
((Gly-Pro-Thr)$_{10}$-Gly-Pro-Cys-Cys)$_3$	13.8	−7.1	−41.5	−11.9	142
Ac-(Gly-Pro-4(R)Hyp)$_8$-Gly-Gly-NH$_2$	44.5	−16.0	−43.8		36
Ac-(Gly-Pro-4(R)Hyp)$_8$-Gly-Gly-NH$_2$	47.3	−19.8		−9.77[d]	58
T1-892	22.9	−15.6	−48.1		96
T1-892(O24A)	18.9	−18.7	−59.7		96
T1-892(P26A)	20.4	−18.3	−57.9		96
HT(I)[e] A	9	−18.6	−55.1		53
HT(I) B	33	−16.6	−47.6		53
HT(I) C	33	−14.7	−41.2		53
HT(I) D	41	−15.6	−43.8		53
HT(IV)[f] A	42	−4.49		−15.2	56
HT(IV) B	30	−4.30		−14.3	56
HT Ac-(Gly-Pro-4(R)Hyp)$_5$-Cys[g]	56.3	−15.8	−48.0	−15.5	44

Tab. 30.1. (continued)

Peptide	T_m (°C)	$\Delta H°_{vH}$ (kJ mol^{-1} tripeptide)	$\Delta S°$ (J mol^{-1} tripeptide K^{-1})	$\Delta H°_{cal}$ (kJ mol^{-1} tripeptide)	Reference
(Ac-(Pro-4(R)Hyp-Gly)$_5$Pro-Cys-Cys-Gly-Gly-Gly-NH$_2$)$_3$	68	−13.8	−40.4	−15.5	45
(Ac-Cys-Cys-Gly-(Pro-4(R)Hyp-Gly)$_5$-Gly-Gly-Gly-NH$_2$)$_3$	58.4	−10.1	−30.5	−14.5	45
GPP*(8)^h	62.5	−3.39	−10.1		162
PP*G(8)	50.7	−3.73	−11.5	−4.25	162
PP*G(7)	44.4	−3.4	−10.7		162
PP*G(6)	39.4	−5.85	−18.7		162
MG^i	16.6	−25.1	−81.3		163
((Gly-Pro-Pro)$_{10}$-Gly-Pro-Cys-Cys)$_3$	82				165
((Gly-Pro-Pro)$_{10}$)$_3$–foldon	66			−10.7	46
	70				165

^{a} The EK peptide is (Pro-4(R)Hyp-Gly)$_4$-Glu-Lys-Gly-(Pro-4(R)Hyp-Gly)$_5$.
^{b} Peptide with a parallel chain arrangement.
^{c} Peptide with an antiparallel chain arrangement.
^{d} Calorimetric data from Ref. [59].
^{e} Heterotrimeric type I collagen peptide containing the collagenase cleavage site.
^{f} Heterotrimeric type IV collagen peptide containing the $\alpha 1 \beta 1$ integrin binding site.
^{g} Heterotrimeric (Gly-Pro-4(R)Hyp)$_5$ with cysteine bridge.
^{h} Dilysine bridged peptide with a collagen type III sequence and a Gly-Pro-4(R)Hyp clamp at the N-terminus.
^{i} Synthetic α1-CB2.
Flp, fluoroproline.

namic data also arise from the facts that the transition curves monitored by CD are measured at a much lower concentration than the calorimetric measurements and that the concentration dependence is unknown in many studies. All measurements of single-chain peptides include the uncertainty of misaligned chains and errors in the fitting procedures. Data from different sources are summarized in Tables 30.1 and 30.2. In view of the many sources of error it is highly recommended to consult the individual publications in order to gain information on reliability.

For the reasons mentioned above, the published thermodynamic data for the triple helix coil transitions of synthetic peptides vary significantly. The commercially available peptides (Pro-Pro-Gly)$_{10}$ and (Pro-4(R)Hyp-Gly)$_{10}$ have been measured in several laboratories. The van't Hoff enthalpy change varies from −7.91 to −18.8 kJ mol^{-1} tripeptide units for (Pro-Pro-Gly)$_{10}$ and from −13.4 to −25.25 kJ mol^{-1} tripeptide units for (Pro-4(R)Hyp-Gly)$_{10}$. The published values for the calorimetric enthalpies of these peptides are more consistent.

Tab. 30.2. Thermodynamic data of the triple helix formation of host-guest peptides in PBS determined by circular dichroism ($\Delta H°_{vH}, \Delta S°$) and differential scanning calorimetry ($\Delta H°_{cal}$). The values are expressed per mol of triple helix.

Peptide	T_m (°C)	$\Delta H°_{vH}$ (kJ mol^{-1})	$\Delta S°$ (J mol^{-1} K^{-1})	$\Delta H°_{cal}$ (kJ mol^{-1})	Reference
HG[a] Gly-Pro-4(R)Flp	43.7			−204	59
HG Gly-4(R)Hyp-Pro	43.0			−204	59
HG Gly-4(R)Hyp-4(R)Hyp	47.3			−217	59
HG Gly-Pro-Pro	45.5			−213	59
HG Gly-Ala-4(R)Hyp	39.9	−423.7	−1214		36
HG Gly-Ala-4(R)Hyp	41.7	−480			58
HG Gly-Leu-4(R)Hyp	39.0	−437.5	−1256		36
HG Gly-Phe-4(R)Hyp	33.5	−514.1	−1549		36
HG Gly-Pro-Ala	38.3	−358.0	−1005		36
HG Gly-Pro-Ala	40.9	−502			58
HG Gly-Pro-Leu	32.7	−514.6	−1549		36
HG Gly-Pro-Leu	31.7	−514			58
HG Gly-Pro-Phe	28.3	−557.3	−1717		36
HG Gly-Pro-Arg	47.2	−610	−1100		62
HG Gly-Pro-Arg pH 2.7	45.5	−560	−1300		62
HG Gly-Pro-Arg pH 12.2	43.1	−390	−1100		62
HG Gly-Pro-Met	42.6	−436			58
HG Gly-Pro-Ile	41.5	−559			58
HG Gly-Pro-Gln	41.3	−559			58
HG Gly-Pro-Val	40.0	−481			58
HG Gly-Pro-Glu	39.7	−630	−1900		62
HG Gly-Pro-Glu pH 2.7	41.9	−570	−1700		62
HG Gly-Pro-Glu	38.5	−640	−1900		62
HG Gly-Pro-Thr	39.7	−647			58
HG Gly-Pro-Cys	37.7	−471			58
HG Gly-Pro-Lys	36.8	−400	−1200		62
HG Gly-Pro-Lys pH 2.7	37.1	−430	−1300		62
HG Gly-Pro-Lys pH 12.2	38.8	−440	−1300		62
HG Gly-Pro-His	35.7	−497			58
HG Gly-Pro-Ser	35.0	−435			58
HG Gly-Pro-Asp	34.0	−776			58
HG Gly-Pro-Asp	30.1	−550	−2000		62
HG Gly-Pro-Asp pH 2.7	33.1	−590	−1800		62
HG Gly-Pro-Asp pH 12.2	30.1	−770	−1700		62
HG Gly-Pro-Gly	32.7	−665			58
HG Gly-Pro-Asn	30.3	−640			58
HG Gly-Pro-Tyr	30.2	−657			58
HG Gly-Pro-Trp	26.1	−670			58
HG Gly-Glu-4(R)Hyp	42.9	−590	−1800		62
HG Gly-Glu-4(R)Hyp pH 2.7	39.7	−470	−1400		62
HG Gly-Glu-4(R)Hyp pH 12.2	40.9	−670	−2000		62
HG Gly-Lys-4(R)Hyp	41.5	−540	−1600		62
HG Gly-Lys-4(R)Hyp pH 2.7	40.4	−510	−1500		62
HG Gly-Lys-4(R)Hyp pH 12.2	38.3	−530	−1600		62

Tab. 30.2. (continued)

Peptide	T_m (°C)	$\Delta H°_{vH}$ (kJ mol^{-1})	$\Delta S°$ (J mol^{-1} K^{-1})	$\Delta H°_{cal}$ (kJ mol^{-1})	Reference
HG Gly-Arg-4(R)Hyp	40.6	−520	−1500		62
HG Gly-Arg-4(R)Hyp pH 2.7	39.4	−470	−1400		62
HG Gly-Arg-4(R)Hyp pH 12.2	38.0	−510	−1500		62
HG Gly-Gln-4(R)Hyp	40.4	−565			58
HG Gly-Asp-4(R)Hyp	40.1	−520	−1500		62
HG Gly-Asp-4(R)Hyp pH 2.7	37.6	−540	−1600		62
HG Gly-Asp-4(R)Hyp pH 12.2	38.0	−560	−1600		62
HG Gly-Val-4(R)Hyp	38.9	−518			58
HG Gly-Met-4(R)Hyp	38.6	−452			58
HG Gly-Ile-4(R)Hyp	38.4	−624			58
HG Gly-Asn-4(R)Hyp	38.3	−502			58
HG Gly-Ser-4(R)Hyp	38.0	−506			58
HG Gly-His-4(R)Hyp	36.5	−580			58
HG Gly-Thr-4(R)Hyp	36.2	−506			58
HG Gly-Cys-4(R)Hyp	36.1	−423			58
HG Gly-Tyr-4(R)Hyp	34.3	−629			58
HG Gly-Gly-4(R)Hyp	33.2	−575			58
HG Gly-Trp-4(R)Hyp	31.9	−593			58
HG Gly-Asp-Lys pH 2.7	26.5	−600	−1000		62
HG Gly-Asp-Lys	30.9	−520	−1600		62
HG Gly-Asp-Lys pH 12.2	29.9	−490	−1500		62
HG Gly-Asp-Arg pH 2.7	33.4	−550	−1700		62
HG Gly-Asp-Arg	37.1	−580	−1700		62
HG Gly-Asp-Arg pH 12.2	34.4	−460	−1300		62
HG Gly-Glu-Lys pH 2.7	29.5	−490	−1500		62
HG Gly-Glu-Lys	35.0	−590	−1800		62
HG Gly-Glu-Lys pH 12.2	33.1	−530	−1600		62
HG Gly-Glu-Arg pH 2.7	37.3	−530	−1600		62
HG Gly-Glu-Arg	40.4	−520	−1500		62
HG Gly-Glu-Arg pH 12.2	39.1	−510	−1500		62
HG Gly-Ala-Ala	29.3	−450	−1400		62
HG Gly-Lys-Asp pH 2.7	30.5	−770	−2400		62
HG Gly-Lys-Asp	35.8	−720	−2200		62
HG Gly-Lys-Asp pH 12.2	30.2	−720	−2300		62
HG Gly-Arg-Asp pH 2.7	28.8	−720	−2300		62
HG Gly-Arg-Asp	35.0	−720	−2200		62
HG Gly-Arg-Asp pH 12.2	31.9	−630	−1900		62
HG Gly-Lys-Glu pH 2.7	36.5	−620	−1900		62
HG Gly-Lys-Glu	35.3	−630	−1900		62
HG Gly-Lys-Glu pH 12.2	31.6	−680	−2100		62
HG Gly-Arg-Glu pH 2.7	35.0	−630	−1900		62
HG Gly-Arg-Glu	33.8	−680	−2100		62
HG Gly-Arg-Glu pH 12.2	32.2	−710	−2200		62
HG Gly-Gly-Phe	19.7	−647	−2093		61
HG Gly-Gly-Leu	23.9	−578	−1800		61
HG Gly-Gly-Ala	25.0	−559	−1758		61

Tab. 30.2. (continued)

Peptide	T_m (°C)	$\Delta H°_{vH}$ (kJ mol^{-1})	$\Delta S°$ (J mol^{-1} K^{-1})	$\Delta H°_{cal}$ (kJ mol^{-1})	Reference
HG Gly-Ala-Leu	27.8	−574	−1758		36
HG Gly-Phe-Ala	23.4	−593	−1884		36
HG Gly-Ala-Phe	20.7	−637	−2051		36

[a] HG refers to the host–guest peptides with the structure Ac(Pro-4(R)Hyp-Gly)$_3$-Gly-Xaa-Yaa-(Pro-4(R)Hyp-Gly)$_4$-Gly-Gly-NH$_2$. The tripeptide sequence given after HG corresponds to Gly-Xaa-Yaa. The peptides were measured in PBS, pH 7, unless indicated otherwise. Flp, fluoroproline.

Generally, the triple helix ⇌ coil transitions of short peptides can be described by the all-or-none model discussed in the previous section. Most peptides for which calorimetric data are available show that the ratio of the van't Hoff enthalpy and the calorimetric enthalpy is close to 1, indicating that the all-or-none mechanism is a good approximation, which is expected for peptides smaller than the cooperative length. However, exceptions have been reported, especially in heterotrimeric peptides [56].

Figure 30.3 shows the triple helix coil transition of Ac-(Gly-Pro-4(R)Hyp)$_{10}$-NH$_2$ measured by CD in water. The heating/cooling rate was 10 °C h^{-1} and the unfolding and refolding curves are shown for three different concentrations.

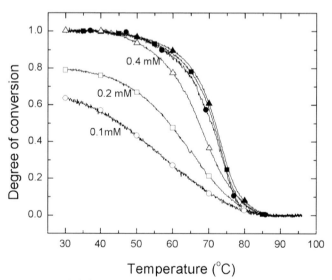

Fig. 30.3. Triple helix–coil transition of Ac-(Gly-Pro-4(R)Hyp)$_{10}$-NH$_2$ in water. The transition was monitored by circular dichroism at 225 nm. Closed symbols are for heating and open symbols for cooling, The heating/cooling rate was 10 °C h^{-1}.

The experimental data can be described by the all-or-none model (Mizuno and Bächinger, unpublished). The T_m increases with increasing concentration, but a concentration-dependent hysteresis is observed during refolding (see Section 30.3.4.4).

The studies with synthetic model peptides confirm the stabilizing role of 4(R)Hyp in the Yaa position of the triple helix. The influence of 4(S)Hyp, 3(S)Hyp, 4(S)Flp, and 4(R)Flp on the stability of the triple helix is entirely based on measurements with synthetic peptides (see Section 30.2.4.7).

The peptides in the host–guest system carry no charges at the ends and the values for the enthalpy and entropy changes are measured over a (Gly-Pro-4(R)Hyp)$_7$ background. The stability of (Pro-4(R)Hyp-Gly)$_{10}$ was shown to be pH dependent. At acidic and basic pH the peptide carries charges at only one end and is more stable than at neutral pH, where charges are present at both ends [57]. The peptides have the general sequence Ac-(Gly-Pro-4(R)Hyp)$_3$-Gly-Xaa-Yaa-(Gly-Pro-4(R)Hyp)$_4$-Gly-Gly-NH$_2$. The thermodynamic data for the guest peptides can be used to predict the stability of triple helices with any sequence [58, 59]. The limitations of this approach are discussed in Section 30.2.4.4. Table 30.2 lists the thermodynamic parameters determined for the host–guest peptides. In contrast to other synthetic peptides and CNBr peptides of type I collagen [60], where the enthalpy change increases with an increasing amount of imino acids, the Gly-Pro-4(R)Hyp containing host peptide has one of the lowest enthalpy change observed. The host–guest studies clearly show that hydrophobic residues do not play a major role in the stabilization of the triple helix, rather they perform important tasks in the assembly of collagen molecules into fibrils [36]. Gly-Gly containing peptides show a strong influence in destabilizing the triple helix, which may play an important role in specific functions by modulating the local triple helical stability [61]. These studies also clarify the dependence of the triple helical stability on the identity, position and ionization state of charged residues [62].

30.2.4.4 Contribution of Different Tripeptide Units to Stability

The high content of imino acids in collagens suggests a crucial role in the stabilization of the triple helix. Indeed, the most stable natural tripeptide unit is Gly-Pro-4(R)Hyp. Heidemann's group used statistical data [63], and published studies of stabilities of synthetic peptides [30, 64, 65] to derive four classes of the most common tripeptide units found in collagens. The first class of tripeptide units, the collagen-typical tripeptides, consists of Gly-Pro-Hyp, Gly-Pro-Ala, Gly-Ala-Hyp, Gly-Glu-Hyp, Gly-Glu-Arg, and Gly-Pro-Lys. The second class consists of tripeptide units often found in clusters with the collagen-typical tripeptides and includes Gly-Leu-Hyp, Gly-Pro-Ser, Gly-Phe-Hyp, Gly-Ala-Ala, Gly-Ala-Arg, Gly-Glu-Hyp, Gly-Asp-Ala, and Gly-Pro-Gln. The third class consists of tripeptide units that do not occur in the collagen-typical clusters and includes Gly-Glu-Ala, Gly-Ala-Lys, Gly-Asp-Arg, Gly-Ala-Asp, Gly-Pro-Ile, Gly-Arg-Hyp, Gly-Glu-Thr, and Gly-Ser-Hyp. The fourth class includes all the rest of the tripeptide units, which do not occur often enough for the statistical analysis.

A sequence-dependent stability profile along the triple helix was generated by assigning numerical values to the classes derived by Heidemann, using a sliding average to simulate cooperativity [66]. While empirical, this method was able to qualitatively explain the occurrence of unfolding intermediates in the triple helix of type V and XI collagens. In contrast to other fibrillar collagens, collagens type V and XI show unfolding intermediates in the triple helix coil transition [29]. A comparison of the sequence-dependent stability profiles indicates that type V and XI collagens lack a very stable region at the N-terminal end of the triple helix, that is present in other fibrillar collagens. The N-terminal region of the triple helix unfolds at lower temperatures in collagen types V and XI, and this leads to stable unfolding intermediates, that consist of the more C-terminal regions of the triple helix.

Brodsky's group used a host–guest system to quantitate the influence of naturally occurring tripeptide units to the triple helical stability [36, 58, 59, 61, 62, 67, 68, 69, 70]. The host–guest peptide system Ac-(Gly-Pro-Hyp)$_3$-Gly-Xaa-Yaa-(Gly-Pro-Hyp)$_4$-Gly-Gly-NH$_2$ is used to study the influence of the Xaa and Yaa residues on the thermal stability of the triple helix. While there are over 400 possible combinations of host–guest peptides, only about 80 appear with a significant frequency in known collagen sequences and only 24 tripeptides have a frequency of higher than 1% [71]. These peptides can be divided into three groups: the Gly-Xaa-Hyp group, the Gly-Pro-Yaa group, and the Gly-Xaa-Yaa group. Not surprisingly, the most stable of the 20 peptides of group 1 is Gly-Pro-Hyp, followed by Gly-Glu-Hyp and Gly-Ala-Hyp. The T_m values range from 47.3 °C for Gly-Pro-Hyp to 31.9 °C for Gly-Trp-Hyp. For group 2, the most stable of the 20 peptides is again Gly-Pro-Hyp, followed by Gly-Pro-Arg and Gly-Pro-Met. The T_m values range from 47.3 °C for Gly-Pro-Hyp to 26.1 °C for Gly-Pro-Trp. A reasonable correlation was found between the observed stabilities and the frequencies of occurrence in fibrillar collagen sequences [58]. Forty-one peptides of group 3 that contained no imino acids were synthesized and their stabilities measured [69]. Theoretical calculations, assuming no interaction between the Xaa and Yaa residues, allow the prediction of the stability of all 361 possible peptides. Sixteen of the 41 peptides show a reasonable agreement between the predicted and the measured stability ($\Delta T_m \leq 2$ °C). Seven peptides had a lower T_m than predicted and 18 had a higher T_m, most likely indicating side-chain interactions in the triple helix. These studies should now allow the prediction of the thermal stability of the triple helix in any Gly-Xaa-Yaa sequence. Unfortunately, this is not the case. The tripeptide units Gly-Hyp-Pro and Gly-3Hyp-Hyp show T_m values in the host-guest system that are comparable to Gly-Pro-Hyp, but neither forms a triple helix as a homopolymer [72]. A similar result was also obtained for the guest tripeptide unit Gly-Pro-fluoroproline [59]. The conclusion one has to draw is that these stabilities are dominated by the very stable tripeptide units at either end of the guest tripeptide unit, and that this system has little predictive power for the stability of Gly-Xaa-Yaa in homopolymers.

30.2.4.5 Crystal and NMR Structures of Triple Helices

The molecular structure of the triple helix was first derived from diffraction patterns of collagen fibrils [74, 75]. These models and data on the diffraction of tail tendon [76] showed a 10/3 symmetry (3.33 residues per turn). The first structure of the synthetic model peptide (Pro-Pro-Gly)$_{10}$ showed a 7/2 symmetry, giving rise to 3.5 residues per turn [77]. In 1994, a true single crystal was grown from a peptide that consisted of 10 repeats of Pro-Hyp-Gly with a single substitution of a glycine residue by an alanine in the middle [78]. This peptide was a triple helical molecule with a length of 87 Å and a diameter of 10 Å. The structure was solved to 1.9 Å and was consistent with the general parameters derived from the fiber diffraction model, but showed a 7/2 symmetry. It was hypothesized that imino acid-rich peptides have a 7/2 symmetry, but imino acid-poor regions adopt a 10/3 helix [78]. This was later confirmed with a synthetic peptide that contained an imino acid poor region [79, 80]. The structure also confirmed the expected hydrogen bond between the GlyNH and C=O of the proline in the Xaa position. In addition, an extensive water network forms hydrogen bonds with all available carbonyl and hydroxyl groups [23, 81].

The structure of the peptides (Pro-Pro-Gly)$_{10}$ [82, 83] and (Pro-Hyp-Gly)$_{10}$ [84] was also determined. The structure of (Pro-Pro-Gly)$_{10}$ obtained from crystals grown in microgravity, which diffracted up to a resolution of 1.3 Å, showed a preferential distribution of proline backbone and side-chain conformations, depending on the position [85, 26]. Proline residues in the Xaa position exhibit an average main chain torsion angle of $-75°$ and a positive side-chain χ_1 (down puckering), while proline residues in the Yaa position were characterized by a significantly smaller main-chain torsion angle of $-60°$ and a negative χ_1 (up puckering). These results were used to explain the stabilizing effect of hydroxyproline in the Yaa position, because hydroxyproline has a strong preference for up puckering. It also would explain the destabilizing effect of 4(R)-hydroxyproline in the Xaa position (see also Section 30.2.4.7).

The structure of (Gly-Pro-Pro)$_{10}$ was also solved in the context of the attached oligomerization domain foldon [86].

NMR studies with synthetic model peptides of the triple helix are difficult because of overlapping resonances of the repetitive sequence and by peak broadening from the shape [87–89]. Isotopic labeling was used to observe specific residues using heteronuclear NMR techniques. Hydrogen exchange studies were used to show that the NH groups of glycine exchanged faster in the imino acid-poor region of a synthetic peptide compared with the Gly-Pro-Hyp region [87]. The NH of threonine and the NH of glycine showed a very slow exchange with deuterium in the galactosyl-containing peptide Ac-(Gly-Pro-Thr(β-Gal))$_{10}$-NH$_2$ [90]. Heteronuclear NMR was used to monitor the folding kinetics of the triple helix [91, 92]. The hydration of the triple helix was also studied by NMR [93]. The hydration shell was found to be kinetically labile with upper limits for water molecule residence times in the nanosecond to subnanosecond range.

30.2.4.6 Conformation of the Randomly Coiled Chains

Little attention has been paid to the conformation of the unfolded chains in considering the stability of the collagen triple helix. Because the stability is determined by the free energy difference of the unfolded and the folded state, changes in the unfolded state can also contribute to the stability of the triple helix. The triple helix requires all peptide bonds to be in the trans conformation, so changes in the *cis* to *trans* ratio of peptide bonds in the unfolded state have a direct influence on the stability (see Section 30.2.4.8). Another example comes from studies with synthetic peptides, in which galactosyl threonine was incorporated [94]. The deep-sea hydrothermal vent worm *Riftia pachyptila* has a cuticle collagen that has a very low content of proline and hydroxyproline residues, but a high T_m of 37 °C [95, 42]. The Yaa position of the triple helical sequence in this cuticle collagen is frequently occupied by threonine. Threonines in the Yaa position were frequently galactosylated with one to three galactosyl units. The peptide Ac-(Gly-Pro-Thr)$_{10}$-NH$_2$ was unable to form a triple helix, whereas the galactosylated version Ac-(Gly-Pro-Thr(β-Gal))$_{10}$-NH$_2$ showed a T_m of 39 °C [94]. The stabilizing effect of galactosyl threonine has been explained by the occlusion of water molecules and by hydrogen bonding [90]. Additionally, galactosylation of the threonine residues also restricts the available conformations in the unfolded state. The end-to-end distance, determined by fluorescence energy transfer, is increased in the peptide Dabcyl-(Gly-Pro-Thr(β-Gal))$_5$-Gly-EDANS when compared with the peptide Dabcyl-(Gly-Pro-Thr)$_5$-Gly-EDANS (Bächinger, unpublished). However, it is difficult to quantitate the contribution of these different effects to the overall stability of the triple helix.

The importance of the restriction of imino acids in the unfolded state for the efficient folding of the triple helix has recently been described [96].

30.2.4.7 Model Studies with Isomers of Hydroxyproline and Fluoroproline

As mentioned before, it was recognized early on that hydroxproline played a crucial role in the stabilization of the triple helix. Synthetic peptides were used to establish that only 4(R)-hydroxyproline stabilized the triple helical structure. 4(S)-Hydroxyproline in the peptides (4(S)Hyp-Pro-Gly)$_{10}$ and (Pro-4(S)Hyp-Gly)$_{10}$ prevented the formation of a triple helix [97]. It was also shown that 4(R)-hydroxyproline in the Xaa position prevented triple helix formation in the peptide (4(R)Hyp-Pro-Gly)$_{10}$ [70]. Recent studies show that 4(R)-hydroxyproline in the Xaa position can lead to an increase in the stability of the triple helix, when the Yaa position is not proline. The peptide Ac-(Gly-Pro-Thr)$_{10}$-NH$_2$ does not form a triple helix in aqueous solution, but when the proline is hydroxylated the peptide Ac-(Gly-4(R)Hyp-Thr)$_{10}$-NH$_2$ does form a stable triple helix [98]. Further studies of peptides with 4(R)-hydroxyproline in the Xaa position showed that valine, but not serine or allo-threonine are able to form stable triple helical peptides, indicating that both the methyl group and the hydroxyl group of threonine, as well as the stereo configuration are important for the stability [73]. It was hypothesized that the methyl group shields the interchain hydrogen bond between the glycine and the Xaa residue from solvent and that the hydroxyl group of threonine and hydroxyproline can form a direct or water-mediated hydrogen bond. In addition, the peptide Ac-(Gly-4(R)Hyp-4(R)Hyp)$_{10}$-NH$_2$ has a T_m that is similar to the that of Ac-(Gly-Pro-4(R)Hyp)$_{10}$-NH$_2$ (Mizuno *et al.*, in press).

A small number of 3(S)-hydroyxyproline residues are present in most collagens in the Xaa position. The occurrence of 3(S)Hyp is much less frequent than that of 4(R)Hyp in the total amino acid content of collagens. In basement membrane collagens, fractions range from 1 [99] to 15 residues per 1000 residues [100]. 3(S)-

Hydroxyproline is also found in other types of collagens, such as type I [100], type V [101, 102], type X [103], and interstitial and cuticle collagens of annelids [105]. The only reported sequences containing 3(S)Hyp are in Gly-3(S)Hyp-4(R)Hyp tripeptide units. Synthetic peptides with 3(S)Hyp in a host–guest system decrease the stability of the triple helix in either the Xaa or the Yaa position [38]. A more severe decrease was observed for 3(S)Hyp in the Yaa position. It was concluded that the inductive effect of the 3-hydroxyl group of 3(S)Hyp decreases the strength of the GlyNH···OC3(S)Hyp hydrogen bond, when 3(S)Hyp is in the Xaa position, and that, when 3(S)Hyp is in the Yaa position, the pyrrolidine ring pucker leads to inappropriate mainchain dihedral angles and steric clashes [38]. When the influence of 3(S)Hyp was studied in homopolymers, the peptide Ac-(Gly-Pro-3(S)Hyp)$_{10}$-NH$_2$ did not form a triple helix [104]. Surprisingly, the peptide Ac-(Gly-3(S)Hyp-4(R)Hyp)$_{10}$-NH$_2$ was also unable to form a triple helix as a homopolymer. Even when foldon was attached to (Gly-3(S)Hyp-4(R)Hyp)$_{10}$, no triple helix formation was observed, ruling out a kinetic difficulty for this peptide to form such a structure [104].

While fluoroproline was used in biosynthetic experiments in the sixties [105, 106, 107], its importance to the stability of the triple helix was established by incorporating 4(R)-fluoroproline (4(R)Flp) into collagen model peptides [25]. Raines and coworkers showed that the substitution of 4(R)Hyp by 4(R)Flp in the peptide (Gly-Pro-4(R)Hyp)$_{10}$ leads to a significant increase in the stability of the triple helix. The authors initially cited an "unappreciated inductive effect" as the reason for this increase in stability, but more importantly, this result showed that an electron withdrawing substituent in the 4(R) position of proline can stabilize the triple helix without an extended network of water molecules [108]. Since this initial discovery, the inductive effect has been characterized in great detail [109, 110, 111]. The *gauche* effect determines the proline ring puckering (up puckering, Cγ-exo) and therefore predetermines the main-chain dihedral angles. In addition, the *trans* to *cis* ratio of the peptide bond preceeding 4(R)Flp increases. Both of these effects increase the stability of the triple helix. Peptides with 4(R)Flp in the Xaa position do not form a triple helix [112, 113]. Peptides with 4(S)Flp in the Yaa position also do not form a triple helix [44, 110]. In contrast to the peptide (4(S)Hyp-Pro-Gly)$_{10}$, which does not form a triple helix, the peptides (4(S)Flp-Pro-Gly)$_n$ form a stable triple helix ($n = 7$ [113]; $n = 10$ [112]). This is an unexpected result, because the *trans* to *cis* ratio of the peptide bond is decreased by 4(S)Flp. On the other hand, the preorganization of the ring puckering should promote triple helix formation. Why then do peptides with 4(S)Hyp in the Xaa position not form a triple helix? It was hypothesized that unfavorable steric interactions occur with 4(S)Hyp that are absent with 4(S)Flp [113].

The substitution of 4(R)Hyp by 4(R)-aminoproline in the peptide (Gly-Pro-4(R)Hyp)$_6$ also increases the stability of the triple helix. The extent of this increase in stability was strongly pH dependent [114].

30.2.4.8 *Cis* ⇌ *trans* Equilibria of Peptide Bonds

The peptide bond shows partial double bond character, as indicated by a shorter distance between the carbonyl carbon and the nitrogen than expected for a single

C–N bond. Consequently, the peptide bond is planar and the flanking Cα atoms can be either in trans ($\omega = 180°$) or cis ($\omega = 0°$) conformation. For peptide bonds preceding residues other than proline, only a small fraction (0.11–0.48%) was found to be in the cis conformation [115]. Peptide bonds preceding proline are much more frequently in the cis conformation, because the energy difference between the two conformations is rather small. Cis contents from 10 to 30% were found in short unstructured peptides [116, 117]. The activation energy for the cis-trans isomerization is high (80 kJ mol^{-1}) and the rate of isomerization is slow (see Section 30.3.2.4).

The collagen triple helix can accommodate only trans peptide bonds, so the ratio of cis to trans peptide bonds in the unfolded state has a direct influence on the stability of the triple helix. As mentioned above, the stereoelectronic effect of 4-substituents of proline residues influences the cis/trans ratio of the peptide bond preceding proline. It was proposed that the conformational stability of the triple helix relies on the change in the cis/trans ratio [110, 118]. These authors measured the $K_{cis/trans}$ for a number of 4-substituted proline residues. Table 30.3 summarizes these results.

There is indeed a good correlation of the observed cis/trans ratios and the thermal stability of homopolymers containing these residues in the Yaa position. However, a quantitation of this effect can be calculated [119]. The observed stability differences in the homopolymeric peptides (Pro-Pro-Gly)$_{10}$ and (Pro-4(R)Hyp-Gly)$_{10}$ are too large to be accounted for solely by the change in the cis/trans ratio of peptide bonds in the unfolded state (Bächinger, unpublished).

30.2.4.9 Interpretations of Stabilities on a Molecular Level

At present it is not possible to derive and quantitate all individual contributions that lead to the stability of the triple helix. It seems likely that a number of small

Tab. 30.3. $K_{cis/trans}$ in model compounds and unfolded type I collagen measured by NMR.

Compound	$K_{cis/trans}$	Reference
Ac-4(S)Hyp-OMe	0.37	Bächinger and Peyton, unpublished
Ac-Pro-OMe	0.16	
Ac-4(R)Hyp-OMe	0.12	
Ac-Pro-OMe	0.217	118
Ac-4(R)Hyp-OMe	0.164	
Ac-4(R)Flp-OMe	0.149	
Ac-4(S)Flp-OMe	0.4	
Ac-4(S)Hyp-OMe	0.417	110
Unfolded type I collagen		
X-Pro	0.19	164
X-Hyp	0.087	

Ac, acetyl; OMe, methyl ester; Flp, fluoroproline; Hyp, hydroxyproline.

contributions from the unfolded and folded state lead ultimately to the stable triple helix.

30.3
Kinetics of Triple Helix Formation

30.3.1
Properties of Collagen Triple Helices that Influence Kinetics

The complexity of the kinetics largely depends on the length of the triple helix to be formed. For short collagens and for short model peptides an all-or-none mechanism is a good approximation. The kinetics of systems with up to 45 tripeptide units per molecule (15 units per chain) have been successfully treated using this approximation [35, 120, 121]. Triple helices of most collagens are however much longer (1000–7200 amino acids) and the extent of cooperativity in triple helix formation is not sufficient for all-or-none transitions in systems of this size. The cooperative length of the triple helix in interstitial collagens with about 1000 tripeptide units was estimated to be about 100 tripeptide units from equilibrium studies [28]. This value was determined for the main triple helix of collagens I, II, and III and may apply to other collagens only in a first approximation because of sequence variations. The value is however consistent with the experimentally observed all-or-none nature of transitions with less than 45 tripeptide units. As expected the kinetics of long triple helices are more complex than those of short naturally occurring or designed model systems.

A second feature that largely influences the kinetics is the presence or absence of a cross-link between the three chains in a collagen molecule. As described in the Introduction most collagens contain nucleation domains (also called registration or oligomerization domains), which are located N-terminal or C-terminal of the triple helical domain. As explained in the first part of this review these domains serve to register the three chains at a side at which triple helix formation is nucleated. They are also involved in the selection of different chains in cases in which the collagen is a heterotrimer. The presence of trimerizing noncollagenous domains (designated as NC-domains or N-propeptides and C-propeptides) was found to be essential for proper folding of the triple helical domain. In addition, many collagen triple helical domains contain disulfide cross-links between their chains. The best-studied example is collagen III in which a disulfide knot of six cysteines (two per chain) connects the three chains at the C-terminus. During the physiological folding process, formation of this knot is most likely dependent on earlier noncovalent interactions between the adjacent NC-domains. After its formation it serves as an ideal registration and nucleation site, thus replacing the action of the NC-domain [122].

We shall first deal with the kinetics of folding from noncross-linked single chains and then turn to collagen triple helices, which are trimerized by either the disulfide knot of collagen III or by NC1-domains with strong trimerization potential. Natural proteins or fragment but also designed collagen-like proteins were employed in these studies.

30.3.2
Folding of Triple Helices from Single Chains

30.3.2.1 Early Work
Historically, all work on collagen folding started with single chains for the simple reason that before about 1980 only collagens from which the NC-domains were removed were available. Key publications of this time are by Piez and Carrillo [123] and Harrington and Rao [124]. In fact, the knowledge about N- and C-terminal NC-domains (frequently named N- and C-propeptides) and their importance developed only at this time and later. Before 1980, some work was also performed with cross-linked collagen chains derived from gelatin [125]. In contrast to the specific links leading to trimers, these chains were more statistically linked by nonreducible cross-links to dimers and higher oligomers.

An intriguing feature of early work in which collagens were refolded from their unfolded denatured chains was the observation that only a small fraction of native trimeric molecules was recovered. This fraction was very small at low temperatures and became higher at refolding temperatures close to the melting temperature [126]. Reaction orders varied between 1 and 2 depending on conditions and the time courses were difficult to interpret by a unique kinetic mechanism. Data indicated formation of products in which collagen triple helices with wrong alignment between the chains were formed. Annealing of these unstable products to native correctly aligned molecules was found to be very slow (for a review see Ref. [127]).

An early breakthrough in the kinetics of collagen triple helix formation from noncross-linked chains was achieved in studies with a short fragment chain of collagen I α1-CB2 [120]. An advantage of this system is its small size (12 tripeptide units per chain, 36 in the triple helix). A reaction order close to 3 was found for α1-CB2 and equilibrium transition curves indicated that the transition was of the all-or-none type with only unfolded chains and triple helices in equilibrium. Because of the relatively low stability of the triple helix formed by α1-CB2 it was difficult to reach a completely folded state at low temperatures. Furthermore, the fragment was isolated from a natural source and was available in small amounts only. Perhaps for these reasons no kinetic constant was derived and measurements of the concentration dependence were performed in a small interval only.

30.3.2.2 Concentration Dependence of the Folding of $(PPG)_{10}$ and $(POG)_{10}$
An extended study on the folding of short chains was performed with the synthetic model peptides $(PPG)_{10}$ and $(POG)_{10}$ [121]. Solutions were first heated to 60 °C ($(PPG)_{10}$) or 70 °C ($(POG)_{10}$) for 5 min to achieve complete unfolding. Time courses of refolding were recorded after fast cooling to 7 °C, at which the triple helix is completely folded according to equilibrium measurements. Kinetics of refolding was found to be extremely concentration dependent as expected for recombination of three chains (Figure 30.4).

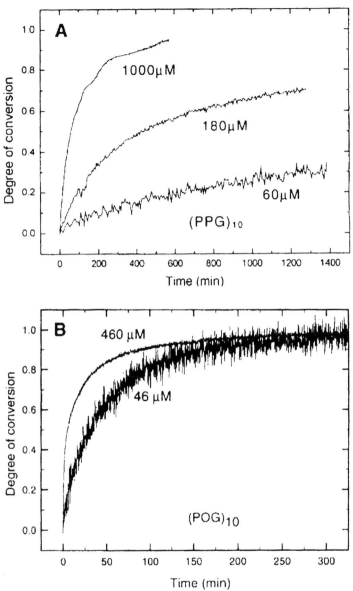

Fig. 30.4. Kinetics of triple helix formation of A) (ProProGly)$_{10}$ ((PPG)$_{10}$)) and B) (ProHypGly)$_{10}$ ((POG)$_{10}$)) at different peptide concentrations. The conversion was monitored at 7 °C after a temperature jump (dead time 2 min) from 60 °C in (A) and 70 °C in (B), temperatures at which the peptides are fully unfolded. The degree of conversion F was calculated by $F = ([\Theta]_{225} - [\Theta]_u)/([\Theta]_f - [\Theta]_u)$, where $[\Theta]_f$ and $[\Theta]_u$ are the ellipticities of the unfolded and folded state. The time courses for 60, 180, and 1000 μM concentrations of (PPG)$_{10}$ demonstrate a high dependence of the folding rate on concentration. For (POG)$_{10}$ the concentration dependence is much smaller and the rates are higher.

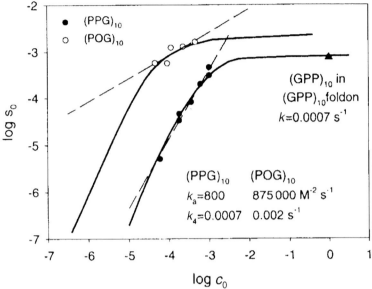

Fig. 30.5. Plot of the logarithm of initial rate of triple helix formation $s_o = (dF/dt)_{t=0}$ as a function of the logarithm of total chain concentration c_o for (ProProGly)$_{10}$ and (ProHypGly)$_{10}$. Broken lines show the best linear fits according to Eq. (2) with the apparent reaction orders of $n = 2.5$ and 1.5 for (ProProGly)$_{10}$ and (ProHypGly)$_{10}$, respectively. Continuous curves show better fits with nonlinear dependencies predicted by mechanism 13. The triangular point indicates the rate constant for (ProProGly)$_{10}$ in (GlyProPro)$_{10}$–foldon. This point was placed at a peptide concentration of 1 M (log $c_o = 0$), which is the estimated intrinsic chain concentration at the junction of the (GlyProPro)$_{10}$ and foldon domain in (GlyProPro)$_{10}$–foldon.

Satisfactory fits of the time courses with theoretical integrated equations for even reaction orders could not be achieved, but fits with reaction orders of 2 were better than for 1 or 3. In agreement with similar observation with host–guest peptides [128], the data suggested that the reaction order changed with the progress of the kinetics. Therefore, initial slopes s_o were determined by careful extrapolations of the initial phases of the time courses (Figure 30.5).

Plotting log s_o as a function of log c_o yielded reaction orders that were 3 at very low concentrations but dropped with increasing concentration. Here c_o is the total concentration of single chains. Interestingly the concentration dependence of s_o for single chains of (PPG)$_{10}$ converged to the value measured for the first-order folding of the peptide cross-linked by foldon or the Cys-knot (see (GPP)$_{10}$–foldon and (GPP)$_{10}$–Cys$_2$ below).

Activation energies were determined from the temperature dependence of the folding kinetics of (PPG)$_{10}$ and (POG)$_{10}$. Values were much lower than for the cross-linked polypeptides (Table 30.4).

30.3 Kinetics of Triple Helix Formation

Tab. 30.4. Rate constants and activation energies of triple helix formation from single chains and trimerized chains at 20 °C.

Protein	k_a (M^{-2} s^{-1})	k_4 (s^{-1})	E_a (kJ mol^{-1})
(ProProGly)$_{10}$	900[a]	–	7
(ProHypGly)$_{10}$	about 10^6*	–	8
(GlyProPro)$_{10}$–foldon	0.00197	54.5	
(GlyProPro)$_{10}$–Cys$_2$	0.00033	52.5	

[a] Experimental values at 7 °C were 800 s^{-1} and 875 000 s^{-1}, respectively.

30.3.2.3 Model Mechanism of the Folding Kinetics

The following mechanism was proposed to explain the high reaction order and its concentration dependence

$$C + 2C \underset{k_2}{\overset{k_1}{\rightleftarrows}} D^* \overset{k_3}{\rightarrow} H^* \overset{k_4}{\rightarrow} H \tag{13}$$

A schematic view of mechanism 13 is presented in Figure 30.6.

In this mechanism a very unstable dimeric nucleus D^* is in fast pre-equilibrium with monomers. Clearly a simultaneous collision of three particles in solution phase is an extremely infrequent event and a direct formation of a triple helical nu-

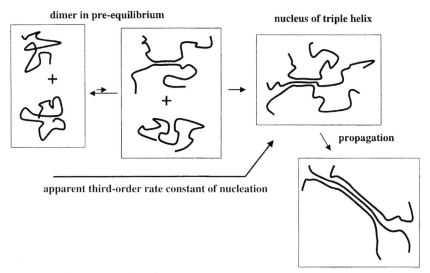

Fig. 30.6. Schematic view of mechanism 13 (see text).

cleus H* can therefore be excluded. D* forms in a bimolecular reaction and H* is formed from it by combination with a third chain.

$$d[D^*]/dt = k_1[C]^2 - k_2[D^*] - k_3[D^*][C] = 0 \tag{14}$$

D* is present only in very small amounts and therefore a steady state equilibrium ($d[D^*]/dt = 0$) is achieved shortly after the start of the reaction. Furthermore, dissociation into monomers is much faster than formation of H* ($k_2 \gg k_3[C]$) and the last term in Eq. (14) can be neglected. It follows that

$$k_1[C]^2 - k_2[D^*] = 0 \quad \text{or} \quad [D^*]/[C]^2 = k_1/k_2 = K \tag{15}$$

The rate of forming H* (neglecting the decay term – $k_4[H^*]$) is

$$d[H^*]/dt = k_3[D^*][C] = Kk_3[C]^3 = k_a[C]^3 \tag{16}$$

It follows that the apparent third order rate constant k_a is the product of the equilibrium constant K and a true second-order rate constant k_3.

It is safe to assume that the nuclei D* and H* do not significantly contribute to the CD signal. This assumption is justified by the high instability and lack of triple helical structure of D* and a very short segment of H* (compared with the final length of H), which acts as a nucleus. Mechanism 13 predicts two limiting cases in which either nucleation or propagation is rate limiting.

At very low concentrations formation of H* is the rate-limiting step and the rate of triple helix formation d[H]/dt equals the rate of nucleus formation as predicted by Eq. (16). In this case, the rate of helix propagation is faster than the rate of nucleation $k_4[H^*] \gg k_a[C]^3$. The experimentally observed degree of conversion F is related to [H] by $F = 3[H]/[C_0]$ in which $[C_0]$ is the total concentration of chains. Consequently the initial rate

$$(dF/dt)_{t=0} = 3k_a[C_0]^2 \tag{17}$$

At sufficiently high concentration nucleation will be faster than propagation and $[H^*] = (1/3)[C_0]$ after an initial phase. Propagation will then be the rate-limiting step of a first-order reaction

$$(dF/dt)_{t=0} = k_4[C_0] \tag{18}$$

Theoretical curves were calculated for mechanism 13 also for intermediate situations by numerical integration [121] and these curves are included in Figure 30.5. The dependencies were calculated with $k_a = 800$ and $875\,000$ M^{-2} s^{-1} and $k_4 = 0.0007$ s^{-1} and 0.002 s^{-1} for (PPG)$_{10}$ and (POG)$_{10}$, respectively. It can be seen that limiting case 1 is nearly reached for (PPG)$_{10}$, whose reaction order is 2.8–3 in the lowest concentration range. The average reaction order for all data was 2.5 and as mentioned an extrapolation to high concentration of 1 M is consistent with the rate constant $k_4 = 0.0007$ s^{-1} determined for (GPP)$_{10}$ in (GPP)$_{10}$–foldon. Folding of (POG)$_{10}$ has an average reaction order of 1.5 and the initial rate plateaus at higher concentrations. It should be mentioned that experimental determinations of reaction orders are possible only in a rather narrow concentration interval, which is defined by the sensitivity of the CD spectrometer and the solubility of the model peptides.

30.3.2.4 Rate Constants of Nucleation and Propagation

The rate constants derived by mechanism 13 are summarized in Table 30.4. The apparent third-order rate constant k_a reflects the finding of the three chains and the nucleation process. According to the model mechanism 13 this includes formation of a dimeric precursor and the first short trimeric helical nucleus. The model is certainly oversimplified and in reality, chains may meet at many sites. Because of the repetitive sequence of the model peptides, precursors and nuclei will be of comparable stability if nucleated at different sites. The apparent rate constant k_a can therefore not be assigned to a single event. It should be noted that wrong nucleations and mismatching would also happen in natural collagens with their long repetitive sequences. This is probably the reason why special noncollagenous nucleation domains have been invented as registration domains.

Interestingly k_a differs by a factor of 1000 between (PPG)$_{10}$ and (POG)$_{10}$. A faster kinetics is expected in the presence of 4-hydroxyprolines in the Y-position because of its stabilizing action (see Section 30.2.4.3). The large effect of hydroxyprolines on the rate of nucleation may be based in a stabilization of the precursor dimer (increase of K) or in an increase of the rate of addition of the third chain (increase of k_3). Activation energies for k_a are much lower than for the rate constant of propagation (Table 30.4). It should be recalled that k_a is the product of an equilibrium constant with a negative temperature dependence and a true kinetic constant, which always increases with increasing temperature. The small activation energy is therefore most likely caused by a compensation of these opposing effects in a similar way as observed for other kinetics with pre-equilibria [129].

Propagation rate constants of (PPG)$_{10}$ and (POG)$_{10}$ differ only by a factor of 6 and are in fair agreement with the values found for the cross-linked trimerized peptides (Table 30.4). It is a long accepted fact that the rate-limiting events in propagation of collagen triple helices are the *cis–trans* isomerization steps of peptide bond preceding prolines or hydroxyprolines [122, 130, 131]. In unfolded chains, a certain fraction of peptide bonds is in *cis* configuration at equilibrium. In the triple helix, only the *trans* configuration can be accommodated. In other prolines containing proteins, *cis–trans* isomerization is also a slow rate-determining step. In collagens, this effect is particularly prominent because of their unusually high content of prolines and hydroxyprolines. *Cis–trans* isomerization steps process an unusually high activation energy, which originates from the need to uncouple the π-electron system of the semi-double bond during the transition from *cis* to *trans*. Rate constants of *cis–trans* isomerization and their activation energies are dependent on amino acids adjacent to prolines [117], but these changes are in a rather narrow range and differences of no more than a factor of 10 in rate constants are expected.

An additional feature that may influence the kinetics is the equilibrium ratio of *cis* to *trans*, which again depends to some extent on the amino acid side chains preceding prolines or hydroxyprolines [117]. The approximate value for the equilibrium constant $K_{\text{cis-trans}}$ is 0.2 for the peptide bond between glycine and prolines. A rate constant $k_{\text{cis-trans}} = 0.003$ s^{-1} at 20 °C was determined for the isomerization of this bond in short peptides [116, 117]. Furthermore, variations of the prolines and hydroxyprolines content in the sequence will lead to differences in the number of *cis–trans* isomerization steps required for folding of a collagen segment. The last point does not apply to the comparison of (PPG)$_{10}$ and (POG)$_{10}$, but is relevant for the folding of natural collagens.

30.3.2.5 Host–guest Peptides and an Alternative Kinetics Model

Numerous studies have been performed with collagen-like model peptides of the design (GPO)$_3$GXY(GPO)$_4$GG. Here a guest tripeptide is housed between stable hydroxyprolines containing host segments. Peptides of this type were mainly employed for studies of relative stability of guest tripeptides with different residues in the X and Y position (see equilibrium part of this review). Kinetic studies were performed comparing guests GPO and GPP [128] in the host. A single GPP interrupting the GPO sequence is not expected to influence the nucleation rate by a large factor. Indeed the rate for the GPP containing peptide was only by a factor of 1.3 slower than that of (GPO)$_8$GG. Apparently, the kinetic data as well as the equilibrium stabilities are mainly determined by the host regions (see equilibrium part). Time courses of refolding could not be fitted with even reactions orders and measurements of concentration dependencies were restricted to two concentrations. A third-order rate constant about 20 000 M^{-2} s^{-1} at 15 °C was derived.

The kinetics of another designed peptide T1-892Y with (GPO)$_4$ at the C-terminus was investigated in greater detail [132]. The N-terminal part of this peptide AcG-PAGPAGPVGPAGARGPA was derived from collagen I. Data were fitted by a mechanism

$$3C_{cis} \underset{k_{tc}}{\overset{k_{ct}}{\rightleftarrows}} 3\ C_{trans} \underset{k_{fr}}{\overset{k_{rf}}{\rightleftarrows}} H$$

in which k_{ct} and k_{tc} are the rate constants of *cis–trans* isomerization and k_{rf} and k_{fr} are the rate constants of folding of residues in *trans* conformation. To improve the fits the mechanism was expanded by a branching reaction at C$_{trans}$. Fitting results supported a nucleation domain composed mainly of the (GPO)$_4$ moiety, which must be in *trans* form before the monomer is competent to initiate triple helix formation. Contrary to the highly positive activation energy of *cis–trans* isomerization, a negative activation energy was found, which was explained by a fast preequilibrium. Contrary to mechanism 13 the mechanism does not contain a dimer precursor and assumes a third-order reaction at any concentration in contrast to experimental findings with other model peptides.

N-terminal triple helix (fragment Col1-3)
```
     Human   (GXY)₁₀ GSOGPOGICESCPT
     Mouse   (GXY)₁₀ GPOGSOGICESCPT
     Bovine  (GXY)₁₀ GSOGPOGICESCPT
```

Central triple helix
```
     Human   (GXY)₃₄₀ GPOGAOGPCCGG
     Mouse   (GXY)₃₄₀ GPOGAOGPCCGG
     Bovine  (GXY)₃₃₉ GPOGAOGPCCGA
```

Fig. 30.7. Sequences of the short N-terminal and the central triple helix in collagen III. In major parts of the GXY repeats, the residues in the X and Y positions are not indicated. Hydroxyprolines residues (O) were only defined by amino acid sequencing of bovine collagen III. In the other sequences proposed by cDNA sequencing, proline residues in the Y position are probably also hydroxylated, and are indicated by O.

The folding of the designed peptide T1-892 was also investigated by NMR spectroscopy [12, 92]. By this technique *cis–trans* isomerization steps of Gly-Pro and Pro-Hyp bonds were monitored. The study identified individual residues involved in *cis–trans* isomerization as the rate-limiting step in triple helix propagation. Interestingly, the authors were able to define the direction of growth. It started in the (Gly-Pro-Hyp)$_4$ part of the peptide, which was defined as a nucleation domain. This finding was incorporated in the above mechanism [132]. Furthermore, a zipper-like folding (see below) was confirmed by the NMR study.

30.3.3
Triple Helix Formation from Linked Chains

Pioneering kinetic work was performed with the short N-terminal and the long central triple helical domains of collagen III, which consists of three identical chains [122, 130, 131]. Both domains are terminated by a disulfide knot. Sequences from three species are compared in Figure 30.7 with the N-terminal parts of the triple helices schematically indicated by GXY repeats [133]. It should be recalled that prolines in Y-positions are hydroxylated to hydroxyproline probably in all chains (Figure 30.7).

Note that the cysteine-containing sequences are different in the two domains. It was however established by molar mass determinations that the three chains are interlinked to trimers in both cases. The three-dimensional structure of the disulfide knots has not been solved yet but two likely models were proposed for the knot terminating the central helix [45, 131].

30.3.3.1   The Short N-terminal Triple Helix of Collagen III in Fragment Col1–3
Fragment Col1–3 consists of the N-terminal propeptide, the short triple helix and a noncollagenous telopeptide. Trimerization by the knot led to an increase in thermal stability and the fragment melts reversibly in an all-or-none type transition

with a midpoint temperature of 53 °C. It refolded in a first-order reaction with a rate constant of 8×10^{-3} s^{-1}, indicating *cis–trans* isomerization to be the rate-determining step. The value matches the later measured values for the propagation rate constants of other systems (Table 30.4). Values are not expected to be identical, because of differences of amino acid composition in the different systems. As a control for the function of the disulfide knot, refolding of Col1–3 was also studied after reductive cleavage. Kinetics was extremely slow and concentration dependent and the transition temperature of the product dropped to 35 °C.

A closer look at the refolding kinetics of Col1–3 trimerized by the disulfide knot revealed that kinetics proceeds in two phases [130]. In experiments with a dead time of 25 s the fast phase remained unresolved. The amplitude of the fast phase was about 50% of the total and rather temperature independent. The amplitude was significantly increased when refolding was started from a nonequilibrium state of the unfolded molecule in which less *cis*-peptide bonds were present than in the equilibrium state. The nonequilibrium state was achieved by refolding from chains, which were unfolded so quickly that most of the *trans* configuration present in the native triple helix was maintained. Experiments closely followed the double jump experiments [134] designed for similar experiments with ribonuclease S. Data are quantitatively analyzed by the zipper model of folding, which will be presented after discussing the kinetic data of the central triple helical domain.

30.3.3.2 Folding of the Central Long Triple Helix of Collagen III

For the central long triple helix the C-terminal propeptide provides a noncovalent link between the chains in the nonprocessed state but even after its removal the chains remain linked by the disulfide knot. The refolding kinetics of this trimerized, purely triple helical domain has been studied in great detail [28, 122]. Contrary to the refolding from noncross-linked chains, the unfolding–refolding process was completely reversible and end-products of refolding at 25 °C were identical to the native molecules as judged by their melting profiles, molecular weights, and sedimentation behavior. Mismatched structures of low stability were formed only at temperatures <15 °C.

The growth of the triple helix was found to proceed from the disulfide knot at the C-terminus at a rather uniform rate in a zipper-like fashion. This was most clearly shown by experiments in which the appearance and disappearance of folding intermediates was monitored directly (Figure 30.8).

The proteolytic assay is based on the ability of the collagen triple helix to resist trypsin digestion whereas the unfolded chains are readily degraded to small fragments [135]. Applying this method it was clearly shown that the folding intermediates all included the disulfide knot. Intermediates with short native triple helices are found at the start of the folding process. They decay and are replaced by longer fragments and finally native molecules with triple helices of full length are formed. In Figure 30.8 this is shown for the chains liberated by disulfide cleavage after digestion by trypsin. Gels were also run under nonreducing conditions [122]. In this

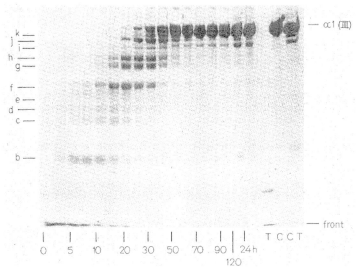

Fig. 30.8. Dodecyl sulfate slab gel electrophoresis of chain fragments which were protected against trypsin digestion by refolding. Type III collagen was denatured at 45 °C for 20 min and renatured at 25 °C for the time interval indicated. After incubation with trypsin at 20 °C for 2 min, sodium dodecyl sulfate was added and the sample was run on 10% polyacrylamide gel under reducing conditions. Eleven trypsin-resistant chain fragments designated by the letters a–k can be clearly distinguished. For comparison, native untreated collagen (C), native trypsin-treated collagen (CT) and trypsin (T) were also run.

case intermediates are not as well resolved because of their three times larger size but the same conclusions can be drawn. Furthermore, identical results were obtained with pN-collagen III, demonstrating that the N-terminal propeptides are not involved in the folding process.

For the zipper-like folding from the C-terminus zero-order kinetics is expected for a large initial fraction of the conversion (see next section). This behavior was experimentally observed for time courses followed by CD (Figure 30.9) [122].

Comparing the kinetics of the long central triple helix with those of the Col1–3 domain and the so-called one-quarter fragment obtained by selective proteolytic cleavage [136], a direct proportionality of the half-times to chain length was observed (Figure 30.9). This observation provides strong additional support to the zipper-like folding.

From the temperature dependence of refolding, activation energies of 85 kJ mol^{-1} were derived for the folding of the central triple helical domains, supporting that the propagation of the zipper is rate limited by *cis–trans* isomerization steps [130]. With the help of a model mechanism, it was concluded that on average, 30 amino acid residues occur in uninterrupted stretches without *cis* peptide bonds. They convert in a fast reaction after each isomerization step. The average rate constant of the isomerization steps was found to be 0.015 s^{-1} at 20 °C.

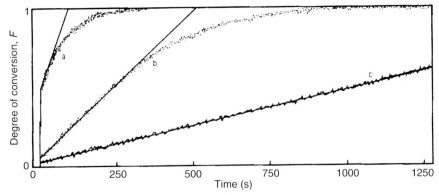

Fig. 30.9. Comparison of the folding kinetics of type III pN-collagen, the quarter fragment of type III collagen, and peptide Col1–3. Refolding of the collagen (c), the quarter fragment (b), and Col1–3 (a) were measured by circular dichroism. The straight lines represent the initial rates.

30.3.3.3 The Zipper Model

The model for the zipper-like folding of collagen triple helices is shown in Figure 30.10.

Here the disulfide knot is called h_0 and tripeptide units in correct helical conformation are called h. In the coiled state tripeptide units (v) with all peptide bonds in *trans* configuration are distinguished from units (w) which contain a *cis* peptide

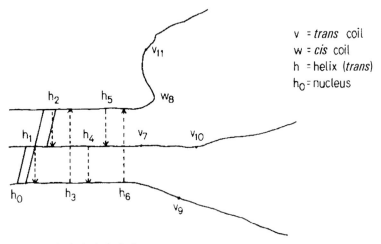

Fig. 30.10. Notation of the collagen-like part of Col1–3 as a linear sequence of letters. The three disulfide bridges are called h_0 and a tripeptide unit while the correct helical conformation is called h. In the coiled state, tripeptide units (v) with all peptide bonds in the *trans* configuration are distinguished from units (w) that contain a *cis* peptide bond.

bond. Tripeptide units are numbered in the way triple helix formation is believed to occur in a sequential way: to an already formed triple helix a tripeptide unit of chain A is added, followed by addition of the closest unit from chain B and then from C. A, B, and C are the designations for the three different chains in collagens. Each addition of a tripeptide unit is defined as a propagation step.

With the above notation, any state during folding can be written as a linear sequence of letters h, v, and w. For example, the partially folded state of Figure 30.10 reads $h_0hhhhhhvwvv\ldots$, with h_0 acting as a permanent nucleus. It was shown by designed models for the triple helix (see following sections) that nucleation is mainly achieved by enforcement of a close neighborhood of residues near to the disulfide knot. The local concentration near the disulfide knot is close to 1 M as estimated from the model of the disulfide knot [131] and from the stability of $(GPP)_{10}$–foldon (see equilibrium part). The studies with designed proteins demonstrated that the disulfide knot may be replaced by any other trimerizing agent without affecting the nucleation potential. It should also be recalled that the disulfide knot (or other trimerizing cross-links) serve the important function of keeping the chains in the correct register and preventing misalignments.

Only the stretch of uninterrupted v units, which starts from h_0 and is terminated by the first w, can convert to the triple helix without *cis–trans* isomerization. Under conditions in which back reactions can be neglected, the relative amplitude of the fast phase (δA_f) or slow phase (δA_s) becomes

$$\left(\frac{\delta A_f}{\delta A_0}\right)_{max} = 1 - \left(\frac{\delta A_s}{\delta A_0}\right) = \frac{\langle i_v \rangle}{n} \tag{19}$$

with

$$\langle i_v \rangle = \sum_{i=1}^{n-1} i p_v^i p_w + n p_v^n = \frac{p_v(1 - p_v^n)}{1 - p_v} \tag{20}$$

Here δA_0 is the total change, $\langle i_v \rangle$ is the average length of uninterrupted v units, and n is the total number of tripeptide units in the triple helix. The probabilities of v and w states, p_v and p_w, are related by the equilibrium constant $K = [v]/[w]$ of *trans* to *cis* isomerization.

$$p_v = \frac{K}{1 + K} \tag{21}$$

It can be shown for large n that $\langle i_v \rangle$ may be approximated by K. For peptide Col1–3 with $n = 43$ (Figure 30.9) the relative amplitude of the fast phase was 0.5 and a value of $K = 25$–30 is derived. Note that this equilibrium constant is an average over regions of variable proline and hydroxyproline content and is much larger than the equilibrium constant for the isomerization of a single X-Pro or X-Hyp bond. Assuming that K is identical for the central triple helix ($n = 1026$) and the quarter fragment ($n = 248$) the relative amplitude is predicted to be inversely proportional to n. This is exactly fulfilled by the experimental data shown in Figure 30.9.

After the initial fast phase, helix propagation will proceed with a rate determined by the rate constant of *cis–trans* isomerization k. Each isomerization step will liberate $(1 + K)$ v segments and the initial rate of the change in degree of helicity F will be

$$\left(\frac{dF}{dt}\right)_{t=0} = k\frac{(1 + K)}{n} = k_{app} \tag{22}$$

From kinetic experiments, (Figure 30.9) it was only possible to determine the apparent rate constant k_{app} from the initial rates of the curves. Assuming a rate constant of $k = 0.015$ s^{-1} as measured for the isomerization of a Gly-Pro bond [137], other data in [116, 117, 130] follow $K = 30$ in close agreement with the value obtained from the amplitude of the fast phase. This implies that on average about every thirtieth tripeptide unit in the coiled chains of collagen III contains a *cis* peptide bond. This value fits well with qualitative estimates from the distribution of proline and hydroxyproline residues.

It was also possible to derive the time course of the entire slow phase (Figure 30.11).

The probability of finding i w segments in a molecule of n coiled segments is given by a binominal distribution with the probabilities for *trans* and *cis* segments defined in Eq. (14)

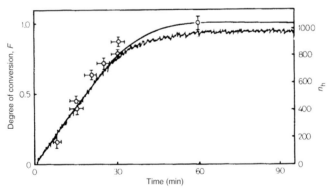

Fig. 30.11. Comparison of the degree of conversion F and length of the chains in triple helical conformation n_h during refolding of type III pN collagen at 25 °C. The experimental curve for the recovery of F with time at 25 °C was taken from Figure 30.9. It is compared with the number of residues in helical conformation n_h (open circles), which was calculated from the length of the trypsin-restistant fragments of Figure 30.8. The experimental time dependencies of n_h and F are compared with theoretical refolding kinetics (drawn-out curve). This was calculated by Eqs (24) and (28) with $k = 0.015$ s^{-1} and $K = 30$.

$$\hat{p}_i^n = \binom{n}{i} p_v^{n-i} p_w^i \tag{23}$$

We consider a species that initially contains i w segments and that converts to a fully helical molecule without w segments. Whenever the growing helix reaches a w segment a *cis* → *trans* isomerization at the h–w junction will lead to a continuation of helix growth and to a species with one w segment less. A *cis* ↔ *trans* isomerization will also occur at other places of the molecule. Since these steps are not coupled to the irreversible helix formation, they are reversible and on average, the same number of w segments will be formed and destroyed. These steps will therefore not contribute to the reaction of interest. As noted above, the rate of helix formation is constant as long as all species contain at least one w segment. This certainly holds true in the beginning of the reaction since $\hat{p}_0^{1018}$ in the binominal distribution (Eq. (23)) is essentially zero. In the later phase of the reaction, however, fully helical species with no h–w junctions will be formed which no longer contribute to the reaction. The approximation $[hw] \approx c_P$ in Eq. (22) is no longer valid and $f_{hw} = [hw]/c_P$ must be calculated as a function of time. Equation (22) becomes

$$\frac{dF}{dt} = \frac{k(1+K)}{n} f_{hw}(t) \tag{24}$$

In order to calculate f_{hw}, the species that initially contains i w segments is designated M_1, the species that is derived from it by disappearance of one w segment M_2, etc. The formation of the final product with no w segments, M_{i+1}, will require i sequential steps which all proceed with the rate constant of *cis* → *trans* isomerization k:

$$\begin{array}{ccccccc} M_1 & \xrightarrow{k} & M_2 & \xrightarrow{k} & \cdots & \xrightarrow{k} & M_r & \xrightarrow{k} & \cdots & \xrightarrow{k} & M_{i+1} \\ i & & i-1 & & & & i-r+1 & & & & 0 \end{array} \tag{25}$$

$i, i-1, i-r+1$ and 0 indicate the number of w segments in the species.
The rate equations are

$$\frac{df_1}{dt} = -kf_1 \tag{26a}$$

$$\frac{df_r}{dt} = -kf_r + kf_{r-1} \tag{26b}$$

$$\frac{df_{r+1}}{dt} = -kf_{r+1} + kf_r \tag{26c}$$

with $1 < r < i+1$.

Here $f_1, f_2 \ldots f_r, f_{r+1} \ldots f_{i+1}$ are the fractions of species in reaction (25). A fraction is defined by $f_r = c_r/c_1^0$ where c_1^0 is the initial concentration of M_1. To find a solution of this set of first-order linear differential equations we introduce

$$f_r = \Phi(t)e^{-kt} \quad \text{and} \quad f_{r+1} = \Psi(t)e^{-kt}$$

it follows that

$$\frac{df_{r+1}}{dt} = -kf_{r+1} + \frac{d\Psi}{dt}e^{-kt}$$

and with Eq. (26c)

$$\frac{d\Psi}{dt} = k\Phi(t)$$

The initial conditions at $t = 0$ are $f_1^0 = 1$ and all other $f_r = 0$. Integration of the first differential Eq. (26a) with this conditions yields

$$f_1 = e^{-kt} \quad \text{and} \quad \Phi(t) = 1$$

It follows that

$$\Psi(t) = kt \quad \text{and} \quad f_2 = kte^{-kt}$$

Stepwise application of this scheme to $r = 3, 4, \ldots r$ yields the general solution

$$f_r = \frac{(kt)^{r-1}}{(r-1)!}e^{-kt} \tag{27}$$

In order to obtain the fraction of molecules which contain at least one w segment we have to take the sum of f_r from $r = 1$ to $r = i$. Since these are fractions of the species present in the initial distribution we have to multiply each sum over f_r by corresponding $\hat{p}_i^n$ and to take the sum over this distribution in order to obtain the total fraction of species with at least one w segment and with one h–w junction

$$f_{hw} = \sum_{i=0}^{i=n} \hat{p}_i^n \sum_{r=1}^{r=i} \frac{(kt)^{r-1}}{(r-1)!}e^{-kt} \tag{28}$$

The rate of helix formation can now be calculated by Eqs (24) and (28) and the time dependence of F can be obtained by numerical integration of Eq. (24) into which f_{hw} Eq. (28) was substituted. A good fit to the experimental time course was obtained with the parameters $k = 0.015$ s^{-1} and $K = 30$ (see Figure 30.11).

30.3.4
Designed Collagen Models with Chains Connected by a Disulfide Knot or by Trimerizing Domains

Natural collagens contain triple helices whose chemical and physical properties are different at different positions of the helix. This property originates from variations of residues in X and Y positions. As described in the equilibrium part stabilities largely depend on the nature of X and Y in short peptides and differences may also result in long triple helices although cooperative influences of neighboring segments may reduce the effect.

Kinetic measurements of designed model proteins with GPP-, GPO-, or other collagen-like repeats, avoid complication caused by sequence heterogeneity. In addition, it is possible to study the influences of cross-linking in a quantitative way. Two methods of cross-linking were applied. In the first one, two Cys residues were added to the ends of the chains either by peptide synthesis or by fusion of the disulfide knot of type III collagen by recombinant technology. The chains were then joined by controlled oxidative coupling. In the second method, noncollagenous domains were fused to either the N- or C-terminus of the model peptide.

30.3.4.1 Disulfide-linked Model Peptides

Cys repeats that differ from those found at the ends of the central helix of collagen III and were different in different chains were placed at the C-termini of model peptides by elaborate peptide chemistry [138]. This strategy was primarily followed to design substrates for matrix metalloproteases (MMPs) and for integrins, which are both structure dependent and recognize triple helices composed of all three chains. The model collagens designed in his way refolded reversibly with first-order kinetics [138]. If reduced they lacked the potential to reform intact triple helices upon reoxidation in the presence of glutathione (L. Moroder, personal communication). Reoxidation was successful, however, when the disulfide sequence GPPGPCCGG of collagen III was used, proving the potential of this sequence for spontaneous formation of a disulfide knot [44–46].

In particular $(GPP)_{10}GPCCGG$ (abbreviated $(GPP)_{10}-Cys_2$) was expressed recombinantly and converted to a trimer in triple helical conformation by reoxidation in the presence of 9 mM oxidized and 0.9 mM reduced glutathione [121]. Data on the stabilization of the $(GPP)_{10}$ triple helix by the cross-link have been reported in the equilibrium part of this review. The kinetics of refolding was monitored after cooling from 70 °C to temperatures between 7 and 35 °C in the presence of 2.5 M GdmCl. The denaturant was added to shift the transition region to a more accessible region, since in plain buffer $(GPP)_{10}-Cys_2$ melts at 90 °C. Kinetics proceeded in a fast, experimentally unresolved phase and a dominant slow phase, which was of first order and concentration independent over a broad range of peptide concentrations. Measurements were performed by CD at the "collagen CD signal" at 221 nm. A large temperature dependence of the CD signal pointed to nonconformation-dependent signal change, which was also kinetically unresolved and subtracted from the time course. The measured first-order kinetic constants are included in Table 30.1 for comparison with the apparent third-order constants for the free uncross-linked chains. For reasons discussed in Ref. [121] activation energies (Table 30.4) were somewhat lower than measured for the *cis–trans* isomerzation of dipeptides [116, 117]. There is no doubt that the slow phase of the refolding reflects the propagation step, whose rate is determined by *cis–trans* isomerization of peptide bonds. The rate constants for the isomerization of individual prolines containing bonds in $(GPP)_{10}$–Cys_2 are unknown and it should be recalled that the overall rate constants are composites of these values.

30.3.4.2 Model Peptides Linked by a Foldon Domain

The foldon domain of T4 phage fibritin consists of three chains with 29 amino acid residues each and forms an obligatory trimer of high stability [46, 86]. Foldon domains were fused to the $(GPP)_{10}$ chains via a GS-linker and the chains were trimerized very efficiently. Care was taken not to unfold the foldon domain in all kinetic measurements performed with $(GPP)_{10}$–foldon. All data therefore correspond to the $(GPP)_{10}$ domains interlinked by an intact foldon trimer.

Results obtained with $(GPP)_{10}$–foldon were very similar to the results obtained with $(GPP)_{10}$–Cys_2 (Table 30.4). The identity of results shows that nucleation of the collagen triple helix is primarily determined by the trimerization at one end and that it does not matter which trimerization domain is applied. It is concluded that a domain like foldon, which was designed to nucleate the folding of fibritin to a trimeric coiled-coil structure, has the same function as the collagen-specific disulfide knot and probably also all other C- or N-terminal propeptide domains of natural collagens. The argument is supported by the observation that propeptide domains differ largely in structure (see Introduction) but do all form trimers.

30.3.4.3 Collagen Triple Helix Formation can be Nucleated at either End

As mentioned earlier, propeptides and disulfide bonds, which are believed to nucleate triple helix folding, are located at the C-terminus of most natural collagens. Exceptions are the membrane-spanning collagens XIII, XVII, XXIII, and XXV, which contain coiled-coil domains at the N-terminal side of the triple helix [139]. Three-stranded coiled-coil domains frequently act as oligomerization domains in extracellular membrane (ECM) proteins [140] and mutation data suggested that the coiled-coil domains nucleate triple helix formation in collagen XIII [141].

In this context, it was of interest to learn whether the kinetics of triple helix formation may be different for nucleation at the N- or C-terminal end. The problem was approached by a comparison of model peptides $(GlyProPro)_{10}$, which were either linked to trimers at the N- or C-terminus [165]. Linkage was either achieved by fusion with a short segment containing the disulfide knot of collagen III or by attachment of the foldon domains. In both cases a stabilization of the triple helix after cross-linking was observed, which was however somewhat less pronounced in the case of N-terminal attachment compared with C-terminal fusion. This difference may be explained by energy differences in the contact regions between the oligomerization domains and the triple helix. In the crystal structure of

(GlyProPro)$_n$–foldon [86] several bad contacts were observed in the contact region between the triple helix and the foldon domain. It was concluded that unfavorable enthalpic interactions may in part counteract the entropic stabilization by cross-linking. The structure of foldon–(GlyProPro)$_n$ was not solved yet but different and probably more unfavorable interactions are anticipated in its contact region. It was also noticed that the insertion of Gly-Ser spacers between the oligomerzation domain and the triple helix is essential. Most importantly, however, the kinetics of triple helix formation was identical for all four model systems (GlyProPro)$_n$–foldon, foldon–(GlyProPro)$_n$, (GlyProPro)$_n$–Cys$_2$ and Cys$_2$–(GlyProPro)$_n$. Also, the activation energies of folding were identical for all four peptides. Rate constants of folding were about 10^{-3} s^{-1} at 20 °C and the activation energy was 50 kJ mol^{-1} [165]. In summary, triple helix formation proceeds with closely similar rates in both directions. For the alignment of strands and nucleation of triple helical folding, oligomerization domains are needed but these may be placed at both ends of the triple helix. Their more frequent occurrence at the C-end is not explained by an easier folding from this side.

30.3.4.4 Hysteresis of Triple Helix Formation

As mentioned in Section 30.2.3, thermal unfolding and refolding of the triple helix is highly rate dependent in the transition region. Consequently, transition curves recorded by heating and by cooling form a hysteresis loop (see Figure 30.1). Hysteresis is very prominent for long natural collagens like collagen III and is also observed under isothermal conditions when unfolding is induced by increasing the concentration of guanidine hydrochloride [28]. In the case of collagen III with an intact disulfide knot at the C-terminus, correct and complete refolding of native molecules is achieved at temperatures 10 or more degrees below the transition temperature (or 0.5 M below the GdmCl midpoint concentration) after short incubation times of a few hours. In these cases, true hysteresis, in which equilibrium values are not reached even after very long waiting times, is limited to the range near the transition temperature. As mentioned in Section 30.3.2.1 and other parts of this review, unlinked single chains refold to misaligned structures at low temperatures. In these cases, an apparent hysteresis is observed, when monitoring CD or other parameters. The nature of this apparent hysteresis is very different from the true hysteresis because refolding products differ from the native parent molecules.

It was noticed that hysteresis was less prominent for short collagen triple helices like the one-quarter fragment of collagen III [28]. More recently, it was found that hysteresis loops are clearly observable also for short model peptides as long as the heating and cooling rates are not too slow. The major difference between native long triple helices and short model peptides is apparently the time dependence. For long triple helices the loops persist even at very slow scan rates whereas for short triple helices equilibrium is achieved more quickly.

For the model systems with linked chains, a simple kinetic hysteresis mechanism is proposed (Boudko, Bächinger, and Engel, in preparation). In the transition region the reciprocal apparent rate constant is

$$k_{app}^{-1} = \frac{1}{k_f + k_u} \qquad (29)$$

in which k_f and k_u are the rate constants of folding and unfolding, respectively. Rate constant k_f is dependent on temperature according to the Arrhenius relation with the high activation energy $E_{a,f}$ of cis–trans isomerization (see Section 30.3.3). The temperature dependence of k_u is much lower and its activation energy $E_{a,v}$ can be calculated from $E_{a,f}$ and the enthalpy of the reaction by $E_{a,v} = E_{a,f} - \Delta H°$. A satisfactory fit of the experimentally observed change of helicity F during heating or cooling is obtained when the rate law

$$\frac{dF}{dt} = k_f(1-F) - k_u F \qquad (30)$$

is integrated with the starting conditions (1) $F = 1$ at $t = 0$ for heating and (2) $F = 0$ at $t = 0$ for cooling. Case 1 yields the time course

$$F = \frac{K}{1+K} + \frac{1}{1+K} e^{-k_{app} t} \qquad (31)$$

and case 2

$$F = \frac{K}{1+K}(1 - e^{-k_{app} t}) \qquad (32)$$

with $K = k_f / k_u$. The hysteresis loops can be constructed by calculation of F after different times at different temperatures. Obviously for infinite times t the equilibrium curve $F = K/(1+K)$ is obtained.

The mechanism of hysteresis for long triple helices of natural collagens is clearly more complicated. They do not fold in an all-or-none reaction like the short proteins as demonstrated by a cooperative length that is about 1/10 of the length of the molecule [28]. A possible mechanism was proposed in Ref. [142].

30.3.5
Influence of cis–trans Isomerase and Chaperones

Catalysis of the folding of many proline-containing proteins was discovered shortly after the discovery of the kidney enzyme peptidyl-prolyl cis-trans isomerase (PPIase) [143]. Three protein families are now known to have PPIase activities: the cyclophilins, the FK506 binding proteins (FKBP) and the parvulins. In vitro experiments showed that cyclophilin has an accelerating effect on the folding of collagen III [144], but the factor was only close to 2. A similar effect was also found for collagen IV [145]. The action of PPIase on collagen was substantiated by the finding that cyclosporin A [146] and to a lesser extent FK506 [147] slows triple helix formation in vivo. FKBP65 had only a small effect on the rate of refolding on type III collagen [148]. PPIases are widely distributed and the *Escherichia coli* variant was also investigated with collagen [149]. It was speculated that the relatively small accelerations may originate from the use of enzymes that are not specific for collagen. A PPIase with an acceleration factor of 100 or more as observed for other proteins has not yet been observed. Such an enzyme may, however, be a necessity for animals living in a cold environment at which *cis–trans* isomerization is very slow, because of its high activation energy.

Numerous other helper proteins have been identified in the context of collagens (for reviews see Refs [150, 151, 152]). A collagen-specific chaperone is HSP47, whose physiological importance is manifested by a severe phenotype of transgenic mice lacking HSP47 [153]. Several very different mechanisms of action were discussed for HSP47, but with the exception of a specific binding to collagen, molecular explanations remain to be explored.

30.3.6
Mutations in Collagen Triple Helices Affect Proper Folding

Mutations in collagen genes cause a number of severe inherited diseases [154, 155]. The best-studied collagen-related genetic disease is osteogenesis imperfecta (brittle bone disease). Point mutations at different positions of the triple helix lead to improperly folded and instable triple helices, disturbed fiber formation, and severe pathological changes of the collagen matrix. Mutations near the C-terminus tend to be more severe than similar mutations near the N-terminus. In view of the steric need of small glycine residues in every third position (see Section 30.1.4), mutations of glycines to residues with larger side chains are particularly disturbing and cause large decreases of transition temperatures. In some cases, even kinks were visualized in such mutated collagens [156]. A delayed triple helix formation of mutant collagen from patients with osteogenesis imperfecta was observed [157]. Hydroxylation of prolines only occurs in the unfolded state and consequently the extent of hydroxylation and other posttranlational modifications was much increased in the mutated collagens, apparently at the N-terminal side of the mutation. Recently model peptides were studied with sequence irregularities designed after important natural mutations [158]. With this approach, it is hoped to gain deeper explanations of how a single-point mutation in the triple helix may cause a global pathological condition.

References

1 MAYNE, R. & BURGESON, R. E. (1987). *Structure and Function of Collagen Types*. Academic Press, Orlando, Fl.
2 OLSEN, B. R. (1991). Collagen biosynthesis. In: *Cell Biology of the Extracellular Matrix* (HAY, E. D., ed.). Plenum Press, New York, pp. 177–212.
3 MIRONOV, A. A., WEIDMAN, P. & LUINI, A. (1997). Variations on the

intracellular transport theme: maturing cisternae and trafficking tubules. *J Cell Biol* **138**, 481–4.

4 BONFANTI, L., MIRONOV, A. A., JR., MARTINEZ-MENARGUEZ, J. A. et al. (1998). Procollagen traverses the Golgi stack without leaving the lumen of cisternae: evidence for cisternal maturation. *Cell* **95**, 993–1003.

5 VUUST, J. & PIEZ, K. A. (1972). A kinetic study of collagen biosynthesis. *J Biol Chem* **247**, 856–62.

6 MILLER, E. J., WOODALL, D. L. & VAIL, M. S. (1973). Biosynthesis of cartilage collagen. Use of pulse labeling to order the cyanogen bromide peptides in the alpha L(II) chain. *J Biol Chem* **248**, 1666–71.

7 MORRIS, N. P., FESSLER, L. I., WEINSTOCK, A. & FESSLER, J. H. (1975). Procollagen assembly and secretion in embryonic chick bone. *J Biol Chem* **250**, 5719–26.

8 KAO, W. W., BERG, R. A. & PROCKOP, D. J. (1977). Kinetics for the secretion of procollagen by freshly isolated tendon cells. *J Biol Chem* **252**, 8391–7.

9 BATEMAN, J. F., LAMANDE, S. R. & RAMSHAW, J. A. M. (1996). Collagen superfamily. In: *Extracellular Matrix* (COMPER, W. D., ed.). Harwood Academic Publishers, Amsterdam, pp. 27–67.

10 ENGEL, J. (1997). Versatile collagens in invertebrates. *Science* **277**, 1785–6.

11 MCALINDEN, A., SMITH, T. A., SANDELL, L. J., FICHEUX, D., PARRY, D. A. & HULMES, D. J. (2003). Alpha-helical coiled-coil oligomerization domains are almost ubiquitous in the collagen superfamily. *J Biol Chem.* **278**, 42200–7.

12 BAUM, J. & BRODSKY, B. (2000) Case study 2: Folding of the collagen triple-helix and its naturally occuring mutants. In: *Frontiers in Molecular Biology: Mechanisms of Protein Folding*, 2nd edn (PAIN, R. H., ed.). Oxford University Press, Oxford.

13 BATEMAN, J. F., LAMANDE, R. & RAMSHAW, J. A. M. (1996) Collagen superfamily. In: *Extracellular Matrix*, Vol. 2 (COMPER, W. D., ed.). Harwood Academic Publishers, Amsterdam, pp. 22–67.

14 LEIKINA, E., MERTTS, M. V., KUZNETSOVA, N. & LEIKIN, S. (2002). Type I collagen is thermally unstable at body temperature. *Proc Natl Acad Sci USA* **99**, 1314–18.

15 PRIVALOV, P. L. (1982). Stability of proteins. Proteins which do not present a single cooperative system. *Adv Protein Chem* **35**, 1–104.

16 BURJANADZE, T. V. (1979). Hydroxyproline content and location in relation to collagen thermal stability. *Biopolymers* **18**, 931–8.

17 BURJANADZE, T. V. & VEIS, A. (1997). A thermodynamic analysis of the contribution of hydroxyproline to the structural stability of the collagen triple helix. *Connect Tissue Res* **36**, 347–65.

18 MILES, C. A., WARDALE, R. J., BIRCH, H. L. & BAILEY, A. J. (1994). Differential scanning calorimetric studies of superficial digital flexor tendon degeneration in the horse. *Equine Vet J* **26**, 291–6.

19 MILES, C. A., BURJANADZE, T. V. & BAILEY, A. J. (1995). The kinetics of the thermal denaturation of collagen in unrestrained rat tail tendon determined by differential scanning calorimetry. *J Mol Biol* **245**, 437–46.

20 BÄCHINGER, H. P. & ENGEL, J. (2001). Thermodynamic vs. kinetic stability of collagen triple helices. *Matrix Biol* **20**, 267–9.

21 KIVIRIKKO, K. I. (1998). Collagen biosynthesis: a mini-review cluster. *Matrix Biol* **16**, 355–6.

22 KIVIRIKKO, K. I. & MYLLYHARJU, J. (1998). Prolyl 4-hydroxylases and their protein disulfide isomerase subunit. *Matrix Biol* **16**, 357–68.

23 BELLA, J., BRODSKY, B. & BERMAN, H. M. (1995). Hydration structure of a collagen peptide. *Structure* **3**, 893–906.

24 PANASIK, N., JR., EBERHARDT, E. S., EDISON, A. S., POWELL, D. R. & RAINES, R. T. (1994). Inductive effects on the structure of proline residues. *Int J Pept Protein Res* **44**, 262–9.

25 HOLMGREN, S. K., TAYLOR, K. M., BRETSCHER, L. E. & RAINES, R. T.

(1998). Code for collagen's stability deciphered [letter]. *Nature* **392**, 666–7.

26 VITAGLIANO, L., BERISIO, R., MASTRANGELO, A., MAZZARELLA, L. & ZAGARI, A. (2001). Preferred proline puckerings in cis and trans peptide groups: implications for collagen stability. *Protein Sci* **10**, 2627–32.

27 VITAGLIANO, L., BERISIO, R., MAZZARELLA, L. & ZAGARI, A. (2001). Structural bases of collagen stabilization induced by proline hydroxylation. *Biopolymers* **58**, 459–64.

28 DAVIS, J. M. & BÄCHINGER, H. P. (1993). Hysteresis in the triple helix-coil transition of type III collagen. *J Biol Chem* **268**, 25965–72.

29 MORRIS, N. P., WATT, S. L., DAVIS, J. M. & BÄCHINGER, H. P. (1990). Unfolding intermediates in the triple helix to coil transition of bovine type XI collagen and human type V collagens alpha 1(2) alpha 2 and alpha 1 alpha 2 alpha 3. *J Biol Chem* **265**, 10081–7.

30 HEIDEMANN, E. & ROTH, W. (1982). Synthesis and Investigation of collagen model peptides. *Adv Polym Sci* **43**, 144–203.

31 SAKAKIBARA, S., KISHIDA, Y., KIKUCHI, Y., SAKAI, R. & KKIUCHI, K. (1968). Synthesis of poly-(L-prolyl-L-prolylglycyl) of defined molecular weights. *Bull Chem Soc Japan* **41**, 1273–80.

32 ROTHE, M., THEYSON, R., STEFFEN, K. D. et al. (1970). Synthesis and conformation of collagen models. *Angew Chem Int Ed Engl* **9**, 535.

33 WEBER, R. W. & NITSCHMANN, H. (1978). Der Einfluss der O-Acetylierung auf das konformative Verhalten des Kollagen-Modellpeptides (L-Pro-L-Hyp-Gly)$_{10}$ und von Gelatine. *Helv Chim Acta* **61**, 701–8.

34 SAKAKIBARA, S., INOUYE, K., SHUDO, K., KISHIDA, Y., KOBAYASHI, Y. & PROCKOP, D. J. (1973). Synthesis of (Pro-Hyp-Gly) n of defined molecular weights. Evidence for the stabilization of collagen triple helix by hydroxypyroline. *Biochim Biophys Acta* **303**, 198–202.

35 SUTO, K. & NODA, H. (1974). Conformational change of the triple-helical structure. IV. Kinetics of the helix-folding of (Pro-Pro-Gly)n (n equals 10, 12, and 15). *Biopolymers* **13**, 2477–88.

36 SHAH, N. K., RAMSHAW, J. A., KIRKPATRICK, A., SHAH, C. & BRODSKY, B. (1996). A host-guest set of triple-helical peptides: stability of Gly-X-Y triplets containing common nonpolar residues. *Biochemistry* **35**, 10262–8.

37 KWAK, J., JEFFERSON, E. A., BHUMRALKAR, M. & GOODMAN, M. (1999). Triple helical stabilities of guest-host collagen mimetic structures. *Bioorg Med Chem* **7**, 153–60.

38 JENKINS, C. L., BRETSCHER, L. E., GUZEI, I. A. & RAINES, R. T. (2003). Effect of 3-hydroxyproline residues on collagen stability. *J Am Chem Soc* **125**, 6422–7.

39 HEIDEMANN, E., NEISS, H. G., KHODADADEH, K., HEYMER, G., SHEIKH, E. M. & SAYGIN, O. (1977). Structural studies of collagen-like sequential polypeptides. *Polymer* **18**, 420–4.

40 FENG, Y., MELACINI, G., TAULANE, J. P. & GOODMAN, M. (1996). Collagen-based structures containing the peptoid residue N-isobutylglycine (Nleu): synthesis and biophysical studies of Gly-Pro-Nleu sequences by circular dichroism, ultraviolet absorbance, and optical rotation. *Biopolymers* **39**, 859–72.

41 FIELDS, C. G., LOVDAHL, C. M., MILES, A. J., HAGEN, V. L. & FIELDS, G. B. (1993). Solid-phase synthesis and stability of triple-helical peptides incorporating native collagen sequences. *Biopolymers* **33**, 1695–707.

42 MANN, K., MECHLING, D. E., BÄCHINGER, H. P., ECKERSKORN, C., GAILL, F. & TIMPL, R. (1996). Glycosylated threonine but not 4-hydroxyproline dominates the triple helix stabilizing positions in the sequence of a hydrothermal vent worm cuticle collagen. *J Mol Biol* **261**, 255–66.

43 MECHLING, D. E. & BÄCHINGER, H. P. (2000). The collagen-like peptide (GER)15GPCCG forms pH-dependent covalently linked triple helical trimers. *J Biol Chem* **275**, 14532–6.

44 BARTH, D., MUSIOL, H. J., SCHUTT, M. et al. (2003). The role of cystine knots in collagen folding and stability, Part I. Conformational properties of (Pro-Hyp-Gly)5 and (Pro-(4S)-FPro-Gly)5 model trimers with an artificial cystine knot. *Chemistry* **9**, 3692–702.

45 BARTH, D., KYRIELEIS, O., FRANK, S., RENNER, C. & MORODER, L. (2003). The role of cystine knots in collagen folding and stability, Part II. Conformational properties of (Pro-Hyp-Gly)n model trimers with N- and C-terminal collagen type III cystine knots. *Chemistry* **9**, 3703–14.

46 FRANK, S., KAMMERER, R. A., MECHLING, D. et al. (2001). Stabilization of short collagen-like triple helices by protein engineering. *J Mol Biol* **308**, 1081–9.

47 KWAK, J., DE CAPUA, A., LOCARDI, E. & GOODMAN, M. (2002). TREN (Tris(2-aminoethyl)amine): an effective scaffold for the assembly of triple helical collagen mimetic structures. *J Am Chem Soc* **124**, 14085–91.

48 TANAKA, Y., SUZUKI, K. & TANAKA, T. (1998). Synthesis and stabilization of amino and carboxy terminal constrained collagenous peptides. *J Pept Res* **51**, 413–19.

49 YU, Y. C., BERNDT, P., TIRRELL, M. & FIELDS, G. B. (1996). Self-assembling amphiphiles for the construction of protein molecular architecture. *J Am Chem Soc* **118**, 12515–20.

50 YU, Y. C., TIRRELL, M. & FIELDS, G. B. (1998). *J Am Chem Soc* **120**, 9979–87.

51 YU, Y. C., ROONTGA, V., DARAGAN, V. A., MAYO, K. H., TIRRELL, M. & FIELDS, G. B. (1999). Structure and dynamics of peptide-amphiphiles incorporating triple-helical proteinlike molecular architecture. *Biochemistry* **38**, 1659–68.

52 KOIDE, T., YUGUCHI, M., KAWAKITA, M. & KONNO, H. (2002). Metal-assisted stabilization and probing of collagenous triple helices. *J Am Chem Soc* **124**, 9388–9.

53 OTTL, J., MUSIOL, H. J. & MORODER, L. (1999). Heterotrimeric collagen peptides containing functional epitopes. Synthesis of single-stranded collagen type I peptides related to the collagenase cleavage site. *J Pept Sci* **5**, 103–10.

54 ENGEL, J., CHEN, H. T., PROCKOP, D. J. & KLUMP, H. (1977). The triple helix in equilibrium with coil conversion of collagen-like polytripeptides in aqueous and nonaqueous solvents. Comparison of the thermodynamic parameters and the binding of water to (L-Pro-L-Pro-Gly)n and (L-Pro-L-Hyp-Gly)n. *Biopolymers* **16**, 601–22.

55 BARTSCH, H.-J. (1982). *Taschenbuch mathematischer Formeln*. Verlag Harri Deutsch, Thun, p. 66.

56 SACCA, B., RENNER, C. & MORODER, L. (2002). The chain register in heterotrimeric collagen peptides affects triple helix stability and folding kinetics. *J Mol Biol* **324**, 309–18.

57 VENUGOPAL, M. G., RAMSHAW, J. A., BRASWELL, E., ZHU, D. & BRODSKY, B. (1994). Electrostatic interactions in collagen-like triple-helical peptides. *Biochemistry* **33**, 7948–56.

58 PERSIKOV, A. V., RAMSHAW, J. A., KIRKPATRICK, A. & BRODSKY, B. (2000). Amino acid propensities for the collagen triple-helix. *Biochemistry* **39**, 14960–7.

59 PERSIKOV, A. V., RAMSHAW, J. A., KIRKPATRICK, A. & BRODSKY, B. (2003). Triple-helix propensity of hydroxyproline and fluoroproline: comparison of host-guest and repeating tripeptide collagen models. *J Am Chem Soc* **125**, 11500–1.

60 SAYGIN, O. & HEIDEMANN, E. (1978). The triple helix-coil transition of cyanogen-bromide peptides of the alpha1-chain of the calf-skin collagen. *Biopolymers* **17**, 511–22.

61 SHAH, N. K., SHARMA, M., KIRKPATRICK, A., RAMSHAW, J. A. & BRODSKY, B. (1997). Gly-Gly-containing triplets of low stability adjacent to a type III collagen epitope. *Biochemistry* **36**, 5878–83.

62 CHAN, V. C., RAMSHAW, J. A., KIRKPATRICK, A., BECK, K. & BRODSKY, B. (1997). Positional preferences of ionizable residues in Gly-X-Y triplets of the collagen triple-helix. *J Biol Chem* **272**, 31441–6.

63 DÖLZ, R. & HEIDEMANN, E. (1986). Influence of different tripeptides on the stability of the collagen triple helix. I. Analysis of the collagen sequence and identification of typical tripeptides. *Biopolymers* **25**, 1069–80.

64 GERMANN, H. P. & HEIDEMANN, E. (1988). A synthetic model of collagen: an experimental investigation of the triple-helix stability. *Biopolymers* **27**, 157–63.

65 THAKUR, S., VADOLAS, D., GERMANN, H. P. & HEIDEMANN, E. (1986). Influence of different tripeptides on the stability of the collagen triple helix. II. An experimental approach with appropriate variations of a trimer model oligotripeptide. *Biopolymers* **25**, 1081–6.

66 BÄCHINGER, H. P. & DAVIS, J. M. (1991). Sequence specific thermal stability of the collagen triple helix. *Int J Biol Macromol* **13**, 152–6.

67 YANG, W., CHAN, V. C., KIRKPATRICK, A., RAMSHAW, J. A. & BRODSKY, B. (1997). Gly-Pro-Arg confers stability similar to Gly-Pro-Hyp in the collagen triple-helix of host-guest peptides. *J Biol Chem* **272**, 28837–40.

68 YANG, W., BATTINENI, M. L. & BRODSKY, B. (1997). Amino acid sequence environment modulates the disruption by osteogenesis imperfecta glycine substitutions in collagen-like peptides. *Biochemistry* **36**, 6930–5.

69 PERSIKOV, A. V., RAMSHAW, J. A., KIRKPATRICK, A. & BRODSKY, B. (2002). Peptide investigations of pairwise interactions in the collagen triple-helix. *J Mol Biol* **316**, 385–94.

70 PERSIKOV, A. V. & BRODSKY, B. (2002). Unstable molecules form stable tissues. *Proc Natl Acad Sci USA* **99**, 1101–3.

71 RAMSHAW, J. A., SHAH, N. K. & BRODSKY, B. (1998). Gly-X-Y tripeptide frequencies in collagen: a context for host-guest triple-helical peptides. *J Struct Biol* **122**, 86–91.

72 INOUYE, K., KOBAYASHI, Y., KYOGOKU, Y., KISHIDA, Y., SAKAKIBARA, S. & PROCKOP, D. J. (1982). Synthesis and physical properties of (hydroxyproline-proline- glycine)10: hydroxyproline in the X-position decreases the melting temperature of the collagen triple helix. *Arch Biochem Biophys* **219**, 198–203.

73 MIZUNO, K., HAYASHI, T. & BÄCHINGER, H. P. (2003). Hydroxylation-induced stabilization of the collagen triple helix: Further characterization of peptides with 4(R)-hydroxyproline in the Xaa position. *J Biol Chem* **278**, 32373–9.

74 RAMACHANDRAN, G. N. & KARTHA, G. (1955). Structure of collagen. *Nature* **176**, 593–5.

75 RICH, A. & CRICK, F. H. (1955). The structure of collagen. *Nature* **176**, 915–16.

76 FRASER, R. D., MACRAE, T. P. & SUZUKI, E. (1979). Chain conformation in the collagen molecule. *J Mol Biol* **129**, 463–81.

77 OKUYAMA, K., ARNOTT, S., TAKAYANAGI, M. & KAKUDO, M. (1981). Crystal and molecular structure of a collagen-like polypeptide (Pro-Pro-Gly)10. *J Mol Biol* **152**, 427–43.

78 BELLA, J., EATON, M., BRODSKY, B. & BERMAN, H. M. (1994). Crystal and molecular structure of a collagen-like peptide at 1.9 Å resolution. *Science* **266**, 75–81.

79 KRAMER, R. Z., BELLA, J., MAYVILLE, P., BRODSKY, B. & BERMAN, H. M. (1999). Sequence dependent conformational variations of collagen triple-helical structure. *Nat Struct Biol* **6**, 454–7.

80 KRAMER, R. Z., BELLA, J., BRODSKY, B. & BERMAN, H. M. (2001). The crystal and molecular structure of a collagen-like peptide with a biologically relevant sequence. *J Mol Biol* **311**, 131–47.

81 BELLA, J., BRODSKY, B. & BERMAN, H. M. (1996). Disrupted collagen architecture in the crystal structure of a triple-helical peptide with a

Gly → Ala substitution. *Connect Tissue Res* **35**, 401–6.

82 KRAMER, R. Z., VITAGLIANO, L., BELLA, J. et al. (1998). X-ray crystallographic determination of a collagen-like peptide with the repeating sequence (Pro-Pro-Gly). *J Mol Biol* **280**, 623–38.

83 NAGARAJAN, V., KAMITORI, S. & OKUYAMA, K. (1998). Crystal structure analysis of collagen model peptide (Pro-pro-Gly)10. *J Biochem (Tokyo)* **124**, 1117–23.

84 NAGARAJAN, V., KAMITORI, S. & OKUYAMA, K. (1999). Structure analysis of a collagen-model peptide with a (Pro-Hyp-Gly) sequence repeat. *J Biochem (Tokyo)* **125**, 310–18.

85 BERISIO, R., VITAGLIANO, L., MAZZARELLA, L. & ZAGARI, A. (2002). Crystal structure of the collagen triple helix model [(Pro-Pro-Gly)(10)](3). *Protein Sci* **11**, 262–70.

86 STETEFELD, J., FRANK, S., JENNY, M. et al. (2003). Collagen stabilization at atomic level: crystal structure of designed (GlyProPro)$_{10}$–foldon. *Structure (Camb)* **11**, 339–46.

87 FAN, P., LI, M. H., BRODSKY, B. & BAUM, J. (1993). Backbone dynamics of (Pro-Hyp-Gly)10 and a designed collagen-like triple-helical peptide by 15N NMR relaxation and hydrogen-exchange measurements. *Biochemistry* **32**, 13299–309.

88 LI, M. H., FAN, P., BRODSKY, B. & BAUM, J. (1993). Two-dimensional NMR assignments and conformation of (Pro-Hyp-Gly)10 and a designed collagen triple-helical peptide. *Biochemistry* **32**, 7377–87.

89 MELACINI, G., FENG, Y. & GOODMAN, M. (1996) Acetyl-terminated and template-assembled collagen-based polypeptides composed of Gly-Pro-Hyp sequences. 3. Conformational analysis by ^{1}H-NMR and molecular modeling studies. *J Am Chem Soc* **118**, 10359–64.

90 BANN, J. G., BÄCHINGER, H. P. & PEYTON, D. H. (2003). Role of carbohydrate in stabilizing the triple-helix in a model for a deep-sea hydrothermal vent worm collagen. *Biochemistry* **42**, 4042–8.

91 LIU, X., SIEGEL, D. L., FAN, P., BRODSKY, B. & BAUM, J. (1996). Direct NMR measurement of folding kinetics of a trimeric peptide. *Biochemistry* **35**, 4306–13.

92 BUEVICH, A. V., DAI, Q. H., LIU, X., BRODSKY, B. & BAUM, J. (2000). Site-specific NMR monitoring of cis-trans isomerization in the folding of the proline-rich collagen triple helix. *Biochemistry* **39**, 4299–308.

93 MELACINI, G., BONVIN, A. M., GOODMAN, M., BOELENS, R. & KAPTEIN, R. (2000). Hydration dynamics of the collagen triple helix by NMR. *J Mol Biol* **300**, 1041–9.

94 BANN, J. G., PEYTON, D. H. & BÄCHINGER, H. P. (2000). Sweet is stable: glycosylation stabilizes collagen. *FEBS Lett* **473**, 237–40.

95 GAILL, F., MANN, K., WIEDEMANN, H., ENGEL, J. & TIMPL, R. (1995). Structural comparison of cuticle and interstitial collagens from annelids living in shallow sea-water and at deep-sea hydrothermal vents. *J Mol Biol* **246**, 284–94.

96 XU, Y., HYDE, T., WANG, X., BHATE, M., BRODSKY, B. & BAUM, J. (2003). NMR and CD spectroscopy show that imino acid restriction of the unfolded state leads to efficient folding. *Biochemistry* **42**, 8696–703.

97 INOUYE, K., SAKAKIBARA, S. & PROCKOP, D. J. (1976). Effects of the stereo-configuration of the hydroxyl group in 4- hydroxyproline on the triple-helical structures formed by homogenous peptides resembling collagen. *Biochim Biophys Acta* **420**, 133–41.

98 BANN, J. G. & BÄCHINGER, H. P. (2000). Glycosylation/hydroxylation-induced stabilization of the collagen triple helix. 4-trans-hydroxyproline in the Xaa position can stabilize the triple helix. *J Biol Chem* **275**, 24466–9.

99 KRESINA, T. F. & MILLER, E. J. (1979). Isolation and characterization of basement membrane collagen from human placental tissue. Evidence for the presence of two genetically distinct collagen chains. *Biochemistry* **18**, 3089–97.

100 KEFALIDES, N. A. (1973). Structure and biosynthesis of basement membranes. *Int Rev Connect Tissue Res* **6**, 63–104.

101 BURGESON, R. E., EL ADLI, F. A., KAITILA, II & HOLLISTER, D. W. (1976). Fetal membrane collagens: identification of two new collagen alpha chains. *Proc Natl Acad Sci USA* **73**, 2579–83.

102 RHODES, R. K. & MILLER, E. J. (1978). Physicochemical characterization and molecular organization of the collagen A and B chains. *Biochemistry* **17**, 3442–8.

103 BOS, K. J., RUCKLIDGE, G. J., DUNBAR, B. & ROBINS, S. P. (1999). Primary structure of the helical domain of porcine collagen X. *Matrix Biol* **18**, 149–53.

104 MIZUNO, K., HAYASHI, T., PEYTON, D. H. & BÄCHINGER, H. P. (2004) The peptides acetyl-(Gly-3(S)Hyp-4(R)Hyp)$_{10}$-NH$_2$ and acetyl-(Gly-Pro-3(S)Hyp)$_{10}$-NH$_2$ do not form a collagen triple helix. *J Biol Chem* **279**, 282–7.

105 BAKERMAN, S., MARTIN, R. L., BURGSTAHLER, A. W. & HAYDEN, J. W. (1966). In vitro studies with fluoroprolines. *Nature* **212**, 849–50.

106 TAKEUCHI, T. & PROCKOP, D. J. (1969). Biosynthesis of abnormal collagens with amino acid analogues. I. Incorporation of L-azetidine-2-carboxylic acid and cis-4-fluoro-L-proline into protocollagen and collagen. *Biochim Biophys Acta* **175**, 142–55.

107 TAKEUCHI, T., ROSENBLOOM, J. & PROCKOP, D. J. (1969). Biosynthesis of abnormal collagens with amino acid analogues. II. Inability of cartilage cells to extrude collagen polypeptides containing L-azetidine-2-carboxylic acid or cis-4-fluoro-L-proline. *Biochim Biophys Acta* **175**, 156–64.

108 ENGEL, J. & PROCKOP, D. J. (1998). Does bound water contribute to the stability of collagen? [letter]. *Matrix Biol* **17**, 679–80.

109 HOLMGREN, S. K., BRETSCHER, L. E., TAYLOR, K. M. & RAINES, R. T. (1999). A hyperstable collagen mimic. *Chem Biol* **6**, 63–70.

110 BRETSCHER, L. E., JENKINS, C. L., TAYLOR, K. M., DERIDER, M. L. & RAINES, R. T. (2001). Conformational stability of collagen relies on a stereoelectronic effect. *J Am Chem Soc* **123**, 777–8.

111 JENKINS, C. L. & RAINES, R. T. (2002). Insights on the conformational stability of collagen. *Nat Prod Rep* **19**, 49–59.

112 DOI, M., NISHI, Y., UCHIYAMA, S., NISHIUCHI, Y., NAKAZAWA, T., OHKUBO, T. & KOBAYASHI, Y. (2003). Characterization of collagen model peptides containing 4-fluoroproline; (4(S)-fluoroproline-pro-gly)10 forms a triple helix, but (4(R)-fluoroproline-pro-gly)10 does not. *J Am Chem Soc* **125**, 9922–3.

113 HODGES, J. A. & RAINES, R. T. (2003). Stereoelectronic effects on collagen stability: the dichotomy of 4-fluoroproline diastereomers. *J Am Chem Soc* **125**, 9262–3.

114 BABU, I. R. & GANESH, K. N. (2001). Enhanced triple helix stability of collagen peptides with 4R-aminoprolyl (Amp) residues: relative roles of electrostatic and hydrogen bonding effects. *J Am Chem Soc* **123**, 2079–80.

115 SCHERER, G., KRAMER, M. L., SCHUTKOWSKI, M. U. R. & FISCHER, G. (1998). Barriers to rotaion of secondary amide peptide bonds. *J Am Chem Soc* **120**, 5568–74.

116 GRATWOHL, C. & WÜTHRICH, K. (1976). The X-Pro peptide bond as an NMR probe for conformational studies of flexible linear peptides. *Biopolymers* **15**, 2025–41.

117 REIMER, U., SCHERER, G., DREWELLO, M., KRUBER, S., SCHUTKOWSKI, M. & FISCHER, G. (1998). Side-chain effects on peptidyl-prolyl cis/trans isomerisation. *J Mol Biol* **279**, 449–60.

118 RAINES, R. T., BRETSCHER, L. E., HOLMGREN, S. K. & TAYLOR, K. M. (1999). The stereoelectronic basis of collagen stability. In: *Peptides for the New Millennium: Proceedings of the 16th American Peptide Symposium* (FIELDS, G. B., TAM, J. P., & BARANY, G., eds). Kluwer Academic, Norwell, MA.

119 KIEFHABER, T., KÖHLER, H. H. & SCHMID, F. X. (1992). Kinetic coupling between protein folding and prolyl isomerization. I. Theoretical models. *J Mol Biol* **224**, 217–29.

120 PIEZ, K. A. & SHERMAN, M. R. (1970). Equilibrium and kinetic studies of the helix-coil transition in alpha 1-CB2, a small peptide from collagen. *Biochemistry* **9**, 4134–40.

121 BOUDKO, S., FRANK, S., KAMMERER, R. A. et al. (2002). Nucleation and propagation of the collagen triple helix in single-chain and trimerized peptides: transition from third to first order kinetics. *J Mol Biol* **317**, 459–70.

122 BÄCHINGER, H. P., BRUCKNER, P., TIMPL, R., PROCKOP, D. J. & ENGEL, J. (1980). Folding mechanism of the triple helix in type-III collagen and type-III pN-collagen. Role of disulfide bridges and peptide bond isomerization. *Eur J Biochem* **106**, 619–32.

123 PIEZ, K. A. & CARRILLO, A. L. (1964). Helix formation by single- and double-chain gelatins from rat skin collagen. *Biochemistry* **155**, 908–14.

124 HARRINGTON, W. F. & RAO, N. V. (1970). Collagen structure in solution. I. Kinetics of helix regeneration in single-chain gelatins. *Biochemistry* **9**, 3714–24.

125 YUAN, L. & VEIS, A. (1973). The characteristics of the intermediates in collagen-fold formation. *Biophys Chem* **1**, 117–24.

126 KÜHN, K., ENGEL, J., ZIMMERMANN, B. & GRASSMANN, W. (1964). Renaturation of soluble collagen. III. Reorganization of native collagen molecules from completely separated units. *Arch Biochem Biophys* **105**, 387–403.

127 ENGEL, J. (1987). Folding and unfolding of collagen triple helices. In: *Advances in Meat Research*, Vol 4. Van Nostrand Reinhold, New York, pp. 145–58.

128 ACKERMAN, M. S., BHATE, M., SHENOY, N., BECK, K., RAMSHAW, J. A. & BRODSKY, B. (1999). Sequence dependence of the folding of collagen-like peptides. Single amino acids affect the rate of triple-helix nucleation. *J Biol Chem* **274**, 7668–73.

129 PÖRSCHKE, D. & EIGEN, M. (1971). Co-operative non-enzymic base recognition. 3. Kinetics of the helix-coil transition of the oligoribouridylic–oligoriboadenylic acid system and of oligoriboadenylic acid alone at acidic pH. *J Mol Biol* **62**, 361–81.

130 BÄCHINGER, H. P., BRUCKNER, P., TIMPL, R. & ENGEL, J. (1978). The role of cis-trans isomerization of peptide bonds in the coil $\rightleftharpoons$ triple helix conversion of collagen. *Eur J Biochem* **90**, 605–13.

131 BRUCKNER, P., BÄCHINGER, H. P., TIMPL, R. & ENGEL, J. (1978). Three conformationally distinct domains in the amino-terminal segment of type III procollagen and its rapid triple helix leads to and comes from coil transition. *Eur J Biochem* **90**, 595–603.

132 XU, Y., BHATE, M. & BRODSKY, B. (2002). Characterization of the nucleation step and folding of a collagen triple-helix peptide. *Biochemistry* **41**, 8143–51.

133 BRANDT, A., GLANVILLE, R. W., HÖRLEIN, D., BRUCKNER, P., TIMPL, R., FIETZEK, P. P. & KUHN, K. (1984). Complete amino acid sequence of the N-terminal extension of calf skin type III procollagen. *Biochem J* **219**, 625–34.

134 BRANDTS, J. F., HALVORSON, H. R. & BRENNAN, M. (1975). Consideration of the Possibility that the slow step in protein denaturation reactions is due to cis-trans isomerism of proline residues. *Biochemistry* **14**, 4953–63.

135 BRUCKNER, P. & PROCKOP, D. J. (1981). Proteolytic enzymes as probes for the triple-helical conformation of procollagen. *Anal Biochem* **110**, 360–8.

136 GROSS, J. & BRUSCHI, A. B. (1971). The pattern of collagen degradation in cultured tadpole tissues. *Dev Biol* **26**, 36–41.

137 CHENG, H. N. & BOVEY, F. A. (1977). Cis-trans equilibrium and kinetic studies of acetyl-L-proline and glycyl-L-proline. *Biopolymers* **16**, 1465–72.

138 FIORI, S., SACCA, B. & MORODER, L. (2002). Structural properties of a

collagenous heterotrimer that mimics the collagenase cleavage site of collagen type I. *J Mol Biol* **319**, 1235–42.
139 FRANZKE, C. W., TASANEN, K., SCHUMANN, H. & BRUCKNER-TUDERMAN, L. (2003). Collagenous transmembrane proteins: collagen XVII as a prototype. *Matrix Biol* **22**, 299–309.
140 ENGEL, J. & KAMMERER, R. A. (2000). What are oligomerization domains good for? *Matrix Biol* **19**, 283–8.
141 SNELLMAN, A., TU, H., VAISANEN, T., KVIST, A. P., HUHTALA, P. & PIHLAJANIEMI, T. (2000). A short sequence in the N-terminal region is required for the trimerization of type XIII collagen and is conserved in other collagenous transmembrane proteins. *EMBO J* **19**, 5051–9.
142 ENGEL, J. & BÄCHINGER, H. P. (2000). Cooperative equilibrium transitions coupled with a slow annealing step explain the sharpness and hysteresis of collagen folding. *Matrix Biol* **19**, 235–44.
143 LANG, K., SCHMID, F. X. & FISCHER, G. (1987). Catalysis of protein folding by prolyl isomerase. *Nature* **329**, 268–70.
144 BÄCHINGER, H. P. (1987). The influence of peptidyl-prolyl cis-trans isomerase on the in vitro folding of type III collagen. *J Biol Chem* **262**, 17144–8.
145 DAVIS, J. M., BOSWELL, B. A. & BÄCHINGER, H. P. (1989). Thermal stability and folding of type IV procollagen and effect of peptidyl-prolyl cis-trans-isomerase on the folding of the triple helix. *J Biol Chem* **264**, 8956–62.
146 STEINMANN, B., BRUCKNER, P. & SUPERTI-FURGA, A. (1991). Cyclosporin A slows collagen triple-helix formation in vivo: indirect evidence for a physiologic role of peptidyl-prolyl cis-trans-isomerase. *J Biol Chem* **266**, 1299–303.
147 BÄCHINGER, H. P., MORRIS, N. P. & DAVIS, J. M. (1993). Thermal stability and folding of the collagen triple helix and the effects of mutations in osteogenesis imperfecta on the triple helix of type I collagen. *Am J Med Genet* **45**, 152–62.
148 ZENG, B., MACDONALD, J. R., BANN, J. G., BECK, K., GAMBEE, J. E., BOSWELL, B. A. & BÄCHINGER, H. P. (1998). Chicken FK506-binding protein, FKBP65, a member of the FKBP family of peptidylprolyl cis-trans isomerases, is only partially inhibited by FK506. *Biochem J* **330**, 109–14.
149 COMPTON, L. A., DAVIS, J. M., MACDONALD, J. R. & BÄCHINGER, H. P. (1992). Structural and functional characterization of Escherichia coli peptidyl-prolyl cis-trans isomerases. *Eur J Biochem* **206**, 927–34.
150 LAMANDE, S. R. & BATEMAN, J. F. (1999). Procollagen folding and assembly: the role of endoplasmic reticulum enzymes and molecular chaperones. *Semin Cell Dev Biol* **10**, 455–64.
151 HENDERSHOT, L. M. & BULLEID, N. J. (2000). Protein-specific chaperones: the role of hsp47 begins to gel. *Curr Biol* **10**, R912–15.
152 TASAB, M., BATTEN, M. R. & BULLEID, N. J. (2000). Hsp47: a molecular chaperone that interacts with and stabilizes correctly-folded procollagen. *EMBO J* **19**, 2204–11.
153 NAGAI, N., HOSOKAWA, M., ITOHARA, S. et al. (2000). Embryonic lethality of molecular chaperone hsp47 knockout mice is associated with defects in collagen biosynthesis. *J Cell Biol* **150**, 1499–506.
154 KUIVANIEMI, H., TROMP, G. & PROCKOP, D. J. (1991). Mutations in collagen genes: causes of rare and some common diseases in humans. *FASEB J* **5**, 2052–60.
155 UITTO, J. & LICHTENSTEIN, J. R. (1976). Defects in the biochemistry of collagen in diseases of connective tissue. *J Invest Dermatol* **66**, 59–79.
156 ENGEL, J. & PROCKOP, D. J. (1991). The zipper-like folding of collagen triple helices and the effects of mutations that disrupt the zipper. *Annu Rev Biophys Biophys Chem* **20**, 137–52.
157 RAGHUNATH, M., BRUCKNER, P. &

Steinmann, B. (1994). Delayed triple helix formation of mutant collagen from patients with osteogenesis imperfecta. *J Mol Biol* **236**, 940–9.

158 Baum, J. & Brodsky, B. (1999). Folding of peptide models of collagen and misfolding in disease. *Curr Opin Struct Biol* **9**, 122–8.

159 Gough, C. A. & Bhatnagar, R. S. (1999). Differential stability of the triple helix of (Pro-Pro-Gly)10 in H2O and D2O: thermodynamic and structural explanations. *J Biomol Struct Dyn* **17**, 481–91.

160 Go, N. & Suezaki, Y. (1973). Letter: Analysis of the helix-coil transition in (Pro-Pro-Gly)n by the all-or-none model. *Biopolymers* **12**, 1927–30.

161 Greiche, Y. & Heidemann, E. (1979). Collagen model peptides with antiparallel structure. *Biopolymers* **18**, 2359–61.

162 Henkel, W., Vogl, T., Echner, H. et al. (1999). Synthesis and folding of native collagen III model peptides. *Biochemistry* **38**, 13610–22.

163 Consonni, R., Zetta, L., Longhi, R., Toma, L., Zanaboni, G. & Tenni, R. (2000). Conformational analysis and stability of collagen peptides by CD and by 1H- and 13C-NMR spectrocopies. *Biopolymers* **53**, 99–111.

164 Sarkar, S. K., Young, P. E., Sullivan, C. E. & Torchia, D. A. (1984). Detection of cis and trans X-Pro peptide bonds in proteins by 13C NMR: application to collagen. *Proc Natl Acad Sci USA* **81**, 4800–3.

165 Frank, S., Boudko, S., Mizuno, K. et al. (2003). Collagen triple helix formation can be unchecked at either end. *J Biol Chem* **278**, 7747–50.

166 Privalov, P. L., Tiktopulo, E. I. & Tischenko, V. M. (1979). Stability and mobility of the collagen structure. *J Mol Biol* **15**, 203–216.

31
Unfolding Induced by Mechanical Force

Jane Clarke and Phil M. Williams

31.1
Introduction

Force spectroscopy has, as yet, been used in very few protein folding laboratories. One has to ask why this is so. There are several possible explanations. First, until relatively recently, only specialized instrumental laboratories had the technical expertise to build instruments. This has been overcome with the advent of commercially available instruments that are relatively simple to use, menu driven, and in one case, where the programming software (Igor) can be adapted for the investigators' convenience – facilitating data collection, collation and analysis. Second, preparing a protein sample is time consuming. With current molecular biology techniques it is possible to clone and express many single domain mutant proteins in the matter of a couple of weeks. But despite the development of versatile cloning systems, the cloning of a polyprotein substrate can take months (and this has to be repeated for each mutant protein you wish to analyze), and after this, a number of simple-to-express, single domain proteins have turned out to be insoluble as polyproteins. Third, the data are complex and time consuming to collect and analyze. Ironically many repeats of the single molecule experiments are required to collect enough data to analyze, and even the very best data can be noisy and are intrinsically unsatisfactory for those of us who are used to collecting many kinetic data points with exquisite accuracy. Worse still, the investigator has to select which data to analyze and which to discard – an anathema to careful experimentalists. Fourth, the data cannot be analyzed by fitting to a simple model, rather a Monte-Carlo or analytical approach has to be used to extract kinetic data.

So, why bother with these experiments? First, there is a biological imperative. It is increasingly apparent that many proteins experience significant force in vivo. In fact, response to mechanical stress has been implicated in a number of signaling pathways. This means that proteins have evolved to resist unfolding when subject to an external force. It has been shown in at least one case that the barrier to unfolding under mechanical stress is not that investigated by traditional unfolding experiments. Only dynamic force experiments can reveal these details of the protein folding landscape. Second, there is an interesting structural biology problem

Protein Folding Handbook. Part I. Edited by J. Buchner and T. Kiefhaber
Copyright © 2005 WILEY-VCH Verlag GmbH & Co. KGaA, Weinheim
ISBN: 3-527-30784-2

of comparison. Why are certain protein folds stronger than others? How does sequence variation affect mechanical stability? Can the effect of mutation be predicted? How are domains assembled in some of the large multidomain natural load-bearing proteins? Finally, forced unfolding experiments can easily be directly compared with computer simulations. The reaction coordinate (N–C length) is known and measurable.

In this chapter we first describe the experiments and summarize the theoretical background for the analysis of these experiments. We describe how the analysis is performed and what kinetic parameters can be obtained and the confidence limits of these parameters. (We are concentrating on atomic force spectroscopy, as techniques that exploit lower loading rates, such as optical tweezers, have yet to be used on single protein domains and the instrumentation is more complex.) We then show how complementary techniques can be used to help our understanding of the data obtained. We end with a case study of a single protein, illustrating how combination of a number of techniques has enabled the forced unfolding mechanism to be examined in detail.

31.2
Experimental Basics

31.2.1
Instrumentation

A number of commercial instruments are available for use in protein folding laboratories. Figure 31.1 shows the basic components of such instruments. The principle components are a stage, to which is attached the protein substrate, a microfabricated cantilever, and a piezoelectric positioner which adjusts the relative position of the cantilever and stage with subnanometer accuracy.

The protein sample is placed onto the stage. In most experimental studies published to date the protein substrate is a long, multimodular molecule. This may be either a natural multimodular protein, such as a portion of a long extracellular matrix or muscle protein or an engineered construct of multiple repeats of single domains (see Section 31.2.2). Attachment of the protein to the stage may be achieved by specific attachment, for example a gold-sulfur linkage with cysteines at the terminus of the protein, or by nonspecific adsorption onto a glass surface. The protein and cantilever are surrounded by solvent, either as a droplet or in an enclosed cell.

Likewise, attachment to the cantilever is generally through nonspecific adsorption to the surface, although cantilevers can be modified to allow specific attachment. The commercial cantilevers used are usually made of silicon nitride, and are $\sim$20–300 μm in length. The key component is an unsharpened tip (radius $\sim$ 50 nm) to which the protein adsorbs. (Sharper tips are available, but this larger, blunter tip gives a better surface for adsorption of the protein.) Cantilevers used in protein unfolding experiments typically have spring constants in the range 10–

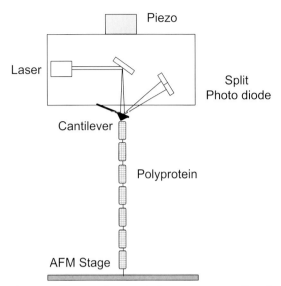

Fig. 31.1. Diagram of an atomic force microscope. The protein molecule is deposited onto the stage and adsorbed to the microfabricated silicon nitride cantilever. The deflection of the cantilever is used to determine the force exerted upon the protein, once the spring constant of the cantilever is known. The separation between stage and cantilever is controlled by the piezo which is capable of repeated cycles of extension and retraction. The photodiode measures the laser light reflected off the back of the cantilever. The deflection of the cantilever is determined from the difference in the output voltages from the two halves.

100 pN nm^{-1}. The actual spring constant has to be measured at the start of each experiment, using methods that are integrated into the software of the commercial instruments. The backs of the cantilevers are reflective and the position of the cantilever is determined by use of a laser reflected onto a split photodiode. Following calibration of the instrument to determine the relationship between photodiode output and bending of the cantilever, the force applied on the protein is determined directly from the deflection of the cantilever. A piezoelectric positioner adjusts the separation of the cantilever tip and the surface.

31.2.2
Sample Preparation

As a tool for protein folding, fragments of large, natural multidomain proteins such as the giant muscle protein titin [1], or the cytoskeleton protein spectrin [2], are of limited use, as they do not report on the unfolding of specific protein domains. Although a few experiments have been reported where a single protein domain has been suspended between tip and stage (see, for example, Ref. [3]), most experiments have used large "polyproteins" which have multiple repeats of a single protein domain cloned in tandem. These large polyproteins have the advantage of holding the protein at a significant distance from the surface before unfolding occurs, so that tip/surface or protein/surface interactions are minimized.

A number of approaches for constructing such repeats have been reported [4–6], but the most versatile are those that use different restriction enzyme sites at the start and end of each domain. One such system is available on request from our laboratory [6]. The first protein to be multimerized in this way was titin I27. It has a poly-His tag at the N-terminus to facilitate affinity purification, and two cysteines at the C-terminus to attach the protein to an AFM stage by a gold-sulfur linkage. This protein can be produced in large amounts as a soluble polyprotein in *Escherichia coli*. The protein can be used after a one-step purification, but we tend to get better results following a second, size exclusion step. I27 is a very versatile protein – it has proved to be a useful "handle" for attaching other proteins, which may not polymerize so easily or where only a single domain of a protein of interest is required [7, 8]. In these circumstances it acts as an internal control.

A few studies have used novel methods to produce multimodular proteins. Yang et al. exploited crystal packing in T4 lysozyme [9]. Cysteines were engineered at contact points and the polyprotein was formed in the crystal. In studies to investigate the effect of attachment point on the effect of force on a protein Fernandez and coworkers used the only known natural polyprotein ubiquitin, linked either via the N- and C-terminus or via specific surface lysine–C-terminus linkages [10]. Brockwell et al. used an elegant, novel lipoic acid–lysine linkage to study the same problem in the protein E2Lip3 [8].

When working with multidomain constructs it is important to know whether the protein is folded in, and how far its properties are changed by, inclusion in a polyprotein. Again, I27 is a "model" protein in this respect; it has the same thermodynamic stability and the same kinetic properties in the polyprotein as does an isolated domain [4]. However, the same is not true of all proteins. Some are stabilized and some destabilized by inclusion in the protein. Remarkably, although traditional equilibrium denaturation and stopped flow kinetic experiments can be performed as easily on the polyprotein as on small individual domains these simple controls are very rarely carried out. We have also been able to show that NMR experiments can be easily undertaken on three-module constructs, so structural integrity can also be verified [7].

31.2.3
Collecting Data

Although there are recent reports of constant force experiments (see, for example, Ref. [11]), most experiments reported to date have used a ramp of force induced by retracting the piezo at constant speed. The cantilever is lowered to the surface repeatedly, picking a protein molecule up at random. Since this is a blind "fishing" step the experimentalist has little control over whether a protein is picked up and no control over the position on the polyprotein where the protein is attached to the cantilever tip. If the protein is adsorbed at too high a concentration onto the substrate then more than one protein molecule is likely to be attached to the tip. Such traces have to be discarded, as they cannot be interpreted. If the protein is too dilute there will be too few pick-ups. It is estimated that where one approach in 10 picks up a protein we can be confident that most data will be from a single molecule [12]. The concentration of protein that results in such a success rate has to be determined empirically.

The tip is retracted at constant speed and when a protein molecule is picked up a force trace is observed. A large number of such traces have to be collected at a number of pulling speeds. The range of experimentally useful pulling speeds allowed by the commercial instruments currently available is on the order of 10 nm s^{-1} to 10 000 nm s^{-1} but in practice, most useful data are collected in the pulling speed range of $\sim$300 to $\sim$3000 nm s^{-1}. Instrumental drift is a problem at low pulling speeds and at very high pulling speeds, viscous drag and cantilever response time introduce error.

31.2.4
Anatomy of a Force Trace

A "typical" force trace is shown and described in Figure 31.2. At the start of the trace there is usually a peak of unpredictable height that reflects tip:surface interactions, deadsorption of the protein from the surface and other nonspecific effects (see arrow on Figure 31.2a). As the cantilever retracts force is applied to the protein. As the force increases unfolded parts of the chain are stretched. At some force one of the protein domains will unfold and this results in a sharp drop in the force trace. The trace does not drop to the baseline, however, as the entropic elasticity of the unfolded protein maintains a force on the cantilever. The force increases again until another domain unfolds, resulting in the signature saw-tooth pattern. The base of subsequent peaks is higher that that of the preceding peak. The spacing between the peaks is regular, reflecting the all-or-none nature of the unfolding events, so that a number of traces can be overlaid. Finally the cantilever is retracted so far that the protein becomes detached. A final "pull off" peak is observed, typically much higher than all preceding peaks and the trace returns to the baseline.

31.2.5
Detecting Intermediates in a Force Trace

If the protein domains unfold without populating intermediates the trace should be simple with a single unfolding event per domain, and the unfolded protein should extend as a featureless smooth curve, fitting to a simple model of an elastic polymer (see Section 31.3.4). However, intermediates have been detected in a few forced unfolding experiments by careful examination of forced unfolding data. The intermediates have been detected either by the presence of two consecutive unfolding events per domain (e.g., Ref. [13]), or from the presence of a "hump" in the unfolding traces [14] (described in more detail in Section 31.7).

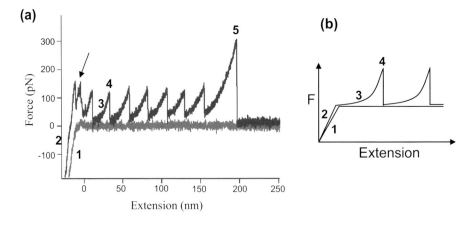

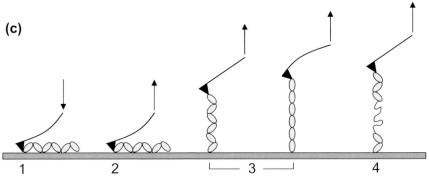

Fig. 31.2. Anatomy of a force trace. Force trace (a) and diagram (b) showing the response of the cantilever during different phases of the AFM approach-retraction cycle. The cartoon in (c) describes the response of the protein and cantilever during the different phases shown in (a). 1) Tip makes contact with surface and is deflected. 2) Protein adsorbs to the cantilever tip. 3) Tip is retracted at a fixed pulling speed from the surface. An entropic restoring force of the protein is generated when it is extended. The cantilever deflects in response to this force. 4) A domain unfolds and extends. The system relaxes. The arrow in (a) shows non-specific adsorption between the tip and/or the protein and the surface.

31.2.6
Analyzing the Force Trace

The unfolding forces can be determined by analysis of the force traces, however, one of the most difficult features of forced protein unfolding experiments is that one has to choose which traces to analyze. As described above, most approach–retract cycles result in no protein being attached to the cantilever at all. Furthermore, a number of the traces where a protein is attached may have large peaks of "noise" at the start of the trace, possibly because the protein was attached to other (possibly unfolded) protein molecules on the surface or there may be more than one molecule attached to the cantilever. It is an essential assumption of all the analysis that the data are collected from a single molecule attached to the tip. It is therefore essential that consistent and rational criteria are applied in choosing which peaks to analyze. This has been discussed elsewhere and a set of criteria have been proposed [15]:

1. Force peaks must be equally spaced and the distance between them must be consistent with the expected contour length of the protein.
2. The trace must include three or more peaks, the last of which is assumed to be the detachment of the protein from the AFM tip.
3. The base of each successive peak should be higher than base of the previous peak.
4. The approach and retraction baselines should overlay, indicating that there was no drift during the course of the pulling experiment.
5. The final baseline should be straight, indicating complete relaxation of the cantilever.

We find that analyzing the trace from the right, at the pull-off peak, towards the left, until the base of the peak touches the baseline, gives us most consistent analysis between investigators and from day to day. Note that all peaks must be counted. The temptation to discard peaks that meet all other criteria on the basis that they are "too high" or "too low" must be resisted – these will not contribute to analysis of the modal forces.

To determine the height of a peak, the full-length final baseline should be fitted to a line and the height of the peak above this line determined. This allows for drift in the instrument to be accounted for. Once these peak heights have been collected, the modal unfolding force can be determined (see Section 31.3.7).

We have found that it is important to collect data with different cantilevers and on different days to minimize error. We estimate that it is necessary to collect at least three full data sets, where each data set contains at least 40–50 force peaks at each pulling speed [15]. A full range of pulling speeds should be used to minimize errors in analysis. However, for some proteins we have found that we need to collect significantly more data to get reliable estimates of the modal unfolding force.

31.3
Analysis of Force Data

31.3.1
Basic Theory behind Dynamic Force Spectroscopy

In a fluid environment molecular structural transformations, such as those of protein unfolding, are driven by Brownian thermal excitation [16]. It is this agitation by the solvent that "kicks" a protein over transition states and provides the activation energy necessary for folding and unfolding. Brownian dynamics therefore sets

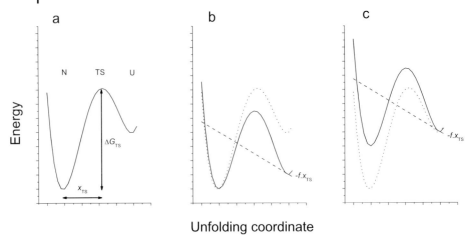

Fig. 31.3. Unfolding energy landscape and the effect of force. a) The rate of unfolding from N is dictated by the magnitude of the free energy relative to the transition state ΔG_{TS} and the shape of the energy potential (encompassed in Eq. (1) as ω_N and ω_{TS}). b) A force f acts on the potential to drop the transition state relative to the native state by the product of the force and the displacement, x_{TS}, of the transition state. Here, force can be considered as having a stabilizing effect on the transition and denatured states. c) Force can be considered as having a destabilizing effect on the native state and unfolding transition states. The effect of force on the unfolding kinetics in (b) and (c) is the same.

the time scale for motions in condensed liquids, which commences at the nanosecond for small ligands and slows with increased molecular size. As shown by Kramers in the 1940s [17], the rate of thermally activated escape from an unfolding potential energy well (Figure 31.3a) can be written as

$$v_0 = \left(\frac{(\omega_N \omega_{TS})^{1/2}}{2\pi\zeta}\right) \exp\left(-\frac{\Delta G_{TS}}{k_B T}\right) \quad (1)$$

where ζ is the frictional coefficient of movement (viscous damping), ω_N and ω_{TS} are the curvatures of the energy potential at the native (N) and transition state (TS) respectively and ΔG_{TS} is the relative energy of TS. Although little is known of the exponential prefactor, one does not require its value to measure kinetics. However, conversion from rates to activation energies is not possible without this.

As shown in Figure 31.3b, force has several effects on an energy landscape. Under a persistent force the landscape is "tilted" in the direction at which the force is applied, and this both lowers and shifts the transition state maximum relative to the potential well of the folded, native state. Alternatively, one can consider force as stabilizing the protein's unfolded state and destabilizing the transition state and the folded state more so (Figure 31.3c). Either way, the net effect on the kinetics is the same. Under an external force potential, therefore, the unfolding rate is increased as force (f) lowers the energy of the unfolding transition state relative to the native state. The amount that the transition state energy has been lowered by is the product of the force applied and the displacement of the state in the direction of force (x_{TS}), and we can re-write Eq. (1) to show this effect:

$$v_u(f) = h\left(\frac{fx_{TS}}{k_B T}\right)\left(\frac{(\omega_N \omega_{TS})^{1/2}}{2\pi\zeta}\right)\exp\left(-\frac{(\Delta G_{TS} - fx_{TS})}{k_B T}\right) \quad (2)$$

where $h(\ )$ describes the movement of the transition state under force [18]. Since the dominant effect of force is the lowering of the transition state, $h(\ )$ can usually be ignored, which gives the rate equation

$$v_u(f) = v_o \exp\left(\frac{fx_{TS}}{k_B T}\right) = v_o \exp\left(\frac{f}{f_\beta}\right) \quad (3)$$

where v_o is the unfolding rate over this TS barrier in the absence of force. Importantly this equation introduces f_β ($= k_B T/x_{TS}$), the *force scale*, commonly used in dynamic force spectroscopy and x_{TS} (or x_u as it is usually referred to in protein unfolding studies) representing the displacement between N and TS. At room temperature, thermal energy $k_B T$ is 4.11×10^{-21} J and over the nanometer length scale of proteins is equal to forces of only a few piconewtons (4.11 pN nm). The catalytic effect of a 100 pN force over this nanometer distance is equivalent to over 14 kcal mol^{-1} (1 kcal mol^{-1} = 6.9 pN nm).

31.3.2
The Ramp of Force Experiment

Virtually all experiments of protein unfolding measured by the atomic force microscope are of the kind where force is increased in time and the maximum force that the protein can withstand recorded. The tip is approached to the surface, protein adheres to the tip, and the tip is then withdrawn from the surface at a constant velocity. The simplest case to consider is where force increases steadily with time and the loading rate on the protein, $r_f = df/dt$, is constant. As we show later, this is usually not true for these experiments but this first approximation provides the introduction to dynamic force spectroscopy. The rate at which force increases is the product of the stiffness of the system (in this first approximation the this is the stiffness of the cantilever alone) and the speed at which the cantilever is withdrawn. Lever stiffness κ_c is of the order of 10–100 pN nm^{-1}, and retract rates v_r can vary from around 10 nm s^{-1} to approaching 10 000 nm s^{-1} (thermal drift and operator patience limiting the low speed and hydrodynamic drag restricting high speeds). AFM loading rates can theoretically range, therefore, between 100 and 1 000 000 pN s^{-1}, although practically achieving more than two to three orders of loading rate is problematic.

As an illustration, we consider the effect of increasing force on a transition state located $x_u = 0.5$ nm along the projection of force. Since at room temperature thermal energy, $k_B T$, is 4.11 pN nm at a loading rate of 1 000 000 pN s^{-1} it takes nearly

10 μs ($k_B T/x_u \cdot r_f = 4.11/0.5/1\,000\,000$) to drop the barrier by 1 $k_B T$. On the time scale of a thermal attempt frequency (see above), therefore, force and the landscape are quasi-static. Experiments of strength adopting a ramp of force can thus normally be considered as a series of many tests under an equally large set of different forces. However, were the rate at which energy is added to the transition state by force ($x_u \cdot r_f/k_B T$) to be comparable to, or exceed, the kinetic prefactor then the dynamics would deviate from that described here.

As unfolding is a kinetic process driven by random fluctuations one must make many measurements to sample the process. In this respect, the measurement of the behavior of a single molecule many times is identical to the measurement of an ensemble of molecules all at once. We can consider each measurement as a sample of this population, the numbers in which follow a simple rate equation. Commencing with all folded proteins, the fraction of the population that are folded under force decays in time as

$$\frac{dS}{dt} = -v_u(t) S(t) + v_f(t)[1 - S(t)] \tag{4}$$

where $v_u(t)$ is the rate at which unfolding occurs at time t, and $v_f(t)$ is the rate of refolding. Since time and force are related through the loading rate r_f, Eq. (4) can be written for an increasing force as

$$\frac{dS}{df} = \frac{1}{r_f}\{-v_u(f) S(f) + v_f(f)[1 - S(f)]\} \tag{5}$$

Until now, we have ignored the refolding term. In fact, since force causes an exponential decrease in folding in the same manner in which it exponentiates unfolding, at forces above a few 10 s of piconewtons refolding can safely be neglected. The master rate equation is simply

$$\frac{dS}{df} = \frac{1}{r_f}\{-v_u(f) S(f)\} \tag{6}$$

This can be integrated to reveal the probability distribution of folded states, as

$$S(f) = \frac{1}{r_f} \exp\left(-\int_0^f (v_u(g)/r_f)\, dg\right) \tag{7}$$

Equation (7) gives the fraction of the ensemble (the probability) that remain folded as the force increases from 0 to f. The probability that a protein will unfold at this force is the product of this, the survivability to force f, and the unfolding rate at this force. The probability of unfolding at f is therefore the product

$$p(f) = \frac{1}{r_f} v_u(f) \exp\left(-\int_0^f (v_u(g)/r_f)\, dg\right) \tag{8}$$

substituting for v_u (Eq. (3)) this can be written as

$$p(f) = \left(\frac{1}{r_f} v_o \exp\left(\frac{f}{f_\beta}\right)\right) \exp\left(-\int_0^f \frac{1}{r_f} v_o \exp\left(\frac{g}{f_\beta}\right) dg\right) \tag{9}$$

The integral of an exponential is unchanged, so taking the limits of force the integral can be solved as

$$p(f) = \left(\frac{1}{r_f} v_o \exp\left(\frac{f}{f_\beta}\right)\right) \exp\left(\frac{1}{r_f}\left(f_\beta v_o - f_\beta v_o \exp\left(\frac{f}{f_\beta}\right)\right)\right) \tag{10}$$

A series of tests of strength using a ramp of force at a constant loading rate will form a distribution given by Eqs (8), (9), and (10). There is not a single force at which a protein will unfold. The most probable unfolding force is that where this distribution is at a maximum, and therefore has a zero gradient. This value is found by finding the force where the first derivative is zero. To solve this, we rely on the property that we can scale the distribution and not affect the location of the maximum, and a useful scaling is the natural logarithm. The logarithm of Eq. (10), expanded for clarity, is

$$\ln[p(f)] = \ln[v_o] + \ln\left[\exp\left(\frac{f}{f_\beta}\right)\right] - \ln[r_f] + \ln\left[\exp\left(\frac{1}{r_f}\left(f_\beta v_o - f_\beta v_o \exp\left(\frac{f}{f_\beta}\right)\right)\right)\right] \tag{11}$$

The logarithms of the exponentials cancel, so

$$\ln[p(f)] = \ln[v_o] + \left(\frac{f}{f_\beta}\right) - \ln[r_f] + \frac{1}{r_f}\left(f_\beta v_o - f_\beta v_o \exp\left(\frac{f}{f_\beta}\right)\right) \tag{12}$$

The maximum of this distribution at the most probable unfolding force is found by differentiation with respect to force. The terms that do not depend of force disappear, leaving the maximum as

$$\partial \ln[p(f)] = \left[\frac{1}{f_\beta} - v_o \exp\left(\frac{f}{f_\beta}\right)\Big/r_f\right]_{f=f^*} = 0 \tag{13}$$

31.3.3
The Golden Equation of DFS

Using Eq. (13) we can find the most probable unfolding force f^* (see Ref. [18])

$$f^* = f_\beta[\ln(r_f) - \ln(f_\beta v_o)] = \frac{k_B T}{x_u} \ln\left[\frac{r_f x_u}{k_B T v_o}\right] \tag{14}$$

Equation (14), the "golden equation" of DFS, has been derived from numerous assumptions (single transition state, stationary in location under force, linear force loading, $x_u \cdot r_f / k_B T \ll 1/t_D$, etc.) but shows elegantly the features of DFS. A plot of f^* against the logarithm of the loading rate has slope f_β, revealing the location (x_u) of the transition state along the force axis. Similar to the *m*-value of chemical unfolding measurements, f_β is the susceptibility of the unfolding kinetics to force. But here this value has a direct physical relationship to geometric coordinates and structure (x_u). Further, the plot cuts the loading rate axis, where f^* is zero, when $r_f = f_\beta v_0$. Since f_β is known one can determine through this extrapolation the force-free unfolding rate across the transition state (v_0).

Equation 14 is derived for the most probable rupture force; the mode of the distribution. The equation does not hold true for the mean force. In fact, the mean of the force distribution (Eq. (8)) is complex to find, involving exponential integrals with no exact solution. Thus it is important to find the mode of the distribution of forces measured and analyze these, and not their mean. Additionally, the mode of the distribution is less affected by outlying forces from nonspecific tip/sample interactions, multiple unfolding events and instrumental noise. As a rule-of-thumb, however, at high forces the mean unfolding force can be crudely approximated by [12]:

$$\bar{f} \sim f_\beta [\ln(r_f) - \ln(f_\beta v_0) - 0.6] \tag{15}$$

and with force scales f_β for proteins approaching 20 pN, mean forces are 10 pN or so lower than the modes.

31.3.4
Nonlinear Loading

The approximation used above, that the stiffness of the system is dominated by the cantilever, is not valid for most protein unfolding measurements. Typically, tandem repeats of the protein are stressed and multiple copies are unfolded in series revealing the characteristic saw-toothed force extension curve. In this arrangement, the mechanical stiffness of the protein polymer may well be less than the cantilever. The stiffness of the system κ_s, with two springs stressed in series (the cantilever with stiffness κ_c and molecular system κ_m) is

$$\frac{1}{\kappa_s} = \frac{1}{\kappa_c} + \frac{1}{\kappa_m} \tag{16}$$

Clearly the molecule has to be stiffer than the lever ($\kappa_m > \kappa_c$) for the approximation that lever stiffness dominates loading. With soft levers of 10 pN nm^{-1} or so this can be true. However, the saw-tooth pattern measured in the AFM experiments derives from the stretching of the unfolded polypeptide chain after each repeat has unfolded. This means that force loading is not constant during protein unfolding experiments. This has two consequences. First data have to be represented as unfolding force vs. retract velocity (v_r) and second, we need to consider the mechanical stiffness of such random-coil polypeptide chains.

31.3.4.1 The Worm-line Chain (WLC)

The WLC is a convenient description of the mechanical properties of a random coil of amino acids, an unfolded protein. The configurational space accessible to a long polymer is considerable and when free in solution the end-to-end distance fluctuates. When this distance is constrained, such as being tethered between an AFM tip and a sample surface, configurational space has been lost and thus the entropy of the system has been reduced. The force required to maintain an extension is a function of this loss in entropy. The polymer behaves as an entropic spring.

The stiffness of an entropic spring is a function of the contour length of the molecule, L, and a persistence length, b. This is the length corresponding to a bending energy of one $k_B T$. It is generally agreed that a value of ≈ 0.4 nm for the persistence length fit the experimental measurements of proteins well. Under low extensions x (where $x \ll L$), the spring is Hookian and exerts a force of [19]:

$$f_{\text{WLC}}(x) \approx \frac{k_B T}{b} \frac{x}{L} \tag{17}$$

As the extension approaches the contour length (the asymptote as the molecule is inextensible), force increases anharmonically as

$$f_{\text{WLC}}(x) \approx \frac{k_B T}{b} \left(1 - \frac{x}{L}\right)^{-2} \tag{18}$$

A convenient equation to show the behaviour of a WLC over all extensions, and one that matches that seen in AFM experiments, is an interpolation between Eqs (17) and (18), as

$$f_{\text{WLC}}(x) \approx \frac{k_B T}{b} \left(\frac{1}{4\left(1 - \frac{x}{L}\right)^2} - \frac{1}{4} + \frac{x}{L} \right) \tag{19}$$

A polymer exerts a force of 150 pN (a typical titin I27 unfolding force) when the chain is extended to around 90% of its contour length. The stiffness of the chain $\kappa_{\text{WLC}}(f)$ at this extension is approximately 10 pN nm^{-1}, and for all but the softest of AFM cantilevers it is the polymer that dominates the loading rate (Eq. (16)). So instead of a steady loading rate, r_f depends on the force and increases nonlinearly in time as $r_f = v_r \kappa_{\text{WLC}}(f)$. Substitution of this loading rate term into Eq. (9) and subsequent solving for f^* gives a good approximation of the unfolding kinetics under WLC polymer loading, as [20]:

$$f^* \approx f_\beta \left[\ln\left(\frac{v_r}{v_\beta}\right) + \ln\left(\frac{f^*}{f_\beta} - \frac{3}{2}\right) + \frac{1}{2} \ln\left(\frac{f^*}{f_\beta}\right) \right]$$

$$v_\beta = \frac{L v_0}{4} \left(\frac{x_u}{b}\right)^{1/2} \tag{20}$$

Equation (20) is transcendental (the force term f^* is expressed in terms of f^*) and therefore can be awkward to solve. Methods to fit this equation to data are explained below. A simpler, although less accurate method of fitting to experimental data is to use the each force value to find the stiffness of the system at the point of unfolding, and then multiply this by the retract velocity v_r to give a loading rate (this is only an estimate as the force distributions are no longer described fully by Eqs (8), (9), and (10) since the loading rate increases during cantilever retraction). The stiffness of the protein chain ($b = 0.4$ nm, $k_B T = 4.11$ pN nm) is approximately

$$\kappa_{\text{WLC}}(f) \approx 4 \frac{k_B T}{bL} \left(\frac{bf}{k_B T} \right)^{3/2} \qquad (21)$$

The rate of loading at the point of unfolding can be estimated, therefore, by calculating the stiffness of the unfolded protein at the unfolding force (Eq. (21)), determining the stiffness of the system (Eq. (16)) and multiplying this by the retract velocity v_r. The modal unfolding force can then be plotted against this loading rate estimate and Eq. (14) used to find a reasonable estimate of the force-free unfolding rate and transition state displacement.

31.3.5
Experiments under Constant Force

An attractive method to remove the problems of the nonlinear loading induced by the dynamics of the polymer is to measure the lifetime of the protein under a constant force [11]. Instrumentally, such an experiment requires the incorporation of a feedback system that modulates the cantilever position to maintain a constant deflection, and hence constant force.

The time for which a protein can withstand a constant force is the reciprocal of its unfolding rate

$$t_u(f) = \frac{1}{v_0} \exp\left(-\frac{f}{f_\beta}\right) \qquad (22)$$

The constant force experiment has the advantage that it should be independent of the system dynamics, such as the nonlinear loading rate introduced by the polymer stiffness, but suffers from two shortcomings. First, analysis assumes that force is kept constant and that the force-feedback system can respond faster than the unfolding rate of the protein under the force applied. Fortunately, this is often the case. However, a greater limitation of this class of experimentation is that control of force relates to exponential changes in lifetime, as seen in Eq. (22). Therefore, as force is lowered, the experiment lasts for exponentially increasing periods of time. Conversely, at high forces, it becomes increasingly difficult to determine lifetime with the necessary accuracy. The advantage of the ramp of force measurement is that the force is a direct measurement of lifetime since these are related through the loading rate. However, since the ramp of force measurement is a continuous collection of constant force measurements, these two regimes can be considered identical. The experiment of constant loading rate can be considered as the ideal and relatively simple technological developments are required to realize this methodology. The potential of the constant force experiment is to measure the kinetics of unfolding under high forces. To measure this, however, requires the ability to apply the force instantaneously (through the use of magnetism or electrostatic attraction for instance) and measure the diminishing lifetime with increasing temporal resolution. Current AFM technology, depending on peizo displacement and at best having kilohertz temporal resolution, is incapable of exploiting this potential.

31.3.6
Effect of Tandem Repeats on Kinetics

The use of tandem repeats of proteins in forced unfolding experiments is beneficial for several reasons as previously discussed; including providing a characteristic signature for a single molecule, extension of the unfolding event away from any interactions of the surface, and increased number of data points per trace. But the presence of multiple proteins affects the unfolding kinetics, and this must be accounted for [19].

For a chain of identical tandem repeats there is no specific order in which they will unfold under force as none is more likely to unfold before another. For a chain of N folded proteins, the probability that any one protein will unfold is N-times higher than the probability of failure of the protein on its own. So the unfolding rate of the first event is N-times the unfolding rate of the constituent protein. For the second event, the number of units that may unfold is $N-1$, and the unfolding rate for this is $N-1$ times one, and so on [21]. The effective unfolding rate of the chain decreases for each peak n of N seen in the experiment, as

$$v_u(f, n, N) = (N - n - 1) v_o \exp\left(\frac{f}{f_\beta}\right) \tag{23}$$

In addition to the unfolding rate decreasing with the event number, the loading rate also decreases as the length of unfolded protein increases with each event. The drop in unfolding rate leads to an increase in the force, whilst the drop in loading rate causes a competing decrease in force. Between the first and second event the fractional change in rate is small (from Nv_o to $(N-1)v_o$) whereas the loading rate changes considerably as the unfolded polymer has increased considerably in length (Eq. (21)). The forces drop over the first events. Conversely, between the second-to-last and last event the unfolding rate halves whereas the fractional change in polymer length is small. The forces increase towards the last events. On average, the net effect is a decrease and then increase in the average rupture force for event number [22].

To analyze accurately the unfolding forces for tandem repeats requires knowledge of the length of repeat being pulled, which can change from test to test, and analysis of individual events based on the each total length, i.e., the 1st, 2nd, and 3rd events of pulls of three repeats, the 1st, 2nd, 3rd, and 4th of pulls of four repeats, etc. A useful observation is that the sixth event in a chain of eight repeats behaves similarly to the mode of all measurements [23]. Fortunately, the effect of changing unfolding rate and polymer length with event number shift the most probable rupture force by only 20 pN or so, and so this effect is usually ignored in all protein folding studies. Titin I27 has been studied in different laboratories using polyproteins with 12, 8, and 5 repeats and the results are essentially the same [4, 5, 24].

31.3.7
Determining the Modal Force

To analyze the kinetics of unfolding using force spectroscopy requires measurement of the full distribution of forces (lifetimes), since unfolding is random. In the inevitable presence of noise, this requires many (several hundred) measurements at each speed. To determine the mode of the distribution (remembering the issues discussed above with the mean) usually requires resorting to a graphics program, binning the data into arbitrary sized bins, and plotting the resulting histogram.

A useful alternative that does not require the use of bins is to estimate a continuous spectrum of forces [25]. Each force measured is assumed to be subject to noise, and assumed to be Gaussian. The largest sources of noise in the force data are from thermal fluctuations of the cantilever, limited sampling rate and hydrodynamic drag. Whilst drag effects are limited by restricting retract velocities to below a few microns per second, and sampling rates are increasing with every new generation of force microscope, thermal noise cannot be easily overcome. The thermal fluctuation of the lever is related to its spring constant κ_c, given as

$$\langle \sigma \rangle = \sqrt{\frac{k_B T}{\kappa_c}} \tag{24}$$

and a cantilever of typical stiffness (30 pN nm^{-1}) fluctuates around 0.3 nm, showing that the standard deviation of the force noise for these soft levers is over 10 pN. The force noise for an AFM is typically between 10 and 20 pN, and we can assume that each force measured is drawn from a Gaussian 10 or 20 pN wide. The sum of all these Gaussians, centered on each force recorded, is calculated and the mode force, where the sum is greatest, found. It is often worthwhile seeing how this estimate of the mode depends on the level of noise chosen, which ideally should be invariant.

The distribution of forces for unfolding over a single transition state has a well-defined form given by Eq. (10). Since the full distribution of forces at a particular rate of loading is given by the two parameters measured, the unfolding rate v_0 and the transition state location x_u, the distribution serves to validate the experimental measurements. Having found the mode, the histogram of recorded forces (and not the artificial Gaussian sum used above) should be plotted and the distribution predicted using Eq. (10) overlaid.

31.3.8
Comparing Behavior

Dynamic force spectroscopy reveals uniquely the presence of multiple transition states, their locations along a reaction coordinate, and the rate at which they are crossed. As shown in Eq. (1), conversion of these rates to a free energy requires knowledge of the exponential prefactor. However, the ratio of two rates does reveal the difference in free energy between the transition states assuming a constant prefactor, as

$$v_1 = \frac{1}{t_D} \exp\left(-\frac{\Delta G_{TS1}}{k_B T}\right)$$

$$v_2 = \frac{1}{t_D} \exp\left(-\frac{\Delta G_{TS2}}{k_B T}\right) \quad (25)$$

$$\frac{v_1}{v_2} = \exp\left(\frac{\Delta G_{TS2} - \Delta G_{TS1}}{k_B T}\right)$$

An assumption of the constant prefactor is one of the difficulties in application of Eq. (25). This is a concern when one is comparing mutant proteins with wild-type. Where the transition state being probed is the same, deformed by a point mutation for example, then this assumption is reasonable. But with such a mutation it is still not assured that the same transition state from the same point along the unfolding coordinate is being probed. Good examples are some of the single point mutants of titin I27 studied by AFM [23, 26]. Many proline mutants, and even conservative mutants such as V86A, have been shown to cause a change in the ground state from which unfolding is measured [23]. Whilst it is believed that the same transition state is being studied, the folded state from which the protein unfolds is different (wild type unfolds from an intermediate whereas the mutants unfold from the native state). Therefore, the ratio of unfolding rates extrapolated from these measurements of the mutant behavior to the wild type does not represent the change in kinetic stability caused by the mutation.

31.3.9
Fitting the Data

The data need to be analyzed to extract the two parameters requires, x_u and the unfolding rate at 0 force v_0. Probably the most obvious method to predict the behavior of a system under load is to perform a "Monte-Carlo" simulation [27]. Here, small intervals of time are sampled and the events that could occur within these windows are chosen at random. Commencing at time $t = 0$, the force on the system is determined by the position of the cantilever ($x = v_r t$) and the mechanical properties of the system (Eq. (19)). This force is used to calculate the probability of an unfolding event within the window of time, as the product of the time slice Δt and Eq. (23). This probability is sampled at random by its comparison to a uniform random number. If the event is chosen (i.e., the random number selected is not more than the probability) then the unfolding event has occurred, this domain is deemed to have unfolded and the length of the unfolded polymer increased by the length of the polypeptide chain forming the domain. Next, the cantilever is moved back by a distance equal to the product of the retract velocity and the time-step, the new force calculated and the process of random sampling continued until all the domains have unfolded.

Monte-Carlo simulations such as these provide full force versus extension traces as they represent a reasonably accurate simulation of the experiment (Figure 31.4). However, for this very reason, the simulation must be repeated many times to sample the stochastic nature of unfolding. Furthermore, whilst capable of predicting a distribution of unfolding forces for given values of v_o and x_u, it is not simple to fit these values to experimental results.

An alternative method of obtaining the full distribution of unfolding forces is to integrate the Master rate equations (for example Eq. (5)) across an increasing force [28]. Unfortunately, the fact that the probability of unfolding increases exponentially in force whilst the probability of surviving to each force decreases exponentially, the equations are so-called stiff, and can be difficult to integrate numerically.

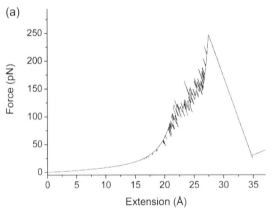

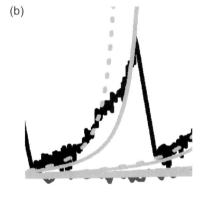

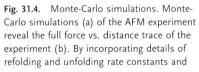

Fig. 31.4. Monte-Carlo simulations. Monte-Carlo simulations (a) of the AFM experiment reveal the full force vs. distance trace of the experiment (b). By incorporating details of refolding and unfolding rate constants and unfolding intermediates fine detail, such as the presence of the force plateau around 100 pN attributed to the population of the force-stabilized unfolding intermediate of I27, can be resolved.

We find the semi-implicit extrapolation method of the Bulirsch-Stoer integration procedure to be the most suitable [23, 29]. Such numerical solutions provide the full unfolding force distribution, an extremely valuable cross-check of the experiment [19]. They do not, however, provide any insight of the individual force vs. extension curve.

Fitting requires predicting the unfolding forces for a set of trial parameters (x_u and v_o) with each measured value at the rates explored, calculating a difference (commonly as the sum of the squares), and changing the trial parameters until this difference is as small as possible. There are many methods available to perform such minimizations, and a useful method is the simplex minimization procedure. Simplex minimization can be slow, but does not require the use of derivatives that are difficult to compute. A good discussion of simplex minimization is given by Press et al. [29]. Simply, in the n parameter space of the fit ($n = 2$ for a single transition state, x_u and v_o) the best result is enclosed by a simplex, a shape of $n + 1$ vertices. For example, all possible values of x_u and v_o can be plotted on a graph, the abscissa representing all the x_u values and the ordinate the v_o values. The smallest simplex that can be drawn at any point on the graph is a triangle ($n + 1 = 3$). Simplex minimization starts with the simplex occupying an arbitrary region of the parameter space, and then walks the simplex through space depending on a series of rules. The rules can include contracting the shape towards the vertex with the best fit, moving the best fitting vertex away from the worst, etc. Minimization is terminated when all the vertices lie close in space and in value. Again, see Press et al. for a much clearer explanation of this and other minimization and optimization methods [29].

A final note is in the assessment of the goodness of fit and the magnitude of errors. A useful method here, again explained in detail by Press et al. is the bootstrap Monte-Carlo method [29]. Having obtained a fit between the measured and the predicted force data, the goodness of the fit can be estimated by the replacement method. The original data is subsampled and padded with copies. About $1/e$ of the data is removed and these points replaced by remaining values. The total number of forces in the set, therefore, remains the same. The x_u and v_o parameters are then fitted to this subset of data, and this procedure repeated many times. The variance in the results of these fittings equals the variance in the original fit. Finally, when measuring and reporting variances in v_o it is important to remember that this value is not linear but exponential, and variances of $\log(v_o)$ and not v_o should be used. For this reason rates are often reported as exponents, as $2 \times 10^{-4 \pm 1.1}$ s^{-1} for example. Since small uncertainties in the experimentally determined x_u lead to large differences in v_o the errors can be large [15, 30].

31.4
Use of Complementary Techniques

Two complementary techniques have been exploited increasingly in the study of protein folding mechanisms – protein engineering Φ-value analysis and computer simulation. Both these techniques can also be exploited in the study of forced unfolding mechanisms.

31.4.1
Protein Engineering

One of the best ways to determine the regions of the protein structure involved in the resistance of a protein domain to force is to make site-specific mutants and to investigate the response of these mutants to force. However, unless we combine force spectroscopy with biophysical data, any information gained is purely qualitative. If mutations in two different regions of the protein have different effects on force it is not possible to say that one region is more structured in the transition state than the other unless the effect of the protein on the stability of the native state has been quantified.

Fortunately, the analysis of mutant proteins as a means of investigating protein folding (and unfolding) mechanisms is well established [31]. Essentially, the effect of the mutation on the stability of the protein and on the stability of the transition state for unfolding is compared:

$$\Phi = \frac{\Delta\Delta G_{D-TS}}{\Delta\Delta G_{D-N}} = 1 - \frac{\Delta\Delta G_{TS-N}}{\Delta\Delta G_{D-N}} \tag{26}$$

where $\Delta\Delta G_{D-N}$ is the change in free energy on mutation (determined from equilibrium denaturation experiments) and $\Delta\Delta G_{D-TS}$ and $\Delta\Delta G_{TS-N}$ are changes in free energy of the transition state (TS), relative to the denatured (D) and native (N) states, respectively, determined from ratios of folding and unfolding rate constants:

$$\Delta\Delta G_{D-TS} = RT \ln\left(\frac{k_f^{wt}}{k_f^{mut}}\right) \quad \text{and} \quad \Delta\Delta G_{\ddagger-N} = -RT \ln\left(\frac{k_u^{wt}}{k_u^{mut}}\right) \tag{27}$$

where k_f and k_u are the folding and unfolding rate constants, and the superscripts wt and mut refer to the rate constants for wild-type and mutant proteins, respectively.

We have shown that this Φ-value analysis can be applied to force experiments, however, they are somewhat more complex, due to the nature of the data collected [30].

31.4.1.1 Choosing Mutants
The assumptions inherent in Φ-value analysis have been clearly described [32]. The three most important criteria are:

1. The mutations should not alter the structure of the native state significantly: To this end, the mutations should be conservative, nondisruptive mutations. They should not distort the structure or add new interactions. The most suitable mutations are simple deletion mutants such as Ile to Val or Val to Ala.

2. The mutations should not alter the stability of the denatured state significantly, or at least the effect on D should be significantly less than on N. For this reason mutations of or to Pro or Gly should be avoided where possible, as should mutants that change the polar/nonpolar nature of the sidechain.
3. The change in stability ($\Delta\Delta G_{D-N}$) should be significant. Where $\Delta\Delta G_{D-N}$ is small Φ cannot be determined with confidence. In AFM experiments this is particularly important since there is much more error in the experimental data. We do not choose to analyze mutants with a $\Delta\Delta G_{D-N} < 2$ kcal mol^{-1}. Where the effect of mutation is less than this it is unlikely that differences between wild type and mutant can be quantified with any confidence.

31.4.1.2 Determining $\Delta\Delta G_{D-N}$

The effect of the mutation on the stability of the native, starting state, relative to the unfolding state has to be determined from equilibrium denaturation experiments. This can be undertaken on either single domains of the protein, where ΔG_{D-N} has been shown to be unaffected by inclusion in a polyprotein, or by equilibrium investigations of the mutant polyprotein itself. Of course, the denatured state in the forced unfolding experiments is not the same as that measured in the equilibrium unfolding studies. One has to make the reasonable assumption that the effect of the mutation in the compact denatured state ensemble will be the same as the effect of the mutation on the stretched denatured state, or that, at least, the difference is significantly less than the effect on the native state. This is most likely to be true where the mutation is conservative. In some conventional kinetic investigations a ΔG has been determined by the ratio of the folding and unfolding rate constants. This cannot be applied to forced unfolding experiments, however. It has been shown to be possible to determine a folding rate constant for some proteins when the force is (very nearly) reduced to zero, but the protein still held on the tip[4]. However, the principle that $\Delta G_{D-N} = RT \ln\left(\dfrac{k_f}{k_u}\right)$, applies only where the folding and unfolding reaction are the reverse of each other, i.e., where the same transition state is being explored. Since it is most unlikely that unfolding in the presence of a directional force will be the reverse of folding in the absence of such a force (and it has been shown that this is not the case for at least one protein) measurements of ΔG from the rate constants derived in AFM experiments is not possible. A final important corollary, however, is that if under force the protein adopts an alternative native state, N^F (or unfolding intermediate, I) that is significantly different in structure to N, then this analysis will not hold.

31.4.1.3 Determining $\Delta\Delta G_{TS-N}$

In principle $\Delta\Delta G_{TS-N}$ can be determined from forced unfolding experiments simply by directly comparing the unfolding rate in the absence of force (v_o, also termed k_f^0) determined from the experimental data. However, as described above, these values are highly sensitive to small changes in the slope of the force vs. log v_r plot. Again, principles well established in protein folding experiments can be applied. This situation is akin to differences in $\Delta\Delta G_{TS-N}$ determined directly from experimental measurements and those extrapolated to 0 M denaturant: small changes in the slope can lead to large errors in v_o. We have shown that where the slope is the same within error, i.e., where x_u can be assumed to be the same for wild-type and mutant proteins a number of methods can be used to determine $\Delta\Delta G_{TS-N}$.

1. Analyze the data for wild type and mutant as described above (Section 31.3.9) but assuming a mean, fixed x_u.
2. Fit the force spectra to a fixed mean slope and then compare the pulling speeds (v_r) of wild type (wt) and mutant (mut) at a fixed force in the range of the experimental data:

$$\Delta\Delta G_{TS-N} = -RT \ln \left(\frac{v_r^{wt}}{v_r^{mut}} \right) \quad (28)$$

3. Fit the force spectra (F vs. $\ln v_r$) to a fixed mean slope (m) and then compare the unfolding force at a fixed pulling speed in the range of the experimental data:

$$\Delta\Delta G_{TS-N} = RT(F^{wt} - F^{mut})/m \quad (29)$$

We have shown that all three methods give the same Φ-value [30].

It is important to note that Φ-value analysis only applies where the unfolding transition that is being measured is the same in wt and mutant proteins. If a mutation changes the folding behaviour so that a different barrier is being explored then this method cannot be used. As an indication of this, the x_u should always be the same for wt and mutant proteins.

31.4.1.4 Interpreting the Φ-values

- Φ = 1: If the mutation is in a region of the protein that is fully structured in the transition state then the mutation will destabilize the transition state to the same extent as the native state. Thus the free energy difference between N and TS will remain the same ($\Delta\Delta G_{TS-N} = 0$) and the force required to unfold the protein will remain the same. In this case Φ = 1.
- Φ = 0: If the mutation is in a region that is fully unstructured in the transition state, then the mutation will have no effect on the stability of TS, while destabilizing N. Thus the difference in free energy between N and TS will be reduced by the same amount as N ($\Delta\Delta G_{TS-N} = \Delta\Delta G_{D-N}$). The unfolding force will be lowered in a manner that can be predicted using Eq. (29).
- 0 < Φ < 1: If the mutation is in a region that is partially structured in TS then the unfolding force will be lowered, but not as significantly as might be predicted from $\Delta\Delta G_{D-N}$. Formally, in traditional protein engineering studies, partial Φ-values can be interpreted in a number of ways. A Φ-value of 0.5, for instance, could indicate that half the free energy of the residue was lost in TS, or it could

mean that the protein unfolded by two pathways, one where the Φ-value was 1 and a second where Φ was 0. In these single molecule experiments, however, these two possibilities can be distinguished. If there are two different pathways then the forced unfolding should show a bi-modal distribution of forces. This has not been seen to date, and so the simpler interpretation of partial loss of contacts can be made.

Given the nature of the AFM experiment the error on Φ is larger than that from traditional unfolding experiments. It is probably not possible to say more than that a region is "fully, or almost fully structured" (Φ close to 1), "fully or almost fully unstructured" (Φ close to 0) or "partly structured" (Φ less than 1 but more than 0). This is one reason why mutants being studied should be significantly less stable than wild type, to allow Φ-values to be assigned with confidence. In traditional Φ-value analysis Φ-values are considered to be too prone to error if $\Delta\Delta G_{D-N} < 0.75$ kcal mol^{-1} (it has recently suggested that this limit should be set significantly higher). We suggest that for mechanical Φ-value analysis $\Delta\Delta G_{D-N}$ should be as large as possible and at least >1.5 kcal mol^{-1}.

31.4.2
Computer Simulation

Since even the most detailed protein engineering experiment can only provide "snapshots" of the unfolding process, simulations are important to understand the effect of force at the atomistic level. Mechanical unfolding experiments are apparently ideal for comparison with simulation. Unlike addition of temperature or denaturant, the perturbation is directional, suggesting that the reaction coordinate (N–C distance) is comparable in simulation and experiment. Furthermore, it is easy to identify in a forced unfolding trajectory, both metastable unfolding intermediates and the position of the transition state barrier. The problem of time scale is, however, just as difficult in simulations of forced unfolding as it is for other unfolding simulations. To increase the rate of unfolding the system has to be perturbed using conditions that are far from experiment. In simulations of thermal unfolding this requires the use of very high temperatures in a simulation. In forced unfolding simulations the rate of application of force (the loading rate) is significantly higher in simulation than experiment. Since theory suggests that different barriers to unfolding will be exposed at different loading rates, it is not possible to assume that the same barrier will be investigated in theory and experiment. Thus it is important that simulation be bench-marked by experiment.

A number of increasingly computationally expensive techniques have been applied to the study of mechanical unfolding and where these can be directly compared to experiment, the simulations seem to perform well. Lattice and off-lattice simulations have been used to describe the effect of force, and other perturbants (denaturant, temperature) on energy landscapes (e.g., [33–36]). The phase diagrams suggest that interesting data may be obtained if force is applied at higher temperatures and/or in the presence of denaturant. Steered molecular dynamics simulations, with and without explicit solvent, have been successful in reproducing a number of experimental features. They have also been successful in describing the order of unfolding events in titin I27 (e.g., [37, 38]). The hierarchy of unfolding of different protein domains has been predicted [39]. Simulations have observed unfolding intermediates in experimental systems where they have later been detected in experiment [13, 14, 37, 40, 41]. (But again, a note of caution is required. In only one case has the intermediate been shown to have experimental properties consistent with the intermediate observed in the simulations [14, 24]. In the other cases, the correct experiments have not yet been performed.) Finally, simulations have been successful in predicting the effect of mutation on the unfolding force [42]. Where simulation and experiment agree, the simulations can be used to understand details of the transitions between experimentally observable states.

Recent advances have reduced the unfolding forces observed in simulations to within an order of magnitude of those observed in experiment. Perhaps the most exciting phase is before us. Unlike previous studies in the field of protein folding where simulations have lagged behind experiment, the opposite is true of simulations of mechanical unfolding experiment. There are far more protein unfolding simulations in the literature than detailed experimental analyzes. The good simulation papers suggest experiments to be done.

31.5
Titin I27: A Case Study

31.5.1
The Protein System

Only one protein has, at the time of writing, been investigated in depth. This study shows how a combination of force spectroscopy, biophysical techniques, protein engineering, and simulation can be used to describe the mechanical unfolding of a single protein in detail. Titin I27 was the first protein made into a polyprotein and thus, the first where domain-specific force characteristics were elucidated [4]. It is one of the most mechanically stable proteins yet studied and remains a paradigm for forced unfolding experiments. The studies described here represent the work of a number of different experimental and theoretical groups.

To understand these experiments the structure of the protein has to be considered [43]. Titin I27 has an I-type immunoglobulin (Ig) fold. It has a beta sandwich structure, with the first sheet comprised of antiparallel strands A, B, E, and D and the second of antiparallel strands C, F, and G with, importantly, the A′ strand running parallel to strand G. In the polyprotein, the N and C termini (A and G strands) are connected by two-residue linkers (encoded by the restriction sites used in the cloning process). The N- and C-termini are found at opposite ends of the protein (Figure 31.5).

The first experiments revealed that the v_0 was $\sim 10^{-4}$ s^{-1} and the $x_u \sim 3$ Å [4]. Since the unfolding rate was close to that observed in standard stopped-flow experi-

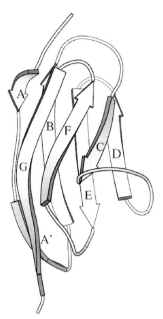

Fig. 31.5. Structure of titin I27. I27 has a beta-sandwich structure with the hydrophobic core enclosed by two antiparallel beta sheets. The strands are labeled A–G. A runs antiparallel to strand B, and A' runs parallel to strand G.

ments it was suggested that the transition state for unfolding under force was the same as that observed in the absence of force. This idea, however, did not hold for long [5, 24].

31.5.2
The Unfolding Intermediate

The first step observed in the simulations of forced unfolding of I27 was the detachment of the A-strand, to form a metastable unfolding intermediate, which lived longer than the native state in the simulation conditions [37, 40]. Careful examination of the experimental force traces revealed the presence of a "hump" in the AFM traces, at around 100 pN, which was attributed to a lengthening due to the detachment of the A-strand in all the domains of the polyprotein [14]. A mutant that disrupted the hydrogen bonding between the A- and B-strands did not show these "humps."

Thus, apparently, at forces around 100 pN, the native state becomes less stable than this intermediate, I. If this explanation were correct, then the observed unfolding force reflects the unfolding of this intermediate, and not the denatured state. Studies of an I27 polyprotein with a conservative mutation V4 to A in the A-strand confirmed this explanation [24]. This mutation destabilized the native state of the protein by ~2.5 kcal mol^{-1}, yet had no effect of the unfolding force, consistent with the intermediate hypothesis, since, if the A-strand is detached in I the V4A mutant would simply lead to I being populated at lower forces as N is destabilized, but I is not. Thus, the experiment and theory agree and the first process, N → I, in the unfolding pathway can be understood, and the unfolding rate constant at 0 force, v_o, and the unfolding distance, x_u, can be assigned to the unfolding of this intermediate. In a traditional protein folding study of I27, no such folding intermediate is observed [44]. A model of this intermediate (I27 with the A-strand deleted) was purified, shown to be stable and, by NMR spectroscopy, to be structurally very similar to the native state. In MD simulations this intermediate had all the properties of the intermediate induced by force [24].

31.5.3
The Transition State

Possibly the most important state along the unfolding pathway is the transition state, the barrier to forced unfolding. Proteins that can withstand significant force have a free energy barrier to unfolding that remains significant under applied force. In order to probe the nature of this barrier protein engineering Φ-value analysis was undertaken [30]. Mutations were made in the A′, B, C′, E, F, and G strands. Most of these mutations, although strongly destabilizing, had no effect on the unfolding force, i.e., had a Φ-value of 1 (Figure 31.6) [42]. Only mutations in the A′ and G strand lowered the unfolding force. The effect of the mutation in the A′-strand (V13A) was to lower the unfolding force by ~25 pN without changing the x_u giving a Φ-value of ~0.5. The mutation in the G-strand (V86A) changed the slope of the force spectrum (Figure 31.6) so that a Φ-value could not be determined (see Section 31.3.8). Since mechanical Φ-values are much harder to determine than in traditional protein engineering studies (the molecular biology is more laborious, the data are harder to collect, and the $\Delta\Delta G_{D-N}$ of each mutant must be significantly larger than traditionally) only eight mutants were made. This can give only a general picture of the transition state. However, on the contrary, the transition state in MD simulations of forced unfolding experiments is easier to identify than in high temperature simulations of unfolding. The transition state is represented by the structures just at the point where extension proceeds rapidly to a fully unfolded structure. Direct comparison was made between the experimental Φ-values and the Φ-values determined from the simulations. The agreement was very good. This means that the structure of the transition state can be directly inferred from the simulation [42].

The transition state structure is essentially the same as the native structure except in the region of the A, A′, and G strands (Figure 31.7). The A-strand is, of course, completely detached, as it is in I. The hydrogen bonding and side-chain interactions between the A′ and G strands are lost, as are interactions between the A′-strand and the E–F loop and between the G-strand and the A′–B loop. Other interactions of the A′ and G strands with the rest of the protein are, however, maintained in this structure. The mechanical strength of the I27 domain lies in the

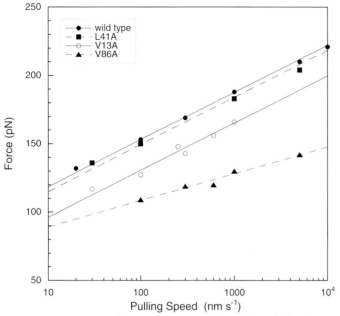

Fig. 31.6. Force spectra of titin I27. Most mutants (such as L41A) have no effect on the unfolding force, nor on the dependence of the unfolding force on the retract speed. V13A lowers the unfolding force but does not affect the pulling speed dependence. V86A both lowers the unfolding force and changes the dependence of the unfolding force on the puling speed. The slope of the force spectrum is $f_\beta \kappa_s = k_B T \kappa_s / x_u$ (i.e., the decrerase in slope for the mutant V86A reflects an increase in x_u).

strength of interactions between the A' and G strands, and in the strength of interactions of these strands with other regions of the protein. (A mutant I27 studied elsewhere had shown that a substitution that made contacts in the BC loop, also lowered the unfolding force, consistent with these results [5].) Importantly, these experiments show that the interactions of the side chains contribute significantly to mechanical strength, and it is not just the hydrogen bonding pattern that determines mechanical strength, as had been suggested in some theoretical studies [38]. Whether it is simply packing interactions, or a more complex effect of shielding of the hydrogen bonds from attack by solvent has yet to be determined. This side chain dependence does, however, seem to be at the root of the relative mechanical stability of different I-band domains of titin [45, 46].

31.5.4
The Relationship Between the Native and Transition States

Thus far theory and experiment had elucidated the structure of the principle states along the pathway and the relative height of the barrier between I and TS (corresponding to $v_0 \sim 10^{-4}$ s^{-1}) and the distance between I and TS ($x_u \approx 3$ Å). However,

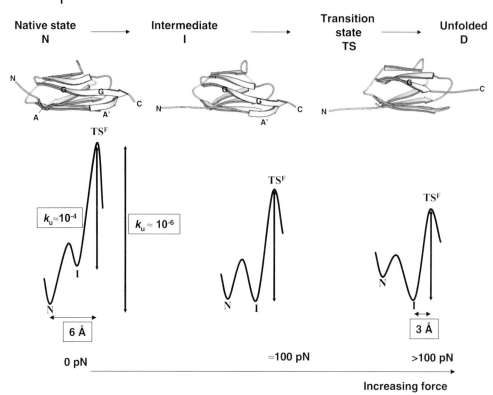

Fig. 31.7. The unfolding of TI I27 in the presence of force. Experiment and simulation combined have allowed the structure of the important states on the forced unfolding pathway to be elucidated. The effect of force on the unfolding landscape and the height and relative position of the energy barriers and the intermediate have been determined. Structures Courtesy of E. Paci.

the relative height of the barrier between N and TS and the distance between these two states was unknown. It is not possible, using AFM, to collect data at pulling speeds low enough to unfold the protein at forces below 100 pN, i.e., when I is not yet populated. However, here again mutant studies were instructive. One of the mutants (V86A) changed the slope of the force spectrum ($x_u \approx 6$ Å) [23]. The same behavior had been seen in a separate study for mutants with prolines inserted into the A'-strand [26]. The anomalous result was that these highly destabilized mutants had, apparently a lower unfolding rate at 0 force (v_0) than wild type. The simplest explanation is that these mutations lowered the stability of I and/or lowered the TS energy to such an extent that the protein unfolded directly from N before I was populated. In this case, v_0 is the unfolding rate directly from N and x_u is the unfolding distance from N to the transition state. By using this as a model and modeling the data for wild type and mutants it was possible to estimate the rate constant between N and TS at 0 force for wild-type I27 ($v_0 \approx 10^{-6}$ s^{-1}) [23].

31.5.5
The Energy Landscape under Force

Using all these results a pathway for the forced unfolding of titin I27 could be constructed (Figure 31.7). It is was also possible to compare the transition state in the absence of force (TS^0, as determined by traditional Φ-value analysis) with that observed in the force experiments (TS^F). The height of the energy barrier between N and TS^0 ($v_o = 5 \times 10^{-4}$ s^{-1}) is >3 kcal mol^{-1} lower than that between N and TS^F ($v_o = 10^{-6}$ s^{-1}). There is also a difference in structure. TS^F is more structured, with higher Φ-values than TS^0. However, the region of the protein that is responsible for mechanical stability (the region that breaks at the transition state) is apparently the same as the region that unfolds early in the denaturant induced unfolding studies, the A' and G strands. When force is applied to the protein energy landscape, the unfolding pathway changes, and the protein unfolds along a pathway occupied by a high-energy barrier, TS^F, that resists unfolding.

31.6
Conclusions – the Future

Forced unfolding of proteins is in its infancy as a technique for studying proteins. Much of the work that has been done has been to demonstrate the power of atomic force microscopy as a tool. However, now forced unfolding is moving beyond a descriptive science towards analysis of the molecular effect of force on protein structure, stability, and unfolding landscapes. The use in the study of proteins that experience force in vivo is obvious. However, it can also be used, combined with simulation, to test our understanding of the molecular forces responsible for protein folding. How, for example, does a protein molecule respond to force applied in different directions? Can we explain the relationship between protein topology and mechanical strength? Altering the solvent conditions (e.g., temperature, pH, salt) may be used to determine the relative importance of different intramolecular forces. We can also look forward to the development of systems allowing proteins to be attached to the beads of optical tweezers, or the biomolecular force probe so that far lower loading rates, probably closer to those experienced in vivo, can be applied. As yet, the single molecule nature of the unfolding process has yet to be exploited, although rare events (misfolding and cooperative unfolding) have been observed [47, 48]. There are moves to combine single molecule fluorescence with single molecule force measurements. We would expect to see significant advances in the field in the next decade.

References

1 RIEF, M., GAUTEL, M., OESTERHELT, F., FERNANDEZ, J. M. & GAUB, H. E. (1997). Reversible unfolding of individual titin immunoglobulin

domains by AFM. *Science* 276, 1109–1112.

2 RIEF, M., PASCUAL, J., SARASTE, M. & GAUB, H. E. (1999). Single molecule force spectroscopy of spectrin repeats: Low unfolding forces in helix bundles. *J. Mol. Biol.* 286, 553–561.

3 WANG, T. & IKAI, A. (1999). Protein stretching III: Force-extension curves of tethered bovine carbonic anhydrase B to the silicon substrate under native, intermediate and denaturing conditions. *Jpn. J. Appl. Phys. 1* 38, 3912–3917.

4 CARRION-VAZQUEZ, M., OBERHAUSER, A. F., FOWLER, S. B., MARSZALEK, P. E., BROEDEL, S. E., CLARKE, J. & FERNANDEZ, J. M. (1999). Mechanical and chemical unfolding of a single protein: A comparison. *Proc. Natl Acad. Sci. USA* 96, 3694–3699.

5 BROCKWELL, D. J., BEDDARD, G. S., CLARKSON, J. et al. (2002). The effect of core destabilization on the mechanical resistance of I27. *Biophys. J.* 83, 458–472.

6 STEWARD, A., TOCA-HERRERA, J. L. & CLARKE, J. (2002). Versatile cloning system for construction of multimeric proteins for use in atomic force microscopy. *Protein Sci.* 11, 2179–2183.

7 BEST, R. B., LI, B., STEWARD, A., DAGGETT, V. & CLARKE, J. (2001). Can non-mechanical proteins withstand force? Stretching barnase by atomic force microscopy and molecular dynamics simulation. *Biophys. J.* 81, 2344–2356.

8 BROCKWELL, D. J., PACI, E., ZINOBER, R. C. et al. (2003). Pulling geometry defines the mechanical resistance of a beta-sheet protein. *Nat. Struct. Biol.* 10, 731–737.

9 YANG, G., CECCONI, C., BAASE, W. A. et al. (2000). Solid-state synthesis and mechanical unfolding of polymers of T4 lysozyme. *Proc. Natl Acad. Sci. USA* 97, 139–144.

10 CARRION-VAZQUEZ, M., LI, H. B., LU, H., MARSZALEK, P. E., OBERHAUSER, A. F. & FERNANDEZ, J. M. (2003). The mechanical stability of ubiquitin is linkage dependent. *Nat. Struct. Biol.* 10, 738–743.

11 OBERHAUSER, A. F., HANSMA, P. K., CARRION-VAZQUEZ, M. & FERNANDEZ, J. M. (2001). Stepwise unfolding of titin under force-clamp atomic force microscopy. *Proc. Natl Acad. Sci. USA* 98, 468–472.

12 WILLIAMS, P. M. (2003). Analytical descriptions of dynamic force spectroscopy: Behaviour of multiple connections. *Anal. Chim. Acta.* 479, 107–115.

13 LENNE, P. F., RAAE, A. J., ALTMANN, S. M., SARASTE, M. & HORBER, J. K. H. (2000). States and transitions during forced unfolding of a single spectrin repeat. *FEBS Lett.* 476, 124–128.

14 MARSZALEK, P. E., LU, H., LI, H. B., CARRION-VAZQUEZ, M., OBERHAUSER, A. F., SCHULTEN, K. & FERNANDEZ, J. M. (1999). Mechanical unfolding intermediates in titin modules. *Nature* 402, 100–103.

15 BEST, R. B., BROCKWELL, D. J., TOCA-HERRERA, J. L. et al. (2003). Force mode atomic force microscopy as a tool for protein folding studies. *Anal. Chim. Acta.* 479.

16 EVANS, E. (2001). Probing the relation between force lifetime and chemistry in single molecular bonds. *Annu. Rev. Biophys. Biomol. Struct.* 30, 105–128.

17 HANGGI, P., TALKNER, P. & BORKOVEC, M. (1990). Reaction-rate theory – 50 Years after Kramers. *Rev. Mod. Phys.* 62, 251–341.

18 EVANS, E. & RITCHIE, K. (1997). Dynamic strength of molecular adhesion bonds. *Biophys. J.* 72, 1541–1555.

19 EVANS, E. & WILLIAMS, P. M. (2002). Dynamic force spectroscopy: I. Single bonds. In *Les Houches Session LXXV. Physics of bio-molecules and cells* (FLYVBJERG, H., JÜLICHER, F., ORMOS, P., & DAVID, F., Eds), pp. 145–186. Berlin: Springer-Verlag.

20 EVANS, E. & RICHIE, K. (1999). Strength of a weak bond connecting flexible polymer chains. *Biophys. J.* 76, 2439–2447.

21 PATEL, A. B., ALLEN, S., DAVIES, M. C., ROBERTS, C. J., TENDLER, S. J. B. & WILLIAMS, P. M. (2004). Influence of architecture on the kinetic stability of

molecular assemblies. *J. Am. Chem. Soc.* 126, 1318–1319.

22 ZINOBER, R. C., BROCKWELL, D. J., BEDDARD, G. S. et al. (2002). Mechanically unfolding proteins: the effect of unfolding history and the supramolecular scaffold. *Protein Sci.* 22, 2759–2765.

23 WILLIAMS, P. M., FOWLER, S. B., BEST, R. B. et al. (2003). Hidden complexity in the mechanical properties of titin. *Nature* 422, 446–449.

24 FOWLER, S. B., BEST, R. B., TOCA-HERRERA, J. L. et al. (2002). Mechanical unfolding of a titin Ig domain: (1) Structure of unfolding intermediate revealed by combining molecular dynamic simulations, NMR and protein engineering. *J. Mol. Biol.* 322, 841–849.

25 WILLIAMS, P. M. A method to analyze bond rupture and protein unfolding forces. Manuscript submitted.

26 LI, H. B., CARRION-VAZQUEZ, M., OBERHAUSER, A. F., MARSZALEK, P. E. & FERNANDEZ, J. M. (2000). Point mutations alter the mechanical stability of immunoglobulin modules. *Nat. Struct. Biol.* 7, 1117–1120.

27 RIEF, M., FERNANDEZ, J. M. & GAUB, H. E. (1998). Elastically coupled two-level systems as a model for biopolymer extensibility. *Phys. Rev. Lett.* 81, 4764–4767.

28 WILLIAMS, P. M. & EVANS, E. (2002). Dynamic force spectroscopy: II. Multiple bonds. In *Les Houches Session LXXV. Physics of bio-molecules and cells* (FLYVBJERG, H., JÜLICHER, F., ORMOS, P., & DAVID, F., Eds), pp. 187–204. Berlin: Springer-Verlag.

29 PRESS, W. H., TEUKOLSKY, S. A., VETTERLING, W. T. & FLANNERY, B. P. (1992). *Numerical recipes in C.* Cambridge: Cambridge University Press.

30 BEST, R. B., FOWLER, S. B., TOCA-HERRERA, J. L. & CLARKE, J. (2002). A simple method for probing the mechanical unfolding pathway of proteins in detail. *Proc. Natl Acad. Sci. USA* 99, 12143–12148.

31 FERSHT, A. R. (1998). *Structure and Mechanism in Protein Science: a Guide to Enzyme Catalysis and Protein Folding.* New York: W.H. Freeman.

32 FERSHT, A. R., MATOUSCHEK, A. & SERRANO, L. (1992). The folding of an enzyme. I. Theory of protein engineering analysis of stability and pathway of protein folding. *J. Mol. Biol.* 224, 771–782.

33 KLIMOV, D. K. & THIRUMALAI, D. (1999). Stretching single-domain proteins: Phase diagram and kinetics of force induced folding. *Proc. Natl Acad. Sci. USA* 96, 6166–6170.

34 KLIMOV, D. K. & THIRUMALAI, D. (2000). Native topology determines force induced unfolding pathways in globular proteins. *Proc. Natl Acad. Sci. USA* 97, 7254–7259.

35 KLIMOV, D. K. & THIRUMALAI, D. (2001). Lattice model studies of force-induced unfolding of proteins. *J. Phys. Chem. B* 105, 6648–6654.

36 CIEPLAK, M., HOANG, T. X. & ROBBINS, M. O. (2001). Thermal unfolding and mechanical unfolding oathways of protein secondary structures. *Proteins* 49, 104–113.

37 LU, H., ISRALEWITZ, B., KRAMMER, A., VOGEL, V. & SCHULTEN, K. (1998). Unfolding of titin immunoglobulin domains by steered molecular dynamics simulation. *Biophys. J.* 75, 662–671.

38 LU, H. & SCHULTEN, K. (2000). The key event in force-induced unfolding of titin's immunoglobulin domains. *Biophys. J.* 79, 51–65.

39 CRAIG, D., GAO, M., SCHULTEN, K. & VOGEL, V. (2004). Tuning the mechanical stability of fibronectin type III modules through sequence variations. *Structure* 12, 21–30.

40 LU, H. & SCHULTEN, K. (1999). Steered molecular dynamics simulation of conformational changes of immunoglobulin domain I27 interpret atomic force microscopy observations. *Chem. Phys.* 247, 141–153.

41 PACI, E. & KARPLUS, M. (2000). Unfolding proteins by external forces and temperature: The importance of topology and energetics. *Proc. Natl Acad. Sci. USA* 97, 6521–6526.

42 Best, R. B., Fowler, S. B., Toca-Herrera, J. L., Steward, A., Paci, E. & Clarke, J. (2003). Mechanical unfolding of a titin Ig domain: (2) Structure of transition state revealed by combining AFM, protein engineering and molecular dynamics simulations. *J. Mol. Biol.* 330, 867–877.

43 Pfuhl, M. & Pastore, A. (1995). Tertiary structure of an immunoglobulin-like domain from the giant muscle protein titin: a new member of the I set. *Structure* 3, 391–401.

44 Fowler, S. B. & Clarke, J. (2001). Mapping the folding pathway of an immunoglobulin domain: Structural detail from phi value analysis and movement of the transition state. *Structure* 9, 355–366.

45 Li, H., Oberhauser, A. F., Fowler, S. B., Clarke, J. & Fernandez, J. M. (2000). Atomic force microscopy reveals the mechanical design of a modular protein. *Proc. Natl Acad. Sci. USA* 92, 6527–6531.

46 Li, H. B., Linke, W. A., Oberhauser, A. F. et al. (2002). Reverse engineering of the giant muscle protein titin. *Nature*, 998–1002.

47 Oberhauser, A. F., Marszalek, P. E., Carrion-Vazquez, M. & Fernandez, J. M. (1999). Single protein misfolding events captured by atomic force microscopy. *Nat. Struct. Biol.* 6, 1025–1028.

48 Law, R., Carl, P., Harper, S., Dalhaimer, P., Speicher, D. W. & Discher, D. E. (2003). Cooperativity in forced unfolding of tandem spectrin repeats. *Biophys. J.* 84, 533–544.

32
Molecular Dynamics Simulations to Study Protein Folding and Unfolding

Amedeo Caflisch and Emanuele Paci

32.1
Introduction

Proteins in solution fold in time scales ranging from microseconds to seconds. A computational approach to folding that should work, in principle, is to use an atom-based model for the potential energy (force field) and to solve the time-discretized Newton equation of motion (molecular dynamics, MD [1]) from a denatured conformer to the native state in the presence of the appropriate solvent. With the available simulation protocols and computing power, such a trajectory would require approximately 10–100 years for a 100-residue protein where the experimental transition to the folded state takes place in about 1 ms. Hence, there is a clear problem related to time scales and sampling (statistical error). On the other hand, we think that current force fields, even in their most detailed and sophisticated versions, i.e., explicit water and accurate treatment of long-range electrostatic effects, are not accurate enough (systematic error) to be able to fold a protein on a computer. In other words, even if one could use a computer 100 times faster than the currently fastest processor to eliminate the time scale problem, most proteins would not fold to the native structure because of the large systematic error and the marginal stability of the folded state typically ranging from 5 to 15 kcal mol^{-1}. Interestingly, only designed peptides of about 20 residues have been folded by MD simulations (see Section 32.2.1) using mainly approximative models of the solvent (see Section 32.3.4). Alternatively, protein unfolding which is a simpler process than folding (e.g., the unfolding rate shows Arrhenius-like temperature dependence whereas folding does not because of the importance of entropy, see Section 32.2.1.2) can be simulated on shorter time scales (1–100 ns) at high temperature or by using a suitable perturbation.

  MD simulations can provide the ultimate detail concerning individual atom motion as a function of time. Hence, future improvements in force fields and simulation protocols will allow specific questions about the folding of proteins to be addressed. The understanding at the atomic level of detail is important for a complicated reaction like protein folding and cannot easily be obtained by experiments. Yet, experimental approaches and results are essential in validating the force fields

Protein Folding Handbook. Part I. Edited by J. Buchner and T. Kiefhaber
Copyright © 2005 WILEY-VCH Verlag GmbH & Co. KGaA, Weinheim
ISBN: 3-527-30784-2

and simulation methods: comparison between simulation and experimental data is *conditio sine qua non* to validate the simulation results and very helpful for improving force fields.

This chapter cannot be comprehensive. Results obtained by using atom-based force fields and MD are presented whereas lattice models [2] as well as off-lattice coarse-grained models (e.g., one interaction center per residue) [3] are not mentioned because of size limitations. It is important to note that the impact of MD simulations of folding and unfolding is increasing thanks to faster computers, more efficient sampling techniques, and more accurate force fields as witnessed by several review articles [1, 4] and books [5–7].

32.2
Molecular Dynamics Simulations of Peptides and Proteins

32.2.1
Folding of Structured Peptides

Several comprehensive review articles on MD simulations of structured peptides have appeared recently [8–10]. Here, we first focus on simulation results obtained in our research group and then discuss the Trp-cage, a model system that has been investigated by others.

32.2.1.1 Reversible Folding and Free Energy Surfaces

β-Sheets The reversible folding of two designed 20-residue sequences, beta3s and ^{D}PG, having the same three-stranded antiparallel β-sheet topology was simulated [11, 12] with an implicit model of the solvent based on the accessible surface area [13]. The solution conformation of beta3s (TWIQNGSTKWYQNGSTKIYT) has been studied by NMR [14]. Nuclear Overhauser enhancement spectroscopy (NOE) and chemical shift data indicate that at 10 °C beta3s populates a single structured form, the expected three-stranded antiparallel β-sheet conformation with turns at Gly6-Ser7 and Gly14-Ser15, (Figure 32.1) in equilibrium with the denatured state. The β-sheet population is 13–31% based on NOE intensities and 30–55% based on the chemical shift data [14]. Furthermore, beta3s was shown to be monomeric in aqueous solution by equilibrium sedimentation and NMR dilution experiments [14].

^{D}PG is a designed amino acid sequence (Ace-VFITS^{D}PGKTYTEV^{D}PG-Orn-KILQ-NH), where ^{D}P are D-prolines and Orn stands for ornithine. Circular dichroism and chemical shift data have provided evidence that ^{D}PG adopts the expected three-stranded antiparallel β-sheet conformation at 24 °C in aqueous solution [15]. Moreover, ^{D}PG was shown to be monomeric by equilibrium sedimentation. Although the percentage of β-sheet population was not estimated, NOE distance restraints indicate that both hairpins are highly populated at 24 °C.

In the MD simulations at 300 K (started from conformations obtained by spon-

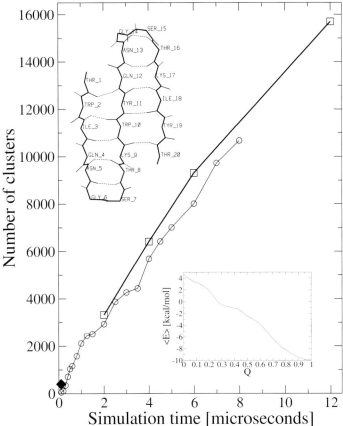

Fig. 32.1. Number of clusters as a function of time. The "leader" clustering procedure was used with a total of 120 000 snapshots saved every 0.1 ns (thick line and square symbols). The clustering algorithm which uses the Cα RMSD values between all pairs of structures was used only for the first 8 μs (80 000 snapshots) because of the computational requirements (thin line and circles). The diamond in the bottom left corner shows the average number of conformers sampled during the folding time which is defined as the average time interval between successive unfolding and refolding events. The insets show a backbone representation of the folded state of beta3s with main chain hydrogen bonds in dashed lines, and the average effective energy as a function of the fraction of native contacts Q which are defined in [11]. Figure from Ref. [33].

taneous folding at 360 K) both peptides satisfy most of the NOE distance restraints (3/26 and 4/44 upper distance violations for beta3s and ^{D}PG, respectively). At a temperature value of 360 K which is above the melting temperature of the model (330 K), a statistically significant sampling of the conformational space was obtained by means of around 50 folding and unfolding events for each peptide [11, 12]. Average effective energy and free energy landscape are similar for both peptides, despite the sequence dissimilarity. Since the average effective energy has a downhill profile at the melting temperature and above it, the free energy barriers are a consequence of the entropic loss involved in the formation of a β-hairpin which represents two-thirds of the chain. The free energy surface of the β-sheet peptides is completely different from the one of a helical peptide of 31 residues, Y(MEARA)$_6$ (see below). For the helical peptide, the folding free energy barrier corresponds to the helix nucleation step, and is much closer to the fully unfolded state than for the β-sheet peptides. This indicates that the native topology determines to a large extent the free energy surface and folding mechanism. On the other hand, the ^{D}PG peptide has a statistically predominant folding pathway with a sequence of events which is the inverse of the one of the most frequent pathway for the beta3s peptide. Hence, the amino acid sequence and specific interactions between different side chains determine the most probable folding route [12].

It is interesting to compare with experimental results on two-state proteins. Despite a sequence identity of only 15%, the 57-residue IgG-binding domains of protein G and protein L have the same native topology. Their folded state is symmetric and consists mainly of two β-hairpins connected such that the resulting four-stranded β-sheet is antiparallel apart from the two central strands which are parallel [16]. The ϕ value analysis (see Section 32.2.3 for a definition of ϕ value) of protein L and protein G indicates that for proteins with symmetric native structure more than one folding pathway may be consistent with the native state topology and the selected route depends on the sequence [16]. Our MD simulation results for the two antiparallel three-stranded β-sheet peptides (whose sequence identity is also 15%) go beyond the experimental findings for protein G and L. The MD trajectories demonstrate the existence of more than one folding pathway for each peptide sequence [12]. Interestingly, Jane Clarke and collaborators [17] have recently provided experimental evidence for two different unfolding pathways using the anomalous kinetic behavior of the 27th immunoglobulin domain (β-sandwich) of the human cardiac muscle protein titin. They have interpreted the upward curvature in the denaturant-dependent unfolding kinetics as due to changes in the flux between transition states on parallel pathways. In the conclusion of their article [17] they leave open the question "whether what is unusual is not the existence of parallel pathways, but the fact that they can be experimentally detected and resolved."

α-Helices Richardson et al. [18] have analyzed the structure and stability of the synthetic peptide Y(MEARA)$_6$ by circular dichroism (CD) and differential scanning calorimetry (DSC). This repetitive sequence was "extracted" from a 60-amino-acid domain of the human CstF-64 polyadenylation factor which contains 12 nearly identical repeats of the consensus motif MEAR(A/G). The CD and DSC data were insensitive to concentration indicating that Y(MEARA)$_6$ is monomeric in solution at concentrations up to 2 mM. The far-UV CD spectrum indicates that the peptide has a helical content of about 65% at 1 °C. The DSC profiles were used to determine an enthalpy difference for helix formation of 0.8 kcal mol^{-1} per amino acid. The length of Y(MEARA)$_6$ makes it difficult to study helix formation by MD simulations with explicit water molecules. Therefore, multiple MD runs were performed with the same implicit solvation model used for the β-sheet peptides [13].

The simulation results indicate that the synthetic peptide Y(MEARA)$_6$ assumes a mainly α-helical structure with a nonnegligible content of π-helix [149]. This is not inconsistent with the currently available experimental evidence [18]. A significant π-helical content was found previously by explicit solvent molecular dynamics simulations of the peptides (AAQAA)$_3$ and (AAKAA)$_3$ [19], which provides further evidence that the π-helical content of Y(MEARA)$_6$ is not an artifact of the approximations inherent to the solvation model.

An exponential decay of the unfolded population is common to both Y(MEARA)$_6$ [149] and the 20-residue three-stranded antiparallel β-sheet [14] previously investigated by MD at the same temperature (360 K) [11]. The free energy surfaces of Y(MEARA)$_6$ and the antiparallel β-sheet peptide differ mainly in the height and location of the folding barrier, which in Y(MEARA)$_6$ is much lower and closer to the fully unfolded state. The main difference between the two types of secondary structure formation consists of the presence of multiple pathways in the α-helix and only two predominant pathways in the three-stranded β-sheet. The helix can nucleate everywhere, with a preference for the C-terminal third of the sequence in Y(MEARA)$_6$. Furthermore, two concomitant nucleation sites far apart in the sequence are possible. Folding of the three-stranded antiparallel β-sheet peptide beta3s started with the formation of most of the side chain contacts and hydrogen bonds between strands 2 and 3, followed by the 1–2 interstrand contacts. The inverse sequence of events, i.e., first formation of 1–2 and then 2–3 contacts was also observed, but less frequently [11].

The free energy barrier seems to have an important entropic component in both helical peptides and antiparallel β-sheets. In an α-helix, it originates from constraining the backbone conformation of three consecutive amino acids before the first helical hydrogen bond can form, while in the antiparallel β-sheet it is due to the constraining of a β-hairpin onto which a third strand can coalesce [11]. Therefore, the folding of the two most common types of secondary structure seems to have similarities (a mainly entropic nucleation barrier and an exponential folding rate) as well as important differences (location of the barrier and multiple vs. two pathways). The similarities are in accord with a plethora of experimental and theoretical evidence [20] while the differences might be a consequence of the fact that Y(MEARA)$_6$ has about 7–9 helical turns whereas the three-stranded antiparallel β-sheet consists of only two "minimal blocks", i.e., two β-hairpins.

32.2.1.2 Non-Arrhenius Temperature Dependence of the Folding Rate

Small molecule reactions show an Arrhenius-like temperature dependence, i.e., faster rates at higher temperatures. Protein folding is a complex reaction involving many degrees of freedom; the folding rate is Arrhenius-like at physiological temperatures, but deviates from Arrhenius behavior at higher temperatures [20].

To quantitatively investigate the kinetics of folding, MD simulations of two model peptides, Ace-(AAQAA)$_3$-NHCH$_3$ (α-helical stable structure) and Ace-V$_5$^{D}PGV$_5$-NH$_2$ (β-hairpin), were performed using the same implicit solvation model [13]. Folding and unfolding at different temperature values were studied by 862 simulations for a total of 4 μs [21]. Different starting conformations (folded and random) were used to obtain a statistically significant sampling of conformational space at each temperature value. An important feature of the folding of both peptides is the negative activation enthalpy at high temperatures. The rate constant for folding initially increases with temperature, goes through a maximum at about T_m, and then decreases [21]. The non-Arrhenius behavior of the folding rate is in accord with experimental data on two mainly alanine α-helical peptides [22, 23], a β-hairpin [24], CI2, and barnase [25], lysozyme [26, 27], and lattice simulation results [28–30]. It has been proposed that the non-Arrhenius profile of the folding rate originates from the temperature dependence of the hydrophobic interaction [31, 32]. The MD simulation results show that a non-Arrhenius behavior can arise at high values of the temperature in a model where all the interactions are temperature independent. This has been found also in lattice simulations [28, 29]. The curvature of the folding rate at high temperature may be a property of a reaction dominated by enthalpy at low temperatures and entropy at high temperatures [30]. The non-Arrhenius behavior for a system where the interactions do not depend on the temperature might be a simple consequence of the temperature dependence of the accessible configuration space. At low temperatures, an increase in temperature makes it easier to jump over the energy barriers, which are rate limiting. However, at very high temperatures, a larger portion of the configuration space becomes accessible, which results in a slowing down of the folding process.

32.2.1.3 Denatured State and Levinthal Paradox

The size of the accessible conformational space and how it depends on the number of residues is not easy to estimate. To investigate the complexity of the denatured state four molecular dynamics runs of beta3s were performed at the melting temperature of the model (330 K) for a total simulation time of 12.6 µs [33]. The simulation length is about two orders of magnitude longer than the average folding or unfolding time (about 85 ns each), which are similar because at the melting temperature the folded and unfolded states are equally populated. The peptide is within 2.5 Å Cα root mean square deviation (RMSD) from the folded conformation about 48% of the time. Figure 32.1 shows the results of a cluster analysis based on Cα RMSD. There are more than 15 000 conformers (cluster centers) and it is evident that a plateau has not been reached within the 12.6 µs of simulation time. However, the number of significantly populated clusters (see Ref. [12] for a detailed description) converges already within 2 µs. Hence, the simulation-length dependence of the total number of clusters is dominated by the small ones. At each simulation interval between an unfolding event and the successive refolding event additional conformations are sampled. More than 90% of the unfolded state conformations are in small clusters (each containing less than 0.1% of the saved snapshots) and the total number of small clusters does not reach a plateau within 12.6 µs. Note that there is also a monotonic growth with simulation time of the number of snapshots in the folded-state cluster. After 12.6 µs (and also within each of the four trajectories) the system has sampled an equilibrium of folded and unfolded states despite a large part of the denatured state ensemble has not yet been explored. In fact, the average folding time converges to a value around 85 ns which shows that the length of each simulation is much larger than the relaxation time of the slowest conformational change. Interestingly, in the average folding time of about 85 ns beta3s visits less than 400 clusters (diamond in Figure 32.1). This is only a small fraction of the total amount of conformers in the denatured state. It is possible to reconcile the fast folding with the large conformational space by analyzing the effective energy, which includes all of the contributions to the free energy except for the configurational entropy of the protein [11, 34]. Fast folding of beta3s is consistent with the monotonically decreasing profile of the effective energy (inset in Figure 32.1). Despite the large number of conformers in the denatured state ensemble, the protein chain efficiently finds its way to the folded state because native-like interactions are on average more stable than nonnative ones.

In conclusion, the unfolded state ensemble at the melting temperature is a large collection of conformers differing among each other, in agreement with previous high temperature molecular dynamics simulations [8, 35]. The energy "bias" which makes fast folding possible does not imply that the unfolded state ensemble is made up of a small number of statistically relevant conformations. The simulations provide further evidence that the number of denatured state conformations is orders of magnitudes larger than the conformers sampled during a folding event. This result also suggests that measurements which imply an average over the unfolded state do not necessarily provide information on the folding mechanism.

32.2.1.4 Folding Events of Trp-cage

Very small proteins are ideal systems to validate force fields and simulation methodology. Neidigh et al. [36] have truncated and mutated a marginally stable 39-residue natural sequence thereby designing a 20-residue peptide, the Trp-cage, that is more than 95% folded in aqueous solution at 280 K. The stability of the Trp-cage is due to the packing of a Trp side chain within three Pro rings and a Tyr side chain. Moreover, the C-terminal half contains four Pro residues which dramatically restrict the conformational space, i.e., entropy, of the unfolded state [36, 37].

Four MD studies have appeared in the 12 months following the publication of the Trp-cage structure [38–41]. All of the simulations were started from the completely extended conformation and used different versions of the AMBER force field and the generalized Born continuum electrostatic solvation model [42]. Two simulations were run with conventional constant temperature MD at 300 K [40] and 325 K [38], a third study used replica exchange MD with a range of temperatures from 250 K to 630 K [41], and in the fourth paper distributed computing simulations at 300 K with full water viscosity were reported [39].

An important problem of the three constant temperature studies is that the Trp-cage seems to fold to a very deep free energy minimum and no unfolding events have been observed [38–40]. Moreover, only one folding event is presented by Simmerling et al. [38] and Chowdhury et al. [40]. The poor statistics does not allow to draw any conclusions on free energy landscapes or on the folding mechanism of the Trp-cage.

Another potential problem is the discrepancy between the most stable state sampled by MD and the NMR conformers. Only in two of the four MD studies NOE distance restraints were measured along the trajectories and about 20% were found to be violated [40, 41]. Moreover, as explicitly stated by the authors, the native state sampled by distributed computing contains a π-helix (instead of the α-helix) and the Trp is not packed correctly in the core [39]. These discrepancies are significant because the Trp-cage has a very small core and a rather rigid C-terminal segment.

32.2.2
Unfolding Simulations of Proteins

32.2.2.1 High-temperature Simulations

Since the early work of Daggett and Levitt [43] and Caflisch and Karplus [44], several other high-temperature simulation studies have been concerned with exploring protein unfolding pathways. Several comprehensive review articles exist on this simulation protocol [45] which has been widely used since. Recent MD simulations at temperatures of 100 °C and 225 °C of a three-helix bundle 61-residue protein, the engrailed homeodomain (En-HD), by Daggett and coworkers [46, 47] have been used to analyze a folding intermediate at atomic level of detail. The unfolding half-life of the En-HD at 100 °C has been extrapolated to be about 7.5 ns, a time scale that can be accessed by MD simulations with explicit water molecules.

Also, unfolding simulations in the presence of explicit urea molecules have shown that the protein (barnase) remains stable at 300 K but unfolds partially at moderately high temperature (360 K) [48]. The results suggested a mechanism for urea induced unfolding due to the interaction of urea with both polar and nonpolar groups of the protein.

32.2.2.2 Biased Unfolding

Because of the limitations on simulation times and height of the barriers to conformational transitions in proteins, a number of methods, alternative to the use of high, nonphysical temperatures, have been proposed to accelerate such transitions by the introduction of an external time-dependent perturbation [49–54]. The perturbation induce the reaction of interest in a reasonable amount of time (the strength of the perturbation is inversely proportional to the available computer time). These methods have been used for studying not only protein unfolding at native or realistic denaturing conditions, but also large conformational changes between known relevant conformers [51]. Their goal is to generate pathways which are realistic, in spite of the several orders of magnitude reduction in the time required for the conformational change. They are not alternatives to methods to compute free energy profiles along defined pathways. The external perturbation is usually applied to a function of the coordinates which is assumed to vary monotonically as the protein goes from the native to the nonnative state of interest. For certain perturbations the unfolding pathways obtained have been shown to depend on the nature of the perturbation and the choice of the reaction coordinate [52];

this is even more the case when the perturbation is strong and the reaction is induced too quickly for the system to relax along the pathway.

A perturbation which is particularly "gentle" since it exploits the intrinsic thermal fluctuations of the system and produces the acceleration by selecting the fluctuations that correspond to the motion along the reaction coordinate has also been used to unfold proteins [55, 56]. This perturbation has been employed, in particular, to expand α-lactalbumin by increasing its radius of gyration starting from the native state, and generate a large number of low-energy conformers that differ in terms of their root mean square deviation, for a given radius of gyration. The resulting structures were relaxed by unbiased simulations and used as models of the molten globule (see Chapter 23) and more unfolded denatured states of α-lactalbumin based on measured radii of gyration obtained from nuclear magnetic resonance experiments [57]. The ensemble of compact nonnative structures agree in their overall properties with experimental data available for the α-lactalbumin molten globule, showing that the native-like fold of the α-domain is preserved and that a considerable proportion of the antiparallel β-sheet in the β-domain is present. This indicated that the lack of hydrogen exchange protection found experimentally for the β-domain [58] is due to rearrangement of the β-sheet involving transient populations of nonnative β-structures in agreement with more recent infrared spectroscopy measurements [59]. The simulations also provide details concerning the ensemble of structures that contribute as the molten globule unfolds and shows, in accord with experimental data [60], that the unfolding is not cooperative, i.e., the various structural elements do not unfold simultaneously.

32.2.2.3 Forced Unfolding

Unfolding by stretching proteins individually has become routinely mainly thanks to the advent of the atomic force microscopy technology [61]. This peculiar way of unfolding proteins opened new perspectives on protein folding studies. Experiments are usually performed on engineered homopolyproteins, and the I27 domain from titin has become the reference system for this type of studies. Experiments measure force-extension profiles, and show typical "saw-tooth" profiles, where peaks are due to the sudden unfolding of individual domains, sequentially in time, causing a drop in the recorded force. These profiles are generally interpreted assuming that the unfolding event is determined by a single barrier which is decreased by the external force. For a detailed description of the experimental techniques and of the most recent results on forced unfolding of single molecules (by atomic force microscopy and optical tweezers) see Chapter 31.

To provide a structural interpretation of the typical saw-tooth-like spectra measured in single molecule stretching experiments, various simulation techniques have been proposed, where detailed all-atom models of proteins are stretched by pulling two atoms apart [62, 63], differing mainly in the way the solvent is treated.

In some cases simulation can effectively explain the force patterns measured (see Ref. [64] for a review). For all the proteins experimentally unfolded by pulling, only a simple saw-tooth pattern has been recorded related to the sudden unfolding when the protein was pulled beyond a certain length, i.e., a simple two-state behavior. Simulations showed a more complex behavior [63] with possible intermediates on the forced unfolding pathways for certain proteins.[1]

Simulation has been used [65] to compare forced unfolding of two protein classes (all-β-sandwich proteins and all-α-helix bundle proteins). In particular, simulations suggested that different proteins should show a significantly different forced unfolding behavior, both within a protein class and for the different classes and dramatic differences between the unfolding induced by high temperature and by external pulling forces. The result was shown to be correlated to the type of perturbation, the folding topologies, the nature of the secondary and tertiary interactions and the relative stability of the various structural elements [65]. Improvements in the AFM technique combined with protein engineering methods have now confirmed (see Chapter 31) that chemical (or thermal) and forced unfolding occur through different pathways and that forced unfolding is related to crossing of a free energy barrier which might not be unique, but might change with force magnitude or upon specific mutations [66].

It should be borne in mind, however, that the forced unfolding of proteins is a nonequilibrium phenomenon strongly dependent on the pulling speed, and, since time scales in simulations and experiments are very different, the respective pathways need not to be the same. Recently, through a combination of experimental analysis and molecular dynamics simulations it has been shown that, in the case of mechanical unfolding, pathways might effectively be the same in a large range of pulling speeds or forces [67, 68], thus providing another demonstration of the robustness of the energy landscape (i.e., the funnel-like shape of the free energy surface sculpted by evolution is not affected by the application of even strong perturbations [69, 70]). In two recent papers [71, 72] it has been shown that proteins resist differently when pulled in different directions. In both cases the experiments have been complemented with simulations, with either explicit or implicit solvents. In both cases the behavior observed experimentally is qualitatively reproduced. This fact strongly suggests, although does not prove it, that in this particular case the forced unfolding mechanisms explored in the simulations is the same as that which determines the experimentally measured force.

Difference between solvation models is discussed in detail in Section 32.3.4 in the context of forced unfolding simulations, the disadvantage of an explicit solvation model [62, 72, 73] relative to an implicit [63, 65, 67, 68, 71] is not only that of being much slower, but also to provide an environment which relaxes slowly relative to the fast unraveling of the protein under force. Moreover, properly hydrating with explicit water a partially extended protein requires a large quantity of water, thus requiring a very large amount of CPU time for a single simulation. Implicit solvent models, on the other hand, allow unfolding to be performed at much lower

---

1) The presence of "late" intermediates on the forced unfolding pathway was first observed [63] in the 10th domain of fibronectin type III from fibronectin (FNfn3). A more complex pattern than equally spaced peaks in the force extension profile was predicted to arise from the presence of a kinetically metastable state. Most recent experimental results (J. Fernandez, personal communication) confirm the behavior predicted by the simulation.

forces (or pulling speeds) and multiple simulations to be used to study the dependence of the results on the initial conditions and/or on the applied force.

32.2.3
Determination of the Transition State Ensemble

The understanding of the folding mechanism has crucially advanced since the development of a method which provide information on the transition state [74] (see Chapter 13). The method allows the structure of the transition state at the level of single residue to be probed by measuring the change in folding and unfolding rates upon mutations. The method provides a so-called ϕ-value for each of the mutated residues which is a measure of the formation of native structure around the residue: a ϕ-value of 1 suggests that the residue is in a native environment at the transition state while a ϕ-value of 0 can be interpreted as a loss of the interactions of the residue at the transition state. Fractional ϕ-values are more difficult to interpret, but have been shown to arise from weakened interactions [75] and not from a mixture of species, some with fully formed and some with fully broken interaction.

As we discussed in Section 32.2.2.1, the use of high temperature makes it possible to observe the unfolding of a protein by MD on a time scale which can be simulated on current computers. Valerie Daggett and collaborators [76] first had the idea of performing a very high-temperature simulation and looking for a sudden change in the structure of the protein along the trajectory, indicating the escape from the native minimum of the free energy surface. The collections of structures around the "jump" were assumed to constitute a sample of the putative transition state. Assuming that the experimental ϕ-values correspond in microscopic terms to fraction of native contacts, they found a good agreement between calculated and experimental values. This approach was initially applied to the protein CI2, a small two-state proteins which has been probably the most thoroughly studied by experimental ϕ-value analysis; it has been subsequently improved and extended to the study of several proteins for which experimental ϕ-values were available (see Ref. [77] for review and other references).

Another related method has been used recently [78] to unfold a protein by high-temperature simulation (srcSH3 in the specific case) and determine a putative transition state by looking for conformations where the difference between calculated and experimental ϕ-values was smallest.

Both methods presented above have the advantage of providing structures extracted from an unfolding trajectory and thus the fast refolding or complete unfolding from these structures (a property of transition states) has been reported [78, 79]. But both approaches only provide few transition state structures, because a long simulation is required to generate each member of the transition state ensemble, while the transition state can be a quite broad ensemble for some proteins [80].

In a recent development, it has been shown that the amount of information that can be obtained from experimental measurements can be expanded further by

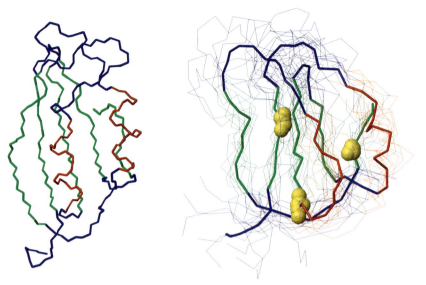

Fig. 32.2. Comparison between the native state structure (left) and the most representative structures of the transition state ensemble of AcP, determined by all-atom molecular dynamics simulations [82]. Native secondary structure elements are show in color (the two α-helices are plotted in red and the β-sheet in green). The three key residues for folding are shown as gold spheres [81, 82]. Figure from Ref. [69].

using the data to build up phenomenological energy functions to bias computer generated trajectories. With this approach (see Section 32.3.2 for a more detailed description of the technique), conformations compatible with experimental data are determined directly during the simulations [81, 82], rather than being obtained from filtering procedures such as those discussed above [78]. The incorporation of experimental data into the energy function creates a minimum in correspondence of the state observed experimentally and therefore allows for a very efficient sampling of conformational space. The transition state for folding of acylphosphatase (see Figure 32.2) was determined in this way [81, 82], showing that the network of interactions that stabilize the transition state is established when a few key residues form their native-like arrangement.

Based on this computational technique, a general approach in which theory and experiments are combined in an iterative manner to provide a detailed description of the transition state ensemble has been recently proposed [83]. In the first iteration, a coarse-grained determination of the transition state ensemble (TSE) is carried out by using a limited set of experimental ϕ-values as constraints in a molecular dynamics simulation. The resulting model of the TSE is used to determine the additional residues whose ϕ-value measurement would provide the most information for refining the TSE. Successive iterations with an increasing number of ϕ-value measurements are carried out until no further changes in the properties of the TSE are detected or there are no additional residues whose ϕ-values can be measured. The method can be also used to find key residues for folding (i.e., those that are most important for the formation of the TSE).

The study of the transition state represents probably the most interesting example of how experiment and molecular dynamics simulations complement each other in understanding and visualizing the folding mechanisms in terms of relevant structures involved. Simulations are performed with approximate force fields and unfolding induced using artificial means (such as high temperature or other perturbations). At this stage in the development of MD simulations, the experiment provides evidence that what is observed in silico is consistent with what happens in vitro. On the other hand, and particularly in the case in which the full ensemble of conformations compatible with the experimental results is generated [82, 83], the simulation suggests further mutations to increase the resolution of the picture of the transition state, and allows detailed hypothesis of the mechanisms, such as the identity and structure of the residues involved in the folding nucleus [150].

32.3
MD Techniques and Protocols

32.3.1
Techniques to Improve Sampling

A thorough sampling of the relevant conformations is required to accurately describe the thermodynamics and kinetics of protein folding. Since the energetic and entropic barriers are higher than the thermal energy at physiological temperature, standard MD techniques often fail to adequately sample the conformational space. As already mentioned in this chapter, even for a small protein it is currently not yet feasible to simulate reversible folding with a high-resolution approach (e.g., MD simulations with an all-atom model). The practical difficulties in performing such brute force simulations have led to several types of computational approaches and/or approximative models to study protein folding. An interesting approach is to unfold starting from the native structure [84–86] but detailed comparison with experiments [47] is mandatory to make sure that the high-temperature sampling does not introduce artifacts. In addition, a number of approaches to enhance sampling of phase space have been introduced [87, 88]. They are based on adaptive umbrella sampling [89], generalized ensembles (e.g., entropic sampling, multicanonical methods, replica exchange methods) [90], modified Hamiltonians [91–93], multiple time steps [94], or combinations thereof.

32.3.1.1 Replica Exchange Molecular Dynamics
Replica exchange is an efficient way to simulate complex systems at low temperature and is the simplest and most general form of simulated tempering [95]. Sugita and Okamoto have been the first to extend the original formulation of replica exchange into an MD-based version (REMD), testing it on the pentapeptide Met-

enkephalin in vacuo [96]. The basic idea of REMD is to simulate different copies (replicas) of the system at the same time but at different temperatures values.

We recently applied a REMD protocol to implicit solvent simulations of a 20-residue three-stranded antiparallel β-sheet peptide (beta3s) [97]. Each replica evolves independently by MD and every 1000 MD steps (2 ps), states i, j with neighbor temperatures are swapped (by velocity rescaling) with a probability $w_{ij} = \exp(-\Delta)$ [96], where $\Delta \equiv (\beta_i - \beta_j)(E_j - E_i)$, $\beta = 1/kT$ and E is the potential energy. During the 1000 MD steps the Berendsen thermostat [98] is used to keep the temperature close to a given value. This rather tight coupling and the length of each MD segment (2 ps) allow the kinetic and potential energy of the system to relax. High temperature simulation segments facilitate the crossing of the energy barriers while the low-temperature ones explore in detail the conformations present in the minimum energy basins. The result of this swapping between different temperatures is that high-temperature replicas help the low-temperature ones to jump across the energy barriers of the system. In the beta3s study eight replicas were used with temperatures between 275 and 465 K [97].

The higher the number of degrees of freedom in the system the more replicas should be used. It is not clear how many replicas should be used if a peptide or protein is simulated with explicit water. The transition probability between two temperatures depends on the overlap of the energy histograms. The histograms' width depends on $1/\sqrt{N}$ (where N is the size of the system). Hence, the number of replicas required to cover a given temperature range increases with the size. Moreover, in order to have a random walk in temperature space (and then a random walk in energy space which enhances the sampling), all the temperature exchanges should occur with the same probability. This probability should be at least of 20–30%. To optimize the efficiency of the method, one should find the best compromise between the number of replicas to be used, the temperature space to cover and the acceptance ratios for temperature exchanges. In the literature there is no clear indication about the selection of temperatures and empirical methods are usually applied (weak point of the method). The choice of the boundary temperatures depends on the system under study. The highest temperature has to be chosen in order to overcome the highest energy barriers (probably higher in explicit water) separating different basins; the lowest temperature to investigate the details of the different basins.

Sanbonmatsu and Garcia have applied REMD to investigate the structure of Met-enkephalin in explicit water [99] and the α-helical stabilization by the arginine side chain which was found to originate from the shielding of main-chain hydrogen bonds [100]. Furthermore, the energy landscape of the C-terminal β-hairpin of protein G in explicit water has been investigated by REMD [101, 102]. Recently, a multiplexed approach with multiple replicas for each temperature level has been applied to large-scale distributed computing of the folding of a 23-residue miniprotein [103]. Starting from a completely extended chain, conformations close to the NMR structures were reached in about 100 trajectories (out of a total of 4000) but no evidence of reversible folding (i.e., several folding and unfolding events in the same trajectory) was presented [103].

32.3.1.2 Methods Based on Path Sampling

A very promising computational method, called transition path sampling (reviewed in Ref. [104]) has been recently used [105] to study the folding of a β-hairpin in explicit solvent. The method allows in principle the study of rare events (such as protein folding) without requiring knowledge of the mechanisms, reaction coordinates, and transition states. Transition path sampling focuses on the sampling not of conformations but of trajectories linking two conformations or regions (possibly basins of attraction) in the conformational space. Other methods focus on building ensemble of paths connecting states; the stochastic path approach [106] and the reaction path method [107] have been also used to study the folding of peptides and small proteins in explicit solvent. The stochastic path ensemble and the reaction path methods introduce a bias in the computed trajectories but allow the exploration of long time scales. All the methods mentioned above are promising but rely on the choice of a somewhat arbitrary initial unfolded conformation beside the final native one.

32.3.2 MD with Restraints

A method to generate structures belonging to the TSE ensemble discussed in Section 32.2.3 consists in performing molecular dynamics simulations restrained with a pseudo-energy function based on the set of experimental ϕ-values. The ϕ-values are interpreted as the fraction of native contacts present in the structures that contribute to the TSE. With this restraint the TSE becomes the most stable state on the potential energy surface rather than being an unstable region, as it is for the true energy function of the protein. This procedure is conceptually related to that used to generate native state structures compatible with measurements from nuclear magnetic resonance (NMR) experiments, in that pseudo-energy terms involving experimental restraints are added to the protein force field [108, 109]. The main difference is that an approach is required to sample a broad state compatible with some experimental restraints, rather than a method to search for an essentially unique native structure.

The method is based on molecular dynamics simulations using an all-atom model of the protein [110, 111] and an implicit model for the solvent [112] with an additional term in the energy function:

$$\rho = \frac{1}{N_\phi} \sum_{i \in E} (\phi_i - \phi_i^{\exp})^2 \tag{1}$$

where E is the list of the N_ϕ available experimental ϕ-values, $\phi_i^{\exp}$. The ϕ_i-value of amino acid i in the conformation at time t is defined as

$$\phi_i(t) = \frac{N_i(t)}{N_i^{\text{nat}}} \tag{2}$$

where $N_i(t)$ is the number of native contacts of i at time t and N_i^{nat} the number of native contacts of i in the native state.

Molecular dynamics simulations are then performed to sample all the possible structures compatible with the restraints. The structures thus generated are not necessarily at the transition state for folding for the potential used. They provide instead a structural model of the experimental transition state, including all possible structures compatible with the restraints derived from the experiment. The experimental information provided by the ϕ-values might not be enough to restrain the sampling to meaningful structures (e.g., when only few mutations have been performed). In such circumstance, other experimentally measured quantities, such as the m-value, which is related to the solvent accessible surface, must be used to restrain the sampling or to a posteriori select meaningful structures.

This type of computational approach relies on the assumption implicit in Eq. (2). This consists in approximating a ϕ-value, measured as a ratio of free energy variations upon mutation, as a ratio of side-chain contacts. A definition based on side-chains is appropriate since experimental ϕ-values are primarily a measure of the loss of side-chain contacts at the transition state, relative to the native state. Although simply counting contacts, rather than calculating their energies, is a crude approximation [113], it has been shown that there is a good correlation between loss of stability and loss of side-chain contacts within about 6 Å on mutation [114]. Also, Shea et al. [115] have found in their model calculations that this approximation for estimating ϕ-values from structures is a good one under certain conditions. A more detailed relation between experimental ϕ-values and atomic contacts could in principle be established by using the energies of the all-atom contacts made by the side chain of the mutated amino acid.

The same approach can be extended to generate the structures corresponding to other unfolded or intermediate states as the site-specific information provided by the experiment is steadily increasing (see Chapters 20 and 21).

32.3.3
Distributed Computing Approach

As mentioned in the introduction, the problem of simulating the folding process of any sequence from a random conformation is mainly a problem of potentials and computer time. Duan and Kollman [116] have showed that a huge effort in parallelizing (on a medium-scale, 256 processors) an MD code and exploiting for several months a several million dollars computer (a Cray T3E) could lead to the simulation of 1 µs of the small protein villin headpiece. Even approaching the typical experimental folding times (which is, however, larger than 1 ms for most proteins), a statistical characterization of the folding process is still impossible in the foreseeable future.

Developing a large-scale parallelization method seems the most viable approach, as the cost of fast CPUs decreases steadily and their performances approach those of much more expensive mainframes. Time being sequential, MD codes are not massively parallelizable in an efficient way. A good scaling is usually obtained for large systems with explicit water and a relatively small number of processors (between 2 and 100, depending on the program and the problem studied). One approach has been proposed that allows the scalability of a MD simulation to be pushed to the level of being able to use efficiently a network of heterogeneous and loosely connected computer [117]. The approach (called distributed computing) exploits the stochastic nature of the folding process. In general protein folding involves the crossing of free energy barriers. The approach is most easily understood assuming that the proteins have a single barrier and a single exponential kinetic (which is the case for a large number of small proteins [118]). The probability that a protein is folded after a time t is $P(t) = 1 - \exp(-kt)$, where k is the folding rate. Thus, for short times, and considering M proteins or independent simulations, the probability of observing a folding event is Mkt. So, if M is large, there is a sizable probability of observing a folding event on simulations much shorter than the time constant of the folding process [119]. The folding rate could then in principle be estimated by running M independent simulations (starting from the completely extended conformation with different random velocities) for a time t and counting the number N of simulations which end up in the folded state as $k = N/(Mt)$. Simulations have been reported where the folding rate estimated in this way (assuming that partial refolding counts as folding) is in good agreement with the experimental one (see, for example, Ref. [39]).

However, it has been argued [120] that even for simple two-state proteins, folding has a series of early conformational steps that lead to lag phases at the beginning of the kinetics. Their presence can bias short simulations toward selecting minor pathways that have fewer or faster lag steps and so miss the major folding pathways. This fact has been clearly observed by comparing equilibrium and fast folding trajectories simulations [121] for a 20-residue three-stranded antiparallel β-sheet peptide (beta3s). It was found that the folding rate is estimated correctly by the distributed computing approach when trajectories longer than a fraction of the equilibrium folding time are considered; in the case of the 20-residue peptide studied within the frictionless implicit solvation model used for the simulations, this time is about 1% of the average folding time at equilibrium. However, careful analysis of the folding trajectories showed that the fastest folding events occur through high-energy pathways, which are unlikely under equilibrium conditions (see Section 32.2.1.1). Along these very fast folding pathways the peptide does not relax within the equilibrium denatured state which is stabilized by the transient presence of both native and nonnative interactions. Instead, collapse and formation of native interactions coincides and, unlike at equilibrium, the formation of the two β-hairpins is nearly simultaneous.

These results demonstrate that the ability to predict the folding rate does not imply that the folding mechanisms are correctly characterized: the fast folding events occur through a pathway that is very unlikely at equilibrium. However, extending the time scale of the short simulations to 10% of the equilibrium folding time, the folding mechanism of the fast folding events becomes almost indistinguishable from equilibrium folding events. It must be stressed that this result is not general but concerns the specific peptide studied; the explicit presence of solvent molecule (and the consequent friction), might decrease the differences between equilibrium and shortest folding events. Unfortunately, this kind of validation of the distributed computing approach is not possible for a generic protein in a realistic solvent, as equilibrium simulations are not feasible.

An alternative method to use many processors simultaneously to access time scales relevant in the folding process by MD simulations has recently been proposed by Settanni et al. [122]. The method is based on parallel MD simulations that are started from the denatured state; trajectories are periodically interrupted, and are restarted only if they approach the transition (or some other target) state. In other words, the method choses trajectories along which a cost function decreases. The effectiveness of such an approach was shown by determining the transition state for folding an SH3 domain using as cost function the deviation between experimental and computed ϕ-values (Eq. (1) in Section 32.3.2). The method can efficiently use a large number of computers simultaneously because simulations are loosely coupled (i.e., only the comparison between final conformations, needed periodically to choose which trajectory to restart, involve communications between CPUs). This method can also be extended to complex nondifferentiable cost functions.

32.3.4
Implicit Solvent Models versus Explicit Water

Incorporating solvent effects in MD and Monte-Carlo simulations is of key importance in quantitatively understanding the chemical and physical properties of biomolecular processes. Accurate electrostatic energies of proteins in an aqueous environment are needed in order to discriminate between native and nonnative conformations. An exact evaluation of electrostatic energies considers the interactions among all possible solute–solute, solute–solvent, and solvent–solvent pairs of charges. However, this is computationally expensive for macromolecules. Continuum dielectric approximations offer a more tractable approach [123–127]. The essential concept in continuum models is to represent the solvent by a high dielectric medium, which eliminates the solvent degrees of freedom, and to describe the macromolecule as a region with a low dielectric constant and a spatial charge distribution. The Poisson equation provides an exact description of such a system. The increase in computation speed for a finite difference solution of the Poisson equation [128–131] with respect to an explicit treatment of the solvent is remarkable but still not enough for effective utilization in computer simulations of macromolecules. The generalized Born (GB) model was introduced to facilitate an efficient evaluation of continuum electrostatic energies [42]. It provides accurate energetics and the most efficient implementations are between five and ten times slower than in vacuo simulations [132–134]. The essential element of the GB approach is the calculation of an effective Born radius for each atom in the system which is a measure of how deeply the atom is buried inside the protein. This information is combined in a heuristic way to obtain a correction to the Coulomb law for each atom pair [42]. For the integration of energy density, necessary to obtain the effective Born radii, both numerical [42, 132, 135] and analytical [134, 136, 137] implementations exist. The former are more accurate but slower than the latter [135]. Moreover, analytical derivatives that are required for MD simulations are not given by numerical implementations.

For efficiency reasons empirical dielectric screening functions are the most common choice in MD simulations with implicit solvent. One kind of solvation model is based on the use of a dielectric function that depends linearly on the distance r between two charges ($\varepsilon(r) = \alpha r$) [138, 139] or has a sigmoidal shape [140, 141]. Although very fast, these options suffer from their inability to discriminate between buried and solvent exposed regions of a macromolecule and are therefore rather inaccurate. A distance and exposure dependent dielectric function has been proposed [142]. Recently, an approach based on the distribution of solute atomic volumes around pairs of charges in a macromolecule has been proposed to calculate the effective dielectric function of proteins in aqueous solution [143].

The simulation results presented in Section 32.2.1 were obtained using an implicit solvent model based on a fast analytical approximation of the solvent accessible surface (SAS) [13] and the CHARMM force field [110]. The former drastically reduces the computational cost with respect to an explicit solvent simulation. The SAS model is based on the approximation proposed by Lazaridis and Karplus [112] for dielectric shielding due to the solvent, and the surface area model for the hydrophobic effect introduced by Eisenberg and McLachlan [144]. Electrostatic screening effects are approximated by a distance-dependent dielectric function and a set of partial charges with neutralized ionic groups [112]. An approximate analytical expression [145] is employed to calculate the SAS because an exact analytical or numerical computation of the SAS is too slow to compete with simulations in explicit solvent. The SAS model is based on the assumptions that most of the solvation energy arises from the first water shell around the protein [144] and that two atomic solvation parameters are sufficient to describe these effects at a qualitative level of accuracy. Within these assumptions, the SAS energy term approximates the solute–solvent interactions (i.e., it should account for the energy of cavity formation, solute–solvent dispersion interactions, and the direct (or Born) solvation of polar groups). The two atomic solvation parameters were optimized by performing 1 ns MD simulations at 300 K on six small proteins [13]. It is important to underline that the structured peptides discussed in Section 32.2.1 were not used for the calibration of the SAS atomic solvation parameters. The SAS model is a good approximation for investigating the folded and denatured state (large ensemble of conformers) of structured peptides. Its limitations, in particular for highly charged peptides and large proteins, have been discussed [13].

The most detailed and physically sound approaches (e.g., explicit solvent and particle mesh Ewald treatment of the long-range electrostatic interactions [146]) are still approximations and might introduces artifacts (see, for example, Ref. [147]). All solvation models, even those computationally most expensive, are approximations and their range of validity is difficult to explore. It is likely that most proteins will unfold fast relative to the experimental time scale if one could afford long (e.g., 100 ns) explicit water MD simulations even at room temperature. Some evidence of this instability has been recently published [148].

32.4
Conclusion

It is a very exciting time for studying protein folding using multidisciplinary approaches rooted in physics, chemistry, and computer science. The time scale gap between folding in vitro and in silico is being continuously reduced and this will bring interesting surprises. We expect an increasing role of MD simulations in the elucidation of protein folding thanks to further improvements in force fields and solvation models.

References

1 KARPLUS, M. & MCCAMMON, J. A. (2002). Molecular dynamics simulations of biomolecules. *Nature Struct. Biol.* **9**, 646–652.

2 DILL, K. A. & CHAN, H. S. (1997). From Levinthal to pathways to funnels. *Nature Struct. Biol.* **4**, 10–19.

3 MIRNY, L. & SHAKHNOVICH, E. (2001). Protein folding theory: From lattice to all-atom models. *Annu. Rev. Biophys. Biomol. Struct.* **30**, 361–396.

4 DAGGETT, V. & FERSHT, A. R. (2003). Is there a unifying mechanism for protein folding? *Trends Biochem. Sci.* **28**, 18–25.

5 CREIGHTON, T. E. (1992). *Protein Folding.* W. H. Freeman & Co., New York.

6 MERZ JR, K. M. & LEGRAND, S. M. (1994). *The Protein Folding Problem and Tertiary Structure Prediction.* Birkhäuser, Boston.

7 PAIN, R. H., ed. (2000). *Mechanisms of Protein Folding.* Oxford University Press, Oxford.

8 SHEA, J. E. & BROOKS III, C. L. (2001). From folding theories to folding proteins: A review and assessment of simulation studies of protein folding and unfolding. *Annu. Rev. Phys. Chem.* **52**, 499–535.

9 GALZITSKAYA, O. V., HIGO, J. & FINKELSTEIN, A. V. (2002). α-helix and β-hairpin folding from experiment, analytical theory and molecular dynamics simulations. *Curr. Protein Pept. Sci.* **3**, 191–200.

10 GNANAKARAN, S., NYMEYER, H., PORTMAN, J., SANBONMATSU, K. Y. & GARCIA, A. E. (2003). Peptide folding simulations. *Curr. Opin. Struct. Biol.* **13**, 168–174.

11 FERRARA, P. & CAFLISCH, A. (2000). Folding simulations of a three-stranded antiparallel β-sheet peptide. *Proc. Natl Acad. Sci. USA* **97**, 10780–10785.

12 FERRARA, P. & CAFLISCH, A. (2001). Native topology or specific interactions: What is more important for protein folding? *J. Mol. Biol.* **306**, 837–850.

13 FERRARA, P., APOSTOLAKIS, J. & CAFLISCH, A. (2002). Evaluation of a fast implicit solvent model for molecular dynamics simulations. *Proteins* **46**, 24–33.

14 DE ALBA, E., SANTORO, J., RICO, M. & JIMENEZ, M. A. (1999). De novo design of a monomeric three-stranded antiparallel β-sheet. *Protein Sci.* **8**, 854–865.

15 SCHENCK, H. L. & GELLMAN, S. H. (1998). Use of a designed triple-stranded antiparallel β-sheet to probe β-sheet cooperativity in aqueous solution. *J. Am. Chem. Soc.* **120**, 4869–4870.

16 MCCALLISTER, E. L., ALM, E. & BAKER, D. (2000). Critical role of β-hairpin formation in protein G folding. *Nature Struct. Biol.* **7**, 669–673.

17 WRIGHT, C. F., LINDORFF-LARSEN, K., RANDLES, L. G. & CLARKE, J. (2003). Parallel protein-unfolding pathways

revealed and mapped. *Nature Struct. Biol.* **10**, 658–662.

18 RICHARDSON, J. M., MCMAHON, K. W., MACDONALD, C. C. & MAKHATADZE, G. I. (1999). MEARA sequence repeat of human CstF-64 polyadenylation factor is helical in solution. A spectroscopic and calorimetric study. *Biochemistry* **38**, 12869–12875.

19 SHIRLEY, W. A. & BROOKS III, C. L. (1997). Curious structure in "canonical" alanine-based peptides. *Proteins* **28**, 59–71.

20 KARPLUS, M. (2000). Aspects of protein reaction dynamics: Deviations from simple behavior. *J. Phys. Chem. B* **104**, 11–27.

21 FERRARA, P., APOSTOLAKIS, J. & CAFLISCH, A. (2000). Thermodynamics and kinetics of folding of two model peptides investigated by molecular dynamics simulations. *J. Phys. Chem. B* **104**, 5000–5010.

22 CLARKE, D. T., DOIG, A. J., STAPLEY, B. J. & JONES, G. R. (1999). The α-helix folds on the millisecond time scale. *Proc. Natl Acad. Sci. USA* **96**, 7232–7237.

23 LEDNEV, I. K., KARNOUP, A. S., SPARROW, M. C. & ASHER, S. A. (1999). α-Helix peptide folding and unfolding activation barriers: A nanosecond UV resonance raman study. *J. Am. Chem. Soc.* **121**, 8074–8086.

24 MUÑOZ, V., THOMPSON, P. A., HOFRICHTER, J. & EATON, W. A. (1997). Folding dynamics and mechanism of β-hairpin formation. *Nature* **390**, 196–199.

25 OLIVEBERG, M., TAN, Y. J. & FERSHT, A. R. (1995). Negative activation enthalpies in the kinetics of protein folding. *Proc. Natl Acad. Sci. USA* **92**, 8926–8929.

26 SEGAWA, S. & SUGIHARA, M. (1984). Characterization of the transition state of lysozyme unfolding. I. Effect. *Biopolymers* **23**, 2473–2488.

27 MATAGNE, A., JAMIN, M., CHUNG, E. W., ROBINSON, C. V., RADFORD, S. E. & DOBSON, C. M. (2000). Thermal unfolding of an intermediate is associated with non-Arrhenius kinetics in the folding of hen lysozyme. *J. Mol. Biol.* **297**, 193–210.

28 KARPLUS, M., CAFLISCH, A., SALI, A. & SHAKHNOVICH, E. (1995). Protein dynamics: From the native to the unfolded state and back again. In *Modelling of Biomolecular Structures and Mechanisms* (PULLMAN, A., JORTNER, J. & PULLMAN, B., eds), pp. 69–84, Kluwer Academic, Dordrecht, The Netherlands.

29 KARPLUS, M. (1997). The Levinthal paradox: Yesterday and today. *Folding Des.* **2**, S69–S75.

30 DOBSON, C. M., SALI, A. & KARPLUS, M. (1998). Protein folding: A perspective from theory and experiment. *Angew. Chem. Int. Ed.* **37**, 868–893.

31 SCALLEY, M. L. & BAKER, D. (1997). Protein folding kinetics exhibit an Arrhenius temperature dependence when corrected for the temperature dependence of protein stability. *Proc. Natl Acad. Sci. USA* **94**, 10636–10640.

32 CHAN, H. S. & DILL, K. A. (1998). Protein folding in the landscape perspective: Chevron plots and non-Arrhenius kinetics. *Proteins* **30**, 2–33.

33 CAVALLI, A., HABERTHÜR, U., PACI, E. & CAFLISCH, A. (2003). Fast protein folding on downhill energy landscape. *Protein Sci.* **12**, 1801–1803.

34 DINNER, A. R., SALI, A., SMITH, L. J., DOBSON, C. M. & KARPLUS, M. (2000). Understanding protein folding via free-energy surfaces from theory and experiment. *Trends Biochem. Sci.* **25**, 331–339.

35 WONG, K. B., CLARKE, J., BOND, C. J. et al. (2000). Towards a complete description of the structural and dynamic properties of the denatured state of barnase and the role of residual structure in folding. *J. Mol. Biol.* **296**, 1257–1282.

36 NEIDIGH, J. W., FESINMEYER, R. M. & ANDERSEN, N. H. (2002). Designing a 20-residue protein. *Nature Struct. Biol.* **9**, 425–430.

37 GELLMAN, S. H. & WOOLFSON, D. N. (2002). Mini-proteins Trp the light

fantastic. *Nature Struct. Biol.* **9**, 408–410.

38 SIMMERLING, C., STROCKBINE, B. & ROITBERG, A. E. (2002). All-atom structure prediction and folding simulations of a stable protein. *J. Am. Chem. Soc.* **124**, 11258–11259.

39 SNOW, C. D., ZAGROVIC, B. & PANDE, V. S. (2002). The Trp cage: Folding kinetics and unfolded state topology via molecular dynamics simulations. *J. Am. Chem. Soc.* **124**, 14548–14549.

40 CHOWDHURY, S., LEE, M. C., XIONG, G. & DUAN, Y. (2003). Ab initio folding simulation of the Trp-cage mini-protein approaches NMR resolution. *J. Mol. Biol.* **327**, 711–717.

41 PITERA, J. W. & SWOPE, W. (2003). Understanding folding and design: replica-exchange simulations of ''Trp-cage'' miniproteins. *Proc. Natl Acad. Sci. USA* **100**, 7587–7592.

42 STILL, W. C., TEMPCZYK, A., HAWLEY, R. C. & HENDRICKSON, T. (1990). Semianalytical treatment of solvation for molecular mechanics and dynamics. *J. Am. Chem. Soc.* **112**, 6127–6129.

43 DAGGETT, V. & LEVITT, M. (1993). Protein unfolding pathways explored through molecular dynamics simulations. *J. Mol. Biol.* **232**, 600–619.

44 CAFLISCH, A. & KARPLUS, M. (1994). Molecular dynamics simulation of protein denaturation: Solvation of the hydrophobic cores and secondary structure of barnase. *Proc. Natl Acad. Sci. USA* **91**, 1746–1750.

45 DAGGETT, V. & FERSHT, A. (2003). Opinion: The present view of the mechanism of protein folding. *Nature Rev. Mol. Cell Biol.* **4**, 497–502.

46 MAYOR, U., JOHNSON, C. M., DAGGETT, V. & FERSHT, A. R. (2000). Protein folding and unfolding in microseconds to nanoseconds by experiment and simulation. *Proc. Natl Acad. Sci. USA* **97**, 13518–13522.

47 MAYOR, U., GUYDOSH, N. R., JOHNSON, C. M. et al. (2003). The complete folding pathway of a protein from nanoseconds to microseconds. *Nature* **421**, 863–867.

48 CAFLISCH, A. & KARPLUS, M. (1999). Structural details of urea binding to barnase: A molecular dynamics analysis. *Structure* **7**, 477–488.

49 HAO, M.-H., PINCUS, M. R., RACHOVSKY, S. & SCHERAGA, H. A. (1993). Unfolding and refolding of the native structure of bovine pancreatic trypsin inhibitor studied by computer simulations. *Biochemistry* **32**, 9614–9631.

50 HARVEY, S. C. & GABB, H. A. (1993). Conformational transition using molecular dynamics with minimum biasing. *Biopolymers* **33**, 1167–1172.

51 SCHLITTER, J., ENGELS, M., KRUGER, P., JACOBY, E. & WOLLMER, A. (1993). Targeted molecular dynamics simulation of conformational change. Application to the TR transition in insulin. *Mol. Simulations* **10**, 291–308.

52 HÜNENBERGER, P. H., MARK, A. E. & VAN GUNSTEREN, W. F. (1995). Computational approaches to study protein unfolding: Hen egg white lysozyme as a case study. *Proteins* **21**, 196–213.

53 FERRARA, P., APOSTOLAKIS, J. & CAFLISCH, A. (2000). Computer simulations of protein folding by targeted molecular dynamics. *Proteins* **39**, 252–260.

54 FERRARA, P., APOSTOLAKIS, J. & CAFLISCH, A. (2000). Targeted molecular dynamics simulations of protein unfolding. *J. Phys. Chem. B* **104**, 4511–4518.

55 PACI, E., SMITH, L. J., DOBSON, C. M. & KARPLUS, M. (2001). Exploration of partially unfolded states of human α-lactalbumin by molecular dynamics simulation. *J. Mol. Biol.* **306**, 329–347.

56 MARCHI, M. & BALLONE, P. (1999). Adiabatic bias mlecular dynamics: A method to navigate the conformational space of complex molecular systems. *J. Chem. Phys.* **110**, 3697–3702.

57 WILKINS, D. K., GRIMSHAW, S. B., RECEVEUR, V., DOBSON, C. M., JONES, J. A. & SMITH, L. J. (1999). Hydrodynamic radii of native and denatured proteins measured by pulse NMR techniques. *Biochemistry* **38**, 16424–16431.

58 KUWAJIMA, K. (1996). The molten globule state of α-lactalbumin. *FASEB J.* **10**, 102–109.

59 TROULLIER, A., REINSTÄDLER, D., DUPONT, Y., NAUMANN, D. & FORGE, V. (2000). Transient nonnative secondary structures during the refolding of α-lactalbumin by infrared spectroscopy. *Nature Struct. Biol.* **7**, 78–86.

60 SCHULMAN, B., KIM, P. S., DOBSON, C. M. & REDFIELD, C. (1997). A residue-specific NMR view of the non-cooperative unfolding of a molten globule. *Nature Struct. Biol.* **4**, 630–634.

61 RIEF, M., GAUTEL, M., OESTERHELT, F., FERNANDEZ, J. M. & GAUB, H. E. (1997). Reversible unfolding of individual titin immunoglobulin domains by AFM. *Science* **276**, 1109–1112.

62 LU, H., ISRALEWITZ, B., KRAMMER, A., VOGEL, V. & SCHULTEN, K. (1998). Unfolding of titin immunoglobulin domains by steered molecular dynamics simulation. *Biophys. J.* **75**, 662–671.

63 PACI, E. & KARPLUS, M. (1999). Forced unfolding of fibronectin type 3 modules: An analysis by biased molecular dynamics simulations. *J. Mol. Biol.* **288**, 441–459.

64 ISRALEWITZ, B., GAO, M. & SCHULTEN, K. (2001). Steered molecular dynamics and mechanical functions of proteins. *Curr. Opin. Struct. Biol.* **11**, 224–230.

65 PACI, E. & KARPLUS, M. (2000). Unfolding proteins by external forces and high temperatures: The importance of topology and energetics. *Proc. Natl Acad. Sci. USA* **97**, 6521–6526.

66 WILLIAMS, P. M., FOWLER, S. B., BEST, R. B. et al. (2003). Hidden complexity in the mechanical properties of titin. *Nature* **422**, 446–449.

67 FOWLER, S., BEST, R. B., TOCA-HERRERA, J. L. et al. (2002). Mechanical unfolding of a titin Ig domain: Structure of unfolding intermediate revealed by combining AFM, molecular dynamics simulations, NMR and protein engineering. *J. Mol. Biol.* **322**, 841–849.

68 BEST, R. B., FOWLER, S., TOCA-HERRERA, J. L., STEWARD, A., PACI, E. & CLARKE, J. (2003). Mechanical unfolding of a titin Ig domain: Structure of transition state revealed by combining atomic force microscopy, protein engineering and molecular dynamics simulations. *J. Mol. Biol.* **330**, 867–877.

69 VENDRUSCOLO, M. & PACI, E. (2003). Protein folding: Bringing theory and experiment closer together. *Curr. Opin. Struct. Biol.* **13**, 82–87.

70 VENDRUSCOLO, M., PACI, E., KARPLUS, M. & DOBSON, C. M. (2003). Structures and relative free energies of partially folded states of proteins. *Proc. Natl Acad. Sci. USA* **100**, 14817–14821.

71 BROCKWELL, D. J., PACI, E., ZINOBER, R. C. et al. (2003). Pulling geometry defines the mechanical resistance of a β-sheet protein. *Nature Struct. Biol.* **10**, 731–737.

72 CARRION-VAZQUEZ, M., LI, H., LU, H., MARSZALEK, P. E., OBERHAUSER, A. F. & FERNANDEZ, J. M. (2003). The mechanical stability of ubiquitin is linkage dependent. *Nature Struct. Biol.* **10**, 738–743.

73 LU, H. & SCHULTEN, K. (2000). The key event in force-induced unfolding of titin's immunoglobulin domains. *Biophys. J.* **79**, 51–65.

74 FERSHT, A. R., MATOUSCHEK, A. & SERRANO, L. (1992). The folding of an enzyme. I. Theory of protein engineering analysis of stability and pathway of protein folding. *J. Mol. Biol.* **224**, 771–782.

75 FERSHT, A. R., ITZHAKI, L. S., ELMASRY, N. F., MATTHEWS, J. M. & OTZEN, D. E. (1994). Single versus parallel pathways of protein folding and fractional structure in the transition state. *Proc. Natl Acad. Sci. USA* **91**, 10426–10429.

76 LI, A. & DAGGETT, V. (1994). Characterization of the transition state of protein unfolding by use of molecular dynamics: Chymotrypsin inhibitor 2. *Proc. Natl Acad. Sci. USA* **91**, 10430–10434.

77 DAGGETT, V. (2002). Molecular dynamics simulations of the protein

unfolding/folding reaction. *Acc. Chem. Res.* **35**, 422–429.

78 GSPONER, J. & CAFLISCH, A. (2002). Molecular dynamics simulations of protein folding from the transition state. *Proc. Natl Acad. Sci. USA* **99**, 6719–6724.

79 DEJONG, D., RILEY, R., ALONSO, D. O. & DAGGETT, V. (2002). Probing the energy landscape of protein folding/unfolding transition states. *J. Mol. Biol.* **319**, 229–242.

80 FERSHT, A. R. (1999). *Structure and Mechanism in Protein Science: A Guide to Enzyme Catalysis and Protein Folding.* W. H. Freeman & Co., New York.

81 VENDRUSCOLO, M., PACI, E., DOBSON, C. M. & KARPLUS, M. (2001). Three key residues form a critical contact network in a protein folding transition state. *Nature* **409**, 641–645.

82 PACI, E., VENDRUSCOLO, M., DOBSON, C. M. & KARPLUS, M. (2002). Determination of a transition state at atomic resolution from protein engineering data. *J. Mol. Biol.* **324**, 151–163.

83 PACI, E., CLARKE, J., STEWARD, A., VENDRUSCOLO, M. & KARPLUS, M. (2003). Self-consistent determination of the transition state for protein folding. Application to a fibronectin type III domain. *Proc. Natl Acad. Sci. USA* **100**, 394–399.

84 DAGGETT, V. & LEVITT, M. (1993). Realistic simulations of native-protein dynamics in solution and beyond. *Annu. Rev. Biophys. Biomol. Struct.* **22**, 353–380.

85 CAFLISCH, A. & KARPLUS, M. (1995). Acid and thermal denaturation of barnase investigated by molecular dynamics simulations. *J. Mol. Biol.* **252**, 672–708.

86 LAZARIDIS, T. & KARPLUS, M. (1997). "New View" of protein folding reconciled with the old through multiple unfolding simulations. *Science* **278**, 1928–1931.

87 FRENKEL, D. & SMIT, B. (1996). *Understanding Molecular Simulation*, 2nd edition, Academic Press, London.

88 BERNE, B. J. & STRAUB, J. E. (1997). Novel methods of sampling phase space in the simulation of biological systems. *Curr. Opin. Struct. Biol.* **7**, 181–189.

89 BARTELS, C. & KARPLUS, M. (1997). Multidimensional adaptive umbrella sampling: Applications to main chain and side chain peptide conformations. *J. Comput. Chem.* **18**, 140–1462.

90 MITSUTAKE, A., SUGITA, Y. & OKAMOTO, Y. (2001). Generalized-ensemble algorithms for molecular simulations of biopolymers. *Biopolymers* **60**, 96–123.

91 WU, X. & WANG, S. (1998). Self-guided molecular dynamics simulation for efficient conformational search. *J. Phys. Chem. B* **102**, 7238–7250.

92 APOSTOLAKIS, J., FERRARA, P. & CAFLISCH, A. (1999). Calculation of conformational transitions and barriers in solvated systems: Application to the alanine dipeptide in water. *J. Chem. Phys.* **110**, 2099–2108.

93 ANDRICIOAEI, I., DINNER, A. R. & KARPLUS, M. (2003). Self-guided enhanced sampling methods for thermodynamic averages. *J. Chem. Phys.* **118**, 1074–1084.

94 SCHLICK, T., BARTH, E. & MANDZIUK, M. (1997). Biomolecular dynamics at long timesteps: bridging the time-scale gap between simulation and experimentation. *Annu. Rev. Biophys. Biomol. Struct.* **26**, 181–222.

95 MARINARI, E. & PARISI, G. (1992). Simulated tempering: A new Monte Carlo scheme. *Europhys. Lett.* **19**, 451–458.

96 SUGITA, Y. & OKAMOTO, Y. (1999). Replica-exchange molecular dynamics method for protein folding. *Chem. Phys. Lett.* **314**, 141–151.

97 RAO, F. & CAFLISCH, A. (2003). Replica exchange molecular dynamics simulations of reversible folding. *J. Chem. Phys.* **119**, 4035–4042.

98 BERENDSEN, H. J. C., POSTMA, J. P. M., VAN GUNSTEREN, W. F., DINOLA, A. & HAAK, J. R. (1984). Molecular dynamics with coupling to an external bath. *J. Chem. Phys.* **81**, 3684–3690.

99 SANBONMATSU, K. Y. & GARCIA, A. E. (2002). Structure of Met-enkephalin in

explicit aqueous solution using replica exchange molecular dynamics. *Proteins* **46**, 225–234.

100 GARCIA, A. E. & SANBONMATSU, K. Y. (2002). Alpha-helical stabilization by side chain shielding of backbone hydrogen bonds. *Proc. Natl Acad. Sci. USA* **99**, 2782–2787.

101 GARCÍA, A. E. & SANBONMATSU, K. Y. (2001). Exploring the energy landscape of a hairpin in explicit solvent. *Proteins* **42**, 345–354.

102 ZHOU, R., BERNE, B. J. & GERMAIN, R. (2001). The free energy landscape for β hairpin folding in explicit water. *Proc. Natl Acad. Sci. USA* **98**, 14931–14936.

103 RHEE, Y. M. & PANDE, V. S. (2003). Multiplexed-replica exchange molecular dynamics method for protein folding simulation. *Biophys. J.* **84**, 775–786.

104 BOLHUIS, P. G., CHANDLER, D., DELLAGO, C. & GEISSLER, P. L. (2002). Transition path sampling: Throwing ropes over rough mountain passes, in the dark. *Annu. Rev. Phys. Chem.* **53**, 291–318.

105 BOLHUIS, P. G. (2003). Transition-path sampling of β-hairpin folding. *Proc. Natl Acad. Sci. USA* **100**, 12129–12134.

106 ELBER, R., MELLER, J. & OLENDER, R. (1999). Stochastic path approach to compute atomically detailed trajectories: Application to the folding of C peptide. *J. Phys. Chem. B*, **103**, 899–911.

107 EASTMAN, P., GRONBECH-JENSEN, N. & DONIACH, S. (2001). Simulation of protein folding by reaction path annealing. *J. Chem. Phys.* **114**, 3823–3841.

108 WÜTHRICH, K. (1989). Protein structure determination in solution by nuclear magnetic resonance spectroscopy. *Science* **243**, 45–50.

109 CLORE, G. M. & SCHWIETERS, C. D. (2002). Theoretical and computational advances in biomolecular NMR spectroscopy. *Curr. Opin. Struct. Biol.* **12**, 146–153.

110 BROOKS, B. R., BRUCCOLERI, R. E., OLAFSON, B. D., STATES, D. J., SWAMINATHAN, S. & KARPLUS, M. (1983). CHARMM: A program for macromolecular energy, minimization and dynamics calculations. *J. Comput. Chem.* **4**, 187–217.

111 NERIA, E., FISCHER, S. & KARPLUS, M. (1996). Simulation of activation free energies in molecular dynamics system. *J. Chem. Phys.* **105**, 1902–1921.

112 LAZARIDIS, T. & KARPLUS, M. (1999). Effective energy function for protein dynamics and thermodynamics. *Proteins* **35**, 133–152.

113 PACI, E., VENDRUSCOLO, M. & KARPLUS, M. (2002). Native and nonnative interactions along protein folding and unfolding pathways. *Proteins* **47**, 379–392.

114 COTA, E., HAMILL, S. J., FOWLER, S. B. & CLARKE, J. (2000). Two proteins with the same structure respond very differently to mutation: The role of plasticity in protein stability. *J. Mol. Biol.* **302**, 713–725.

115 SHEA, J.-E., ONUCHIC, J. N. & BROOKS III, C. L. (1999). Exploring the origins of topological frustration: Design of a minimally frustrated model of fragment B of protein A. *Proc. Natl Acad. Sci. USA* **96**, 12512–12517.

116 DUAN, Y. & KOLLMAN, P. A. (1998). Pathways to a protein folding intermediate observed in a 1-microsecond simulation in aqueous solution. *Science* **282**, 740–744.

117 SHIRTS, M. & PANDE, V. (2000). COMPUTING: Screen savers of the world unite! *Science* **290**, 1903–1904.

118 JACKSON, S. E. (1998). How do small single-domain proteins fold? *Folding Des.* **3**, R81–R91.

119 PANDE, V. S., BAKER, I., CHAPMAN, J. et al. (2003). Atomistic protein folding simulations on the submillisecond time scale using worldwide distributed computing. *Biopolymers* **68**, 91–109.

120 FERSHT, A. R. (2002). On the simulation of protein folding by short time scale molecular dynamics and distributed computing. *Proc. Natl Acad. Sci. USA* **99**, 14122–14125.

121 PACI, E., CAVALLI, A., VENDRUSCOLO, M. & CAFLISCH, A. (2003). Analysis of

122 SETTANNI, G., GSPONER, J. & CAFLISCH, A. (2004). Formation of the folding nucleus of an SH3 domain investigated by loosely coupled molecular dynamics simulations. *Biophys. J.* **86**, 1691–1701.

the distributed computing approach applied to the folding of a small β peptide. *Proc. Natl Acad. Sci. USA* **100**, 8217–8222.

123 ROUX, B. & SIMONSON, T. (1999). Implicit solvent models. *Biophys. Chem.* **78**, 1–20.

124 GILSON, M. K. (1995). Theory of electrostatic interactions in macromolecules. *Curr. Opin. Struct. Biol.* **5**, 216–223.

125 TOMASI, J. & PERSICO, M. (1994). Molecular interactions in solution: An overview of methods based on continuous distributions of the solvent. *Chem. Rev.* **94**, 2027–2094.

126 CRAMER, C. J. & TRUHLAR, D. G. (1999). Implicit solvation models: Equilibria, structure, spectra, and dynamics. *Chem. Rev.* **99**, 2161–2200.

127 OROZCO, M. & LUQUE, F. J. (2000). Theoretical methods for the description of the solvent effect in biomolecular systems. *Chem. Rev.* **100**, 4187–4226.

128 WARWICKER, J. & WATSON, H. C. (1982). Calculation of the electric potential in the active site cleft due to α-helix dipoles. *J. Mol. Biol.* **157**, 671–679.

129 GILSON, M. K. & HONIG, B. H. (1988). Energetics of charge-charge interactions in proteins. *Proteins* **3**, 32–52.

130 BASHFORD, D. & KARPLUS, M. (1990). pKa's of ionizable groups in proteins: Atomic detail from a continuum electrostatic model. *Biochemistry* **29**, 10219–10225.

131 DAVIS, M. E., MADURA, J. D., LUTY, B. A. & MCCAMMON, J. A. (1991). Electrostatics and diffusion of molecules in solution – simulations with the University-of-Houston-brownian dynamics program. *Comput. Phys. Comm.* **62**, 187–197.

132 SCARSI, M., APOSTOLAKIS, J. & CAFLISCH, A. (1997). Continuum electrostatic energies of macromolecules in aqueous solutions. *J. Phys. Chem. B* **101**, 8098–8106.

133 BASHFORD, D. & CASE, D. A. (2000). Generalized Born models of macromolecular solvation effects. *Annu. Rev. Phys. Chem.* **51**, 129–152.

134 LEE, M. S., FEIG, M., SALSBURY, F. R. & BROOKS III, C. L. (2003). New analytic approximation to the standard molecular volume definition and its application to generalized Born calculations. *J. Comput. Chem.* **24**, 1348–1356.

135 LEE, M. S., SALSBURY, F. R. & BROOKS III, C. L. (2002). Novel generalized Born methods. *J. Chem. Phys.* **116**, 10606–10614.

136 QIU, D., SHENKIN, P. S., HOLLINGER, F. P. & STILL, W. C. (1997). The GB/SA continuum model for solvation. A fast analytical method for the calculation of approximate Born radii. *J. Phys. Chem. A* **101**, 3005–3014.

137 DOMINY, B. N. & BROOKS III, C. L. (1999). Development of a generalized Born model parametrization for proteins and nucleic acids. *J. Phys. Chem. B* **103**, 3765–3773.

138 WARSHEL, A. & LEVITT, M. (1976). Theoretical studies of enzymic reactions: dielectric, electrostatic and steric stabilization of the carbonium ion in the reaction of lysozyme. *J. Mol. Biol.* **103**, 227–249.

139 GELIN, B. R. & KARPLUS, M. (1979). Side-chain torsional potentials: effect of dipeptide, protein, and solvent environment. *Biochemistry* **18**, 1256–1268.

140 MEHLER, E. L. (1990). Comparison of dielectric response models for simulating electrostatic effects in proteins. *Protein Eng.* **3**, 415–417.

141 WANG, L., HINGERTY, B. E., SRINIVASAN, A. R., OLSON, W. K. & BROYDE, S. (2002). Accurate representation of B-DNA double helical structure with implicit solvent and counterions. *Biophys. J.* **83**, 382–406.

142 MALLIK, B., MASUNOV, A. & LAZARIDIS, T. (2002). Distance and exposure dependent effective dielectric

function. *J. Comput. Chem.* **23**, 1090–1099.

143 HABERTHÜR, U., MAJEUX, N., WERNER, P. & CAFLISCH, A. (2003). Efficient evaluation of the effective dielectric function of a macromolecule in aqueous solution. *J. Comput. Chem.* **24**, 1936–1949.

144 EISENBERG, D. & MCLACHLAN, A. D. (1986). Solvation energy in protein folding and binding. *Nature* **319**, 199–203.

145 HASEL, W., HENDRICKSON, T. F. & STILL, W. C. (1988). A rapid approximation to the solvent accessible surface areas of atoms. *Tetrahedron Comput. Methodol.* **1**, 103–116.

146 DARDEN, T. A., YORK, D. M. & PEDERSEN, L. (1993). Particle mesh Ewald: An N log(N) method for computing Ewald sums. *J. Chem. Phys.* **98**, 10089–10092.

147 WEBER, W., HUNENBERGER, P. H. & MCCAMMON, J. A. (2000). Molecular dynamics simulations of a polyalanine octapeptide under Ewald boundary conditions: Influence of artificial periodicity on peptide conformation. *J. Phys. Chem. B* **104**, 3668–3675.

148 FAN, H. & MARK, A. E. (2003). Relative stability of protein structures determined by X-ray crystallography or NMR spectroscopy: a molecular dynamics simulation study. *Proteins* **53**, 111–120.

149 HILTPOLD, A., FERRARA, P., GSPONER, J. & CAFLISCH, A. (2000). Free energy surface of the helical peptide Y(MEARA)6. *J. Phys. Chem. B* **104**, 10080–10086.

150 LINDORFF-LARSEN, K., VENDRUSCOLO, M., PACI, E. & DOBSON, C. M. (2004). Transition states for protein folding have native topologies despite high structural variability. *Nature Struct. Mol. Biol* **11**, 443–449.

33
Molecular Dynamics Simulations of Proteins and Peptides: Problems, Achievements, and Perspectives

Paul Tavan, Heiko Carstens, and Gerald Mathias

33.1
Introduction

When viewed from the standpoint of theoretical physics, proteins are extremely complex materials. They are inhomogeneous, non-isotropic, and exhibit dynamic processes covering many spatial and temporal scales (instead of only one or two), that is, they do not offer any of the nice features that make other materials readily accessible to simplified or coarse-grained descriptions. Therefore, theoretical approaches towards protein and peptide dynamics have to be based on microscopic descriptions. Here, atomistic molecular dynamics (MD) simulations, which are based on molecular mechanics (MM) force fields, currently represent the standard [1]. It should be noted, however, that these MM methods represent a first level of coarse graining, because they try to account for the electronic degrees of freedom mediating the interactions between the atoms by simplified parametric descriptions coded into standard MM force fields (for a recent review see Ponder and Case [2]).

In this article we want to address the question, to what extent can MM-MD simulations currently or in the near future contribute to the understanding of protein folding? Although simulations have been published (for a review see Ref. [3]) that allegedly describe folding processes of peptides or small proteins, there are also serious arguments (e.g., in Ref. [2]) that the available MM force fields may not be accurate enough for qualitatively correct descriptions. This would imply that the published simulations, instead of providing a "virtual reality" of folding processes, actually represent "real artifacts." Thus, it is worthwhile addressing the issue of accuracy.

Because there are many excellent textbooks and review articles on MD simulation methods for liquids [4] and proteins (Ref. [2] and references therein), there is no point in duplicating this material. Instead, we have chosen to outline our own perspective of the issue, which has been and still is guiding our longstanding efforts to develop and improve the required computational methods. We start with a sketch of the basic physics of protein structure and dynamics. This sketch will

Protein Folding Handbook. Part I. Edited by J. Buchner and T. Kiefhaber
Copyright © 2005 WILEY-VCH Verlag GmbH & Co. KGaA, Weinheim
ISBN: 3-527-30784-2

serve to identify key challenges that have to be mastered by any attempt aiming at a theoretical description of these complicated processes.

33.2
Basic Physics of Protein Structure and Dynamics

The molecular machinery of life, designed by Darwinian selection in the process of evolution, essentially consists of proteins. Like all machines, these nano-machines are composed of rigid parts that define a specific three-dimensional structure and flexible parts that enable the particular functional dynamics for which the respective protein has been designed. Because proteins are polymer molecules, "flexibility" here implies a liquid state, like the one found in polymer melts. Therefore, under physiological thermodynamic conditions, proteins are generally found in a mixed state close to the melting transition between the solid and liquid phases.

The solid "secondary" substructures of proteins, for example the α-helices and β-sheets, are shaped by attractive interactions between strong electric dipoles, which are attached to the locally stiff and planar peptide groups within the polypeptide chains (Figure 33.1 illustrates the dipole–dipole interactions within a β-sheet). Here, the specific sequence of amino acid residues selects the respective secondary structures or, alternatively, precludes the formation of such rigid structures by inducing a local melting of the backbone into a flexible loop or random coil. There-

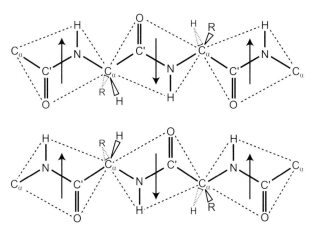

Fig. 33.1. Organization of the dipole–dipole interactions within a β-sheet. As explained by the resonance structures shown in Figure 33.3, the peptide groups form rigid platelets (indicated by the dashed lines) and exhibit strong electric dipole moments (arrows). Such dipoles attract each other, if their arrangement is parallel and axial ($\rightarrow\rightarrow$) or antiparallel and equatorial ($\uparrow\downarrow$). The structural stability of β-sheets is caused by a dipolar organization integrating both arrangements in an optimized fashion. Here, in contrast to α-helices, which exclusively exhibit the ($\rightarrow\rightarrow$)-arrangement, the local dipoles do not add up to a macroscopic dipole.

fore, the specific combination and three-dimensional organization of solid and liquid building blocks into the complex structure and dynamics characteristic for proteins is predetermined by their amino acid sequence.

Considering the thermodynamic state, proteins resemble a mixture of ice and liquid water, which, likewise, is characterized by a competition between enthalpic dipolar binding interactions attempting to generate solid order and temperature-driven entropic unbinding interactions attempting to generate liquid disorder. Nevertheless, there is a distinct difference concerning the complexity of the system: whereas a water–ice mixture homogeneously consists of only one molecular species, proteins are made up of a specific sequence of at least the 20 different residues coded in the genes, not counting all the additional cofactors, which also contribute to the complexity of protein structure and dynamics.

33.2.1
Protein Electrostatics

When attempts are made at a theoretical description, a first difficulty becomes apparent here. The dipole–dipole interactions (commonly called hydrogen bonds), which provide the main contribution to the folding of the polypeptide chains into specific three-dimensional structures, provide binding energies not much larger than the thermal energy $k_B T$ and it is only the combination of many different dipole–dipole interactions that generates the stability of a specific structure. Therefore, the precise description of the many weak electrostatic interactions within a protein is of key importance, since these interactions determine whether particular solid structures are formed or not. This instance is only a specific manifestation of the general fact that electrostatic interactions are dominant in shaping protein structure and dynamics [5]. In contrast, the weaker van der Waals interactions are less important as they are nonspecifically acting between all chemically nonbonded atoms in a similar way.

To obtain a quantitative estimate, consider a hypothetical protein with $N = 100$ residues, whose native conformation lies the typical energy value of $\Delta E = 20$ kJ mol^{-1} below either a compact misfolded conformation or the huge variety of unfolded ones. Then the acceptable error per residue allowing us still to identify the correct conformation is approximately $\Delta E/N^{1/2}$ or 2 kJ mol^{-1} (see Ref. [2] and references therein), where the errors are assumed to be randomly distributed. Systematic errors in the description of the electrostatics will reduce the acceptable error down to $\Delta E/N$ or 0.2 kJ mol^{-1}. Thus, great care has to be taken in the description of intraprotein electrostatics if one wants to generate a "virtual reality." Its quantitatively correct description constitutes the first key problem for theoretical approaches.

33.2.2
Relaxation Times and Spatial Scales

Besides the strong variations of the physicochemical (predominantly electrostatic) properties among the residues, there is a second important difference between a water–ice equilibrium and a protein, which adds to the indicated complexity. In liquid water the molecules are free to move individually, which confines the relaxation time of water in response to an external perturbation (e.g., to a change of the electric field) to the scale of a few picoseconds. Furthermore, in pure water at room temperature spatially ordered structures are found only up to a distance of about 1.5 nm from a given water molecule; beyond this distance water starts to behave like a dielectric continuum [6]. In contrast, in a protein the motions of the molecular components are strongly constrained by the polymeric connectivity and by solid substructures. These constraints lead to relaxation times that are larger than those in liquid water by up to 12 orders of magnitude or more.

Motions that change the conformation of a protein proceed through thermally activated rotations around the covalent single bonds linking the Cα atoms to the adjacent peptide groups of the backbone. At room temperature such rotations take at least a few 10–100 ps, which is why a protein can start to statistically sample its available conformational space only at time scales above 1 ns. Here, the number of atoms involved in a conformational transition defines a spatial scale and the time required for this transition increases rapidly with the size of the spatial scale. Up to now it is not yet clear whether the resulting time scales of conformational transitions represent a well-ordered hierarchy, within which distinct time scales can be associated to specific processes, or a continuum extending from nanoseconds to seconds. Whereas the former alternative would immediately suggest coarse-grained approaches to protein dynamics, the latter would make them difficult to construct. But whichever of these alternative pictures may be true, the vastly different temporal and spatial scales of molecular motions in proteins generate the second key problem for theoretical descriptions by simulation techniques.

33.2.3
Solvent Environment

Up to this point we have looked at the problem of describing the native structure and dynamics of proteins as if they could be understood as isolated objects. Assuming this hypothesis we have identified two key problems with which any attempt of a realistic description is confronted. The first is posed by the high accuracy required for the modeling of the electrostatics within a protein and the second by the huge span of time scales that have to be covered. But the issue is even more complicated, since proteins cannot be isolated from their surroundings.

Proteins acquire their native structure and functional dynamics solely within their physiological environment, which in the case of soluble proteins is characterized by certain ranges of temperature, pressure, pH, and concentrations of ions and of other molecular components within the aqueous solvent. Although these ranges may vary from protein to protein, there are specific physiological ranges for each of them. Therefore, any attempt to provide a quantitative description for the dynamic processes of protein folding and function has to properly include the environment at its physiological thermodynamic state.

33.2.4
Water

The key component of the environment is liquid water, whose adequate representation in computer simulations represents a challenge on its own (see Ref. [7] for a critical review and Ref. [6] for recent results). A water molecule is nearly spherical and very small (approximately the size of an oxygen atom), it carries large electrical dipole and quadrupole moments, and is strongly polarizable by electric fields generated, for example, by other water molecules, ions, or polar groups in its surroundings.

The large dipole moment causes the unusually large dielectric constant of 78, by which liquid water scales down electrostatic interactions. For proteins and peptides this dielectric mean-field property of water is important, because they are usually polyvalent ions and therefore generate a reaction field by polarization of the aqueous environment, which modifies the electrostatic interactions within the protein. The proper description of these very long-ranged effects thus represents the first problem specifically posed by MD simulations of protein–solvent systems.

The sizable quadrupole moment of the water molecule shapes the detailed structures of the hydration shells of solute molecules and, in particular, also those of proteins. If, as suggested by Brooks [8], water molecules actively support secondary structure formation, for example, by bridging nascent β-sheets, then an accurate modeling of the quadrupole moment is required for the understanding of the detailed folding pathways.

The strong electronic polarizability of the water molecules poses yet another challenge. As illustrated by the computational result in Figure 33.2, due to polar-

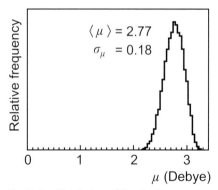

Fig. 33.2. Distribution of the dipole moment of a water molecule dissolved in pure water as obtained from a hybrid MD simulation combining density functional theory (DFT) for the given water molecule with an MM description of the aqueous environment. The DFT description of an isolated water molecule yields a dipole moment of 1.92 D, which is very close to the experimental value of 1.85 D. Thus, the induced dipole moment determined by the calculations is 0.85 D, that is about 40% of the gas-phase value. In liquid water the fluctuations of the dipole moment generated by the electronic polarization are 6.5% as measured by $\sigma_\mu/\langle\mu\rangle$. For technical details and explanations see Ref. [9].

ization the dipole moment increases from a value of 1.85 D for a water molecule isolated in the gas-phase to a value of about 2.6–3.0 D in the liquid phase [7, 9, 10]. Correspondingly the dipole moment of a water molecule will sizably change when it moves from a hydration shell of an ion through bulk water towards a nonpolar surface of a protein.

33.2.5
Polarizability of the Peptide Groups and of Other Protein Components

The polarization-induced changes of the dipole moments of the water molecules sketched above are important in the hydrophobic interactions [11], which are commonly believed to provide a key driving force in protein folding (see Ref. [12] for a discussion). Here, particularly the dipole–induced-dipole interactions, for example, between ordered hydration shells and nonpolar surfaces, are expected to provide large contributions. Therefore, not only the electronic polarizability of the water molecules, but also that of the various protein components has to be included for a quantitative description of the hydrophobic effect.

Furthermore, the electronic polarizability of the components can stabilize local structures in proteins. For instance, about 50% of the binding energy between a positive charge and an aromatic ring is due to polarization (as shown in Ref. [2] for the representative example of potassium bound to benzene). Particularly the polarizability of the peptide groups can be expected to decide about the relative stabilities of the secondary structure motifs in proteins. This claim is proven by the following simple argument:

We have stated further above (and illustrated in Figure 33.1 for the case of a β-sheet) that the dipole–dipole interactions within a protein, particularly those between the peptide groups of the backbone, shape the secondary structures of proteins. The peptide groups, in addition to exhibiting strong dipole moments, are highly polarizable as one can deduce from the resonance structures depicted in Figure 33.3. In response to the electric field generated by the dipole moments of neighboring peptide groups, with which a peptide group interacts in a given secondary structure motif (cf. Figure 33.1), an induced dipole moment will be added to the static dipole moment of the given group, and will correspondingly strengthen or weaken its dipole–dipole interactions (H-bonds). Here, the size and direction of the induced dipole moment will be determined by the spatial distribution and orientation of the neighboring dipoles and partial charges, that is, predominantly by the surrounding secondary structure.

As is also apparent from Figure 33.3, the electronic polarization will concomitantly change the force constants and equilibrium lengths of the chemical bonds (O=C–N–H) within the peptide groups. This effect is apparent in the vibrational spectra of proteins, which are dominated by the normal modes of the peptide groups giving rise to so-called amide bands. The shape of these broad bands is a signature of the differences in polarization experienced by the various peptide groups and, therefore, can be used for an approximate determination of secondary

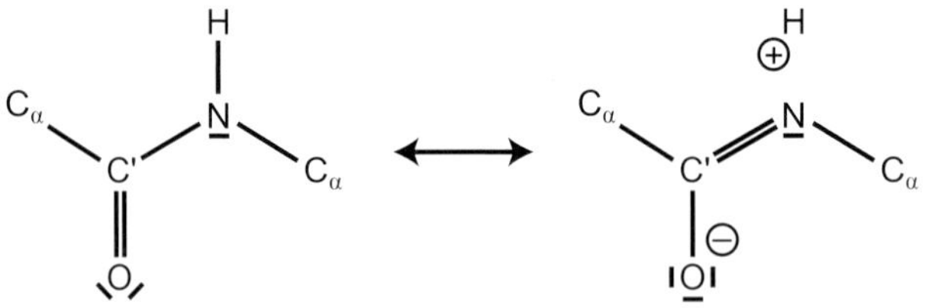

Fig. 33.3. π-electron resonance structures of a peptide group explaining its strong dipole moment and polarizability. An external electric field can modify the relative weights, by which these resonance structures contribute to the electronic wave function. Correspondingly, it will sizably change the dipole moment and the force constants.

structure content (see Ref. [13] for further information). Peaks of the so-called amide-I band, for instance, which belongs to the C=O stretching vibrations of the backbone, are found in the spectral range between 1620 and 1700 cm^{-1}, indicating that the polarization can change the associated C=O force constants by 5%. Conversely, (i) the fact that the large and specific frequency shifts of the amide modes can indicate the secondary structure environment of the corresponding peptide groups, and (ii) the resonance argument in Figure 33.3 uniquely demonstrate that the electronic polarization will entail sizably different strengths ($\geq 5\%$) of the dipolar interactions ("H-bonds") within the various secondary structure motifs. This proves our above conjecture that the electronic polarizability provides a differential contribution to the relative stabilities of secondary structure motifs.

As a result, theoretical descriptions of protein dynamics have to include the electronic polarizability. In addition to this general insight, the above considerations have led us also to quantitative estimates as to how much the polarizability contributes to the electrostatics. We found numbers ranging between 50% (charge on top of an aromatic ring) and 5% (peptide groups in different secondary structure motifs). These contributions systematically and differentially change the electrostatic interactions within protein–solvent systems and do not represent random variations. This instance is obvious if one compares the interaction of a dipolar group with a charged and a nonpolar group, respectively. Furthermore, it has been demonstrated above for the dipolar interactions within secondary structure motifs. Below we will refer to the thus established 5% lower bound of the polarizability contribution to protein electrostatics for purposes of accuracy estimates.

Finally, we would like to add that enzymatic catalysis in proteins, which is a prominent feature of these materials, is essentially an effect of polarizability. Here, a substrate is noncovalently attached in a specific orientation to a binding pocket and a highly structured electric field, which is generated by a particular arrangement of charged, polar and nonpolar residues, exerts a polarizing strain on its charge distribution such that a particular chemical bond either breaks or is formed. This example underlines the decisive contribution of the electronic polarizability to protein structure, dynamics and function.

33.3
State of the Art

By considering the basic physics of protein–solvent systems we have identified key challenges that have to be met by MD-based theoretical descriptions of protein and peptide dynamics. The majority of these problems are posed by (i) the complex electrostatics in such systems and comprise (a) the long range of these interactions, (b) their shielding by the surrounding solvent, and (c) their local variability, which is caused by differential electronic polarization in these inhomogeneous and nonisotropic materials. Further challenges derive from the need (ii) to properly account for the thermodynamic conditions and (iii) to cover the relevant time scales.

During the past 25 years, since the first MD simulation of a protein was reported [14], the true complexities of the challenge have become more and more apparent. By now some of the problems have been solved, whereas others are still awaiting a solution. Quite naïvely, the first MD simulation dealt with a protein embedded in vacuum. However, the importance of the aqueous environment and the correct choice of the thermodynamic conditions soon became apparent [15, 16].

33.3.1
Control of Thermodynamic Conditions

In the theory of liquids and long before the history of protein simulations began, MD simulation systems representing a well-defined thermodynamic ensemble had been set up by filling periodic boxes with MM models of the material of interest (cf. the first simulation of liquid water by Rahman and Stillinger [17]). Whereas the temperature (or internal energy) is trivially accessible in any MD approach, up to now the pressure (or volume) can be controlled solely (cf. Mathias et al. [18] for a discussion) through the use of periodic boundary conditions (PBCs). This control can be executed safely and poses no problems.

33.3.2
Long-range Electrostatics

PBCs generate an unbounded simulation system of finite size with the topology of a three-dimensional torus. Therefore, they create the problem of artificial self-interactions whenever the interactions among the simulated particles are long ranged as is the case for electrostatics. To avoid such self-interactions one has to follow the minimum image convention (MIC), which dictates that all interactions beyond the MIC-distance $R_{\mathrm{MIC}} = L/2$ should be neglected, where L is the size of the periodic simulation box [4]. On the other hand, a corresponding truncation of the long-ranged electrostatic interactions in the so-called straight-cutoff (SC) approach leads to serious structural and energetic artifacts, even if the simulated system consists solely of pure water [18–20]. For solutions of polyvalent ions, like proteins or nucleic acids, the SC-artifacts become even worse.

Nevertheless, up to about a decade ago MD simulations of pure water and protein–solvent systems used to apply the SC approach with small cutoff distances $R_C \approx 1$ nm (i.e., $R_C \ll R_{MIC}$) mainly for computational reasons: In an N-particle system the number of long-range interactions scales like N^2 and, therefore, seems to imply an intractable computational effort. However, in the sequel, the unacceptably large errors connected with this crude approximation forced investigators to ignore the MIC and to apply lattice sum (LS) methods [21–23] to the computation of the long-range electrostatics. LS methods take advantage of the periodicity of the simulated system and can evaluate the correspondingly periodic electrostatic potential at a computational effort scaling with $N \log N$. On the other hand, the neglect of the MIC now introduces periodicity artifacts into the description of non-periodic systems such as liquids and proteins in solution [24]. However, for large simulation systems with $L > 6$ nm in the case of pure water and $L \gg 2d$ in the case of a solvated protein of diameter d, the periodicity artifacts should become negligible or small, respectively [6, 18].

Today, we can even give absolute error bounds for the accuracy of the electrostatics computation in MD simulations. This is due to the recent development of a new algorithmic approach, which avoids the use of a periodic potential and, nevertheless, provides an accurate and computationally efficient alternative to the LS methods [18]. Therefore, it allows us to estimate, for example, the size of the LS artifacts and, in combination with LS results, to actually measure the SC errors. Because a comparably accurate alternative to LS had been lacking before, such accuracy measurements had been impossible. Figure 33.4 sketches the basic concept of this recent multiple-scale approach, which combines toroidal boundary conditions with a reaction field correction (TBC/RF).

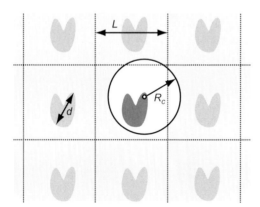

Fig. 33.4. Concept of the TBC/RF approach to the computation of long-range electrostatics in MD simulations of proteins in solution [18]. Each atom of the periodic simulation system of size L containing a protein of diameter d in solution is surrounded by a sphere of radius $R_C = R_{MIC} = L/2$. The polarization of the dielectric outside of that sphere generates a reaction field acting on the central charge, whose interactions with the charges within the sphere are rapidly calculated by fast, hierarchical, and structure-adapted multipole expansions. Here, the applied SAMM algorithm [25, 26] scales linearly with N.

To provide an example for such error bounds, we consider the PBC/LS, TBC/RF, and SC simulations on large water systems ($L = 8$–12 nm, $N = 34\,000$–$120\,000$), which have been executed with two different programs (GROMACS [27], EGO-MMII [28]) under otherwise identical conditions (see Refs [6, 18]). The enthalpies of vaporization ΔH per water molecule calculated by PBC/LS and TBC/RF showed relative deviations of at most $3/1000$, that is 0.1 kJ mol^{-1}. This deviation is smaller by a factor of 2 than the acceptable systematic error per residue estimated in Section 33.1.1 and by a factor of 20 below the acceptable random error. In contrast, the ΔH value calculated by SC using a cutoff radius $R_C \approx 4$ nm (system size $L = 8$ nm) deviated by 1.2 kJ mol^{-1} from the PBC/LS and TBC/RF values, which is larger by a factor of 6 than the acceptable systematic error introduced in Section 33.1.1.

This example quantitatively proves that SC simulations grossly miss the accuracy required for protein structure prediction in purely dipolar systems, whereas here the two other methods are acceptable. For solutions containing ions and particularly proteins similar comparisons of PBC/LS and TBC/RF treatments will soon provide clues as to which measures can be used to avoid computational artifacts and to achieve the required accuracy in these more complicated cases. As a result, the longstanding problem of how the long-range electrostatics of protein–solvent systems can be treated in MD simulations with sufficient accuracy and manageable computational effort will find a final answer in the near future. Furthermore, we can safely state that the remaining accuracy problems connected with long-range electrostatics computation are much smaller than those associated with the MM force fields to which we will now turn our attention.

33.3.3
Polarizability

As shown in Sections 33.1.4–33.1.5, many components of a protein–solvent system exhibit a significant electronic polarizability and, therefore, their electrostatic signatures, as measured by their multipole moments, will strongly vary with their position within these complex and inhomogeneous systems. In spite of the decisive part played by the electronic polarizability in protein structure, dynamics, and function, all available MM force fields still use static partial charges to model the electrostatic properties of the molecules (see, however, the ongoing development reported by Kaminski et al. [29]). Thus, they try to account for the electronic polarizability in a mean field fashion, which is invalid in inhomogeneous and nonisotropic materials. The examples discussed in Sections 33.1.4–33.1.5 suggest a lower bound of 5% for the size of the relative errors, which the neglect of polarization introduces into the description of the electrostatics.

To convert this relative error into an absolute number, we have conducted a sample TBC/RF simulation for bovine pancreatic trypsin inhibitor (BPTI) [30] in water at a temperature of 300 K and a pressure of 1 bar using the set-up described in Ref. [31]. Here, a periodic simulation box of 8 nm side length, filled with 17 329 rigid TIP3P water molecules [32] (cf. Figure 33.5) and a CHARMM22 model [33] of BPTI ($N_{res} = 58$ amino acid residues), was equilibrated for 1 ns. We took this equilibrated system as the starting point for a 100 ps simulation and evaluated the average total electrostatic energy $E_{e,\text{tot}}$ of the protein atoms. Less than 1% of all electrostatic interactions between the protein charges occur within a given residue. Due to the so-called 1–3 exclusion commonly used in MM force fields [33] they act at comparable distances like the interactions with charges residing at neighboring residues or solvent molecules. Therefore, we estimate the intra-residue contribution to $E_{e,\text{tot}}$ to be sizably smaller than at most 20% of $E_{e,\text{tot}}$. As a result, the expression

$$E_{e,\text{res}} \approx 0.8 \times E_{e,\text{tot}}/N_{\text{res}}$$

should represent a lower limit for the average electrostatic energy $E_{e,\text{res}}$ per residue. Inserting the computed number, we find $E_{e,\text{res}} \approx 160$ kJ Mol^{-1}. With the 5% lower bound for the systematic error, which is expected to be caused by the neglect of polarizability, we thus find a lower limit to the "polarizability error" of 8 kJ mol^{-1} per residue. This polarizability error is by a factor 4 larger than the acceptable random error and exceeds the acceptable systematic error, which should be more relevant here, by a factor of 40.

Therefore, the neglect of the electronic polarizability is the key drawback of the available MM force fields and, up to now, prevents quantitative MD descriptions of protein structure. Furthermore, it is the reason why the parameterization of such force fields is confronted with insurmountable difficulties. As outlined in Section 33.1.5, the electronic polarizability of the peptide groups provides a differential contribution to the relative stabilities of secondary structure motifs. As a result, the common nonpolarizable MM force fields can either describe the stability of α-helices or that of β-sheets at an acceptable accuracy, but never both. Correspondingly and as illustrated in Ponder and Case [2] by sample simulations, such force fields either favor α-helical structures (e.g., AMBER ff94 [34]) or β-sheets (e.g., OPLS-AA [35]), or else fail concerning intermediate structures (e.g., CHARMM22 [33]).

33.3.4
Higher Multipole Moments of the Molecular Components

For computational reasons, MM force fields apply the partial charge approximation, according to which the charge distribution within the molecular components of a simulation system is modeled by charges located at the positions of the nuclei. Although this approximation should be sufficiently accurate in most instances (if the problem of the missing polarizability is put aside), it will cause problems concerning the description of local structures, whenever "electron lone pairs" occur at oxygen atoms [29]. Important examples for this chemical motif are the peptide C=O groups (cf. Figure 33.3) and the water molecule. At the small oxygen atoms the lone pairs cause large higher multipole moments, whose directional signature cannot be properly represented within the partial charge approximation. Although the electric field generated by these multipole moments rapidly decays beyond the

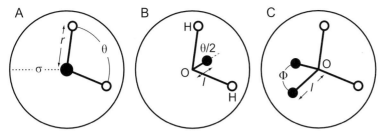

Fig. 33.5. Geometries of three different nonpolarizable MM models for water. Black circles indicate the positions of negative and open circles those of the positive partial charges. The large circle indicates the van der Waals radius. A) TIP3P, B) TIP4P (both Jorgensen et al. [33]), C) TIP5P [36].

first shell of neighboring atoms, it steers the directions of the H-bonds within that shell and, thus, shapes the corresponding local structures. The consequences of this subtle effect are yet to be explored for proteins. For water, however, whose experimental and theoretical characterization has always been and still is at a much higher stage than that of proteins, the importance of this effect is well known.

33.3.5
MM Models of Water

In an excellent review, Guillot [7] has recently given "a reappraisal of what we have learnt during three decades of computer simulations on water." Within the partial charge approximation, MM models of water comprise three charges at the positions of its three atoms ("three point models"). By symmetry and neutrality of the molecule, in this case three parameters (one charge, bond angle, and bond length) suffice to specify the charge distribution (cf. Figure 33.5A). However, these simple three-point models, which are commonly used in MD simulations of protein–water systems, do not correctly account for the higher multipole moments and, therefore, miss essential features of the correlation functions in pure water (see, for example, Ref. [6]). Because of this defect and of the missing polarizability, they cannot be expected to provide an adequate description of the hydration shell structures surrounding proteins.

For an improved description of the higher multipole moments, four- and five-point models have been suggested positioning partial charges away from the positions of the nuclei. Figure 33.5 compares the geometries of correspondingly extended models with that of a simple three-point model. In addition, a series of polarizable models using "fluctuating charges" or inducible dipoles have been proposed. Guillot gives an impressive table listing about 50 different MM models of water and, subsequently, classifies the performance of these models concerning the reproduction of experimentally established properties in simulations. Although Guillot mainly discusses pure water, he also shortly comments on the topic of transferability, which is the unsolved problem of how to construct an MM water model performing equally well for all kinds of solute molecules. Summarizing his review he arrives at the conclusion that "one has a taste of incompletion, if one considers that not a single water model available in the literature is able to reproduce with a great accuracy all the water properties. Despite many efforts to improve this situation very few significant progress can be asserted." In his view this is due to "the extreme complexity of the water force field in the details, and their influence on the macroscopic properties of the condensed phase." Nevertheless, being a committed scientist he does not suggest that we should quit attempts at an MM description but instead adds several suggestions as to how one should account for the polarizability and for other details in a better way.

33.3.6
Complexity of Protein–Solvent Systems and Consequences for MM-MD

The state of the MM-MD descriptions of pure water sketched above is sobering when one turns to protein–solvent systems, whose complexity exceeds that of pure water by orders of magnitude. This enormous increase of complexity is caused only in part by the much larger spatial and temporal scales characterizing protein dynamics (cf. Section 33.1.2). More important is the combinatorial complexity, which is posed by the fact that one has to parameterize the interactions between a large variety of chemically different components (water, peptide groups, amino acid residues, ions, etc.) into a single effective energy function in a balanced way. If already in the case of a one-component liquid like water subtle details of the MM models determine the macroscopic properties, a difficulty which, up to now, has prevented the construction of a model capable of explaining all known properties, then one may ask whether the MM approach will ever live up to its promises concerning proteins at all. This critical question gains support by our above estimate concerning the size of the errors introduced by the neglect of the electronic polarizability characteristic for the common force fields. Our conservative estimate suggests that nowadays not even a single and most basic property of soluble proteins, namely their structure, can be reliably predicted by extended MM-MD simulations.

33.3.7
What about Successes of MD Methods?

However, this disastrous result on the accuracy of current MM-MD simulations does not imply that their results are always completely wrong. Instead it may sometimes happen that numbers computed by MD for certain observables surprisingly agree with experimental data within a few per cent. Examples are the peptide simulations presented by Daura et al. [37], which have shown a mixture of peptide conformations compatible with NMR data, the simulation of AFM experiments on the rupture force required to pull a biotin substrate out of its binding pocket in streptavidin [38], or the simulation of the light-induced relaxation dynamics in small cyclic peptides [39] (see Section 33.3 for a more detailed discussion of this example). In such cases, the apparent "success" of an MD simulation may be due (i) to the choice of an observable that happens to be extremely stable with respect to the sizable inaccuracies of the MM force field, or (ii) to the fortunate cancellation of many different errors [40]. Therefore, one cannot readily conclude from such favorable examples that the MD approach towards biomolecular dynamics has "come of age" [41]. Instead, one must expect that observables computed by unrestricted MD simulations differ either quantitatively from experimental data by 10–50% or are qualitatively wrong.

Furthermore, the deficiencies of the MM force fields do not imply that results of MD simulations that are restricted by experimental data are wrong. On the contrary, this type of simulation [42] has led to the greatest success of the MD method and is largely responsible for its widespread application on proteins. In this experimentally restricted MD approach the chemical expertise that is coded into the various parameters of the MM force field (providing standard values for the bond lengths, bond angles, dihedral angles, for the elasticity of these internal coordinates etc.) is combined with a large set of experimental constraints (e.g., from X-ray diffraction or more-dimensional NMR). Shortly after their suggestion in 1987, the corresponding MD-based "annealing simulations" have become the standard tool in the computation of protein and peptide structures from X-ray and NMR data and have acquired a key role in structural biology, because they have enabled the automation of protein structure refinement. In this context, the deficiencies of the MM force fields play a minor role as they are largely corrected by the experimental constraints. Note that the protein electrostatics and its shielding by the solvent are usually ignored in these experimentally restricted simulations, because here one neither wants nor has to deal with the associated complexities.

Finally, another type of restricted MD simulation also does not suffer too much from the deficiencies of the MM force fields. These are the so-called QM/MM hybrid methods, in which a small fragment of a simulation system is treated by quantum mechanics (QM) and the large remainder by MM (the basic concepts of this approach were described in a seminal paper by Warshel and Levitt in 1976 [43]). Here, the attention is restricted to the properties of the QM fragment, which are calculated as exactly as possible by numerical solution of the electronic Schrödinger equation. The electronic polarization of the QM fragment is obtained by importing the electric field generated by the MM environment as an external perturbation into the QM Hamiltonian. If this perturbation changes a certain property of the QM fragment by 10%, then even 10% errors in the external field calculated by MM will entail only errors of 1% in the QM result.

To provide an example we consider the vibrational spectra of molecules. Due to the development [44, 45] and widespread accessibility [46] of density functional theory (DFT) highly accurate computations of intramolecular force fields and of associated gas-phase vibrational spectra have become feasible at a moderate computational effort. On average, the spectral positions of the computed lines deviate by only about 1.5% from the observations, which is sufficient for a safe assignment of observed bands to normal modes [47, 48]. Note that calculated molecular structures are more exact by orders of magnitude than the spectra, because the molecu-

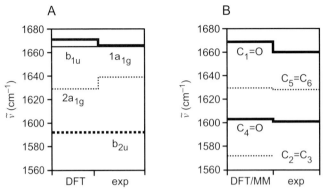

Fig. 33.6. Comparison of DFT (A) and DFT/MM (B) computations with experimental data for the vibrational spectra of the C=O (solid lines) and C=C (dashed lines) modes of quinone molecules in the gas-phase and in a protein, respectively. The DFT results pertain to p-benzoquinone [48] and the DFT/MM results to the ubiquinone Q_A in the reaction center of Rb. sphaeroides [49]. Both computational results agree very well with the observations. In particular, the DFT/MM treatment correctly predicts the sizable 60 cm^{-1} red shift of one of the C=O modes, which is caused by the protein environment through electrostatic polarization of the quinone.

lar structure is much less sensitive to inaccuracies of the QM treatment than the intramolecular force field.

In condensed phase, the computation of vibrational spectra [49, 50] has recently been enabled by the development of a DFT/MM hybrid method [9]. The perturbation argument given further above entails the expectation that DFT/MM can describe condensed-phase vibrational spectra at an accuracy comparable to the one previously achieved by DFT for the gas-phase spectra. This expectation is confirmed by the results shown in Figure 33.6 and by the infrared spectra of p-bezoquinone on water, which have been calculated from a QM/MM hybrid MD trajectory [50]. Thus, although the common nonpolarizable MM force fields are not accurate enough for unrestricted MD simulations of biomolecules, they are valuable in more restricted settings.

33.3.8
Accessible Time Scales and Accuracy Issues

Concerning MM-MD simulations of protein or peptide folding, the insufficient accuracy of MM force fields has the following consequence: If we had computer power enough to simulate the folding dynamics of a small protein or peptide in solution for milliseconds or seconds starting from a random coil conformation, we would have to expect at a high probability that the conformation predicted by the simulation has nothing to do with the native conformation, that is, represents a "real artifact."

In this respect it is, in a sense, fortunate that computational limitations have previously restricted MM-MD simulations of protein–solvent systems to the time scale of a few nanoseconds. Because simulations usually start from a structure determined by X-ray crystallography or by more-dimensional NMR, and because in proteins small-scale conformational transitions only begin to occur at this time scale, the actual metastability of a MM-MD protein model is likely to become apparent only in much more extended simulations than in the ones that are currently feasible. At the accessible nanosecond time scale, the large-scale and slow relaxation processes changing a metastable structure are simply "frozen out," and protein simulations are strongly restricted by the accessible time scales. Structural stabilities observed in short-time simulations are due to this restriction and cannot be attributed to the particular quality of the force field. Nevertheless, the apparent stability of protein structures even in grossly erroneous MM-MD simulations (like in those applying the SC truncation to the long-range electrostatics) may have led scientists to believe in the validity of MD results. For the field of MD simulations on protein dynamics this was fortunate, because it has created a certain public acceptance (cf. e.g., Ref. [41]) and has attracted researchers. On the other hand it was also quite unfortunate, because the urgent necessity to improve the MD methods to meet the challenges outlined in Section 33.1 remained an issue of concern only in a small community of methods developers.

However, the speed of computation has been increasing by a factor of 2 each year during the past decades and, according to the plans of the computer industry, will continue to exhibit this type of exponential growth within the coming decade. For the available MM force fields this progress would imply that MD simulations spanning about 0.5 ms become feasible, because the current technology already allows series of MD simulations spanning 10 ns each for protein–solvent systems comprising 10^4–10^5 atoms. Therefore, the deficiencies of the current MM force fields are becoming increasingly apparent in unrestricted long-time simulations through changes of supposedly stable protein or peptide conformations. This experience will drive the efforts to construct better force fields, which must be more accurate by at least one order of magnitude, although such an improvement necessarily will imply a larger computational effort and, therefore, will impede the tractability of large proteins at extended time scales. Here, the accurate and computationally efficient inclusion of the electronic polarizability will be of key importance.

33.3.9
Continuum Solvent Models

The above discussion on the accessible time scales was based on the assumption that a microscopic model of the aqueous environment is employed to account for the important solvent–protein interactions. This approach represents the state of the art and, at the same time, an obstacle in accessing more extended time scales, because in the corresponding simulation systems 90% of the atoms must belong to the solvent (for an explanation see Section 33.2.2 and Mathias et al. [18]). Thus, one simulates mainly a liquid solvent slightly polluted by protein atoms, and most of the computer time is spent on the calculation of the forces acting on the solvent molecules.

It is tempting to get rid of this enormous computational effort by applying so-called implicit solvent models. Here, one tries to replace the microscopic modeling of the solvent by mean field representations of its interactions with the solute protein, which are mainly the electrostatic free energy of solvation and the hydrophobic effect caused by the solvent. The hydrophobic effect is usually included by an energy term proportional to the surface of the solute. Much more complicated is the computation of the electrostatic free energy of solvation. Here one has to find solutions to the electrostatics problem of an irregularly shaped cavity filled with point charges (the solute) and embedded in a dielectric and possibly ionic continuum (the solvent). Standard numerical methods to solve the corresponding Poisson and Poisson-Boltzmann equations are much too expensive to be used in MD simulations and have many other drawbacks (see Egwolf and Tavan [31] for a discussion).

To circumvent the problem of actually having to solve a partial differential equation (PDE) at each MD time step, the so-called generalized Born methods (GB) have been suggested (for a review see Bashford and Case [51]). GB methods introduce screening functions that are supposed to describe the solvent-induced shielding of the electrostatic interactions within the solute and are empirically parameterized using sets of sample molecules. The required parameters then add to the combinatorial complexity of the force field parameterization.

The GB approximation enables the computation of impressive MD trajectories covering several hundred microseconds [52]. However, not surprisingly, the resulting free energy landscapes drastically differ from those obtained with explicit solvent [53, 54]. Thus, the GB methods apparently oversimplify the complicated electrostatics problem for the sake of computational efficiency and, therefore, fail in structure prediction.

A physically correct method to compute the free energy of solvation and, in particular, the forces on the solute atoms in implicit solvent MD simulations has recently been developed by Egwolf and Tavan [31, 55]. In this analytical "one-parameter" approach the reaction field arising from the polarization of the implicit solvent continuum is represented by Gaussian dipole densities localized at the atoms of the solute. These dipole densities are determined by a set of coupled equations, which can be solved numerically by a self-consistent iterative procedure. The required computational effort is comparable to that of introducing the electronic polarizability into MM force fields. This effort is by many orders of magnitude smaller than that of standard numerical methods for the solution of PDEs.

The resulting reaction field forces agree very well with those of explicit solvent simulations, and the solvation free energies match closely those obtained by the standard numerical methods [31]. However, this progress is only a first step towards MD simulations with implicit solvent. For its use in MD, one has to correct the violations of the reaction principle resulting from the reaction field forces. In addition, the surface term covering the hydrophobic forces has to be parameterized in order to match the free energy landscapes found in explicit solvent simulations.

33.3.10
Are there Further Problems beyond Electrostatics and Structure Prediction?

Up to this point, our analysis has been focused on the complex electrostatics in protein–solvent systems and the accuracy problems that are caused by the electrostatics in the construction and numerical solution of MM-MD models. In this discussion we have taken the task of protein structure prediction as our yardstick to judge the accuracy of the available force fields and computational techniques. If MD is supposed to provide a contribution to protein folding or to the related problem of structure-based drug design, this yardstick is certainly relevant. However, there are many more possible fields of application for MD simulation than just these two. Each of them has its own accuracy requirements pertaining to different parts of the force field, as has been illustrated by the various examples of restricted MD-based methods given in Section 33.2.7.

If MD is taken seriously as a method for generating a virtual reality of the conformational dynamics that enables, for instance, energy-driven processes of protein function, then additional aspects of the MM force fields become relevant. Because this conformational dynamics involves torsions around chemical bonds and requires evasive actions to circumvent steric hindrances in these strongly constrained macromolecules, the potential functions mapping the mechanical elasticity and flexibility must be adequately represented for correct descriptions of the kinetics. Here, these so-called "bonded" potential functions must be balanced with the van der Waals potentials of the various atoms and with the modeling of the electrostatics. To illustrate the delicate dependence of simulation descriptions on these details of the force field parameterization, we will now return to one of the examples mentioned in Section 33.2.7.

33.4
Conformational Dynamics of a Light-switchable Model Peptide

An important obstacle impeding the improvement of MM force fields is the lack of sufficiently specific and detailed experimental data on the conformational dynamics in proteins and peptides. Structural methods like X-ray or NMR render only information on the equilibrium fluctuations or on the equilibrium conformational ensemble, respectively. In X-ray, a structurally and temporally resolved monitoring of relaxation processes may become accessible through the Laue technique, but this approach is still in its infancy. On the other hand, time-resolved spectroscopic methods for the visible and infrared spectral regions are well-established and, with the recent development of two-dimensional correlation techniques [56], promising perspectives are opening up.

To become useful for the evaluation of MM force fields, these techniques must be applied to small peptide–solvent systems of strongly restricted complexity, because, otherwise, the multitude of the many weak interactions characteristic for large proteins in solution prevents the unique identification of force field deficien-

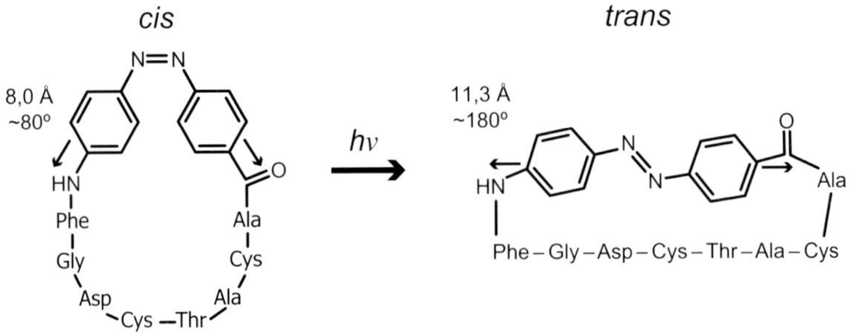

Fig. 33.7. Chemical structures of a small cyclic peptide (cAPB) and scheme of the light-induced photoisomerization, which changes the isomeric state of the azobenzene dye integrated into the backbone of the peptide from *cis* to *trans*. The isomerization widens the angle of the stiff linkage between the chromophore and the peptide from about 80° to 180° and elongates the chromophore by 3.3 Å. The strain exerted by the chromophore after photoisomerization forces the peptide to relax from an ensemble of helical and loop-like conformations into a stretched β-sheet (see Figure 33.9 for structural details).

cies. Furthermore, the dynamical processes monitored spectroscopically should proceed rapidly on the time scale of at most nanoseconds, because this time scale is already accessible to MD simulations.

Adopting this view, Spörlein et al. [39] have recently presented an integrated approach towards the understanding of peptide conformational dynamics, in which femtosecond time-resolved spectroscopy on model peptides with built-in light switches has been combined with MD simulation of the light-triggered motions. It has been applied to monitor the light-induced relaxation dynamics occurring on subnanosecond time scales in a peptide that was backbone-cyclized with an azobenzene derivative as optical switch and spectroscopic probe.

Figure 33.7 depicts the chemical structure of this molecule (cAPB) and schematically indicates the changes of geometry that are induced by the photoisomerization of the azobenzene dye. The femtosecond spectra [39] allows us to clearly distinguish and characterize the subpicosecond photoisomerization of the chromophore, the subsequent dissipation of vibrational energy and the subnanosecond conformational relaxation of the peptide. It was interesting to see to what extent these processes can be described by MM-MD simulation.

33.4.1
Computational Methods

To enable the MD simulation of the photochemical *cis–trans* isomerization of the chromophore and of the resulting peptide relaxations, an MM model potential has been constructed (see Refs [39] and [57] for details), which drives the chromophore along an inversion coordinate at one of the central nitrogen atoms from *cis* to *trans* and, concomitantly, deposits the energy of 260 kJ mol^{-1}, which is equal to that of the absorbed photon, into the cAPB molecule. This approach serves to

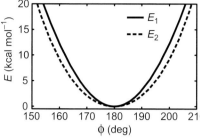

Fig. 33.8. Two different MM potentials obtained from fits to DFT potential curves for the torsion of the chromophore around the central N=N double bond by a dihedral angle Φ [57]. The stiff potential E_1 results from a fit of a cosine to the DFT potential curve in the vicinity of the *trans*-configuration (Φ = 180°) and the softer potential E_2 from a fit of a Fourier expansion aiming at an improved description of the DFT potential curve in a wider range of torsion angles Φ ∈ [70°, 300°].

model the photoexcitation of the dye, its relaxation on the excited-state surface, and its ballistic crossing back to the potential energy surface of the electronic ground state through a conical intersection. To describe the subsequent dissipation of the absorbed energy into the solvent and the relaxation of the peptide from the initial *cis*-ensemble of conformations to the target *trans*-conformation, MM parameters for the chromophore, the peptide and the surrounding solvent were required.

Apart from a slight modification [57], the peptide parameters were adopted from the CHARMM22 force field [33]. Like in the experimental set-up, dimethyl-sulfoxid (DMSO) was used as the solvent. For this purpose an all-atom, fully flexible DMSO model was designed by DFT/MM techniques [9], was characterized by TBC/RF-MD simulations, and was compared with other MM models known in the literature [57]. Likewise, the MM parameters of the chromophore were determined by series of DFT calculations. Figure 33.8 illustrates the subtle differences that may result from such a procedure of MM parameter calculation, taking the stiffness of the central N=N bond with respect to torsions as a typical example. The stiffer potential E_1 has been employed in Spörlein et al. [39] and the consequences of choosing the slightly softer potential E_2 will be discussed further below.

A larger number of MD simulations, each spanning 1–10 ns (some even up to 40 ns), have been carried out to characterize the equilibrium conformational ensembles in the *cis* and *trans* states as well as the kinetics and pathways of their photo-induced conversion [57]. The simulation system was periodic, shaped as a rhombic dodecahedron (inner diameter of 54 Å), and was initially filled with 960 DMSO molecules. The best 10 NMR solutions [58] for the structure of the cAPB peptide were taken as starting structures for the simulations of the photoisomerization. They were placed into the simulation system, overlapping DMSO molecules were removed, and the solvent was allowed to adjust in a 100 ps MD simulation at 300 K to the presence of the rigid peptide. Then the geometry restraints initially imposed on the peptide were slowly removed during 50 ps, the system was equilibrated for another 100 ps at 300 K and 1 bar, it was heated to 500 K for

50 ps, and cooled within 200 ps to 300 K. In this way 10 different starting structures were obtained for the 1 ns simulations of the *cis–trans* isomerization, which was initiated by suddenly turning on the model potential driving the inversion reaction. To obtain estimates for the equilibrium conformational ensembles in the *cis* and *trans* states, 10 ns MD simulations were carried out at 500 K starting from the best NMR structures modified by the equilibration procedure outlined above. At this elevated temperature a large number of conformational transitions occur in the peptide during the simulated 10 ns. In all simulations the TBC/RF approach (cf. Section 33.2.2) was used with a dielectric constant of 45.8 modeling the DMSO continuum at large distances (for further details see Ref. [57]).

Due to the lack of space, the methods applied for the statistical analysis of the simulation data cannot be presented in detail here. Instead a few remarks must suffice: The 10 ns trajectories at 500 K have been analyzed by a new clustering tool, which is based on the computation of an analytical model of the configuration space density sampled by the simulation. The model is a Gaussian mixture [59], enables the construction of the free energy landscape and allows the identification of a hierarchy of conformational substates. The associated minima of the free energy are then characterized by prototypical conformations [57]. Concerning the characterization of the photo-induced relaxation dynamics, we will restrict our analysis to the temporal evolution of the total energy left in the cAPB molecule after photoisomerization and will omit structural aspects.

33.4.2
Results and Discussion

Figure 33.9 shows the prototypical conformations of *cis*- and *trans*-cAPB in DMSO at 500 K and 1 bar, which have been obtained from the 10 ns simulations by the procedures sketched in the preceding section. Whereas *cis*-cAPB exhibits a series

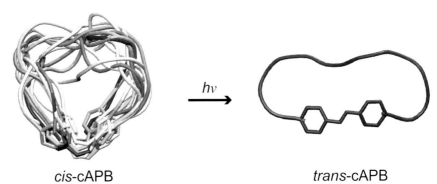

cis-cAPB   *trans*-cAPB

Fig. 33.9. Conformational ensembles in the *cis* and *trans* states of the cAPB peptide as obtained from 10 ns simulations at 500 K. Each backbone structure represents a different local minimum of the free energy landscape. The gray-scale indicates the depth of the respective minimum. Whereas there is only one such minimum in *trans*, there are many in *cis* (see Ref. [57] for details).

of different conformations, which are dynamically related by thermal flips around the dihedral angles at the Cα atoms of the backbone, there is only a single *trans*-cAPB conformer. Although the depicted conformers derive from high-temperature simulations, they are compatible [57] with the NOE-restraints observed by NMR [58]. This agreement may not be taken as a big surprise, because the conformational flexibility of peptide is sizably restricted by the strong covalent linkages within the cyclic structure. Apparently, in the given case these restrictions suffice to compensate those inaccuracies of the force field, which involve weaker dipolar interactions or the neglect of the electronic polarizability.

Figure 33.9 illustrates the process that is driven by the photoisomerization. The photons are absorbed by an ensemble of *cis*-conformers, and the geometry change of the chromophore forces the peptide backbone to relax through flips around the bonds at the Cα atoms towards the *trans*-conformation. In this conformation, the structure of the peptide is that of a β-sheet, up to turns at the covalent linkage with the chromophore. Therefore, the process may be considered as protein folding "en miniature."

As far as the photo-induced relaxation dynamics is concerned, Figure 33.10 illustrates a surprising and, as we will argue below, also somewhat frustrating result, reported previously in Ref. [39]. The kinetics of energy relaxation calculated by MD for the *cis–trans* photoisomerization of cAPB agrees nearly perfectly with that of a corresponding spectroscopic observable! As far as the match within the first picosecond is concerned, this agreement simply reflects the careful design of the MM model potential driving the ballistic photoisomerization and, therefore, is what one should expect, if the assumption of an inversion reaction is correct.

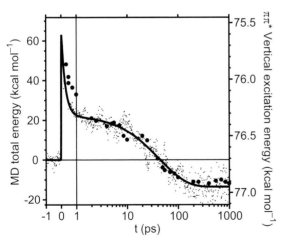

Fig. 33.10. Comparison of the energy relaxation kinetics following *cis–trans* isomerization as observed by femtosecond spectroscopy (black dots, monitoring the spectral position of the main absorption band) and as calculated by MD (black pixels and fit to the MD trajectories, monitoring the total energy of the cAPB molecule) on a linear-logarithmic time scale [39].

From a global fit of all spectroscopic data, covering a wide spectral range, four time constants have been obtained for the relaxation processes, and the analysis of the spectra has allowed to assign elementary processes to these kinetics [39]. Thus, according to spectroscopy, the ballistic isomerization proceeds within 230 fs, a second relaxation channel from the excited state surface to the electronic ground state (not included in the MM model by construction) takes 3 ps, the dissipation of heat from the chromophore into the solvent and the peptide chain proceeds within 15 ps, and a 50 ps kinetics is assigned to conformational transitions in the peptide. As shown in Ref. [60] further conformational transitions in the peptide occur above the time scale of 1 ns until the relaxation process is complete.

The limited statistics of the MD data shown in Figure 33.10 allows us to determine two time constants within the first nanosecond. These time constants are 280 fs for the isomerization of the chromophore and 45 ps for the conformational relaxation of the peptide form the *cis*-ensemble towards the *trans*-conformation. The corresponding two-exponential decay is depicted by the solid fit curve in Figure 33.10. Due to the proximity of the cooling kinetics (15 ps) to that of the conformational dynamics (50 ps) these processes cannot be distinguished in the MD data on cAPB. According to the simulations, after 1 ns the *trans*-conformation has not yet been reached in most trajectories. Thus, within the first nanosecond the relaxation kinetics of the simulated ensemble quantitatively agrees with the observations to the extent, to which the limited MD statistics allows the identification of kinetic constants.

Further above we have stated that this result has been somewhat frustrating for us. Representing the theory group that had been involved in the design of the model system and of its experimental characterization from the very beginning about a decade ago, we had hoped that our integrated approach [39] would provide hints as to how one has to improve the MD methods and MM force fields. This hope had been a key driving force in the original set-up of this interdisciplinary approach. Instead, the results have merely validated our MD methods, raising the question whether (i) the integrated approach is sensitive enough to the details of the MM force field or (ii) our skepticism with respect to the quality of MD descriptions is justified.

To check these questions we have conducted a series of simulations in which certain details of the applied MM force fields were changed one by one [57]. From the huge pile of results collected in this way we have selected a typical example which provides enough evidence to answer both questions. In this example we have changed a single model potential of the MM force field, that is, we have replaced the stiff torsion potential E_1 by the slightly weaker potential E_2 (see Figure 33.8 for plots of these function and Section 33.3.1 for further explanation). One can expect that this potential is important for transmitting the mechanical strain from the chromophore to the peptide after *cis–trans* isomerization. If it is stiff (E_1), then the strain is strong and the relaxation should be fast, and if it is weaker (E_2) the relaxation should become slowed down.

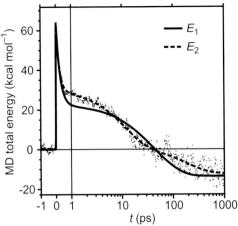

Fig. 33.11. Comparison of the energy relaxation kinetics calculated by MD for two slightly different force fields: The original model potential E_1 is replaced by E_2 (cf. Figure 33.8). The fit curve belonging to E_1 has been taken from Figure 33.10. The dashed fit curve to the 10 MD trajectories (pixels) calculated with E_2 exhibits three kinetic constants. Besides showing a 190 fs ballistic isomerization (exp: 230 fs), it separates an 11 ps heat dissipation (exp: 15 ps) from a 190 ps conformational relaxation (exp: 50 ps).

The above expectation is confirmed by the MD results for the softer potential E_2 shown in Figure 33.11 and by the kinetic data listed in the associated caption. In the case of E_2, the prolonged time scale (190 ps) of the conformational relaxation, which is caused by the slightly softer spring, has enabled us to additionally identify the kinetics of heat dissipation (11 ps) in the data. The results demonstrate that a small change of a single model potential can change the kinetic time constant calculated for the conformational relaxation from 45 ps to 190 ps, that is, by a factor of 4.

This example answers the two questions voiced above, because it provides evidence that (i) our integrated approach is sensitive enough to detect small changes in a force field and (ii) that our skepticism with respect to the quality of MM force fields is well-justified. The latter is apparent from the fact that both fit procedures for obtaining a torsion potential from DFT calculations, the one that has led to E_1, and the one yielding E_2, represent a priori valid choices. It was actually a matter of luck that we happened to choose the right one for the first few simulations published in Ref. [39]. As we now definitely know [57], only a series of further lucky choices in the force field used for our original simulations enabled the surprisingly perfect match. On the other hand, the perfect match obtained in our first attempt to describes the light-induced relaxation in the cAPB peptide also demonstrates that, in principle, MD simulations can give quantitative descriptions. In this sense it gives credit to the method and justifies further efforts aiming at the improvement of the force fields and the extension of time scales.

Summary

As of today, MM force fields are not yet accurate enough to enable a reliable prediction of protein and peptide structures by MD simulation. Here the key deficiency is the neglect of the electronic polarizability in the MM force fields. However, MM-MD is valid in restricted settings, for instance, when complemented by experimental constraints, when applied to systems of reduced complexity, or in the restricted context of QM/MM hybrid calculations. In particular, the development of the DFT/MM methods has generated key tools that are required for the improvement of the MM force fields, because they allow highly accurate computations of intramolecular force fields of molecules polarized by a condensed phase environment. The latter has been demonstrated by computations of IR spectra in condensed phase, whose accurate description critically depends on the quality of the computed force field. For the validation of future MM force fields more specific experimental characterizations of the materials are necessary. Thus, the improvement of the computational methods has to rely on close cooperation between scientists working in experimental and theoretical biophysics, chemistry, biochemistry, and beyond.

Acknowledgments

Financial support by the Volkswagen-Stiftung (Project I/73 224) and by the Deutsche Forschungsgemeinschaft (SFB533/C1) is gratefully acknowledged. The authors would like to thank W. Zinth, F. Siebert, L. Moroder, C. Renner, and J. Wachtveitl for fruitful discussions.

References

1 M. Karplus, J. A. McCammon, Nature Struct. Biol. 2002, 9, 646–652.
2 J. W. Ponder, D. A. Case, Adv. Protein Chem. 2003, 66, 27–85.
3 S. Gnanakaran et al., Current Opinion in Struct. Biol. 2003, 13, 168–174.
4 M. P. Allen, D. J. Tildesley, Computer Simulations of Liquids. Clarendon, Oxford, 1987.
5 A. Warshel, S. T. Russel, Q. Rev. Biophys. 1984, 17, 283–422.
6 G. Mathias, P. Tavan, J. Chem. Phys. 2004, 120, 4393–4403.
7 B. Guillot, J. Mol. Liquids 2002, 101, 219–260.
8 C. L. Brooks, Acc. Chem. Res. 2002, 35, 447–454.
9 M. Eichinger, P. Tavan, J. Hutter, M. Parrinello, J. Chem. Phys. 1999, 110, 10452–10467.
10 K. Laasonen, M. Sprik, M. Parrinello, R. Car, J. Chem. Phys. 1993, 99, 9080–9089.
11 C. Tanford, The Hydrophobic Effect: Formation of Micelles and Biological Membranes. Wiley, New York, 1980.
12 P. L. Privalov, in Protein Folding, Ed: T. E. Creighton, Freeman, New York, 1992.
13 F. Siebert, Meth. Enzymol. 1995, 246, 501–526.
14 J. A. McCammon, B. R. Gelin, M. Karplus, Nature 1977, 267, 585–590.
15 W. F. van Gunsteren, H. J. C.

Berendsen, *J. Mol. Biol.* **1984**, *176*, 559–564.
16 M. Levitt, R. Sharon, *Proc. Natl Acad. Sci. USA* **1988**, *85*, 7557–7561.
17 A. Rahman, F. H. Stillinger, *J. Chem. Phys.* **1971**, *55*, 3336–3359.
18 G. Mathias, B. Egwolf, M. Nonella, P. Tavan, *J. Chem. Phys.* **2003**, *118*, 10847–10860.
19 I. G. Tironi, R. Sperb, P. E. Smith, W. F. van Gunsteren, *J. Chem. Phys.* **1995**, *102*, 5451–5459.
20 P. H. Hünenberger, W. F. van Gunsteren, *J. Chem. Phys.* **1998**, *108*, 6117–6134.
21 T. A. Darden, D. York, L. Pedersen, *J. Chem. Phys.* **1993**, *98*, 10089–10092.
22 U. Essmann et al., *J. Chem. Phys.* **1995**, *103*, 8577–8593.
23 B. A. Luty, I. G. Tironi, W. F. van Gunsteren, *J. Chem. Phys.* **1995**, *103*, 3014–3021.
24 P. H. Hünenberger, J. A. McCammon, *Biophys. Chem.* **1999**, *78*, 69–88.
25 C. Niedermeier, P. Tavan, *J. Chem. Phys.* **1994**, *101*, 734–748.
26 C. Niedermeier, P. Tavan, *Mol. Simul.* **1996**, *17*, 57–66.
27 E. Lindahl, B. Hess, D. van der Spoel, *J. Mol. Mod.* **2001**, *7*, 306–317.
28 G. Mathias et al., *EGO-MMII Users Guide*. Lehrstuhl für BioMolekulare Optik, LMU München, Oettingenstrasse 67, D-80538 München, unpublished.
29 G. A. Kaminski et al., *J. Comput. Chem.* **2002**, *23*, 1515–1531.
30 M. Marquart et al., *Acta Crystallogr. Sec.* **1983**, *39*, 480–490.
31 B. Egwolf, P. Tavan, *J. Chem. Phys.* **2003**, *118*, 2039–2056.
32 L. Jorgensen et al., *J. Chem. Phys.* **1983**, *79*, 926–935.
33 A. D. MacKerell et al., *J. Phys. Chem. B* **1998**, *102*, 3586–3616.
34 W. D. Cornell et al., *J. Am. Chem. Soc.* **1995**, *115*, 9620–9631.
35 G. A. Kaminski, R. A. Friesner, J. Tirado-Rives, W. L. Jorgensen, *J. Phys. Chem. B* **2001**, *105*, 6474–6487.
36 M. W. Mahoney, W. L. Jorgensen, *J. Chem. Phys.* **2000**, *112*, 8910–8922.
37 X. Daura et al., *J. Am. Chem. Soc.* **2001**, *123*, 2393–2404.
38 H. Grubmüller, B. Heymann, P. Tavan, *Science* **1996**, *271*, 997–999.
39 S. Spörlein et al., *Proc. Natl Acad. Sci. USA* **2002**, *99*, 7998–8002.
40 W. F. van Gunsteren, A. E. Mark, *J. Chem. Phys.* **1998**, *108*, 6109–6116.
41 H. J. C. Berendsen, *Science* **1996**, *271*, 954–955.
42 A. T. Brünger et al., *Science* **1987**, *235*, 1049–1053.
43 A. Warshel and M. Levitt, *J. Mol. Biol* **1976**, *103*, 227–249.
44 P. Hohenberg, W. Kohn, *Phys. Rev B* **1964**, *136*, 864–870.
45 W. Kohn, L. J. Sham, *Phys. Rev. A* **1965**, *140*, 1133–1138.
46 M. J. Frisch et al., *Gaussian98*, Gaussian, Inc., Pittsburgh PA, 1998.
47 J. Neugebauer, B. A. Hess, *J. Chem. Phys.* **2003**, *118*, 7215–7225.
48 M. Nonella, P. Tavan, *Chem. Phys.* **1995**, *199*, 19–32.
49 M. Nonella, G. Mathias, M. Eichinger, P. Tavan, *J. Phys. Chem. B* **2003**, *107*, 316–322.
50 M. Nonella, G. Mathias, P. Tavan, *J. Phys. Chem. A* **2003**, *107*, 8638–8647.
51 D. Bashford, D. A. Case, *Annu. Rev. Phys. Chem.* **2000**, *51*, 129–152.
52 C. D. Snow, N. Nguyen, V. S. Pande, M. Gruebele, *Nature* **2002**, *420*, 102–106.
53 R. H. Zhou, B. J. Berne, *Proc. Natl Acad. Sci. USA* **2002**, *99*, 12777–12782.
54 H. Nymeyer, A. E. Garcia, *Proc. Natl Acad. Sci. USA* **2003**, *100*, 13934–13939.
55 B. Egwolf, P. Tavan, *J. Chem. Phys.* **2004**, *120*, 2056–2068.
56 S. Woutersen, P. Hamm, *J. Phys. Chem. B* **2000**, *104*, 11316–11320.
57 H. Carstens, Dissertation, Ludwig-Maximilians-Universität München, 2004.
58 C. Renner et al., *Biopolymers* **2000**, *54*, 489–500.
59 M. Kloppenburg, P. Tavan, *Phys. Rev. E* **1997**, *55*, 2089–2092.
60 S. Spörlein, Dissertation, Ludwig-Maximilians-Universität München, 2001.